American Academy of Orthopaedic Surgeons
American Academy of Family Physicians
American Academy of Pediatrics

Musculoskeletal Medicine

Edited by
Joseph Bernstein, MD, MS

Staff

Mark Wieting,
Vice President, Educational Programs

Marilyn L. Fox, PhD,
Director, Department of Publications

Lynne Roby Shindoll,
Managing Editor

Keith Huff,
Senior Editor

Mary Steermann,
Manager, Production and Archives

David Stanley,
Assistant Production Manager

Sophie Tosta,
Assistant Production Manager

Michael Bujewski,
Database Coordinator

Dena Lozano,
Desktop Publishing Assistant

Karen Danca,
Production Assistant

Laura Andujar,
Publications Assistant

Published by the
American Academy of Orthopaedic Surgeons
6300 North River Road
Rosemont, IL 60018

First Edition

The material presented in *Musculoskeletal Medicine* has been made available by the American Academy of Orthopaedic Surgeons for educational purposes only. This material is not intended to present the only, or necessarily best, methods or procedures for the medical situations discussed, but rather is intended to represent an approach, view, statement, or opinion of the author(s) or producer(s), which may be helpful to others who face similar situations.

Some drugs or medical devices demonstrated in Academy courses or described in Academy print or electronic publications have not been cleared by the Food and Drug Administration (FDA) or have been cleared for specific uses only. The FDA has stated that it is the responsibility of the physician to determine the FDA clearance status of each drug or device he or she wishes to use in clinical practice.

The U.S. FDA has expressed concern about potential serious patient care issues involved with the use of polymethylmethacrylate (PMMA) bone cement in spine. A physician might insert the PMMA bone cement into vertebrae by various procedures, including vertebroplasty and kyphoplasty.

PMMA bone cement is considered a device for FDA purposes. In October 1999, the FDA reclassified PMMA bone cement as a Class II device for its intended use "in arthroplastic procedures of the hip, knee, and other joints for the fixation of polymer or metallic prosthetic implants to living bone." The use of a device for other than its FDA-cleared indication is an off-label use. Physicians may use a device off-label if they believe, in their best medical judgment, that its use is appropriate for a particular patient (eg, tumors).

The use of PMMA bone cement in the spine is described in Academy educational courses, videotapes, and publications for educational purposes only. As is the Academy's policy regarding all of its educational offerings, the fact that the use of PMMA bone cement in the spine is discussed does not constitute and Academy endorsement of this use.

Furthermore, any statements about commercial products are solely the opinion(s) of the author(s) and do not represent an Academy endorsement or evaluation of these products. These statements may not be used in advertising or for any commercial purpose.

ISBN 0-89203-294-4

Acknowledgments
The figures for Shoulder and Arm, Hand and Wrist, Knee and Leg, Foot and Ankle, Spine, and Special Tests in Chapter 26 (Physical Examination) have been reproduced from Greene WB (ed): *Essentials of Musculoskeletal Care*, ed 2. Rosemont, IL, American Academy of Orthopaedic Surgeons, 2001.

The photographs for Hip and Thigh in Chapter 26 (Physical Examination) were taken by Cheryl Fort, Fort Photography, Inc.

Anatomy illustrations by Scott Thorn Barrows, CMI, FAMI.

American Academy of Orthopaedic Surgeons Editorial Board

American Academy of Family Physicians Review Board

Allen Fongemie, MD
Health Partners, Inc.
St. Paul, Minnesota

Martha B. Johns, MD, MPH, FACPM
Instructor
Department of Family Medicine
Department of Preventive Medicine
University of Colorado School of Medicine
Denver, Colorado
Physician
Wardenburg Health Center
University of Colorado at Boulder
Boulder, Colorado

Michael W. McShan, MD, PhD
Diplomate of American Board of Emergency Medicine
Kilgore, Texas

American Academy of Pediatrics Review Board

David D. Aronsson, MD
Department of Orthopaedics and Rehabilitation
University of Vermont, College of Medicine
Burlington, Vermont

Peter D. Pizzutillo, MD
Orthopedic Department
St. Christopher's Hospital
Philadelphia, Pennsylvania

John F. Sarwark, MD
Professor of Orthopaedic Surgery
Northwestern University Medical School
Division of Pediatric Orthopaedic Surgery
Children's Memorial Hospital
Chicago, Illinois

Contributors

Jaimo Ahn, MD, PhD
Resident Physician
Department of Orthopaedic Surgery
University of Pennsylvania School of Medicine
Philadelphia, Pennsylvania

Donald S. Bae, MD
Harvard Combined Orthopaedic Residency Program
Massachusetts General Hospital
Boston, Massachusetts

Marshall L. Balk, MD
Clinical Assistant Professor
Department of Orthopaedic Surgery
University of Pittsburgh School of Medicine
Attending Physician
Western Pennsylvania Hand Center
Pittsburgh, Pennsylvania

Holly J. Benjamin, MD
Assistant Professor of Clinical Pediatrics and Surgery
Director of Primary Care Sports Medicine
University of Chicago
Chicago, Illinois

Joseph Bernstein, MD, MS
University of Pennsylvania School of Medicine
Veterans' Hospital
Philadelphia, Pennsylvania

Philip E. Blazar, MD
Assistant Professor, Harvard Medical School
Department of Orthopaedic Surgery
Brigham and Women's Hospital
Boston, Massachusetts

Michael Brage, MD
Assistant Professor of Clinical Orthopaedic Surgery
University of California, San Diego
San Diego, California

Joseph A. Buckwalter, MS, MD
Professor and Chair Orthopaedic Surgery
University of Iowa
Iowa City, Iowa

Douglas C. Burton, MD
Assistant Professor
Section of Orthopaedic Surgery
University of Kansas Medical Center
Kansas City, Kansas

John T. Campbell, MD
Assistant Professor
Johns Hopkins University School of Medicine
Department of Orthopaedic Surgery
Johns Hopkins Bayview Medical Center
Baltimore, Maryland

Hans L. Carlson, MD
Assistant Professor of Physical Medicine and Rehabilitation
Department of Orthopaedics and Rehabilitation
Oregon Health and Science University
Portland, Oregon

Nels L. Carlson, MD
Assistant Professor
Department of Orthopaedics and Rehabilitation
Oregon Health and Science University
Portland, Oregon

Creg A. Carpenter, MD
Chief Orthopedic Adult Reconstruction
Orthopedic Surgery
Wilford Hall Medical Center
Lackland Air Force Base, Texas

Raymond M. Carroll, MD
Clinical Instructor
Department of Orthopaedic Surgery
Georgetown University Medical Center
Washington, DC

Peter Foster Cronholm, MD
Clinical Instructor
Family Practice and Community Medicine
University of Pennsylvania School of Medicine
Philadelphia, Pennsylvania

Contributors

Nicholas A. DiNubile, MD
Orthopaedic Consultant,
Philadelphia 76ers Basketball
Clinical Assistant Professor
Department of Orthopaedic Surgery
University of Pennsylvania School of Medicine
Philadelphia, Pennsylvania

Robert K. Eastlack, MD
Resident Orthopaedic Surgery
University of California, San Diego
San Diego, California

John L. Esterhai, Jr, MD
Professor of Orthopaedic Surgery
University of Pennsylvania School of Medicine
Philadelphia, Pennsylvania

Worth W. Everett, MD
Clinical Instructor of Emergency Medicine
University of Pennsylvania School of Medicine
Philadelphia, Pennsylvania

Sarah Yuxing Fan, MD
University of Pennsylvania School of Medicine
Philadelphia, Pennsylvania

Kevin B. Freedman, MD
Co-Director, Section of Sports Medicine
Department of Orthopaedic Surgery
and Rehabilitation
Loyola University Medical Center
Maywood, Illinois

Andrew A. Freiberg, MD
Chief, Hip and Implant Unit
Massachusetts General Hospital
Boston, Massachusetts

Meral Gunay-Aygun, MD
Clinical Geneticist, Division of Medical Genetics
Assistant Professor of Medical Genetics in Pathology
Northeastern Ohio University College of Medicine
Children's Hospital Medical Center of Akron
Akron, Ohio

Kevin M. Guskiewicz, PhD, ATC-L
Associate Professor
Exercise and Sport Science, Orthopaedics
University of North Carolina
Chapel Hill, North Carolina

Robert A. Hart, MD
Director, Division of Spine Surgery
Assistant Professor
Department of Orthopaedics and Rehabilitation
Oregon Health and Science University
Portland, Oregon

James M. Hartford, MD
Assistant Professor
Department of Orthopaedic Surgery
University of Kentucky
Lexington, Kentucky

David J. Hill, PT
Clinical Operations Director
NovaCare Rehabilitation
Philadelphia, Pennsylvania

Michael H. Huo, MD
Associate Professor
University of Kansas School of Medicine
Kansas City, Kansas

Mark R. Hutchinson, MD
Associate Professor of Orthopaedics and Sports
Medicine
Director of Sports Medicine Services
Department of Orthopaedics
University of Illinois at Chicago
Chicago, Illinois

Kerwyn C. Jones, MS, MD
Children's Orthopaedic Surgery Associates
Children's Hospital Medical Center of Akron
Akron, Ohio

Rosemary D. Laird, MD, MHSA
Assistant Professor of Medicine
Internal Medicine
University of Kansas
Kansas City, Kansas

Contributors

Jeffrey N. Lawton, MD
Hand and Upper Extremity Surgery
Department of Orthopaedic Surgery
The Cleveland Clinic Foundation
Cleveland, Ohio

Arabella I. Leet, MD
Assistant Professor
Johns Hopkins University
Division of Pediatric Orthopedics
Baltimore, Maryland

Seth S. Leopold, MD
Associate Professor
Department of Orthopaedics and Sports Medicine
University of Washington Medical Center
Seattle, Washington

Gregory N. Lervick, MD
Orthopaedic Surgeon
Department of Orthopaedics
Park Nicollett Clinic
St. Louis Park, Minnesota

William N. Levine, MD
Director, Sports Medicine
Associate Director, Center for Shoulder, Elbow, and Sports Medicine
Orthopaedic Surgery
Columbia-Presbyterian Medical Center
New York, New York

Carol B. Lindsley, MD
Professor and Chair
Department of Pediatrics
University of Kansas
Kansas City, Kansas

Herbert B. Lindsley, MD
Professor of Internal Medicine
Department of Internal Medicine
University of Kansas Medical Center
Kansas City, Kansas

Robert Lohman, MD
Assistant Professor
Surgery, Section of Plastic Surgery
University of Chicago
Chicago, Illinois

Catherine E. Lucasey, RN, MA
Nurse, Hiatt Osteoporosis Center
Department of Medicine
University of Kansas School of Medicine
Kansas City, Kansas

Barbara P. Lukert, MD
Professor of Medicine
Division of Metabolism, Endocrinology, and Genetics
Department of Internal Medicine
University of Kansas School of Medicine
Kansas City, Kansas

Scott D. Mair, MD
Assistant Professor
Section of Sports Medicine
University of Kentucky
Lexington, Kentucky

Henry J. Mankin, MD
Edith M. Ashley Distinguished Professor Orthopaedics
Harvard Medical School
Orthopaedic Oncology
Massachusetts General Hospital
Boston, Massachusetts

Jennifer Marler, MD
Assistant Professor of Surgery and Biomedical Engineering
University of Cincinnati
Attending Surgeon, Division of Plastic Surgery
Cincinnati Children's Hospital Medical Center
Cincinnati, Ohio

Joel L. Mayerson, MD
Chief, Division of Orthopedic Oncology
Assistant Professor of Orthopedic Surgery
Medical College of Ohio at Toledo
Toledo, Ohio

Contributors

Samir Mehta, MD
Department of Orthopaedic Surgery
University of Pennsylvania
Philadelphia, Pennsylvania

Jason J. Mickels, MD
Resident, Orthopaedic Surgery
Department of Surgery
University of Kansas Medical Center
Kansas City, Kansas

R. Alden Milam, IV, MD
Department of Orthopaedic Surgery
Case Western Reserve University School of Medicine
Cleveland, Ohio

Giuseppe Militello, MD
Department of Dermatology
University of Pennsylvania School of Medicine
Philadelphia, Pennsylvania

Van C. Mow, PhD
Stanley Dicker Professor of Biomedical Engineering
Chair, Department of Biomedical Engineering
Columbia University
New York, New York

Martha M. Murray, MD
Instructor, Harvard Medical School
Department of Orthopaedic Surgery
Brigham and Women's Hospital
Boston, Massachusetts

James A. Onate, MA, ATC-L
Research Associate
Sports Medicine Research Laboratory
Exercise and Sport Science
University of North Carolina at Chapel Hill
Chapel Hill, North Carolina

Craig S. Phillips, MD
Assistant Professor of Surgery
Reconstructive Hand and Upper-Extremity Surgery
The University of Chicago Hospitals
Chicago, Illinois

Brigham B. Redd, MD
Captain, United States Army Medical Corps
William Beumont Army Medical Center
Orthopaedic Surgery Service
El Paso, Texas

Donald Resnick, MD
Professor of Radiology
Chief, Osteoradiology Section
Department of Radiology, University of California
San Diego and Veterans Affairs Medical Center
San Diego, California

Sally K. Rigler, MD, MPH
Assistant Professor of Medicine
Department of Internal Medicine
University of Kansas School of Medicine
Kansas City, Kansas

Haynes Robinson, MD
Director of Medical Genetics in Pathology
Division of Medical Genetics
Children's Hospital Medical Center of Akron
Akron, Ohio

Philip E. Rosen, MD
Associate Professor, Orthopedic Surgery
Department of Orthopedic Surgery
Louisiana State University
New Orleans, Louisiana

Jonathan Scherl, MD
Englewood Knee and Sports Medicine
Englewood, New Jersey

Susan A. Scherl, MD
Assistant Professor of Pediatric Orthopaedics
Department of Orthopaedics and Rehabilitation
University of Nebraska
Omaha, Nebraska

Sonya Shortkroff, PhD
Instructor, Harvard Medical School
Department of Orthopedic Surgery
Brigham and Women's Hospital
Boston, Massachusetts

Contributors

Jeffrey D. Stone, MD
Division of Hand, Upper Extremity and Microvascular Surgery
Florida Orthopaedic Institute
Tampa, Florida

Vishwas R. Talwalkar, MD
Assistant Professor, Division of Orthopaedic Surgery
University of Kentucky and Shriners Hospital for Children
Lexington, Kentucky

Kimberly Templeton, MD
Associate Professor of Orthopaedic Surgery
Section of Orthopaedic Surgery
University of Kansas Medical Center
Kansas City, Kansas

Daphne J. Theodorou, MD
Musculoskeletal Radiologist
Department of Radiology, University of California
San Diego and Veterans Affairs Medical Center
San Diego, California

Stavroula J. Theodorou, MD
Musculoskeletal Radiologist
Department of Radiology, University of California
San Diego and Veterans Affairs Medical Center
San Diego, California

Brian C. Toolan, MD
Assistant Professor
Department of Surgery, Section of Orthopaedic Surgery and Rehabilitation Medicine
The University of Chicago
Chicago, Illinois

Carole S. Vetter, MD
Assistant Professor Orthopaedic Surgery
Medical College of Wisconsin
Milwaukee, Wisconsin

Dennis S. Weiner, MD
Chairman, Department of Pediatric Orthopaedic Surgery
Children's Hospital Medical Center of Akron
Professor of Orthopaedics
Northeastern Ohio University College of Medicine
Akron, Ohio

Brent B. Wiesel, MD
Department of Orthopaedic Surgery
University of Pennsylvania School of Medicine
Philadelphia, Pennsylvania

Robert H. Wilson, MD
Assistant Professor
Division of Orthopaedic Surgery
Howard University
Washington, DC

Robert J. Winn, MD
Clinical Instructor
Department of Family Medicine
Thomas Jefferson University
Philadelphia, Pennsylvania

A.J. Yates, Jr, MD
Associate Professor
Department of Orthopaedics
University of Rochester Medical Center
Rochester, New York

Reviewers

Edward Abraham, MD
Professor and Department Head
Department of Orthopaedics
University of Illinois at Chicago Medical School
Chicago, Illinois

Louis M. Adler, MD
New England Orthopedic Surgeons
Springfield, Massachusetts

Frederick M. Azar, MD
Campbell Clinic
Germantown, Tennessee

Pedro Beredjiklian, MD
Assistant Professor of Orthopaedic Surgery
University of Pennsylvania School of Medicine
Philadelphia, Pennsylvania

Martin I. Boyer, MD
Department of Orthopaedic Surgery
Washington University School of Medicine
St. Louis, Missouri

Joseph A. Buckwalter, MS, MD
Professor and Chair
Department of Orthopaedic Surgery
Orthopaedic Oncology
University of Iowa Hospitals and Clinics
Iowa City, Iowa

Richard H. Gelberman, MD
Department of Orthopaedic Surgery
Washington University School of Medicine
St. Louis, Missouri

Sherwin Ho, MD
Orthopaedic Surgery
University of Chicago
Chicago, Illinois

Craig L. Israelite, MD
Assistant Professor of Orthopaedic Surgery
University of Pennsylvania School of Medicine
Philadelphia, Pennsylvania

Douglas W. Jackson, MD
Medical Director, Orthopaedic Research Institute
Memorial Medical Center
Long Beach, California

Alan M. Levine, MD
Director, Alvin and Lois Lapidus Cancer Institute
Head, Division of Orthopedic Oncology at Sinai Hospital
Clinical Professor
Orthopaedic Surgery and Oncology
University of Maryland School of Medicine
Baltimore, Maryland

Courtland Lewis, MD
Orthopaedic Associates of Hartford
Hartford, Connecticut

Josh Miller, MD, PhD
Orthopaedic Research Laboratories
University of Michigan
Ann Arbor, Michigan

Jeremy K. Souder
University of Pennsylvania School of Medicine
Philadelphia, Pennsylvania

Andrew Urquhart, MD
Orthopaedic Surgery
University of Michigan
Ann Arbor, Michigan

Brian Victoroff, MD
University Orthopaedic Associates, Inc.
Department of Orthopaedic Surgery
Case Western Reserve University
Cleveland, Ohio

To my chairmen,
Carl Brighton, Joe Iannotti, Marvin Steinberg,
and Bob Fitzgerald, who got me started;
and
To my family,
Kirsten, James, Jillian, and Jacob,
who keep me going.

Acknowledgments

Samuel Johnson once said, "no man but a blockhead ever wrote except for money." Samuel Johnson was wrong. This book is the product of the volunteer efforts of dozens of writers—not one blockhead among them, but none paid a penny either. Needless to say, without their hard work, this book would not have been possible. I am grateful to them for the quality of their contributions and the grace with which they allowed me and the section editors to alter their prose to fit the larger purpose of the book.

The professional editing staff at the AAOS deserves special mention and thanks as well. Unlike the contributing writers, they are paid—that's what the word "professional" means, after all—but to none of them was this simply a job. Lynne Shindoll, an early champion of this project, lent her considerable organizational skills and fine taste to the composition of the book throughout. Keith Huff, our senior editor, has pored over each word in this book at least five times. Through each iteration, he added rigor and clarity. If holes remain, it is only because Keith did not have a sixth pass through the book; the faults are mine. Mary Steermann, Dave Stanley, Sophie Tosta, Mike Bujewski, and Dena Lozano maintained the standards of AAOS book publishing—a high compliment indeed. I also thank Marilyn Fox, PhD, and Drs. Martin Boyer, Doug Jackson, Gary Friedlaender, and Alan Levine for subtle help at vital times.

My approach to teaching, writing, and the subject of musculoskeletal medicine in general is based in large part on the lessons of my teachers over the last 20 years. I can identify the gifts I received from the following professors: Dan Alonso, Andrew Cooper, John Cuckler, Bill DeLong, John Dormans, John Duda, John Esterhai, Bill Frayer, Ugo Gagliardi, Wallace Gray, Wilbur Hagamen, Joe Iannotti, Fred Kaplan, Joe Lane, Paul Lotke, Powers Peterson, Alex Sapega, and Marvin Steinberg. I am sure there are others whose mark appears in this book, and I apologize for not naming them.

Some of my best teachers have been students, colleagues, and family members, including Bruce Abramson, Jaimo Ahn, Lou Adler, Anita Bernstein, Sally Bernstein, Pedro Beredjiklian, Dan Jacob, Kevin Freedman, Gerhard Holt, and Tim Johnson. Their influence is strongly felt.

I appreciate the enthusiastic students of the University of Pennsylvania School of Medicine, who have been willing subjects as I experimented with various teaching methods.

The Veterans' Hospital has been a consistent source of support. My debt is to John Esterhai and Bob Fitzgerald for making medical student education part of my mission there.

Last, I thank you, the reader, for your indulgence in what I hope will be seen as a work in progress. In order to help us plan subsequent editions, please send the enclosed comment card to *Musculoskeletal Medicine*, AAOS Publications Department, 6300 River Road, Rosemont, IL, 60018 (shindoll@aaos.org). You may also send your comments and criticisms to me directly at orthodoc@post.harvard.edu.

Joseph Bernstein, MD, MS

Table of Contents

Section 1: Biology

Section Editors: Kerwyn C. Jones, MD, MS, and Martha M. Murray, MD

Section 2: Anatomy

Section Editor: Phillip E. Blazar, MD

Section 3: Disorders

Section Editors: Susan A. Scherl, MD, and Kimberly Templeton, MD

Section 4: Clinical Evaluation

Section Editors: Kevin B. Freedman, MD, and Robert A. Hart, MD

Section 5: Management

Section Editor: Marshall L. Balk, MD

Preface

The idea for this book took root on a chilly Sunday in April 1990 as I was nearing the end of my internship. On this particular day, I got a call from a medical resident working in the emergency department asking me to see a patient with a "fibia" fracture. The patient, it turned out, had only an ankle sprain, a common and not particularly thought-provoking injury. What got me thinking, though, was that the doctor who called me—a sharp guy, who today is probably chief of cardiology somewhere—did not know the names of the long bones.

Years later, Dr. Kevin Freedman and I administered a basic competency examination in musculoskeletal medicine to all of the interns at our institution. Using a passing threshold set by directors of orthopaedic surgery and internal medicine residencies, we found that my colleague in the emergency department was not alone in his lack of mastery: a vast majority of the interns failed the test. Further work showed that nearly half of American medical schools did not require a course in musculoskeletal medicine. A theme emerged: medical students were allowed to graduate without sufficient instruction in musculoskeletal medicine, and, not surprisingly, many of them did not know enough. This book was written to help solve that problem.

Musculoskeletal Medicine is a book for students. It attempts to present an overview and to help place concurrent observations, readings, and clinical experiences in perspective. It aims to serve as a springboard to further study and application. There are gaps in this book, no doubt, but they are unavoidable. To make things manageable, some detail had to be sacrificed. (The Argentinean writer Jorge Luis Borges described a library that contains everything—a flawed library, of course, because it was the size of the universe itself.) I assume that all readers own—or at least have access to—an atlas of anatomy and good textbooks of pathology, physiology, internal medicine, and general surgery. My hope is that you will read this book, but not only this book.

Musculoskeletal Medicine addresses a topic of great importance. Obviously, there is an element of professional pride at work when I say that, but it is a fair claim nonetheless. The simplest reason is the large number of people with musculoskeletal conditions. Granted, few people die of bone and joint diseases, yet many people do not enjoy life to the fullest because of them. Thus, for students who went to medical school to help people, mastering musculoskeletal medicine will be an essential task.

Mastering musculoskeletal medicine will be easy once you get the hang of it. I say this from the perspective of one who nearly failed out of medical school. My performance in biochemistry as a first-year student set ignominious records that probably still stand. But fortunately I had the chance later to work with teachers who not only taught the material well but also taught effective methods of learning. I hope you will find that this book carries that tradition forward. Some of the material presented may be a bit simplified—too simplified for experts in the particular field. To them, I apologize. To everyone else, I invite you to use the fundamentals presented here to acquaint yourself with this interesting and important field of medicine. Of course, I hope you won't stop with only the basics if your interests take you further. But please, at the very least, learn the names of the long bones!

—Joseph Bernstein, MD, MS

National Bone and Joint Decade, 2002–2011

By the President of the United States of America

A Proclamation

Living a life free from daily bone pain or joint discomfort is something most people take for granted. Our bones, joints, and connective tissues are the structure upon which all other systems of the body depend. They give us strength, mobility, protection, and stability. And they permit us to perform a great variety of physical activities that shape our daily lives.

Our musculoskeletal structure is a complex system of tissue and bone that is regularly subjected to trauma, metabolic and genetic processes, and the gradual wear and tear of an active life. When these bones and tissues become damaged or diseased, they can create chronic conditions that may seriously impede and sometimes permanently affect one's health and well-being.

In the United States, musculoskeletal disorders are a leading cause of physical disability. Conditions such as osteoporosis, osteoarthritis, rheumatoid arthritis, back pain, spinal disorders, and fractures, also affect hundreds of millions of people around the world. And many children suffer from crippling bone and joint diseases and deformities, impeding normal development and preventing them from experiencing a full and healthy life.

The incidence of musculoskeletal conditions will increase as the average age of our population increases. And our culture's increasing emphasis on physical activity, while important to society's overall well-being, will also increase the stress factors on bones and joints. Ensuing disorders, if left untreated, could result in significant pain and suffering that would affect employment, well-being, and healthcare costs.

National Bone and Joint Decade, 2002-2011, envisions a series of international initiatives among physicians, health professionals, patients, and communities, working together to raise awareness about musculoskeletal disorders and promoting research and development into therapies, preventative measures, and cures for these disorders. Advances in the prevention, diagnosis, treatment, and research of musculoskeletal conditions will greatly enhance the quality of life of our aging population.

The National Institutes of Health, the National Institute of Arthritis and Musculoskeletal and Skin Diseases, and other Federal agencies support many bone and joint studies. Industry and private professional and voluntary agencies support other initiatives. This work involves scientists examining the possible genetic causes of bone and joint diseases and studying how hormones, growth factors, and drugs regulate the skeleton. Other researchers are studying bone density, quality, and metabolism, and other ways to increase the longevity of joint replacements for those whose daily activities have become painful, difficult, or even impossible. These research efforts can help relieve pain and suffering and give countless children and adults the opportunity for a better life.

Thanks to the hard work of these dedicated researchers, we have made great progress in understanding and treating musculoskeletal disorders. I commend their efforts and encourage them to pursue diligently further research that will help those suffering from these disorders. And I hope that all Americans will learn more about musculoskeletal problems, their long- and short-term effects, and the therapies and treatments available to help them.

NOW, THEREFORE, I, GEORGE W. BUSH, President of the United States of America, by virtue of the authority vested in me by the Constitution and laws of the United States, do hereby proclaim the years 2002–2011, as National Bone and Joint Decade. I call upon the people of the United States to observe the decade with appropriate programs and activities; and I call upon the medical community to pursue research in this important area.

IN WITNESS WHEREOF, I have hereunto set my hand this twenty-first day of March, in the year of our Lord two thousand two, and of the Independence of the United States of America the two hundred and twenty-sixth.

Foreword

I left the hospital one evening and walked through a narrow passageway toward the parking garage. A large trash dumpster was blocking my path. As I squeezed by, I looked inside the dumpster and saw, to my astonishment, hundreds of old medical textbooks. I telephoned the hospital librarian at home to inquire about this mysterious and troubling disposal and was informed that "there is simply no room in the library for old textbooks, and besides, they are hopelessly out of date." "But they're textbooks!" I exclaimed. "If you want them, you may take them," she said. "They are going to the dumps."

A textbook is a fossil of knowledge, and it is important not to be mesmerized by the apparent certainty of the knowledge it contains. Some of that knowledge will eventually be proven wrong, and much of the rest will appear quaint, if not downright foolish, over time. An old textbook is a subtle reminder that today's textbooks, although seemingly fresh and new, will eventually become relics of the past.

If much of the knowledge within a textbook will eventually become obsolete, does that mean we should not bother learning it? No indeed. As Banquo, one of the generals of the King's armies, said in Act I of Shakespeare's *Macbeth*, "If you can look into the seeds of time, and say which grain will grow and which will not, speak."

If every student of science and medicine who read a textbook at the turn of the last century had been content with the knowledge it contained, we would not now have a polio vaccine, total joint replacement, antibiotics, recombinant growth hormone, bone morphogenetic protein, or the promise of stem cell therapy and gene therapy—treatments that have revolutionized the field of musculoskelatal medicine. Textbooks, clearly, are not the last word.

The first textbook of musculoskeletal medicine, one might say, was written by Nicolas Andry in 1740. Nicolas Andry was born in Lyon, France, and became a professor of theology. At the age of 32, he left the clergy to study medicine. He became Professor of Medicine and eventually Dean of the Faculty of Medicine in Paris. As Dean, he launched a vitriolic attack on barber surgeons, and through that farsighted but unpopular move, lost the Deanship.

Andry died at the age of 84. It was a project taken up in his 83rd year that left an indelible mark on the history of medicine. He wrote, in that last year of his life, a book, the title of which, *Orthopaedia* is engraved in our vocabulary. "As to the title," he wrote, "I have formed it of two Greek works, *orthos* which signifies straight, free from deformity, and *paedios*, a child. Out of these two words, I have compounded that of *orthopaedia* to express in one term the design I propose, which is to teach the different method of preventing and correcting the deformities of children." Thus, in the very title and opening paragraphs of his prophetic little textbook, he planted a seed for the future. From that seed, a crooked little tree of great fame has grown, as we shall see.

Many of the treatments that Andry proposed were, in fact, weak and devoid of value. But the book offers much more. In the middle, we find a simple drawing that is so charming and advanced, so far reaching, that we can forgive the author of any transgression on our language or our thought.

On the basis of one illustration, that of a charming crooked little tree, our knowledge takes root and our imagination soars. Andry uses the metaphor of a crooked little tree to illustrate a critical point. He writes: "The same method must be used for recovering the shape of the leg, as is used for making straight the crooked trunk of a young tree." With those words, he set the world of science and medicine on a search for a more profound understanding of how physical factors affect living systems.

By the mid-1800s, many scientists and physicians had written on the relationship between bone morphology and applied load. In 1892, the physician Julius Wolff of Berlin, wrote a treatise entitled *The Law of Bone Transformation*. Wolff summarized 30 years of experiments and proposed a new theory that described how every change in the function of a bone is followed by certain definite changes in morphology. Wolff's work was so immediately popular that by the time of his death in 1902, the scientific community referred to the effect of load on bone mass as Wolff's law. The architect, Lewis Sullivan, familiar with Wolff's work, adopted this principle in his famous theorem: "Form follows function."

Theories abound on the biological effects of physical stimuli—positing roles for prostaglandins, electricity, calmodulin, integrins, and more. But the transduction of a mechanical signal to a biological message remains a mystery even today.

A little more than a decade ago, the scientific world gained a fugitive glimpse of how a physical strain is converted into a molecular response. The article appeared in the journal *Cell* with a cover caption reading, "Plant Growth Response To Touch." In the accompanying article, the authors described how touch induces the expression of four calmodulin-related genes in plants. This work suggested that calcium ions in calmodulin were involved in the transduction of mechanical signals and provided some of the first clues at the molecular level of how a complex living organism responds to a mechanical stimulus in its environment.

The opportunities provided by our molecular tools will undoubtedly allow us to decipher the puzzle further. We are certain to gain a better understanding of the factors controlling bone morphology: the genetic program, the humoral factors, the applied load, and most importantly, the complex molecular signaling pathways that provide the integration. When we do, we may be able to accomplish what Nicolas Andry envisioned in the preface to his famous little book a quarter of a millennium ago: the art of correcting and preventing the deformities in children.

Leaving Andry, a brief stop in time takes us back about 2,000 years, to the Isle of Cos in ancient Greece. There, we find Hippocrates, the Father of Medicine, remembered today for his Oath, but an accomplished scholar of musculoskeletal conditions as well. His method of reducing shoulder dislocations is still employed today. From our visit with Hippocrates, we take a major leap back to a time about 20,000 years ago. We find ourselves staring at the brilliant white calcite walls of the Hall of Bulls in a cave in South-central France near the present day village of Lascaux. There, on a grand scale, a Paleolithic artist painted images of the animals that he or she knew. Among the images is a depiction of an early human figure in what one might imagine to be a supremely traumatic context. So, when did musculoskeletal medicine begin?

Our next stop back in time requires a truly giant leap—65 millions years. For a moment, imagine yourself on the plains of what is now South Dakota with a very old creature by the name of Sue. Sue has sustained an open fracture of her left fibula and many closed fractures of the right ribs in a terrible accident. Sue is a dinosaur, and not just any dinosaur, but one of the largest Tyrannosaurus rex dinosaurs to have ever lived. Sue has osteomyelitis of the left fibula and multiple healing fractures of the right ribs.

In reality, Sue stands motionless before us in the Field Museum of Natural History in Chicago, on the same continent where she roamed 65 million years ago. What tickles us is that long ago, Sue knew how to heal a fracture, and her body knew how to react to a bone infection. The scars of her skeleton are a testimonial to that undeniable fact. Here we are, the beneficiaries of the most advanced educational systems that have ever existed on the planet, and we still do not know exactly how fractures heal or how the body fights a bone infection! But Sue knew because she did it!

Science is a recent invention of human beings, a method of inquiry that provides a structured discipline to our curiosity. Science can answer questions, but it cannot ask them. As students of medicine, questions may occur to us in the clinic today (or in a virtual clinic from 65 million years ago that we can visit tomorrow in a museum courtyard in Chicago). As William Osler, the famous nineteenth century physician said, "Clinics are laboratories, laboratories of the highest order."

The clinic—the laboratory of the highest order—will be where you can learn things that assuredly will pass the test of time: the skills of physical diagnosis. The importance of physical diagnosis was best exemplified at a high-powered scientific conference that I attended several years ago on the molecular genetics of the skeleton. A young molecular biologist, who had been working on the developmental biology of the skeleton, dazzled the audience with her beautiful work on a genetically engineered mouse in which she had successfully deleted both alleles of a gene not previously known to be involved in the formation or maintenance of the skeleton. The scientist mesmerized the audience with her methods of molecular biology. At the end of the talk, however, all of the questions that followed (not some of them, but all of them) pertained to the clinical findings (or phenotype) of the mouse. None of the questions pertained to the molecular genetics or to the experimental methods. It was as if we were in a mouse clinic and curious mouse doctors were asking questions about the physical examination of the interesting mouse patient.

As we begin to mine the bounties of the human genome, our skills of physical diagnosis will become even more important. They will have to become more important as we struggle to integrate the emerging knowledge of genetic variation into a useful clinical context and as we strive to understand the often subtle and nebulous boundaries of health and disease. So instead of being relegated to obsolescence, like yesterday's textbooks, our observational and integrative skills of physical diagnosis will likely become even more important over time. As Albert Einstein said, "Concern for man himself and his fate must always form the chief interest of all endeavors. Never forget this in the midst of diagrams and equations."

Let's review our travels back in time. Our visit to the apartment of Dr Nicolas Andry let us eavesdrop on the creation of the field of orthopaedics. Our meeting with Hippocrates suggested that there were doctors of musculoskeletal medicine long before Andry. Our encounter with the caveman suggested that musculoskeletal concerns were present at the dawn of human history. Our visit with Sue taught us that reality precedes knowledge. Our lecture at the molecular genetics conference inspired us that the skills of physical diagnosis may become even more relevant over time.

The next story teaches us that one never knows when the future will intrude to question the present or possibly just how unnoticeable the future may be when it first appears.

The time was 30 years ago. The place was a classroom in Baltimore. I was a second-year medical student struggling to learn microbiology. One morning, the professor teaching the course strayed from the outline to tell us about the laboratory experiments he had been conducting on some obscure enzymes in bacteria. He mentioned that these enzymes cleaved the DNA in predictable patterns, and then he used them to cut DNA into predictable pieces. He then showed us a primitive map from another organism using the enzymatic tools that he had isolated from various strains of bacteria. None of the students could see much immediate use for this. After all, we were studying to be physicians to care for patients. What possibly could be the use of some obscure concoction of enzymes that could chop up DNA in a test tube?

Our professor's comments provoked little interest, and we were all relieved when he got back on track and followed the textbook that he had assigned us to read. There was no reference to the professor's curious digression on the final exam, and we heard no further mention of his arcane experiments during the remainder of our time in medical school. We all passed the microbiology exam, entered the clinics, and graduated as medical doctors soon thereafter.

Several years later, I turned on the television to listen to the news and saw the face of my microbiology professor prominently displayed. With that, the TV news anchorman announced that the microbiologist had just been awarded the Nobel Prize in medicine. I was truly shocked! For what could he have possibly won the Nobel Prize in medicine? Surely, it was not for teaching a microbiology class to medical students!

The next morning, I bought a newspaper and read the citation of the Nobel Prize Committee about my professor's work on restriction endonucleases, those curious little enzymes from bacteria that he had told us about years before. Still the importance and implication of the message eluded me.

Fifteen years later, as a medical faculty member, I was having lunch with a medical student and listening to him explain the evolution of molecular biology during the past 2 decades. I knew very little about molecular biology and felt like Rip Van Winkle awaking from a long sleep. One of the highlights that the student mentioned was the discovery of restriction endonucleases and their ability to cleave DNA in predictable patterns. After about 5 minutes of discussion, the lights went on! It became obvious that little could be accomplished in molecular biology without that discovery. Somehow it had bypassed me in medical school, throughout my residency, and throughout my early years of practice. Somehow it had bypassed all of us who sat in the classroom that day in medical school. But it did not bypass the Nobel Prize Committee or the next generation of medical students.

On that morning in medical school 30 years ago, I had no idea that I was witnessing the birth of a new field of human inquiry—a field that would change the world. It was the one topic in the entire microbiology course that our professor took directly from his lab notebook and brought to the classroom. It was the only topic that he discussed with us in the entire microbiology course that was not in the textbook, and it was one of the only topics that he ever discussed with us that would have any direct or long-lasting importance to our lives and the lives of our patients. It is in the textbooks now. Even more importantly, there are people alive today (who otherwise would not be) because of it.

I leave you with ten untextbook-like thoughts:

1. Clinics are laboratories—laboratories of the highest order.

2. Physical diagnosis skills are timeless. Learn them, practice them, and teach them to others.

3. The greatest discoveries are made not within a field but at the boundaries of a field and

within other fields. Step outside and glimpse the future.

4. One never knows when the future will intrude to question the present. When the discussion strays from the text, pay attention.

5. Don't expect to learn everything, but try hard to learn something. Keep asking questions.

6. If you want to be a top practitioner of musculoskeletal medicine, try first to be a knowledgeable doctor. A broad knowledge of internal medicine, pharmacology, pediatrics, radiology, neurology, epidemiology, genetics, molecular biology, and psychiatry will also be relevant—along with many other fields not mentioned. Likewise, subscribe to a general medical journal and read it.

7. Knowledge alone is not enough. Caring is part of the cure.

8. Learn to communicate and take time to communicate well. When the patient returns home from a visit to your clinic or office, the family will most certainly ask: "What did the doctor say?"

9. There isn't a condition known to man that surgery can't make worse.

10. Old textbooks in a dumpster may be a sign of progress.

—Frederick S. Kaplan, MD
Isaac and Rose Nassau Professor of Orthopaedic Molecular Medicine
University of Pennsylvania School of Medicine
Hospital of the University of Pennsylvania
Philadelphia, Pennsylvania

Section Editors
Kerwyn C. Jones, MS, MD
Martha M. Murray, MD

Section 1

Biology

Contributing Authors
Joseph Bernstein, MD, MS
Joseph A. Buckwalter, MS, MD
Creg A. Carpenter, MD
Andrew A. Freiberg, MD
Henry J. Mankin, MD
Jennifer Marler, MD
Joel L. Mayerson, MD
R. Alden Milam, IV, MD
Van C. Mow, PhD
Martha M. Murray, MD
Sonya Shortkroff, PhD

Biology

Approaching basic science can be daunting. Without clinical experience to offer perspective, the novice student must acquire, assimilate, and apply information within a near vacuum. And yet it must be done. As the word "basic" implies, this is a body of fundamental knowledge that must be mastered before you move on to advanced topics. So how can one effectively master basic science?

The best first step is reading. Some educators recommend reading a little every day; others advocate reading in continuous thematic chunks. Whichever method you choose, the key is to make reading a constant part of your learning schedule, even when clinical responsibilities dominate your days. An excellent first source is a textbook that provides a good general overview supported by illustrative details. This should then be supplemented with more specialized texts and journal articles, particularly reviews.

Reading is only the first step; not all will be clear after the first pass through the material or even a second. The goal of reading is to give yourself a "prepared mind" (as Louis Pasteur put it), to make you ready for the real learning encounter: taking care of patients. As you progress through your clinical training, form connections between patient scenarios and basic scientific causes. After examining patients, reinforce your clinical observations by rereading basic science material. You may, for example, see a patient with osteoarthritis of the knee. Observe that he is slightly bowlegged (a varus deformity), and note that his radiograph shows a white line in the tibia near the joint on the medial side, a finding that indicates subchondral sclerosis. Your reading will remind you that radiographs record density and that this white line therefore indicates an area of additional bone formation. Your basic science reading will also remind you of Wolff's law: that bone grows in response to mechanical stress. Consequently, by recognizing the connection between varus alignment, abnormally high loads, and new bone formation, you can infer that the appropriate treatment for isolated medial compartment arthritis may be to unload that stress with a brace or surgical realignment of the bone perhaps. As this example illustrates, clinical observations reinforce basic science reading, which in turn helps explain the clinical findings.

How much should you memorize? It's hard to say. Because medicine is a practical art, you must collect facts. But go beyond that. Strive to discern relationships among facts and thereby identify patterns. For a useful and satisfying mental exercise, write a list of four or five topics on a blank sheet of paper and attempt to establish meaningful relationships among the entities based on scientific or clinical principles. Try, for instance, this list: osteoporosis, stress fractures, compartment syndrome, and long bone fractures. After reading, you may see the connection between osteoporosis and stress fractures (both are characterized by perturbations of bone remodeling); stress fractures and compartment syndrome (both are conditions that can cause leg pain); compartment syndrome and long bone fractures (compartment syndrome being a dreaded complication of fractures, typically from bleeding into the soft tissues); and last, fractures and osteoporosis (osteoporosis is a prime cause of fractures, especially of the hip, wrist, and spine). Challenging yourself to make these connections will enhance your comprehension.

You may ask yourself the relevance of the basic science. A glib response is that your enlightenment begins when relevance is no longer in question. But until that point, it is fair to wonder whether the study of organs, cells, and molecules has much to do with clinical practice. The short answer is that almost every major improvement in prevention, diagnosis, and treatment has come from the translation of basic science into clinical medicine. So tackle the basic science section of this book with gusto. Even if you don't fancy yourself a scientist, the basic science you study today may well be the basis of the clinical science you apply later in your career.

—Jaimo Ahn, PhD

—Joseph Bernstein, MD, MS

Chapter 1

Bone

Bone is living, dynamic tissue that affects three main functions of the body: skeletal homeostasis (providing structure to house the internal organs as well as a system of pulleys and levers to move the body), mineral homeostasis (storing and releasing ions), and hematopoiesis (accommodating the machinery of blood cell formation). As living tissue, bone is in a state of continuous flux and renewal; osteoclasts resorb bone, and osteoblasts create new bone. This process allows ions to be released and stored. It also repairs areas of structural damage.

Most bone forms according to the cartilage model, a process called endochondral ossification. The process of bone growth and development, therefore, is one of transformation from a cartilage skeleton in utero to one made entirely of bone in the adult. The pediatric skeleton contains remnants of this cartilage skeleton, the cartilaginous growth plate or physis. The physis is active metabolically and allows the pediatric skeleton to grow longitudinally; however, it is also structurally weaker than bone, which makes the pediatric skeleton susceptible to a class of diseases and injuries not seen in the adult.

Bone Formation

Bone forms by one of two mechanisms: endochondral or intramembranous ossification. Endochondral ossification is the most common form, producing the long bones; "long" refers to the longitudinal orientation of the bone. (Even the distal phalanx of the fifth toe is a long bone.) The flat bones, such as the skull and clavicle, form via intramembranous ossification. The key distinction between the two processes is that endochondral ossification employs a cartilage template whereas intramembranous ossification does not.

Morphogenesis, or the development of the skeleton in utero, begins in the fourth week of development, at which time limb buds form in the embryo. When the fetus reaches the sixth week of development, mesenchymal cells (cells from the middle of the three germ cell layers) appear within the central region of the limb buds. These cells then differentiate into cartilage-forming chondroblasts, which elaborate a cartilage template of the bony skeleton. After these cells die, blood vessels migrate into the mass of matrix, bringing the osteoblasts, or bone-forming cells. Osteoblasts synthesize and secrete a collagen-based material called osteoid; this organic bone matrix then becomes calcified.

A region of calcified cartilage called the primary center of ossification forms midway between the two ends of the long bone. Calcification begins here and continues longitudinally toward the ends of the long bone, a process called interstitial growth, which takes place entirely within the cartilage model.

As interstitial growth continues, the primary ossification center enlarges, migrating toward but never ultimately reaching the ends of the bone. At the edge of the advancing primary ossification center, a specialized structure, the physis, or growth plate, forms (Fig. 1). The physis remains cartilaginous throughout adolescence to allow longitudinal growth of the maturing skeleton.

Just beyond the physeal growth plate, in the regions of the long bones farthest away from the primary ossification center, lies the chondroepiphysis. This area will become a secondary center of ossification. Some bones, such as the proximal and distal humeri, have several secondary centers of ossification that appear in a sequential manner during childhood and then fuse just before skeletal maturity to yield the adult form of the bone.

The primary and secondary ossification centers on either side of the physis define the following regions of the formed bone: the diaphysis (shaft), representing the primary center; the metaphysis, the flared region just

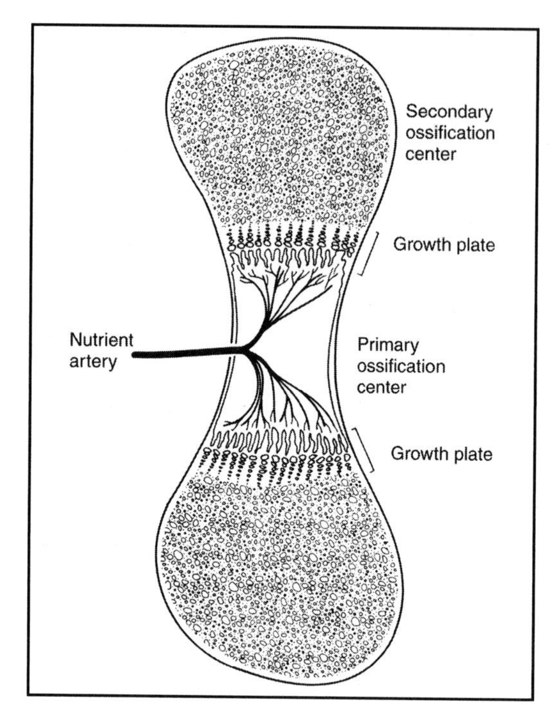

Figure 1
The primary ossification center stands at the midportion of the developing bone. Toward each end lies a growth plate (physis) and beyond it, a secondary ossification center will develop.

(Reproduced with permission from Brighton CT: The growth plate. Orthop Clin North Am 1984;15:571.)

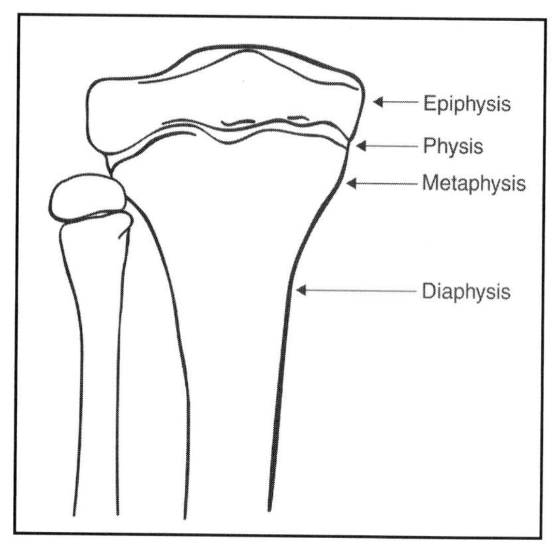

Figure 2
Schematic representation of the regions of the long bones in the tibia.

below the physis; and the epiphysis, which is located across the physis just below the articular cartilage lining the joint (Fig. 2). Within the diaphysis, a central core called the medullary canal is hollowed out by resorption, creating a cavity for the future home of the blood-forming cells.

As interstitial growth lengthens the bone, a parallel process of appositional growth widens the bone. Appositional growth takes place under the periosteum, a sleeve of connective tissue surrounding the shaft of the bone. The periosteum is a thick and distinct membrane in children but remains present in an attenuated form throughout life. It is the source of cells used for fracture healing and the source of some new bone formation.

Endochondral ossification continues long after the skeleton ceases longitudinal growth. Instead of stopping the healing process at the scar formation stage (as most tissues are apt to do), the repair of bone continues via endochondral ossification, which converts cartilaginous scar into new bone. Thus, it is quite possible that many years after a long bone fracture, no structural evidence of the original injury exists.

Growth Plate Anatomy and Function

At the physis, chondrocytes proliferate, mature, and secrete extracellular matrix, which eventually ossifies. The physis is polar: moving away from the epiphysis and joint surface toward the metaphysis, there is increasing maturation of the cells. Just below the epiphysis is a zone of immature resting cells; just above the metaphysis, calcification takes place in the setting of programmed cell death (apoptosis) of the hypertrophic chondrocytes. New bone formation thus occurs at the point farthest from the articular surface, at the junction between the physis and the metaphysis. This region is named the zone of provisional calcification.

The physis itself is divided into three major zones: the reserve zone, the proliferative zone, and the hypertrophic zone (Fig. 3). Each one of these zones has a distinct histology directly related to its function. The zones are ordered such that metabolic activity and maturation increases the farther the cells are located from the joint surface.

Closest to the surface (ie, nearest the epiphysis) is the reserve zone. The cells in this zone produce cartilaginous matrix, primarily in the form of type II collagen, which is used for eventual ossification into bone. These cells do not actively divide, nor are they very metabolically active; accordingly, they have the poorest blood supply.

Below the reserve zone, moving toward the metaphysis, is the proliferative zone. Within this zone, the cells are stacked in columns. These columns of cells synthesize proteoglycans, thereby contributing to the extracellular matrix. This region, characterized by synthesis and cell division, has the most extensive blood supply within the growth plate. This blood supply provides nutrition, of course, but also allows hormonal signals to reach their targets in the growth plate effectively.

The third zone of the growth plate, the hypertrophic zone, lies closest to the calcified bone of the metaphysis. The cells in this zone are unusually large and plump; hence, the name "hypertrophic." This zone of the growth plate is highly active metabolically, even though it has a poor blood supply. It therefore relies on anaerobic metabolism and uses stored glycogen as its source of energy. It is also the region that participates in mineralization of the cartilage. Calcium is stored in the cells in the upper levels of the hypertrophic zones. In the lower levels of the hypertrophic zone, these cells liberate their calcium in order to assist in matrix mineralization.[1]

The actual process of matrix mineralization occurs at the interface between the hypertrophic zone and the metaphysis: the zone of provisional calcification. Within this area, vascular invasion allows osteoblasts to arrive and replace the calcified cartilage with bone. This bone is a primitive, less-organized form called woven bone. In time, this tissue will be replaced by mature lamellar bone via the process of bone remodeling. The distinction between woven and lamellar bone is one of material orientation: the fibers of woven bone are haphazard, whereas lamellar bone aligns the structure in the direction of load.

Blood does not flow easily through the physis; the intramedullary blood supply does not reach the epiphysis or secondary centers of ossification. Accordingly, in children, whose growth plates are open, the main blood supply to the epiphysis is a direct epiphyseal artery. This artery loses its prominence once the growth plates close at skeletal maturity.

Numerous circulating hormones affect growth plate activity. Thyroid hormone (thyroxine), growth hormone, parathyroid hormone (PTH), calcitonin, and testosterone are among the hormones used to regulate growth plate physiology. These hormones can stimulate matrix synthesis, cell division, and calcification (and thus growth plate closure). The precise method and zone of action of these hormones are beyond the scope of this text. Even without entirely understanding their mechanisms of action, however, it is clear that abnormalities of these hormones can significantly affect the development of the human skeleton.[2]

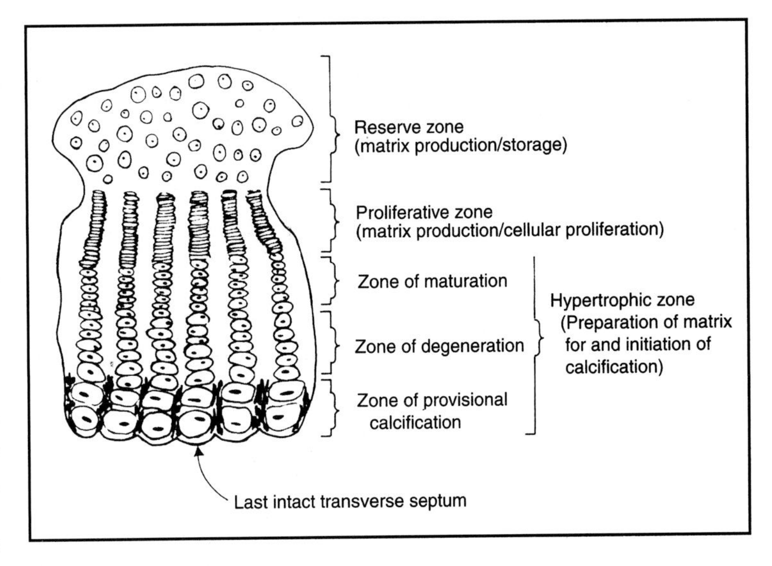

Figure 3
The zones of the growth plate. Note that the epiphysis is above the reserve zone and the metaphysis is below the hypertrophic zone.

(Reproduced with permission from Brighton CT: Structure and function of the growth plate. Clin Orthop 1978;136:24.)

Bone Cells

Three types of cells, osteoblasts, osteocytes, and osteoclasts, are essential to the development of bone and its dynamic remodeling (ie, the process of bone resorption and deposition that releases and deposits ions, adapts the bone to new loads, and repairs microscopic damage). All three types of cells can be seen in any given area of bone, and many metabolic bone diseases are caused by loss of regulation, resulting in disorders of bone formation or remodeling.

Osteoblasts

Osteoblasts are bone-forming cells. Although this is a useful definition for the purpose of discussion, osteoblasts do not actually form complete bone. Rather, they synthesize the organic (ie, nonmineral) component of bone, the primary element of which is osteoid (a protein matrix of type I collagen). Osteo-

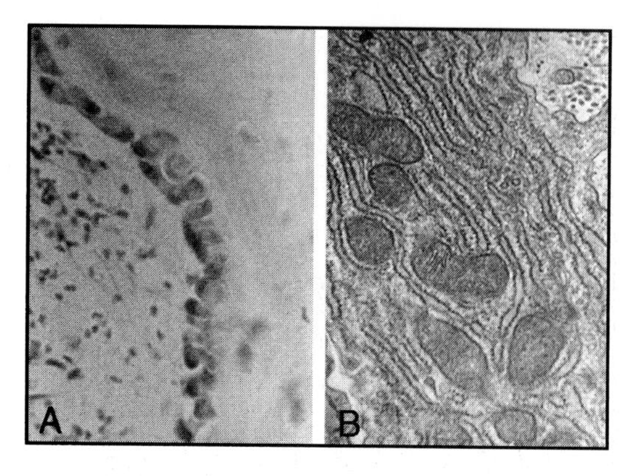

Figure 4
Light **(A)** and electron **(B)** microscope views of the osteoblast.

(Reproduced from Bostrom MPG, Boskey A, Kaufman JK, Einhorn TA: Form and function of bone, in Buckwalter JA, Einhorn TA, Simon SR (eds): ***Orthopaedic Basic Science: Biology and Biomechanics of the Musculoskeletal System****, ed 2. Rosemont, IL, American Academy of Orthopaedic Surgeons, 2000, p 323.)*

calcin and bone sialoproteins are other organic products of osteoblasts. Through these products, the osteoblasts regulate the mineralization of osteoid. Bone, therefore, is a composite of organic products of cellular synthesis and mineral deposited by extracellular matrix calcification.[3,4]

Because osteoblasts are metabolically active, they typically contain a large Golgi apparatus (Fig. 4). This organelle is responsible for secreting type I collagen, the prime component of the bone matrix. Osteoblasts also contain a large nucleus that is usually oriented away from the osteoid-producing side of the cell. In addition to collagen, osteoblasts also produce an enzyme called alkaline phosphatase. The precise function of alkaline phosphatase is still unknown; however, based on the observation that patients who are producing bone (ie, those with fractures or bone-forming tumors) have increased serum levels of alkaline phosphatase, researchers reasonably conclude that it plays some role in the mineralization of osteoid. Alkaline phosphatase is therefore a useful clinical marker of bone formation activity.

Osteoblasts are derived from the mesenchymal cell line. Once a mesenchymal cell becomes committed to the bone cell lineage, it is known as a preosteoblast. These preosteoblasts become osteoblasts when properly stimulated. The precise signals responsible are not known, but substances shown to have a role in the regulation of the differentiation of this cell line include bone morphogenetic proteins, growth factors, interleukins, insulin-derived growth factor, and platelet-derived growth factor.

The organic matrix secreted by osteoblasts is primarily type I collagen. Approximately 10% of the mass of the osteoid is composed of other matrix proteins and growth factors. Mineralization of the matrix occurs via a controlled process that has not been completely identified. It is known, however, that calcium must be locally concentrated (to form precipitates) and that energy is required to initiate the process. Some of the noncollagenous matrix proteins apparently promote or inhibit mineralization and modulate crystal size.

Osteocytes

Osteocytes, the cells of established bone, are derived from osteoblasts. In fact, an osteocyte is simply an osteoblast that has surrounded itself with bone matrix. Of course, the morphology of the cell changes: because the osteocyte is less metabolically active, it contains fewer organelles, and the nucleus occupies a greater proportion of the entire cell body. Although osteocytes are, by definition, surrounded by osteoid, they are able to communicate with other distant osteocytes through long cell processes traveling through the canaliculi of bone. This network of osteocytes can regulate its metabolic environment and respond to signals applied to the surface of the bone.

Osteoclasts

Whereas osteoblasts and osteocytes are responsible for the production of bone, osteoclasts break it down. Osteoclasts are large cells, derived from stem cells in the bone marrow that give rise to the monocyte-macrophage cell line. Macrophages and osteoclasts, both formed by the fusion of monocyte precursors, are multinucleated. When examined under light microscopy, osteoclasts are distinguished easily from other bone cells by their nuclear morphology (Fig. 5) and ruffled borders. This ruffled border increases the surface area of the cell in contact with bone, and its presence indicates an "active" osteoclast.

Osteoclasts reside in pits on the surface of bone called Howship's lacunae. The stimulus for osteoclasts to begin bone resorption is provided by osteoblasts. In response to signals for bone resorption (such as PTH), os-

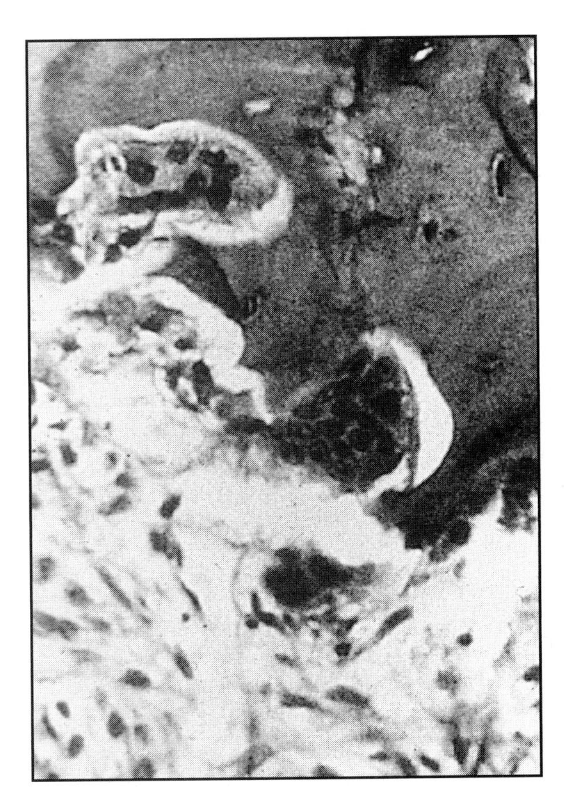

Figure 5
Light photomicrograph of a multinucleated osteoclast with a resorption (Howship's) lacuna.

(Reproduced from Bostrom MPG, Boskey A, Kaufman JK, Einhorn TA: Form and function of bone, in Buckwalter JA, Einhorn TA, Simon SR (eds): ***Orthopaedic Basic Science: Biology and Biomechanics of the Musculoskeletal System****, ed 2. Rosemont, IL, American Academy of Orthopaedic Surgeons, 2000, p 324.)*

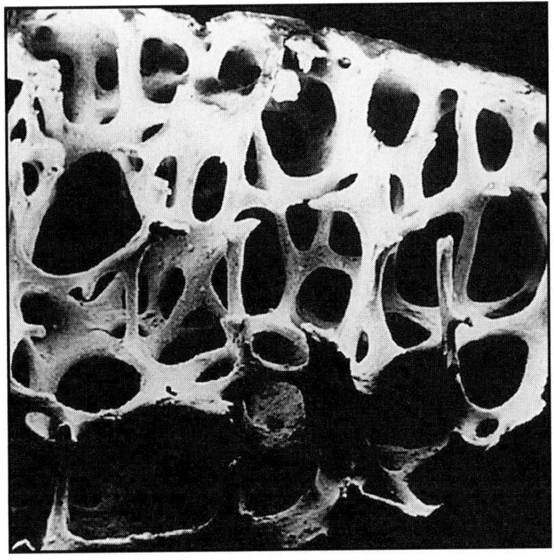

Figure 6
Trabecular bone is characterized by a three-dimensional lattice of bony spicules with considerable space among them.

(Reproduced with permission from Gibson JL: The mechanical behavior of cancellous bone. ***J Biomech*** *1985;18:317-328.)*

teoblasts secrete neutral proteases that expose the surface of the bone. Osteoclasts then work on this surface by attaching to the bone by means of receptors known as integrins. Their ruffled borders then secrete proteolytic enzymes and hydrogen ions (acidifying the local environment) that degrade the bone matrix. In this acidic environment, calcium hydroxyapatite crystals, which make up the mineral content of bone, become more soluble and are released into the circulation.

Normal Histology and Composition

Bone is efficient. While forming and remodeling, mass is kept to a minimum. Bone is divided into two structural types: cortical (bark-like) and trabecular. Cortical bone is dense, with its collagen aligned in the direction of applied forces; the cortex of bone is found at the periphery and is responsible for skeletal homeostasis. Trabecular bone is characterized by lattices of bone spicules, which are also aligned in the direction of load but with considerably more empty space than cortical bone. This type of bone, therefore, has a large surface-to-volume ratio and as such is best able to answer the demands of mineral homeostasis (as noted above, osteoclasts, which release mineral ions, require a surface on which to operate). Thus, trabecular bone is the area most likely affected by metabolic bone diseases. Trabecular bone also contributes to efficient mechanical support by orienting its spicules in the direction of load (Fig. 6).

Cortical bone, although dense, has its own microanatomy. The basic unit of cortical bone is the osteon. An osteon is a central canal housing a vascular channel surrounded by rings of lamellar bone (Fig. 7). Osteons communicate with each other via side-to-side canals. Osteons allow even cortical bone to remodel by maintaining the tissue's blood supply; they are likely to play a mechanical role as well, such as stopping the propagation of cracks within the cortex.

Skeletal Homeostasis

The resorption and removal of bone by osteoclasts shapes bones during development and continues throughout life. In the healthy state, bone resorption is coupled with new bone deposition, and the paired process is called bone remodeling. Bone remodeling in the adult frees mineral from the skeleton and

Figure 7
A bone osteon has a central core surrounded by concentric rings of lamellar bone.

*(Reproduced from Bostrom MPG, Boskey A, Kaufman JK, Einhorn TA: Form and function of bone, in Buckwalter JA, Einhorn TA, Simon SR (eds): **Orthopaedic Basic Science: Biology and Biomechanics of the Musculoskeletal System**, ed 2. Rosemont, IL, American Academy of Orthopaedic Surgeons, 2000, p 321.)*

removes microscopic areas of damaged bone. Thus, bone remodeling contributes to both mineral homeostasis and skeletal homeostasis, although mineral homeostasis takes precedence: insufficient amounts of calcium can cause cardiac arrhythmias and sudden death, whereas the transient loss of bone structure, however important, does not typically cause rapid and devastating consequences.

Bone remodeling is initiated by both chemical and mechanical signals. The primary chemical signal is PTH, which is mobilized in response to low serum calcium. Estrogen, too, is an important chemical signal. It helps ensure that the actions of osteoclasts are coupled with those of osteoblasts. When an estrogen shortage occurs, the activities of osteoclasts may exceed those of osteoblasts, resulting in decreased skeletal mass over time. This is the phenomenon that underlies perimenopausal osteoporosis. Testosterone also affects bone metabolism via osteoblast function, but its role is less clear than the role of estrogen.

Skeletal signals in the form of physical stresses can also stimulate bone remodeling. Bone is deposited in areas where load bearing is high, a process known as **Wolff's law**. Bone deposition may be mediated by direct stress receptors on the bone cells or by the electrical potentials created by deforming (ie, bending) the bone under load bearing. The converse of Wolff's law also applies to bone remodeling: bone will be resorbed in areas not subject to stress. Accordingly, disuse will lead to significant bone loss. This has practical relevance. For example, a bone that is casted for treatment of a fracture will be weaker than normal, even in the areas that were not fractured.

Bone remodeling requires biologic substrate (material) and energy (power). Thus, bone remodeling takes place only in areas with an adequate blood supply. If an area of bone **infarcts** (ie, is cut off from circulation and dies), then that area will not remodel. This occurs in osteonecrosis, a process in which the inability to repair subchondral bone eventually leads to fracture and collapse within the joint.

Hematopoiesis

Hematopoiesis, the process of forming blood cells, normally occurs in tandem with the processes of skeletal and mineral homeostasis and not in opposition to them. Nonetheless, diseases of hematopoiesis may impact bone health. Specifically, hematologic malignancies are often found in bone. Also, because the bone is part of the circulatory system (albeit a place where circulation moves sluggishly), metastatic tumor deposits are often found in bone as well. Likewise, circulating bacteria may lodge in the bone causing osteomyelitis.

Mineral Homeostasis

Calcium

Calcium is a key mineral that contributes to the structural strength of the bone, but it has a myriad of other functions in the body as well, including participating in enzyme activation, nerve transmission, and muscle contraction.[5] In addition, it functions as an intracellular messenger and as a cofactor in the coagulation cascade. Accordingly, the precise concentration of active (ionized) calcium must be tightly regulated.

Nearly all (99%) of the 1,000 g of calcium held in the body is stored within the bone in the form of a crystalline calcium phosphate called hydroxyapatite ($Ca_{10}(PO_4)_6(OH)_2$). Apatite crystals are 5×80 nm or smaller. In bone, there may be impurities of the crystal, in which other ions, such as potassium, sodium, or magnesium, replace the calcium; these substitutions can affect material properties of the bone.

Only 1% of the body's calcium is found outside the bone, primarily in the body fluids. Within cells, the concentration of ionized calcium is very low, 0.1% that of the extracellular fluid, as free calcium could form precipitates and kill the cell. In the extracellular fluid, approximately 50% of the calcium is protein bound (ie, attached primarily to albumin) or complexed to negatively charged ions, such as citrate. The other 50% is ionized. It is this ionized fraction that is biologically active.

Calcium enters the body through the intestines and is reabsorbed through the renal tubule. Both processes control the level of circulating calcium ions. However, because dissolution of bone matrix can free calcium for other physiologic needs, much of the body's minute-to-minute calcium requirements are supplied via bone resorption. Nonetheless, the intestinal absorption of dietary calcium is needed to replenish skeletal stores.

In a typical 24-hour period, 500 mg of calcium are released by the bone via osteoclast resorption, and 500 mg are deposited back into the bone via osteoblast-mediated bone formation. In those 24 hours, approximately 10,000 mg of calcium are delivered to the kidneys, and nearly all of it (98%) is retained, with only 200 mg excreted in the urine. This 200-mg loss must be replaced by calcium in the diet; however, 1,000 mg must be ingested—not 200 mg—because intestinal absorption is imperfect and 800 mg are excreted in the stool. Thus, at the point of balance, 1,000 mg of calcium are ingested and 1,000 mg are excreted. The exchange between the bone and the plasma is similarly balanced (Fig. 8).

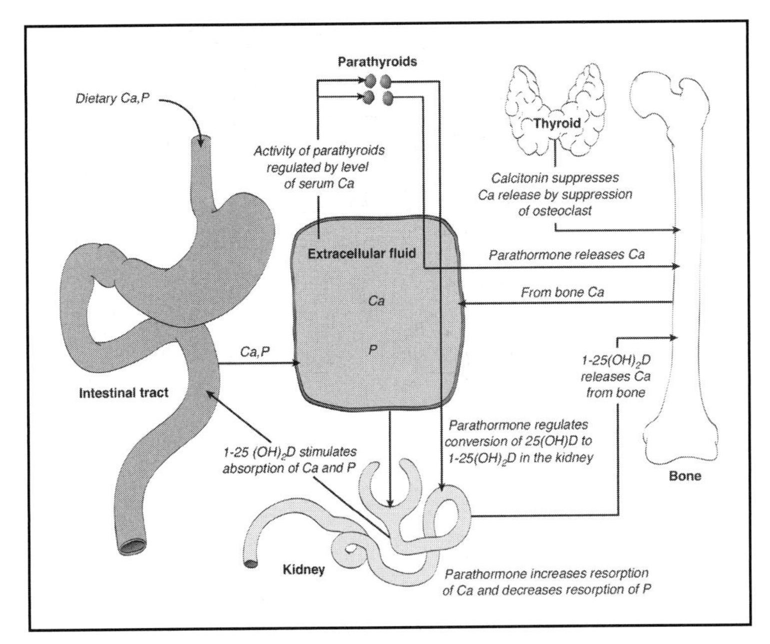

Figure 8
Calcium balance.

*(Adapted with permission from Bullough PG: **Atlas of Orthopedic Pathology**. Hampshire, England, Gower Press, 1992, p 7.5.)*

Phosphorus

Like calcium, phosphorus is stored in the body primarily in the bone, although nearly 100 g are present in the extracellular fluid, approximately 10 times the amount of extraskeletal calcium. In the body, phosphorus is found mainly in the form of phosphate ions, which play a key role in energy generation. The phosphorylation of adenosine is the primary energy exchange system in the cell. Phosphorous molecules, in the form of phospholipids, are plentiful in cell membranes as well.

The average daily intake of phosphorus is approximately 800 to 1,000 mg in the typical American diet, 60% of which is absorbed in the intestine. This rate of absorption can be increased to 90% via the action of vitamin D, if necessary; urinary excretion also helps maintain phosphorus balance. Bone remodeling can also modulate the serum concentration of phosphorus.

Regulation of Calcium Flux

The three points of calcium flux, the intestine, the kidneys, and the bone, are all regulated and can modify their normal exchange rates in response to metabolic demands. Should a drop in the concentration of serum calcium occur—or, more specifically, a drop in the concentration of biologically active, ionized calcium—the kidneys can retain a greater percentage of the serum calcium they receive for filtration. Similarly, bone can also assist in regulating serum calcium by mobilizing skeletal calcium through the resorption of mineralized bone matrix.

Mediators of Mineral Balance

The primary circulating factors that affect calcium balance are PTH, vitamin D, calcitonin, and calcium itself (Fig. 8).

Parathyroid hormone

PTH is synthesized by the parathyroid glands and stored within the glands themselves. In response to low concentrations of serum calcium, PTH is released, and the synthesis of additional supplies is initiated. PTH exerts its effects on calcium and phosphate metab-

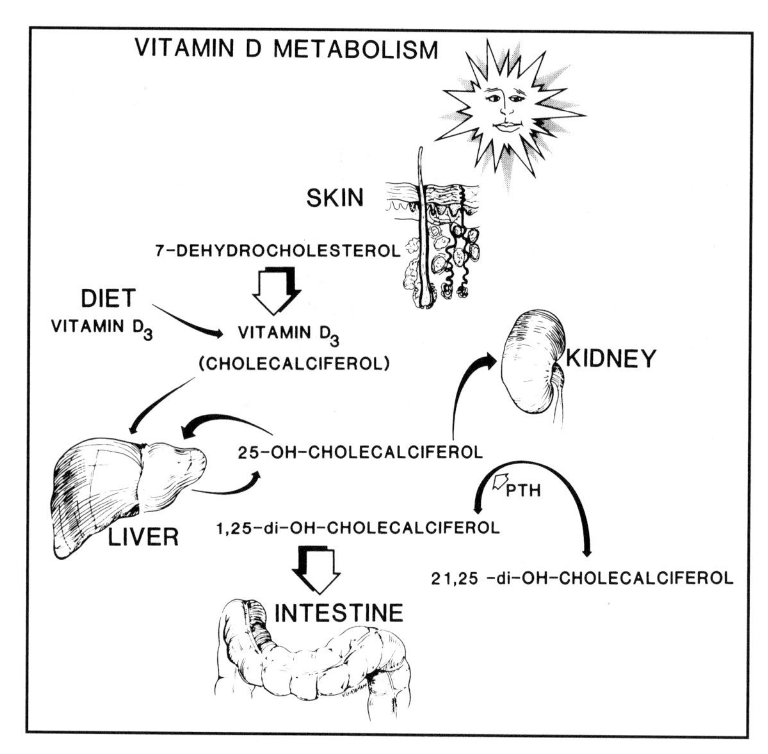

Figure 9
Vitamin D synthesis. Note that both the liver and the kidney hydroxylate inactive forms of cholecalciferol; the kidney's hydroxylation is under PTH control.

(Reproduced from Bostrom MPG, Boskey A, Kaufman JK, Einhorn TA: Form and function of bone, in Buckwalter JA, Einhorn TA, Simon SR (eds): ***Orthopaedic Basic Science: Biology and Biomechanics of the Musculoskeletal System****, ed 2. Rosemont, IL, American Academy of Orthopaedic Surgeons, 2000, p 354.)*

olism in the bones and the kidneys. In bone, PTH stimulates the release of calcium and phosphate in response to low serum calcium levels. Paradoxically, it is not the bone-resorbing osteoclasts but the bone-forming osteoblasts that have receptors for PTH. Osteoblasts, in turn, signal osteoclasts (via means not completely understood) to increase bone resorption and mobilize calcium and phosphate stores. In the kidneys, PTH stimulates the resorption of calcium, decreases phosphate resorption, and promotes the hydroxylation of inactive vitamin D into its biologically active form.

Vitamin D

Vitamin D is a steroid hormone; as such, it binds to the nucleus of a cell and modifies gene expression and protein synthesis. The function of vitamin D in mineral metabolism is to increase the serum concentration of calcium. The body obtains vitamin D directly from diet or through primary synthesis (Fig. 9). The endogenic synthesis of vitamin D begins in the skin when the skin is exposed to sunlight. Sunlight modifies 7-dehydrocholesterol to form the base (inactive) compound of vitamin D_3 (cholecalciferol), which is then transported to the liver where it is hydroxylated. This hydroxylation occurs at what is referred to in chemistry nomenclature as position 25; thus, the molecule at this point in the process is called 25-OH cholecalciferol.

25-OH cholecalciferol is then transported to the kidneys, where it is hydroxylated again. Hydroxylation can occur at either position 1 or position 24; the former is the active form, the latter inactive. The "decision" regarding which position will be hydroxylated depends on the availability of circulating vitamin D and PTH. Low levels of vitamin D or high levels of PTH stimulate hydroxylation at position 1, yielding the active form of the vitamin, 1,25 OH_2 vitamin D. Likewise, high levels of vitamin D create a feedback loop of sorts and increase the formation of inactive 24,25 $(OH)_2$ vitamin D.

Bone, the intestine, and the kidneys are the target end organs of vitamin D. Like all steroid hormones, vitamin D arrives at these target organs via the bloodstream. Each organ has specific receptors for vitamin D.[6] In the bone, vitamin D stimulates the release of calcium and phosphate. In the small intestine, vitamin D promotes the synthesis of proteins that transport dietary calcium into the circulation. In the kidney, it increases calcium resorption.

Calcitonin

Calcitonin is a peptide that is produced and secreted by the chief cells of the thyroid gland. This peptide has a direct effect on osteoclasts by decreasing their size and inhibiting their ability to resorb bone. Calcitonin is directly regulated by serum calcium level. When serum calcium levels are high, the chief cells respond by producing and secreting more calcitonin in order to decrease bone breakdown and thereby decrease the mobilization of calcium. Calcitonin has been used to treat several diseases in which osteoclast resorption of bone is uncontrolled. It also may have a role in decreasing the long-term effects of osteoporosis.

Research and New Directions During the Bone and Joint Decade (2002-2011)

Skeletal Formation

Current investigations attempt to identify the genetic, biochemical, and biomechanical signals that cause the skeleton to be formed, take the shape it does, and stop the formation process when complete. These basic biologic studies add to our knowledge, of course, but they may also suggest why diseases of bone loss or dysfunction occur.

Fracture Healing

Fracture healing is known to recapitulate some of the steps of normal bone formation. Studies of the cellular machinery of bone formation are ongoing and may reveal how fracture healing can be accelerated or enhanced.

Regulation of Bone Metabolism

Osteoporosis is a disease in which bone formation is not coupled perfectly with bone resorption. Ongoing laboratory studies, such as those involving menopause and estrogen loss, attempt to further regulate this process to ensure that bone is not lost from the skeleton even as some natural mediators are altered.

Key Terms

Alkaline phosphatase An enzyme produced by osteoblasts that is believed to play a role in the mineralization of bone

Apoptosis Programmed cell death

Bone remodeling A process that couples bone resorption by osteoclasts with deposition by osteoblasts (new bone cells)

Chondroblasts The cells that form cartilage

Cortical bone Dense bone, literally "bark," that is responsible for skeletal homeostasis

Diaphysis The shaft of a long bone

Endochondral ossification The formation of bone within a cartilage model

Epiphysis The region of bone across the physis, just below the articular cartilage

Hematopoiesis The process of forming blood cells

Hypertrophic zone A layer of large, plump cells in the physis that assists in mineralization of the cartilage

Infarct An area of tissue that is cut off from its blood supply, becomes ischemic, and dies

Lamellar bone Mature, well-organized form of cortical bone

Medullary canal The relatively hollow central core of a long bone that houses blood-forming cells

Metaphysis The region of bone just below the physis

Mineral homeostasis The function of bone responsible for the storing and releasing of ions, principally calcium

Morphogenesis The development of the skeleton in utero

Osteoblasts The cells that synthesize the organic component of bone; also thought of as the bone-forming cells

Osteoclasts Large cells that resorb bone matrix when activated

Osteocytes The cells of established bone

Periosteum A sleeve of connective tissue that surrounds the shaft of the bone and contributes to fracture healing

Physis The growth plate

Proliferative zone A layer of stacked cells in the physis that synthesizes proteoglycans for the cartilaginous matrix; also has the most extensive blood supply within the physis

Reserve zone A layer of cells in the physis that produces the cartilaginous matrix for ossification into bone

Skeletal homeostasis The function of bone that supplies structural support and movement for the body

Trabecular bone Bone characterized by its "lattice" appearance and is responsible for mineral homeostasis

Wolff's law A law that states that the growth and remodeling of bone is influenced and modulated by mechanical stresses

Woven bone Primitive, less-organized form of cortical bone

References

1. Zaleske DJ: Cartilage and bone development. *Instr Course Lect* 1998;47:461-468.
2. Iannotti JP: Growth plate physiology and pathology. *Orthop Clin North Am* 1990;21:1-17.
3. Tonna EA, Cronkite EP: The periosteum: Autoradiographic studies on cellular proliferation and transformation utilizing tritiated thymidine. *Clin Orthop* 1963;30:218-232.
4. Baron R, Ravesloot JH, Neff L, et al: Cellular and molecular biology of the osteoclast, in Noda M (ed): *Cellular and Molecular Biology of Bone.* San Diego, CA, Academic Press, 1993, pp 445-495.
5. Boden SD, Kaplan FS: Calcium homeostasis. *Orthop Clin North Am* 1990;21:31-42.
6. Walters MR: Newly identified actions of the vitamin D endocrine system. *Endocr Rev* 1992;13:719-764.

Chapter 2

Articular Cartilage

Articular cartilage, the resilient durable tissue that forms the opposing articulating surfaces of synovial joints, provides these surfaces with the low friction, lubrication, and wear characteristics that make possible smooth painless movement.[1] It also absorbs mechanical load and spreads these applied loads onto subchondral bone.[2] No synthetic or reparative material performs as well as a natural joint surface. Injuries or diseases of articular cartilage cause pain and loss of mobility for more people than disorders of any other musculoskeletal tissue. This chapter reviews current understanding of how the unique structure and composition of articular cartilage give the tissue its remarkable mechanical properties and durability. It also introduces the subjects of articular cartilage injury and repair.

Physiologic Function

The articular cartilage of synovial joints is subject to high loads applied repetitively for many decades. Thus, the structural molecules, including collagens, proteoglycans, and other molecules of articular cartilage, must be organized into a strong, fatigue-resistant, and tough, solid matrix. This matrix must be capable of sustaining the high stresses and strains developed within the tissue from these loads. In terms of material behavior, this solid matrix is porous, permeable, and very soft.[2,3] Water resides in the microscopic pores, and application of loads to the tissue forces the water through the porous-permeable solid matrix. Thus, articular cartilage is a biphasic material, composed of solid and fluid phases.[2,3]

Normal Histology and Composition

Unlike most tissues, articular cartilage does not have blood vessels, nerves, or lymphatics. It consists of a highly organized **extracellular matrix (ECM)** with a sparse population of highly specialized cells (chondrocytes) distributed throughout the tissue.[1] The primary components of the ECM are water, proteoglycans, and collagens, with other proteins and glycoproteins present in smaller amounts.

The structure and composition of the articular cartilage vary throughout its depth (Fig. 1), from the articular surface to the subchondral bone. These differences include cell shape and volume, collagen fibril diameter and orientation, proteoglycan concentration and water content. The cartilage can be divided into four zones: the superficial tangential zone; the middle or transitional zone; the deep zone; and the calcified zone.

The **superficial tangential zone** forms the smooth, nearly frictionless gliding surface. The thin collagen fibrils there are arranged parallel to the surface, and the chondrocytes are elongated with the long axis parallel to the surface. Here the proteoglycan content is at its lowest concentration and the water content at its highest. The **middle (transitional) zone** contains collagen fibers with less apparent organization, and the chondrocytes have a more rounded appearance. The **deep zone** contains the highest concentration of proteoglycans and the lowest water content. The collagen fibers are organized vertical to the joint surface, and the chondrocytes are arranged in a columnar fashion. The deepest layer, the **calcified zone**, separates hyaline cartilage from the subchondral bone. Histologic staining with hematoxylin and eosin shows a wavy bluish line, called the **tidemark**, which separates the deep zone from the calcified zone.

In addition to these articular surface-to-bone zonal distinctions, the ECM is divided into pericellular, territorial, or interterritorial regions (Fig. 2). The pericellular matrix is a thin layer that completely surrounds each chondrocyte. It contains proteoglycans and other noncollagenous matrix components. This pericellular matrix serves as an important biomechanical buffer between the terri-

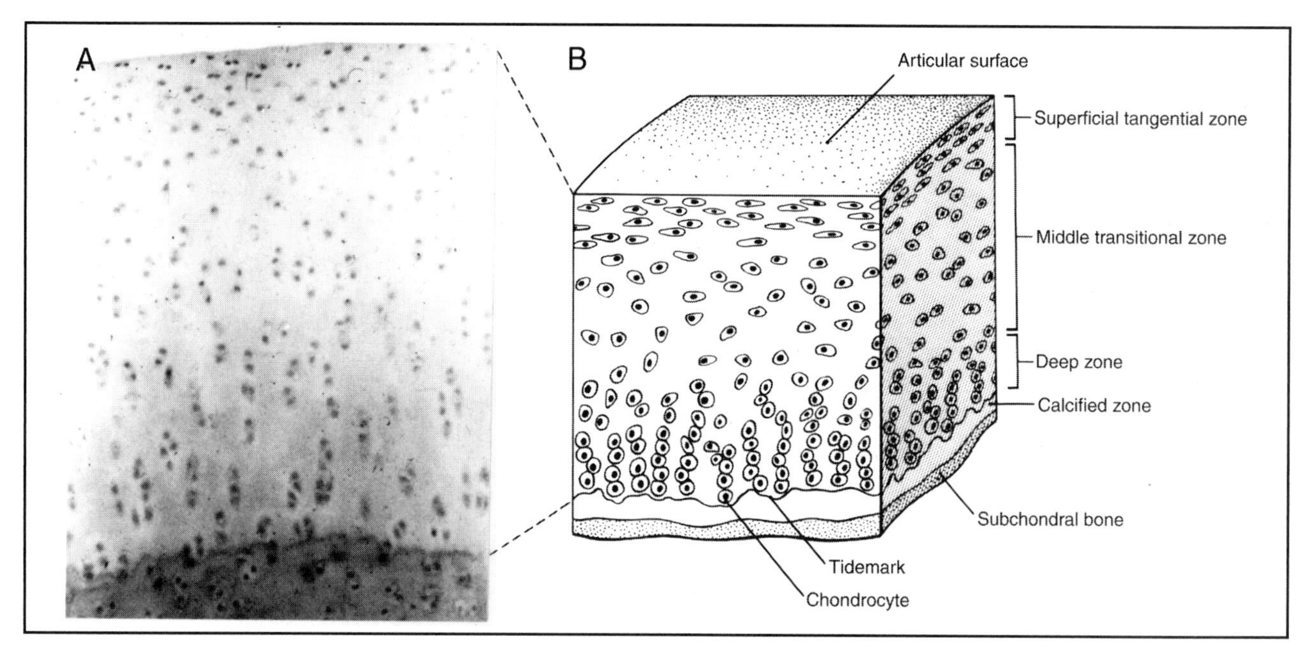

Figure 1
A, Histologic section of normal adult articular cartilage showing even Safranin O staining and distribution of chondrocytes. **B**, Schematic diagram of chondrocyte organization in the three major zones of the uncalcified cartilage, the tidemark, and the subchondral bone.

(Reproduced with permission from Mow VC, Proctor CS, Kelly MA: Biomechanics of articular cartilage, in Nordin M, Frankel VH (eds): ***Basic Biomechanics of the Musculoskeletal System****, ed 2. Philadelphia, PA, Lea & Febiger, 1989, pp 31–57.)*

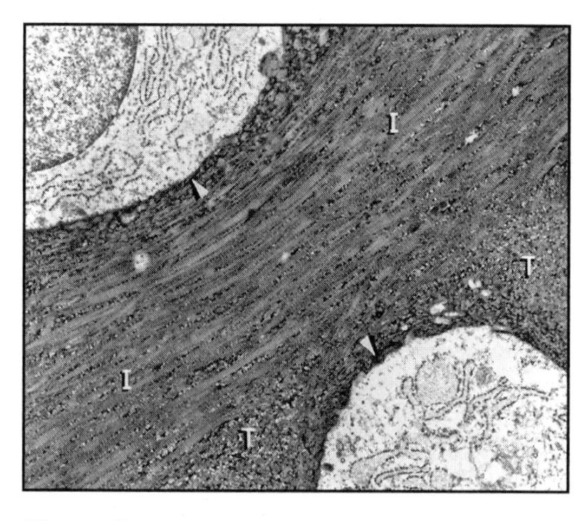

Figure 2
Electron microscopic view (×8700) of mature rabbit articular cartilage from the medial femoral condyle. Micrograph shows cytoskeletal elements, and pericellular matrix (arrowhead), territorial matrix (T), and interterritorial matrix (I). The pericellular matrix lacks cross-striated collagen fibrils, whereas the territorial matrix has a fine fibrillar collagen network. The collagen of the interterritorial matrix has coarser fibers, and they tend to lie parallel to each other.

(Reproduced with permission from Buckwalter JA, Hunziker EB: Articular cartilage biology and morphology, in Mow VC, Ratcliffe A (eds): ***Structure and Function of Articular Cartilage****. Boca Raton, FL, CRC Press, 1993.)*

torial matrix and the cell. The territorial matrix surrounds the pericellular matrix, and it is characterized by thin collagen fibrils that form a fibrillar network that is distinct from the surrounding interterritorial matrix. The interterritorial matrix is the largest of the matrix regions and contributes most of the material properties of the articular cartilage. It encompasses all of the matrix between the territorial matrices of the individual cells or clusters of cells and contains large collagen fibers and most of the proteoglycans.

Chondrocytes

The formation and maintenance of articular cartilage depends on chondrocytes.[1,4,5] These are derived from mesenchymal cells, which differentiate during skeletal morphogenesis. During skeletal growth, these cells generate a large amount of ECM, and in mature tissue, they maintain this matrix. They are metabolically active and respond to a variety of environmental stimuli, including growth factors, interleukins, and pharmaceutical agents; matrix molecules; mechanical loads; electrical potential and currents; and hydrostatic and osmotic pressure changes.

Extracellular Matrix

Because the chondrocytes of articular cartilage occupy only a small proportion of the total volume of the tissue, the material properties of cartilage are determined primarily by the ECM.[6] Normal cartilage has water contents ranging from 65% to 80% of its total wet weight. The remaining wet weight of the tissue is accounted for principally by two major classes of structural macromolecular materials: collagens and proteoglycans. Several other classes of molecules, including lipids, phospholipids, proteins, and glycoproteins, make up the remaining portion of the ECM.

Water

Water is the most abundant component of normal articular cartilage, making up from 65% to 80% of the wet weight of the tissue.[6,7] A small percentage of this water is contained in the intracellular space, about 30% is associated with the intrafibrillar space within the collagen, and the remainder is contained in the ECM. Water content varies throughout cartilage, decreasing in concentration from approximately 80% at the surface to 65% in the deep zone. Most of the water can be moved through the ECM by compressing the solid matrix. Frictional resistance against this flow through the ECM is very high, and this resistance to water flow within the ECM is the basis of articular cartilage's ability to cushion very high joint loads.

Collagen

Collagen is a triple helix protein that is the major structural macromolecule of the ECM.[8,9] There are at least 15 distinct collagen types composed of at least 29 genetically distinct chains. All members of the collagen family contain a characteristic triple-helical structure that may constitute most of the length of the molecule. Over 50% of the dry weight of articular cartilage consists of collagen. The major cartilage collagen, which represents 90% to 95% of the total, is known as type II. Articular cartilage collagens provide the tissue's tensile properties and serve to immobilize the proteoglycans within the ECM. Collagen fibers in cartilage are generally thinner than those seen in tendon or bone, and this may be, in part, a function of their interaction with the relatively large amount of proteoglycan in this tissue.

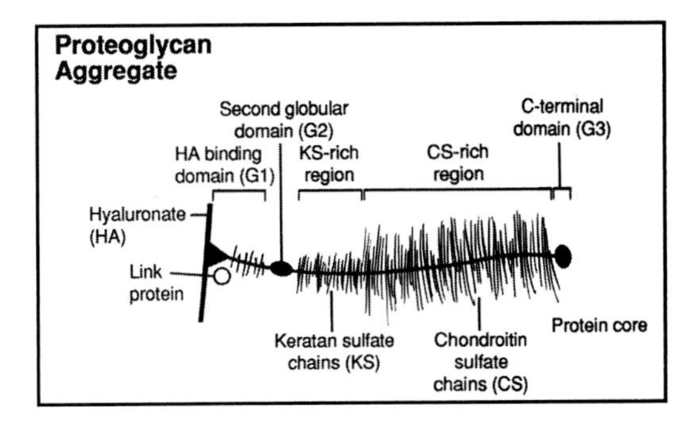

Figure 3
Schematic diagram of a proteoglycan. The protein core has several globular domains (G1, G2, and G3), with other regions containing the keratan sulfate and chondroitin sulfate glycosaminoglycan chains. The N-terminal G1 domain is able to bind specifically to hyaluronate. This binding is stabilized by link protein.

(Reproduced from Mankin HJ, Mow VC, Buckwalter JA, Iannotti JP, Ratcliffe A: Articular cartilage structure, composition, and function, in Buckwalter JA, Einhorn TA, Simon SR (eds): ***Orthopaedic Basic Science: Biology and Biomechanics of the Musculoskeletal System****, ed 2. Rosemont, IL, American Academy of Orthopaedic Surgeons, 2000, pp 443-470.)*

Proteoglycans

Proteoglycans are complex macromolecules that consist of a protein core with covalently bound polysaccharide (glycosaminoglycan) chains (Fig. 3). Glycosaminoglycans consist of long-chain, unbranched, repeating disaccharide units. Three major types have been found in cartilage: (1) chondroitin sulfate (found in two isomeric forms, chondroitin 4-sulfate and chondroitin 6-sulfate); (2) keratan sulfate; and (3) dermatan sulfate. The chondroitin sulfates are the most prevalent glycosaminoglycans in cartilage. They account for 55% to 90% of the total population, depending principally on the age of the subject. Each chain is composed of 25 to 30 repeating disaccharide units. The keratan sulfate constituent of articular cartilage resides primarily within the large, aggregating proteoglycan. Hyaluronate is also a glycosaminoglycan, but unlike those described above, it is not sulfated. A further distinguishing feature of hyaluronate is that it is not bound to a protein core, and, therefore, is not part of a proteoglycan.

Approximately 80% to 90% of all proteoglycans in cartilage are of the large, aggregating type, called aggrecans (Fig. 4). They consist of a long, extended protein core with up to 100 chondroitin sulfate and 50 keratan sulfate glycosaminoglycan chains covalently

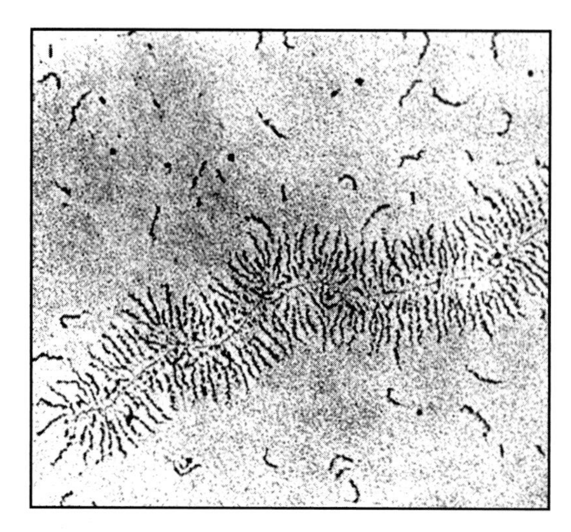

Figure 4
Electron micrograph of bovine articular cartilage proteoglycan aggregates from skeletally immature calf. The aggregates consist of a central hyaluronic acid filament and multiple attached monomers.

(Reproduced with permission from Buckwalter JA, Kuettner KE, Thonar EJ: Age-related changes in articular cartilage proteoglycans: Electron microscopic studies. J Orthop Res *1985;3:251–257.)*

bound to the protein core. In young individuals, the concentration of keratan sulfate is relatively low, and chondroitin 4-sulfate is the predominant form of chondroitin sulfate. With increasing age, the concentration of keratan sulfate increases, and chondroitin 6-sulfate becomes the predominant isomeric form of chondroitin sulfate.

Metabolism

Despite the lack of a blood supply, articular cartilage chondrocytes have a high level of metabolic activity[1] (Fig. 5). Chondrocytes synthesize and assemble the cartilaginous matrix components and direct their distribution within the tissue. These synthetic and assembly processes involve synthesis of proteins, synthesis of glycosaminoglycan chains and their addition to the appropriate protein cores, and secretion of the completed molecules into the ECM. All of these actions take place under avascular and, at times, anaerobic conditions, with considerable variation in local mechanical, electrical and physicochemical states. In addition, the chondrocyte directs internal ECM remodeling by regulating an elaborate series of degradative enzymes.

The maintenance of a normal ECM depends on three factors: (1) the ability of the chondrocytes to balance the rates of synthesis of matrix components, (2) the component's appropriate incorporation into the matrix, and (3) the component's degradation and release rate from the cartilage. The cells do this by responding to mechanical, electrical, and chemical stimuli in the local environments. Soluble mediators (eg, growth factors and interleukins), and changes in matrix composition, mechanical loads, and electric fields all influence the metabolic activities of the chondrocytes. The response of the chondrocytes usually will maintain a stable matrix. However, in some cases, the response of the cells can lead to a change of matrix composition and ultrastructural organization, and eventually to cartilage degeneration.

Mechanical Signal Transduction

Because articular cartilage is an aneural tissue, the nerve impulses that regulate many body processes cannot provide information to chondrocytes. Cellular and humoral immune responses also are not likely to occur in cartilage, because both monocytes and immunoglobulins tend to be excluded from the tissue by virtue of their size. However, the cells derive considerable information from the mechanical stresses and strains that act on their membranes as a result of physical forces applied to the tissue.[10-12] Joint motion and loading appear to serve as principal stimuli to the cartilage cells. How the chondrocytes sense their mechanical environment and convert the information received to changes in gene expression is unknown, although integrins (molecules that span the plasma membrane of the chondrocytes and are connected to the intracellular cytoplasm) are likely to be involved.

Effects of Joint Motion and Loading

Joint motion and loading are required to maintain the normal composition, structure, and mechanical properties of adult articular cartilage.[1,13] When the intensity or frequency of loading exceeds or falls below these necessary levels, the balance between synthesis and degradation will be altered, and changes in the composition and microstructure of cartilage will follow.

Prolonged reduction of joint motion caused by rigid immobilization leads to degeneration of articular cartilage. These changes result, in part, because normal nutritive transport to cartilage from the synovial fluid by means of diffusion is diminished when the fluid does not move. In addition, the mechanical properties of articular cartilage will be compromised with immobilization. These biochemical and biomechanical changes are, at least in part, reversible if the joint is allowed to move. A better understanding of the deleterious effects of joint immobilization has led to treatments of joint injuries that allow patients at least some joint motion while the injuries are healing.

Excessive joint loading also may affect articular cartilage. Catabolic effects can be induced by a single-impact or repetitive trauma and may serve as the initiating factor for progressive degenerative changes. However, regular joint use, including running, has not been shown to cause joint damage in normal joints.[14]

The specific mechanisms by which joint loading influences chondrocyte function remain unknown, although various mechanical, physicochemical, and electrical transduction mechanisms have been proposed.[10,12] Matrix deformation produces fluid and ion flow, which may facilitate chondrocyte nutrition. Deformation and fluid flow lead to changes in the local charge density within the matrix, resulting in an electric potential that may serve as cellular signals.

Development and Aging

Unlike many other tissues, immature articular cartilage differs considerably from adult articular cartilage. On gross inspection, the cartilage from skeletally immature individuals appears blue-white, presumably because of the presence of vascular structures in the underlying immature bone, and is relatively thick. The thickness reflects the two tasks of the cartilage mass: to serve as a cartilaginous articular surface for the joint and to be a source of endochondral ossification of the underlying bone. Immature articular cartilage is also considerably more cellular than the adult tissue.

With increasing age, the cartilage undergoes changes in matrix organization, mechanical properties, and cell function. All of these changes increase the risk of tissue degeneration.[15,16] The tensile strength of the superficial cartilage layer decreases and perhaps most important, the ability of the chondrocytes to maintain and restore the tissue diminishes.[17,18]

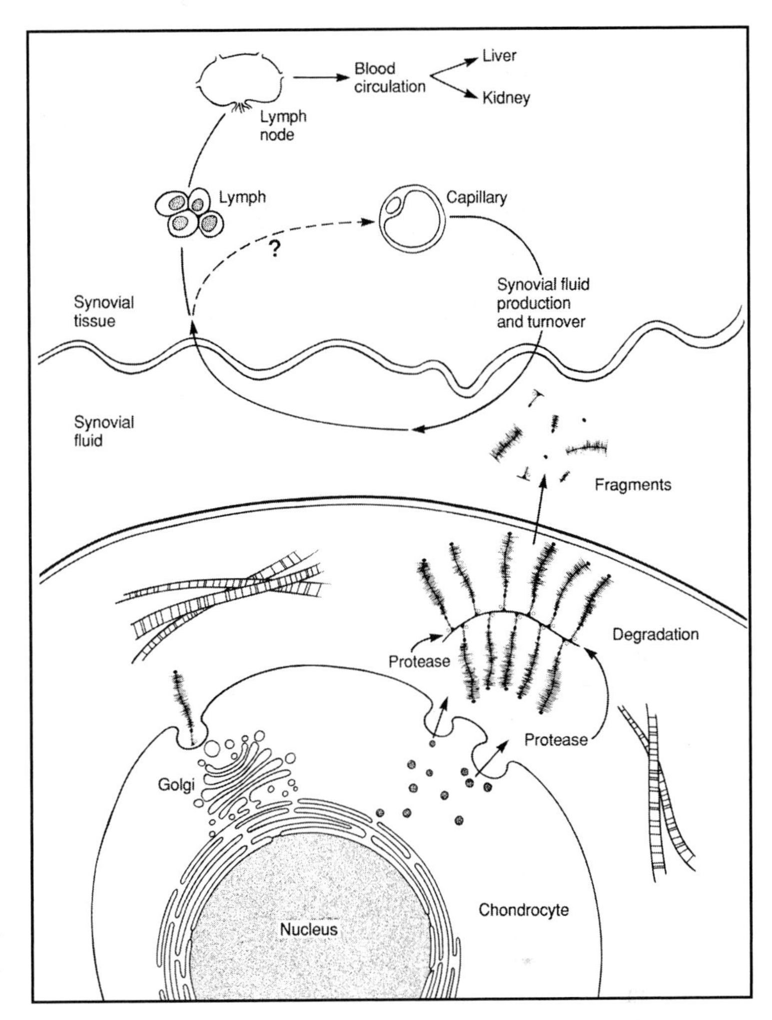

Figure 5

Schematic diagram of the metabolic events controlling the proteoglycans in cartilage. The chondrocytes synthesize and secrete aggrecan, link protein and hyaluronate, and they become incorporated into functional aggregates. Enzymes released by the cells break down the proteoglycan aggregates. The fragments are released from the matrix into the synovial fluid, and from there the fragments are taken up by the lymphatics and moved into the circulating blood.

(Reproduced from Mankin HJ, Mow VC, Buckwalter JA, Iannotti JP, Ratcliffe A: Articular cartilage structure, composition, and function, in Buckwalter JA, Einhorn TA, Simon SR (eds): ***Orthopaedic Basic Science: Biology and Biomechanics of the Musculoskeletal System,*** *ed 2. Rosemont, IL, American Academy of Orthopaedic Surgeons, 2000, pp 443-470.)*

Pathophysiology

Injury and Repair

Although articular cartilage can withstand many decades of joint use, excessive force can disrupt the tissue. Articular cartilage is metabolically active, yet it has a limited capacity to replace damaged or lost cells and matrix.[19] Factors responsible for this limited response include a paucity of native cells and a lack of blood vessels to import more. Because of the unique characteristics of articular cartilage, the repair responses to superfi-

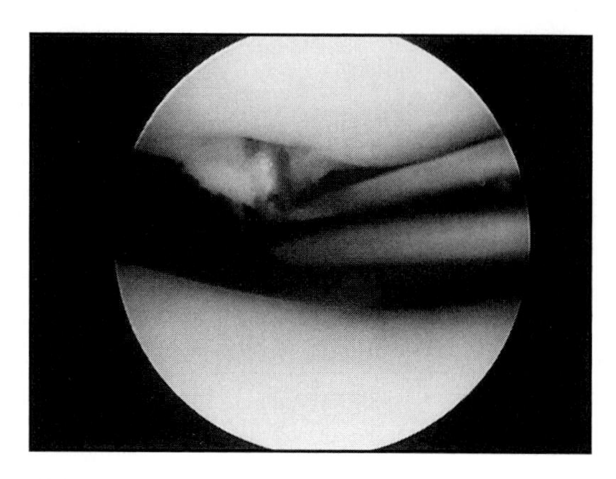

Figure 6
Arthroscopic image of a partial-thickness lesion of the articular cartilage of the patella.

(Reproduced from Boden BP, Pearsall AW, Garrett WE Jr, Feagin JA: Patellofemoral instability: Evaluation and management. J Am Acad Orthop Surg *1997;5:47-57.)*

cial injuries and injuries that damage subchondral bone as well as articular cartilage differ considerably.[19]

Superficial Articular Cartilage Injuries
High-level force applied to articular surfaces can cause articular cartilage ruptures or tears that do not extend into the underlying bone (Fig. 6). Articular cartilage injuries that do not cross the tidemark generally do not heal.[19] Why do superficial injuries behave this way? First, these lesions do not cause hemorrhage or initiate an inflammatory response. Also, fibrin clots rarely form on exposed surfaces of normal cartilage. Chondrocytes near the injury may proliferate and synthesize new matrix, but they do not migrate into the lesion. The new matrix they produce remains in the immediate region of the chondrocytes, and their proliferative and synthetic activity fails to provide new tissue to repair the damage. By increasing the load on adjacent tissue, large articular surface defects may also cause degeneration of previously normal regions of the articular surface.

Deep Articular Cartilage and Subchondral Bone Injuries
Injuries that disrupt articular cartilage and the underlying subchondral bone are referred to as osteochondral fractures. Mechanical injury that disrupts bone as well as articular cartilage causes hemorrhage, fibrin clot formation, and inflammation.[19] Soon after creation of the osteochondral defect, a fibrin clot fills the injury site and inflammatory cells migrate into the clot. Injury to bone and subsequent clot formation release growth factors and proteins that influence multiple cell functions, including migration, proliferation, and differentiation. Bone matrix contains a number of growth factors, and platelets release at least two important growth factors: platelet-derived growth factor and transforming growth factor-β.

Cells within the bony portion of the osteochondral defect form immature bone, fibrous tissue, and cartilage with a hyaline matrix. This bone formation restores the original volume of injured subchondral bone but rarely, if ever, progresses into the chondral portion of the defect. In general, by 6 months after injury, the subchondral bone defect has been repaired with a tissue that is primarily bone but also contains some regions of fibrous tissue and hyaline cartilage. In contrast, the chondral defect is not repaired completely and does not restore the normal structure, composition, or mechanical properties of an articular surface.

In most injuries, the chondral repair tissue begins to show evidence of fibrillation and loss of chondrocytes and hyaline matrix in less than 1 year. However, the fate of cartilage repair tissue is not always progressive deterioration. Occasionally, the repair tissue persists and appears to function satisfactorily as an articular surface for a prolonged period of time. The reasons why some repair tissue persists for a prolonged period of time while most repair tissue deteriorates remain unknown. Because repair tissue can provide a functional surface in some instances, many orthopaedic surgeons will drill, puncture, or abrade below the tidemark and stimulate bleeding of the bone in an attempt to restore the articular surface. Of course, the repair tissue is fibrocartilage, not hyaline cartilage. Fibrocartilage is a form of cartilage that is less resilient, less smooth, and less suited to load bearing.

Osteoarthritis
Osteoarthritis (OA), also referred to as degenerative joint disease, is characterized by a generally progressive loss of articular cartilage accompanied by attempted repair of articular cartilage, remodeling, and sclerosis of subchondral bone, and in many instances, the formation of subchondral bone cysts and osteophytes[19-24] (Fig. 7). In addition to the histopathologic changes in the synovial joint,

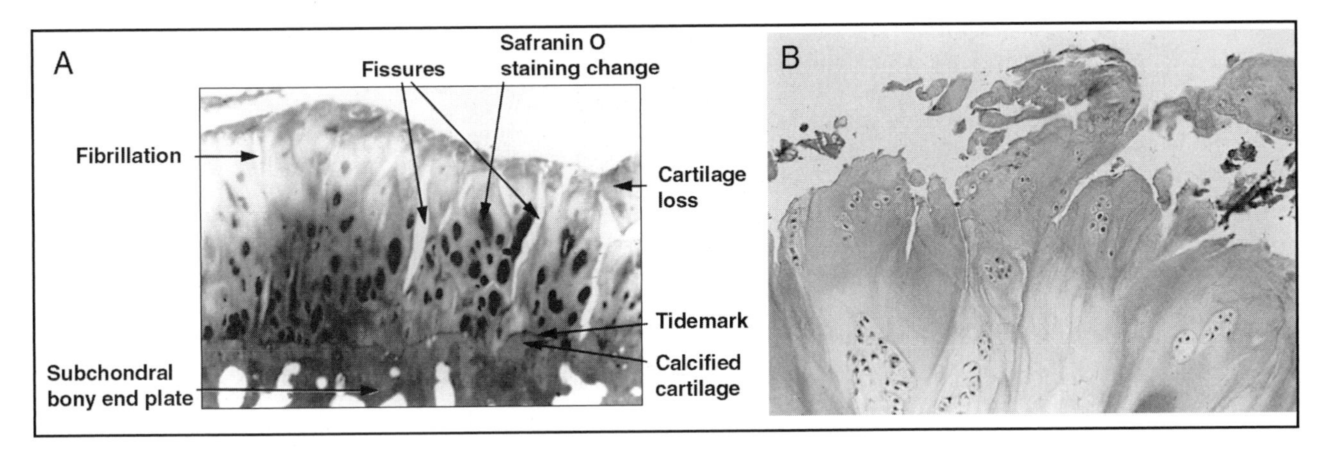

Figure 7
A, Histologic section of osteoarthritic cartilage from a humeral head removed at surgery for total shoulder arthroplasty. Note the significant fibrillation, vertical cleft formation, the tidemark, and the subchondral bony end plate. **B**, Another view of surface fibrillation showing vertical cleft formation and widespread large necrotic regions of the tissue devoid of cells. Clusters of cells, common in osteoarthritic tissues, are also seen.

(Reproduced from Mankin HJ, Mow VC, Buckwalter JA: Articular cartilage repair and osteoarthritis, in Buckwalter JA, Einhorn TA, Simon SR (eds): ***Orthopaedic Basic Science: Biology and Biomechanics of the Musculoskeletal System****, ed 2. Rosemont, IL, American Academy of Orthopaedic Surgeons, 2000, pp 471-488.)*

diagnosis of OA requires the presence of symptoms and signs that may include joint pain, restricted motion, crepitus with motion, joint effusions, and deformity. The structural changes characteristic of OA include fissuring and focal erosive cartilage lesions, cartilage loss and destruction, subchondral bone sclerosis, and cyst and large osteophyte formation at the joint margins.

Unlike the destruction seen in joint diseases with a major inflammatory component, OA consists of a sequence of cell and matrix changes that result in loss of articular cartilage structure and function, accompanied by cartilage repair and bone remodeling reactions. Because of the repair and remodeling reactions, the degeneration of the articular surface in OA is not uniformly progressive,[24] and the rate of joint degeneration varies among individuals and among joints. Occasionally, it occurs rapidly, but in most joints it progresses slowly over many years. OA may stabilize or even improve spontaneously with at least partial restoration of the articular surface and a decrease in symptoms.

OA has no single cause, but by a variety of means reaches a common end stage. Despite many years of research, considerable speculation still exists concerning various factors that may contribute to the initiation and perpetuation of the disorder. OA develops most commonly in the absence of a specific known cause, a condition referred to as primary or idiopathic OA.[19] Less frequently, it develops as a result of joint injuries, infections, or a variety of hereditary, developmental, metabolic, and neurologic disorders, a group of conditions referred to as secondary OA.[19] The age of onset of secondary OA depends on the underlying cause; thus, it may develop in young adults and even children as well as the elderly. In contrast, a strong association exists between primary OA and age. Despite this strong association and the widespread view that OA results from wear-and-tear processes, the relationships among joint use, aging, and joint degeneration remain uncertain. Furthermore, the changes observed in articular cartilage from the normal joints of older individuals differ from those observed in OA, and normal lifelong joint use has not been shown to cause degeneration.[14] Thus, OA is not simply the result of mechanical wear from joint use; the biologic processes of tissue maintenance and remodeling certainly play important roles in the continued normal function of articular cartilage with age.

Research and New Directions During the Bone and Joint Decade (2002-2011)

Role of ECM

An improved understanding of the biomechanics of articular cartilage has clarified how the components of the ECM provide a durable, resilient articular surface. Based on these advances and the understanding of articular cartilage biology, researchers are now exploring the complex relationships between chondrocytes and their ECM. In particular, they are learning how molecules in the ECM influence chondrocyte function and how the chondrocytes are continually remodeling and maintaining their ECM.

Mechanical Signs

Wolff's law notes that bone forms in response to mechanical load. A similar law has not been articulated for cartilage, but new insights from studies of mechanical signal transduction in articular cartilage help explain how loading of joint surfaces influences chondrocyte function. Ultimately, the composition, molecular structure, and mechanical properties of the articular cartilage matrix may be shown to be dependent on the mechanical signals presented.

Growth Factors

Exciting new advances in the use of growth factors that stimulate chondrocyte function are leading to new strategies for facilitating maintenance and repair of articular cartilage. Although cartilage normally does not heal, it is hoped that the addition of various factors may help induce such repair.

Transplantation

Developments in transplantation and the ability to transplant articular chondrocytes have shown that this technique is clinically feasible and has great potential for promoting the resurfacing of damaged synovial joints. These techniques are still in their infancy and research to improve them is ongoing.

Medications

Although most medications for OA aim to provide symptomatic relief, new research into the mechanisms of articular cartilage degradation suggests that it may be possible to develop medications that will inhibit matrix degradation. Such medications would truly modify the disease and not simply palliate it.

Key Terms

Aggrecans An aggregating form of proteoglycan composed of many glycosaminoglycan chains bound to a protein core

Articular cartilage The tissue that forms the opposing surfaces of synovial joints

Calcified zone The fourth zone and deepest layer of articular cartilage; separates hyaline cartilage from the subchondral bone

Collagen A triple helix protein that is the major structural macromolecule of the ECM of articular cartilage; found also in bone, tendon, and ligament

Deep zone The third zone of articular cartilage in which the collagen fibers are organized vertical to the joint surface, and the chondrocytes are arranged in a columnar fashion

Extracellular matrix (ECM) A complex structural entity surrounding and supporting cells that are found within tissues; the primary components of ECM in cartilage are water, proteoglycans, and collagens, with other proteins and glycoproteins present in smaller amounts

Glycosaminoglycans Polysaccharides consisting of long-chain, unbranched, repeating disaccharide units, such as keratan sulfate and chondroitin sulfate

Middle (transitional) zone The middle zone of articular cartilage in which collagen fibers are less organized and the chondrocytes have a more rounded appearance

Osteochondral fractures Injuries that disrupt articular cartilage and the underlying subchondral bone

Proteoglycans Complex macromolecules that consist of a protein core with covalently bound polysaccharide (glycosaminoglycan) chains

Superficial tangential zone The smooth, nearly frictionless gliding surface of articular cartilage in which thin collagen fibrils are arranged parallel to the surface

Tidemark A wavy bluish line visible on histologic staining with hematoxylin and eosin that signifies the border between the deep zone and the zone of calcified cartilage

References

1. Buckwalter JA, Mankin HJ: Articular cartilage: Part I. Tissue design and chondrocyte-matrix interactions. *J Bone Joint Surg Am* 1997;79:600-611.
2. Mow VC, Ateshian GA: Lubrication and wear of diarthrodial joints, in Mow VC, Hayes WC (eds): *Basic Orthopaedic Biomechanics,* ed 2. Philadelphia, PA, Lippincott-Raven, 1997, pp 275-315.
3. Setton LA, Zhu W, Mow VC: The biphasic poroviscoelastic behavior of articular cartilage: Role of the surface zone in governing the compressive behavior. *J Biomech* 1993;26:581-592.
4. Aydelotte MB, Schumacher BL, Kuettner KE: Heterogeneity of articular chondrocytes, in Kuettner KE, Schleyerbach R, Peyron JG, Hascall VC (eds): *Articular Cartilage and Osteoarthritis.* New York, NY, Raven Press, 1992, pp 237-249.
5. Stockwell RA: (ed): *Biology of Cartilage Cells.* Cambridge, England, Cambridge University Press, 1979.
6. Mankin HJ, Thrasher AZ: Water content and binding in normal and osteoarthritic human cartilage. *J Bone Joint Surg Am* 1975;57:76-80.
7. Mankin HJ: The water of articular cartilage, in Simon WH (ed): *The Human Joint in Health and Disease.* Philadelphia, PA, University of Pennsylvania Press, 1978, pp 37-42.
8. Eyre DR: Collagen structure and function in articular cartilage: Metabolic changes in the development of osteoarthritis, in Kuettner KE, Goldberg VM (eds): *Osteoarthritic Disorders.* Rosemont, IL, American Academy of Orthopaedic Surgeons, 1995, pp 219-227.
9. Eyre DR, Wu JJ, Woods P: Cartilage-specific collagens: Structural studies, in Kuettner KE, Schleyerbach R, Peyron JG, Hascall VC (eds): *Articular Cartilage and Osteoarthritis.* New York, NY, Raven Press, 1992, pp 119-131.
10. Buschmann MD, Gluzband YA, Grodzinsky AJ, et al: Mechanical compression modulates matrix biosynthesis in chondrocyte/agarose culture. *J Cell Sci* 1995;108:1497-1508.
11. Gray ML, Pizzanelli AM, Grodzinsky AJ, et al: Mechanical and physiochemical determinants of chondrocyte biosynthetic response. *J Orthop Res* 1988;6:777-792.
12. Kim YJ, Sah RL, Grodzinsky AJ, et al: Mechanical regulation of cartilage biosynthetic behavior: Physical stimuli. *Arch Biochem Biophys* 1994;311:1-12.
13. Buckwalter JA: Activity vs. rest in the treatment of bone, soft tissue and joint injuries. *Iowa Orthop J* 1995;15:29-42.
14. Buckwalter JA, Lane NE: Athletics and osteoarthritis. *Am J Sports Med* 1997;25:873-881.
15. Buckwalter JA, Woo SL, Goldberg VM, et al: Soft-tissue aging and musculoskeletal function. *J Bone Joint Surg Am* 1993;75:1533-1548.
16. Roughley PJ: Articular cartilage: Matrix changes with aging, in Buckwalter JA, Goldberg VM, Woo SL-Y (eds): *Musculoskeletal Soft-Tissue Aging: Impact on Mobility.* Rosemont, IL, American Academy of Orthopaedic Surgeons, 1993, pp 151-164.
17. Martin JA, Ellerbroek SM, Buckwalter JA: Age-related decline in chondrocyte response to insulin-like growth factor-I: The role of growth factor binding proteins. *J Orthop Res* 1997;15:491-498.
18. Martin JA, Buckwalter JA: Telomere erosion and senescence in human articular cartilage chondrocytes. *J Gerontol A Biol Sci Med Sci* 2001;56:B172-B179.
19. Buckwalter JA, Mankin HJ: Articular cartilage: Part II. Degeneration and osteoarthrosis, repair, regeneration, and transplantation. *J Bone Joint Surg Am* 1997;79:612-632.
20. Dieppe P: The classification and diagnosis of osteoarthritis, in Kuettner KE, Goldberg VM (eds): *Osteoarthritic Disorders.* Rosemont, IL, American Academy of Orthopaedic Surgeons, 1995, pp 5-12.
21. Felson DT: The epidemiology of osteoarthritis: Prevalence and risk factors, in Kuettner KE, Goldberg VM (eds): *Osteoarthritic Disorders.* Rosemont, IL, American Academy of Orthopaedic Surgeons, 1995, pp 13-24.
22. Schiller AL: Pathology of osteoarthritis, in Kuettner KE, Goldberg VM (eds): *Osteoarthritic Disorders.* Rosemont, IL, American Academy of Orthopaedic Surgeons, 1995, pp 95-101.
23. Buckwalter JA, Martin J: Degenerative joint disease. *Clin Symp* 1995;47:2-32.
24. Buckwalter JA, Mow VC, Hunziker EB: Concepts of cartilage repair in osteoarthritis, in Moskowitz RW (ed): *Osteoarthritis: Diagnosis and Medical/Surgical Management,* ed 3. Philadelphia, PA, WB Saunders, 2001, pp 101-114.

Chapter 3

Synovial Tissue

Synovial tissue is found in joints, tendon sheaths, and bursae and allows smooth motion. Synovial tissue is organized in a membranous structure called the synovium. This chapter describes the gross anatomy, histology, and function of synovium, on which the joints, tendons, and bursae depend for motion and metabolism. Synovial fluid, which is produced by the synovium and functions as the conduit for the synovium's complex role, will also be discussed.

Both inflammatory and noninflammatory conditions affect the synovium. The discussion of normal synovium and synovial fluid will lay the groundwork for understanding these conditions as well as synovial fluid collections, termed effusions, which occur in inflammatory and noninflammatory states. An understanding of normal synovium is also required to appreciate uncommon diseases, such as pigmented villonodular synovitis (PVNS) and synovial chondromatosis, that are unique to synovial tissue. The second half of this chapter focuses on the pathologic conditions that can affect the synovium.

Physiologic Function

Synovial joints have two opposing surfaces covered with articular cartilage that are attached at their periphery by the joint capsule. The synovium is the lining tissue of the capsule and originates at the periphery of the articular cartilage on both sides of the joint, coating the capsular structures. The capsule, along with its synovial lining, creates a pouch that contains synovial fluid secreted by the synovium. All intra-articular ligaments, such as the anterior cruciate ligament in the knee and ligamentum teres in the hip, are also surrounded by synovium. Intra-articular tendons, such as tendons of the long head of the biceps in the shoulder and the popliteus in the knee are also covered by synovial tissue. The unique structure of synovial joints permits the sliding of the articular surfaces with less friction than is created by ice rubbing on ice.

Synovium is also found lining bursae and tendon sheaths. Similar to its function in the synovial joint, the synovium secretes fluid and allows the smooth passage of tendons through their sheath and decreases friction between a bony prominence and overlying skin or tendon.

Normal Histology and Anatomy

Histology

Synovial cells are classically defined as type A or type B synoviocytes. Type A synoviocytes are macrophage-like and derived from the hematopoietic system. These cells contain multiple lysosomes and play a phagocytic role in removing particulate debris from the joint. Type B synoviocytes are fibroblastic in nature and derived from the mesenchyme. These cells also secrete the hyaluronic acid present in synovial fluid and in the matrix surrounding synovial cells.

Gross Anatomy

The number of cell layers that form the synovium varies but usually ranges between two and four cell layers thick (Fig. 1). Synoviocytes that cover structures subject to pressure (such as tendons and ligaments) may be widely separated from one another. Loose cellular connections make the synovial lining semipermeable. The topography of the synovial lining consists of multiple villi and microvilli, which results in a huge surface area. The surface area of the synovial tissue in the knee alone is 100 m^2.

A rich vascular network, adipose tissue, and cells such as fibroblasts, macrophages, and lymphocytes, support the synovium. This subsynovial layer becomes more fibrous with dense bands of collagen as it merges with the underlying joint capsule. Diffusion is essential to the synovium's role in articular nutrition, lubrication, and the removal of debris.

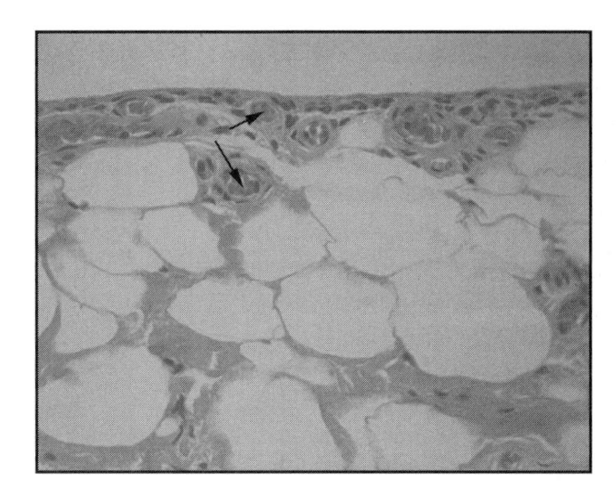

Figure 1
Normal synovium consisting of two cell layers of synoviocytes overlying fibrofatty supportive stroma. Note the presence of blood-filled capillaries (arrows) immediately beneath the synoviocytes.

(Reproduced with permission from Carpenter CA, Rosenberg AE, Freiberg AA: Synovial conditions of the knee, in JJ Callahan, AG Rosenberg, HE Rubash, PT Simonian, TA Wickiewicz (eds): ***The Adult Knee****. Philadelphia, PA, Lippincott-Williams & Wilkins, in press.)*

The large surface area and semipermeable nature of the synovium are essential for efficient delivery of nutrients and removal of waste products from the articular chondrocytes. The synovial fluid provides a conduit for this exchange of materials. The articular surface does not have a blood supply or lymphatic system and thus depends on the synovial fluid to deliver and remove metabolites and to lubricate the articular interface to allow frictionless motion.

Synovial Fluid

Normal joints contain only a small amount of clear, straw-colored synovial fluid. For example, the adult knee normally contains only 2 mL. The fluid is formed by the filtration of capillary plasma through a sieve of hyaluronate molecules. Most small molecules pass from the subsynovial capillaries through the synovium by diffusion. The concentration of electrolytes and glucose in nondiseased synovial fluid is the same as in plasma.

Because synovial fluid is created by filtration, it contains proteins in concentrations that are inversely proportional to their molecular weight. Large molecules such as α-2-macroglobulin and immunoglobulins are virtually excluded. Fibrinogen is also excluded; therefore, synovial fluid will not clot. Normal synovial fluid also contains very few white blood cells and virtually no neutrophils. The viscosity of synovial fluid parallels the concentration of hyaluronic acid. Normal synovial fluid has a high concentration of hyaluronate and therefore high viscosity.

Development and Aging

In the embryo, joints develop from the primitive mesenchyme located between the moving components of the maturing skeleton. The synovium originates from a specialized area known as the interzonal mesenchyme.[1] The synovium is not considered to be a true membrane because it lacks the tight intercellular junctions (desmosomes) and associated epithelial structures that separate cells from the vascular network. Instead, the synovium consists of modified connective tissue cells loosely arranged in a bed of hyaluronic acid and collagen. The synovium persists throughout life.

Pathophysiology

Synovitis

Synovitis is a nonspecific term that simply means inflammation of the synovial lining. Numerous conditions cause or are associated with synovitis. The pathologic characteristics of the synovitis vary depending on the condition. Most conditions that cause synovitis can be categorized as either inflammatory or noninflammatory joint disease (Table 1).

There are at least 10 conditions that cause inflammatory joint disease. An acute inflammatory reaction characterizes these diseases. Inflammatory arthritides can be further classified as monoarticular (single joint) or polyarticular (multiple joints).

Bacterial Septic Arthritis

This condition is the most commonly encountered type of monoarticular inflammatory arthritis. The hallmark on histologic examination of the synovium is many polymorphonuclear leukocytes (Fig. 2). Many physicians employ a clinical-decision rule that holds that any effusion with a polymorphonuclear leukocyte concentration above 50,000/mm^3 represents an infection until proven otherwise. The release of proteolytic enzymes by the inflammatory infiltrate will lead to rapid cartilage destruction if this condition is not identified early and treated.

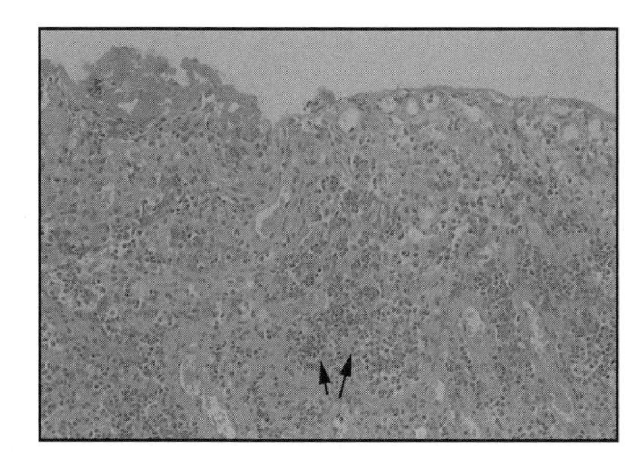

Figure 2
Synovium in septic arthritis. Note polymorphonuclear leukocyte (arrows) infiltration of the synovium and subsynovial tissue.

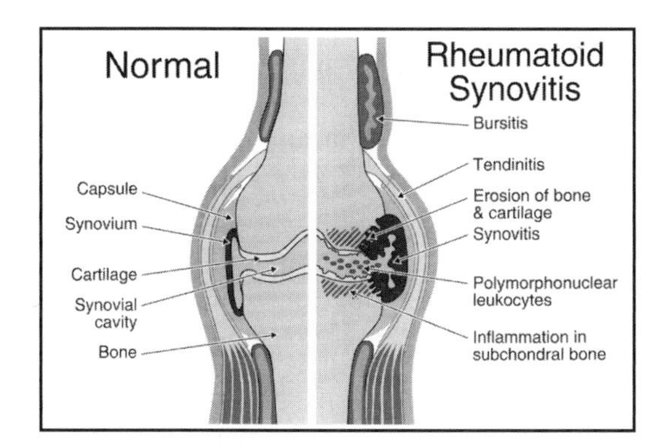

Figure 3
Schematic illustration of a diarthrodial joint and tissues affected by inflammatory processes, using the knee joint as an example. The left side shows structures in the healthy joint; the right side illustrates the widespread involvement of all joint tissues.

(Reproduced from Recklies AD, Poole AR, Banerjee S, et al: Pathophysiologic aspects of inflammation in diarthrodial joints, in Buckwalter JA, Einhorn TA, Simon SR (eds): ***Orthopaedic Basic Science: Biology and Biomechanics of the Musculoskeletal System****, ed 2. Rosemont, IL, American Academy of Orthopaedic Surgeons, 2000, pp 489-530.)*

Rheumatoid Arthritis

Representative of the polyarticular noninfectious inflammatory arthritides, rheumatoid arthritis (RA) is a systemic disease in which synovitis is the essential component involved in symptom development and joint destruction (Fig. 3). The etiology of RA is not entirely understood, but infectious, autoimmune, and hereditary causes are suspected. Over time, the deposition of antigen-antibody complexes within the synovium causes an inflammatory reaction that perpetuates a vicious cycle that results in destruction of the joint. Although RA is a systemic disease, the principal symptoms affect the synovial joints in a symmetric fashion.[2] The disease initially affects the small joints of the hand and wrist, followed by the larger joints as the disease progresses.

Histologically, the early stages of RA are manifest as edema and chronic inflammation in the synovium. Hypertrophy and hyperplasia of the synovial cells produce thickening of the synovial membrane. The subsynovial connective tissue is expanded by a proliferation of capillaries and venules and a cellular infiltrate composed of many lymphocytes, plasma cells, and macrophages. The lymphocytes frequently form cuffs around the blood vessels or become arranged into follicles with central germinal centers (Fig. 4). These changes cause the synovium to appear dense with a papillary architecture. Neutrophils are rarely seen within the synovium because they quickly exit the capillaries, traverse the subsynovium, and pass between the synoviocytes into the synovial fluid.

As RA progresses, subsynovial fibroblasts proliferate and deposit collagen fibers. Eventually, the synovial surface becomes focally coated by organizing fibrin. Some of the aggregates of fibrin float freely in the joint. The fibrin coat also makes the synovial surface sticky, allowing it to adhere to the periphery of the articular surface forming the pannus. The synovium continues to proliferate lead-

Table 1
Conditions Causing Synovitis

Inflammatory Joint Diseases	Noninflammatory Joint Diseases
Rheumatoid arthritis	Osteoarthritis
Juvenile rheumatoid arthritis	Traumatic arthritis
Collagen vascular disease	Charcot arthropathy
Scleroderma	Osteonecrosis
Systemic lupus erythematosus	Osteochondritis dissecans
Septic arthritis	
Crystalline arthropathies	
Gout	
Pseudogout	
Psoriasis	
Reiter's syndrome	
Ankylosing spondylitis	
Enteropathic arthritis	
Rheumatic fever	

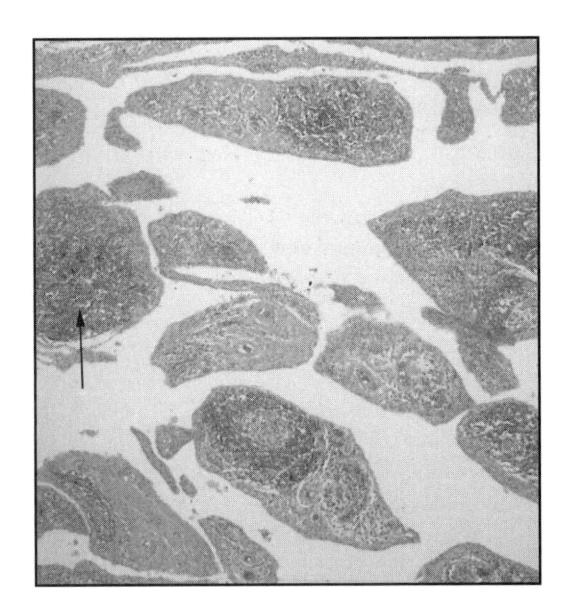

Figure 4
Rheumatoid synovium with papillary transformation of the synovium by inflammatory infiltrate (arrow).

(Reproduced with permission from Carpenter CA, Rosenberg AE, Freiberg AA: Synovial conditions of the knee, in Callahan, Rosenberg, Rubash, Simonian, Wickiewicz (eds): **The Adult Knee***. Philadelphia, PA, Lippincott-Williams & Wilkins, in press.)*

ing to villous fronds of inflamed tissue filling the joint. In the later stages of RA, there is complete destruction of the articular cartilage, creating the "burnt out" joint, which may be fused by either fibrous or bony ankylosis.[3]

Osteoarthritis
This condition (also called degenerative joint disease) is a common, slowly progressive noninflammatory disorder of the joints. Osteoarthritis (OA) is associated with aging, but no true etiology is known. Mechanical wear, trauma, and genetics play a role in cartilage loss. Single or multiple joints may be affected.

Unlike RA, in which the cartilage is destroyed by synovitis and proteolytic enzymes, OA appears to be limited (at least initially) to the cartilage. The subsequent synovitis is a response to articular cartilage injury and loss. The synovium develops a villous pattern with hyperplasia of the synovial membrane. The subsynovium typically lacks significant cell infiltrate, but a chronic inflammatory infiltrate may be seen in the synovium. As cartilage destruction occurs, small fragments will be shed into the synovial fluid, which may then be deposited in the synovium.

Effusions
Inflammatory and noninflammatory synovitis alters both the synovium and the composition of synovial fluid. In the synovial fluid of inflammatory joint diseases, the lysosomal enzymes released break down the hyaluronic acid "sieve" barrier. The loss of the hyaluronic sieve allows large proteins to enter the synovial fluid. This protein-rich fluid can form clots resulting from the presence of fibrinogen. A decreased concentration of hyaluronic acid within the synovial fluid also reduces its viscosity.

In inflammatory conditions, leukocytes pass freely across the synovium and into the synovial fluid. As the white blood cell count within the synovial fluid increases, its clarity decreases. Fluid that contains large numbers of crystals also appears cloudy on gross examination and can be confused with the fluid of infection. In cases of infection or severe inflammation, impaired glucose delivery and increased metabolism by synovial components combine to decrease the glucose level in the synovial fluid relative to plasma concentration; Table 2 summarizes the characteristics of synovial fluid in different types of effusions. Note that parameters suggested in Table 2 are only rough guidelines, as the microscopic and macroscopic characteristics of synovial fluids vary widely depending on the condition. Host response, underlying disease states, and medications such as steroids can greatly alter the composition of synovial fluids.

Aspiration of the joint (arthrocentesis) hones the differential diagnosis. If there is a history of trauma, blood in the joint (hemarthrosis) suggests a ligament tear, fracture, retinacular tear (ie, patellar dislocation), or peripheral meniscal tear. If there is no history of trauma, a hemarthrosis suggests a clotting disease, hemangioma, or PVNS.

Benign Synovial Tumors
Neoplasms of the synovium are uncommon and can involve the synovial lining of joints, bursae, and tendons. Most tumors of the synovium are benign and mimic the cells and tissue types that compose the articular and periarticular tissues. Hemangiomas and lipomas are examples of benign tumors that can affect the synovium and mimic the native cell types.[4-6] In addition, pigmented villonodular synovitis and synovial chondromatosis are two uncommon yet benign diseases that are unique to the synovium.

Table 2
General Characteristics of Synovial Fluid in Different Types of Effusions

Characteristics	Normal	Inflammatory Conditions	Noninflammatory Conditions	Septic
Volume	1 to 4 mL	Increased	Increased	Increased
Color	Clear to pale yellow	Cloudy, yellow/green	Straw	White/yellow
Clarity	Transparent	Opaque	Transparent	Opaque
Viscosity	Very high	Low	High	Very low
White blood cell count (per mm^3)	Few	10,000 to 20,000	500	50,000 or more
Predominant white blood cell type	Monocytes	Neutrophils	Monocytes	Neutrophils
Glucose (relative to serum)	Equal	Low	Equal	0 to Low
Crystals	Negative	Positive in gout and CPPD	Negative	Negative
Cultures	Negative	Negative	Negative	Positive

CPPD = crystalline
(Reproduced with permission from Carpenter CA, Rosenberg AE, Freiberg AA: Synovial conditions of the knee, in Callahan, Rosenberg, Rubash, Simonian, Wickiewicz (eds): ***The Adult Knee.*** *Philadelphia, PA, Lippincott-Williams & Wilkins, in press.)*

Pigmented villonodular synovitis is a proliferative process of the synovial membrane that most commonly affects the knee and tendon sheaths of the fingers. It is also found less commonly in the hip, ankle, toe, or wrist.[7,8] The pathogenesis of PVNS remains unclear; inflammatory, metabolic, and neoplastic causes have been proposed.

The onset of PVNS typically is insidious with slow progression of symptoms, including pain, warmth, swelling, stiffness, and sometimes a palpable mass. Men may be slightly more affected than women, with most cases occurring in the third or fourth decade of life.

Plain radiographs can be normal or reveal a soft-tissue density mass with osseous erosions, cysts, and loss of joint space. PVNS has low-signal intensity on T1- and T2-weighted MRI scans because of hemosiderin deposition within the lesion. MRI scans show the lesion throughout the synovium, but the lesion may extend beyond the joint bursa.[9]

In the diffuse form of PVNS, most of the synovium is affected, yet some areas remain unaffected. This form of PVNS is characterized by many villi that coalesce and fuse into nodules that stud the synovial surface. These nodules range in size from less than a millimeter to several centimeters and are red-brown to yellow. The localized form of PVNS is characterized by a single, pedunculated or sessile red-brown or yellow nodule that is attached to the synovium. Microscopically, both forms have identical morphologic features. The subsynovium is expanded by a proliferation of medium-sized polyhedral cells with central nuclei and eosinophilic cytoplasm. Scattered throughout are hemosiderin-laden histiocytes, foamy macrophages, and multinucleated giant cells. The stroma is collagenous and frequently focally sclerotic. The lesion can also invade adjacent bones. Aspiration of the joint often reveals large effusions composed of a brownish, bloody fluid.

Primary synovial chondromatosis is characterized by growth of hyaline cartilage nodules within the synovial lining of a joint, bursa, or tendon sheath.[10,11] It is unclear whether these nodules are a result of a metaplastic or neoplastic process.

Patients between the ages of 30 and 50 years are most commonly affected, with men affected twice as often as women. There is no known cause. The knee joint is most frequently involved, and it is usually unilateral in distribution. Common symptoms include pain, swelling, and mechanical symptoms such as locking or giving way. Physical examination may reveal an effusion, a palpable mass, tenderness, soft-tissue swelling, muscle atrophy, and crepitus.

Once the cartilaginous nodules form, they may detach and become loose bodies.

These nodules differ from those that may develop secondary to trauma or degenerative joint disease. These conditions produce joint debris that sticks to the synovium, a process called secondary synovial chondromatosis. The nodules may be loose or attached to the synovial lining and range in size from a few millimeters to several centimeters. Primary and secondary forms of synovial chondromatosis are distinguished histologically. In the primary form of the disease, the nodules of cartilage form de novo. Histologic sections show nodules of hyaline cartilage, which are at most moderately cellular and contain plump chondrocytes (Fig. 5). The chondrocytes may develop from resting fibroblasts, which undergo metaplasia and begin secreting chondroid matrix. The cartilage nodules may calcify or undergo enchondral ossification with the formation of mature lamellar bone.

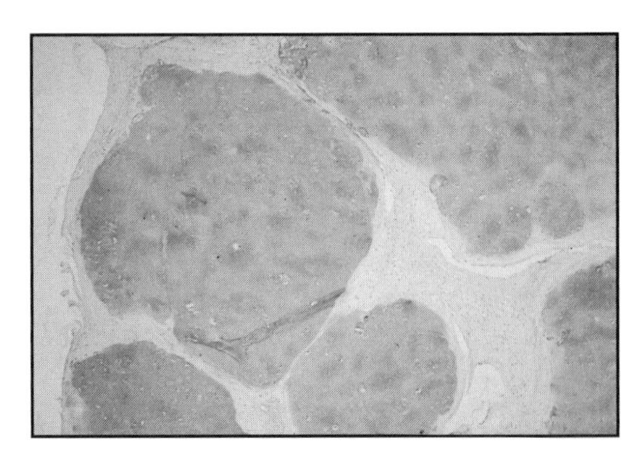

Figure 5
Synovial chondromatosis with multiple nodules of hyaline cartilage.

Research and New Directions During the Bone and Joint Decade (2002-2011)

Synovial Supplementation

One of the most active synovium research frontiers involves direct intra-articular supplementation of synovial products (hyaluronan or its derivative, hylan). It has been observed that synovial fluid viscosity is decreased in osteoarthritis. Injection of synovial fluid supplements (Synvisc [hylan G-F20] or Hyalgan [hyaluronan]) has been shown to be clinically effective. Research is ongoing to determine the mechanism of action of these products and to produce compounds that are both longer lasting and less likely to induce local inflammatory reactions with repeated treatments.[12]

Etiology of RA

RA continues to be a topic of intense investigation. The exact instigator of synovial inflammation is still under debate, although recent epidemiologic and laboratory data suggest a noninfectious etiology.[13] Investigation of cellular activities and interactions now include those of chondrocytes, osteoclasts, fibroblasts, and the neuroendocrine system.[14,15] The focus of such research is not only on elucidation of the basic biology of inflammatory joint diseases but on application of this knowledge to creating novel strategies for preventing and managing joint inflammation.

Key Terms

Arthrocentesis Aspiration of the joint

Capsule Part of the synovial joint that, along with the lining, creates a pouch in which synovial fluid is secreted

Desmosomes A small body that forms the site of attachment between cells

Hemarthrosis The presence of blood in the joint

Monoarticular Affecting a single joint

Pannus A proliferation of synovium beginning at the periphery of the joint surface as seen in rheumatoid arthritis

Pigmented villonodular synovitis A proliferative process of the synovial membrane of unknown etiology

Polyarticular Affecting multiple joints

Primary synovial chondromatosis A condition in which hyaline cartilage nodules grow within the synovial lining of a joint, bursa, or tendon sheath

Synovial fluid The straw-colored fluid in the joint that is formed by filtration of capillary plasma

Synovial joints Joints that have two opposing surfaces covered with articular cartilage and are attached at their periphery by the joint capsule

Synoviocytes Cells that form the synovial membrane, remove debris, and secrete hyaluronic acid

Synovitis A condition characterized by inflammation of the synovial lining

Synovium A membrane-like lining of the synovial joints, tendon sheaths, and bursae

Type A synoviocytes A type of synovial cell that plays a role in removing particulate debris from the joint

Type B synoviocytes A type of synovial cell that secretes hyaluronic acid

References

1. Sledge CB: Structure, development, and function of joints. *Orthop Clin North Am* 1975;6:619-628.
2. Arnett FC, Edworthy SM, Bloch DA, et al: The American Rheumatism Association 1987 revised criteria for the classification of rheumatoid arthritis. *Arthritis Rheum* 1988;31:315-324.
3. Robertsson O, Knutson K, Lewold S, et al: Knee arthroplasty in rheumatoid arthritis: A report from the Swedish Knee Arthroplasty Register on 4,381 primary operations 1985-1995. *Acta Orthop Scand* 1997;68:545-553.
4. Coventry MB, Harrison EG Jr., Martin JF: Benign synovial tumors of the knee: A diagnostic problem. *J Bone Joint Surg Am* 1966;48:1350-1358.
5. Jaffe HL: (ed): *Tumors and Tumorous Conditions of the Bones and Joints*. Philadelphia, PA, Lea & Febiger, 1958.
6. Moon NF: Synovial hemangioma of the knee joint: A review of previously reported cases and inclusion of two new cases. *Clin Orthop* 1973;90:183-190.
7. Jaffe HL, Lichtenstein L, Sutro CJ: Pigmented villonodular synovitis, bursitis, and tenosynovitis: A discussion of the synovial and bursal equivalents of the tenosynovial lesion commonly denoted as xanthoma, xanthogranuloma, giant cell tumor or myeloplaxoma of the tendon sheath, with some consideration of this tendon sheath lesion itself. *Arch Pathol* 1941;31:731-765.
8. Granowitz SP, D'Antonio J, Mankin H: The pathogenesis and long-term end results of pigmented villonodular synovitis. *Clin Orthop* 1976;114:335-351.
9. Giudici MA, Moser RP Jr., Kransdorf MJ: Cartilaginous bone tumors. *Radiol Clin North Am* 1993;31:237-259.
10. Jeffreys TE: Synovial chondromatosis. *J Bone Joint Surg Br* 1967;49:530-534.
11. McCarthy EF, Dorfman HD: Primary synovial chondromatosis: An ultrastructural study. *Clin Orthop* 1982;168:178-186.
12. Leopold SS, Warme WJ, Pettis PD, et al: Increased frequency of acute local reaction to intra-articular hylan GF-20 (Synvisc) in patients receiving more than one course of treatment. *J Bone Joint Surg Am* 2002;84:1619-1623.
13. Silman AJ, Pearson JE: Epidemiology and genetics of rheumatoid arthritis. *Arthritis Res* 2002;4(suppl 3):S265-S272.
14. Choy EH, Panayi GS: Cytokine pathways and joint inflammation in rheumatoid arthritis. *N Engl J Med* 2001;344:907-916.
15. Straub RH, Cutolo M: Involvement of the hypothalamic—pituitary—adrenal/gonadal axis and the peripheral nervous system in rheumatoid arthritis: Viewpoint based on a systemic pathogenetic role. *Arthritis Rheum* 2001;44:493-507.

Chapter 4

Ligaments

Skeletal ligaments are highly organized, fibrous tissues that connect bone to bone. Some ligaments are large and easily seen or felt; others are small and subtle. All share the task of protecting the joints from instability and allowing normal motion to occur with minimal resistance.

The orientation of a ligament relative to the plane of the joint it crosses determines its mechanical function. For example, the anterior talofibular ligament, the structure most commonly injured in an ankle sprain, attaches the distal fibula to the lateral side of the talus, sloping somewhat anteriorly as it courses toward its distal attachment. Thus, it resists inversion of the ankle joint, especially when the ankle is plantar flexed (the forefoot pointed slightly toward the floor).

Physiologic Function

Ligaments are like ropes in that they offer little resistance to compression, but they are strong in tension. A structure such as a ligament, whose mechanical properties depend on the orientation of the force applied, is anisotropic. To stabilize the joint, most ligaments work in pairs. In the ankle, the lateral talofibular ligament is balanced by the deltoid ligament on the medial side.

A ligament's resistance to tension differs slightly compared with that of a tendon, the tissue that attaches a muscle to bone. A tendon is uniformly stiff, and it does not elongate much when pulled. This stiffness ensures that the entire tug of the muscle is used to move the joint and that no force is wasted by simply elongating the tendon. A ligament does, however, have some built-in laxity in response to low tension forces. This lower stiffness allows the joint to withstand small deforming forces without damage—just as a tree may bend and not break in response to wind. At higher tension, however, the ligament becomes stiffer, thus keeping the joint stable. The structural properties of tissue can be described by measuring the relationship between stress and strain (Fig. 1). Stress is the deforming force, defined as the amount of load per unit of cross-sectional area of the tissue. Strain is defined as the amount of tissue elongation divided by the original tissue length—a percentage of the deformation. When a material is stressed, a given strain will be observed.

Consider, for example, an individual walking over rocky terrain. Whenever weight is disproportionately borne on the medial side of the foot, an inversion force is placed on the lateral ankle. When the forces are small, the anterior talofibular ligament stretches in response to the load and springs back to its normal length once the force

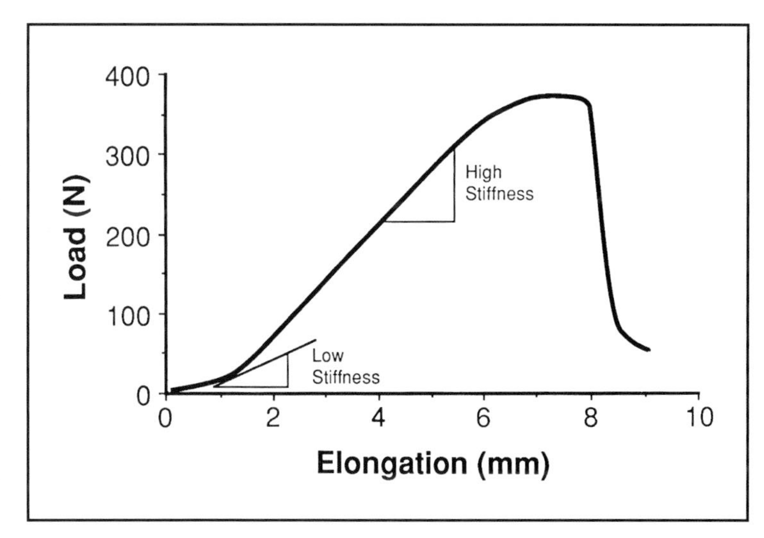

Figure 1
The mechanical response of the tissue can be illustrated by plotting a load versus elongation curve or a stress versus strain curve. The slope of this curve defines the stiffness of the tissue. The load-elongation curve shown in the graph is typical of that of ligaments, with an initial region of low stiffness (the toe region) and a second region of high stiffness (linear region).

(Reproduced from Woo SLY, An KN, Frank CB, et al: Anatomy, biology, and biomechanics of tendon and ligament, in Buckwalter JA, Einhorn TA, Simon SR (eds): ***Orthopaedic Basic Science: Biology and Biomechanics of the Mulsculoskeletal System****, ed 2. Rosemont, IL, American Academy of Orthopaedic Surgeons, 2000, pp 581-616.)*

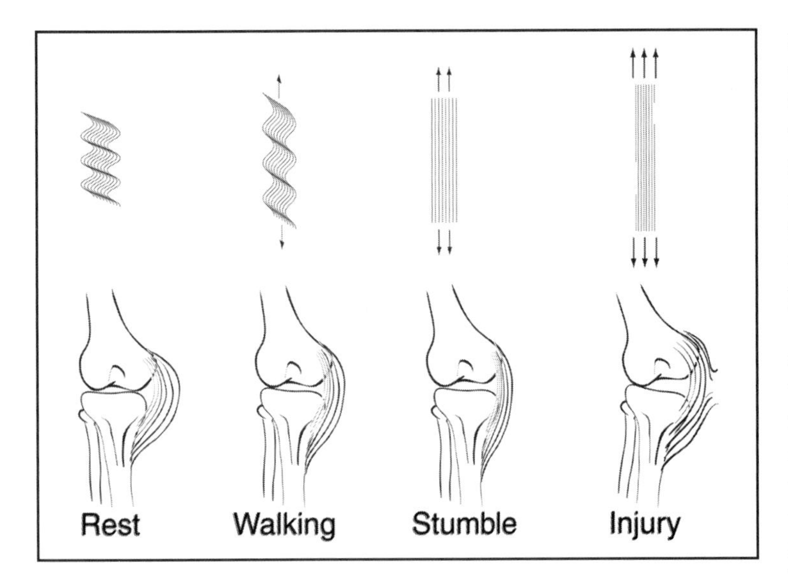

Figure 2
Schematic representation of the elongation of the medial collateral ligament as progressively stronger valgus stresses are applied to the knee. Initially, the wavy collagen fibers straighten, but as stronger forces are placed across the knee, the fibers within the ligament begin to rupture, resulting in ligament injury.

abates. When the forces are increased and the joint threatens to give way, the ligament becomes taut and barely elongates at all. Of course, when the force exceeds the ligament's own strength, such as that produced when a jumping basketball player lands unevenly on another player's foot, the ligament will tear and no longer stabilize the joint.

It is worth noting that the extent of ligament disruption does not always predict the degree of clinical instability that an individual will perceive. There are instances in which the lateral ankle ligaments are completely torn and yet there may not be much instability, as the individual can use his or her muscles (the peroneus longus and brevis) to actively stabilize the joint. Conversely, in some cases, even once the torn ligament heals, the individual may still sense subtle instability. This perceived instability may occur because the nerves within the ligament that sense movement and displacement can be damaged in a severe sprain. When these nerves are injured, the individual may lose proprioception and report a poor sense of control, even if instability is not objectively confirmed.

Normal Histology and Composition

The gross inspection of a ligament reveals band-like connective tissue, with little suggestion of its microscopic complexity. In fact, skeletal ligaments are intricate amalgams of extracellular matrix proteins that are maintained by a diverse population of cells. Ligaments are typically composed of longitudinally oriented fascicles, 20 to 400 μm in diameter, that course between bony insertion sites. The fascicles consist of densely organized collagen fiber bundles that are approximately 20 μm wide.

When viewed along the longitudinal axis of the fascicles, an organized, sinusoidal waveform can be appreciated. This pattern is referred to as the crimp. The longitudinal orientation of fibers and the crimp within them produce the unique structural properties of ligament, namely, anisotropy and increased stiffness in response to increased load. The longitudinal orientation of the fascicles produces anisotropy: the ligament on the whole behaves like a rope because its elemental constituents are molecular ropes. The crimp is responsible for the nonlinear stiffness of ligament.[1] As noted above, the stiffness of a ligament increases with increasing tension. This change in stiffness occurs because some of the fibrils within the crimp region are bent like a spring when the ligament is not under any tension. When tension is first applied, the crimp elongates (Fig. 2). After the crimp is maximally stretched, however, any additional displacement requires stretching of the individual collagen fibrils themselves.[2]

Ligament cells are typically oriented along the longitudinal axis of the tissue. These cells have been distinguished histologically on the basis of nuclear shape and the presence of a lacunar space. Three principal cell types have been described: fusiform, ovoid, and spheroid. Fusiform cells are spindle-shaped and have been noted to be intimately related to the crimped collagen fibers. Ovoid and spheroid cells are typically found in columns, with amorphous extracellular matrix in the pericellular space. The specific function of these types of cells is unknown.

The vascular supply of ligaments typically originates in the blood vessels near the joint surfaces. Bundles of individual collagen fibrils within the ligament are surrounded by a vascularized layer of loose connective tissue known as the endoligament. The surface of the entire ligament is also covered with a vascularized layer called the epiligament. The vascularity of ligaments is not necessarily constant throughout their entire length. In the anterior cruciate ligament (ACL) of the knee, for example, there is greater vascularity at the femoral end, with vessels seen predominantly in the epiligament and endoligament. Few vessels are seen in the fibers themselves. The cells within the fibers depend primarily on diffusion of nutrients from the surrounding vascularized tissues.

Nerve fibers are typically located near the vessels, also predominantly in the epiligament. Four types of cells have been identified in the ligaments of the knee, specifically Ruffini receptors, pacinian receptors, Golgi receptors, and free unmyelinated nerve endings. The neurosensory role of these receptors has been supported by studies demonstrating somatosensory evoked potentials with mechanical stimulation of the human ACL,[3] as well as changes in electromyographic function after disruption of the ligament.

Mechanoreceptors in ligament play an important role in the body's proprioception by helping to provide the sense of where the body is in space. In an injured ligament, this function can be disrupted, causing disability that is seemingly out of proportion to the objective skeletal injury. When the ligament has healed but these nerve pathways have not yet been reestablished, an individual not staring directly at his or her foot may have a poor sense of where the foot is relative to the ground. This subtle clumsiness may be reported as instability: an individual will say that he or she simply does not "trust" the ankle even if an examiner cannot document objective instability. It may take a while for proprioception to return—long after the ligament is healed—but it may be restored more quickly with physical therapy and rehabilitation.

Ligament attaches to bone in two ways. In the direct attachment pattern, four histologic zones are seen[4] (Fig. 3). The first zone contains the ligament itself, the second zone contains nonmineralized fibrocartilage, the third zone contains mineralized fibrocartilage, and the fourth zone contains bone. Some ligamentous fibers pass directly into bone without an intervening layer of fibrocartilage. In this so-called indirect attachment, the fibers of the ligament blend with those of the periosteal layer.[5]

Collagen makes up 70% to 80% of the dry weight of ligaments. There are at least 15 distinct types of collagen, all of which are macromolecular triple helices composed of three amino acid alpha chains. All alpha chains are composed of repeating triplets of polypeptides. The most common sequence begins with the small amino acid glycine, followed typically by proline and hydroxyproline. Hydroxylysine (with or without a sugar attached) is seen frequently as well. Because of the geometry of the proline molecule itself, each alpha chain is wound into a left-handed helix. These in turn intermingle into one long, right-handed triple helix. This spiral configuration gives collagen the distinct ability to resist tensile forces.

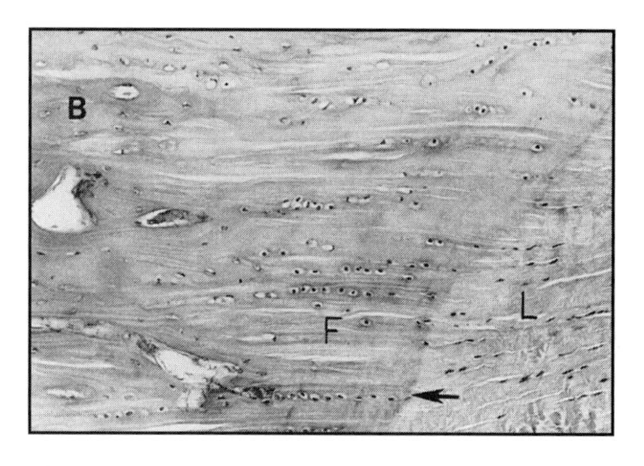

Figure 3
The four zones of the direct insertion of the medial collateral ligament into the femur. The deep fibers of the ligament (first zone) (L) pass through a well-defined area (arrow) combining a zone of uncalcified fibrocartilage and a zone of calcified fibrocartilage (second and third zones) (F) before inserting into the bone (fourth zone) (B).

(Reproduced with permission from Woo SLY, Gomez MA, Sites TJ, et al: The biomechanical and morphological changes in the medial collateral ligament of the rabbit after immobilization and remobilization. J Bone Joint Surg Am *1987;69:1200-1211.)*

The type of collagen is defined by the molecular constituents of the amino acid chains as well as the presence of specific nonhelical domains. More than 90% of the collagen found in ligament is fibrillar type I collagen, with most of the remainder thought to be type III. Type II is found in articular cartilage and is more suited to resist compression. Fibrillar type I collagen is produced intracellularly and modified extracellularly by cleavage. Once modified extracellularly, these molecules then aggregate into microfibrils. A microfibril is composed of a staggered array of adjacent collagen molecules, each offset by a quarter of its length. This allows many oppositely charged amino acids to align and bond, giving the fibril additional tensile strength (Fig. 4).

Collagen is synthesized and degraded continuously, with a half-life of 300 to 500 days.

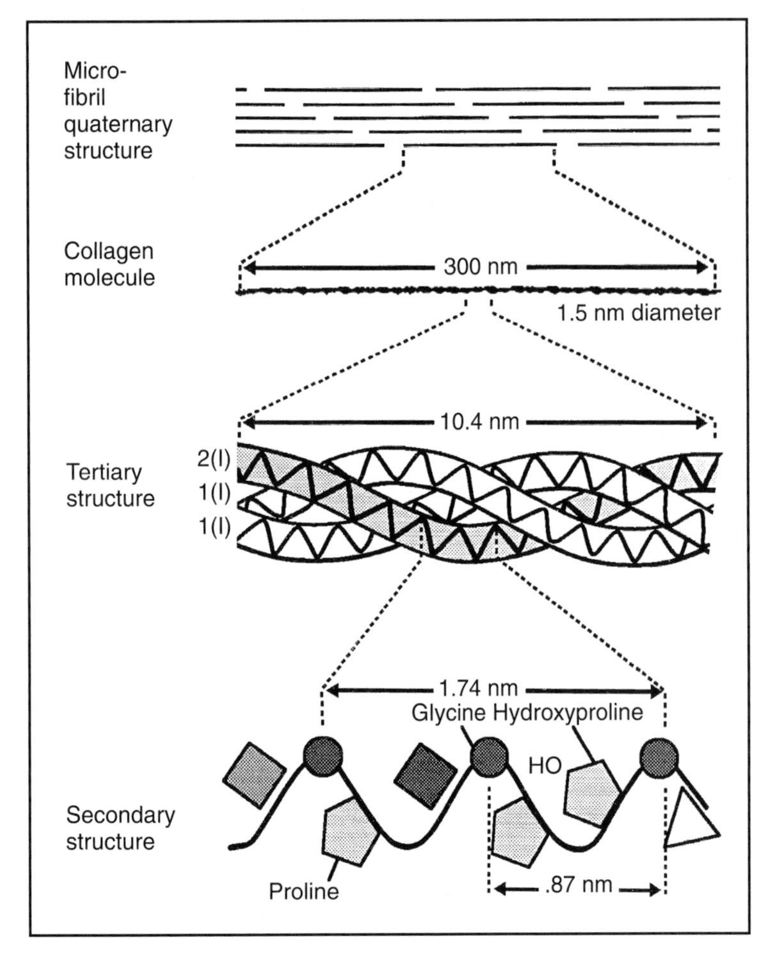

Figure 4

Schematic representation of the structural organization of collagen into the microfibril. The secondary structure is a single helix, and the tertiary structure is a triple helix.

(Reproduced from Woo SLY, An KN, Frank CB, et al: Anatomy, biology, and biomechanics of tendon and ligament, in Buckwalter JA, Einhorn TA, Simon SR (eds): ***Orthopaedic Basic Science: Biology and Biomechanics of the Musculoskeletal System****, ed 2. Rosemont, IL, American Academy of Orthopaedic Surgeons, 2000, pp 581-616.)*

Consequently, after a period of years, an individual essentially rebuilds his or her native ligaments. This turnover is akin to the remodeling of bone; however, in contrast to the process in bone, the factors regulating collagen remodeling have yet to be elucidated.

Proteoglycans, such as chondroitin 4-sulfate and dermatan sulfate, make up less than 1% of the total dry weight of ligaments. Proteoglycan molecules are polar and thus hydrophilic, binding water within the ligament. Ligaments also contain elastin, fibronectin, and other glycoproteins. Up to 70% of the wet weight of ligaments is water, which is both structurally bound to collagen and freely associated with the interfibrillar gel.

Development and Aging

Ligaments develop in the periarticular mesenchymal tissue, with a layer of synovium separating them from the joint spaces. Development of the ACL, for example, begins with fibroblasts aligning between the femoral and tibial attachment sites. Collagen fibrils are deposited between the cells, parallel to the longitudinal axis of the ligament. The organelles of the fetal and young adult rat ligament cells contain much rough endoplasmic reticulum and a prominent Golgi apparatus, which suggests that these are actively synthesizing cells. As the cells deposit more and more collagen, the bundles of matrix push the cells apart, thus forming the mature ligament with a lower density of cells and higher collagen content.

Studies of growth in rabbit medial collateral ligaments (MCLs) have demonstrated consistent elongation of the ligament, in contrast to the focal centers of growth located at the physes (ie, growth plates) in long bones.[6] At the site where ligament inserts into bone, bone cells divide rapidly, and the collagen of the ligament is incorporated into the bone. This active bone remodeling at the insertion site allows the ligament to move in concert with the longitudinal growth of the bone. Therefore, the ligament remains attached to the bone's metaphyseal region.

Changes in cell content and ligament structure as a function of aging are not yet well known in humans. In animal models, however, it has been shown that the stiffness of the MCL of the knee reaches a maximum at maturity and then gradually declines.[7] It is not yet known whether this phenomenon is present in all ligaments or in all species. The strength of the attachment of ligament to bone is also suspected to vary with age; adults tend to rupture their ACL in its midsubstance, whereas in skeletally immature patients, the injury more commonly occurs in a bony avulsion of the attachment site of the ligament.

Pathophysiology

Injury and Repair

Damage to a ligament is technically termed a sprain. Sprains are graded according to the amount of gross displacement of the joint appreciated on clinical examination and microscopic findings. A grade 1 sprain is associated with pain but no joint instability on clinical examination. This represents intrafibrous injury to the ligament. A grade 2 sprain occurs when some but not all fibers have been torn; because some fibers are still in continuity, the joint is only minimally unstable when the ligament is stressed. A grade 3 sprain is a complete tear of the ligament. Sprains can be caused both by contact or noncontact mechanisms (such as twisting); in fact, any mechanism that produces tension on the ligament can cause injury. Ligament injuries are common, accounting for 25% to 40% of all knee injuries,[8,9] and they are present, by definition, in all joint dislocations.

Ligaments can be broadly divided into those that heal and those that do not, although the propensity for healing is more apt to be a function of the local environment rather than the ligament tissue itself. Ligaments with potential to heal, such as the MCL of the knee, do so by progressing through three characteristic stages: inflammation, proliferation, and remodeling.[10,11] The inflammatory phase begins within hours of the injury and predominates throughout the first 7 to 10 days after the injury.[12,13] This phase is characterized by the presence of necrosis, inflammatory cells, and the growth of neovascular tissue on the surface of the ligament remnant.[13] Corresponding injury to local capillaries causes hematoma formation and signals the continued migration of inflammatory cells. Because inflammation initiates the healing response, use of medications that inhibit inflammation may impede the repair.

The proliferative phase is characterized by an increase in cell density in the ligament remnants with ovoid cells and gradual neovascularization of the entire ligament remnant.[12,13] New collagen formation—the provisional scar—and cell migration from the torn ligament ends into the granulation tissue are also seen during this phase.[12] Thus, the space between the torn ends of the original ligament is filled with a matrix containing disorganized proteins and many cells. In the rabbit MCL, this phase has started by 10 days and peaks at 21 days after rupture.[13]

The transition to the remodeling phase is characterized by a decrease in the proliferative fibroblast response and an increase in matrix alignment.[12,13] Thus, there is less cellularity and vascularity and denser collagen within the scar. Over time, this collagen becomes more aligned with the natural tensile forces of the given joint. In the rabbit MCL, this phase starts at 21 days and continues for several years.[13]

The progression from inflammation to proliferation to remodeling also has been reported in other dense, connective tissues, such as tendon and skin. The product of this process is a functional fibrovascular scar. Through the process of remodeling, this scar becomes like the original, uninjured tissue—but not perfectly so. Although remodeling begins as early as 6 weeks after injury, the final reorganization of the tissue (and thus recreation of normal mechanical properties) is not complete for up to 1 year after injury.

Some ligaments, especially intra-articular ligaments such as the ACL, fail to heal after rupture, even when they are sutured back together. It is not clear whether this lack of clinical healing is caused by an inability to initiate a healing response, an inhibition of the response, or simply a lack of nutrition to support healing. It is likely that the failure of these tissues to heal is the result of a combination of factors.

Because of their healing potential, most injuries to ligaments found outside the joint, or extra-articular ligaments, can be treated successfully with immobilization or primary repair. Other treatments, such as ligament reconstruction, are used for intra-articular ligament injuries. For example, the ACL, which is an intra-articular ligament, is commonly treated by removal of the torn ligament and replacement with a tendon graft that is secured in bony tunnels. This graft is only an approximation of the normal ligament, as it is currently impossible to recreate the complex geometry of the native ACL with an artificial ligament. Moreover, it is probable that some of the neurologic function (ie, proprioception) is lost when the ligament is torn.

Even those ligaments that heal require well-conceived treatment. It is known that immobilization can lead to joint stiffness. This is because without the appropriate signals provided by normal mechanical stress, the healing tissue fails to orient itself in the correct direction. This helter-skelter formation of scar causes adhesions. Prolonged immobilization also weakens intact ligaments, as well as their insertion into bone. These processes are reversible, but reversing them takes time.

Systemic Diseases

The Ehlers-Danlos syndromes are a heterogeneous group of syndromes that cause laxity and weakness of ligaments, skin, and blood vessels. There are at least nine clinical and genetic subtypes, all of which are caused by mutations in fibrillar collagen genes and genes that modify fibrillar collagen. For example, Ehlers-Danlos type I is an autosomal dominant disorder caused by a mutation in the gene that is coding for collagen type V, which is important for forming collagen I fibrils. This defect in fibril formation leads to decreased ligament stiffness and joint laxity.

Marfan's syndrome is an autosomal dominant disease caused by a defect in the gene that is coding for fibrillin. Fibrillin is a large glycoprotein that is a structural component of elastin-containing microfibrils. The gene defect results in decreased fibrillin formation and a structurally weakened elastin. As elastin is one of the structural proteins in ligaments, clinical manifestations of the disease include ligament laxity, loose joints, and other connective tissue problems, such as aortic dilation.

Research and New Directions During the Bone and Joint Decade (2002-2011)

Function and Physiologic Response of Various Cells

Although many cells (including vascular cells, nerve cells, myofibroblasts, and fibroblasts) have been identified in ligaments, the contribution of each cell type to normal ligament function and healing after injury is not completely known. It is thought that better characterization of the normal responses of these cells and the signals that modulate them can allow us to augment the healing process.

Growth Factors and Tissue Engineering

Reconstruction of the ACL, while providing functional stability, does not prevent the premature onset of osteoarthritic changes after a knee injury.[14] This may be because reconstruction allows only an approximation of the native ligament. Scientists are therefore seeking new treatment methods to promote healing of torn intra-articular ligaments to preserve innervation and the precise ligament geometry.[15]

Gene Therapy

Injured ligaments heal with tissue that is often mechanically inferior to the original tissue, probably because the collagen of the repair tissue is not identical to the original. Investigators are studying the possibility of using gene therapy to direct the healing response to reproduce the original, native collagen structure.

Key Terms

Anisotropic Having unlike mechanical properties in different directions; that is, the mechanical properties depend on the orientation of the applied force

Crimp An organized, sinusoidal waveform of ligament fascicles that provides minimal stiffness in response to light stress but increased stiffness in response to increased stress

Ehlers-Danlos syndromes A heterogeneous group of syndromes that cause laxity and weakness of ligaments, skin, and blood vessels

Elastin One of the structural proteins in ligaments

Endoligament The vascularized layer of loose connective tissue that surrounds bundles of individual collagen fibrils within a ligament

Epiligament The vascularized layer that covers the surface of the entire ligament

Fascicles Densely organized fiber bundles in muscle and ligament

Fibrillin A large glycoprotein that is a structural component of elastin-containing microfibrils

Hematoma A collection of blood resulting from injury

Ligaments Highly organized, fibrous tissues that connect bone to bone

References

1. Dorlot JM, Sidi MAB, Tremblay GM, et al: Load elongation behaviour of the canine anterior cruciate ligament. *J Biomech Eng* 1980;102:190-193.
2. Diamant J, Keller A, Baer E, et al: Collagen: Ultrastructure and its relation to mechanical properties as a function of ageing. *Proc R Soc Lond B Biol Sci* 1972;180:293-315.
3. Pitman MI, Nainzadeh N, Menche D, et al: The intraoperative evaluation of the neurosensory function of the anterior cruciate ligament in humans using somatosensory evoked potentials. *Arthroscopy* 1992;8:442-447.
4. Cooper RR, Misol S: Tendon and ligament insertion: A light and electron microscopic study. *J Bone Joint Surg Am* 1970;52:1-20.
5. Laros GS, Tipton CM, Cooper RR: Influence of physical activity on ligament insertions in the knees of dogs. *J Bone Joint Surg Am* 1971;53:275-286.
6. Muller P, Dahners LE: A study of ligamentous growth. *Clin Orthop* 1988;229:274-277.
7. Woo SL, Ohland KJ, Weiss JA: Aging and sex-related changes in the biomechanical properties of the rabbit medial collateral ligament. *Mech Ageing Dev* 1990;56:129-142.
8. DeHaven KE, Lintner DM: Athletic injuries: Comparison by age, sport, and gender. *Am J Sports Med* 1986;14:218-224.
9. Praemer A, Furner S, Rice DP: (eds): *Musculoskeletal Conditions in the United States*, ed 2. Rosemont, IL, American Academy of Orthopaedic Surgeons, 1999, p 170.
10. Andriacchi T, Sabiston P, DeHaven K, et al: Ligament: Injury and repair, in Woo SLY, Buckwalter JA (eds): *Injury and Repair of the Musculoskeletal Soft Tissues*. Park Ridge, IL, American Academy of Orthopaedic Surgeons, 1988, pp 103-128.
11. Arnoczky SP: Physiologic principles of ligament injuries and healing, in Scott WN (ed): *Ligament and Extensor Mechanism Injuries of the Knee: Diagnosis and Treatment*. St Louis, MO, Mosby-Year Book, 1991, pp 67-81.
12. Jack EA: Experimental rupture of the medial collateral ligament of the knee. *J Bone Joint Surg Br* 1950;32:396-402.
13. Frank C, Amiel D, Akeson WH: Healing of the medial collateral ligament of the knee: A morphological and biochemical assessment in rabbits. *Acta Orthop Scand* 1983;54:917-923.
14. Daniel DM, Stone ML, Dobson BE, et al: Fate of the ACL-injured patient: A prospective outcome study. *Am J Sports Med* 1994;22:632-644.
15. Murray MM, Martin SD, Spector M: Migration of cells from human anterior cruciate ligament explants into collagen-glycosaminoglycan scaffolds. *J Orthop Res* 2000;18:557-564.

Chapter 5

Meniscus and Intervertebral Disks

Once thought to be vestigial and functionless, the menisci are now believed to play a critical role in knee biomechanics. The meniscus cushions the proximal tibia from impact, distributes the load of weight bearing over an area of maximal size, helps stabilize the joint against anterior displacement, helps lubricate the joint, and assists with proprioception (Fig. 1). Loss of the menisci through trauma, degeneration, or surgical excision has been shown to lead to articular cartilage damage and development of osteoarthritis.[1,2]

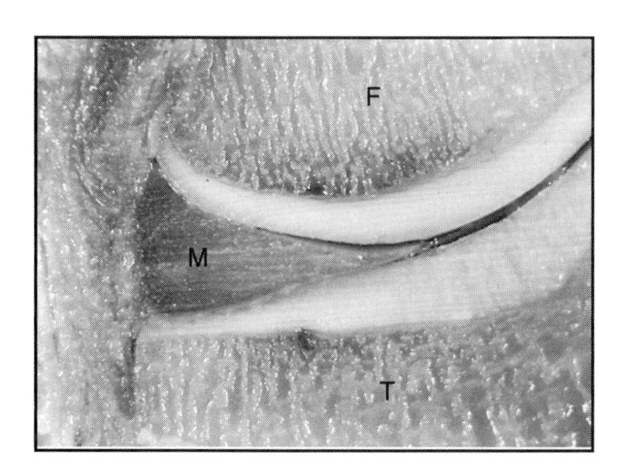

Figure 1
The medial compartment of the knee showing the articulation of the menisci (M) with the condyles of the femur (F) and tibia (T). As shown, the meniscus increases contact between the bones and therefore distributes stress.

(Reproduced with permission from Warren RF, Arnoczky SP, Wickiewicz TL: Anatomy of the knee, in Nicholas JA, Hershman EB (eds): ***The Lower Extremity and Spine in Sports Medicine****. St. Louis, MO, CV Mosby, 1986, pp 657-694.)*

Meniscal injuries represent the most common intra-articular knee problems treated by physicians.[3] An understanding of the gross and microscopic architecture of the menisci is integral to appreciating their biomechanical properties and function. Knowledge of the basic science of meniscal cartilage is also helpful when deciding the appropriate treatment of an injured meniscus.

Physiologic Function

The menisci play an important role in knee biomechanics. During walking, the knee joint experiences forces two to four times body weight; during running and high-impact activities, the forces increase dramatically. The primary function of the menisci is to distribute these loads and protect the articular cartilage. The menisci also provide passive stability, lubrication, and proprioception to the knee (Fig. 2).

Shock Absorption

Macroscopically, the meniscus has a fibrous, sponge-like structure that is filled with a gel of proteoglycans and water. This morphology allows the meniscus to function as a viscoelastic structure, dampening the load generated during weight bearing. (A viscoelastic structure is one whose mechanical properties depend on the rate of loading.) Initial loading of the meniscus results in deformation of the fibrous structure and flow of the gel through the fibers. The resistance to flow of the gel provides energy dissipation, or shock absorption, with compressive loading of the meniscus, as is seen during running or

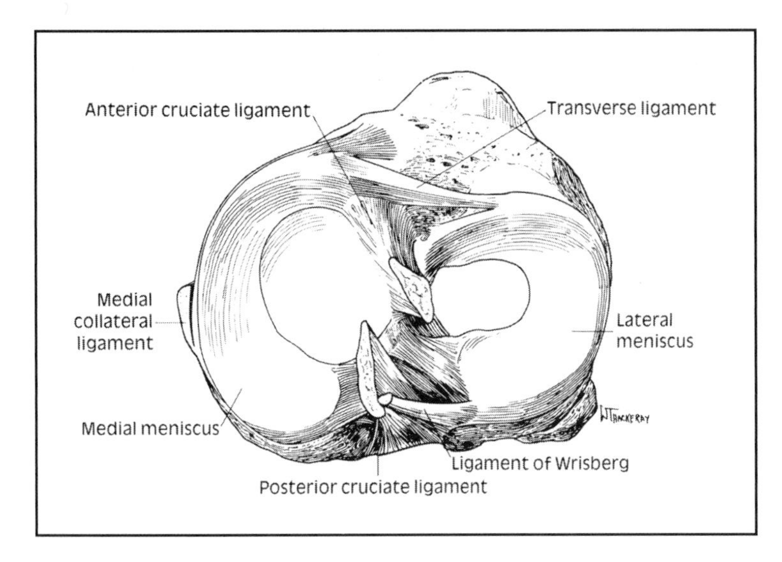

Figure 2
Cross-section of a tibial plateau showing the shape and attachments of the medial and lateral menisci.

(Reproduced with permission from Warren RF, Arnoczky SP, Wickiewcz TL: Anatomy of the knee, in Nicholas JA, Hershman EB (eds): ***The Lower Extremity and Spine in Sports Medicine****. St. Louis, MO, CV Mosby, 1986, pp 657-694.)*

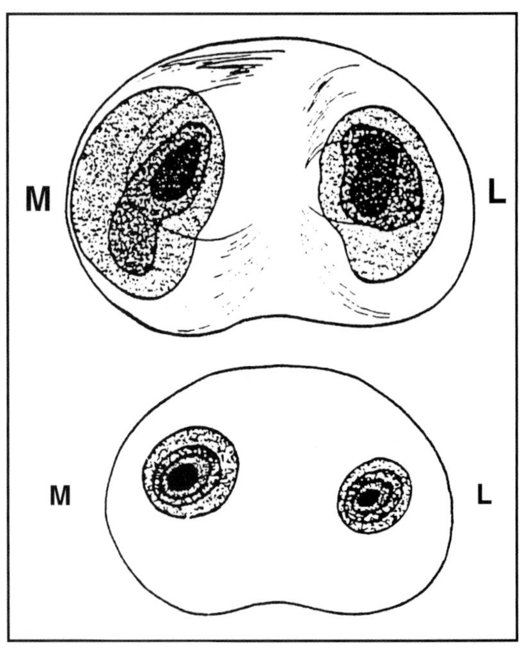

Figure 3
Schematic representation of a tibial plateau showing contact area and contact stress in the intact knee (top) and meniscectomized knee (bottom). M = medial, L = lateral. Note the focal concentration of stress when the menisci are absent.

(Reproduced with permission from Fukubayashi T, Kurosawa H: The contact area and pressure distribution pattern of the knee: A study of normal and osteoarthrotic knee joints. ***Acta Orthop Scand*** *1980;51:871-879.)*

jumping. A normal knee has 20% more shock-absorbing capacity than a knee that has had its meniscus removed.

Load Transmission

The triangular cross-sectional anatomy of the meniscus allows maximal contact between the rounded end of the distal femur and the relatively flat tibial plateau. This lowers the peak stresses in the articular cartilage and subchondral bone (Fig. 3). Removal of the medial meniscus has been shown to decrease the contact area between the femoral condyle and tibial plateau by 60% and more than double the stresses imparted on the articular cartilage leading to damage and degeneration.

In full extension, the medial and lateral menisci transmit 50% and 75% of the total compartmental load across the knee joint, respectively. With 90° of knee flexion, up to 85% of the total load is transmitted through the menisci. The menisci are stiffer near the anterior and posterior attachments (or horns) and have less stiffness modulus in their midsections. The anterior and posterior horns also have the greatest degree of parallel fibers, which suggests that the tensile properties of the meniscus are influenced by the arrangement of collagen fibers.

Passive Joint Stability

The anterior cruciate ligament (ACL) of the knee is the main source of resistance to anterior tibial subluxation. When the ACL is injured, secondary restraints, such as the medial meniscus, become important. The medial meniscus provides a passive restraint to anterior tibial translation in the ACL-deficient knee by blocking the femur from gliding too far off the tibia. Also, its wedged shape helps in rotational and varus stability through its space-filling effect within the knee joint. The lateral meniscus does not offer the same passive restraint to anterior tibial translation, even in the ACL-deficient knee, because it is more mobile.

Joint Lubrication

The menisci reduce friction between the femur and tibia by increasing congruency and maintaining a constant layer of synovial fluid between their gliding articular surfaces. Fluid is mechanically pumped or squeezed into and out of the menisci and articular cartilage, providing nutrients and removal of waste for the fibrochondrocytes and chondrocytes, respectively.

Proprioception

Nerve fibers, both myelinated and unmyelinated, have been identified throughout the entire meniscus.[4,5] The fibers originate from the perimeniscal and synovial tissue of the knee and radiate into the periphery of the meniscus. Many of the fibers are seen to accompany the vascular supply of the tissue. Three types of mechanoreceptors have been identified within the medial meniscus: (1) Ruffini endings, (2) Golgi tendon organs, and (3) pacinian corpuscles. Nerve endings, along with the nerve fibers, are found in the greatest concentration in the anterior and posterior horns of the menisci. The proprioceptive function of the menisci has been inferred from the finding of the three different types of mechanoreceptors. They are thought to trigger a proprioceptive reflex and contribute to the functional stability of the knee.

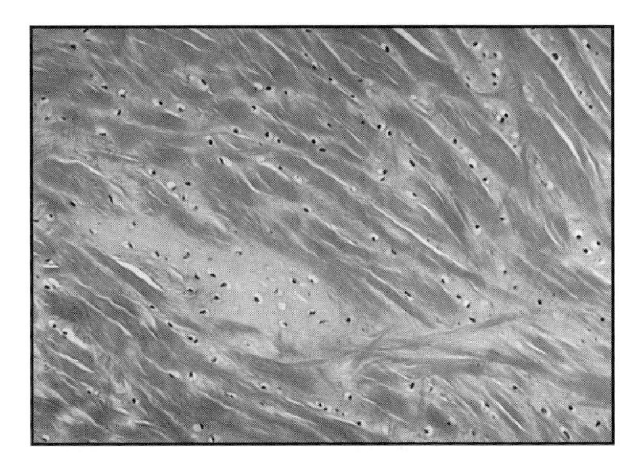

Figure 4
Photomicrograph of a longitudinal section of a human meniscus. (Hematoxylin and eosin, magnification ×100.)

(Reproduced from Arnoczky SP, McDevit CA: The meniscus: Structure, function, repair, and placement, in Buckwalter JA, Einhorn TA, Simon SR (eds): ***Orthopaedic Basic Science: Biology and Biomechanics of the Musculoskeletal System****, ed 2. Rosemont, IL, American Academy of Orthopaedic Surgeons, 2000, pp 531-545.)*

Normal Histology and Anatomy

Histology

The microstructure of the meniscus consists of water, cells, collagen, proteoglycans, and glycoproteins. The organization of these basic building blocks gives the menisci their biomechanical properties. Two major types of cells have been identified and classified in the meniscus. The first is a cell that is found in the peripheral meniscus and is spindle-shaped and fibroblast-like (Fig. 4). In the inner, avascular zones of the meniscus, the cells are ovoid or polygonal. These cells are called fibrochondrocytes as they are able to synthesize fibrous extracellular proteins (including type I collagen), yet have the rounded appearance of chondrocytes. Both cell types have abundant endoplasmic reticulum and Golgi complexes, with minimal mitochondria, suggesting a dependence on anaerobic metabolism. It is unclear whether these two cell types are actually distinct or whether they are the same cell type with modulation in phenotypic expression as a result of environmental influences.

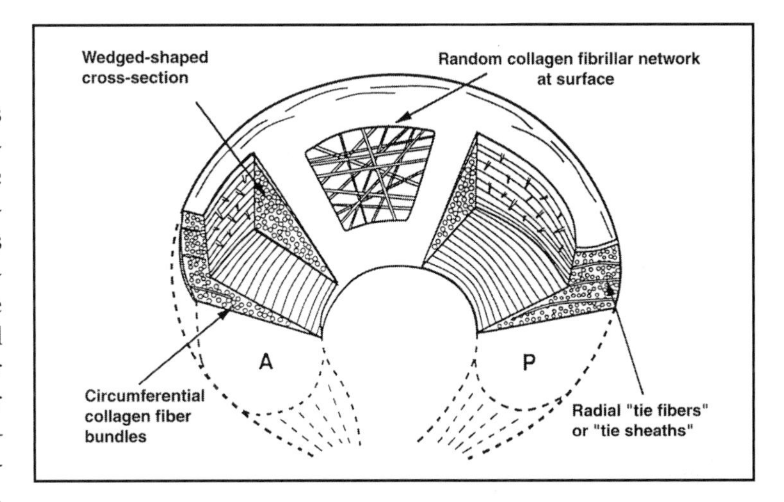

Figure 5
Collagen fiber ultrastructure of the meniscus. A = anterior. P = posterior.

(Reproduced with permission from Mow VC, Ratcliffe A, Chern KY, Kelley MA: Structure and function relationships of the menisci of the knee, in Mow VC, Arnoczky SP, Jackson DW (eds): ***Knee Meniscus: Basic and Clinical Foundations****. New York, NY Raven Press, 1992, pp 37-57.)*

Collagen represents up to 75% of the dry weight of the menisci. More than 90% of the collagen is fibrillar type I, with types II, III, V, and VI making up the remainder.[6] Proteoglycans such as aggrecan, chondroitin 6-sulfate, and chondroitin 4-sulfate contribute 1% to 2% of the dry weight of the menisci. The menisci also contain elastin, fibronectin, and other glycoproteins. As much as 70% of the wet weight of the meniscus is water, which is both structurally bound to collagen and freely associated with the interfibrillar proteoglycans.

The arrangement of the collagen fibers within the wedge-shaped meniscus is adapted for load transmission (Fig. 5). Most of the type I collagen fibers are arranged in circles to withstand the circumferential tension (technically termed "hoop stress"), which the meniscus develops during normal loading. Without these fibers, the meniscus would be extruded from the joint space. Other fibers are oriented radially and act to tie the circumferential fibers together to resist longitudinal splitting of the meniscus. This structural design converts the axially directed load into a tensile stress, which the circumferentially oriented collagen fibers are well suited to bear.

Gross Anatomy

The knee menisci are crescent-shaped disks of fibrocartilage interposed between the femoral condyles and the tibial plateaus. The word meniscus means "little moon" in Greek—when viewed from above, the meniscus has a crescent moon shape. Triangular or wedge shaped in cross section, the menisci increase the congruency within the knee joint, filling the gap between the convex distal femur and flat proximal tibia. The menisci are broadest at their peripheral fixed margin, measuring approximately 5 mm thick, and taper to a thin, free, central edge, creating a shallow cup to hold the round condyles of the femur.

Medial Meniscus The semicircular medial meniscus of an adult male is approximately 3.5 cm in diameter. It covers 60% of the medial tibial plateau articular cartilage and is less mobile than the lateral meniscus. The transverse intermeniscal ligament connects the anterior horns of the medial and lateral menisci. An additional point of anterior attachment to the plateau can be found in front of the ACL in the region of the intercondylar fossa. Along the entire periphery of the meniscus, the coronary ligament attaches the meniscus to the joint capsule and the tibial plateau. At the body of the medial meniscus, the attachment to the capsule is reinforced by the deep fibers of the medial collateral ligament. These fibers, called the posterior oblique ligament, tether the meniscus and allow it to function in a role similar to that of a block placed behind the tire of a car to prevent rolling. By blocking the rolling of the femoral condyle, the meniscus helps resist anterior tibial subluxation. The posterior horn is attached to the posterior intercondylar fossa between the lateral meniscus and the posterior cruciate ligament.

Lateral Meniscus Up to 80% of the lateral tibial plateau articular cartilage is covered by the circular lateral meniscus. Anteriorly, it is attached to the tibia behind the attachment of the ACL. Along its peripheral margin the lateral meniscus is loosely attached to the joint capsule by the coronary ligament. The absence of an attachment of the meniscus to the lateral collateral ligament (LCL) allows passage of the popliteal tendon between the lateral meniscus and LCL—the popliteal hiatus. Behind the intercondylar eminence and anterior to the posterior horn of the medial meniscus the posterior horn of the lateral meniscus is attached to the tibia. In 50% of people, the posterior horn of the lateral meniscus receives extra stability from the meniscofemoral ligaments of Humphry and Wrisberg.

Knowledge of the meniscal attachments helps in understanding the differential mobility of the medial and lateral menisci. With knee flexion and extension, the menisci are able to slide relative to the articular surface of the tibial plateau. Through three-dimensional MRI, the less constrained lateral meniscus has been shown to have an average excursion of 11 mm, with the more extensively fixed medial meniscus averaging only 5 mm of excursion (Fig. 6). The more limited range of motion of the medial meniscus allows it to serve as an anterior stabilizer but also makes it more vulnerable to tearing and injury. In both menisci, the anterior horns are more mobile than the wider posterior horns, rendering the posterior horns more susceptible to injury.

Vascular Anatomy

Most of the blood supply to the menisci originates from the superior and inferior branches of the medial and lateral geniculate arteries, which are fed, in turn, by the popliteal artery. These vessels give branches that supply the capillary bed of the perimeniscal tissue. The perimeniscal vessels are oriented in a circumferential pattern with radial branches extending into the meniscus along its periphery (Fig. 7). These vessels penetrate the outer 10% to 30% of the adult meniscus, a region referred to as the "red zone" of the meniscus. The inner 70% to 90% of the cartilage is avascular and referred to as the "white zone." The posterolateral aspect of the lateral meniscus at the popliteal hiatus is devoid of vascular branches entering its peripheral margin.[7]

Synovial branches of the middle genicular artery also supply the meniscal horns, but these have limited penetration within the meniscus and end in capillary loops. The cells in the white zone derive their nutritional supply from a combination of diffusion and fluid flow induced by joint loading and motion. These meniscal zones are crucial in predicting the success of meniscal healing and repair after injury. Repairs within the red zone, as might be expected, are more apt to heal than repairs within the inner or white region of the meniscus. Of course, a healing response can begin only if there is the potential for vascular ingrowth, so repairs within the white zone are apt to fail unless a blood supply is provided.

Development and Aging

The menisci form during the first 8 weeks of gestation as condensations and differentiation of mesenchymal cells. As the fetus develops, collagenous fibers appear and are oriented in a circumferential pattern, with the meniscus morphology and relationship to the rest of the knee being established by the 14th week of development. The organization and concentration of collagen fibers increase into early adulthood, then remains constant for the next 50 years before decreasing.

During the initial formation of the menisci, there is an abundance of cells and vascularity. With neonatal development, both the number of cells and extent of vascularity decrease. At birth, the vascular network extends from the peripheral perimeniscal capillary plexus to the free central margin of the meniscus. Over the first few months of life, the vascularity decreases such that by the ninth month the inner third of the meniscus is avascular. This process continues so that by adulthood only the peripheral 10% to 30% has a vascular supply. A similar phenomenon occurs with the cells of the meniscus. Replete with cells at birth, the meniscus becomes progressively less cellular with age. These changes are believed to be induced by the increased joint motion and weight bearing seen in early child development.

The proteoglycan content also shifts with age, with an increase in chondroitin 6-sulfate and a decrease in chondroitin 4-sulfate. The water content does not change. The percentage of noncollagenous proteins also declines from 20% at birth to 10% in individuals older than 50 years. A gradual discoloration is seen in the menisci with the normal aging process.

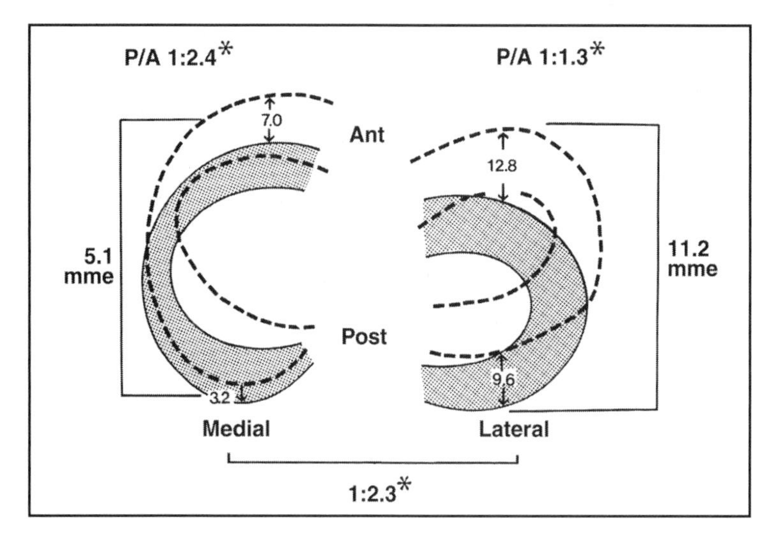

Figure 6

Schematic representation of mean meniscal excursion (mme) along the tibial plateau. The ratio of posterior to anterior translation (P/A) was significant (*$P < 0.05$). The medial meniscus moves less than half as much as the lateral meniscus, making it a better stabilizer but more prone to injury.

(Reproduced with permission from Thompson WO, Thaete FL, Fu FH, Dye SF: Tibial meniscal dynamics using three-dimensional reconstruction of magnetic resonance images. ***Am J Sports Med*** *1991;19:210-215.)*

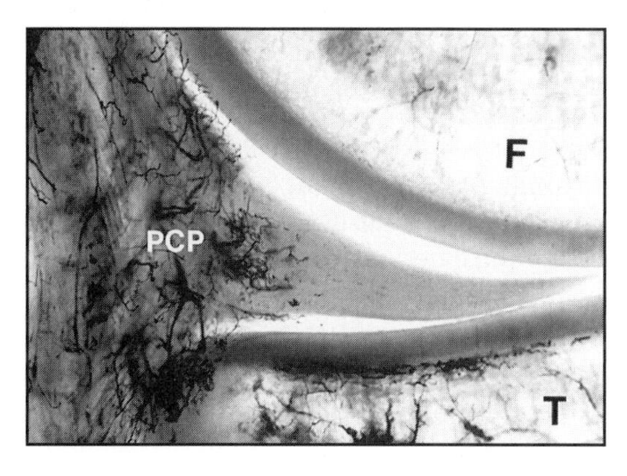

Figure 7

A 5-mm thick frontal section of the medial compartment of the knee is shown after vascular perfusion with India ink and tissue clearing. Branching radial vessels from the perimeniscal capillary plexus (PCP) can be seen penetrating the peripheral border of the medial meniscus. The central regions are avascular. F = femur, T = tibia.

(Reproduced with permission from Arnoczky SP, Warren RF: Microvasculature of the human meniscus. ***Am J Sports Med*** *1982; 10:90-95.)*

Pathophysiology

The location of the tear (in the vascular or avascular zone) plays a large role in the de-

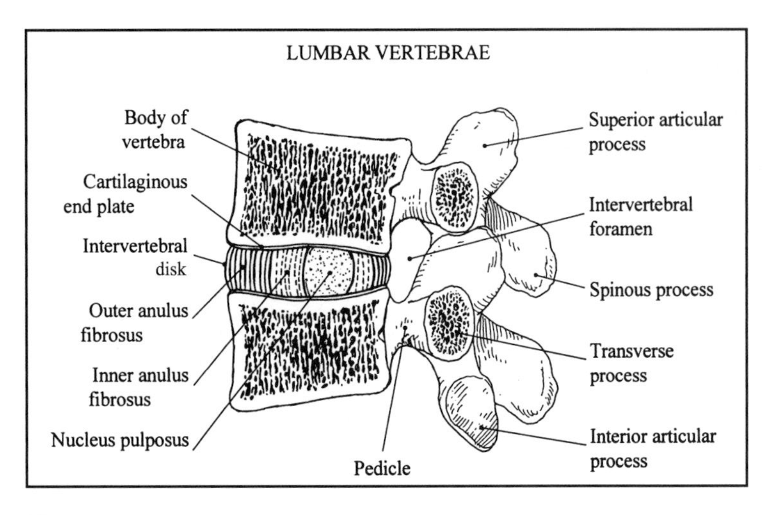

Figure 8

Sagittal section view of two vertebral bodies and an intervertebral disk. The three regions of the disk are shown: cartilaginous end plate, outer anulus fibrosus, inner anulus fibrosus, and nucleus pulposus. The posterior articular and spinous processes and the articular surface of a facet joint are also shown.

(Reproduced with permission from Ashton-Miller JA, Schultz AB: Biomechanics of the human spine, in Mow VC, Hayes WC (eds): ***Basic Orthopaedic Biomechanics****. Philadelphia, PA, Lippincott-Raven, 1997, pp 353-393.)*

cision whether to repair or excise a torn piece of meniscus. Vascularized tears are located in the periphery of the meniscus and have a functional blood supply that can provide fibrin clot and cells to populate the scar area. In animal models, radial tears, which extend from the avascular inner edge of the meniscus to the blood-rich synovium and capsule, heal with fibrovascular scar.[7] In humans, tears in this region may have potential to heal but need to be stabilized surgically first. Often, these are repaired with sutures or bioresorbable arrows to immobilize the torn area and allow healing to occur.

Tears fully within the inner edge of the meniscus are avascular and have been shown to be incapable of healing even when they are stabilized with simple repair. Therefore, these tears are typically treated with removal of the torn tissue. The tears in the middle of the meniscus are sometimes capable of healing, and therefore repair with additional techniques to stimulate healing (such as the creation of "vascular access channels" or the placement of fibrin clot into the tear before repair) is attempted, especially in young patients.

One of the goals of treatment is to preserve as much of the meniscal tissue as possible. As noted, the meniscus serves as a shock absorber of the knee. Complete excision of the meniscus results in a decrease in the load transmission area of approximately 50%; not surprisingly, the long-term results of total meniscectomy demonstrate an acceleration of osteoarthritis of the knee. However, when the torn tissue is so damaged that it can no longer function and, moreover, can create symptoms of catching or pain, meniscal resection may be the best option.

Techniques of meniscal replacement with allograft or synthetic materials are also being developed. Allograft menisci are fastened to the knee capsule with sutures. It is thought that the initially acellular allograft is repopulated with cells, at least in the superficial regions of the allograft.[8] Allograft transplantation is complicated by an immune response directed against the allograft meniscus, as shown in histologic studies of biopsies of previously implanted allograft menisci.

Intervertebral Disks

Physiologic Function

The intervertebral disk stabilizes the spine, helps maintain its alignment, allows motion between vertebral levels, absorbs energy, and distributes loads applied to the spine. The intervertebral disks of the spine are similar to the menisci of the knee in that both are viscoelastic shock absorbers. Like the meniscus, the disk is composed primarily of extracellular matrix, containing only few cells within it.

Normal Histology and Anatomy

Histology

The disks have an outer ring of fibrocartilage, the anulus fibrosus, and within it, a gelatinous material called the nucleus pulposus (Fig. 8). The organic matrix characteristic of each zone is distinct. In the anulus fibrosus, there is an outer layer of approximately 90 concentric lamellae of type I collagen fibers; deep within that is a region of a less dense, type II collagenous matrix. This inner zone is less organized than the outer zone of the anulus fibrosus. The most superficial anterior fibers of the anulus fibrosus blend with the anterior longitudinal ligament, whereas the most superficial posterior fibers blend with the posterior anterior longitudinal ligament. Protrusion of the nucleus pulposus through tears in the anulus fibrosus is called a herniated disk. Because the anterior anulus fibrosus is thicker than its posterior counterparts, posterior herniations are more com-

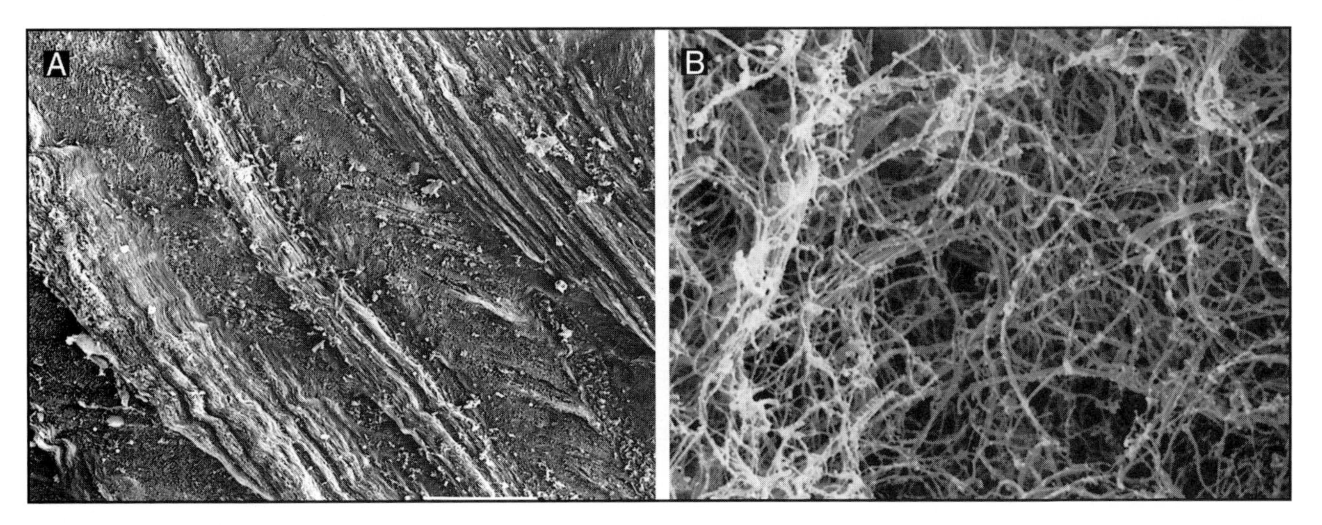

Figure 9
Micrographs showing the arrangement of the collagen fibrils in the outer anulus fibrosus (**A**) and the central nucleus pulposus (**B**). Note the tightly packed, highly oriented lamellae of collagen fibrils in the outer anulus fibrosus and the loose, almost random pattern of collagen fibrils in the nucleus pulposus.

(Reproduced with permission from Buckwalter JA: The fine structure of human intervertebral disc, in White AA III, Gordon SL (eds): ***American Academy of Orthopaedic Surgeons Symposium on Idiopathic Low Back Pain****. St. Louis, MO, CV Mosby, 1982, pp 108-143.)*

mon. When this herniation compresses a spinal nerve root, pain that radiates down the leg (known as sciatica) can result.

In contrast to the collagen-rich anulus fibrosus, the nucleus pulposus is predominantly proteoglycan (Fig. 9). It is the interaction between large proteoglycan molecules and water that gives the nucleus pulposus its resistance to compression. The cells populating the anulus fibrosus are like fibroblasts, while those in the nucleus pulposus are more chondrocytic in appearance and have synthetic characteristics.

Gross Anatomy
The intervertebral disk is the primary articulation between the vertebral bodies of two adjacent vertebrae. The spine is composed of 23 such disks that unite the 24 vertebrae of the human spine to form four regions (cervical, thoracic, lumbar, and sacrum). The disks increase in height and diameter from the cervical to the lumbar spine (Fig. 10).

The anterior aspects of the cervical and lumbar disks are thicker than the posterior aspects and thus produce the normal lordosis seen in those regions. By contrast, the thoracic disks are of uniform height from anterior to posterior; normal thoracic kyphosis is caused by the shape of the thoracic vertebral body itself. The thoracic disks are thinner than in the other regions of the spine, a property that may restrict mobility of this region of the spine. The outer edges of the disks are attached to the vertebral bodies by small collagen fibers called Sharpey's fibers.

Vascular Anatomy
Like the meniscus, the disk has a relatively poor blood supply and what exists enters from the periphery. There is no direct vessel to the central disk. Small vessels penetrate the outer ring of the anulus fibrosus. Nutrition of the inner disk, thus, is by diffusion. The nerve supply to the disk comes from the sinu-vertebral nerve and the gray ramus communicans. Like the blood vessels, free nerve endings do not penetrate deeply.

Growth and Development
Changes in disk microstructure, composition, size, and vascular supply occur throughout growth and development. Even at birth, the regions of the intervertebral disk are distinct. The end plate has discrete hyaline cartilage that separates the vertebral body from the disk. The outer anulus fibrosus is composed of dense circumferential lamellae of collagen that penetrate the vertebral bodies of adjacent vertebrae. Small blood vessels may be found adjacent to the cartilaginous end plates; occasionally, blood vessels penetrate the inner anulus fibrosus. Numerous free nerve endings lie on and within the most peripheral layers of the anulus fibrosus.

During skeletal growth, disk volume and

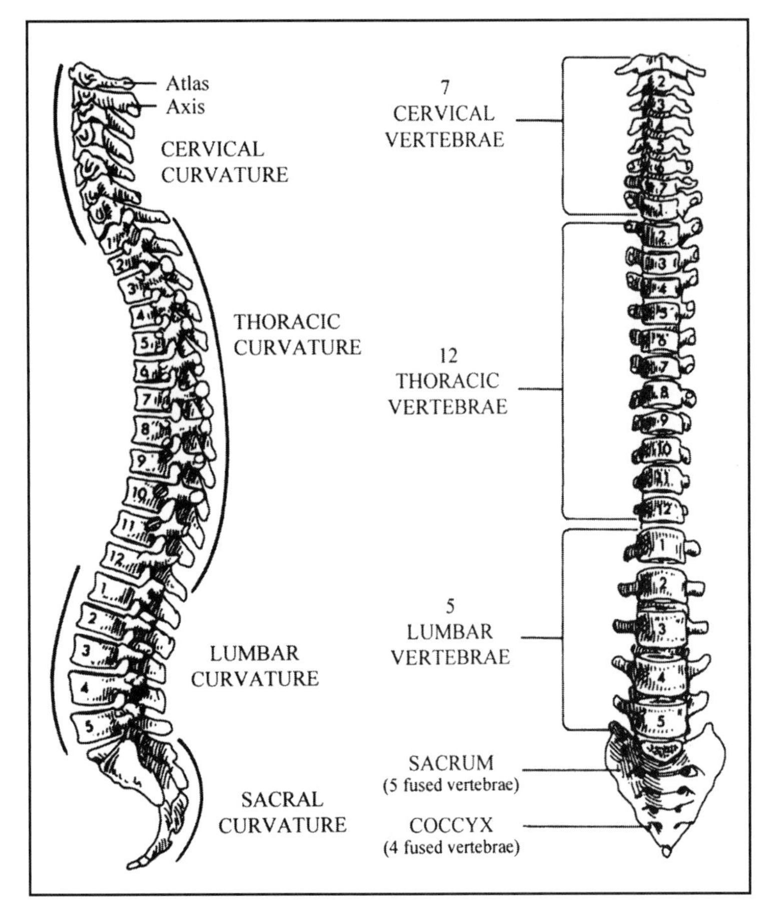

Figure 10

Drawing of the human spine illustrating the differences in the sizes of the vertebral bodies and intervertebral disks.

(Reproduced with permission from Ashton-Miller JA, Schultz AB: Biomechanics of the human spine, in Mow VC, Hayes WC (eds): ***Basic Orthopaedic Biomechanics****. Philadelphia, PA, Lippincott-Raven, 1997, pp 353-393.)*

diameter increase several-fold, thus increasing the distance from the peripheral vessels and the central portions of the disk. Additionally, the three vessels of the end plate gradually disappear, leaving scars in the cartilaginous end plate, and the peripheral blood vessels of the anulus fibrosus become smaller and less numerous. As a result, the disk becomes avascular. The concentration of cells decrease as the nucleus pulposus becomes more dense and fibrous. By early adulthood, the embryonic cells have disappeared completely, leaving behind chondocyte-like cells. The high water content of the nucleus pulposus also decreases with age as the proportion of proteoglycans that do not bind water progressively increases and the size of the aggrecan molecules decreases. This change in proteoglycans occurs in all regions of the disk but is greatest in the nucleus pulposus. Throughout all regions of the disk, the diameter and the variability of collagen fibrils increases.

Pathophysiology

The spinal disk changes throughout life (beginning soon after birth), but these changes accelerate after skeletal maturity, at approximately age 20 years. Once skeletal maturity is achieved, all intervertebral disks undergo progressive alterations in volume, shape, microstructure, and composition. As such, these changes can decrease motion, adversely affect the mechanics of the spine, and lead to the two most common clinical disorders of the axial skeleton: herniated nucleus pulposus and degenerative diseases of the spine.

The most extensive changes occur in the nucleus pulposus, where there is a sharp decline in the number of viable cells and concentrations of proteoglycans that bind water. The aggregate proteoglycans also become fragmented. Accompanying these changes are increases in the concentration of collagens and noncollagen proteins as dense granular material accumulates throughout the extracellular matrix. As a result, the nucleus pulposus becomes firm and white rather than soft and translucent.

With aging, the size of the outer anulus fibrosus remains constant, while the inner anulus fibrosus increases in size. This growth comes at the expense of the nucleus pulposus, which becomes more fibrotic. Myxomatous degeneration also occurs in parts of the anulus fibrosus. Fissures and cracks appear, some extending from the periphery to the central portion of the disk.

Over time, the outer lamellae of the outer anulus fibrosus become stiff fibrocartilage. The height of the disk may decline, and prominent fissures and clefts form in the center. These changes affect mobility, alter alignment, and change the loads applied to the facet joints, paraspinous muscles, and the spinal ligaments. In sum, age-related changes following skeletal maturity decrease structural integrity and contribute to changes in disk volume and shape, which, in turn, increase the probability of mechanical failure leading to disk herniation.

Research and New Directions During the Bone and Joint Decade (2002-2011)

Function and Physiologic Response of Various Cells

The fibrochondrocyte is thought to be the major cell type in the meniscus. However, many other types of cells have also been identified in menisci, including nerve cells, vascular cells, and synovial cells. The role of these cells in meniscal function and repair are currently under investigation.

Gene Therapy

The use of recombinant adenoviral vectors to transfer genes into meniscal fibrochondrocytes has been successfully demonstrated in vitro.[9] Although further understanding of the genetic basis of the healing response is still in the preliminary stage, the potential exists for future transfer of genetic material that codes for cytokines. This enhances the healing response and accelerates the repair and regeneration of the meniscus after injury.

Key Terms

Fibrochondrocytes Cells that are able to synthesize fibrous extracellular proteins and have the rounded appearance of chondrocytes

Intervertebral disk The structure located between two moving vertebrae that stabilizes the spine, helps maintain its alignment, allows motion between vertebral levels, absorbs energy, and distributes loads applied to the spine

Meniscus A crescent-shaped disk of fibrocartilage located between the femoral condyle and the tibial plateau

Sharpey's fibers The small collagen fibers that attach tendon to bone; in the spine they connect the outer edges of intervertebral disks to the vertebral bodies

Viscoelastic structure A structure whose mechanical properties depend on the rate of loading

References

1. McNicholas MJ, Rowley DI, McGurty D, et al: Total meniscectomy in adolescence: A thirty-year follow-up. *J Bone Joint Surg Br* 2000;82:217-221.
2. Cox JS, Nye CE, Schaefer WW, et al: The degenerative effects of partial and total resection of the medial meniscus in dogs' knees. *Clin Orthop* 1975;109:178-183.
3. Praemer A, Furner S, Rice DP: (eds): *Musculoskeletal Conditions in the United States,* ed 2. Rosemont, IL, American Academy of Orthopaedic Surgeons, 1999, p 170 (Appendix A: Table B).
4. Day B, Mackenzie WG, Shim SS, et al: The vascular and nerve supply of the human meniscus. *Arthroscopy* 1985;1:58-62.
5. O'Connor BL, McConnaughey JS: The structure and innervation of cat knee menisci, and their relation to a "sensory hypothesis" of meniscal function. *Am J Anat* 1978;153:431-442.
6. Aronoczky S, Adams M, DeHaven K, et al: Meniscus, in Woo SLY, Buckwalter JA (eds): *Injury and Repair of the Musculoskeletal Soft Tissues.* Park Ridge, IL, American Academy of Orthopaedic Surgeons, 1988, pp 483-537.
7. Arnoczky SP, Warren RF: The microvasculature of the meniscus and its response to injury: An experimental study in the dog. *Am J Sports Med* 1983;11:131-141.
8. Rodeo SA, Seneviratne A, Suzuki K, et al: Histological analysis of human meniscal allografts: A preliminary report. *J Bone Joint Surg Am* 2000;82:1071-1082.
9. Goto H, Shuler FD, Niyibizi C, et al: Gene therapy for meniscal injury: Enhanced synthesis of proteoglycan and collagen by meniscal cells transduced with a TGFbeta(1) gene. *Osteoarthritis Cartilage* 2000;8:266-271.

Chapter 6

Skeletal Muscle

Skeletal muscle is the single largest tissue mass in the body, constituting 40% to 45% of the dry body weight. Muscles attach to the bones and produce movements or exert static forces. They can be connected directly to bone or insert on a bone by means of a tendon, which is a specialized type of connective tissue. The area of interface between a skeletal muscle and its tendon is called the musculotendinous junction.

Skeletal muscles may take a variety of forms, from the slender sartorius to the broad, fan-shaped pectoralis major. However, their histologic architecture remains the same. Each muscle is composed of muscle fibers called myofibers (Fig. 1). Each of these fibers is essentially one large multinucleated cell created from the fusion of many other cells. An individual muscle fiber typically spans part of the muscle, though it may run its entire length.

A framework of connective tissue supports muscle. Each fiber is surrounded by an endomysium. Groups of fibers are surrounded by a perimysium, resulting in fascicles that are often large enough to be visible to the naked eye. The entire muscle is enveloped by an epimysium. This architectural arrangement supports integrated motion among the fibers.

Tendons allow the force generated by the muscle to be transferred into motion by connecting the skeletal muscle to the bony skeleton.

Physiologic Function

Mechanics

Muscle contractions can be described as isotonic, isometric, or isokinetic. Isotonic contraction occurs, for example, when a biceps curl is performed with a free weight. The tension in the muscle is constant, but its length changes throughout the range of motion. The magnitude of isotonic contraction is a measure of dynamic strength. Isometric contraction occurs when a person pushes against a wall with his or her arms. Tension is generated, but muscle length remains unchanged. The magnitude of isometric contraction is a measure of static strength. Finally, isokinetic contraction signifies muscle tension generated by muscle contraction at a constant velocity over a full range of motion. The magnitude of isokinetic contraction is also a measure of dynamic strength. Muscle contraction can also be described as concentric or eccentric. Concentric contraction oc-

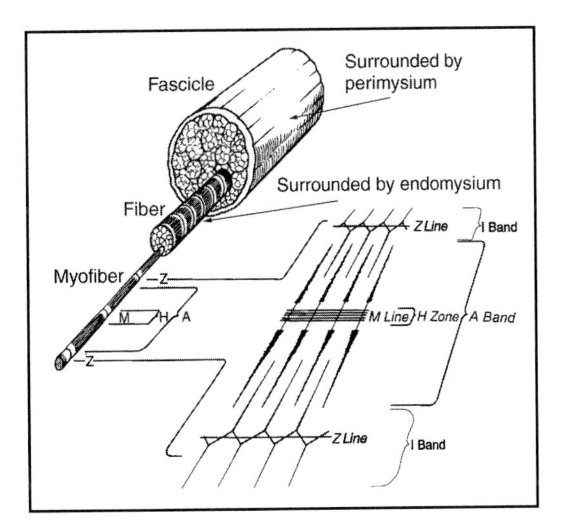

Figure 1
Schematic representation of the structural design of human muscle. The structure of the bands within a myofiber is shown in Figure 2.

(Reproduced from Garrett WE Jr, Best TM: Anatomy, physiology, and mechanics of the skeletal muscle, in Buckwalter JA, Einhorn TA, Simon SR (eds): ***Orthopaedic Basic Science: Biology and Biomechanics of the Musculoskeletal System****, ed 2. Rosemont, IL, American Academy of Orthopaedic Surgeons, 2000, pp 683-716.)*

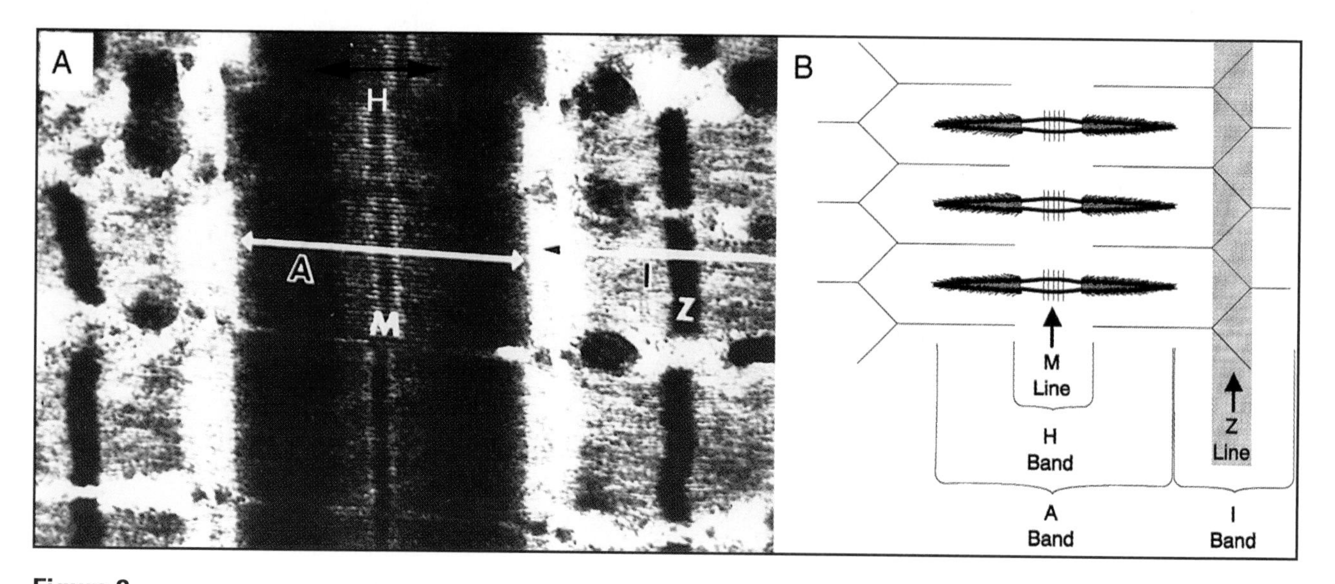

Figure 2

A, Electron micrograph of skeletal muscle illustrating the striated, banded appearance. A = A-band, M = M-line, H = H-band, I = I-band, and Z = Z-line. **B**, The basic functional unit of skeletal muscle, the sarcomere.

(Reproduced from Garrett WE Jr, Best TM: Anatomy, physiology, and mechanics of the skeletal muscle, in Buckwalter JA, Einhorn TA, Simon SR (eds): ***Orthopaedic Basic Science: Biology and Biomechanics of the Musculoskeletal System****, ed 2. Rosemont, IL, American Academy of Orthopaedic Surgeons, 2000, pp 683-716.)*

curs when the muscle is shortened as it contracts. Eccentric contraction occurs when tension is generated even though the muscle is elongated; the muscle power is used to decelerate the joint. Injuries occur more commonly with eccentric contractions.

Contraction of a muscle occurs as a result of the coordinated shortening of its component myofibers. For the contraction to occur, a highly organized intracellular apparatus consisting of long, slender myofibrils must function properly. Myofibrils are composed of repeating units called sarcomeres, which are the fundamental components of the contractile apparatus. Sarcomeres are made up of thick (myosin) and thin (actin) filaments in an intricate arrangement of bands and lines that allows these structures to slide past each other (Fig. 2). Activation of the muscle fiber causes the myosin heads (globular regions of the myosin molecules) to bind to actin. This results in a change in the shape of the proteins, drawing the thin filament a short distance (approximately 10 nm) past the thick filament. Then the linkages between actin and myosin break and reform (requiring adenosine triphosphate [ATP], pulling the filaments past each other with a ratchet-like action.

Muscle contraction is stimulated by the motor nerve endings. A motor neuron in the spinal cord generates an electrical impulse, or action potential, that travels down its axon. The nerve enters the muscle and contacts an individual myofiber. This specialized communication point is called the motor end plate, or neuromuscular junction (Fig. 3). Here, the nerve forms a synapse with the muscle, a specialized site at which an electrical signal is transmitted chemically across a cleft to produce a similar electrical impulse on the other side of the gap.[1] The presence of such gaps allows for the pharmacologic modification of what is essentially an electrical process.

As the axon approaches the muscle at the synapse, it loses its myelin sheath, and the entire terminal axon is covered by a Schwann cell, a specialized support cell that encases nerve fibers. The nerve terminal covers a region of the muscle where the membrane systems of the nerve and the muscle weave to maximize the area of contact between them. These folds are called synaptic folds, and the space between them, called the synaptic cleft, is 50 nm across.

The chemical mediator involved in neuromuscular synaptic transmission is acetylcholine, which is stored in the presynaptic axon in membrane-bound sacs called synaptic vesicles. The electrical impulse (action potential) produces an influx of extracellular

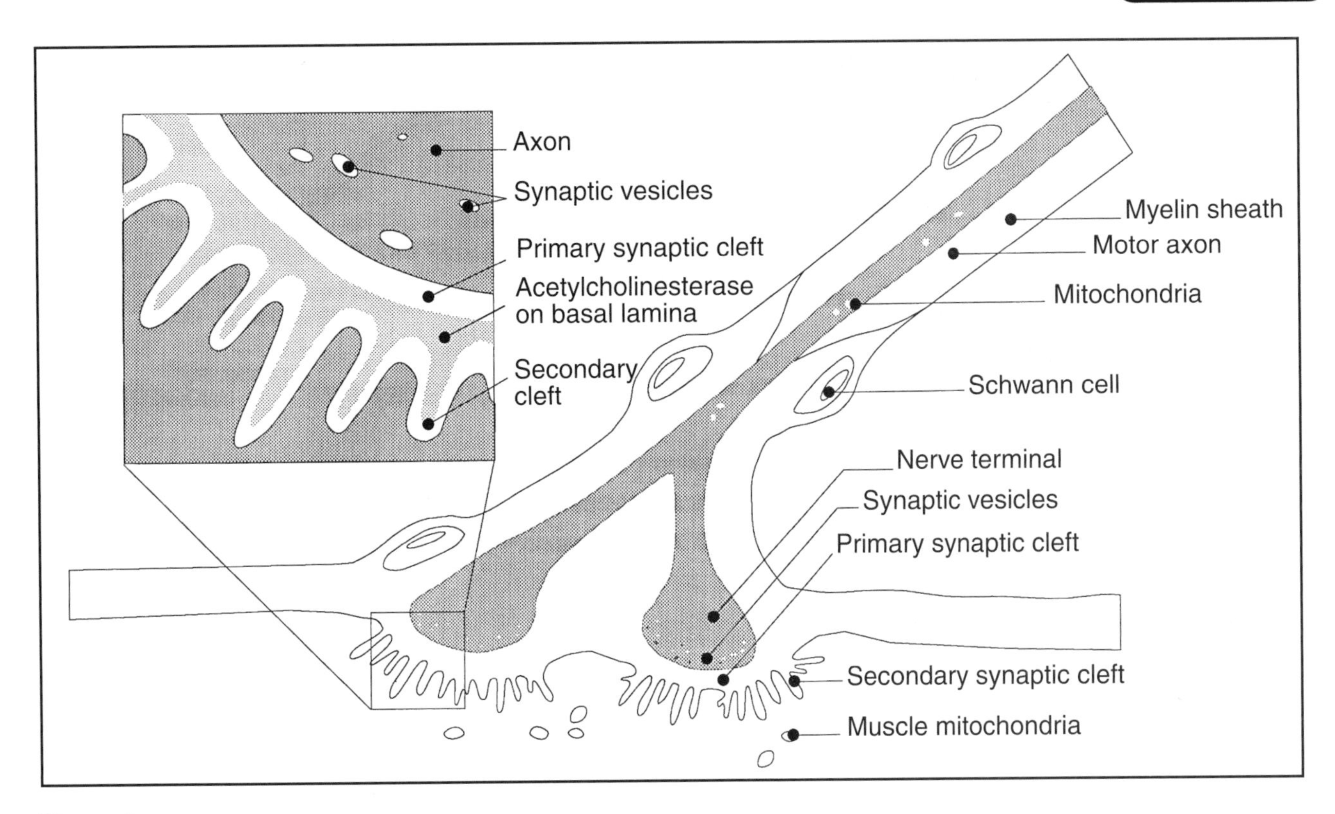

Figure 3
Schematic representation of the motor end plate.

(Reproduced from Garrett WE Jr, Best TM: Anatomy, physiology, and mechanics of the skeletal muscle, in Buckwalter JA, Einhorn TA, Simon SR (eds): ***Orthopaedic Basic Science: Biology and Biomechanics of the Musculoskeletal System****, ed 2. Rosemont, IL, American Academy of Orthopaedic Surgeons, 2000, pp 683-716.)*

calcium into the presynaptic terminal, which leads to fusion of the vesicle with the adjacent membrane and then the release of acetylcholine. The acetylcholine diffuses across the synaptic cleft and then binds to postsynaptic acetylcholine receptors, changing the membrane potential and generating an action potential in the muscle. Acetylcholine is deactivated by the enzyme acetylcholinesterase, which is also located in the synaptic cleft. The action potential reaches the interior of the muscle fiber through a membrane system that includes transverse tubules, directed perpendicularly to the axis of the fiber, and cisternae of the sarcoplasmic reticulum (SR), directed parallel to the axis of the fiber (Fig. 4). The SR is the system of membranes responsible for transmission of the electrical signal from one muscle cell to the next. An action potential moving over the surface of the fiber passes down the transverse tubules and causes Ca^{+2} release from the outer vesicles of the SR.

Each muscle fiber is contacted by a single nerve terminal. Collectively, a single motor axon and all the myofibers it contacts constitute a **motor unit**. The number of muscle fibers within a motor unit and the number of

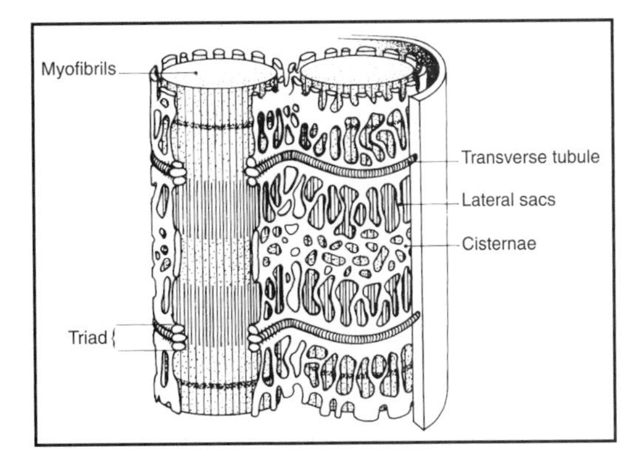

Figure 4
Schematic representation of the sarcoplasmic reticulum.

(Reproduced from Garrett WE Jr, Best TM: Anatomy, physiology, and mechanics of the skeletal muscle, in Buckwalter JA, Einhorn TA, Simon SR (eds): ***Orthopaedic Basic Science: Biology and Biomechanics of the Musculoskeletal System****, ed 2. Rosemont, IL, American Academy of Orthopaedic Surgeons, 2000, pp 683-716.)*

Table 1
Characteristics of Human Skeletal Muscle Fiber Types

	Type I (Sustained Slow)	Type IIA (Intermediate)	Type IIB (Fast Fatigable)
Other names	Red, slow twitch (ST) Slow oxidative (SO)	White, fast twitch (FT) Fast oxidative glycolytic (FOG)	Fast glycolytic (FG)
Speed of contraction	Slow	Fast	Fast
Strength of contraction	Low	High	High
Fatigability	Fatigue-resistant	Fatigable	Most fatigable
Aerobic capacity	High	Medium	Low
Anaerobic capacity	Low	Medium	High
Motor unit size	Small	Larger	Largest
Capillary density	High	High	Low

(Reproduced from Garrett WE Jr, Best TM: Anatomy, physiology, and mechanics of the skeletal muscle, in Buckwalter JA, Einhorn TA, Simon SR (eds): ***Orthopaedic Basic Science: Biology and Biomechanics of the Musculoskeletal System****, ed 2. Rosemont, IL, American Academy of Orthopaedic Surgeons, 2000, pp 683-716.)*

motor units within a given muscle vary considerably. The number of myofibers within a motor unit determines the precision with which the muscle can be used. A muscle under fine motor control may contain only 10 fibers per unit, whereas muscle under gross motor control may contain more than 1,000 fibers.[2,3]

The types of fibers within motor units also differ, providing an additional level of physiologic flexibility. At least three types of fibers have been identified in human muscle and are classified according to the different isoforms of myosin adenosinetriphosphatase (ATPase) they contain (Table 1). ATPase hydrolyzes ATP during muscle contraction, as myosin and actin slide over one another. The various types of ATPase work at different speeds, and thus the type of ATPase correlates closely with the intrinsic speed of muscle shortening. Essentially, these may be described as sustained slow fibers (type I), intermediate fibers (type IIA), and fast fatigable fibers (type IIB).[4] Slow fibers use aerobic metabolism and thus depend on adequate perfusion and oxygenation.

Proprioception

Proprioception is the ability of the body to sense its position in space. Muscles contribute to proprioception. Sensory nerve endings include those on muscle spindles, which are sensitive to length, and in the Golgi tendon organs, which are sensitive to tension. There is also a variety of free endings in the muscle, some of which are involved in sensations of pain. These nerve endings transmit information to the central nervous system, including the motor neurons that innervate the muscle. Through reflexes and higher order processing, postural adjustments are made and the body "senses" the positions of the limbs in space.

Normal Histology and Composition

Each muscle fiber contains multiple nuclei that lie immediately beneath the sarcolemma, which is its plasma membrane. A small proportion of the nuclei at the periphery of the myofiber are stem cells, also called satellite cells, which can repopulate damaged fibers after injury.[5] The cytoplasm, called sarcoplasm, is similar to that of other cells. It contains a cellular matrix, organelles, and a variety of other molecules. Among the organelles, the Golgi apparatus and mitochondria are abundant and lie close to the nuclei. The sarcoplasmic reticulum is a continuous branching network of membrane, which is a specialized form of endoplasmic reticulum that is unique to muscle. Glycogen, lipid droplets, and myoglobin are among other cytoplasmic components.

Other cell types within muscle include fibroblasts, endothelial and smooth muscle cells constituting blood vessels, and Schwann cells around the sheath of nerve axons.

The cells in tendons are called tenocytes. These cells are typical fibroblasts with long cytoplasmic extensions. The cell density of tendons is similar to that of ligaments but lower than that of other tissues such as bone marrow or liver, conferring mechanical strength to these structures.

External to the sarcolemma is a basement membrane that merges with the extracellular matrix (ECM) to form the endomysium. The basement membrane is rich in protein and carbohydrate components, including collagen, laminin, fibronectin, and a variety of glycoproteins.

The ECM of tendons is composed of dense, parallel bundles of collagen fibers. These bundles are oriented along the line of tension between muscle and bone insertion for maximal transmission of load, and they have less crimp than the collagen bundles in ligaments. The collagen is almost 95% type I, with the remainder primarily type III collagen and proteoglycans. Tendons generally attach to bone via a specialized direct insertion site that has four zones: tendon, fibrocartilage, mineralized fibrocartilage, and bone. Sharpey's fibers are collagen bundles that extend from the tendon or periosteum into the bone.

Blood vessels run parallel to the axis of myofiber in the connective tissue, often with several capillaries around each myofiber. They are arranged with enough redundancy to permit changes in length during the contraction-extension cycle of a muscle.

Development and Aging

Skeletal muscle cells develop from a mesodermal cell population called myoblasts. These spindle-shaped cells divide and fuse to form long, multinucleated tubes, called myotubes, that differentiate into muscle fibers. During this period, many contractile proteins appear, some of which exist in embryonic isoforms. By the seventh week of gestation, distinct muscle and tendinous structures can be identified. Further steps in muscle differentiation include the production of structural and metabolic proteins, organization of the ECM, and innervation.

Classic developmental studies demonstrate that the formation of neuromuscular junctions is a mutually inductive event; neurons induce postsynaptic differentiation in myofibers, and myofibers induce presynaptic differentiation in motor axon terminals. More recent experiments indicate that Schwann cells, which also surround axon terminals, play an active role in the formation and maintenance of the neuromuscular junction.

In childhood, both muscle fibers and tendons increase in length. The region of the myotendinous junction appears to be the site at which these fibers lengthen, rather than near the muscle belly.[6] Sarcomeres remain constant in size; additional units are added to the fiber during growth. In adulthood, the primary mechanism by which the musculotendinous unit lengthens is elongation of the muscle belly.

During aging, there is loss of muscle mass and strength, termed sarcopenia. It is perhaps, unfortunately, a universal process. The mechanisms are complex, with both myogenic and neural components.[7] It appears that regeneration of muscle fibers in older individuals proceeds more slowly, a result of biochemical, hormonal, and cytokine-related changes. Aging may also express its effects via the relative loss of testosterone, which occurs with advancing age. Testosterone increases protein synthesis in muscle and decreases its breakdown.

Pathophysiology

Muscular Dystrophy

Muscular dystrophies (also called myopathies) are noninflammatory inherited disorders of progressive muscle weakness. There are several types, distinguished by their inheritance patterns. Duchenne's muscular dystrophy is an X-linked recessive disorder of young boys that is characterized by clumsy walking, decreasing motor skills, lumbar lordosis, and muscle weakness. The hip extensors are usually affected first. Muscle biopsy demonstrates foci of necrosis and connective tissue infiltration, with absence of the dystrophin protein and elevated creatine phosphokinase (CPK) on DNA testing. These individuals typically lose independent ambulation by age 10 and become wheelchair dependent by age 15 years. They usually die of cardiorespiratory complications before age 20 years. Becker's muscular dystrophy has a similar but less severe pathophysiology. Affected individuals also have an abnormal absence of dystrophin and histologic evidence of necrosis with connective tissue infiltration. However, they can often live beyond age 20 years without respiratory support. Other rare types of muscular dystrophy are seen in older individuals. Facioscapulohumeral muscular dystrophy is an autosomal dominant disorder seen in individuals between 6 and 20 years of age. They have facial muscle abnormalities, normal CPK, and winging of the scapula. Limb-girdle muscular dystrophy is an autosomal recessive disorder diagnosed in individuals between 10 and 30 years of age and is characterized by pelvic or shoulder girdle involvement and elevated CPK.

Muscle Injuries

Muscle injury can occur by a variety of mechanisms ranging from ischemia to direct injury by crush, laceration, or excessive force. Injured muscle undergoes processes of degeneration and regeneration.[8] When muscle fibers are damaged, an inflammatory process ensues, followed by removal of debris by macrophages.[9] New fibers appear within the connective tissue framework and are believed to be generated from a population of satellite cells that exist in a quiescent state within the original muscle syncytium. Synchronous processes that affect the functional recovery of muscle are connective tissue formation (fibrosis) and revascularization. Fibrosis can be extensive enough to interfere with muscle regeneration, preventing fibers from shortening or lengthening fully.

Forms of muscle injury include lacerations, contusions, and strains. Lacerations typically result from direct trauma from a sharp object. Recovery of function requires reorganization of devitalized tissue, muscle regeneration across the injury site, and reinnervation of denervated myofibers; it is usually partial rather than total.

Contusions usually result from blunt injury, occurring frequently in motor vehicle accidents and sports injuries. A contusion leads to an inflammatory process and hematoma, with subsequent muscle regeneration and scar formation. Severe blunt injury to muscle may result in heterotopic bone formation (synthesis of bone where it is not normally present) within the muscle, referred to as myositis ossificans.[10]

Muscle strains are indirect injuries caused by excessive tension on a muscle rather than by direct trauma. A strain may result in either a complete or an incomplete muscle tear. The muscle fibers can avulse from the tendon at the myotendinous junction. However, failure usually occurs in the muscle fiber within several millimeters of the junction. The terminal sarcomeres near the myotendinous junction are stiffer than the middle sarcomeres of a muscle fiber. The injury to muscle occurs within this region of relative stiffness.[11] Among the most frequently injured muscles are the hamstrings, the rectus femoris, and the gastrocnemius; these muscles cross at least two joints and as such may be subject to more stretch. In addition, the architecture in these muscles demonstrates an extensive length of the myotendinous junction.

Tendon Injuries

Tendon injuries can result from a sudden tensile force or laceration. Injuries that result in large gaps in the tendon, such as occurs after laceration of the finger flexor tendons, require surgical repair and suturing to hold the ends together during healing. Tendons respond to injury in a way that is similar to that of skin and other connective tissues. An initial phase of inflammation is followed by a reparative phase in which surrounding cells enter the site and produce a collagen scar, after which remodeling occurs over time. Controversy still exists as to whether the invading cells come from within the tendon (intrinsic cells) or from the surrounding tissues (extrinsic cells). The strength of the healing tendon is significantly lower than the uninjured tendon in the initial phase of healing, and strength does not increase until 3 weeks after injury. With improved techniques of tendon suture repair, early motion is now possible after tendon lacerations. This movement is thought to improve the functional outcome by minimizing adhesion formation.

Research and New Directions During the Bone and Joint Decade (2002-2011)

Repair of Skeletal Muscle

While there is some understanding of how skeletal muscle regenerates, it is incomplete. Topics under investigation include the molecular and cellular biology of satellite cells and the roles of blood vessel formation and innervation. This information provides a basis for developing strategies to augment muscle regeneration.[12]

Activity and Functional Recovery of Muscle

The influences of activity on functional recovery of muscle after injury are only beginning to be identified. These include the effects of immobilization and passive and active stimulation of myofibers. These insights are relevant to age-associated changes in skeletal muscle as well as the changes seen in muscle during space flight.

Gene Therapy and Tissue Engineering

The ability to retrieve and genetically alter stem cells (satellite cells) in individuals of all ages makes muscle an ideal tissue for both gene therapy and tissue engineering. A number of investigations have already focused on the possibility of using dystrophin-expressing stem cells to restore muscle function in individuals with muscular dystrophy. Tissue engineering research has focused on the use of satellite cells in both myocardial and striated muscle reanimation and the development of three-dimensional engineered replacement muscle tissues.

Key Terms

Acetylcholine A key chemical mediator involved in neuromuscular synaptic transmission

Action potential An electrical impulse generated by neurons

Concentric contraction The shortening of a muscle during activation

Dynamic strength The magnitude of isotonic or isokinetic contraction

Eccentric contraction The lengthening of a muscle during activation

Endomysium The connective tissue surrounding a muscle cell

Epimysium The connective tissue surrounding the entire muscle

Isokinetic Literally "same speed"; when applied to muscle action, it implies constant velocity of shortening

Isometric Literally "same length"; when applied to muscle action, it implies that the muscle length is held constant

Isotonic When applied to muscle action, it implies that the load is constant

Motor end plate (neuromuscular junction) The synapse between a motor neuron and a muscle fiber

Motor unit The motor nerve axon and the myofibers with which it contacts

Musculotendinous junction The area of interface between a skeletal muscle and its tendon

Myoblasts The embryonic cells that develop into skeletal muscle cells

Myofibers The fibers that constitute a muscle

Perimysium The connective tissue surrounding a fascicle

Sarcolemma Muscle-cell membrane and its associated basement membrane

Sarcomeres The fundamental components of the contracting unit of the myofibril

Sarcopenia The loss of muscle mass and strength as a result of aging

Sarcoplasmic reticulum A continuous branching network of membrane, which is a specialized form of endoplasmic reticulum unique to muscle

Schwann cell A specialized support cell that encases nerve fibers

Static strength The magnitude of isometric contraction

Stem cells Cells with the unlimited ability of self-renewal and regeneration; serve to regenerate tissue

Synapse A specialized site at which an electrical signal is transmitted chemically across a junction to produce a similar electrical impulse on the opposite side

Tenocytes The cells in tendons

References

1. Huxley AF: Muscular contraction. *J Physiol* 1974;243:1-43.
2. Burke RE: Motor unit properties and selective involvement in movement. *Exerc Sport Sci Rev* 1975;3:31-81.
3. Buchthal F, Schmalbruch H: Motor unit of mammalian muscle. *Physiol Rev* 1980;60:90-142.
4. Close RI: Dynamic properties of mammalian skeletal muscles. *Physiol Rev* 1972;52:129-197.
5. Garrett WE Jr., Seaber AV, Boswick J, et al: Recovery of skeletal muscle after laceration and repair. *J Hand Surg Am* 1984;9:683-692.
6. Griffin GE, Williams PE, Goldspink G: Region of longitudinal growth in striated muscle fibres. *Nat New Biol* 1971;232:28-29.
7. Roubenoff R: Origins and clinical relevance of sarcopenia. *Can J Appl Physiol* 2001;26:78-89.
8. Nikolaou PK, Macdonald BL, Glisson RR, et al: Biomechanical and histological evaluation of muscle after controlled strain injury. *Am J Sports Med* 1987;15:9-14.
9. Tidball JG: Inflammatory cell response to acute muscle injury. *Med Sci Sports Exerc* 1995;27:1022-1032.
10. King JB: Post-traumatic ectopic calcification in the muscles of athletes: A review. *Br J Sports Med* 1998;32:287-290.
11. Tidball JG: Myotendinous junction: Morphological changes and mechanical failure associated with muscle cell atrophy. *Exp Mol Pathol* 1984;40:1-12.
12. Marler JJ: Skeletal muscle, in Atala A, Lanza RP (eds): *Methods of Tissue Engineering*. San Diego, CA, Academic Press, 2001.

Chapter 7

Inflammation

Inflammation is characterized by erythema, warmth, pain, and edema. Acute inflammation generally occurs in response to an injury or introduction of foreign material at a specific site and is an important part of wound healing. Chronic inflammation is usually associated with a systemic disease process, such as rheumatoid arthritis (RA), and itself can be a major source of disability. When inflammation affects the joints of the musculoskeletal system, it most often affects the synovial membrane, resulting in a condition called synovitis. Inflammation may also affect tendons or tendon sheaths (tendinitis); it also can occur extra-articularly in bursae (bursitis).

While acute inflammation is often a healthful response and usually is self-limiting, chronic inflammation, as in arthritic disease, can be harmful. For example, inflammation can destroy the cartilage that cushions the bone. Progressive cartilage destruction leads to joint instability and loss of function. Because of these harmful effects, inflammation should be inhibited in certain circumstances. Knowing when and how to modulate inflammation demands an understanding of the cellular, biochemical, and molecular aspects of inflammatory processes and the regimens currently available to treat them.

This chapter describes the inflammatory process, how inflammation is triggered, how chronic inflammation affects the joints, and how it is ameliorated through endogenous and exogenous agents. In addition, the pathophysiology of several types of arthritis that are manifested through the inflammatory process are described, along with current trends in research aimed at targeting the molecular and immunologic pathways of inflammation.

The Inflammatory Process

In response to a trigger, such as an injury or an antigen, cells in affected tissue produce signals to initiate the infiltration of white blood cells (such as monocytes, granulocytes, and lymphocytes) to the site. Once these cells move from the blood vessels into the tissue, they attack microorganisms and bacteria and ingest senescent cells and cellular debris in a process called **phagocytosis**. Along with the resident tissue cells, these cells produce inflammatory mediators that result in vasodilation, edema, and cell proliferation.

This inflammatory response occurs through a sequence of metabolic events that involves five types of agents: (1) **cytokines**, such as the interleukins, which induce cellular production of chemokines and growth factors; (2) **chemokines**, which attract or recruit cells to the site; (3) **arachidonic acid** metabolites, such as prostaglandins and leukotrienes, which induce enzyme production and activation; (4) **growth factors**, which stimulate cell proliferation; and (5) catabolic enzymes, such as **matrix metalloproteinase** (MMP), which degrade the extracellular matrix (ECM). The cascade is perpetuated by the cellular response to the autocrine, paracrine, and endocrine effects of these same agents (Fig. 1).

When a cytokine binds to its specific receptor on the cell membrane, the receptor mediates activation of protein kinase-C, which by intracellular phosphorylation produces transcription factors, such as activation protein-1 (AP-1) and nuclear factor κB (NFκB). Upon phosphorylation, the NFκB complex is cleaved and passes into the nucleus where it binds to the promoter region of genes responsible for expression of inflammatory proteins[1] (Fig. 2). An AP-1 binding site is located on the promoter for MMP genes. In conjunction with other sites on the promoter, binding of AP-1 to the AP-1 binding site leads to transcriptional activation of the MMP genes and subsequent production of MMP as proenzymes.[2-4]

Activation of AP-1 and NFκB stimulates the

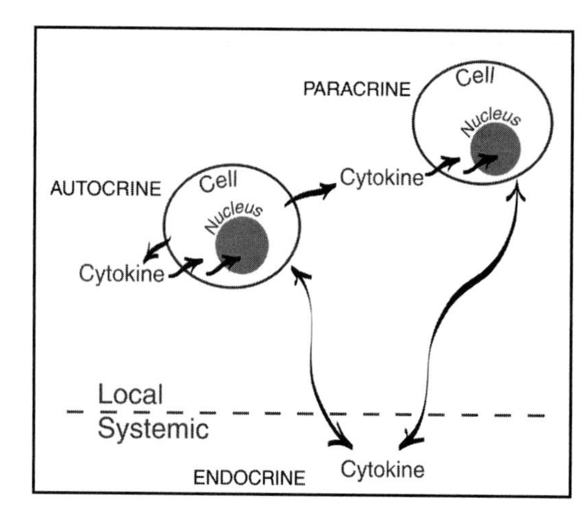

Figure 1
Schematic diagram of the types of cytokine interactions. Autocrine activity involves the cytokine-producing cell itself. Paracrine activity impacts neighboring cells, and endocrine activity involves the cellular release of cytokines into the vascular system, affecting cells remotely.

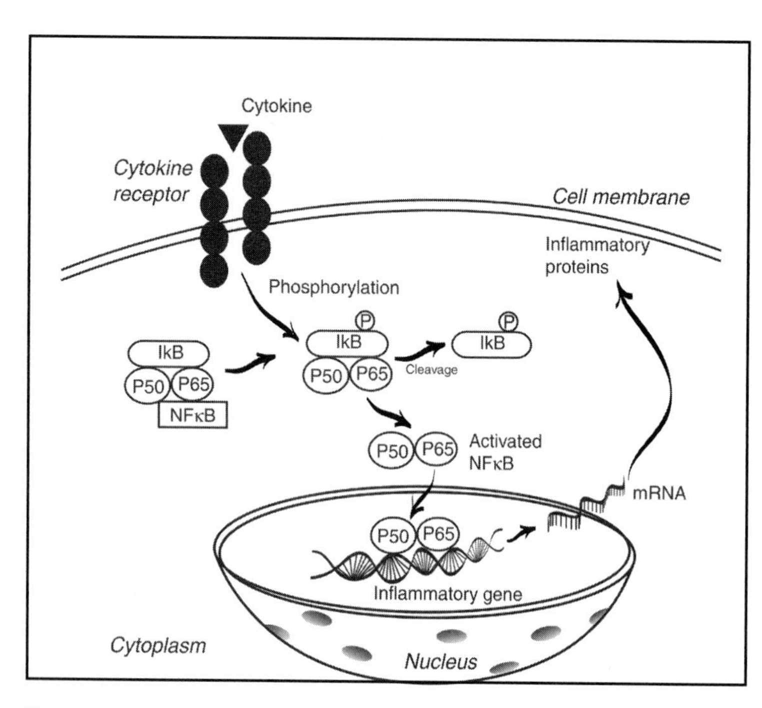

Figure 2
Schematic representation of a cytokine signal transduction pathway that results in release of inflammatory proteins. When a cytokine binds to its receptor, NFκB in the cell cytoplasm is activated by C-kinase phosphorylation and subsequent cleavage of IκB. The activated NFκB then translocates to the nucleus where it binds to an inflammatory gene and signals production of messenger RNA that codes for certain inflammatory proteins.

(Adapted with permission from Barnes PJ, Karin M: Nuclear factor-κB: A pivotal transcription factor in chronic inflammatory diseases. ***N Engl J Med*** *1997;336:1066-1071.)*

synthesis of chemokines and adhesion molecules that promote recruitment and infiltration of mononuclear cells. This activation also produces cytokines and growth factors that perpetuate the immune response and proenzymes such as MMPs that degrade matrix proteins. In addition, enzymes that initiate the arachidonic acid cascade and production of eicosanoids are produced. For example, tumor necrosis factor-α (TNF-α) and interleukin (IL)-6, both of which are cytokines associated with inflammation, have been shown to stimulate the messenger RNA responsible for production of inflammatory cytokines through activation of NFκB and AP-1. Thus, the presence of a cytokine can promote synthesis of itself and other cytokines.

Another cytokine, IL-1, also has been shown to activate MMP. When activated, MMP enzymes degrade the ECM in the vicinity of the cell. Degradative products of the ECM derived from MMP activity can stimulate chemokines, which then recruit phagocytic cells to the site. Infiltration of leukocytes produces additional cytokines, and the process continues until the inducing agent is eliminated.

Immune-Mediated Inflammation

The immune response to an antigen can be classified as humoral (antibody based) or cell mediated. Both processes require lymphocytes and antigen-presenting cells (APCs), which may be lymphocytes, macrophages, or dendritic cells. Lymphocytes are described as B cells or T cells, a naming convention based on the site of their original differentiation—bone marrow (B cells) or thymus (T cells).[5]

Humoral immunity occurs through B cells, which produce soluble, membrane-bound immunoglobulin (antibody) for a specific antigen. Cell-mediated immunity occurs through T cells that recognize an antigen when it is bound to a major histocompatibility complex (MHC) molecule, also termed a human leukocyte antigen (HLA). Thus, B cells are frequently the APCs to the T-cell receptors. Binding of the APCs with the T-cell receptors stimulates division of T cells into one of two types of helper cells (Th1 or Th2) and secretion of numerous cytokines that induce an inflammatory response (Fig. 3).

There are two classes of MHC molecules (class I and II), both of which contain a large number of alleles (variable regions). Although MHC molecules bind antigens, they differ from B-cell receptors in that MHC molecules are expressed by a number of cell types. They also lack the specificity of the antibodies produced by B cells. Several chronic inflammatory diseases, such as RA, are associated with distinct alleles that may confer genetic predisposition to the disease. That the inflammation persists or recurs in these chronic diseases indicates that attempts to achieve homeostasis are impeded, possibly by a recurrent trigger.

Triggers of Inflammation

Inflammation may be initiated by either endogenous or exogenous factors. It can be an acute response to trauma (as in a ligament tear) or a chronic response (as in autoimmune diseases such as RA). Infection and crystalline deposits also can provoke an inflammatory response that will persist until the underlying cause is eliminated.

When trauma is the cause, inflammation promotes healing. As wounds heal, metabolic activity consumes oxygen, resulting in a hypoxic environment that stimulates proliferation of fibroblasts and production of blood vessels (angiogenesis) at the site. Anoxia leads to increased levels of enzymes such as cathepsin, which provide a means to remove dead tissue and debris.[6] The process is reversed with angiogenesis and tissue reperfusion; thus begins the next stage of wound healing and a return to homeostasis.

Viral or bacterial infection is also a trigger for inflammation, as it promotes an immune response that recruits leukocytes to destroy the antigen. At the same time, this response increases cytokine production and vascular permeability. When the infection is controlled, the acute inflammation subsides. However, with chronic autoimmune disease, the infection may produce bacterial or viral compounds that alter T-cell recognition and the tissue response.

With diseases such as RA or systemic lupus erythematosus (SLE), the trigger may be a T-cell–mediated immune response to an autoantigen such as an epitope of type II collagen[7] or a nucleoprotein, respectively.[8,9] Because a specific antigen has not been identified as the initiator of either disease, both must be considered to be multifactorial processes of unknown etiology.

Crystal-induced arthritis, namely gout and pseudogout, is an acute process that may be treated with medication but may also develop into a chronic synovitis or recurrent acute inflammation. With gout, the trigger is deposition of the endogenously produced microcrystalline compound monosodium urate within the joint. With pseudogout, the agent typically is calcium pyrophosphate dihydrate. Hydroxyapatite and calcium oxalate can also initiate inflammation. These crystals form nodes or tophi that act as irritants,

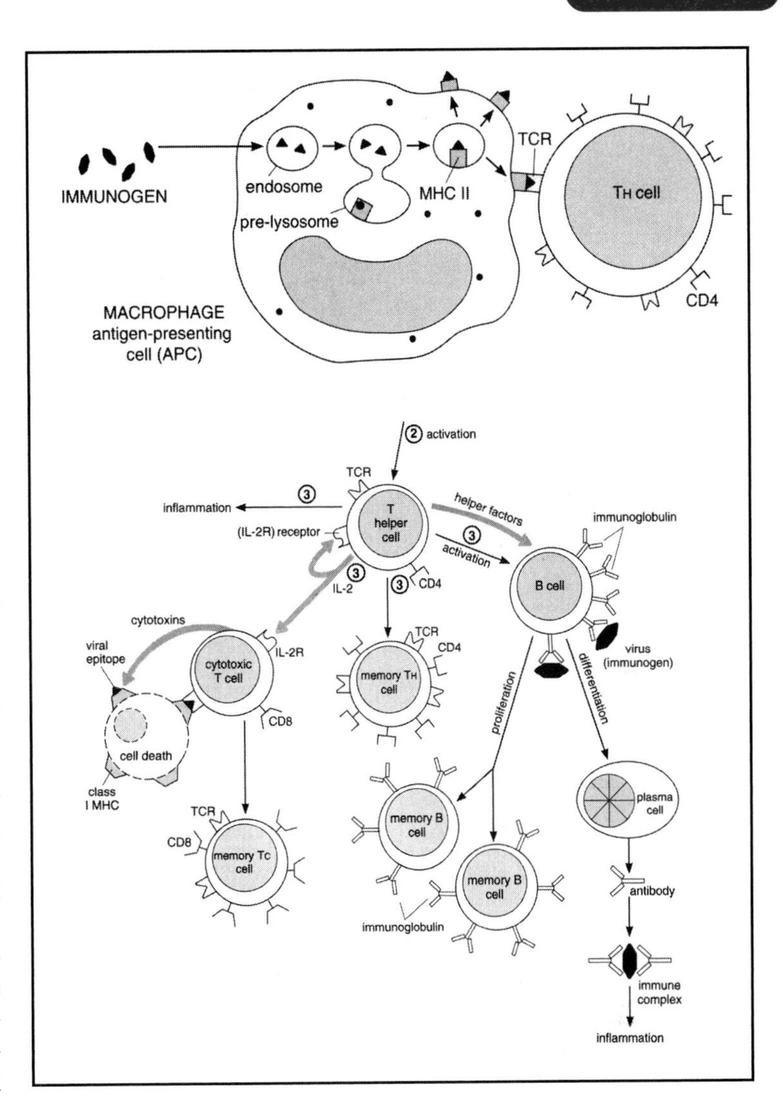

Figure 3

Stimulation of the immune response involves internalization of an antigen by macrophages. The antigen is then bound to a receptor (MHC II) and expressed on the cell surface. The CD4-expressing helper T cells are activated by receptor interaction with the antigen (2) if it is recognized as foreign. The activated CD4 cell (3) elaborates factors that stimulate B cells to express antibody and also inflammatory cytokines. B cells can differentiate into antibody- producing plasma cells or remain as memory cells. Similarly, the sensitized T cells can expand by proliferation and mediate cytotoxicity or be retained as memory cells.

(Reproduced with permission from Stites DP, Terr AI, Parslow TG: ***Medical Immunology****, ed 8. Norwalk, CT, Appleton & Lange, 1994, pp 43-44.)*

causing the adjacent tissue to become inflamed.[10]

Mediators of Inflammation

Inflammatory mediators often play important roles in the normal cell, regulating the synthesis and turnover of ECM, for example. Accordingly, blocking their production can have adverse effects on normal cell physiology. At best, inflammation can be controlled by modulating the production of these mediators.

Eicosanoids

Arachidonic acid is produced by cleavage of a membrane-bound phospholipid, which sets into motion the arachidonic acid cascade and synthesis of many different carbon fatty acid derivatives. The cascade includes two principal arms: (1) the cyclooxygenase pathway, which leads to production of prostaglandins and thromboxanes, and (2) the lipoxygenase pathway, which produces leukotrienes and lipoxins. These compounds have varying effects. Some, such as leukotriene B_4 (LTB_4) and prostaglandin E_2 (PGE_2), can stimulate cell infiltration (and thus be considered proinflammatory). Others such as lipoxin A_4 can inhibit cell recruitment (and thus be considered anti-inflammatory).

While many prostaglandins are inflammatory mediators, the one that plays a pivotal role in inflammation, and consequently has been studied in greatest depth, is PGE_2. This eicosanoid is found at elevated levels in the synovial fluid of inflamed joints and mediates the edema associated with joint inflammation. Results of recent studies describe both proinflammatory and anti-inflammatory roles for PGE_2: it interacts with LTB_4 to recruit cells, and it activates lipoxin A_4 to resolve inflammation.

Cytokines

Interleukins

The interleukins are a class of low molecular weight proteins that modulate cell activity and regulate the immune system. Some interleukins are synthesized preferentially by certain cell types, while others appear to be produced by a wide variety of cells. For instance, IL-2 is released by T cells almost exclusively, whereas IL-1 is produced by macrophages, fibroblasts, and lymphocytes. Even tissue-specific cells, such as chondrocytes, can produce IL-1. At least 19 interleukins have been identified, and each has a specific role in cell and tissue maintenance.[11-14] While the mechanisms of action of each of the interleukins are still being investigated, it is clear that they are the regulators of the immune system (Tables 1 and 2). Because of their roles in cartilage and bone metabolism, IL-1 and IL-6 are of particular interest to musculoskeletal medicine. IL-1 has been shown to both decrease synthesis and increase degradation of cartilage ECM components, and IL-1 and IL-6 stimulate the differentiation and recruitment of osteoclasts, the cells responsible for bone resorption.

Tumor Necrosis Factor-α

This cytokine derives its name from its ability to lyse tumors. Its name may be misleading because TNF-α has been shown to have a much broader range of activity: it plays a major role in inflammation and in the immune response.[12,13] As it appears early in the inflammation cascade, TNF-α is considered a possible initiator of further cytokine activity. Increased TNF-α induces and augments IL-1 activity; inhibitors of TNF-α decrease IL-1 activity. For this reason, TNF-α has become the target cytokine for treatment of chronic inflammatory diseases. Several approaches to decrease production of this cytokine are currently being investigated.

Growth Factors

Several growth factors participate in the inflammatory process and in the degradative stages of arthritis. Vascular endothelial growth factor and basic fibroblast growth factor are two important agents; both stimulate endothelial proliferation and increase vascular permeability and angiogenesis.[15,16] Both factors are produced by fibroblasts and macrophages in the vicinity of blood vessels and capillaries within the inflamed tissue.

Modification of the Mediators of Inflammation

Endogenous Inhibitors

To regulate the inflammatory response, numerous endogenous agents have been employed. IL-4 and IL-10 inhibit cytokine production. PGE_2 can either stimulate or inhibit inflammation indirectly. Several other agents are competitive inhibitors of inflammation.[13] Soluble TNF-α receptor proteins cleaved from the cell membrane, for example, can bind with cytokine in the extracellular fluid. This binding decreases the concentration of

Table 1
Proinflammatory Cytokines and Their Role in the Pathogenesis of Arthritis

Cytokine	Major Cellular Source	Major Targets and Biologic Effects
TNF-α	Monocytes Macrophages T lymphocytes	Monocytes, synovial macrophages, fibroblasts, chondrocytes, endothelial cells Stimulation of proinflammatory cytokine and chemokine synthesis Activation of granulocytes Increase MHC class II expression Secretion of MMPs leading to cartilage matrix degradation
IL-1	Monocytes Macrophages Many cell types	Monocytes, synovial macrophages, fibroblasts, chondrocytes, endothelial cells Inhibition of matrix synthesis in chondrocytes Secretion of MMPs leading to matrix degradation Stimulation of proinflammatory cytokine and chemokine synthesis Fibroblast proliferation T-cell proliferation
IL-6	Activated T cells (Th2) Many cell types (induced by IL-1 or TNF-α)	Stimulation of acute phase protein synthesis in liver B-cell proliferation and differentiation T-cell proliferation Differentiation of hematopoietic precursor cells Differentiation and maturation of osteoclasts (induction of MMP-inhibitor, TIMP-1)
TGF-β	Many cell types	Immune suppression (inhibition of B- and T-cell proliferation) Monocyte chemotaxis Differentiation of mesenchymal and epithelial cells (chondrogenesis) Anabolic for cartilage • stimulation of matrix synthesis • reduced production of MMPs • increased production of proteinase inhibitors

TIMP = tissue inhibitors of matrix metalloproteinases

(Reproduced from Recklies AD, Poole AR, Banerjee S, et al: Pathophysiologic aspects of inflammation in diarthrodial joints, in Buckwalter JA, Einhorn TA, Simon SR (eds): ***Orthopaedic Basic Science: Biology and Biomechanics of the Musculoskeletal System****, ed 2. Rosemont, IL, American Academy of Orthopaedic Surgeons, 2000, pp 489-530.)*

TNF-α available for attachment to the cell membrane receptor. Similarly, interleukin-1 receptor antagonist (IL-1ra) protein prevents binding of IL-1 to its receptor by competing for the binding site.

Tissue inhibitors of metalloproteinases are a class of proteins that block activation of the MMPs and prevent enzymatic degradation of the ECM. Transforming growth factor-β (TGF-β) is an agent that plays an important role in decreasing the destructive enzymes of the inflammatory cascade. It also promotes collagen formation, increases MMP inhibitors, and modulates IL-1. With chronic inflammatory disease, however, some or all of these inhibitory agents are produced in low concentrations or are themselves inhibited. Augmenting their effectiveness offers a means by which medical intervention can decrease or slow the effects of chronic inflammation.

Anti-Inflammatory Medications

Joint inflammation has been treated successfully with nonsteroidal anti-inflammatory drugs (NSAIDs), including acetylsalicylic acid (aspirin), naproxen, flurbiprofen, diclofenac, and indomethacin. These compounds inhibit the arachidonic acid cascade at the cyclooxygenase pathway (Fig. 4).

Table 2
Major Cytokines Involved in Immune Regulation

Cytokine	Cellular Source	Biologic Effects
IL-2	Activated T cells (Th1)	Clonal proliferation of T cells Proliferation of B cells
IFN-γ	Activated T cells (Th1)	Induction of MHC class II expression on monocytes, connective tissue cells (fibroblasts, chondrocytes), and endothelial cells Increased expression of MHC class I molecules
IL-4	Activated T cells (Th2)	B-cell proliferation and class switching of immunoglobulin synthesis Inhibition of production of proinflammatory cytokines by monocytes
IL-5	Activated T cells (Th2)	Growth and differentiation of eosinophils
IL-12	Activated macrophages	Development of the Th1 response
IL-10	Activated T cells (Th2) Macrophages	Inhibition of synthesis of proinflammatory cytokines by T cells and macrophages
IL-15	Activated macrophages Connective tissue cells?	T-cell proliferation (similar to IL-2)
IL-17	Activated memory T cells	Proinflammatory; stimulation of cytokine secretion (IL-6, IL-8, MCP1, granulocyte-stimulating factor) and PGE_2 in epithelial, endothelial, and fibroblastic cells

IFN-α = interferon-α, MCP1 = monocyte chemotactic protein 1

(Reproduced from Recklies AD, Poole AR, Banerjee S, et al: Pathophysiologic aspects of inflammation in diarthrodial joints, in Buckwalter JA, Einhorn TA, Simon SR (eds): ***Orthopaedic Basic Science: Biology and Biomechanics of the Musculoskeletal System****, ed 2. Rosemont, IL, American Academy of Orthopaedic Surgeons, 2000, pp 489-530.)*

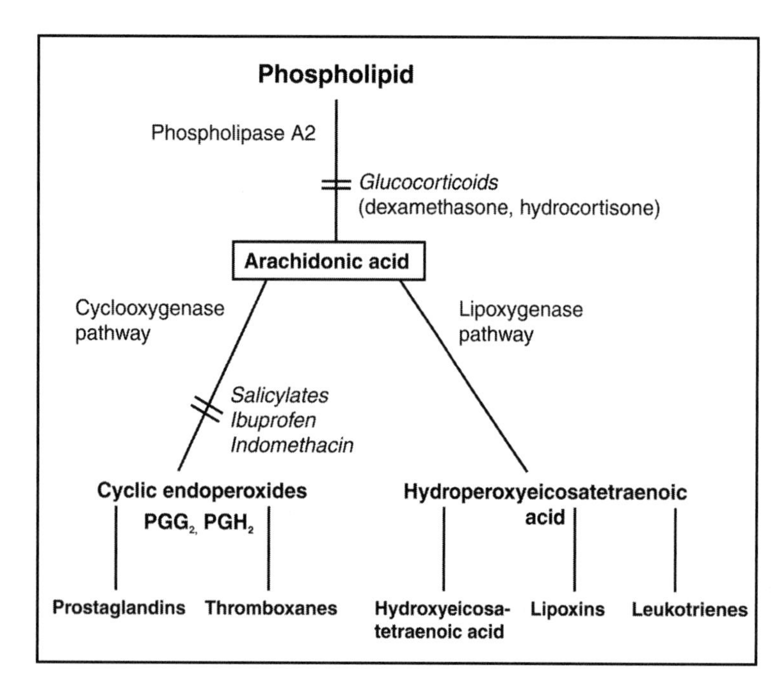

Figure 4
The arachidonic acid cascade and points of inhibition by NSAIDs (the cyclooxygenase pathway) and glucocorticoids (the conversion of phospholipid into arachidonic acid).

Currently, more than 20 NSAIDs are commonly used to treat acute and chronic inflammation; each varies in its biologic half-life and ability to inhibit the enzymes cyclooxygenase 1 (COX-1) and 2 (COX-2). COX-2 is thought to be an induced enzyme that catalyzes the production of prostaglandins in the inflammatory response. COX-1 is active (constitutive) in normal physiology and produces prostaglandins responsible for the homeostasis of many cells and tissues in the body. Therefore, the newer NSAIDs attempt to selectively inhibit COX-2 while preserving the activity of COX-1.

NSAIDs affect one arm of the arachidonic acid cascade: prostaglandin synthesis. Corticosteroids inhibit both prostaglandin synthesis and production of leukotrienes. They can be used locally, as an intra-articular injection, but they are also often administered orally for systemic management of chronic inflammatory diseases. Pharmaceutical steroids are synthetic analogs of cortisol, the body's endogenous steroid. They mediate the inflammatory processes and inhibit the im-

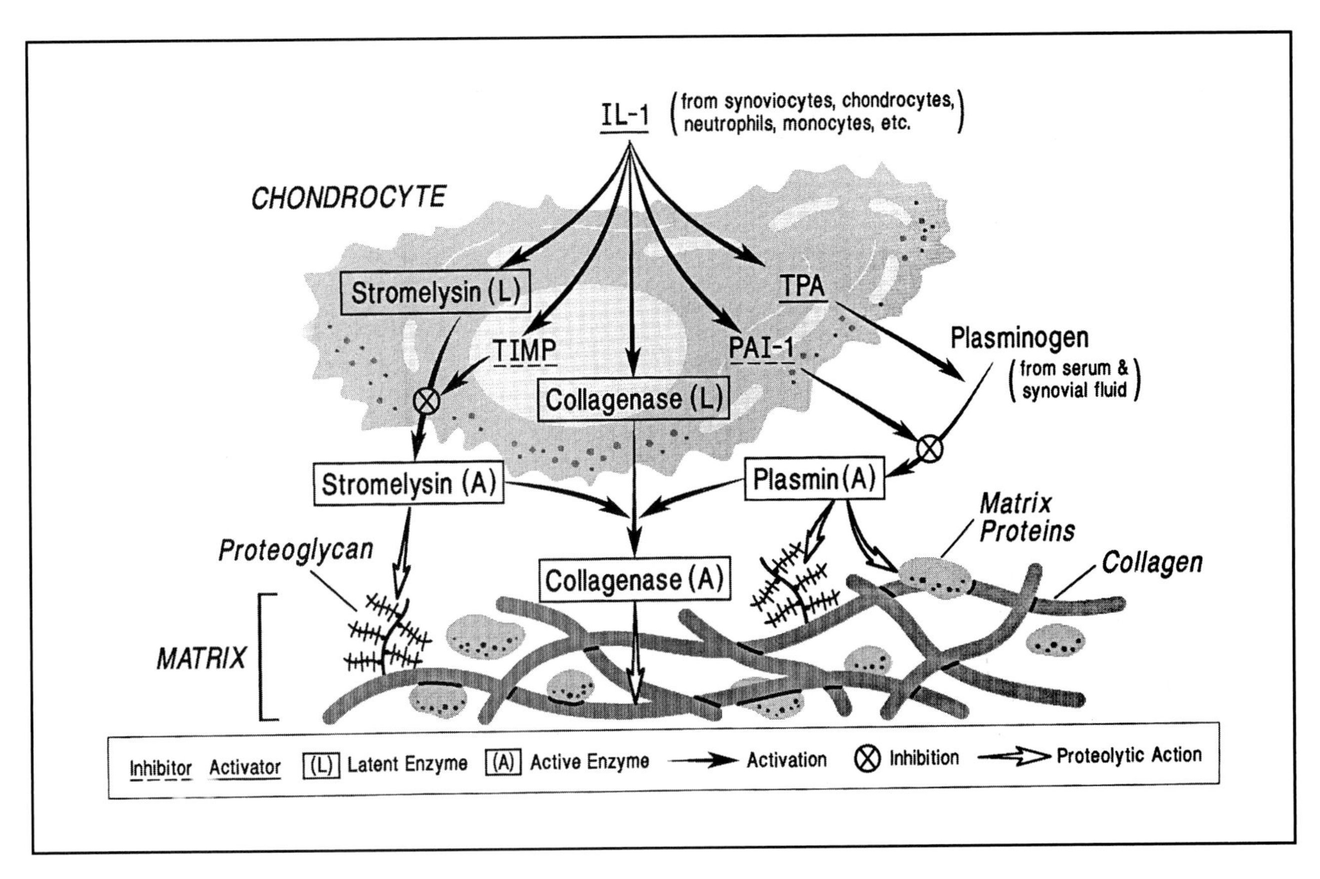

Figure 5
Cytokine-mediated cartilage matrix destruction. TIMP = tissue inhibitor of metalloproteinases. TPA = tissue-type plasminogen activator. PAI-1 = plasminogen activator inhibitor-1.

(Reproduced from Mankin HJ, Mow VC, Buckwalter JA: Articular cartilage repair and osteoarthritis, in Buckwalter JA, Einhorn TA, Simon SR (eds): ***Orthopaedic Basic Science: Biology and Biomechanics of the Musculoskeletal System****, ed 2. Rosemont, IL, American Academy of Orthopaedic Surgeons, 2000, pp 471-488.)*

mune response through cellular actions at the transcriptional, translational, and post-translational levels. As the steroid binds to the glucocorticoid receptor on the cell membrane, human heat shock protein is released and the receptor complex moves to the cell nucleus. There it inhibits transcription of enzymes and cytokines involved in the inflammatory process.[1] Long-term use of corticosteroids can result in numerous adverse effects, including increased susceptibility to infection, osteoporosis, osteonecrosis, mood changes, depression, insomnia, cataracts, and glaucoma. Hypertension and a number of other endocrine and metabolic symptoms such as weight gain, glucose intolerance, and suppression of the hypothalamic-pituitary-adrenal axis are also seen with their long-term use. These endocrine effects can result in adrenal insufficiency and occasionally diabetes mellitus.[17]

Effects of Inflammation on Tissue

Chronic inflammation affects two principal areas in the joints—the synovium and the cartilage. In the early stages of disease, inflammation primarily affects the synovial lining. As disease progresses, changes in cartilage occur, culminating in chondrocyte death and complete cartilage destruction by disintegration of its ECM.

Synovium

Inflammation causes a large number of leukocytes to infiltrate the synovium, which dramatically increases the numbers of macrophages and B and T lymphocytes. The thickness of the synovial tissue increases as does that of the intimal layer, which may increase from one to four cells to several hundred.[18] As the tissue volume increases, the synovial lining extends into the joint space and onto cartilaginous surfaces. To accommodate this increase in tissue volume, neovascularization occurs both below and within the thickened intimal layer. Because the synovium is the source of synovial fluid, an increase in volume with a concurrent increase in vascular permeability leads to increased fluid volume within the joint space (joint effusion).[7]

Cartilage

The effects of inflammation on cartilage appear to be secondary to the process itself and may even be independent of the initial inflammation. Several studies have demonstrated that in chronic arthritic diseases, cartilage degradation continues even when the inflammation has been suppressed.[19] By decreasing proteoglycan synthesis by the chondrocytes and increasing synthesis and activation of the MMPs (collagenase, aggrecanase, and stromelysin), inflammation destroys the organization of the ECM in cartilage (Fig. 5). Formation of pannus on the cartilage surface establishes sites of synovial invasion into the ECM through enzymatic degradation by the MMPs. With continual release and activation of the MMPs, the cartilage completely disintegrates and the joint becomes unstable.

Key Terms

Arachidonic acid The substance that is a precursor of inflammatory mediators such as prostaglandins and leukotrienes

Chemokines An agent of the inflammatory process that attracts or recruits cells to the site

Cyclooxygenase pathway One arm of the arachidonic acid cascade that leads to production of prostaglandins and thromboxanes

Cytokines Proteins produced by a cell to modulate the actions of other cells; also known as messenger proteins

Growth factors The molecules that stimulate cell growth or activation

Lipoxygenase pathway One arm of the arachidonic acid cascade that leads to production of leukotrienes and lipoxins

Major histocompatibility complex (MHC) A cluster of genes important in immune recognition and signaling between cells of the immune system; also called human leukocyte antigen

Matrix metalloproteinases (MMP) Agents of the inflammatory process that degrade the extracellular matrix

Phagocytosis The process by which white blood cells ingest debris or microorganisms

Research and New Directions During the Bone and Joint Decade (2002-2011)

Anti-TNF and TNF-Receptor Proteins

Because TNF-α seems to be a pivotal cytokine, several agents have been developed that can slow the inflammatory cascade by decreasing the available endogenous TNF. Three drugs—TNF-receptor (TNF-R) p55 Fc fusion protein (Lenercept), TNF-R p75 Fc fusion protein (Enbrel, Etanercept), and a chimeric monoclonal antibody to TNF-α (Infliximab)—interfere with TNF-α activity by competing with the endogenous TNF-α receptor. In preliminary clinical trials, all three have resulted in decreased swelling and decreased serum C-reactive protein levels.[20] Infliximab has been shown to decrease IL-6 and increase IL-10 serum levels up to 12 weeks after a single intravenous infusion.[21] While these treatments appear promising, they have associated problems and adverse effects. In most cases, adverse reactions such as headaches, injection site irritation, and infections were minor. However, there were two reported incidences of drug-induced SLE, presumably from development of autoantibodies. Furthermore, multiple doses demonstrated a decrease in the length of benefit, which suggests immunogenicity.[13]

Interleukins

Several approaches to affect inflammation through the interleukins have been proposed.[13] Since both IL-10 and IL-4 have been shown to slow the inflammatory response, a systemic increase in the levels of these cytokines should affect the cascade. However, clinical trials with recombinant IL-10 and IL-4 found these inhibitory agents to be ineffective, possibly because of their short biologic half-life. Another approach is to inhibit IL-6 by introduction of an antibody to the IL-6 receptor, similar to the use of TNF-α receptor antibody. Although this agent has not yet been tested in human trials, it has shown promise in an animal model of RA. Third, the use of recombinant IL-1ra has shown some promise in the treatment of RA, especially in its ability to decrease the rate of disease progression. Unfortunately, the half-life of the agent is extremely short (6 hours), and the dose required is quite high. Studies have demonstrated that a 10- to 1,000-fold excess of IL-1ra is required to effectively block IL-1.[13] In an attempt to provide high doses that are regularly dispensed at the site, gene therapy has been investigated.

Gene Therapy

Stimulation of endogenous inhibitors of the inflammatory cytokines by manipulation of the genes or genetic regulators in synovial fibroblasts may not only decrease inflammation but also may have protective effects on the cartilage. IL-1ra is currently being used in gene therapy. With this novel form of treatment, synovial cells from the patient are grown in culture and using a vector, the complementary DNA of the gene responsible for production of IL-1ra is transferred to the synoviocytes. The transfected cells are then injected into the joint in which the upregulated protein should be expressed. An initial study identified expression of the protein in injected joints 1 week after injection with transfected cells.[22] Long-term trials are necessary to determine if the transfected cells continue to produce IL-1ra and demonstrate efficacy. Although further study is required to fully understand the mechanisms and to assess the long-term benefits of cytokine-targeted therapy, this approach may be the most promising strategy for treating chronic inflammatory disease. The use of autogenous transfected cells opens the field to the feasibility of targeting multiple inflammatory mediators in the local environment and possibly providing effective chondroprotection as well as decreasing the inflammatory response. Finally, by manipulating the abnormal or mutated gene(s) involved in the predisposition to these diseases, it may be possible to arrest the joint destruction and even to prevent the disease process itself.

References

1. Barnes PJ, Karin M: Nuclear factor-kappaB: A pivotal transcription factor in chronic inflammatory diseases. *N Engl J Med* 1997;336:1066-1071.
2. Benbow U, Brinckerhoff CE: The AP-1 site and MMP gene regulation: What is all the fuss about? *Matrix Biol* 1997;15:519-526.
3. Crawford HC, Matrisian LM: Mechanisms controlling the transcription of matrix metalloproteinase genes in normal and neoplastic cells. *Enzyme Protein* 1996;49:20-37.
4. He C: Molecular mechanism of transcriptional activation of human gelatinase B by proximal promoter. *Cancer Lett* 1996;106:185-191.
5. Delves PJ, Roitt IM: The immune system: First of two parts. *N Engl J Med* 2000;343:37-49.
6. Anderson GR, Volpe CM, Russo CA, et al: The anoxic fibroblast response is an early-stage wound healing program. *J Surg Res* 1995;59:666-674.
7. Andersson EC, Hansen BE, Jacobsen H, et al: Definition of MHC and T cell receptor contacts in the HLA-DR4 restricted immunodominant epitope in type II collagen and characterization of collagen-induced arthritis in HLA-DR4 and human CD4 transgenic mice. *Proc Natl Acad Sci USA* 1998;95:7574-7579.
8. Shoenfeld Y, Alarcon-Segovia D, Buskila M, et al: Frontiers of SLE: Review of the 5th International Congress of Systemic Lupus Erythematosus, Cancun, Mexico, April 20-25, 1998. *Semin Arthritis Rheum* 1999;29:112-130.
9. Maddison PJ: Autoantibodies in SLE: Disease associations. *Adv Exp Med Biol* 1999;455:141-145.
10. Agudelo CA, Wise CM: Gout: Diagnosis, pathogenesis, and clinical manifestations. *Curr Opin Rheumatol* 2001;13:234-239.
11. Delves PJ, Roitt IM: The immune system: Second of two parts. *N Engl J Med* 2000;343:108-117.
12. Burger D: Cell contact interactions in rheumatology, The Kennedy Institute for Rheumatology, London, UK, 1-2 June 2000. *Arthritis Res* 2000;2:472-476.
13. Choy EH, Panayi GS: Cytokine pathways and joint inflammation in rheumatoid arthritis. *N Engl J Med* 2001;344:907-916.
14. Gallagher G, Dickensheets H, Eskdale J, et al: Cloning, expression and initial characterization of interleukin-19 (IL- 19): A novel homologue of human interleukin-10 (IL-10). *Genes Immun* 2000;1:442-450.
15. Sunderkotter C, Steinbrink K, Goebeler M, et al: Macrophages and angiogenesis. *J Leukoc Biol* 1994;55:410-422.
16. Yanagisawa K, Hamada K, Gotoh M, et al: Vascular endothelial growth factor (VEGF) expression in the subacromial bursa is increased in patients with impingement syndrome. *J Orthop Res* 2001;19:448-455.
17. Stein CM, Pincus T: Glucocorticoids, in Kelley WN, Harris ED Jr, Ruddy S, Sledge CB (eds): *Textbook of Rheumatology*, ed 5. Philadelphia, PA, WB Saunders, 1997, pp 787-803.
18. Bresnihan B, Cunnane G, Youssef P, et al: Microscopic measurement of synovial membrane inflammation in rheumatoid arthritis: Proposals for the evaluation of tissue samples by quantitative analysis. *Br J Rheumatol* 1998;37:636-642.
19. Fassbender HG: What destroys the joint in rheumatoid arthritis? *Arch Orthop Trauma Surg* 1998;117:2-7.
20. Feldmann M, Charles P, Taylor P, et al: Biological insights from clinical trials with anti-TNF therapy. *Springer Semin Immunopathol* 1998;20:211-228.
21. Ohshima S, Saeki Y, Mima T, et al: Long-term follow-up of the changes in circulating cytokines, soluble cytokine receptors, and white blood cell subset counts in patients with rheumatoid arthritis (RA) after monoclonal anti-TNF alpha antibody therapy. *J Clin Immunol* 1999;19:305-313.
22. Evans CH, Robbins PD, Ghivizzani SC, et al: Abstract: Results from the first human clinical trial of gene therapy for arthritis. *Arthritis Rheum* 1999;42(suppl 9):S170.

Section Editor
Philip E. Blazar, MD

Section 2

Anatomy

Contributing Authors
Joseph Bernstein, MD, MS
Philip E. Blazar, MD
John T. Campbell, MD
James M. Hartford, MD
Jeffrey N. Lawton, MD
Scott D. Mair, MD
Vishwas R. Talwalkar, MD

Anatomy

The English word *anatomy* is derived from the Greek *anatome*, meaning dissection; and for many generations of students, the primary center of anatomy instruction was the dissection laboratory. Dissection will no doubt continue to be part of medical school training, if for no reason beyond tradition: the transformation from college senior to freshman medical student begins with the pickling of hands in cadaver preservative. Nevertheless, anatomy instruction must go beyond that because a semester or two of dissection usually provides too much distracting detail for those students not entering the field of surgery and too little necessary detail for those who do.

Students certainly can augment their dissections using modern pedagogical tools such as the visible human project (*http://www.nlm.nih.gov/research/visible/visible_human.html*) or by studying MRI and CT scans. But perhaps the best method is a simple one: learning to draw uncomplicated anatomic sketches from memory.

Over 500 years ago, Leonardo da Vinci correctly noted that all muscles work by pulling, not pushing. Therefore, to know a muscle's origin and insertion is to know its function. Drawing anatomic sketches will help you grasp these relationships and, consequently, understand the workings of the musculoskeletal system.

Each of the illustrations in the section that follows also appears as a plate at the end of the chapter. On these plates, a few simple bold lines have been added to highlight the essential anatomic relationships. Begin by tracing these lines. After tracing a few times, try to recreate the drawings from memory or from a different perspective.

You may think that sketching seemingly naïve cartoons requires glossing over important details. That may be. Yet this method ensures active focus on the big picture—how the whole comprises its parts—and will provide a solid foundation upon which additional detail can later be layered. Moreover, after drawing from memory you can compare your sketches to the originals and determine the extent to which you have mastered the material.

Beyond these sketches, strive to master surface anatomy. A thorough knowledge of surface anatomy will enable you to correlate your understanding of structure and function to a given patient's presentation. As such, familiarity with surface anatomy will bring your sketches to life. You can certainly learn surface anatomy from cadaveric dissection, but you should also try to learn it from every patient encounter and even from examination of your own surface anatomy.

Throughout medical school, each professor will try to convince you that his or her area of specialty is the most important. For musculoskeletal problems, this issue was resolved long ago: anatomy is the essential discipline. It is also the field of medicine in which the answers are least likely to change from generation to generation. Many of your basic science classes are no doubt vastly different than the ones I took 20 years ago. The anatomy you study, however, is essentially the same corpus of knowledge that was available to da Vinci and all who took an interest in anatomy before him. So learn anatomy well, and reinforce and demonstrate your learning by drawing. If you do so, mastery of musculoskeletal medicine is within your reach.

—Joseph Bernstein, MD, MS

Chapter 8

Anatomy

Shoulder and Arm

Bones and Joints

The shoulder comprises three bones (the clavicle, scapula, and humerus) and three joints (the glenohumeral, the acromioclavicular [AC], and the scapulothoracic) (Fig. 1). The glenohumeral joint is the primary articulation of the shoulder, the point where the head of the humerus meets the glenoid. The AC and scapulothoracic joints attach the trunk to the scapula, with the arm appended to it.

The clavicle (collarbone) is an S-shaped bone that serves primarily as a strut to which muscle is attached. It is directly under the skin and is thus prone to fracture. The scapula (shoulder blade) is a thin, flat bone with several named prominences, one of which, the coracoid process, projects anteriorly off of the scapula, anterior and medial to the glenohumeral joint. Just medial to the coracoid lies the brachial plexus and the subclavian artery and vein. The spine of the scapula, another important prominence, runs along the posterior surface of the bone. The spine terminates in the acromion process. The acromion lies approximately 1 cm above the humeral head, creating a space for the rotator cuff tendons. Laterally, the scapula widens to form the glenoid fossa (shoulder socket). The glenoid fossa is smaller than the humeral head, such that only about one third of the humeral head contacts the glenoid fossa. This joint accordingly lacks the static stability of the hip; however, the decreased stability permits far more motion.

The proximal portion of the humerus is the humeral head. It has an articular surface covered with hyaline cartilage and two tuberosities that serve as the attachment site of the rotator cuff. The greater tuberosity is on the superior lateral aspect of the humeral head and serves as the attachment site of three of the four rotator cuff tendons: the supraspinatus, the infraspinatus, and the teres minor. The lesser tuberosity is medial to the greater; the remaining rotator cuff tendon, the subscapularis, attaches to it. Between the tuberosities lies the bicipital groove, housing the long head of the biceps tendon as it makes its way from its scapular origin down the front of the arm.

The AC joint is the only articulation between the clavicle and the scapula. The arm and scapula are suspended from the clavicle and held in place by strong coracoclavicular ligaments, which prevent inferior displacement of the arm. When the AC joint is dislocated (shoulder separation), it appears that the clavicle has moved superiorly, but actually the arm sags inferiorly. The scapulothoracic joint is not a true (cartilage-lined) joint but simply the muscular attachment of the scapula to the thorax. Approximately 60° of elevation of the arm takes place at this "joint." There are more than a dozen muscles that

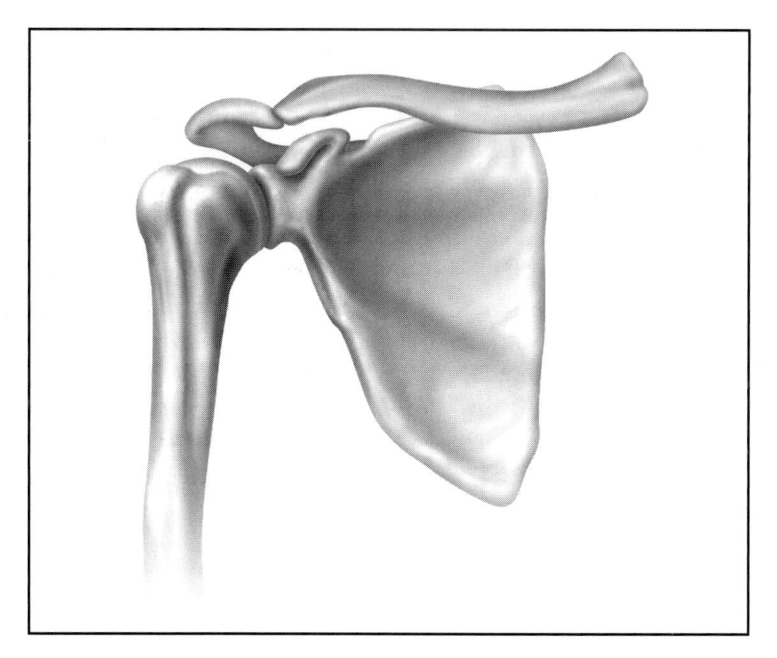

Figure 1
The bones of the shoulder: the clavicle, humerus, and scapula. The clavicle articulates with the scapula at the acromioclavicular joint. The humerus articulates with the scapula at the glenohumeral joint.

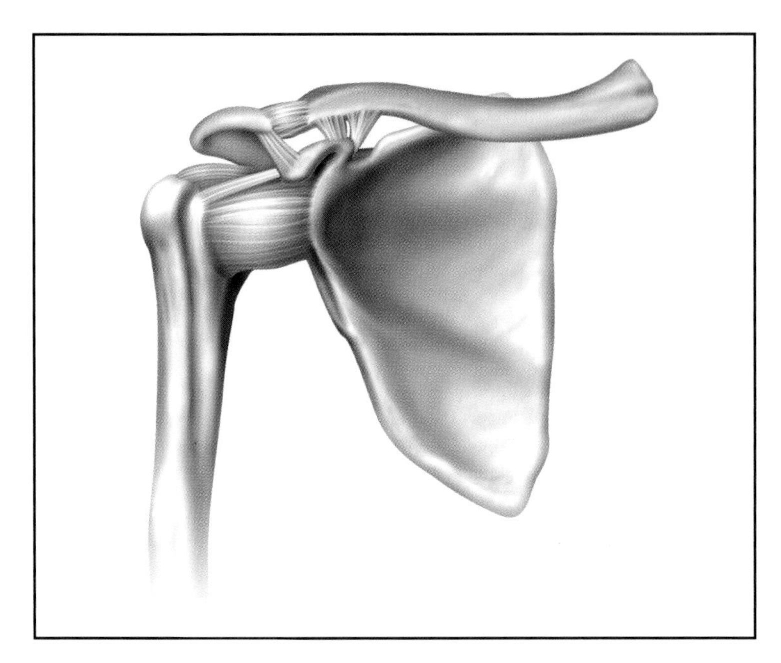

Figure 2
The ligaments of the shoulder: the capsule (and confluent glenohumeral ligaments) and the acromioclavicular, coracoclavicular, and coracoacromial ligaments.

attach to and move the scapula, each of which contributes to the scapula's normal motion.

The glenohumeral joint is a variant of a ball-and-socket articulation. In fact, it is closer in shape to a golf ball sitting on a tee. This configuration allows for the remarkable range of motion of the shoulder but provides little inherent stability. The glenoid surface is deepened by the glenoid labrum, a fibrocartilage ring that surrounds the glenoid articular cartilage. At the top of the labrum, the long head of the biceps tendon attaches. Stability of the glenohumeral joint is provided by both static and dynamic means. Structures that provide static stability include the labrum, the capsule that runs from the glenoid to the humerus, and the thickened areas of the capsule known as the glenohumeral ligaments. The rotator cuff muscles and the biceps provide dynamic stability.

Ligaments and Soft Tissues

The glenohumeral ligaments, which are thickened portions of the capsule, limit excessive translation and rotation of the humeral head on the glenoid during motion of the arm (Fig. 2). The largest and most important glenohumeral ligament is the inferior glenohumeral ligament complex. This complex is shaped like a hammock in which the inferior humeral head rests. It attaches to the glenoid labrum around the lower half of the "clock face." This structure is the primary stabilizer of the shoulder when the arm is abducted. It prevents excessive anterior, posterior, and inferior translation.

The middle and superior glenohumeral ligaments are less important than the inferior glenohumeral ligament and provide stability when the arm is not abducted. The tightness of the shoulder capsule and ligaments varies among individuals. In some people, generalized laxity of these structures leads to excessive translation of the shoulder during activities, with resultant shoulder pain.

The acromion is held in correct relation to the clavicle primarily by the coracoclavicular ligaments. The coracoid and the acromion are both connected to the scapula; thus, when the coracoid is held a fixed distance from the clavicle, the acromion will also be held in place. There are two coracoclavicular ligaments: the conoid and the trapezoid. There is also an acromioclavicular ligament, but this is a minor stabilizer. The coracoacromial ligament attaches the coracoid to the acromion. This clearly does not stabilize those two bones—as noted, they are both processes of the scapula. Rather, this ligament most likely serves to help contain the humeral head.

Muscles

Muscles of the shoulder can be divided into three groups: those that control scapulothoracic joint motion, those that produce motion at the glenohumeral joint, and those that cross more than one joint.

The trapezius muscle is the largest and most superficial of the muscles of the scapulothoracic joint. The primary function of the trapezius is to retract the scapula. The rhomboid muscles also function as scapular retractors. The serratus anterior protracts the scapula and is very important in stabilizing the scapula during forward elevation of the arm.

The deltoid muscle powers glenohumeral abduction and flexion and extension. It is divided into three parts: anterior, which originates from the clavicle; middle, which arises from the lateral acromion; and posterior, which attaches to the spine of the scapula. Intuitively, you can see that the anterior deltoid is a forward flexor of the arm, the middle deltoid is an abductor, and the posterior deltoid powers extension. Paralysis of the deltoid caused by an axillary nerve injury is a devastating injury. In many cases, function of the glenohumeral joint is nearly completely lost.

The rotator cuff is made up of four separate muscles that blend at their insertions and form a cuff that envelopes the humeral head (Figs. 3 and 4). While these muscles provide some power for motion, their primary purpose is to stabilize the humeral head against the glenoid fossa. The supraspinatus muscle lies above the scapular spine and attaches along with the infraspinatus muscle on the greater tuberosity. The supraspinatus is active in any motion involving elevation of the humerus, working to stabilize the glenohumeral joint. The infraspinatus also stabilizes the humeral head; in addition, it accounts for up to 60% of external rotation strength at the shoulder, the remainder of which is provided by the teres minor below it. The subscapularis muscle is the only muscle attaching to the lesser tuberosity and as such serves as an internal rotator. It also helps stabilize the glenohumeral joint and prevents anterior subluxation of the shoulder.

The pectoralis major muscle, inserting on the lateral edge of the bicipital groove of the proximal humerus, serves as a powerful adductor and internal rotator of the glenohumeral joint. The latissimus dorsi muscle also provides adduction and internal rotation. The biceps brachii crosses both the shoulder and elbow joints, but its primary action is supination and flexion at the elbow (Fig. 5). The triceps brachii inserts on the olecranon process of the ulna at the posterior elbow, and its predominant function is extension of the arm and forearm (Fig. 6).

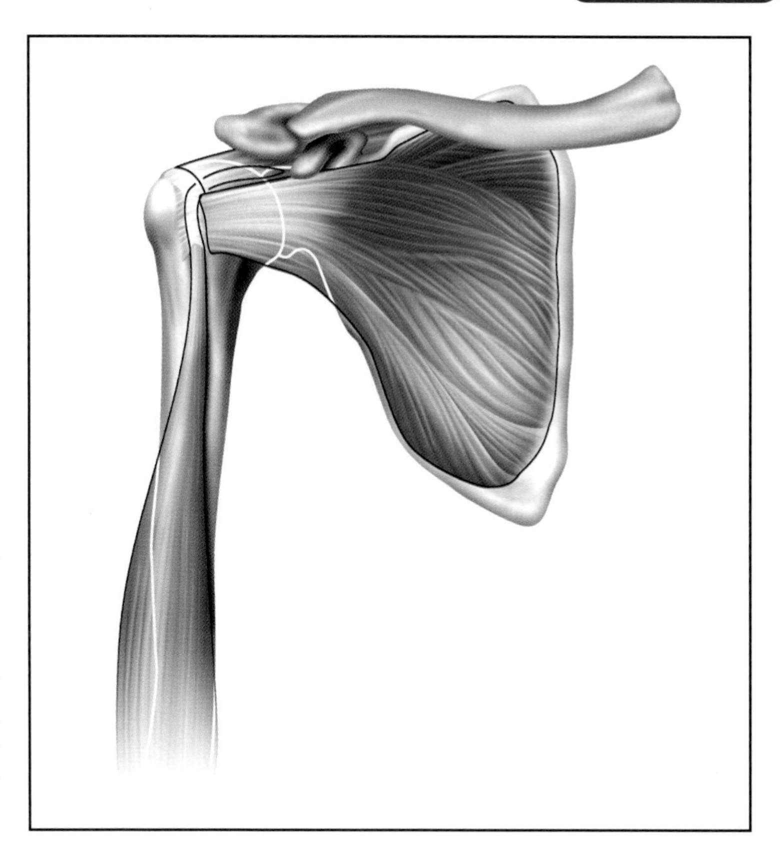

Figure 3
Anterior view of the rotator cuff, showing the subscapularis and supraspinatus. The long head of the biceps is also shown.

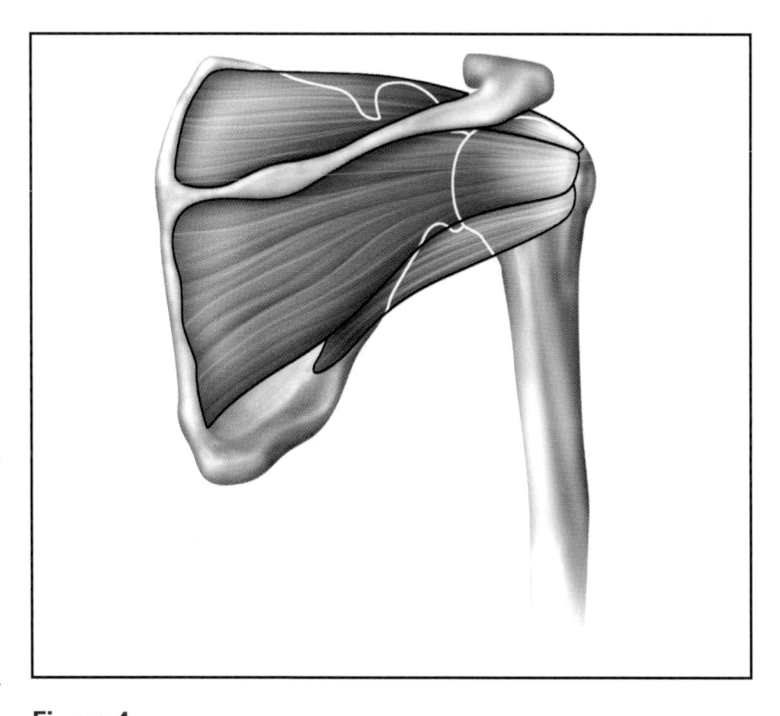

Figure 4
Posterior view of the rotator cuff, showing the supraspinatus, infraspinatus, and teres minor. The course of the infraspinatus and teres minor allows these muscles to serve as external rotators.

Nerves and Blood Vessels

All of the important nerves to the shoulder girdle are branches of the brachial plexus, with one exception: cranial nerve XI, also known as the spinal accessory nerve, which supplies the trapezius muscle. Paralysis of the trapezius results in protraction and downward rotation of the scapula, with significant destabilization of the scapulothoracic joint. The brachial plexus is formed from the nerve roots of C5-T1 and terminates in five main branches (Fig. 7).

Two of the major terminal branches of the plexus, the median nerve and the ulnar nerve, supply muscles distal to the elbow. Another, the radial nerve, supplies both muscles of the shoulder and those originating

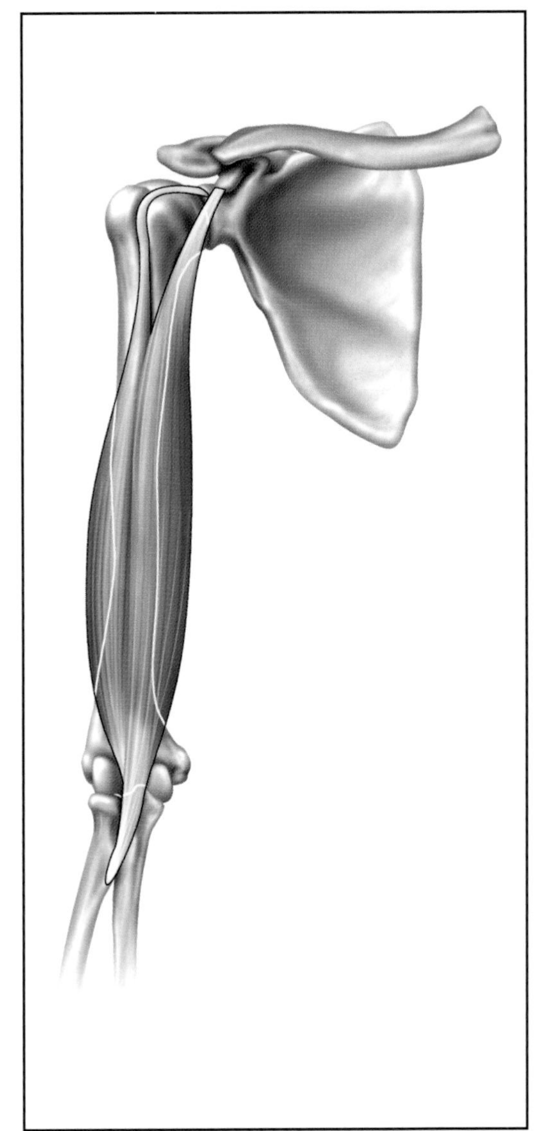

Figure 5
The biceps originates from the coracoid and above the glenoid and inserts across the elbow. It therefore flexes the elbow, supinates the forearm, and flexes the shoulder.

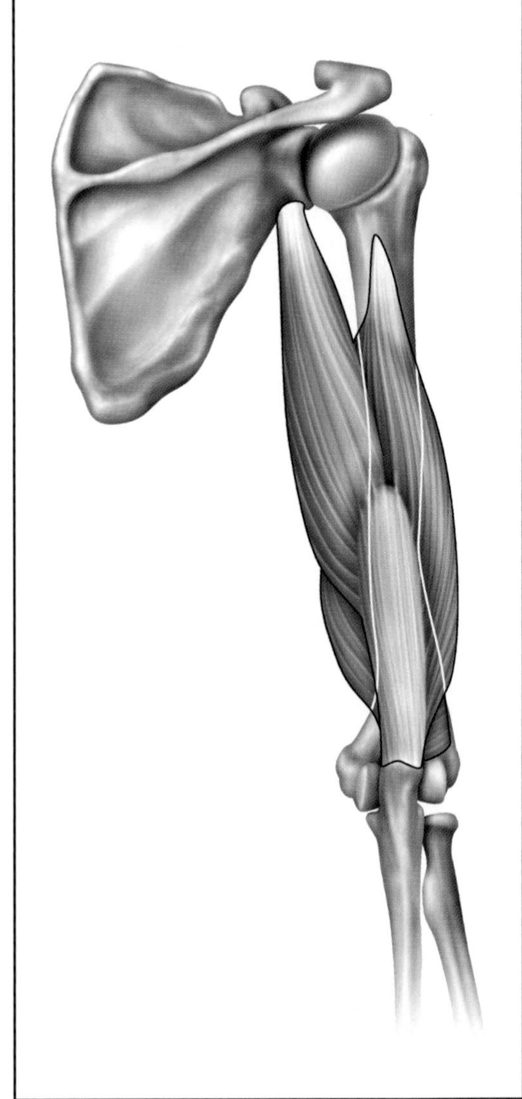

Figure 6
The triceps lies on the posterior aspect of the arm and inserts on the ulna. Its main function is to extend the elbow.

below the elbow. The other two, the axillary nerve and the musculocutaneous nerve, supply muscles housed in the shoulder and arm. The axillary nerve branches from the brachial plexus and runs inferior to the glenohumeral capsule as it passes toward the back of the shoulder. The axillary nerve supplies the teres minor and the deltoid muscles. Given its proximity to the shoulder joint, this is the nerve most at risk for injury in fractures and dislocations. The musculocutaneous nerve also courses close to the shoulder and is at risk for iatrogenic injury during surgery. This nerve supplies the biceps muscle. The radial nerve innervates the triceps, a muscle that extends the humerus at the glenohumeral joint, but the main targets of this nerve are in the forearm. Although the radial nerve does not have much function in the upper arm, it is prone to injury there because it spirals around the humerus to emerge on the lateral side of the arm. As such, it may be stretched or torn with a humeral shaft fracture.

The suprascapular nerve innervates two muscles of the rotator cuff, the supraspinatus and infraspinatus. This nerve may become trapped as it enters the supraspinatus or as it courses around the spine of the scapula to reach the infraspinatus. This compres-

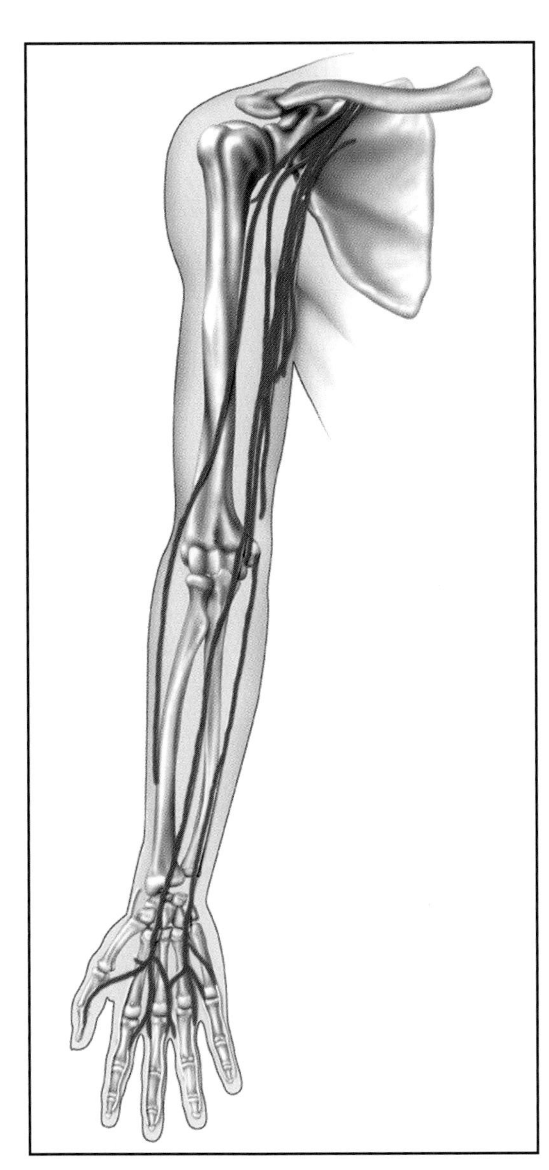

Figure 7
The brachial plexus. Note that this is a highly simplified view to show basic anatomic relationships. A precise wiring diagram can be found in many anatomy textbooks.

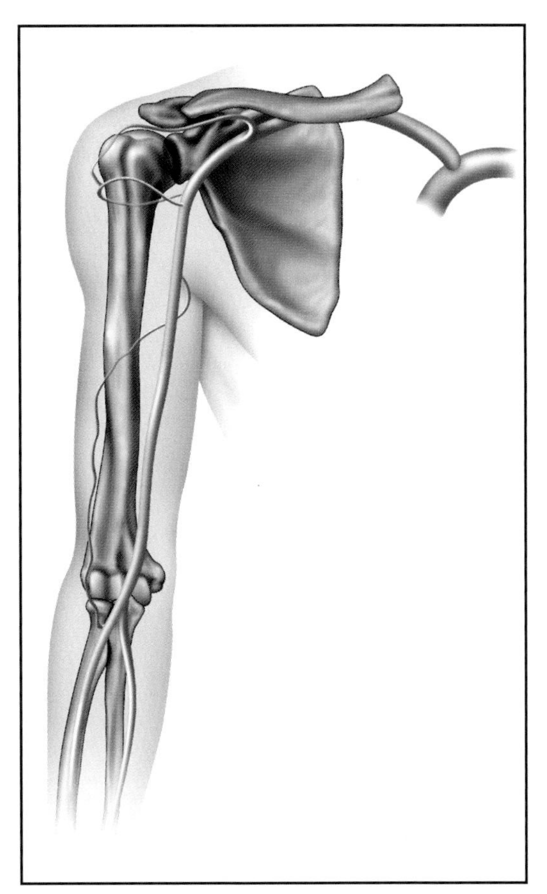

Figure 8
The arterial blood supply of the upper extremity.

sion may result in a motor palsy that can mimic a rotator cuff tear. The long thoracic nerve innervates the serratus anterior muscle, which can be prone to stretch injury. This can cause winging of the scapula with elevation of the arm and is generally accompanied by significant pain and functional deficit.

The subclavian artery becomes the axillary artery at the outer border of the first rib. Six branches extend from the axillary artery, all of which provide rich collateral circulation to much of the shoulder. Unfortunately, the humeral head does not have extensive collateral circulation because it receives its blood supply predominantly from one vessel, the anterior humeral circumflex. If this supply is disrupted, as may happen with a displaced proximal humerus fracture, there is a high risk of development of osteonecrosis of the humeral head. The axillary artery is renamed the brachial artery as it runs behind the teres major. The brachial artery branches into the profunda brachii and then continues to the elbow (Fig. 8).

Elbow and Forearm

Bones and Joints

The elbow is the junction of the distal humerus and the two bones of the forearm, the radius and ulna (Fig. 9). Unlike the knee, at which the femur makes contact with only one of the two distal bones (the tibia but not the fibula), in the elbow joint, the humerus makes contact with bone at the ulna (a joint built for elbow flexion and extension) and the radius (a joint that allows forearm rotation).

The distal humerus flares medially and laterally, forming two condyles. The con-

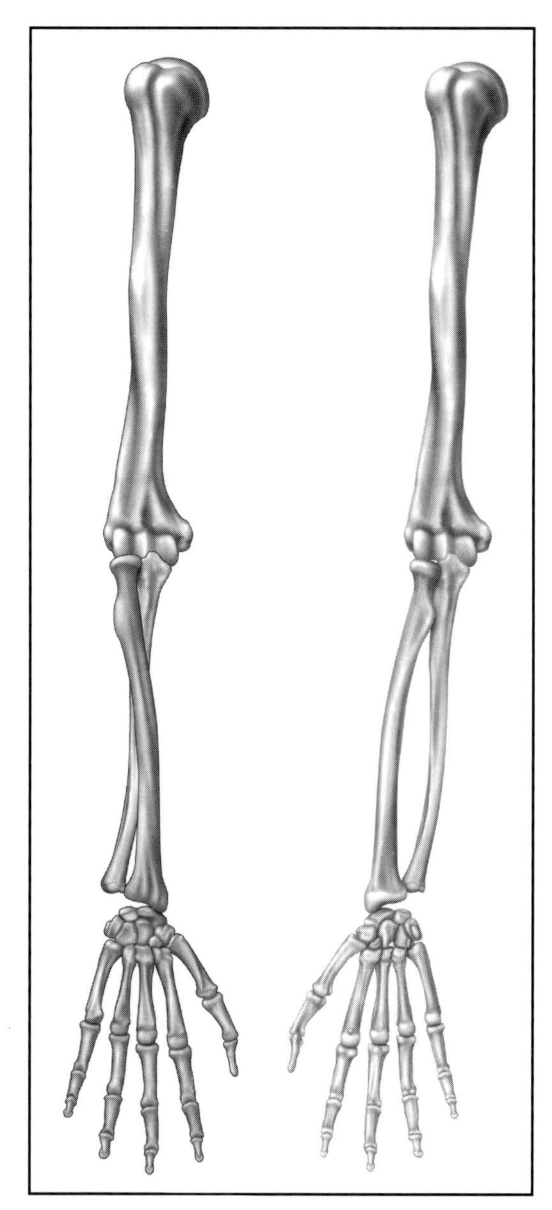

Figure 9
The elbow is a hinge joint but also allows for pronation **(left)** and supination **(right)**.

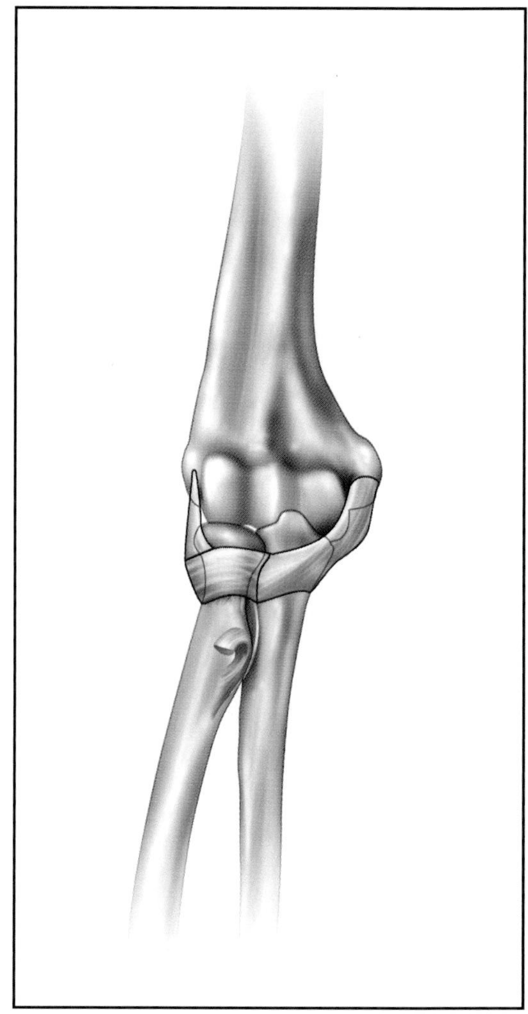

Figure 10
The ligaments of the elbow.

dyles are covered with articular cartilage and serve as the contact points for the joint. The smaller prominences just proximal to the joint are the epicondyles. These are not covered with cartilage but rather serve as the points of origin for many forearm muscles. The medial epicondyle is the site of attachment of the flexor-pronator muscle group of the forearm as well as the ulnar collateral ligament. The lateral epicondyle is the origin of the extensor muscles of the forearm.

The elbow is made up of two joints. The first, the humeroulnar joint, is hinge-like and allows flexion and extension. This joint is formed by the trochlea on the medial surface of the distal humerus and the proximal ulna. The ulna has a coronoid process anteriorly and an olecranon process posteriorly; these grasp the humeral trochlea front and back. Lateral to the trochlea is the second joint in the elbow, mating the capitellum of the humerus to the radial head. The radial head is discoid and the capitellum is hemispherical, allowing pronation and supination of the forearm. Since the radius is intimately bound to the ulna, a so-called proximal radioulnar joint is formed.

Ligaments and Soft Tissues

The bony congruity between the trochlea and the ulna provides significant stability to the elbow. This joint is nonetheless supplemented by the ligaments and capsule (Fig. 10). The ulnar (or medial) collateral ligament is

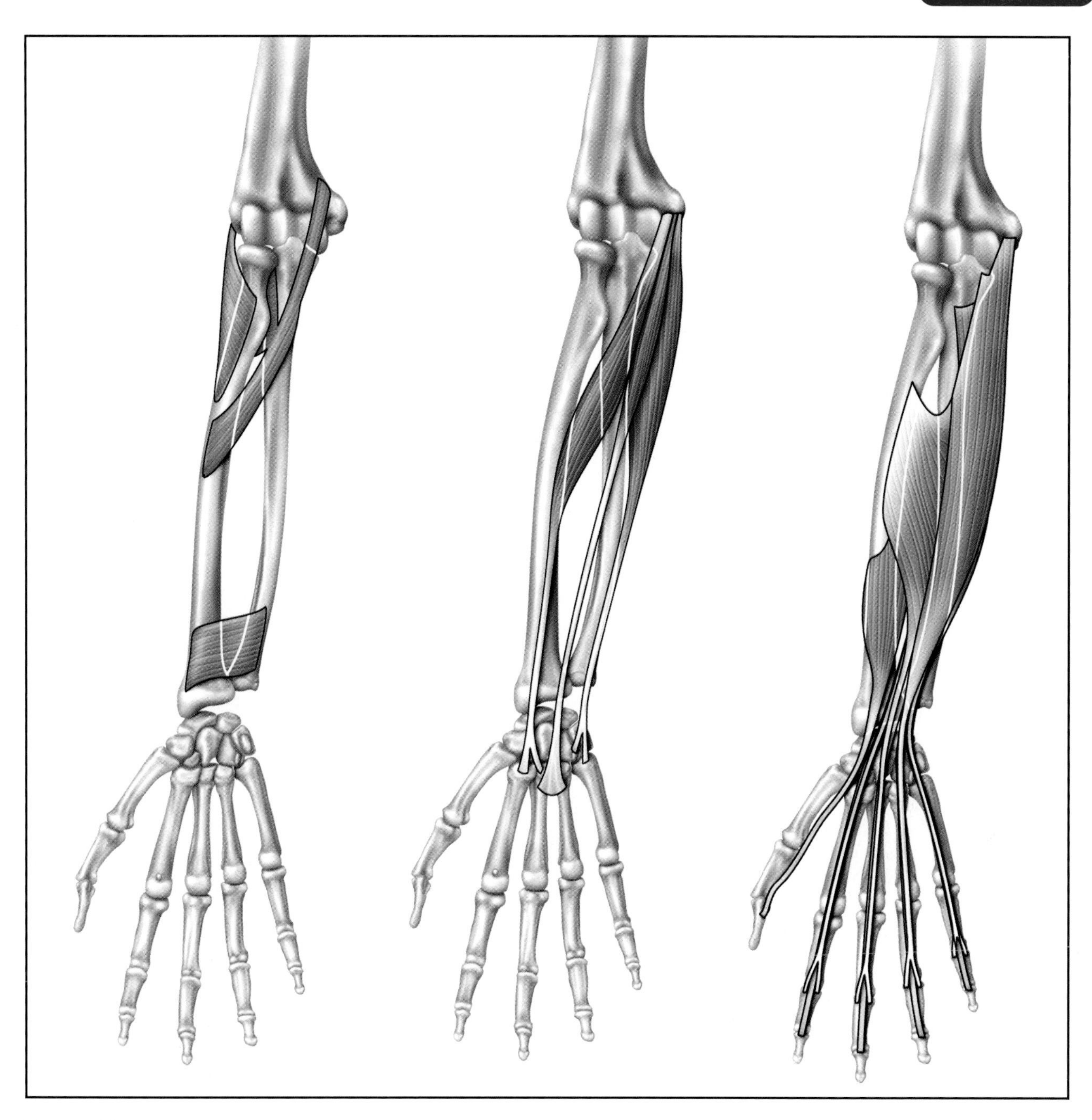

Figure 11
Left, The rotators of the forearm: the pronator teres, pronator quadratus, and supinator. Because the bones here are shown in supination, the supinator is at its shortest length. **Center**, The flexors of the wrist: the flexor carpi radialis, flexor carpi ulnaris, and palmaris longus. **Right**, The flexors of the digits: the flexor pollicis longus and flexor digitorum superficialis. The flexor digitorum profundus lies under the superficialis and is obscured in this view until it inserts on the distal phalanx.

the primary stabilizer of the medial side of the elbow. This ligament is particularly important in the throwing athlete, who places significant forces on the elbow when cocking the arm. The radial (or lateral) collateral ligament supports the lateral side of the elbow.

The joint between the radial head and the capitellum lacks bony stability. This joint is therefore stabilized by the annular ligament. This structure, as its name implies, is a ring that surrounds the radial neck just distal to the radial head. It attaches the radius to the ulna—but since the ulna is firmly attached to the humerus, it also stabilizes the radius to the humerus.

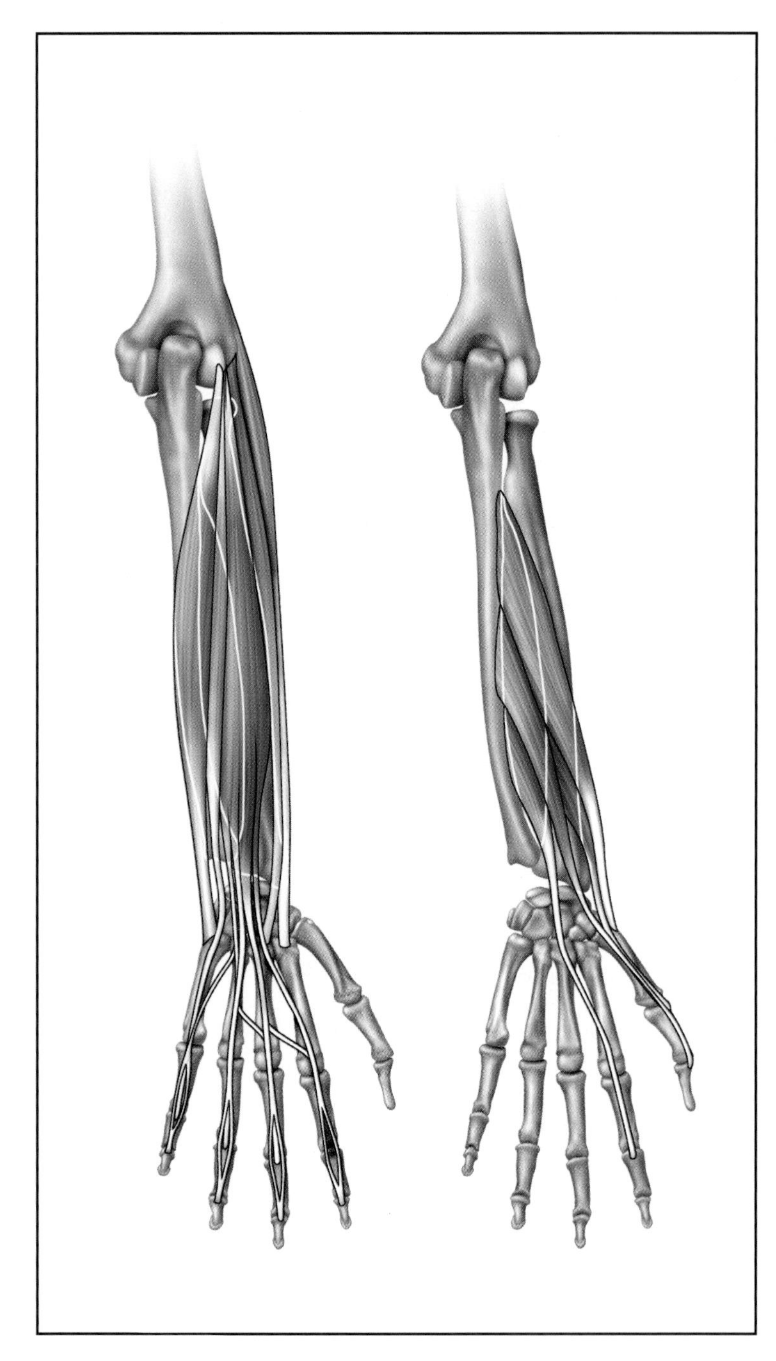

Figure 12
Left, The extensor muscles of the forearm. This figure shows the extensor carpi radialis longus, extensor carpi radialis brevis, extensor carpi ulnaris, extensor digitorum, and extensor digiti minimi. The second extensor to the index finger and the muscles to the thumb are obscured on this view and shown on the right. **Right**, The extensor muscles of the forearm, deep layer. The abductor pollicis longus, extensor pollicis longus and brevis, and extensor indicis proprius.

Muscles

Most of the muscles that cross the elbow joint originate from one of the two epicondyles. Muscles on the dorsum of the forearm originate from the lateral epicondyle, whereas those on the volar surface originate from the medial epicondyle. The volar muscles are primarily flexors (Fig. 11) and the dorsal muscles are extensors (Fig. 12). There are three muscles that cross the elbow but originate proximal to the epicondyle. The biceps originates from two points on the scapula and proximal humerus and courses on the anterior surface of the arm. Its tendon crosses the anterior aspect of the elbow joint and inserts on the proximal radius. As such, it both flexes the elbow and supinates the forearm. The brachialis muscle originates from the anterior humerus and inserts on the proximal ulna, providing elbow flexion. The triceps brachii originates from both the scapula and the posterior humerus, courses posterior to the elbow joint, and inserts on the olecranon process of the ulna, providing elbow extension.

The medial epicondyle is the origin of the "superficial" flexor muscle group. This group includes the pronator teres, the flexor carpi radialis, the palmaris longus, the flexor carpi ulnaris, and the flexor digitorum superficialis, all of which provide some stability to the medial elbow. The pronator teres allows for forearm pronation and elbow flexion. The others flex the wrist, with the exception of the flexor digitorum superficialis, whose action is on the proximal interphalangeal (finger) joint. The median nerve innervates all of the superficial flexor muscles, except for the flexor carpi ulnaris, which is supplied by the ulnar nerve.

The "deep" flexors of the forearm originate in the medial forearm, most of them from the ulna. This group includes the flexor digitorum profundus, the flexor pollicis longus, and the pronator quadratus, all of which are supplied by the anterior interosseous branch of the median nerve, except for the ulnar half of the flexor digitorum profundus, which is supplied by the ulnar nerve.

From the lateral epicondyle arise the extensors of the wrist and hand. The superficial group includes the brachioradialis, the extensor carpi radialis longus, and the extensor carpi radialis brevis. These three muscles are known as the mobile wad of three—a name that makes sense if you pinch and shake them. The brachioradialis attaches to the distal radius and flexes the elbow; the extensor carpi radialis longus and the extensor carpi radialis brevis insert on the index and middle metacarpals, respectively, and extend the wrist. All receive innervation from the radial nerve. The lateral epicondyle is the point of origin for three other superficial extensors: the extensor digitorum, the extensor digiti minimi, and the extensor carpi ulnaris. The extensor carpi ulnaris attaches to the fifth metacarpal and extends the wrist. (The insertions of the finger extensors are somewhat complex and are discussed in the hand section below.) The posterior interosseous nerve, a branch of the radial nerve, innervates these three superficial extensor muscles as well as the deep extensors below them. The deep extensors include the supinator, three muscles to the thumb (the abductor pollicis longus, the extensor pollicis longus, and the extensor pollicis brevis), and the extensor indicis proprius.

Nerves and Blood Vessels

The median nerve crosses the anterior elbow, superficial to the brachialis muscle and medial to the brachial artery. The median nerve supplies the superficial flexors and travels through the carpal tunnel into the hand. In the forearm, the median nerve gives off the anterior interosseous nerve, which supplies all of the deep flexors, except the ulnar half of the flexor digitorum profundus. The ulnar nerve crosses the elbow joint just posterior to the medial epicondyle, a common site of nerve compression and irritation. When the medial elbow is hit, the superficial placement of the ulnar nerve causes it to absorb some of that energy, resulting in paresthesias (colloquially termed "hitting the funny bone").

In the forearm, the ulnar nerve supplies the flexor carpi ulnaris and the ulnar half of the flexor digitorum profundus. Its main targets are in the hand, however. The ulnar nerve reaches the hand by way of Guyon's canal at the wrist. It supplies the hypothenar muscles, most of the intrinsics, and the adductor pollicis. The radial nerve crosses the elbow anterior to the lateral epicondyle, where it gives off the posterior interosseous nerve. The radial nerve itself supplies the mobile wad of three. The posterior interosseous branch of the radial nerve feeds all extensor muscles distal to the mobile wad of three.

The brachial artery travels on the anterior surface of the humerus. It crosses the elbow anteriorly, where, at the level of the neck of the radius, it bifurcates into the ulnar and radial arteries. It also gives off recurrent branches that course back toward the arm both medially and laterally. The radial artery runs on the dorsum of the forearm under the brachioradialis muscle and terminates in the deep palmar arch in the hand. The ulnar artery runs on the volar surface under the superficial flexor muscles and terminates in the hand as the superficial palmar arch.

Hand and Wrist

Bone and Joints

The radius and ulna are the bones of the forearm; they connect the hand to the arm. Proximally, the main articulation is between the ulna and the distal humerus, forming the hinge of the elbow. The radius functions primarily to allow pronation and supination. At the wrist, the roles are reversed. Here, the distal radius provides the primary articulation with the carpal bones of the hand, with the ulna participating mainly in pronation and supination. The radius accepts approximately 80% of the weight transfer from the hand across the wrist. Distal to the radius and the ulna at the wrist are the carpal bones, which are organized into two rows. The proximal row consists of the scaphoid, the lunate, the triquetrum, and the pisiform. The distal row includes the trapezium, the trapezoid, the capitate, and the hamate.

The radial and the ulnar styloids are palpable subcutaneous landmarks at the wrist. The distal pole of the scaphoid is palpable on the palmar aspect of the wrist. The scaphoid is also palpable in the "anatomic snuff box," the space between the extensor pollicis brevis and longus tendons to the thumb. The pisiform and the hook of the hamate are palpable on the volar ulnar surface of the palm.

Each finger of the hand is composed of a metacarpal bone and three phalanges, with the exception of the thumb (Fig. 13). The thumb has a proximal and a distal phalanx articulating at a single interphalangeal joint, while the remaining fingers have proximal, middle, and distal phalanges that form a proximal interphalangeal (PIP) joint and a distal interphalangeal (DIP) joint.

Ligaments and Soft Tissues

The shaft of the radius is attached to the shaft of the ulna by an interosseous membrane. This membrane, technically a ligament, helps transfer forces from the radius (which supports the hand) to the ulna, which

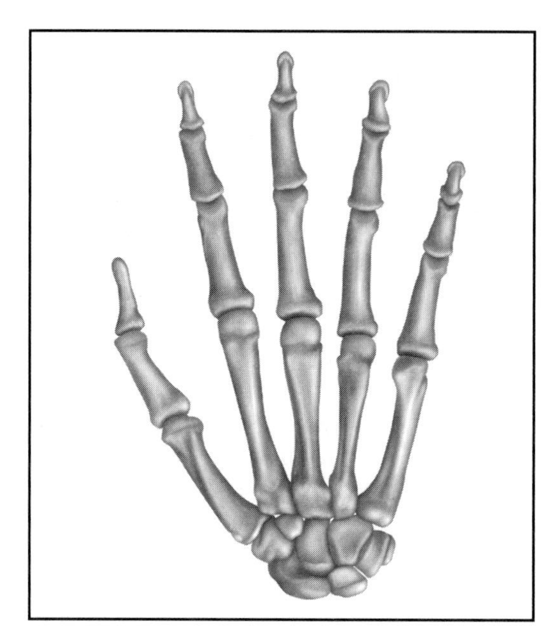

Figure 13
The bones of the wrist and hand. A more detailed description of the carpal bones can be found in many anatomy textbooks.

has the primary articulation with the humerus. Distally, the radius is attached to the proximal row of carpal bones by strong volar ligaments. The radius is attached to the ulna by the dorsal and volar radioulnar ligaments of the triangular fibrocartilage complex. The proximal and the distal rows of carpal bones are connected via a joint capsule that allows for both flexion/extension and radial/ulnar deviation of the hand. The metacarpophalangeal joints as well as the interphalangeal joints are stabilized by joint capsules and collateral ligaments. The ulnar collateral ligament of the thumb is especially susceptible to sprains (causing a skier's or gamekeeper's thumb).

There are two important transverse ligaments at the wrist. On the dorsal surface, there is an extensor retinaculum. This pulley-like structure tethers the extensor tendons close to the bones of the wrist, even as the muscle contracts. The extensor retinaculum houses six separate synovial sheaths, creating six extensor compartments. There is also a transverse covering to the volar surface, the main component of which is the transverse carpal ligament. This is functionally a "flexor retinaculum" and forms the roof of the carpal canal. This canal (or tunnel) houses nine flexor tendons and the median nerve as they pass into the hand from the forearm

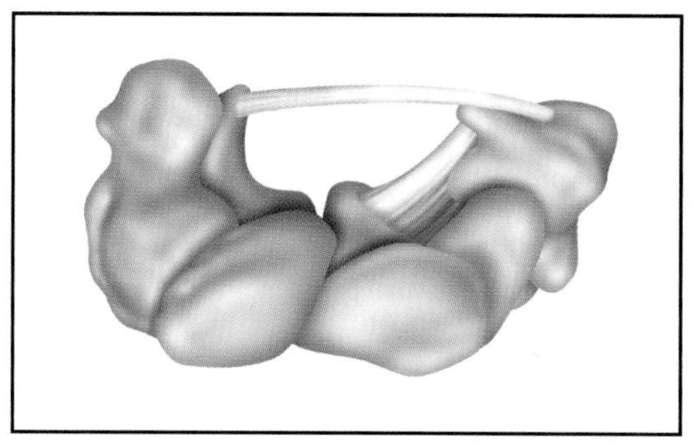

Figure 14
The carpal tunnel, shown here in cross section, is formed by the transverse carpal ligament above and the carpal bones below. The carpal tunnel contains the median nerve and nine flexor tendons.

(Fig. 14). Compression of the median nerve at the wrist causes carpal tunnel syndrome.

Muscles

The muscles of the hand can be categorized into two groups: intrinsics and extrinsics. Intrinsic muscles reside exclusively in the hand itself. The extrinsic muscles have tendons that attach in the hand but their muscle bellies reside in the forearm. The thenar group of intrinsics powers the thumb. This group includes the abductor pollicis brevis, the opponens pollicis, the flexor pollicis brevis, and the adductor pollicis. The first three are supplied by the median nerve in the hand after it passes under the transverse carpal ligament and are responsible for opposition. A similar (but functionally less important) group of muscles, the hypothenar muscles, attach to the little finger.

The final constituents of the intrinsic group are the lumbricals and the interosseous muscles. The dorsal and volar (palmar) interosseous muscles originate from the metacarpals and insert on the proximal phalanges. The dorsal interosseous muscles abduct the fingers; the volar group adducts the fingers. The intrinsics also have a key role in finger flexion and extension. To understand this role, the anatomy of the extrinsics, the flexor and extensor tendons of the long fingers, must first be understood (Fig. 15).

The tendons of the finger flexor muscles pass into the wrist via the carpal tunnel. They then travel to the fingers bound by fibrous digital sheaths, an intricate pulley system that both nourishes the tendons and prevents "bowstringing" when the muscles contract.

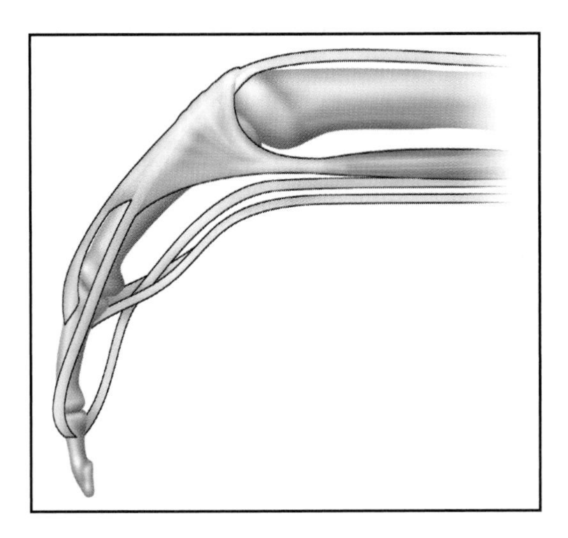

Figure 15
Insertion of the two flexor and common extensor tendons and the course of the lumbrical muscle. Note that the lumbrical originates on the volar surface and inserts onto the extensor tendon; it thus flexes the metacarpophalangeal joint and extends the interphalangeal joints. The interosseous muscles that abduct and adduct the digits are not shown in this view.

The superficialis attaches to the middle phalanx, and the profundus attaches to the distal phalanx. The superficialis is, as its name implies, superficial; thus, to allow the profundus (ie, deep) flexor to reach the distal phalanx, it divides into two slips, and the profundus tendon passes between them (Fig 11, *right*). The superficialis attaches to the middle phalanx and flexes the PIP joint. The profundus, attaching to the distal phalanx, flexes the DIP joint.

Note that there is no extrinsic flexor tendon that attaches directly to the proximal phalanx. Flexion of the metacarpophalangeal joint, rather, is powered by the intrinsics: the interosseous muscles, which attach to the proximal phalanx; and the lumbricals, which blend into the common extensor tendon on the dorsal surface. The lumbricals originate from the flexor digitorum profundus tendon, cross the finger on its radial border, and attach dorsally into the common extensor tendon. The lumbricals, thus, power metacarpophalangeal flexion as well as extension of the interphalangeal joints—especially when the metacarpophalangeal joint is flexed. The combined function of the intrinsics and the extrinsic flexor-extensor tendons gives a smooth and coordinated composite motion to the digits.

Nerves and Blood Vessels

The three major nerves of the hand are the radial, median, and ulnar nerves. The radial nerve, through its posterior interosseus branch, provides innervation to extensors of the fingers. The radial nerve also provides sensation on the radial aspect of the dorsum of the hand. The median nerve provides sensation to the thumb and index and long fingers, as well as to the radial half of the ring finger. It provides the motor innervation of all of the forearm flexor muscles, with the exception of the flexor carpi ulnaris and the ulnar aspect of the flexor digitorum profundus (that part of the muscle going to the ring and little fingers), which are supplied by the ulnar nerve. The ulnar nerve most importantly provides all of the motor innervation of most of the intrinsic muscles of the hand. The ulnar nerve also provides sensation to the ulnar aspect of the palm, to the little finger, and to the ulnar aspect of the ring finger. The lumbricals to the index and long finger and the thenar muscles, with the exception of the adductor pollicis, are supplied by the median nerve.

The radial artery enters the hand dorsally (although the pulse is more easily palpated on the volar side) and terminates in the deep palmar arch. The ulnar artery enters the hand through Guyon's canal, at which point it is very close to the ulnar nerve. It terminates in the superficial palmar arch farther distal in the hand. In most people, these two arches communicate. This allows a patient with an injury to one of the two main arteries to maintain viability of the hand and fingers. The palmar arches supply the common digital arteries that further bifurcate into proper digital arteries. Each finger has a radial and ulnar artery that run in the volar aspect of the finger adjacent to the flexor sheath with the digital nerves.

Hip and Thigh

Bones and Joints

The hip joint is the articulation of the femur within the acetabulum of the pelvis. The hip spans the pelvis and the proximal femur. The proximal femur contains the femoral head, which is covered with articular cartilage; the femoral neck, which connects the head to the shaft; and two bony prominences, the greater trochanter and the lesser trochanter (Fig. 16). The greater and lesser trochanters serve as attachment points for the hip abductors and

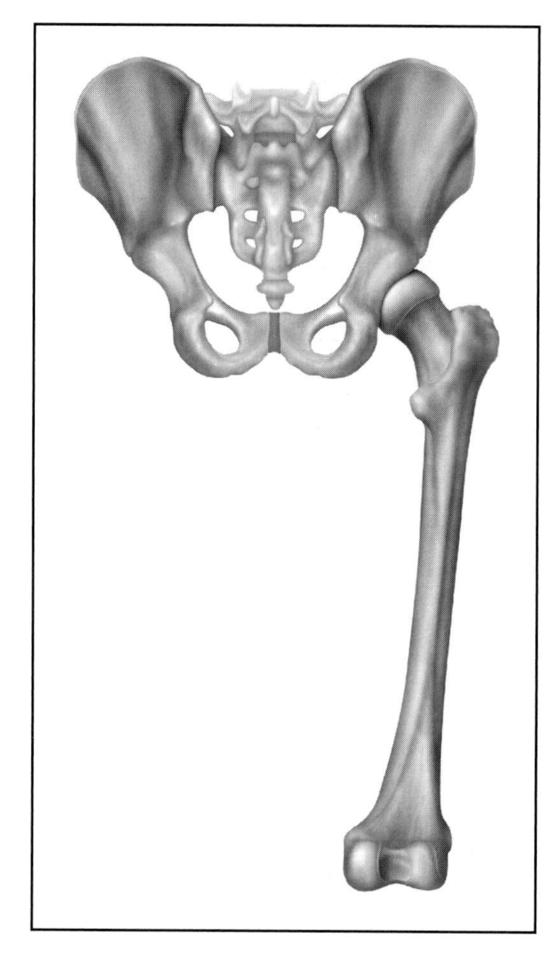

Figure 16
The bones of the hip: the pelvis and femur. Contrast the inherent bony stability of this ball-and-socket joint with that of the shoulder.

flexors, respectively. The ridge between the trochanters is called the intertrochanteric region. The intertrochanteric region and the femoral neck are the two most common sites of hip fracture in older people. Femoral neck fractures are especially troubling, not only because of their lower propensity to heal but also because the blood supply to the femoral head, which courses along the neck, is prone to disruption with such fractures. Disruption of the blood supply can cause ischemia, resulting in death of the femoral head.

The femoral neck is angulated approximately 135° with respect to the shaft of the femur. In addition, the femoral neck points anteriorly approximately 15°. As a result, when a femur rests on a flat surface, the femoral head is elevated off that surface as the femoral neck inclines up out of the true anteroposterior plane. Increased anteversion causes a compensatory internal rotation of the femur (which is needed to keep the head seated in the acetabulum) and therefore abnormal alignment of the feet or legs.

The pelvis contains a right and left hemipelvis, each of which is composed of three bones: the ilium above the hip joint, the ischium behind and below the hip joint, and the pubis medial to the hip joint. These bones unite at the center of the acetabulum. The pubis and ischium also meet to surround the obturator foramen. Congruity between the femoral head and the acetabular socket depends on contact as the fetus develops; accordingly, children who have developmental or congenital dislocation of the hip may have malformed femoral heads and acetabuli.

Each of the three bones of the pelvis is typically palpable. The ischial tuberosity is the bony prominence on which we sit. The iliac crest is normally felt right below the beltline laterally. The junction of the right and left pubic bones, the symphysis pubis, is palpable inferior to the umbilicus. The entire proximal femur is deeply enveloped within a sleeve of muscles; thus, with the exception of the greater trochanter laterally, it is often not directly palpable.

Ligaments

The hip joint is a true ball-and-socket joint and therefore inherently stable. Nonetheless, it is supplemented with ligamentous attachments. Three anterior ligaments, one from each of the pelvic bones to the femur, comprise the hip capsule. In addition, a ligamentum teres attaches from the deepest base of the acetabulum to the femoral head directly. The pelvic bones themselves are attached to the sacrum by anterior and posterior sacroiliac ligaments. The right and left sides of the pelvis are attached anteriorly at the symphysis pubis by strong ligaments.

The acetabular socket is deepened by the labrum, a lip of cartilage similar to the meniscus of the knee and the labrum of the shoulder. It is a dense fibrocartilagenous ring that surrounds the rim of the acetabulum. Because the labrum increases the depth of the acetabulum, it enhances the stability of the hip joint.

Muscles

Muscles of the hip are best considered in terms of functional groups, including the flexors, the extensors, the abductors, the adductors, and the external rotators. There are

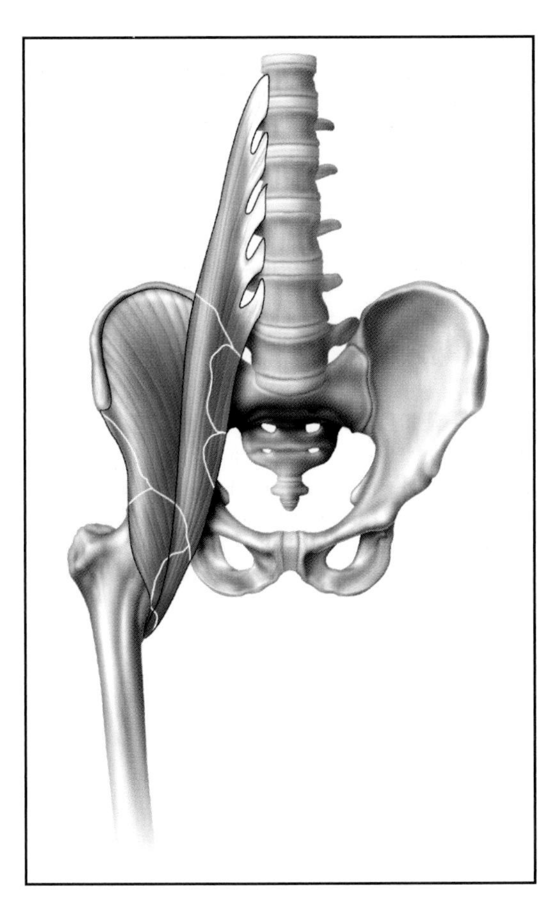

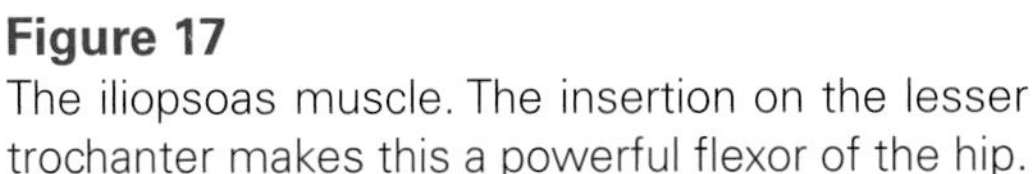

Figure 17
The iliopsoas muscle. The insertion on the lesser trochanter makes this a powerful flexor of the hip.

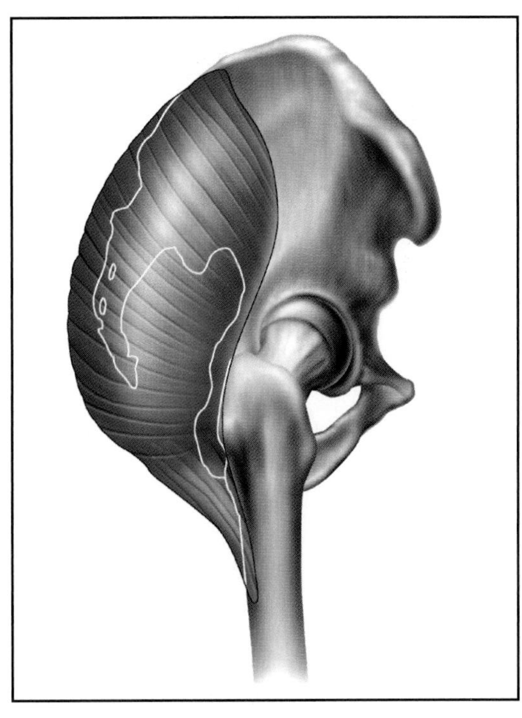

Figure 18
The gluteus maximus as seen on a profile view. This powerful muscle extends the hip joint.

no true internal rotators of the hip joint.

The strongest hip flexor is the iliopsoas, a muscle formed by the fusion of the iliacus and psoas muscles (Fig. 17). These muscles originate in the pelvis and unite to form a common tendon that inserts on the lesser trochanter. The rectus femoris, another flexor, originates on the anterior-inferior iliac spine of the pelvis and courses down the femur, uniting with the vastus muscles to form the quadriceps tendon. The quadriceps ultimately inserts on the tibia via the patellar tendon. It thus extends the knee as well. The final flexor is the sartorius, the longest muscle in the body. It originates from the anterosuperior iliac spine and inserts with the gracilis and semitendinosus on the anteromedial tibia to form the pes anserinus. Because it reaches the tibia behind the axis of the knee (in contrast to the rectus, which is located in front of the axis throughout), the sartorius is a flexor of the knee. The femoral nerve powers all of the hip flexors.

Hip extension is performed primarily by the gluteus maximus and the hamstring muscles. The gluteus maximus arises from the outer surface of the ilium and inserts on the posterior femur (Fig. 18). It also has a common insertion with the tensor fascia lata muscle onto the iliotibial band. The hamstring muscles, namely, the biceps femoris, the semitendinosus, and the semimembranosus, all originate from the ischial tuberosity. They cross both the hip and knee joints, and insert on the tibia and fibular head (Fig. 19). The gluteus maximus receives innervation from the inferior gluteal nerve; the hamstrings are all powered by the tibial nerve, with the exception of the short head of the biceps femoris, which receives its innervation from the peroneal part of the sciatic nerve.

The abductors of the hip joint are the gluteus medius and the gluteus minimus (Fig. 20). Hip abduction refers to the motion of pulling the leg away from the plane of the body; but in day-to-day life, the more important function of the hip abductors is to isometrically resist adduction. That is, these muscles keep the pelvis horizontal when the contralateral leg is not touching the ground. The gluteus medius and minimus arise from the ilium and insert onto the greater tro-

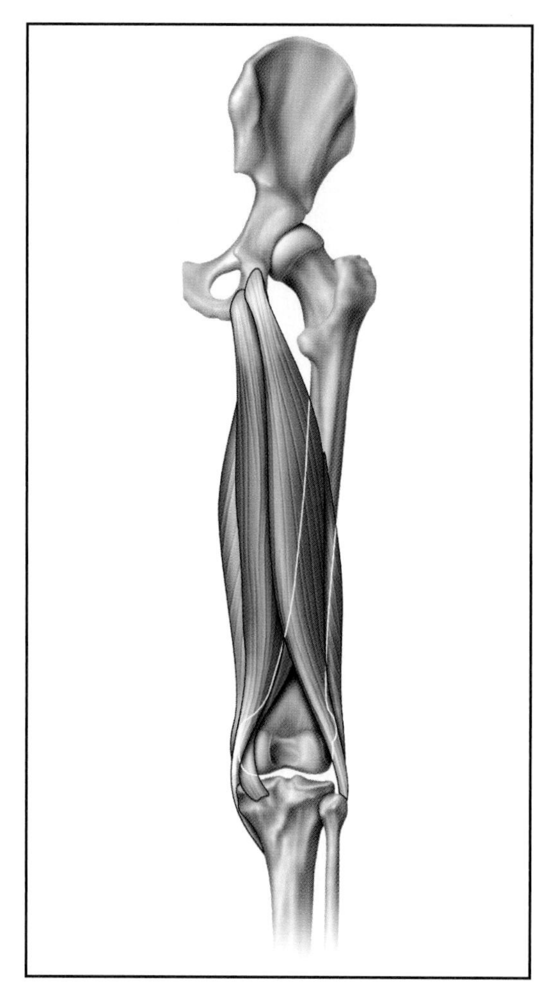

Figure 19
The hamstring muscles: the semimembranosus and semitendinosus on the medial side and the biceps femoris on the lateral side. Note that the biceps inserts on the fibula and the semitendinosus courses medially to insert on the anterior tibia.

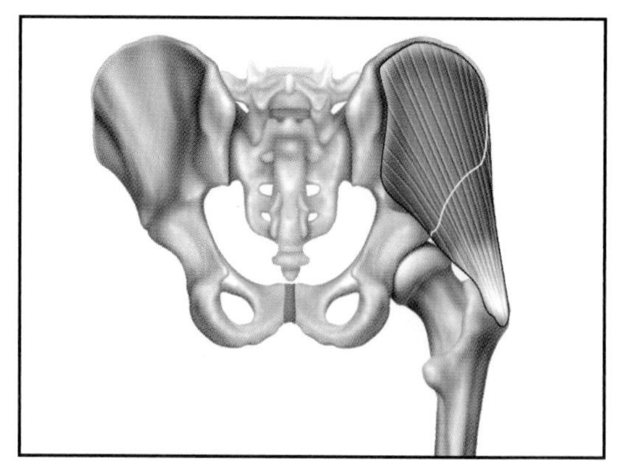

Figure 20
The hip abductors (in this view, the gluteus medius obscures the gluteus minimus underneath it). One essential function of the abductors (which can be inferred from this drawing showing only one femur) is to hold the pelvis level when the contralateral foot is off the ground.

chanter; both are supplied by the superior gluteal nerve.

The hip adductors originate from the pelvis and insert on the femoral shaft (Fig. 21). They are powered by the obturator nerve. This group includes the adductor longus, the adductor brevis, the adductor magnus, and the gracilis. The gracilis, unlike the other muscles in this group, inserts on the anteromedial tibia near the sartorius and semitendinosus.

The external rotators are six short muscles that course behind the joint and insert on the medial aspect of the greater trochanter. One of these, the piriformis muscle, may irritate the sciatic nerve as it exits the sciatic foramen, causing "piriformis syndrome." The piriformis muscle also serves as

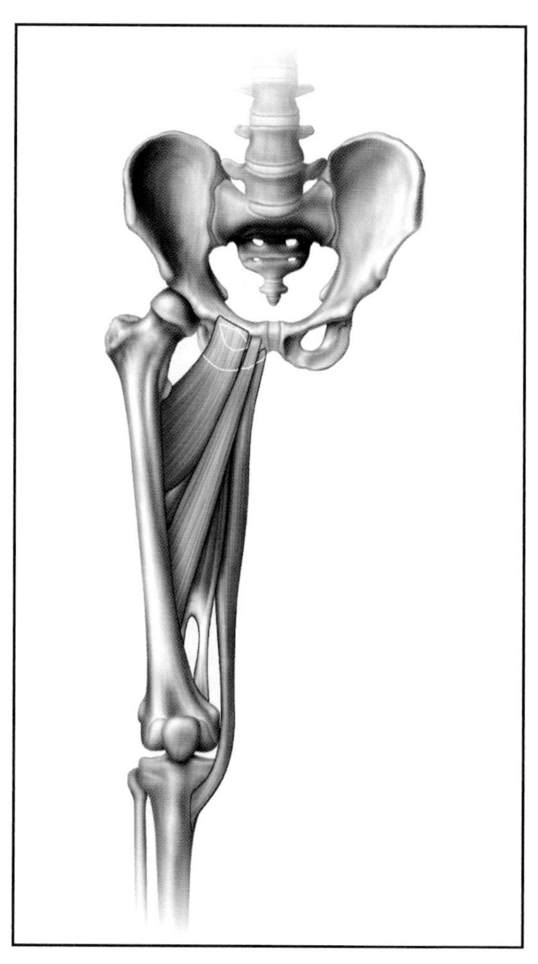

Figure 21
The hip adductors. The adductor magnus is the most powerful of these. Note the hiatus above the medial femoral condyle. The femoral artery passes through here on its course to the popliteal fossa.

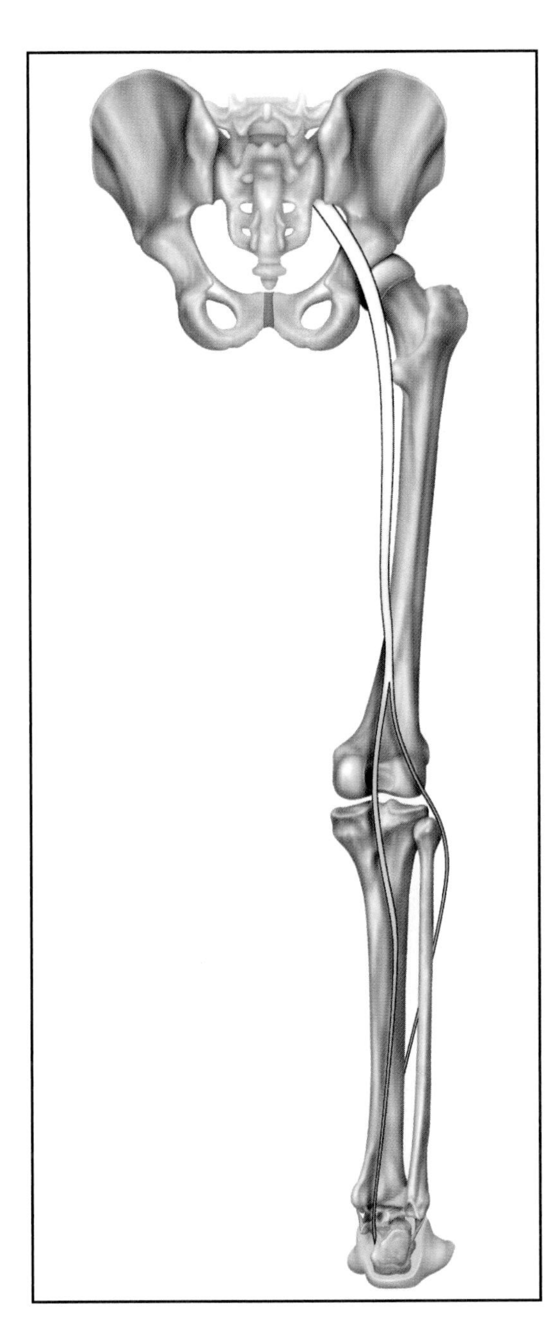

Figure 22
The sciatic nerve.

a surgical landmark in the sciatic foramen above which the superior gluteal nerve and vessels can be found.

Nerves and Blood Vessels

The major nerves of the hip region all come from the lumbosacral plexus. The femoral nerve is composed of branches from L2, L3, and L4. The sciatic nerve has both a tibial branch and a peroneal branch (Fig. 22). These travel through the hip region together but split at the popliteal fossa into distinct nerves.

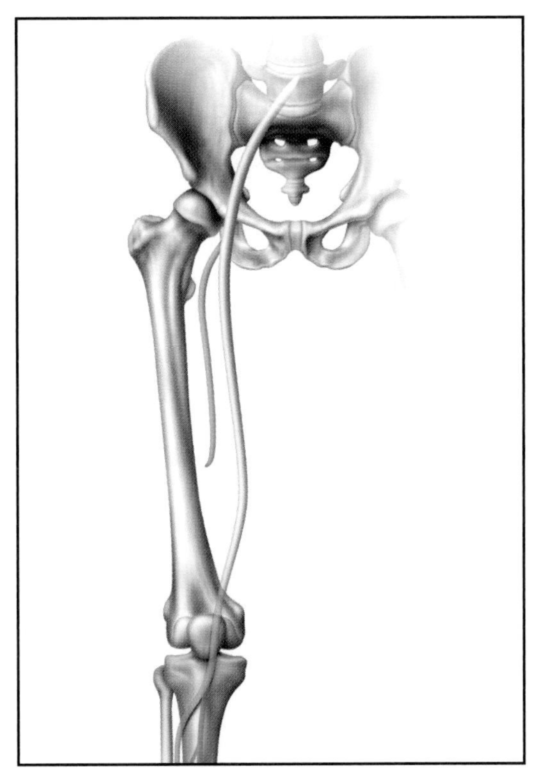

Figure 23
The blood supply to the lower extremity comes from the external iliac artery. The femoral artery moves to the posterior side through the adductor hiatus and is renamed the popliteal artery.

The superior and inferior gluteal nerves are also branches from the posterior division of the sacral plexus.

Blood vessels enter the pelvis as branches of the aorta. The aorta branches into left and right common iliac arteries at a point located approximately at the level of the L4 vertebral body. These in turn divide into the internal and external iliac arteries at the level of the sacrum. The internal iliac provides branches to supply both the superior and inferior gluteal arteries. The external iliac crosses under the inguinal ligament and is renamed the femoral artery (Fig. 23). The femoral artery gives off the profunda femoris and then continues on to the knee as the popliteal artery. Two other important branches of the femoral artery are the medial and lateral femoral circumflex arteries. The medial femoral circumflex artery provides the majority of blood to the femoral head. It travels up the femoral neck and is at risk for disruption with femoral neck fractures. The femoral artery pulse is readily palpable at the groin. The femoral nerve is lateral to the artery and the vein

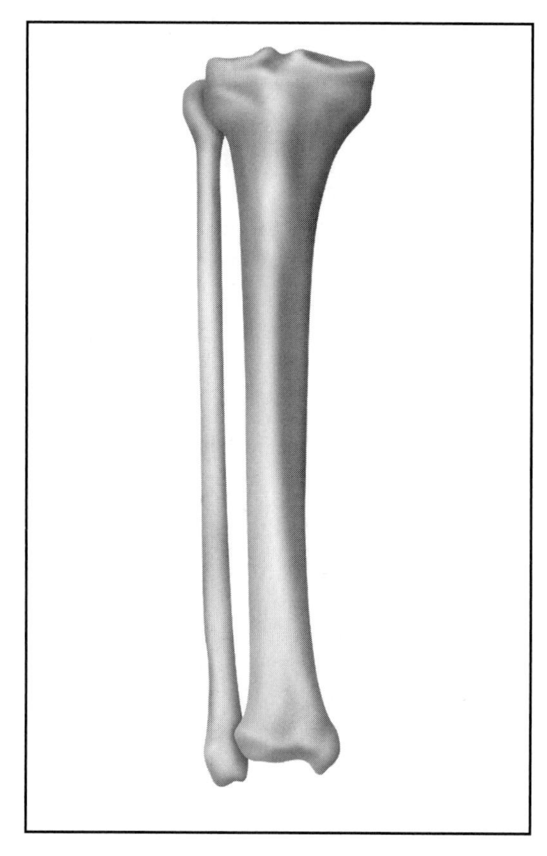

Figure 24
The bones of the leg: the tibia and fibula. The fibula does not contact the femur (as shown in Figure 25), but it is an important part of the ankle joint.

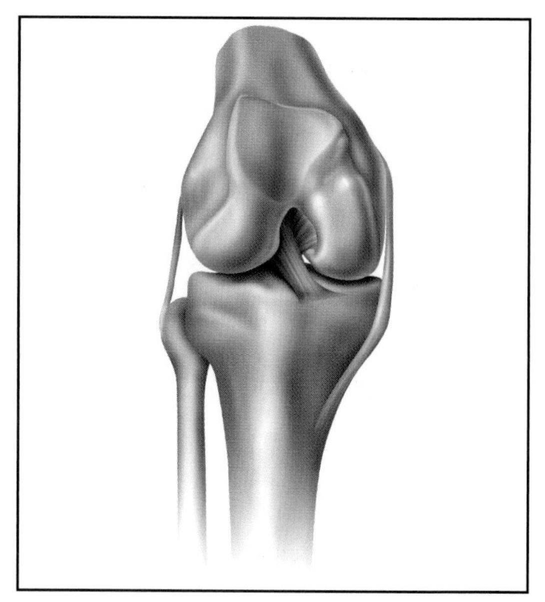

Figure 25
The collateral and cruciate ligaments of the knee. The cruciates reside within the notch between the femoral condyles: the posterior cruciate originates from the medial side and courses laterally toward its attachment on the tibia; the anterior cruciate originates on the lateral side and courses medially. These ligaments thus form a cross (hence the name "cruciate") as they pass each other.

medial to it; thus, you can draw blood or obtain venous access via the groin by palpating the pulse and placing a needle or cannula medial to it.

Knee and Leg

Bones and Joints

There are four bones joined at the knee, although no one bone touches all three of the others. The femur from above and the tibia from below articulate to form a joint that approximates a hinge. The patella, a sesamoid bone within the quadriceps tendon, articulates with the femur to create a joint that is similar to a pulley mechanism. The tibia articulates with the second leg bone, the fibula, in a joint that allows a small degree of rotation (Figs. 24 and 25).

The ends of the femur and tibia are covered with articular cartilage, as is the surface of the patella. In between the condyles is a space called the intracondylar notch, which houses the cruciate ligaments. The femur does not articulate with the fibula. The distal ends of the femur, the condyles, can be thought of as two wheels that roll and glide along the relatively flat surface of the tibia. As the knee flexes, the femur rolls posteriorly. As the knee is straightened, the lateral plateau of the tibia reaches full extension before the medial side does; thus, at terminal extension the tibia rotates externally (with the medial plateau continuing to extend and the lateral plateau remaining motionless). These last few degrees of motion ("screw-home movement") allow the knee to lock in full extension. As a result, the inability to fully extend the knee (a "flexion contracture") can lead to an abnormal gait.

The function of the patella is to hold the quadriceps farther anterior to the central hinge of the knee, thus increasing the moment arm of extension. Knee extension can be considered a rotational motion around the center of the knee as seen on the lateral view. This torque can be increased by either pulling harder or increasing the distance between the line of pull and the axis of rotation. Increasing the distance between the line of pull of the quadriceps and the center of the knee is the goal of the patella. In fact, it is able to increase the torque by as much as 30%. The patella is consequently subject to high joint reactive forces—many times a person's body weight, in fact.

Ligaments and Soft Tissues

The knee, unlike the hip, does not have much bony congruity to make it stable; and, unlike the shoulder, it relies less on muscles to hold it in place. Rather, it uses ligaments to secure the joint. The main ligaments are the anterior cruciate ligament (ACL), the posterior cruciate ligament (PCL), the medial collateral ligament (MCL), and the lateral collateral ligament (LCL).

The MCL is long and broad, extending from the femur to the tibial metaphysis. It prevents the loss of contact between the femur and tibia when the knee is hit from the outside. Since most people have a slight knock-knee (valgus) alignment, the MCL also prevents the joint from gapping during simple weight bearing. The LCL is much smaller (reflecting, perhaps, the reduced likelihood of being hit from the inside) and attaches to the fibula. It is the fibular attachments to the tibia, in turn, that complete the link, stabilizing the lateral side of the knee.

The cruciate ligaments are found between the condyles in the intracondylar notch (Fig. 26). The ACL resists anterior subluxation of the tibia, and the PCL resists posterior motion. The orientation of these ligaments is primarily vertical. This permits the knee to move through a wide arc of motion. In fact, you can think of the tibia as suspended from the femur by the cruciates and the hinge motion of the knee as the swinging that such suspension allows.

When the knee is flexed, the patella is stabilized by bone congruity: it sits firmly in the groove of the femoral trochlea. At full extension, there is no bony contact, and stability is achieved by the medial and lateral retinacula (ligaments that hold the patella to the femur) and the balanced tension of the muscles of the quadriceps. If the pull of the quadriceps is excessively lateral (as may be the case with a knock-knee deformity), the patella may be unstable. At the very least, this lateral pull can lead to imperfect tracking between the articular surfaces, with greater contact on the lateral facet of the patella. This increased focal contact will increase pressure and, in turn, can lead to cartilage breakdown and pain.

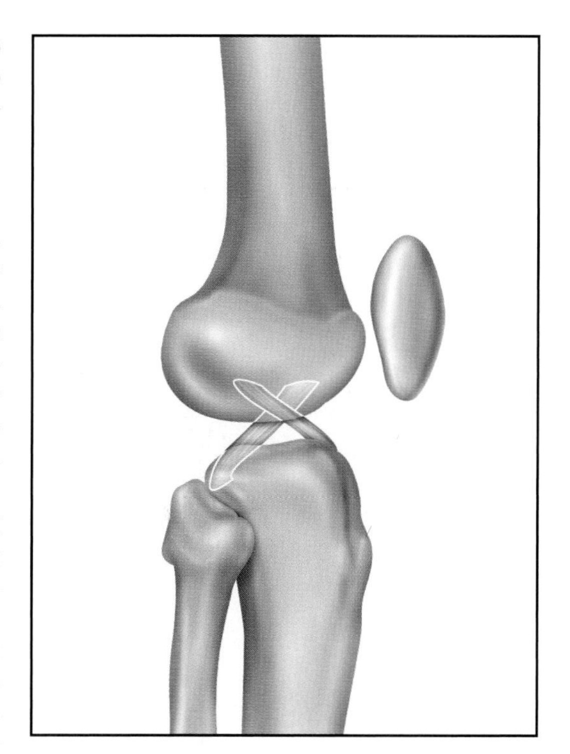

Figure 26
Sagittal view of a right knee with the femur rendered transparent to show the cruciates in the center of the knee.

The intra-articular space of the knee is lined with synovium, which produces lubricating and nourishing fluid. This space is bounded by a thin capsule that holds this fluid around the knee and prevents it from dissipating into the soft tissues. The extent of the pouch is at least a few inches superior to the patella, allowing a large space for injecting or aspirating the knee, if needed.

There are also two menisci (singular "meniscus"; Greek for "little moon") in each knee, crescent-shaped cartilages that rest on the tibial plateaus (Fig. 27). They are progressively thicker toward the periphery. Consequently, a sagittal or coronal slice (as seen in MRI, for example) looks like a wedge. The function of the meniscus is threefold: (1) it is a cushion that functions as a shock absorber; (2) it creates a greater contact area between the femur and tibia, allowing the force of weight bearing to be dissipated across a larger surface area; and (3) it helps stabilize the knee, much in the way a brick behind a back tire prevents a car from rolling while changing the front tire. Loss of the meniscus (as with an irreparable tear) may increase pressure in the knee along with a sense of instability.

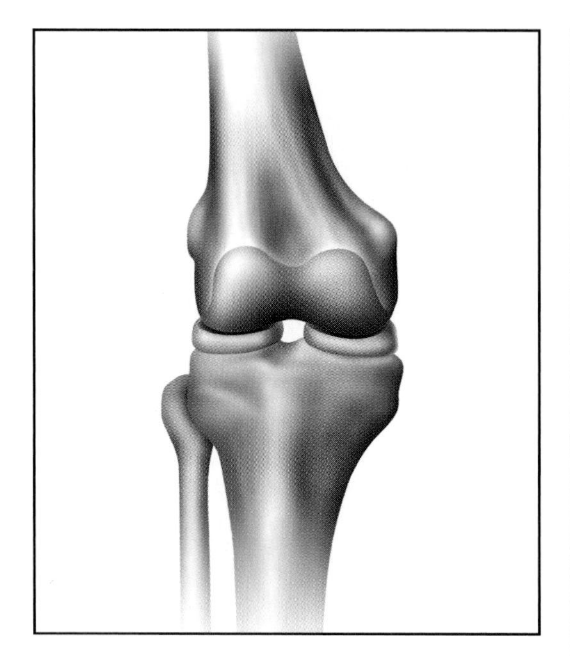

Figure 27
The medial and lateral menisci lie on the tibia and cushion the knee joint.

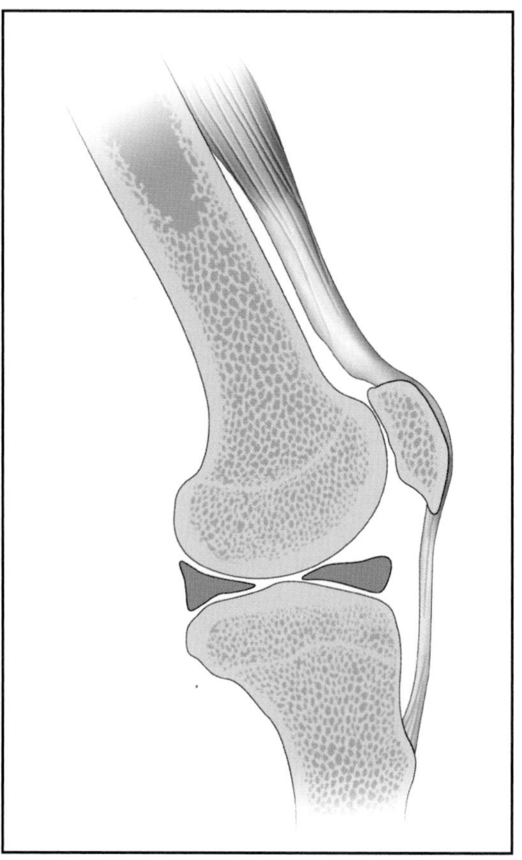

Figure 28
A profile view of the knee, illustrating how the patella increases the distance between the quadriceps and the center of the knee. This lengthens the moment arm of extension, thus adding leverage. The menisci are also shown.

Muscles

Extension is powered by the quadriceps (innervated, primarily, by the L4 nerve root contribution to the femoral nerve) (Fig. 28). The quadriceps, as the name implies, is composed of four muscles: three vastus muscles and the rectus femoris (Fig. 29). The hamstrings, which power flexion, also have four parts: the two heads of the biceps, which attach laterally on the fibula; and the medial group, the semitendinosus and the semimembranosus, which attach on the tibia. The gastrocnemius, which spans the knee posteriorly, attaches to the femur and also powers knee flexion. The muscles of the knee can also help stabilize the joint, especially when the cruciates are injured: the quadriceps has a slight anterior pull and the hamstrings and gastrocnemius pull posteriorly.

Although the biceps attaches to the fibular head and the semitendinosus attaches to the anterior tibia, in terms of flexion function, the hamstrings behave as if all were attached to the posterior tibia. Understanding the actual attachments of the tendons becomes important when examining patients whose tendons are inflamed or injured.

One tendon crossing the knee, the iliotibial band, is not easily categorized. This is the tendon of the tensor fascia lata muscle, and it receives some fibers from the gluteus maximus. The band courses down the side of the leg and attaches on the anterolateral aspect of the tibia. When the knee is flexed beyond 30°, the band passes posterior to the axis of the knee and serves as a flexor. When the knee is flexed less than 30°, however, the band passes anterior to the axis and serves as an extensor.

The muscles of the leg are housed in four distinct fascial compartments: two posterior (one deep and one superficial), one lateral, and one anterior (Fig. 30). The superficial posterior compartment of the leg contains the gastrocnemius, soleus, and plantaris muscles, all of which are innervated by branches of the tibial nerve. The gastrocnemius muscle is composed of medial and lateral heads that arise from the posterior distal femur (Fig. 31). Its fibers cross the knee joint (and thus flex it) and insert into the superior portion of the Achilles tendon. The soleus muscle arises from the proximal tibia and fib-

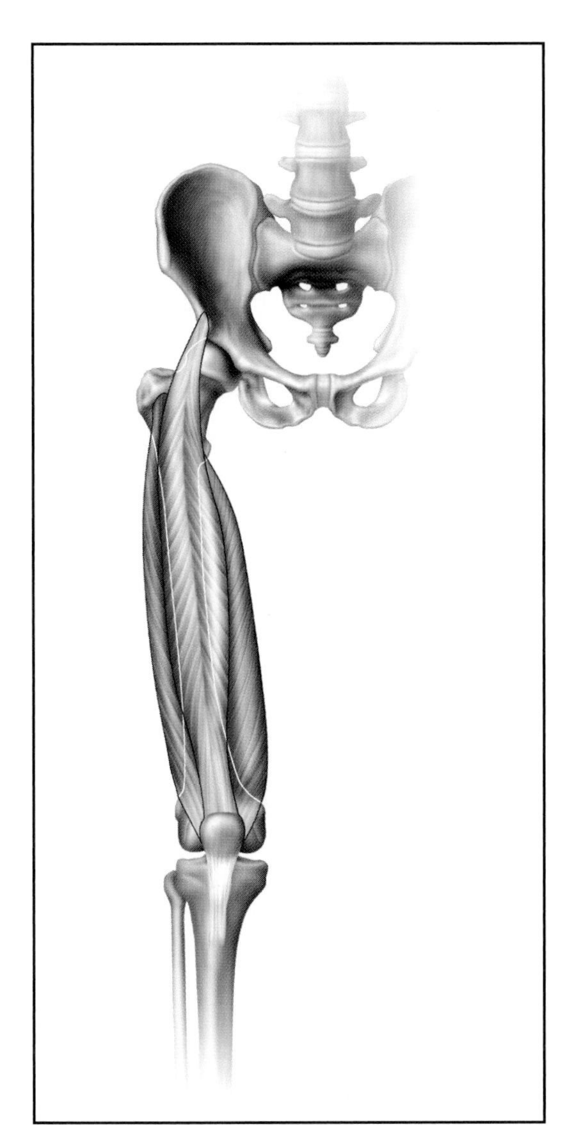

Figure 29
The quadriceps muscles attach to the patella, which in turn attaches to the proximal tibia.

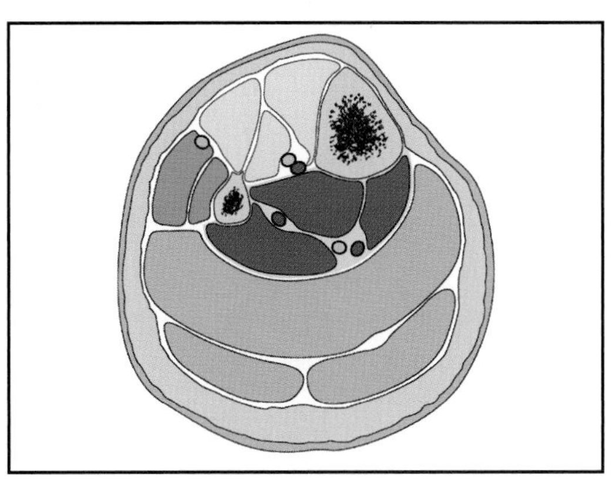

Figure 30
A cross-sectional view of the four compartments of the leg. Note that the lateral compartment holds the superficial peroneal nerve but no artery. The superficial posterior compartment, which contains the gastrocnemius and soleus muscles, has no major neurovascular structures.

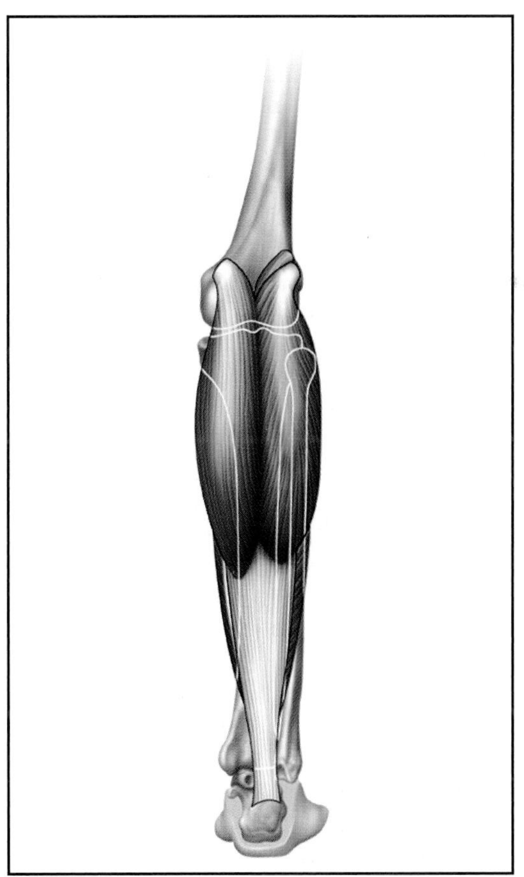

Figure 31
The gastrocnemius and soleus muscles combine to form the Achilles tendon. Both cross the ankle joint. In this view, the soleus is obscured by the gastrocnemius. Note that only the gastrocnemius flexes the knee joint; the soleus does not cross the knee and serves only as an ankle plantar flexor.

ula and inserts more distally into the Achilles tendon. These two muscles provide the vast majority of plantar flexion strength to the ankle. The plantaris is a small vestigial muscle whose strength contribution is minimal.

The deep posterior compartment contains the posterior tibialis, flexor digitorum longus, and flexor hallucis longus muscles (Fig. 32). These muscles arise from the posterior tibia, fibula, and interosseous membrane; are innervated by the tibial nerve; and course behind the medial malleolus. The posterior tibialis inserts on the navicular.

The anterior compartment contains the anterior tibialis, extensor hallucis longus, extensor digitorum longus, and peroneus tertius muscles (Fig. 33). Innervated by the deep

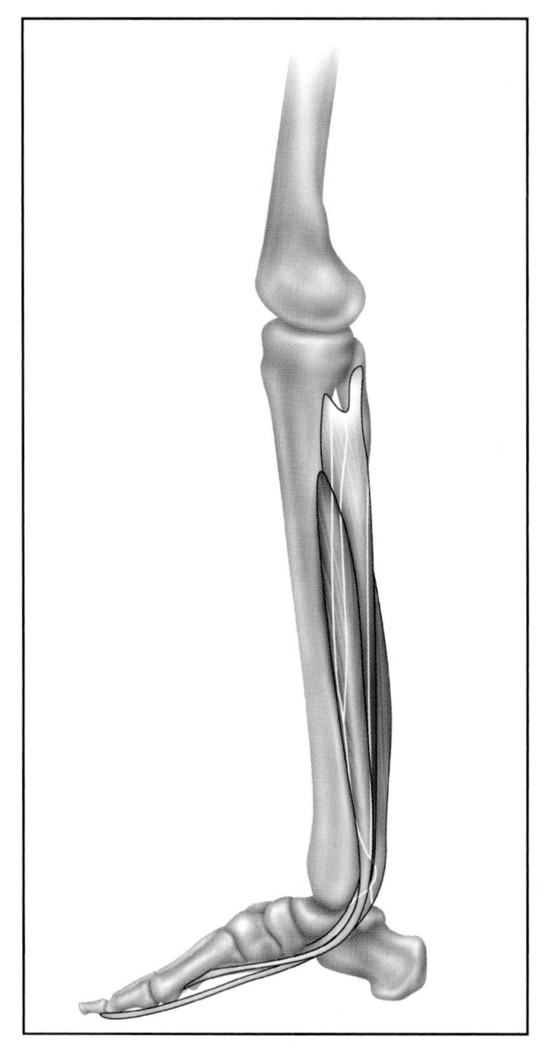

Figure 32
The flexor hallucis longus, flexor digitorum longus, and posterior tibialis. These muscles of the deep posterior compartment course behind the medial malleolus at the ankle.

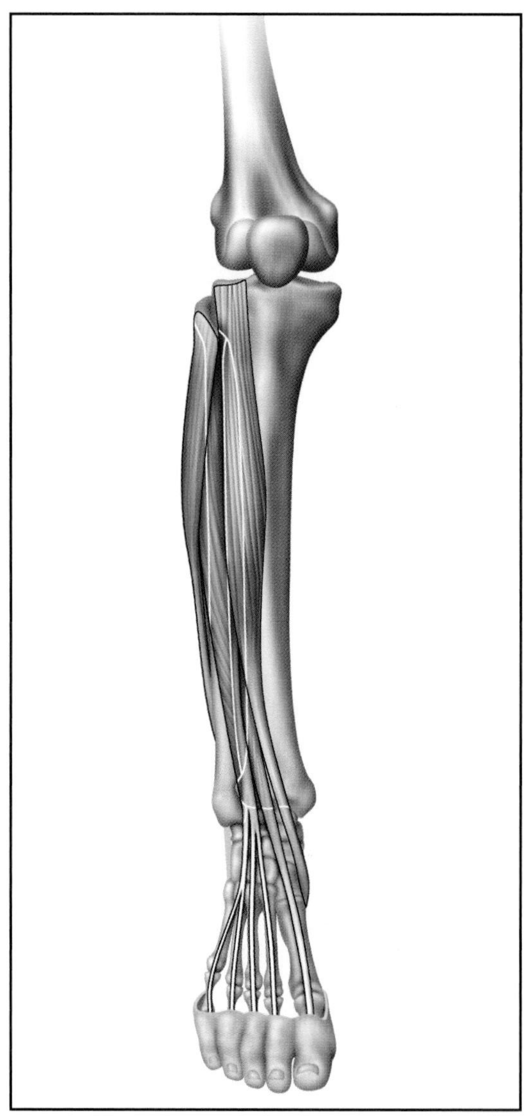

Figure 33
The anterior tibialis, extensor hallucis longus, and extensor digitorum longus are muscles of the anterior compartment and cross the ankle along its dorsal surface.

peroneal nerve, these muscles act as extensors of the ankle during the swing phase of gait. They arise from the anterior tibia and interosseous membrane of the leg and enter the ankle under the extensor retinaculum. Injuries to the peroneal nerve, therefore, result in a loss of active ankle dorsiflexion, a functional deficit known as a footdrop.

The lateral compartment of the leg contains the peroneus longus and peroneus brevis muscles (Fig. 34). These muscles are innervated by the superficial peroneal nerve and act to evert the ankle and protect against inversion sprains. They arise from the fibula, and their tendons cross the ankle directly posterior to the lateral malleolus.

Nerves and Blood Vessels

Nerves and blood vessels are important components of the knee; however, for the most part, they are transient—on their way to perform important tasks in the leg and foot. The nerves and blood vessels are located in back of the knee in the diamond-shaped popliteal fossa. The popliteal artery is an extension of the femoral artery, which changes its name as it dives posteriorly through the adductor canal. The popliteal artery gives off the genicular arteries, which supply the knee, and continues to the leg as the anterior and posterior tibial arteries. The posterior tibial ar-

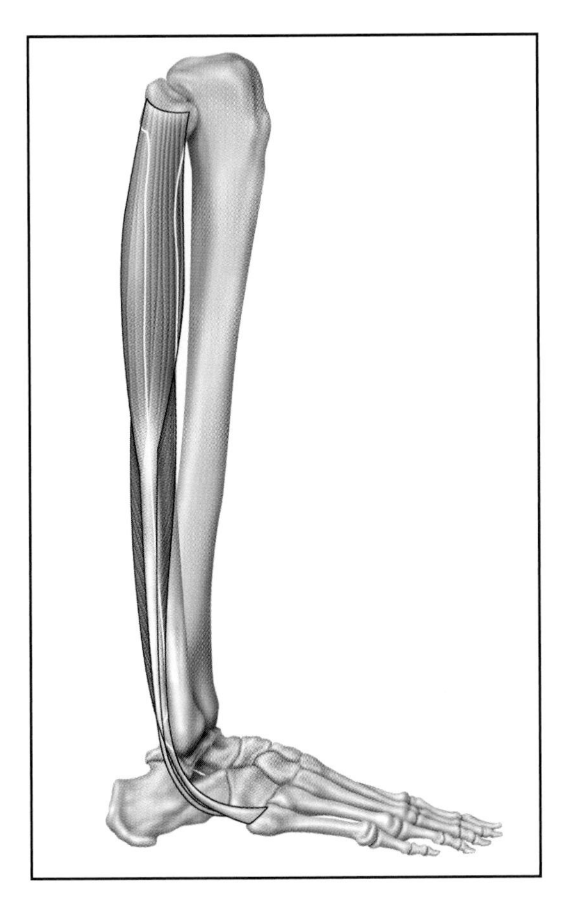

Figure 34
The peroneus longus and peroneus brevis muscles are muscles of the lateral compartment and cross the ankle behind the fibula. The brevis inserts on the fifth metatarsal but the longus courses under the foot to insert in the first metatarsal.

tery splits and gives off the peroneal artery. The popliteal artery is at risk for damage during a knee dislocation because it is fairly well tethered within the fossa.

The nerves also pass through the popliteal fossa. In the thigh, the sciatic nerve splits into separate tibial and peroneal branches. These enter the fossa distinctly, with the peroneal laterally placed. The tibial nerve remains posterior in the calf throughout. The peroneal wraps around the fibular neck where it is subject to injury during fracture or dislocation. This nerve then splits into the deep peroneal nerve, which supplies the anterior muscles of the leg, and the superficial peroneal nerve, which supplies the muscles of the lateral compartment (the peroneus longus and brevis).

The vessels and nerves continue into the leg in distinct fascial compartments. This arrangement is somewhat counterintuitive because there are four compartments but only three nerves and arteries (the superficial posterior compartment has neither a vessel nor an artery). In addition, the lateral compartment has its own nerve (the superficial peroneal) but no blood vessel. The vessel that feeds it, the peroneal artery, actually resides in the deep posterior compartment.

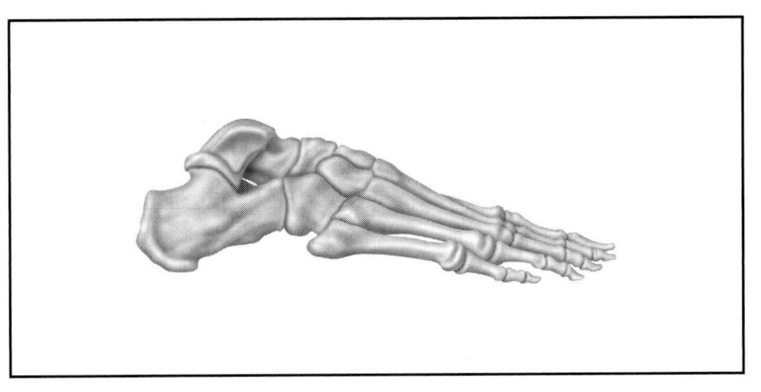

Figure 35
The bones of the foot as seen on a lateral view.

Foot and Ankle

Bone and Joints

The ankle joint is the articulation of the leg bones (the tibia and fibula) and the talus (the superior-most bone of the foot). This joint is formed as the tibial shaft flares out distally to form the medial malleolus. This malleolus serves as a medial buttress of the ankle and as the proximal point of origin of the deltoid ligament. The distal articular surface of the tibia (the plafond) is not perfectly flat but instead has a slight central ridge that corresponds to a central depression in the talar dome. This shape increases the congruity and stability of the joint. The fibula also flares out distally to form the lateral malleolus, the anchor point for the lateral ankle ligaments and a buttress preventing lateral displacement of the talus. The distal tibia and fibula thus form a "mortise," an inverted U, which contains the dome of the talus. The talar dome accordingly comprises the floor of the ankle joint and articulates with the tibial plafond superiorly, the medial malleolus medially, and the lateral malleolus laterally.

The architecture of the foot can be divided into three parts: the hindfoot, the midfoot, and the forefoot (Figs. 35 and 36). The hindfoot consists of the talus and calcaneus and serves as the link between the ankle and the remainder of the foot. The undersurface

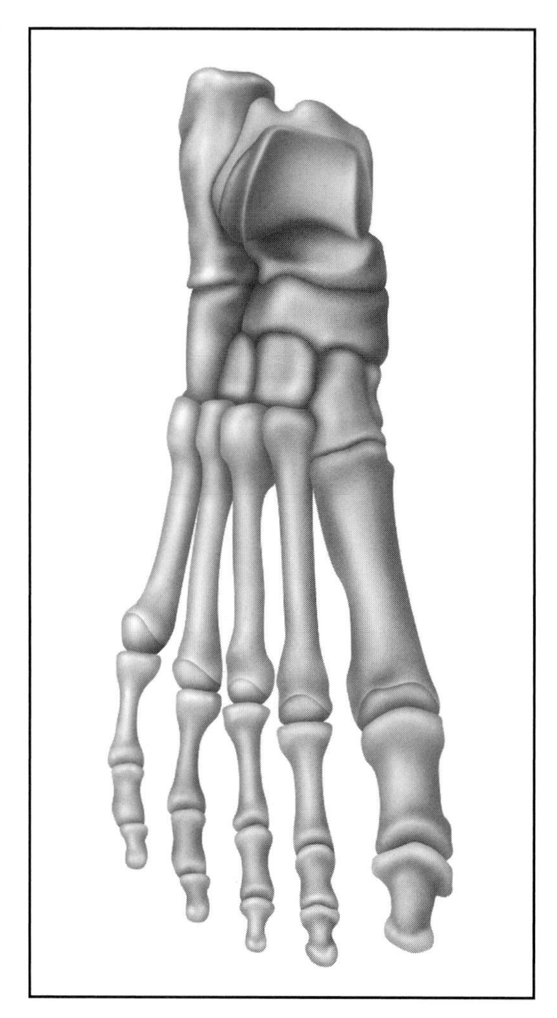

Figure 36
The bones of the foot as seen from above.

of the talar body consists of articular cartilage that contacts the calcaneus to form the subtalar joint. Much of the talus is covered by articular cartilage with no direct muscle attachments and few areas for the entry of blood vessels into the bone. In the presence of severe fractures or dislocations, this limited vascular supply predisposes the talus to osteonecrosis. The body of the calcaneus is its largest portion and bears large compressive loads during weight bearing. The sustentaculum tali, a dense bony projection off the medial side of the calcaneal body, helps support the talus.

The midfoot is made up of the tarsal bones, namely, the navicular; the cuboid; and the medial, middle, and lateral cuneiforms. The midfoot meets the hindfoot at the talonavicular and calcaneocuboid joints, which collectively are known as the transverse tarsal joint, or Chopart's joint. The distal extent of the midfoot is the tarsometatarsal joint, or the Lisfranc joint. Medially, the navicular serves as the insertion for the posterior tibialis tendon. The cuboid, the most lateral bone, has a shallow groove on its lateral side in which the peroneus longus tendon is housed as it turns from the lateral side of the foot toward the plantar aspect.

The forefoot bones include the metatarsals, the phalanges, and two accessory bones beneath the hallux, the medial and lateral sesamoids. The metatarsal bones link the midfoot with the toes. The great toe, or hallux, has two phalanges (proximal and distal) and one interphalangeal joint. The four lesser toes each have three phalanges (proximal, middle, and distal), with two intervening joints, the proximal interphalangeal (PIP) joint and the distal interphalangeal (DIP) joint. The phalanges are short tubular bones. The concave proximal phalanx base articulates with the metatarsal head.

The sesamoid bones beneath the metatarsophalangeal joint of the great toe have a function similar to that of the patella and increase the mechanical advantage of the muscles that flex that metatarsophalangeal joint. Because of their position at the ball of the foot, the sesamoids undergo substantial mechanical force and are often the site of overuse conditions or injuries.

Ligaments

An extensive array of ligaments stabilize the ankle (Fig. 37). Medially, the deltoid ligament resists eversion (tilting up of the lateral foot) as well as rotation of the talus within the mortise. The lateral ankle ligaments resist inversion of the ankle joint (tilting up of the medial foot). The lateral ligament complex consists of the anterior talofibular, calcaneofibular, and posterior talofibular ligaments (Fig. 38). The calcaneofibular ligament resists varus tilting of the joint when the ankle is in the neutral position (neither plantar flexed nor dorsiflexed). When the ankle is in the neutral position, the anterior talofibular ligament resists anterior subluxation of the talus. In plantar flexion, the anterior talofibular ligament resists inversion because in that position the ligament is oriented more vertically.

The syndesmotic ligaments (syndesmosis) stabilize the tibia to the fibula above the level of the ankle joint. This complex is composed of four major structures that resist separation (or diastasis) of the tibia from the fibula.

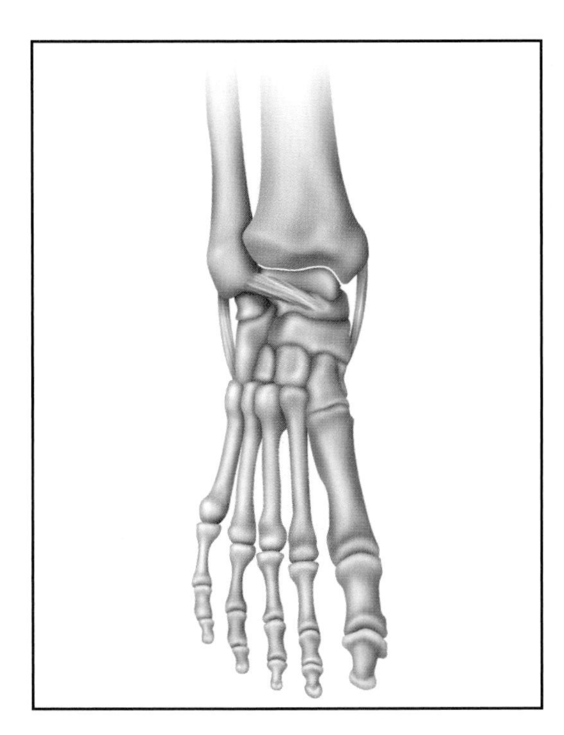

Figure 37
Frontal view of the ankle joint. The deltoid ligament lies medially and the talofibular and calcaneofibular ligaments are lateral. The syndesmosis is not shown.

The subtalar joint is stabilized primarily by ligaments directly between the talus and calcaneus. The motions at the subtalar joint are inversion and eversion. The midfoot demonstrates a complex variety of motion, allowing the foot to accommodate uneven surfaces or terrain. These motions include dorsiflexion, plantar flexion, abduction, adduction, and rotation. The metatarsophalangeal joints in the forefoot are stabilized by medial and lateral collateral ligaments, a weak dorsal joint capsule, and a strong plantar ligament complex (the "plantar plate"). The PIP and DIP joints have a similar ligamentous structure that, along with the bony anatomy, provides stability.

Muscles

Muscles of the foot and ankle are divided into two groups: intrinsic muscles, which have their bellies in the foot; and extrinsic muscles, which are based in the compartments of calf, with only their tendons crossing the ankle. The muscles of the superficial posterior compartment (gastrocnemius and soleus) attach to the calcaneus and are powerful plantar flexors of the ankle. The muscles of the deep posterior compartment cross the ankle behind the medial malleolus deep to the flexor retinaculum. The posterior tibialis inverts the ankle and provides dynamic support of the arch of the foot. The flexor hallucis longus and flexor digitorum longus tendons insert into the distal phalanges of the great toe and lesser toes; flex the toes; and, to a small degree, contribute to ankle plantar flexion.

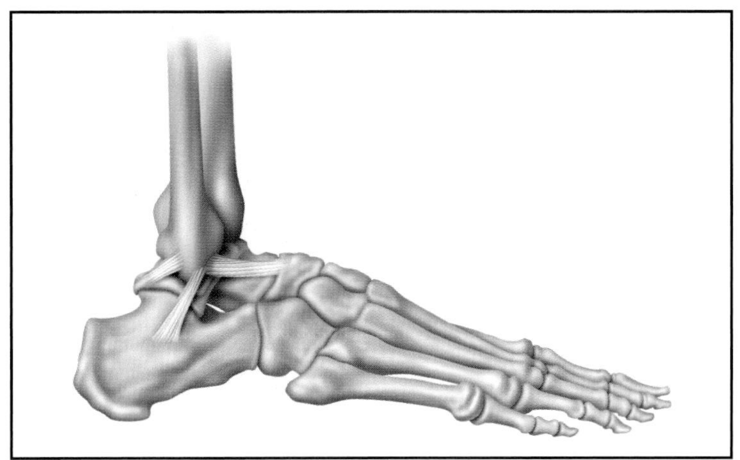

Figure 38
The talofibular and calcaneofibular ligaments.

The anterior compartment muscles (tibialis anterior, extensor hallucis longus, and extensor digitorum longus) are extensors of the ankle. They cross the ankle anteriorly and pass deep to the extensor retinaculum to exit onto the dorsum of the foot. The tibialis anterior tendon inserts on the medial cuneiform and the base of the first metatarsal and is the primary extensor of the ankle. The extensor hallucis longus and extensor digitorum longus extend the toes.

The lateral compartment of the leg contains the peroneus longus and peroneus brevis muscles, which are important dynamic stabilizers of the lateral ankle and protect against inversion sprains. These two muscles cross the ankle behind the lateral malleolus in a shallow groove that is covered by the peroneal retinaculum. The peroneus brevis tendon inserts on the base of the fifth metatarsal, whereas the peroneus longus tendon curves around a groove in the cuboid and travels deep into the plantar aspect of the foot to insert on the base of the first metatarsal.

The intrinsic muscles are housed in the foot itself. The extensor hallucis brevis and extensor digitorum brevis muscles arise from the lateral hindfoot, and their tendons travel across the dorsum of the foot. The extensor

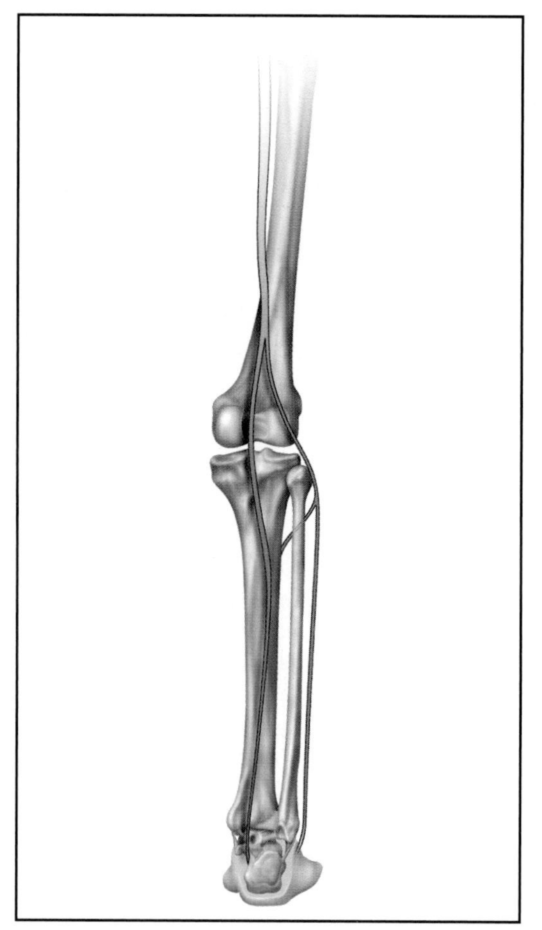

Figure 39
Nerve supply to the foot and ankle. Note the course of the peroneal nerve around the fibula. The tibial nerve enters the foot behind the medial malleolus. The sural and saphenous nerves are not shown.

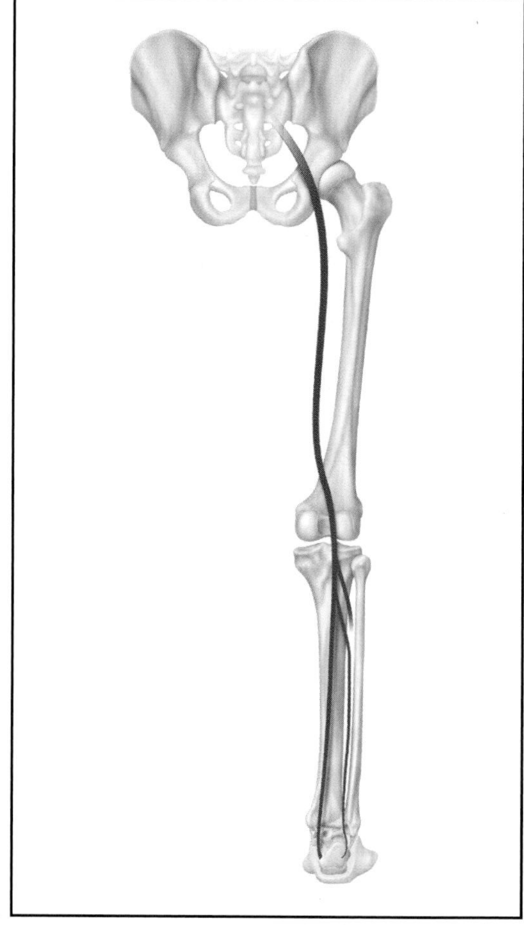

Figure 40
The blood supply to the foot. Note the trifurcation of the popliteal artery just distal to the knee joint.

hallucis brevis inserts on the proximal phalanx of the great toe and extends the metatarsophalangeal joint. The extensor digitorum brevis tendons extend the metatarsophalangeal joint and are innervated by a branch from the deep peroneal nerve.

On the plantar side, the plantar fascia originates from the calcaneus and inserts on the plantar plate of the metatarsophalangeal joint, which helps support the longitudinal arch of the foot. Branches of the tibial nerve innervate the plantar intrinsic foot muscles. These muscles, which abduct the toes and assist with flexion and extension, are similar in function to the intrinsic muscles of the hand.

Nerves and Blood Vessels

Five nerves supply the lower legs, ankles, and feet: the tibial, saphenous, sural, superficial peroneal, and deep peroneal nerves (Fig. 39). The saphenous and sural nerves are purely sensory nerves, whereas the other three are mixed sensory and motor nerves. At the level of the ankle, the tibial nerve travels posterior to the medial malleolus and runs deep to the flexor retinaculum in the tarsal tunnel. It then trifurcates into the medial calcaneal, medial plantar, and lateral plantar nerve branches. High in the calf, the common peroneal nerve divides into two branches, the superficial and deep peroneal nerves. The deep peroneal nerve runs within the anterior compartment of the leg with the anterior tibial vessels and innervates the muscles there. It sends a sensory branch onto the dorsum of the foot with the dorsalis pedis artery, which terminates in the web space between the great toe and the second toe. The superficial peroneal nerve runs within the lateral muscle compartment of the leg, innervating the peroneus longus and brevis muscles. The nerve continues distally to supply sensation to most of the dorsum of the foot.

The ankle and foot have three major arteries providing vascular inflow: the posterior tibial, anterior tibial, and peroneal arteries (Fig. 40). These three branches arise in the proximal leg at the "trifurcation" of the popliteal artery. (This is not a true trifurcation as the anterior tibial splits off first.) The posterior tibial artery pulse is palpable posterior to the medial malleolus. The anterior tibial artery crosses the ankle and courses on the dorsum of the foot, at which point it is renamed the dorsalis pedis artery. The peroneal artery runs within the deep posterior compartment but supplies the peroneal muscles and the lateral foot and ankle. As in the hand, these vessels have extensive anastomoses; therefore, collateral flow is often sufficient when an isolated arterial injury occurs.

Spine

Bones and Joints

The spine is composed of 24 vertebrae that are organized into four regions: cervical, thoracic, lumbar, and sacral (Fig. 41). There are seven cervical vertebrae. Two of these are distinctly shaped: C1, the atlas (which holds up the globe of the head) and C2, the axis. C3 through C7 are more similar in shape to the remaining vertebrae below. The 12 thoracic vertebrae give off 12 pairs of ribs, each emanating from a single thoracic vertebral body. The first seven ribs articulate with the sternum. The eighth, ninth, and tenth ribs do not reach all the way around to the sternum directly but form a common cartilage bar that angles up toward the inferior border of the sternum. The eleventh and twelfth ribs are not attached to anything anteriorly; thus, they are called "floating ribs." The five lumbar vertebrae attach the thoracic region to the sacrum. The sacrum itself represents the fusion of five spinal elements into one bone.

When viewed in profile, the overall configuration of the spine places C1 directly above the sacrum; however, each region is curved. The cervical and lumbar regions demonstrate lordosis (arching back), whereas the thoracic spine is kyphotic (hunched). The three curves offset one another, which balances the spine overall. When viewed in the coronal (anteroposterior) plane, the spine should appear perfectly straight. Deviations to either the right or left represent scoliosis (the clinical entity termed "scoliosis" is a three-dimensional deformity, with a rotational component as well.)

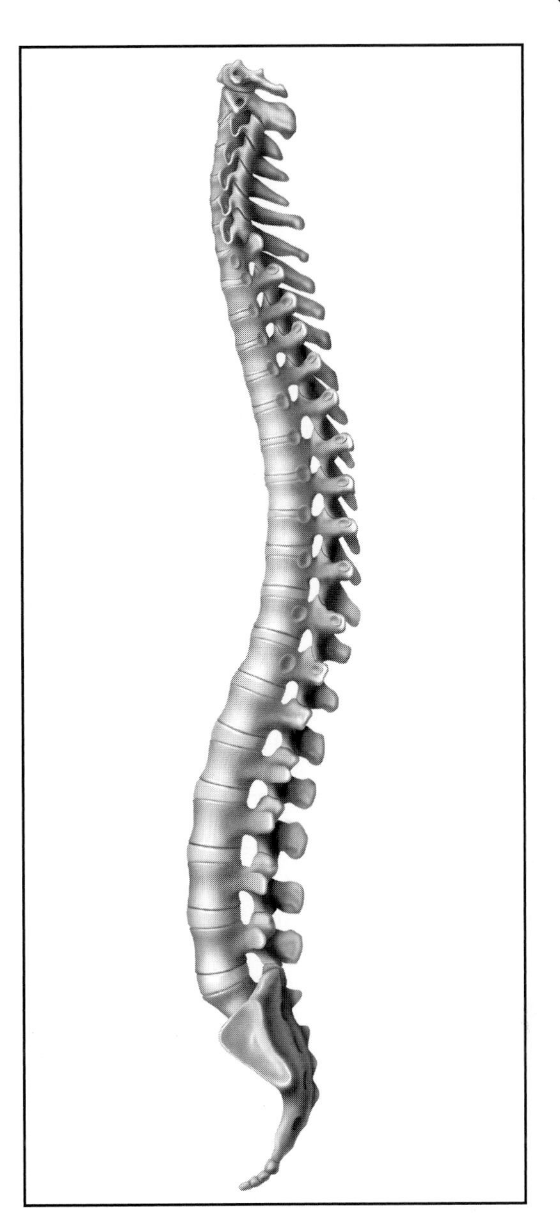

Figure 41
Sagittal view of the spine. Note the lordosis of the cervical and lumbar regions and kyphosis of the thoracic spine.

The morphology of the vertebrae varies by region. For example, the body of a lumbar vertebra viewed from above is anterior (Fig. 42). Projecting posteriorly off the body are left and right pedicles. From there, the transverse processes emanate laterally and the articular processes aim in the inferior and superior directions. Posterior to the transverse processes are the right and left laminae, which come together at the spinous process. The vertebrae of the lumbar region are

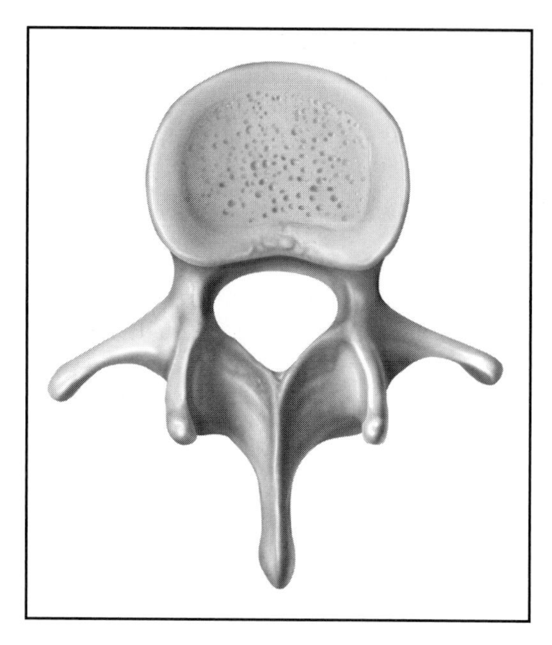

Figure 42
Axial view of a lumbar vertebral body. The space surrounded by the vertebral body, pedicles, and laminae houses the spinal cord and cauda equina.

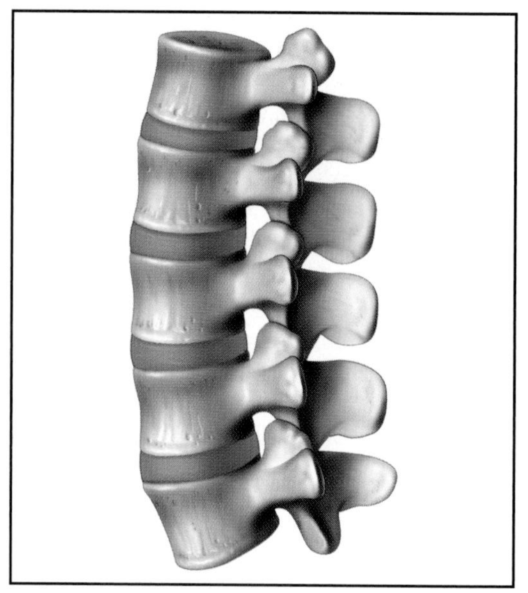

Figure 43
Lateral view of the lumbar spine showing the intervertebral disks, the neural foramina, and the posterior elements.

stacked one on top of the other, with disks cushioning the bodies (Fig. 43). The inferior articular process of the vertebra above contacts the superior articular process of the one below, forming the facet joint. (Note that the articular processes are named not in reference to their position around the neural foramen but according to their location on their own vertebral body.) The superior and inferior articular processes of a single vertebra are connected by a piece of bone called the pars intra-articularis, which is commonly subject to stress fracture. When the articular processes contact each other, a space called the neural foramen is defined by the pedicles above and below it. It is through this space that the nerve roots exit.

The standard cervical vertebrae, C3-C7, differ somewhat from the organization of vertebrae in the lumbar region. Unlike the vertebrae of the lumbar region, there are holes (foramina) in the transverse processes of cervical vertebrae that allow passage of the vertebral artery. In addition, the bodies are smaller and more square. The atlas (C1) is even more unusual in that it does not have a body; it is more like a bony ring supporting the skull. The axis below (C2) can be considered to have taken this body, as it has a superior post-like process called the dens (also known as the odontoid) that projects into the ring of C1. The facet joints of the cervical spine are also more horizontally oriented than those of the lumbar spine, allowing slippage with trauma more easily than in the caudal (distal) segments.

The main distinction between the thoracic vertebrae and those above and below is that the thoracic region attaches to the rib cage and is less mobile. Therefore, it is at the junction of the thoracic and the (more flexible) lumbar region that traumatic hyperflexion (as caused by a motor vehicle accident, for example) inflicts the most damage.

The sacrum, a triangular bone that is the result of the fusion of five sacral vertebrae, is the link between the spine and the lower limbs. The sacrum forms three important articulations. The first is an oblique junction with the last lumbar vertebra, L5. Congenital or developmental defects of this articulation may cause instability between L5 and S1 and anterior translation of L5 relative to S1 (spondylolisthesis). The sacrum also forms two lateral articulations with the iliac bones of the pelvis. The sacroiliac joints are stabilized by strong anterior and posterior ligaments. The sacrum is also a site of attachment of the gluteal and pelvic floor muscles. Four pairs of anterior and posterior (dorsal) foramina allow the sacral nerve roots to exit. A central longitudinal canal houses the sacral nerve roots. The coccyx is a highly variable vestigial

Figure 44
The ligaments of the lumbar spine. The anterior and posterior longitudinal ligaments lie on the ventral and dorsal surfaces of the body. The ligamentum flavum is adjacent to the laminae. The supraspinous ligament connects the spinous process.

remnant of a tail that lies at the end of the sacrum. It becomes mobile in pregnancy to allow delivery and fuses to the sacrum in later adulthood. It also serves as a site of muscular and ligamentous attachments.

Ligaments and Soft Tissues

The front and back of the vertebral bodies are bound to those above and below by the anterior and posterior longitudinal ligaments, which prevent hyperextension and hyperflexion, respectively (Fig. 44). The laminae are connected by the ligamentum flavum. There is also a supraspinous ligament that connects the posterior-most aspects of the spinous processes.

There are specialized ligaments at the junction of the skull and the upper cervical spine. The dens of C2 is held within C1 by a transverse ligament. There are also vertically oriented fibers, which join the transverse ligament to form the cruciform ligament. This extends to the skull and helps stabilize the C1 articulation. The alar ligaments, right and left, connect the dens to the occiput. Accordingly, the upper cervical spine depends on ligaments for stability to a greater extent than the rest of the spine below; conditions that destroy connective tissue at synovial joints (such as rheumatoid arthritis) can lead to instability in this region.

In between the vertebral bodies lie the intravertebral disks, which make up approximately 25% of the height of the spine. The disks are essentially shock absorbers that are composed of a fibrous outer ring (the anulus fibrosus) around a soft inner core (the nucleus pulposus). The nucleus pulposus is made of a gel with a high concentration of proteoglycans; this component attracts and binds water and thereby provides cushioning. The anulus, a ring of about 90 sheets of collagen fibers laminated together, holds the nucleus in place. The most superficial fibers of the outer anulus blend with the anterior and posterior longitudinal ligaments. The anulus is thicker on its anterior aspect, which may explain why the disk usually herniates posteriorly. There are also end plates above and below each disk composed of hyaline cartilage that separates the disk from the adjacent vertebral bodies.

Muscles

Muscles around the spine are essential for postural stability and movement of the trunk. Muscles that attach directly to the bony processes of the spine and those that attach remotely, such as the abdominal wall musculature, can stabilize or move the spinal column. If the spinal column were stripped of its musculature, it would easily collapse. (The names of all of the paraspinal muscles and their attachments can be found in most anatomy texts.) The primary purpose of the muscles around the cervical spine is to position the head, while those attaching to the thoracic and lumbar regions control trunk position. One muscle worth knowing by name is the sternocleidomastoid, which connects the head to the trunk. A contracture of this muscle may lead to lateral head tilt and rotation (torticollis). Muscles that do not directly attach to the spine, such as the rectus abdominis, can still play an important role in stabilizing the spine. For this reason, abdominal strengthening is a key feature of physical therapy for back pain.

Nerves and Blood Vessels

The spinal cord runs from the foramen magnum to the L1 vertebral level, where it terminates in the conus medullaris. The cauda equina (a collection of nerve roots that has not yet exited the canal) continues distally within the dural sac. Nerve roots emanate from the spinal cord at every level (Fig. 45). Within the cervical spine, roots are named by the body below their point of exit. Thus, the C7 root exits from the neural foramen

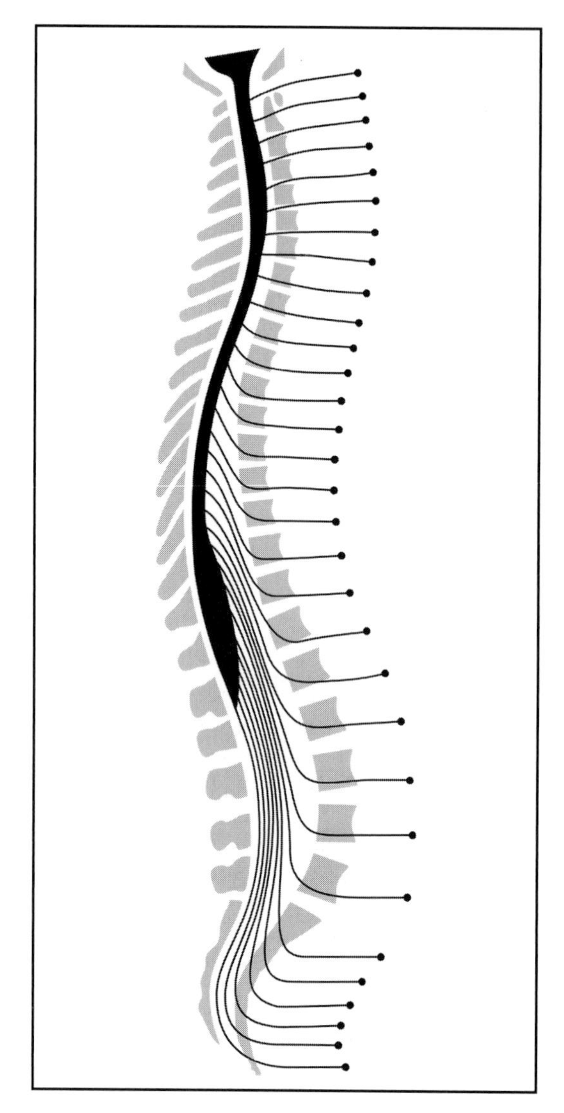

Figure 45
The spinal cord and segmental roots.

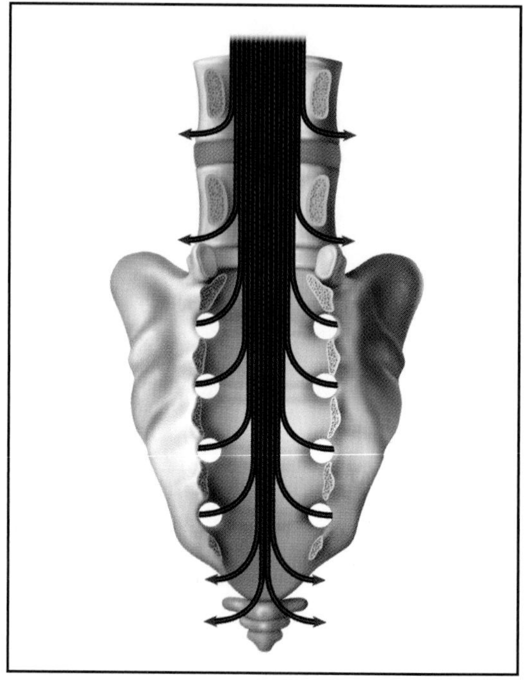

Figure 46
AP view of the nerve root exiting under the pedicles. Note that a herniated disk at L4-5 will contact the L5 nerve root unless it is a very lateral herniation, in which case it may compress the L4 nerve root.

above the body of C7. In the rest of the spine, the roots are named according to the body above. For example, the L5 root exits below the L5 vertebral body (Fig. 46). The root that exits below C7 does not adhere to this nomenclature and is arbitrarily named C8, even though there is no C8 vertebral body.

Blood is supplied to the spinal column via a complex network of arteries and veins with countless variations and connections. In general, the arterial supply is provided by the cervical and vertebral arteries in the neck, the segmental branches of the aorta in the thoracic and lumbar regions, and the sacral arteries in the sacral region. The paraspinal soft tissues also receive their blood supply from branches of these vessels. The arterial blood supply for the pelvis and surrounding structures is provided by the iliac vessels and their branches. The venous drainage of the spine and pelvis roughly mirrors the arterial tree.

One anterior and two posterior spinal arteries provide vascular inflow to the spinal cord. The anterior spinal artery is formed superiorly by branches of the vertebral arteries. More distally, this vessel is supplied by radicular branches from segmental vessels along the length of the spinal cord. The anterior spinal artery supplies the anterior two thirds of the spinal cord. The posterior spinal arteries form in the cervical spine as branches of the vertebral system and are also fed by radicular vessels. The posterior vessels supply the posterior one third of the spinal cord. Although the anterior and posterior spinal vessels traverse the length of the spinal cord, the majority of the vascular supply below the cervical spine comes from the segmental radicular arteries. The area of the spinal cord with the least collateral circulation, the midthoracic region, is more susceptible to injury from insufficient blood flow. This "watershed area" may suffer permanent damage in situations in which the blood supply is disrupted or arterial blood pressure is diminished.

Anatomy Plates

The following illustrations are modifications of the anatomy figures in chapter 8. Some of the artistic detail has been eliminated, and a few bold lines have been added. What remains, it is hoped, are drawings that embody essential anatomic relationships. Anatomy is visual, not semantic. Therefore, instead of memorizing anatomic descriptions, try to visualize the course of the various anatomic structures. These simple drawings may be very helpful in this regard. I suggest that you begin by tracing the line drawing. (You can print each drawing from the enclosed CD.) After tracing a few times, try to recreate the drawings from memory. Next, to determine the extent to which you have mastered the material, compare your work from memory with the original.

There are fewer than 50 drawings here, which clearly represent only a small portion of all of musculoskeletal anatomy. You may find it beneficial to create your own drawings from other anatomy books. I welcome submissions for consideration in subsequent editions. Enjoy.

—Joseph Bernstein, MD, MS

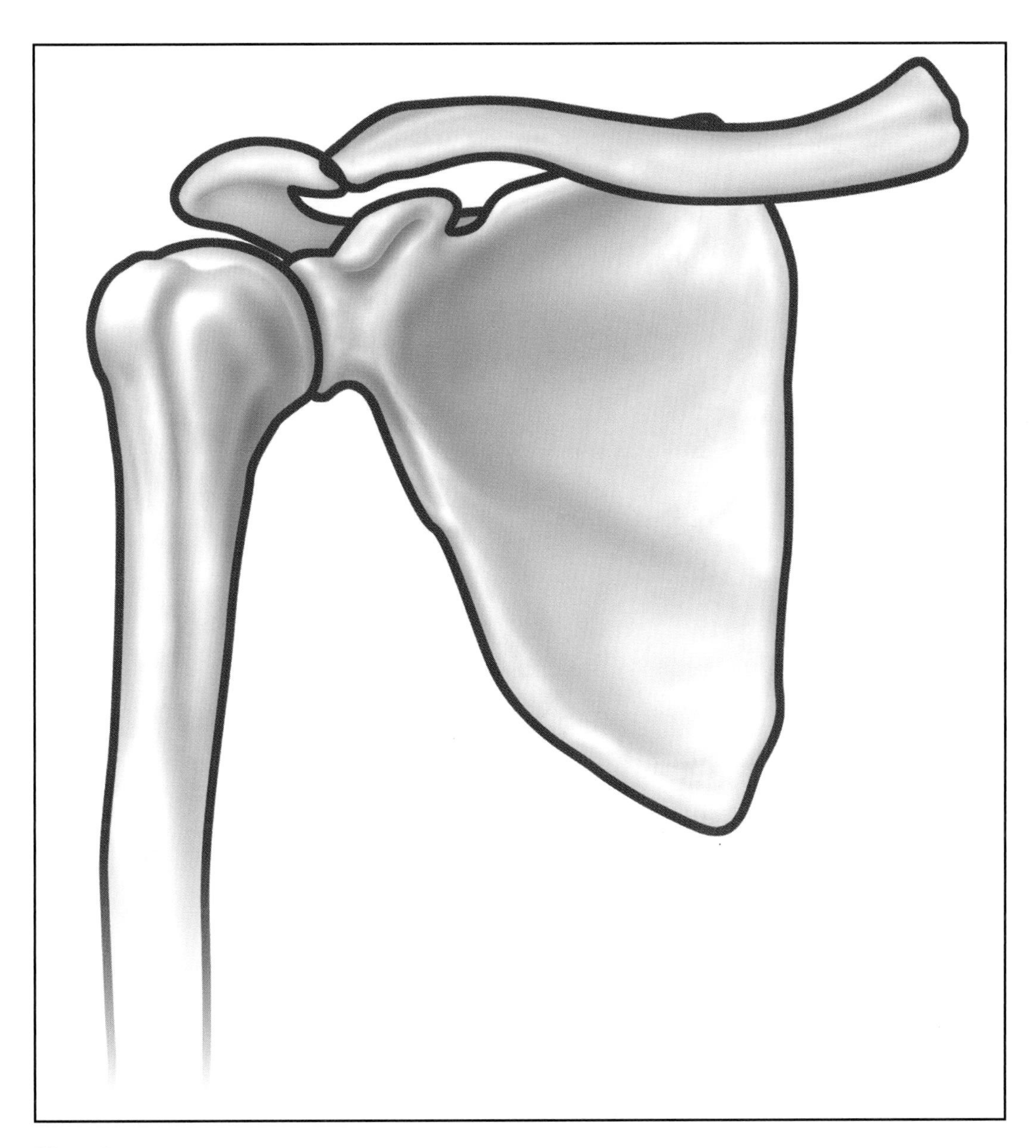

Plate 1
Draw the bones of the shoulder.

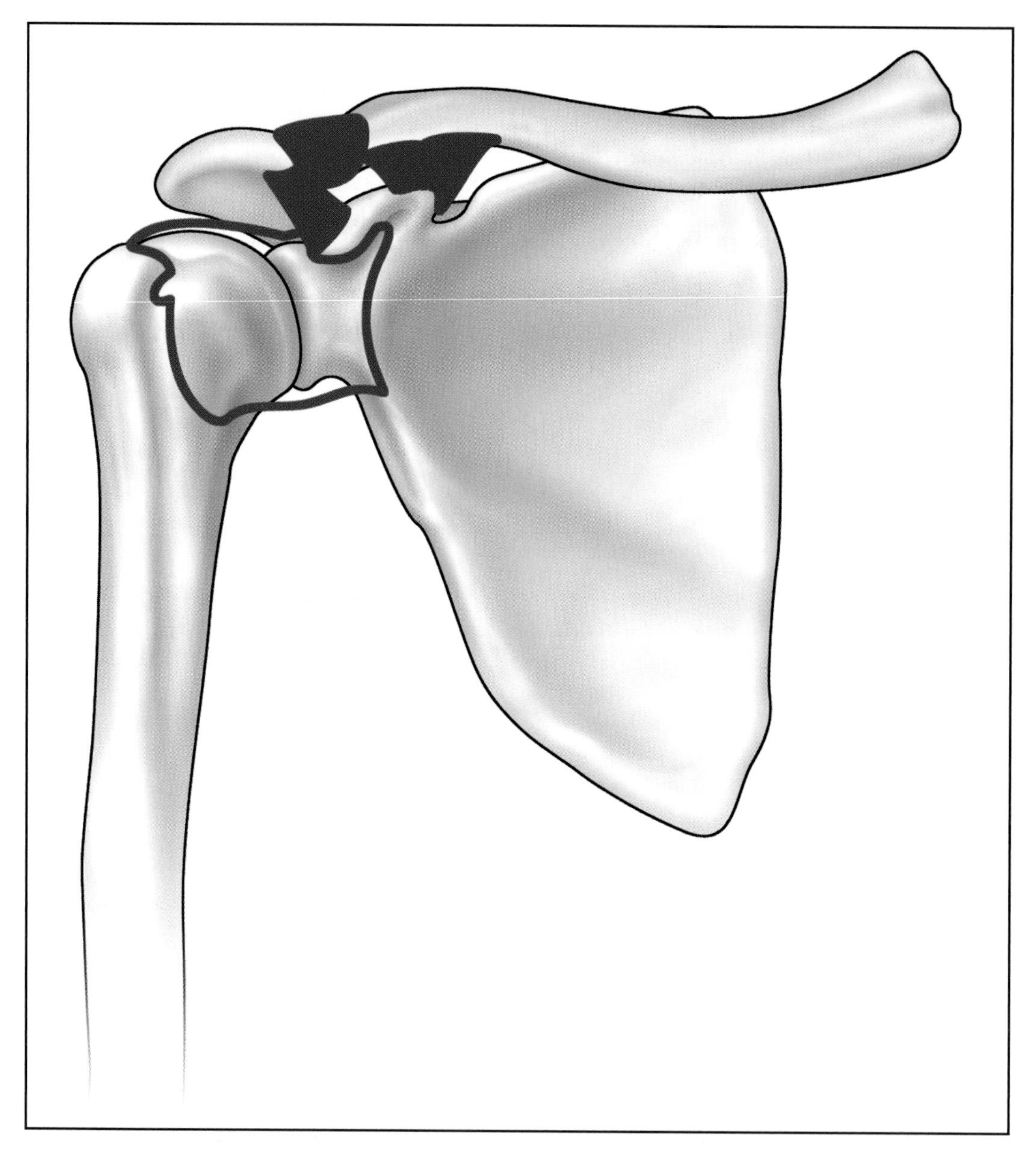

Plate 2
Draw the shoulder capsule and the acromioclavicular, coracoclavicular, and coracoacromial ligaments.

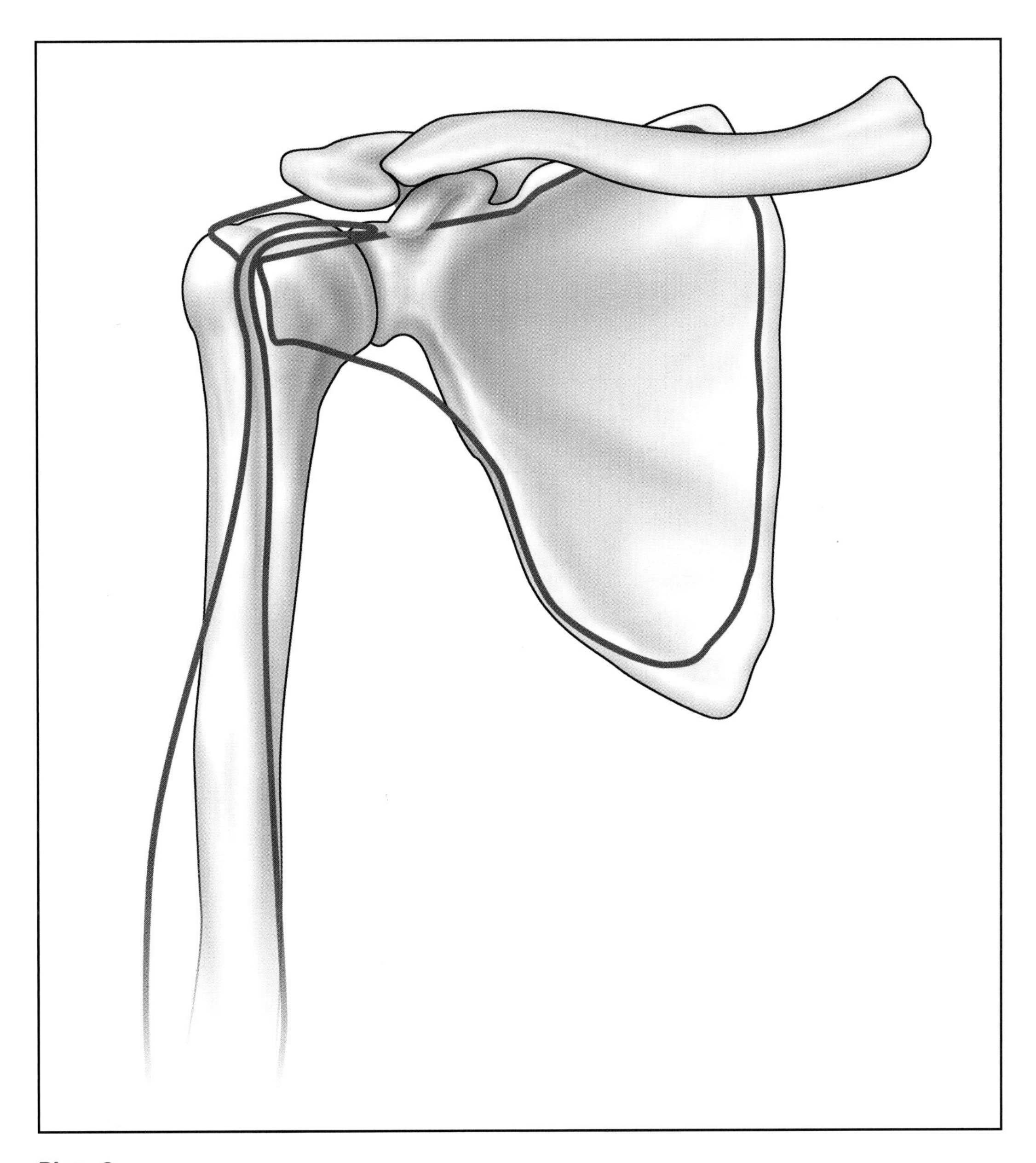

Plate 3
Draw an anterior view of the shoulder showing the subscapularis and supraspinatus. Note the course of the long head of the biceps tendon, which may be considered a functional part of the cuff.

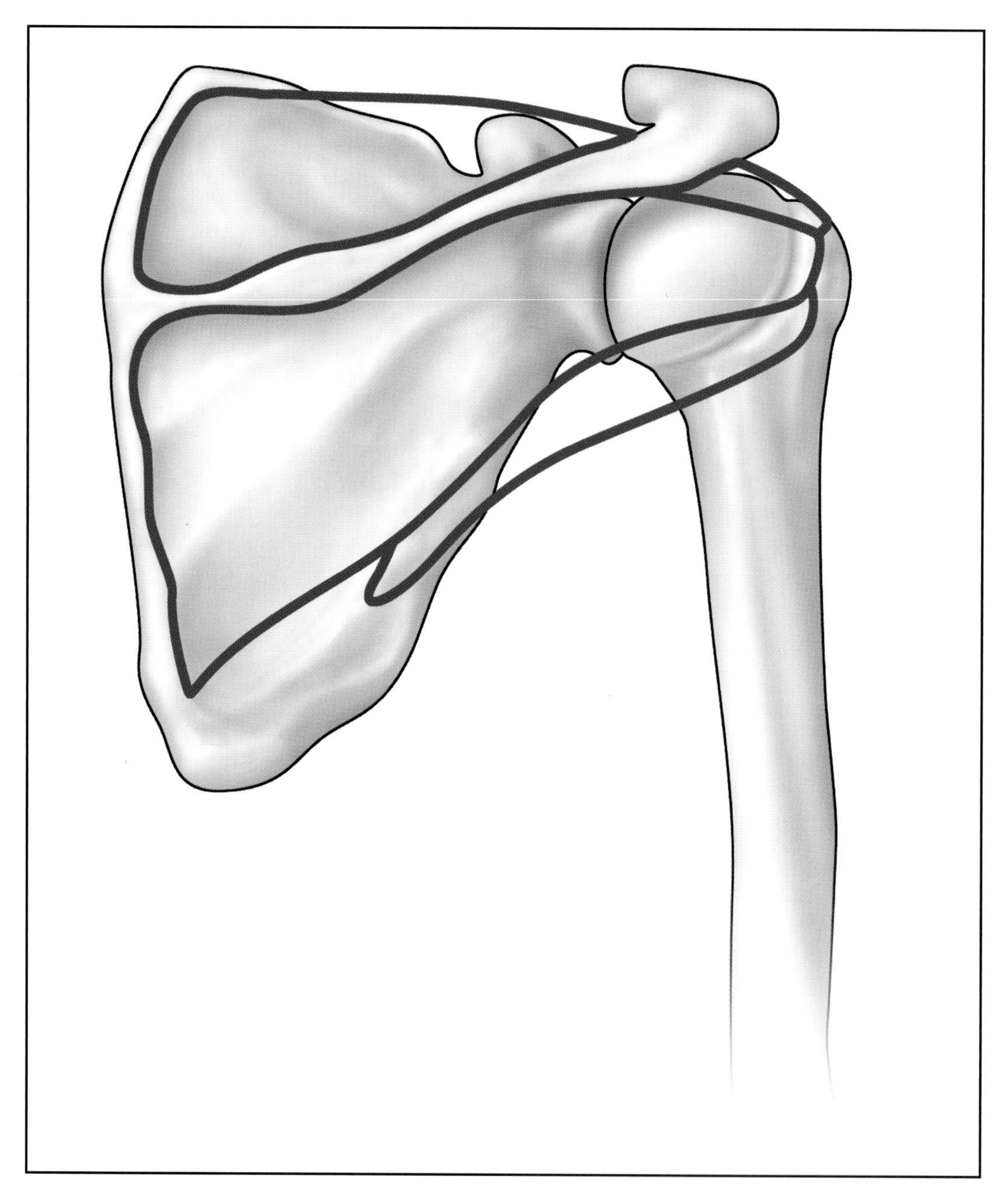

Plate 4
Draw a posterior view of the shoulder with the supraspinatus, infraspinatus, and teres minor.

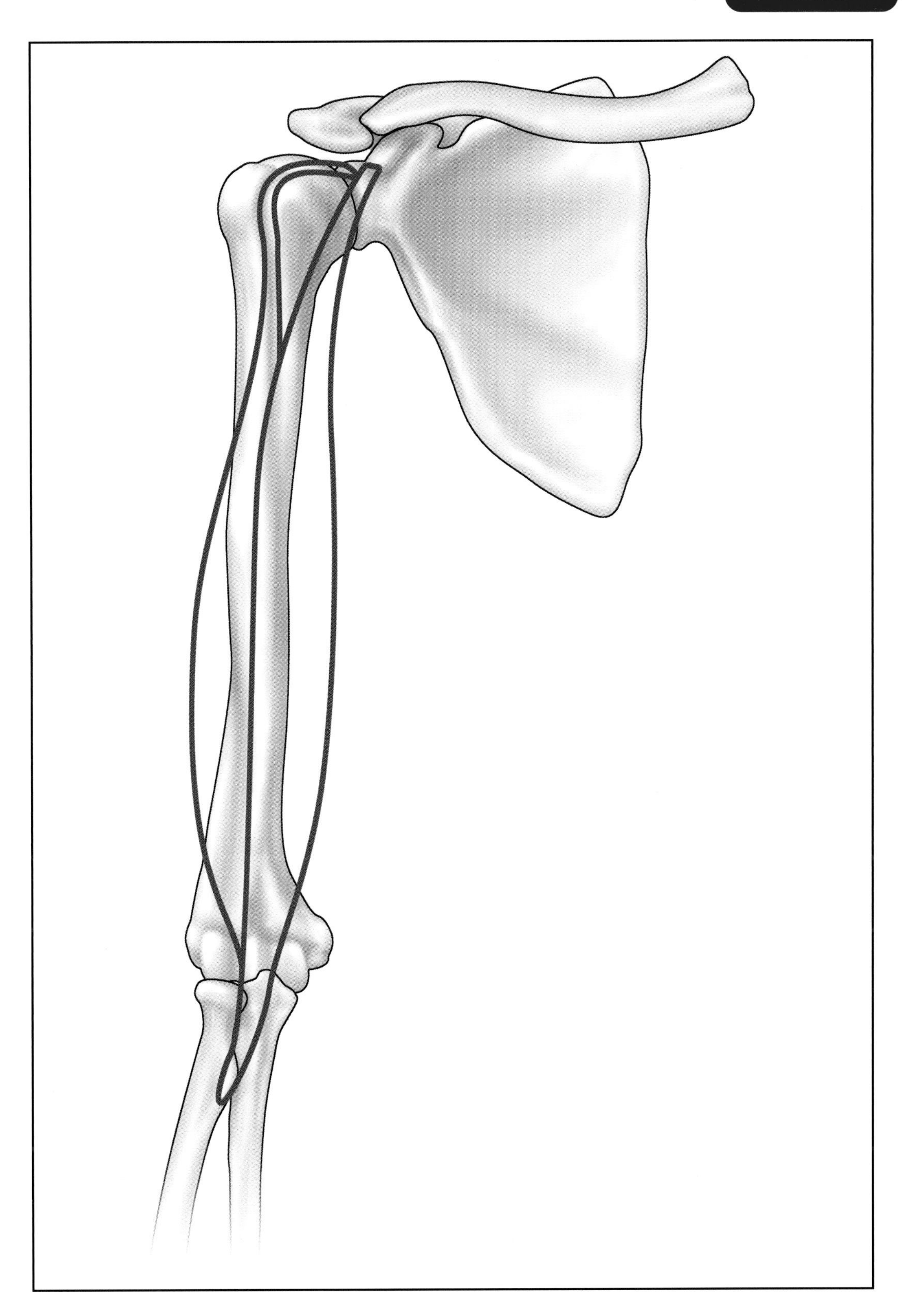

Plate 5
Draw the course of the biceps. Note the origin from the coracoid and above the glenoid and the insertion on radius, which allows for supination.

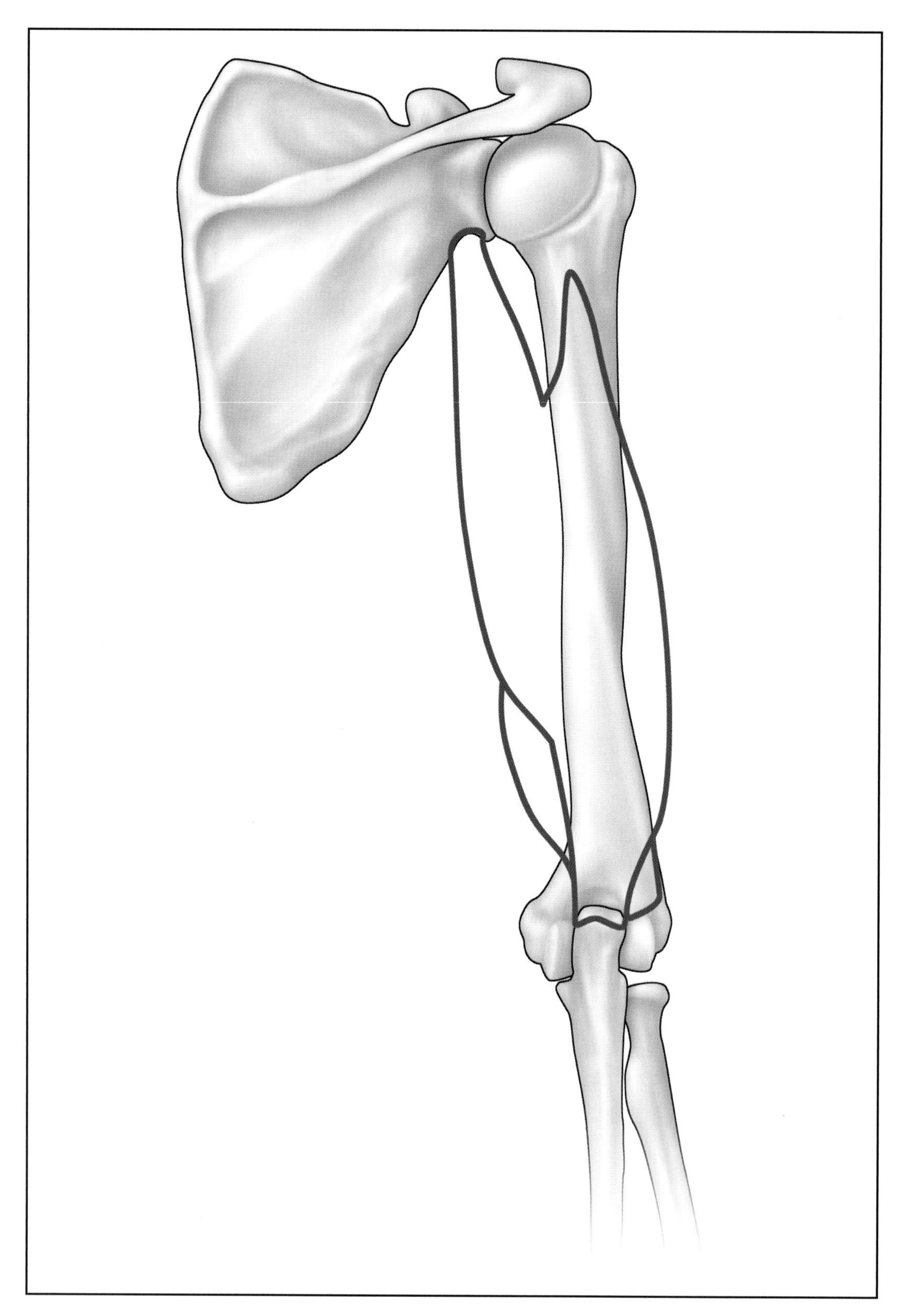

Plate 6
Draw the course of the triceps.

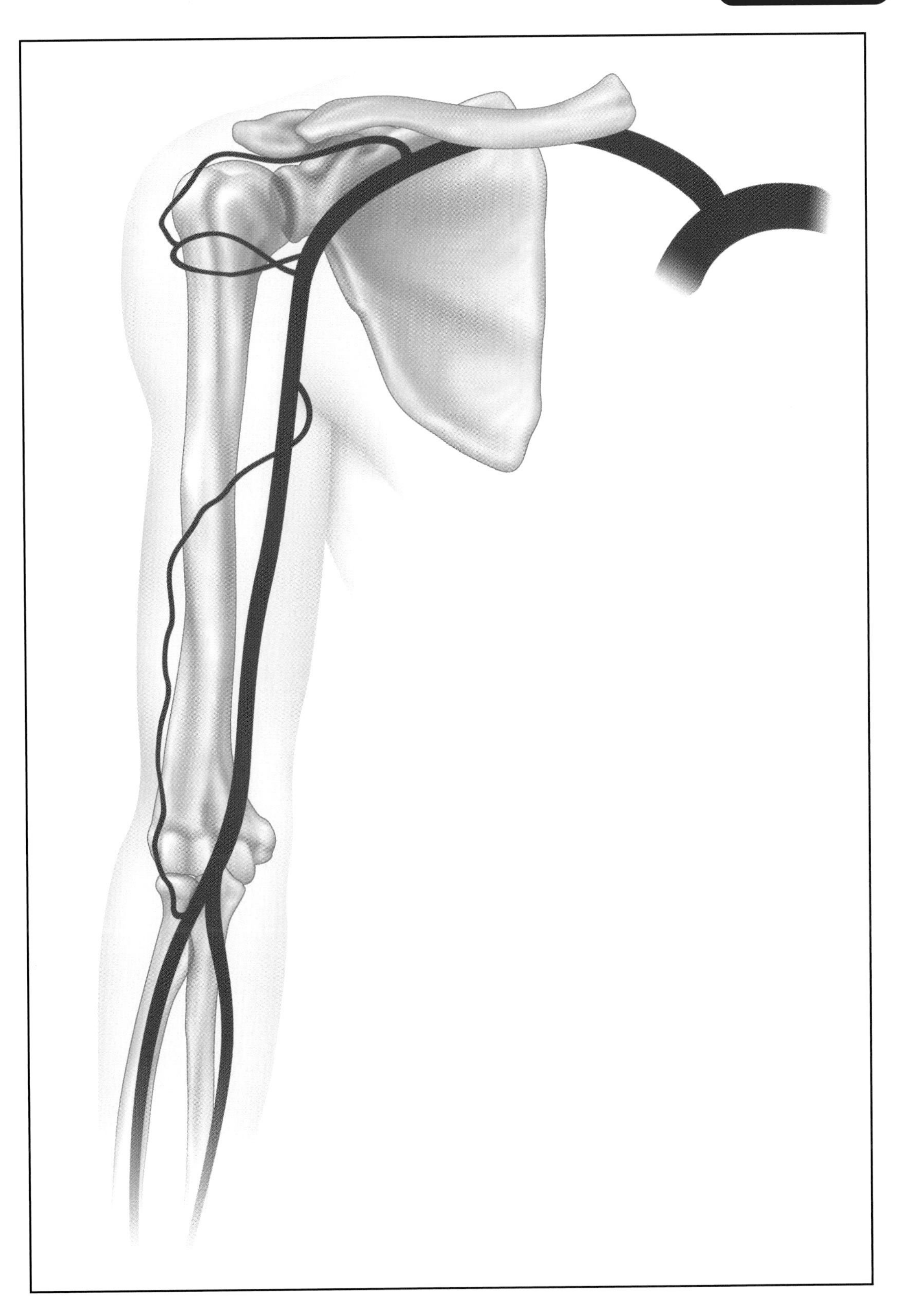

Plate 7
Draw the arterial tree of the arm.

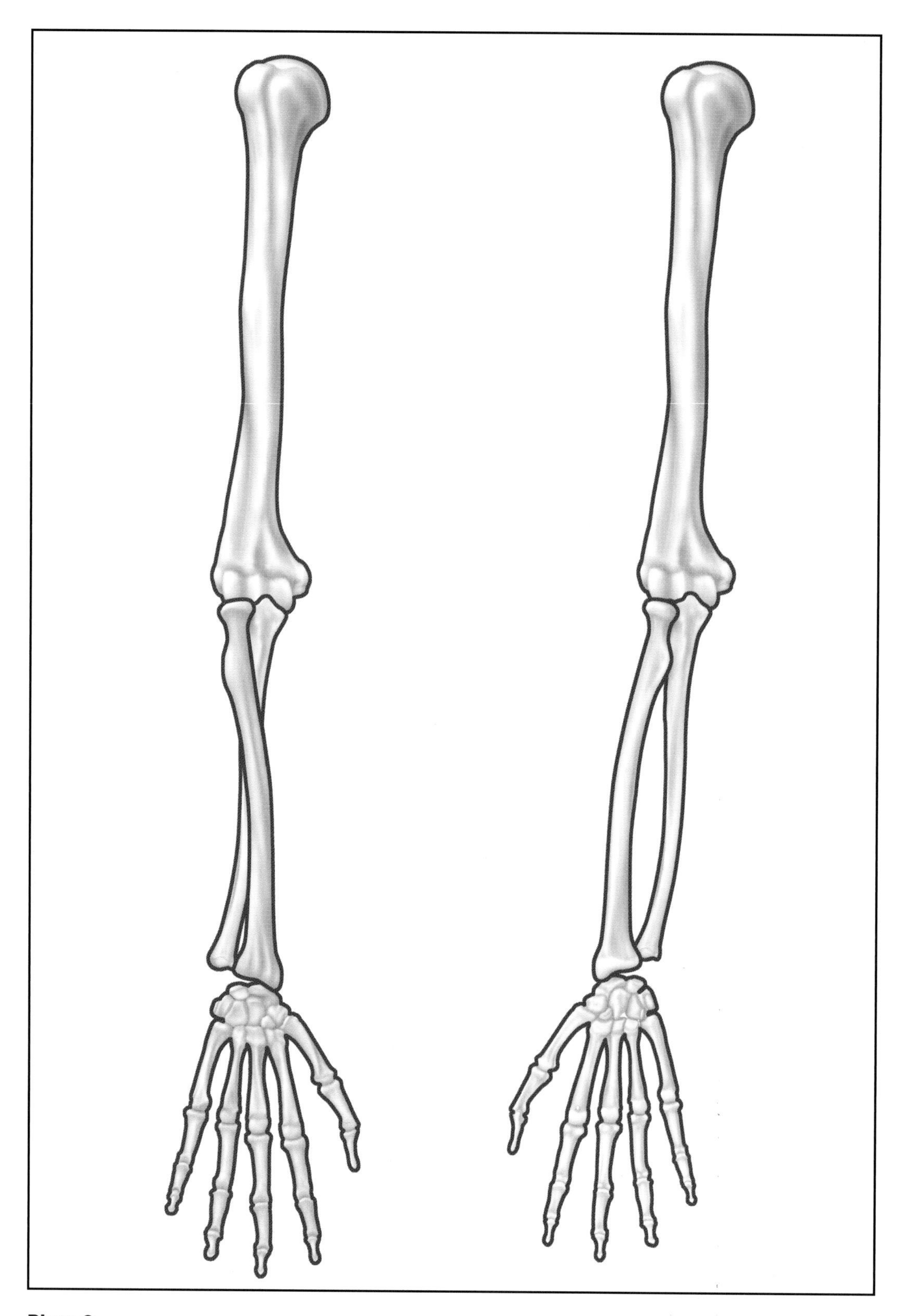

Plate 8
Draw the bones of the elbow. Note the points of articulation between the humerus and ulna (which allow flexion and extension) and humerus and radius (which allow pronation **[left]** and supination **[right]**).

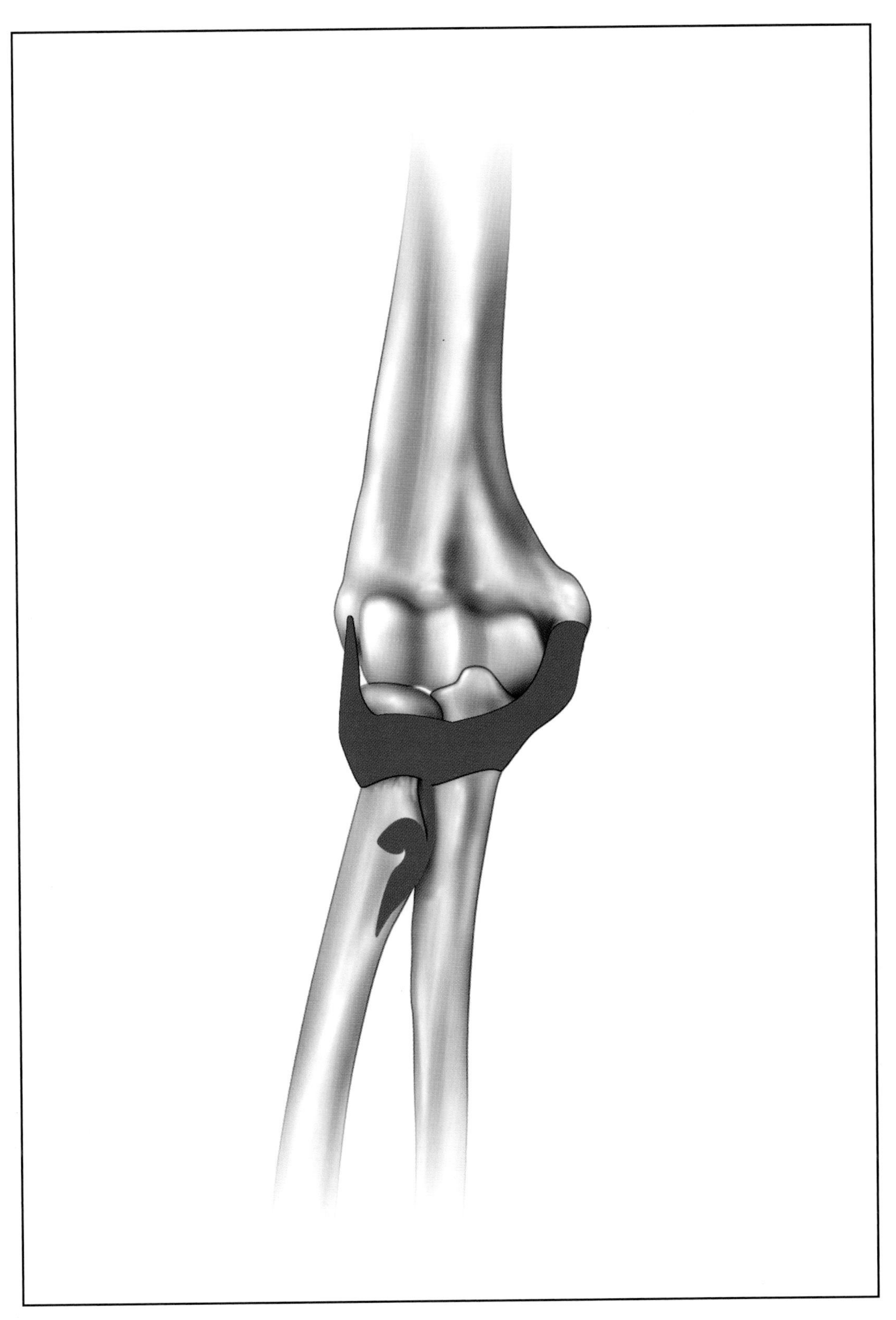

Plate 9
Draw the ligaments of the elbow: the medial (ulnar) collateral, radial collateral, and annular ligaments. Also shown is the biceps tendon as it inserts on the radius.

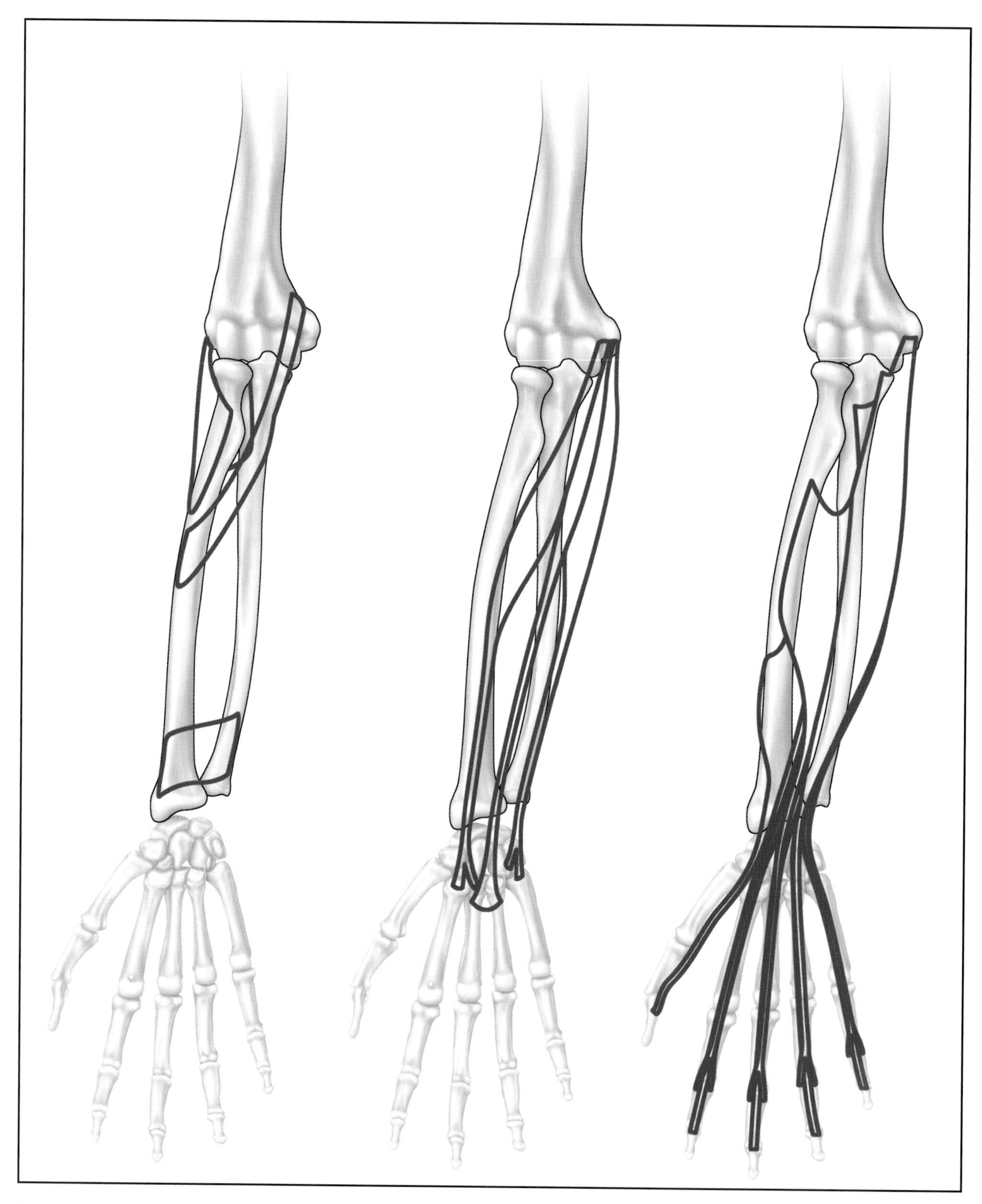

Plate 10

Left, (Based on Fig. 11, *left*) Draw the rotators of the forearm: the pronator teres, pronator quadratus, and supinator. **Center,** (Based on Fig. 11, *center*.) Draw the flexors of the wrist: the flexor carpi radialis, flexor carpi ulnaris, and palmaris longus. **Right,** (Based on Fig. 11, *right*.) Draw the flexors of the digits: the flexor pollicis longus and flexor digitorum superficialis. The flexor digitorum profundus lies under the superficialis and is obscured in this view until it inserts distally.

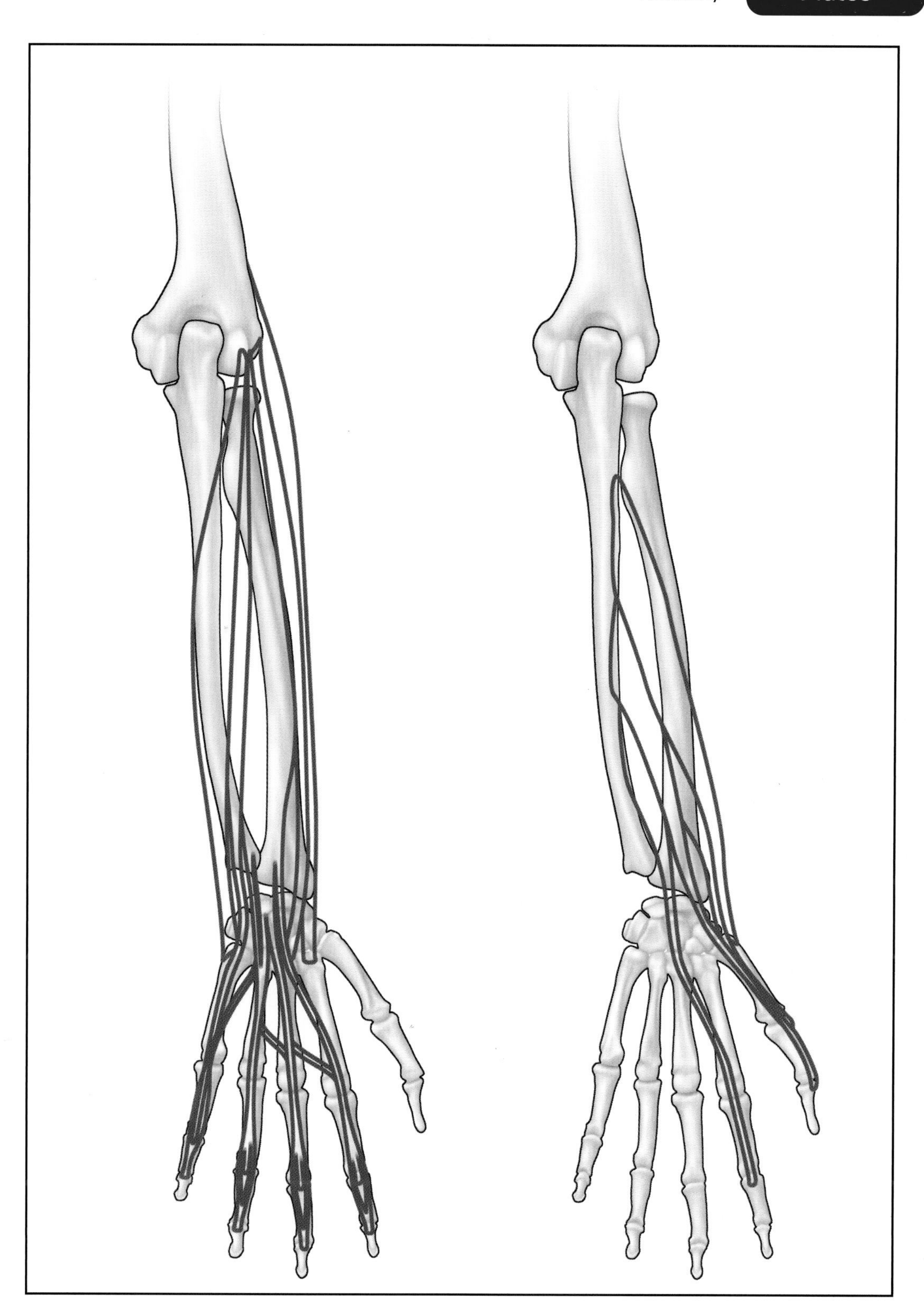

Plate 11

Left, (Based on Fig. 12, *left*.) Draw the extensor muscles of the forearm: the extensor carpi radialis longus, extensor carpi radialis brevis, extensor carpi ulnaris, extensor digitorum, and extensor digiti minimi. **Right**, (Based on Fig. 12, *right*.) Draw the deep extensor muscles of the forearm (not seen on the plate to the left). These include the abductor pollicis longus, extensor pollicis longus and brevis, and extensor indicis proprius.

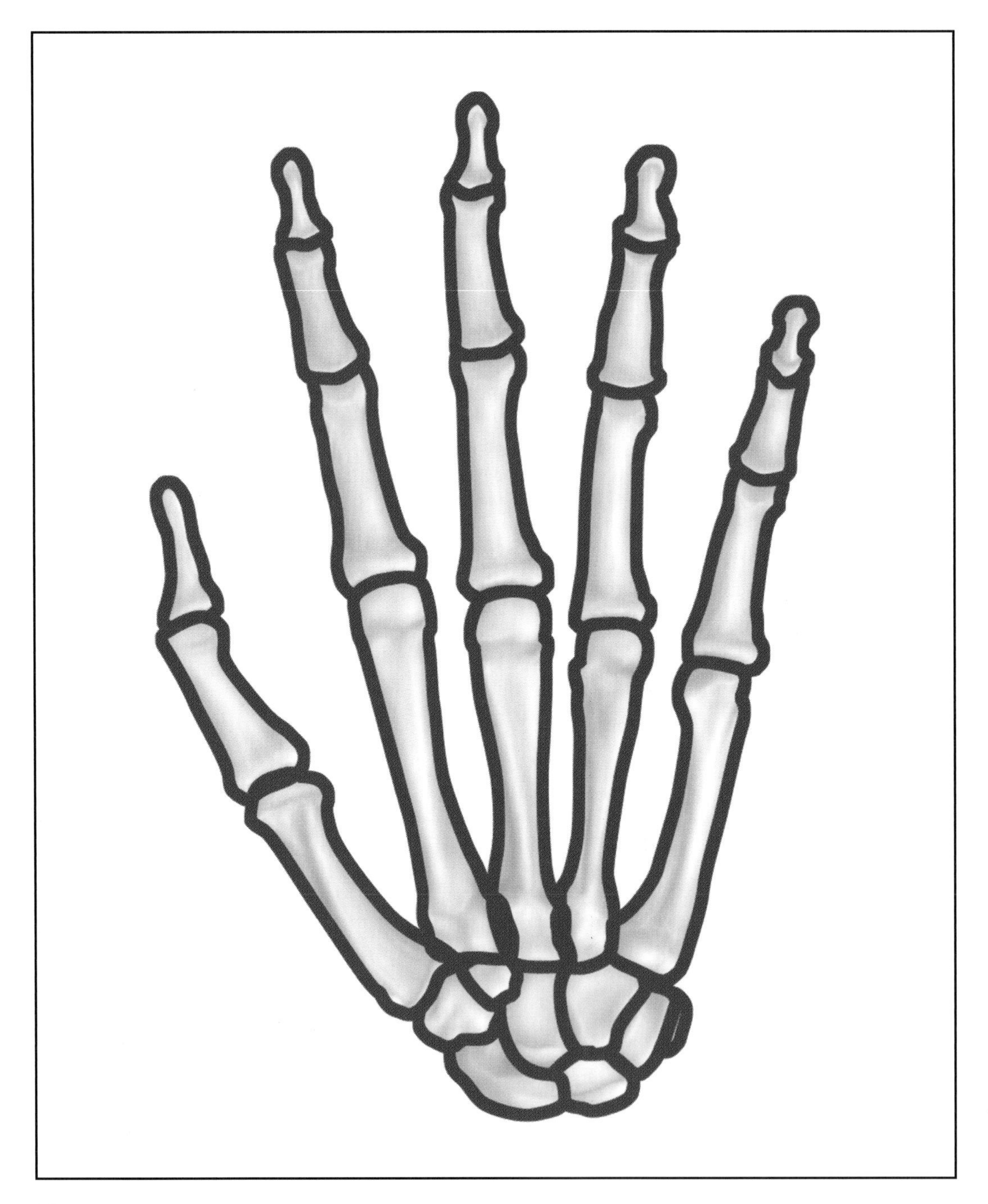

Plate 12
Draw the bones of the hand. In what is probably a gross oversimplification, you may draw the carpal bones as one big block and refine that view as you learn more.

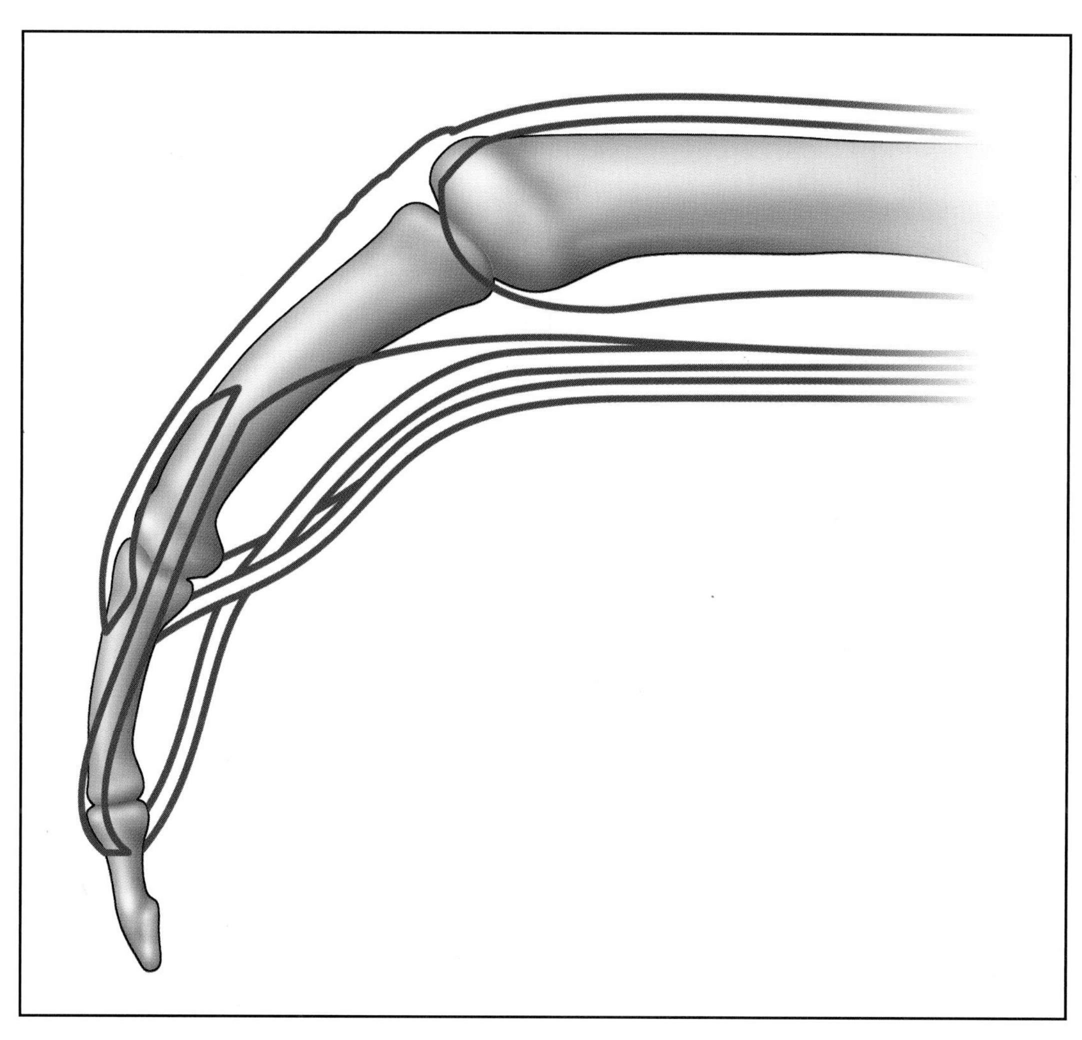

Plate 13
Draw a profile of the finger, noting the insertion of the two flexor tendons, common extensor tendon, and lumbrical. The flexor tendons are shown here without their pulleys. (Obviously, these course closer to the bone.)

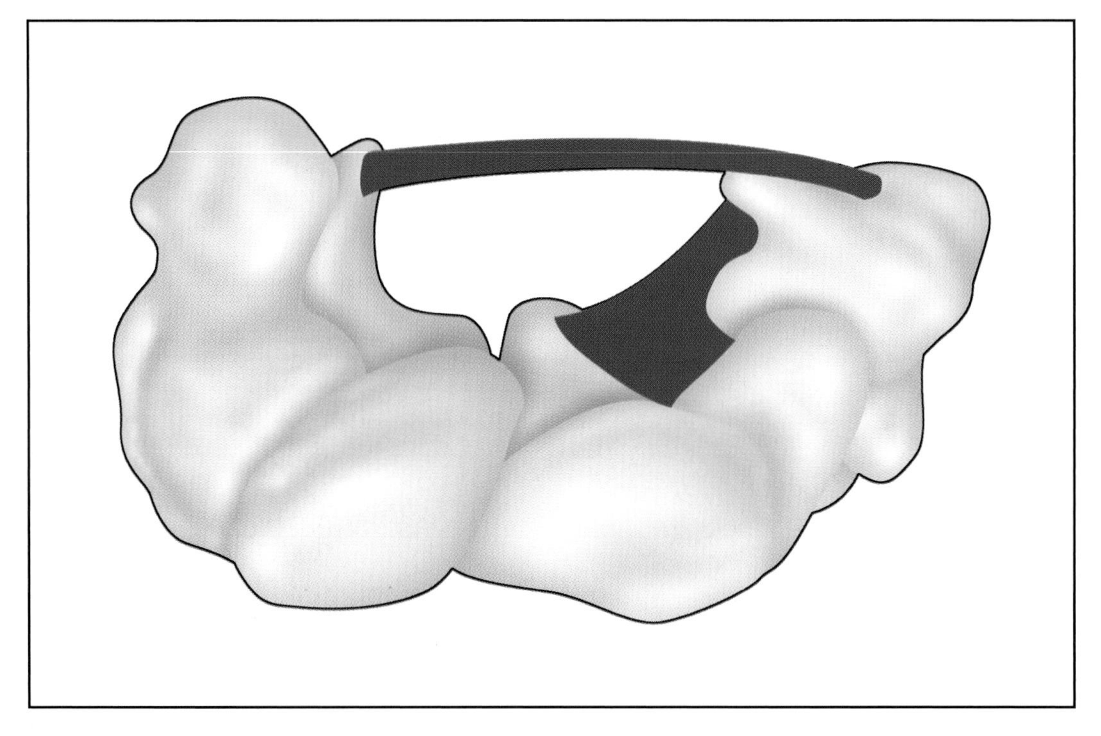

Plate 14

Draw a cross section of the carpal tunnel. Note that the contents include the flexor tendons and median nerve.

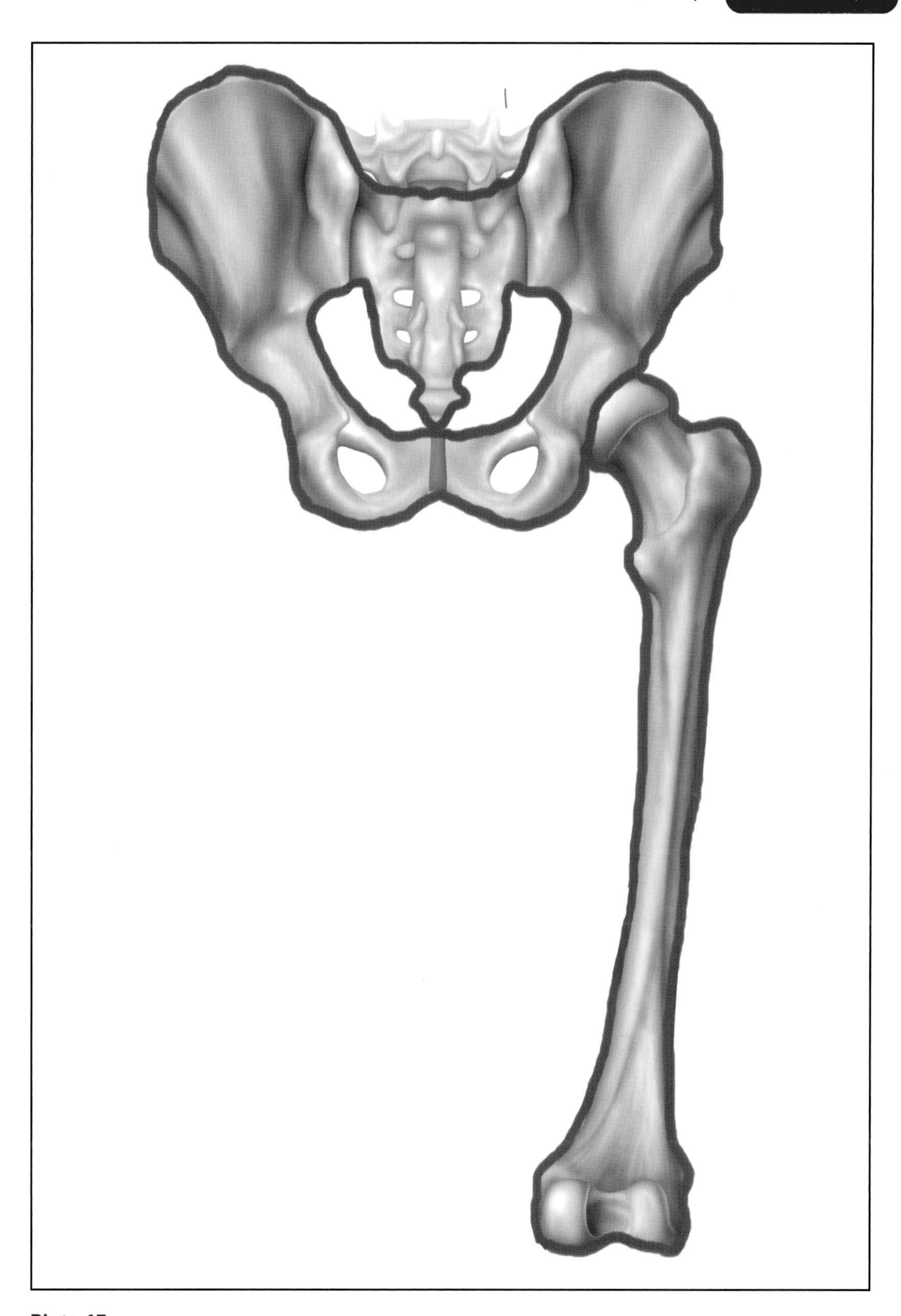

Plate 15
Draw a femur and pelvis. Note the femoral head, femoral neck, and trochanters.

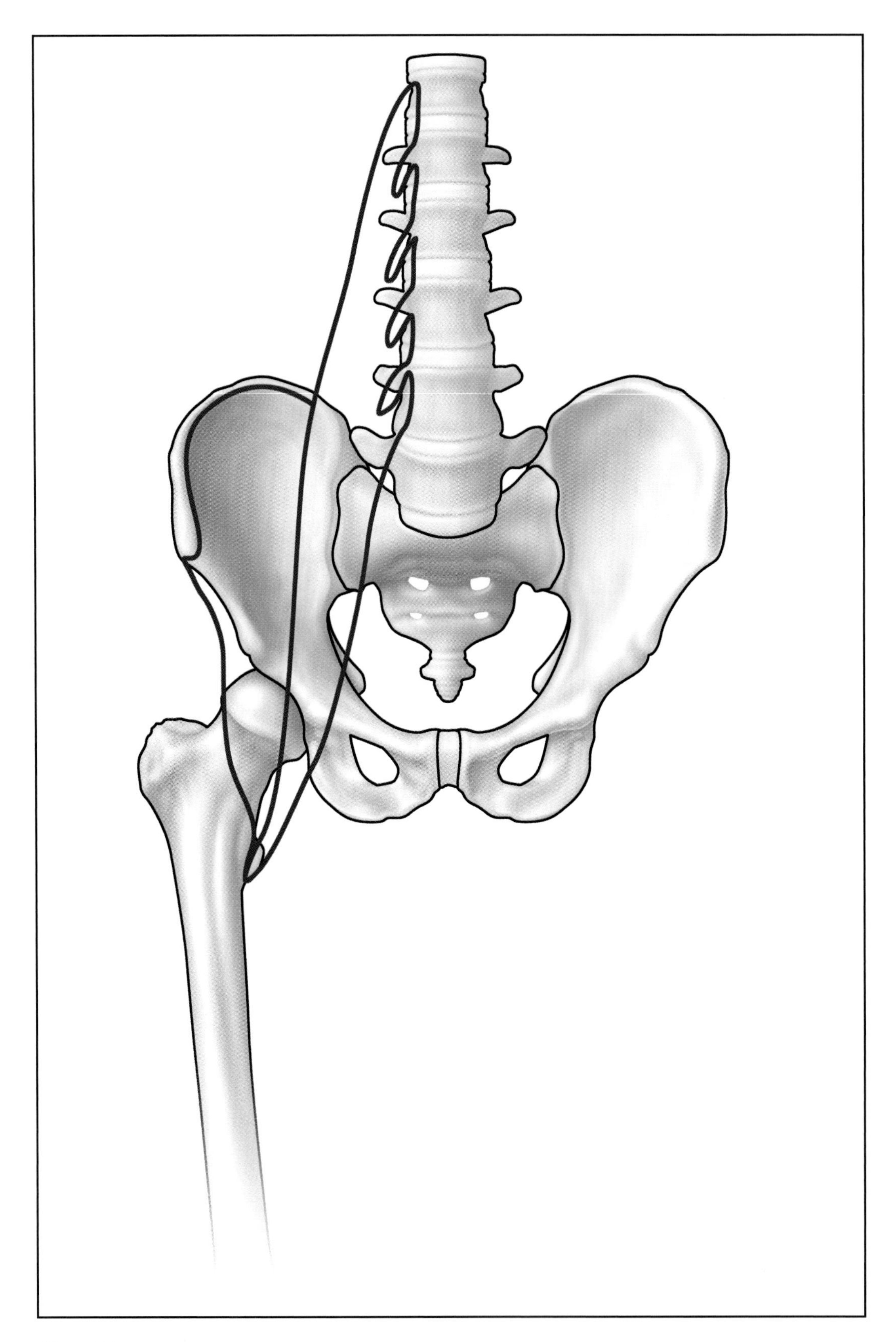

Plate 16
Draw the iliopsoas muscle group. The iliac and psoas muscles actually originate separately but insert together on the lesser trochanter.

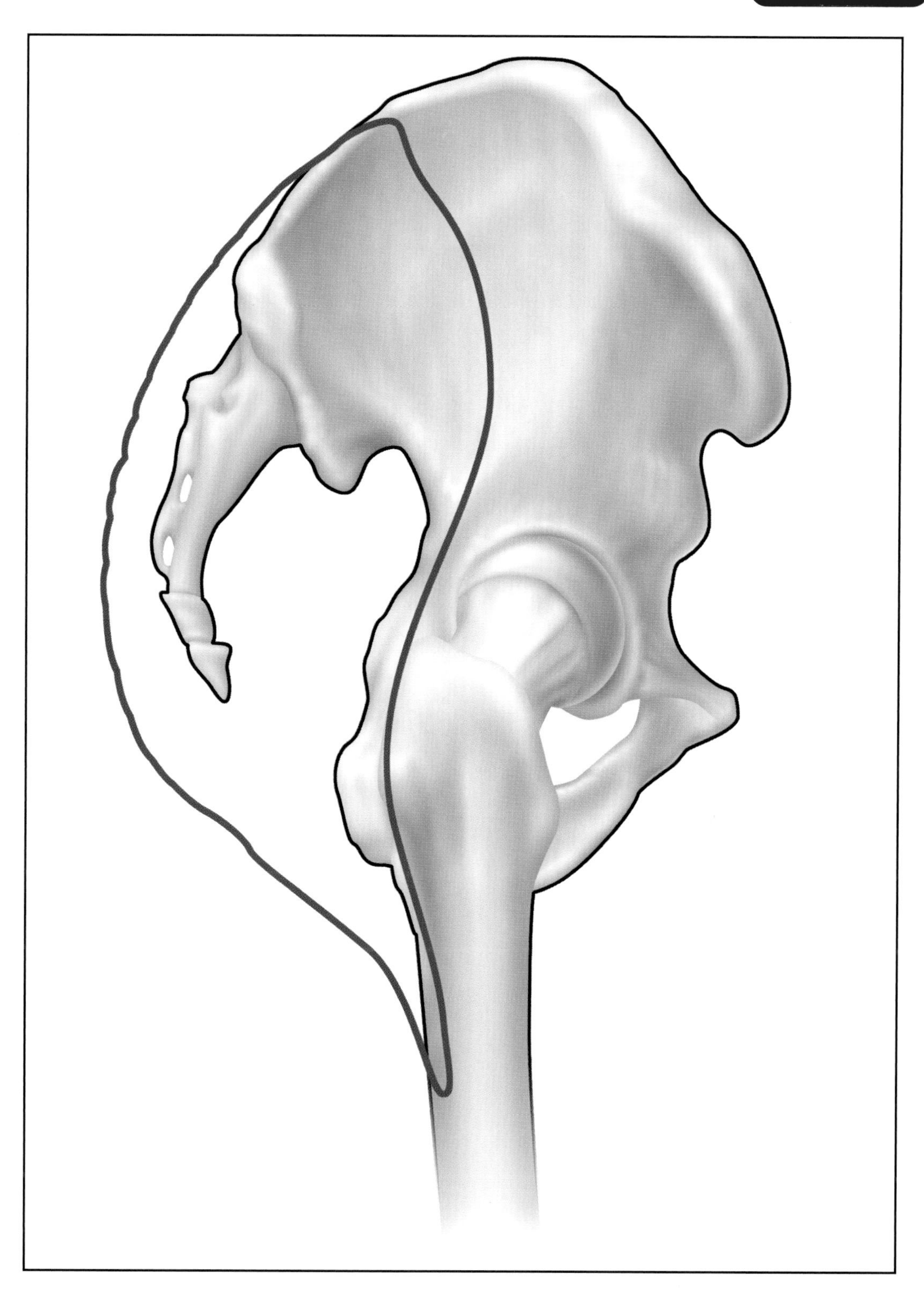

Plate 17
Draw a lateral view of the gluteus maximus.

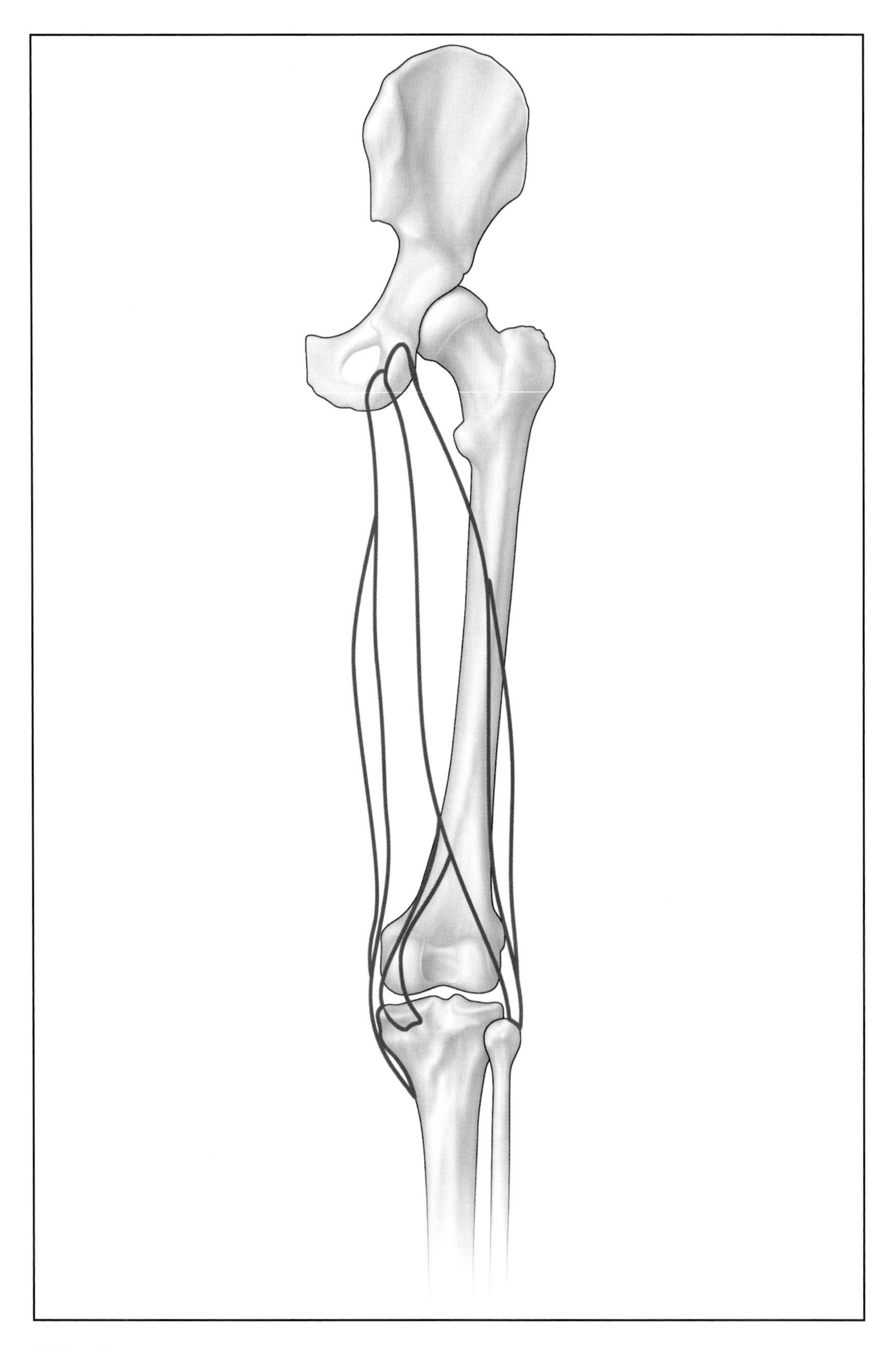

Plate 18
Draw the hamstrings. Note that the semitendinosus courses medially to insert on the anterior surface of the tibia.

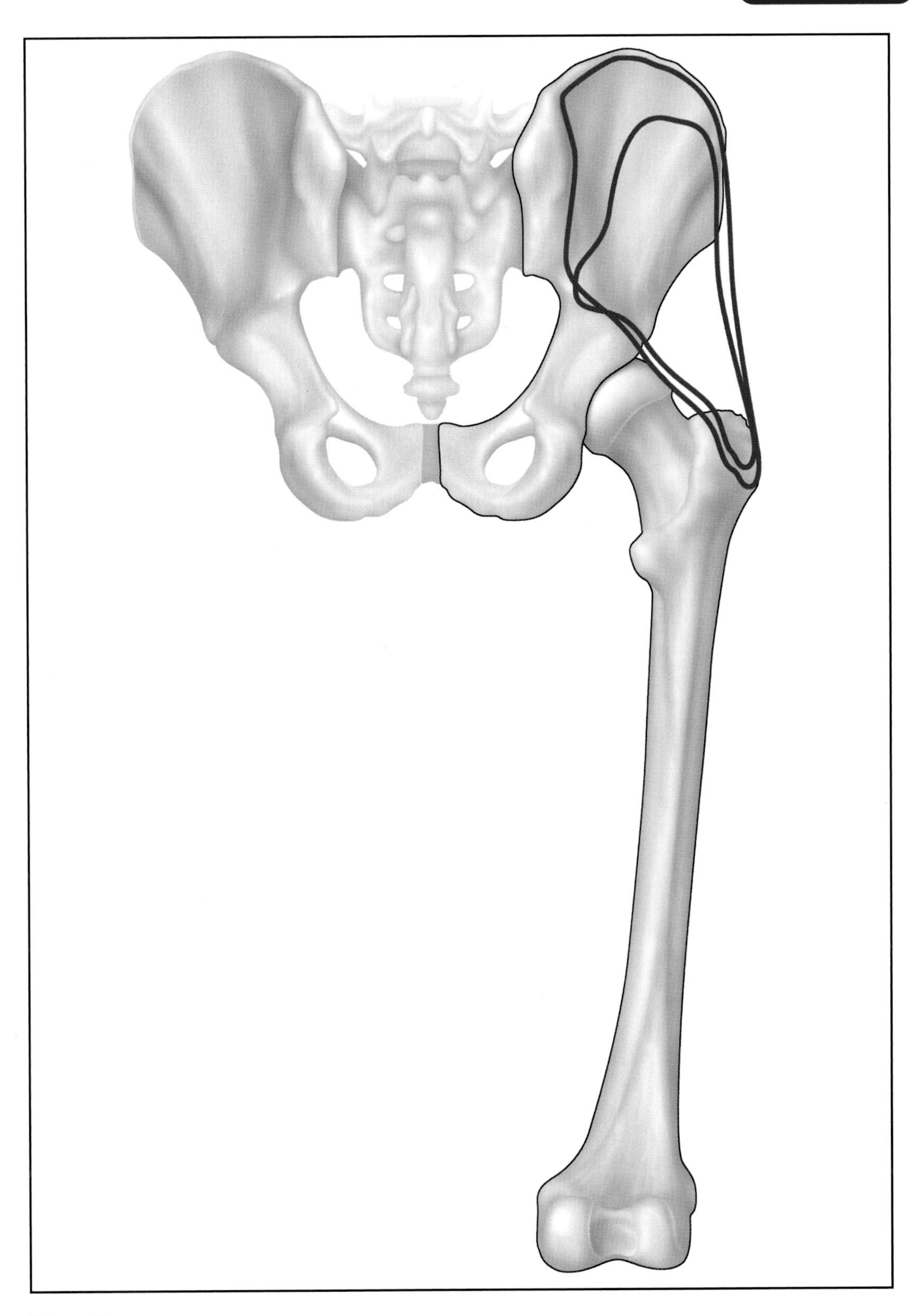

Plate 19
Draw the abductors of the hip: the gluteus minimus and gluteus medius. Note that the gluteus minimus is deep to the medius. If you draw the pelvis tilted downward slightly in the opposite direction, you will see how the abductors work to maintain a level pelvis while standing on one leg.

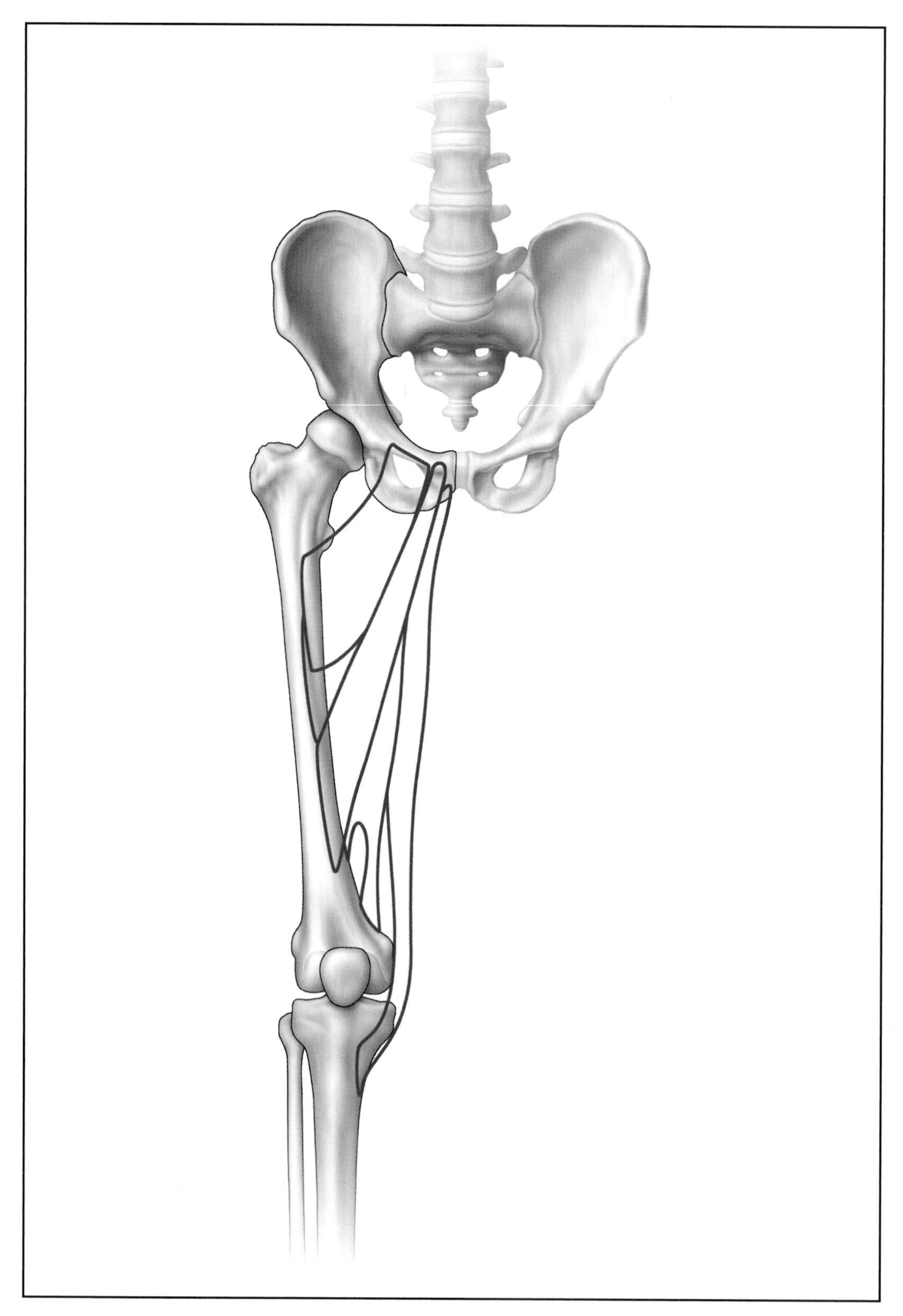

Plate 20
Draw the course of the adductor group, and show the hiatus in the adductor magnus, which allows passage of the femoral artery. Note that the gracilis attaches on the tibia.

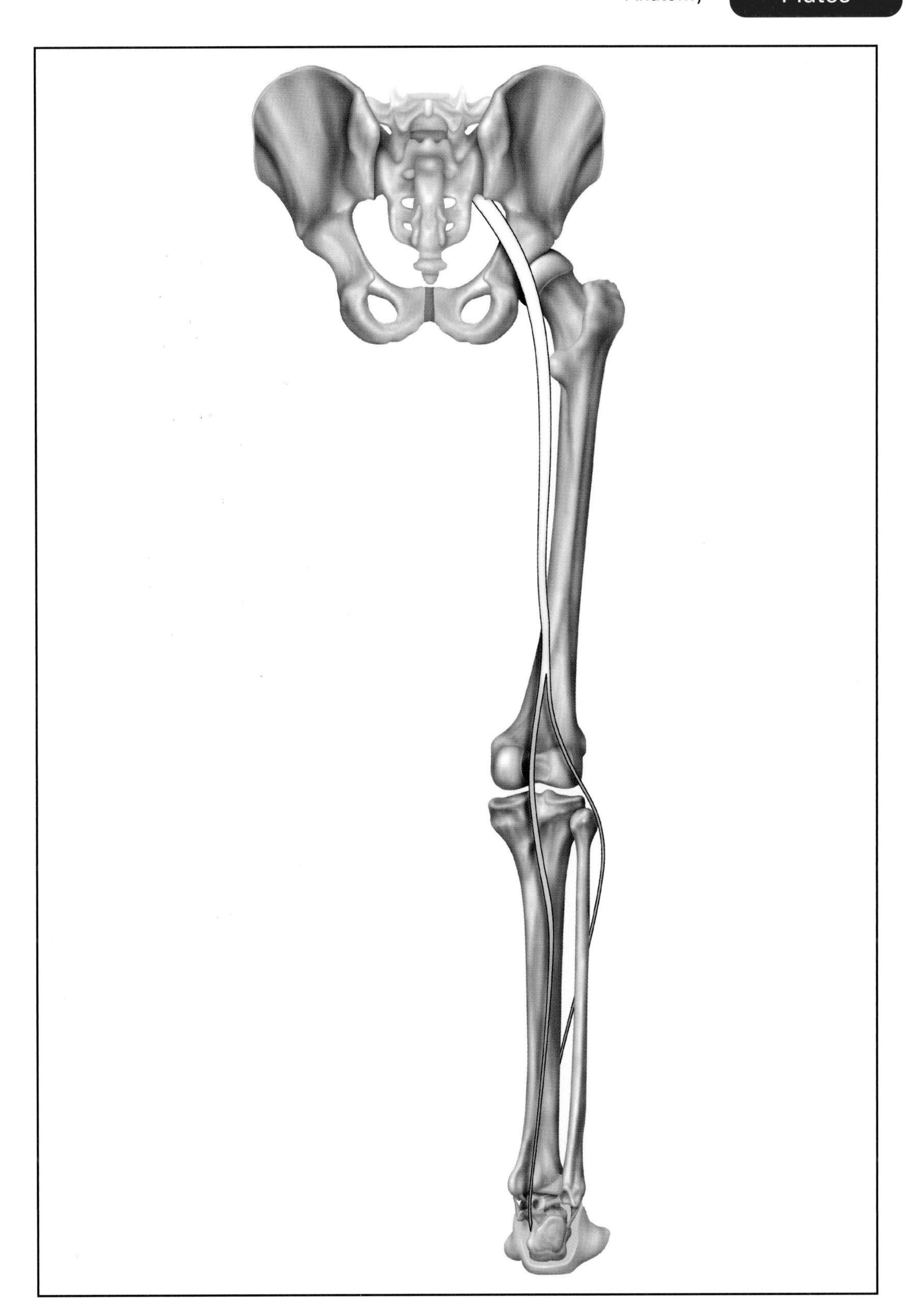

Plate 21
Draw the course of the sciatic nerve as seen from the rear.

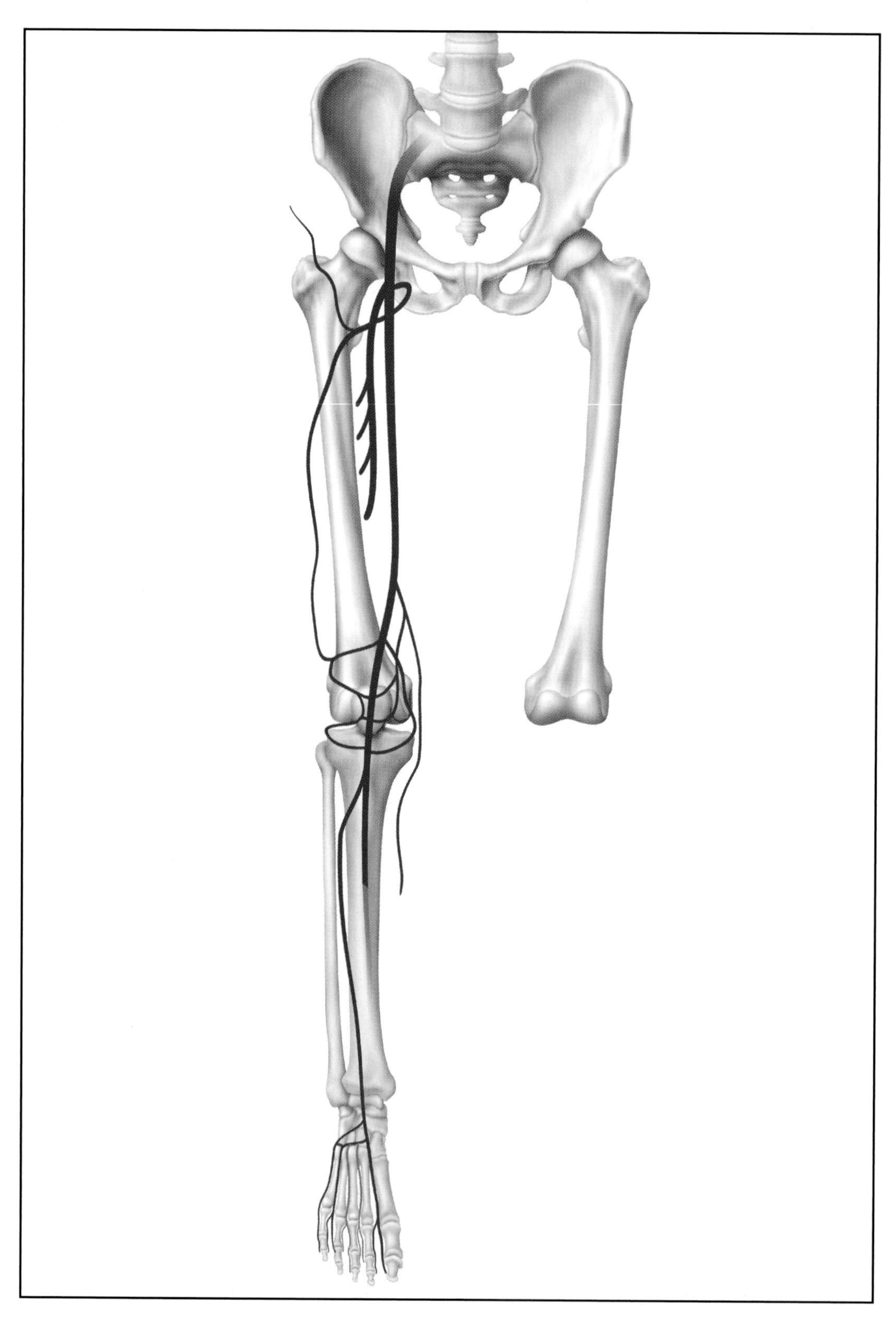

Plate 22

Blood supply to the lower extremity. This plate shows some vessels not seen in the figures including the lateral femoral circumflex near the hip, the perforating branches of the profunda femoris in the thigh, and the genicular arteries near the knee. Also shown is the continuation of the anterior tibial artery (cut off on plate 39) as it becomes the dorsalis pedis in the foot.

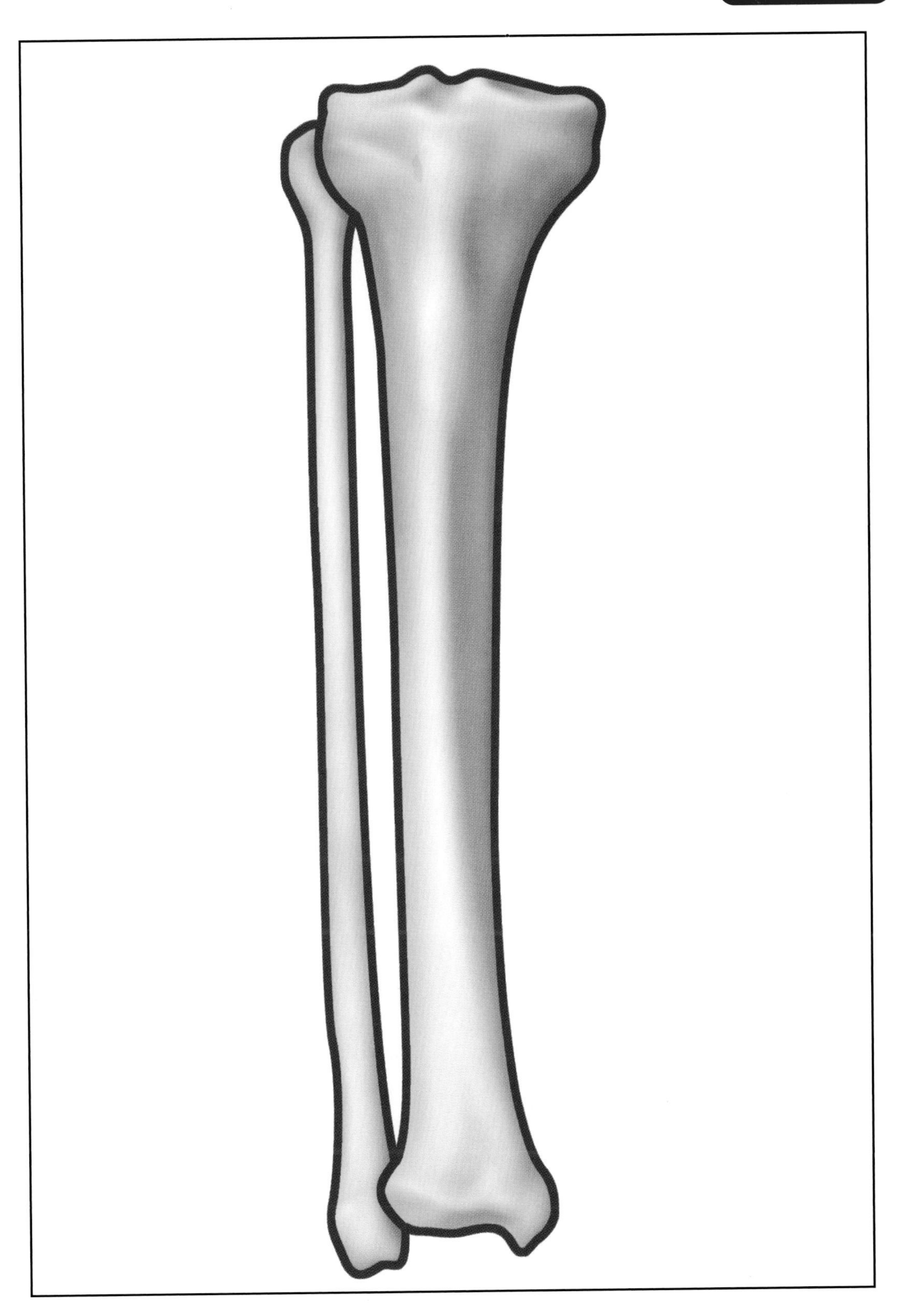

Plate 23
Draw the tibia and fibula.

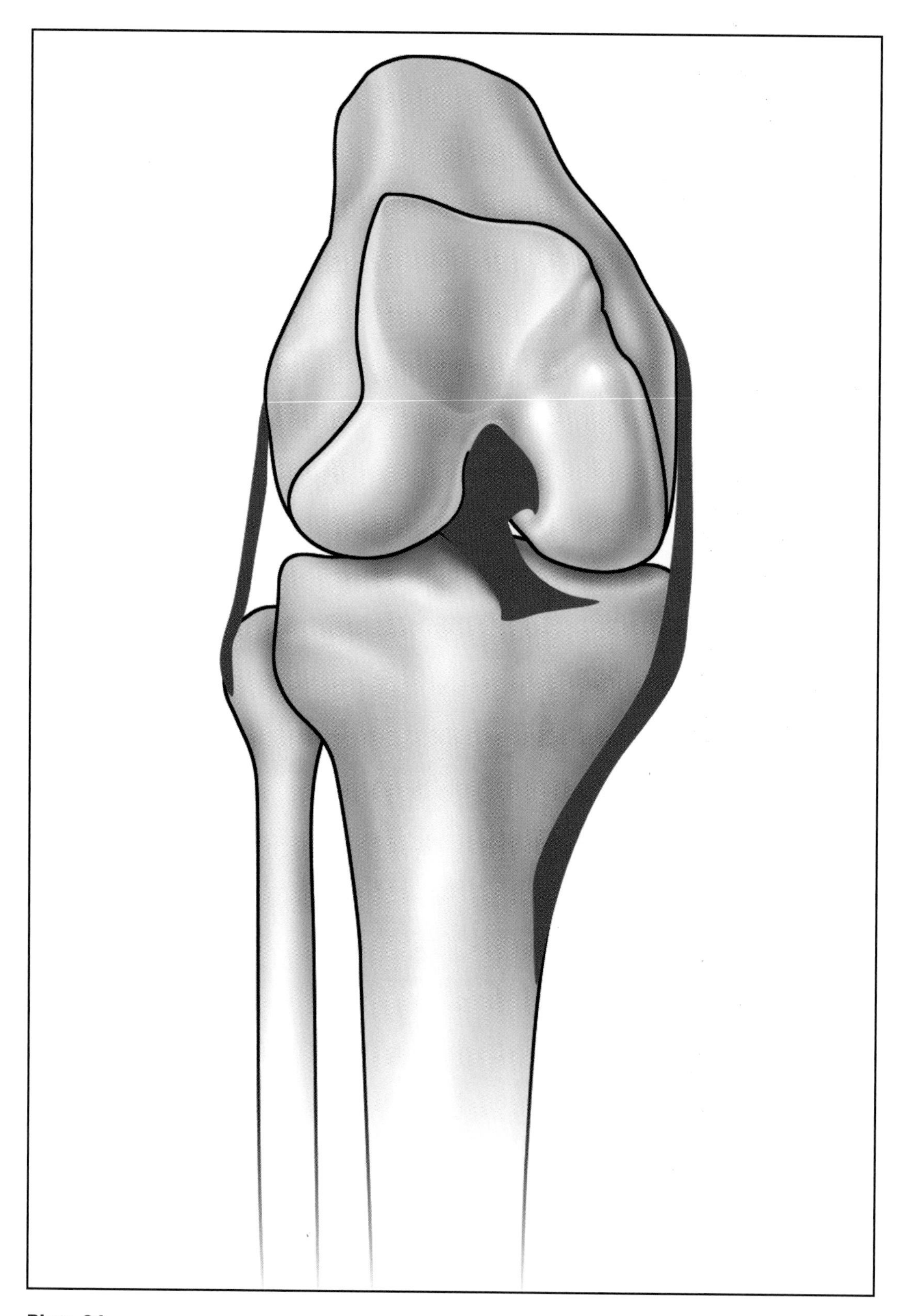

Plate 24
Draw the collateral and cruciate ligaments of the knee. This view should help explain the functional role of the collateral ligaments.

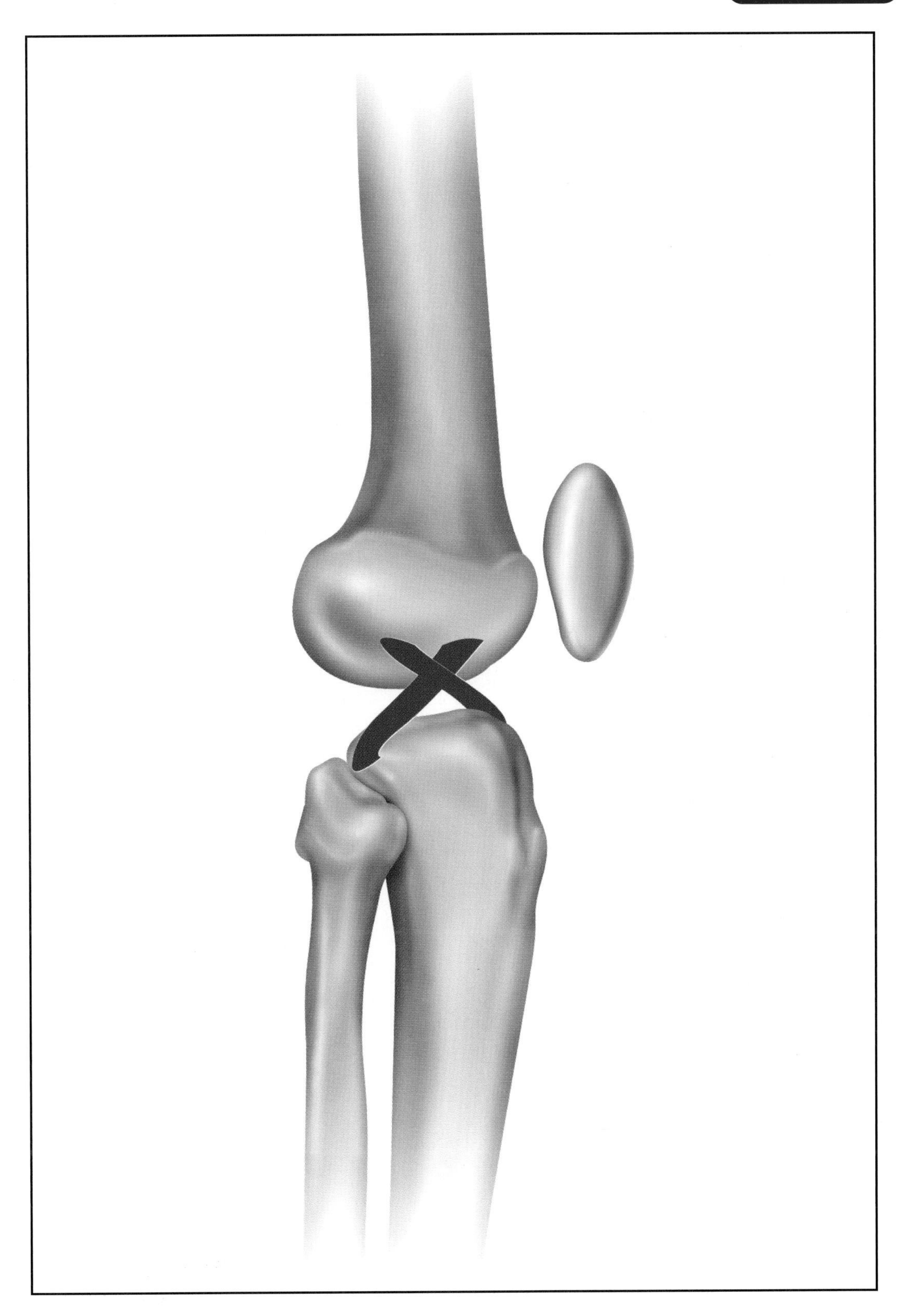

Plate 25
Draw the cruciate ligaments of the knee. This view should help explain how these ligaments resist anterior and posterior displacement of the tibia.

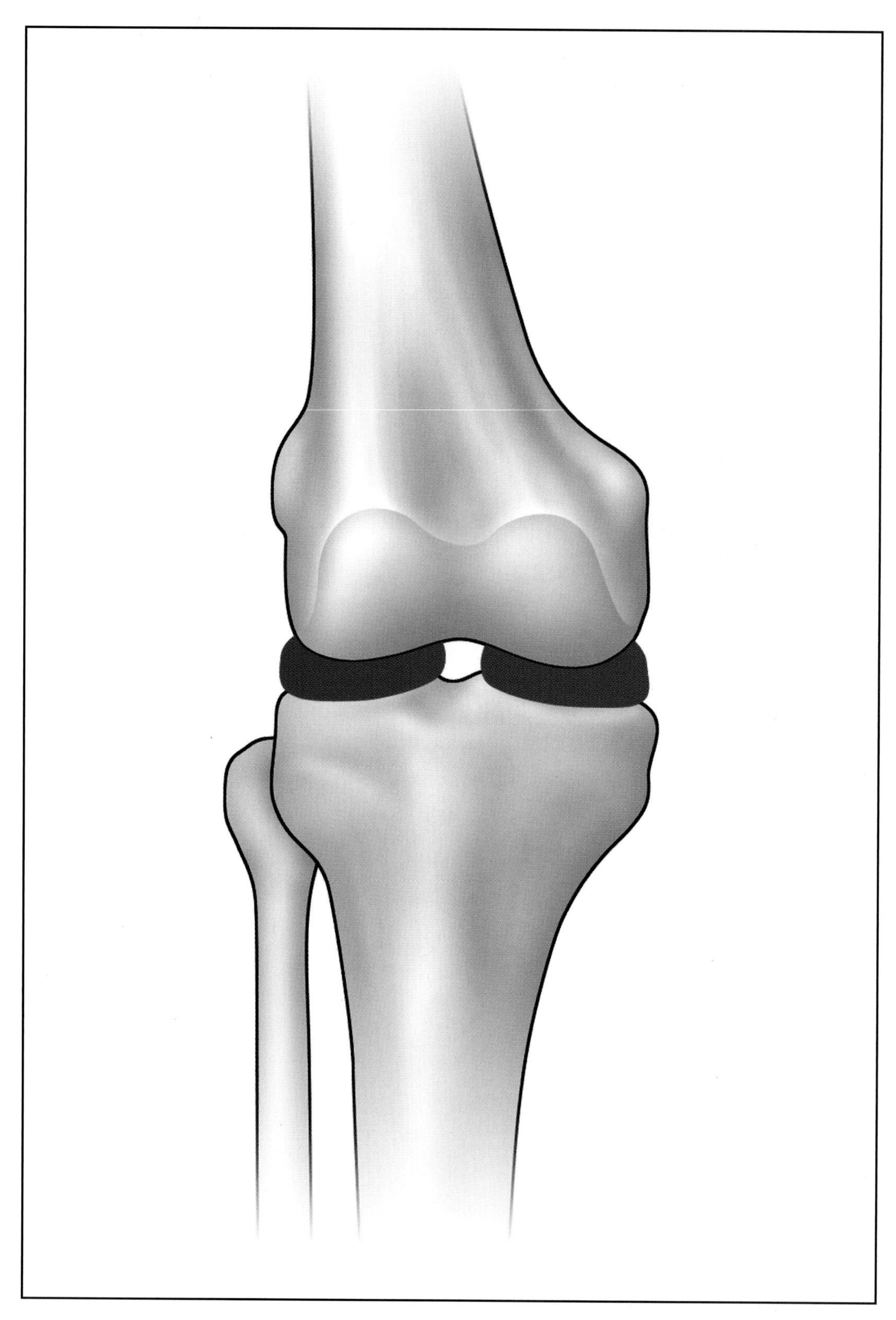

Plate 26
Draw the menisci as seen from the front.

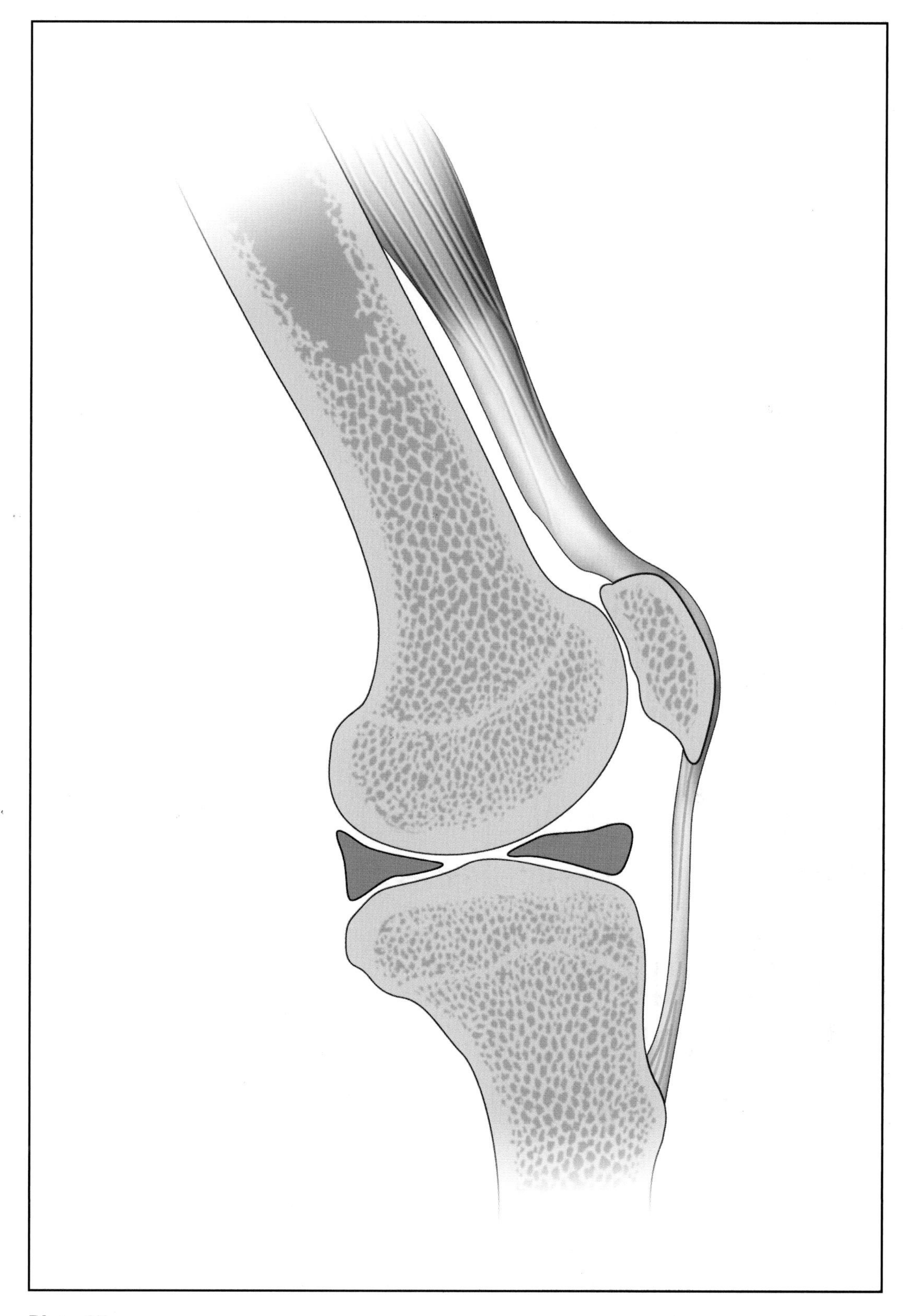

Plate 27

Draw a profile view of the knee showing the quadriceps and menisci. Note that the quadriceps generates torque around the center of the knee and the menisci help stabilize the knee (like a door jamb).

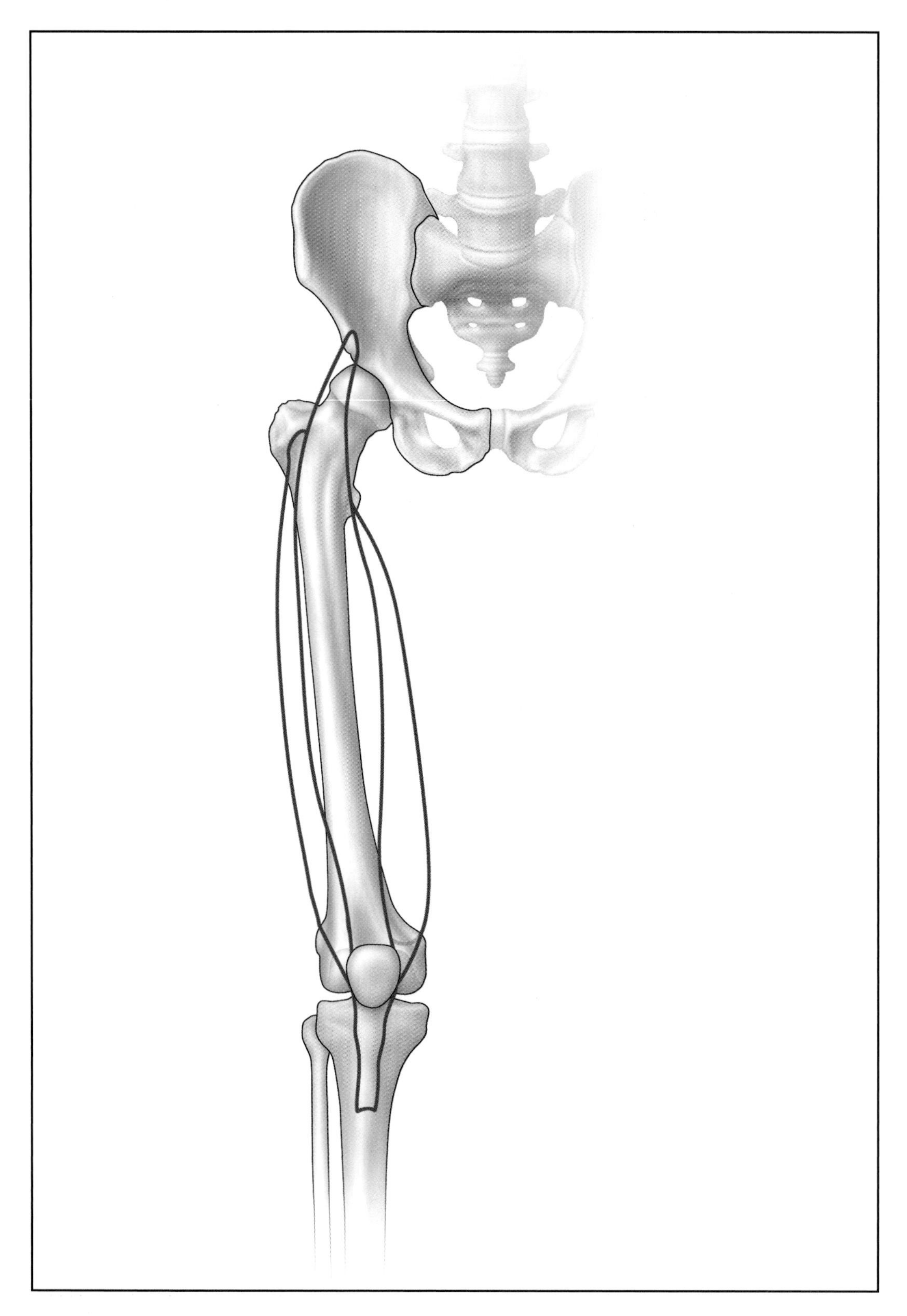

Plate 28
Draw the quadriceps muscle group. Note that the rectus femoris actually originates above the hip and thus flexes it.

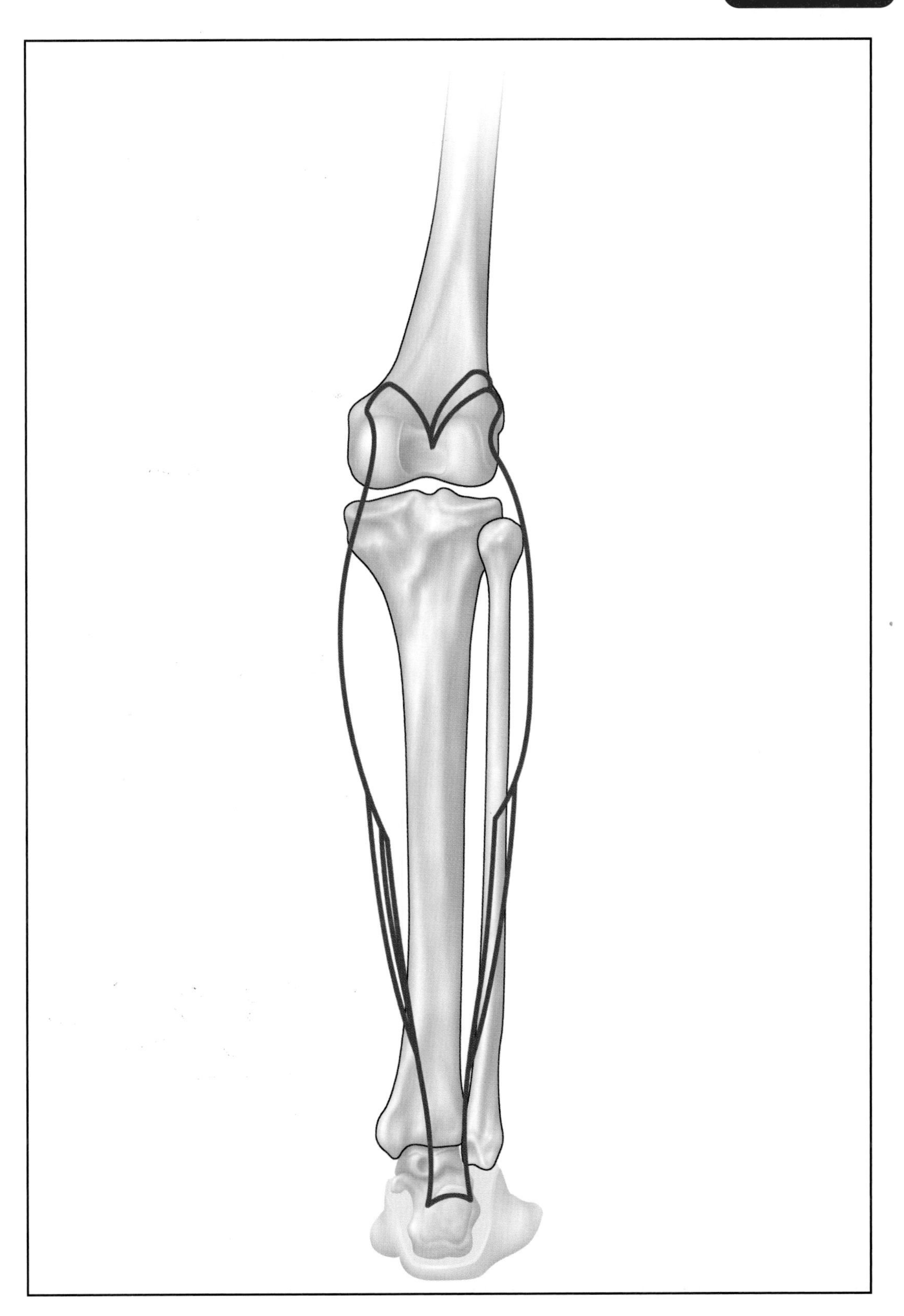

Plate 29
Draw the posterior calf, illustrating the gastrocnemius-soleus complex.

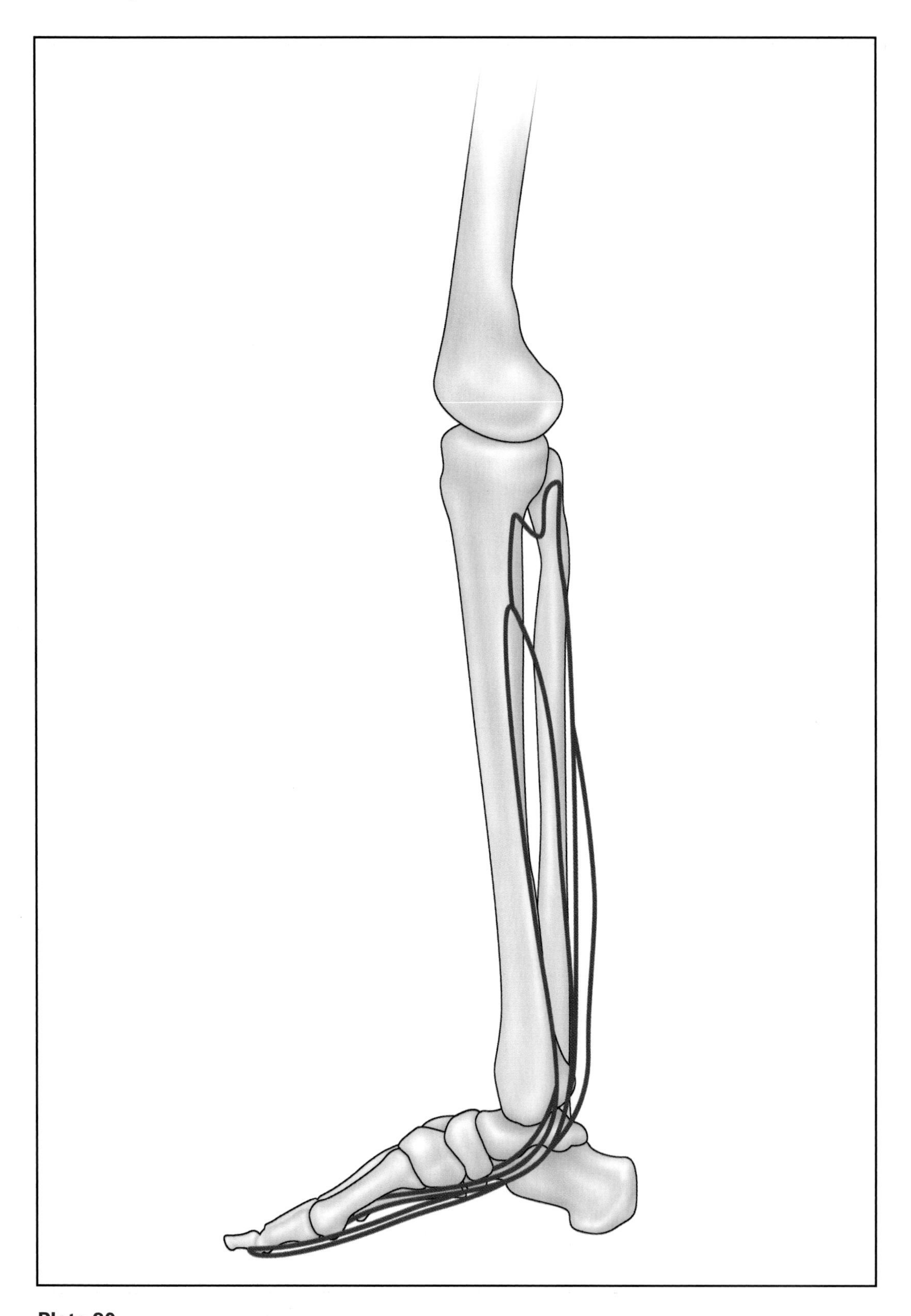

Plate 30
Draw a profile view of the calf, showing the flexor hallucis longus, flexor digitorum longus, and posterior tibialis.

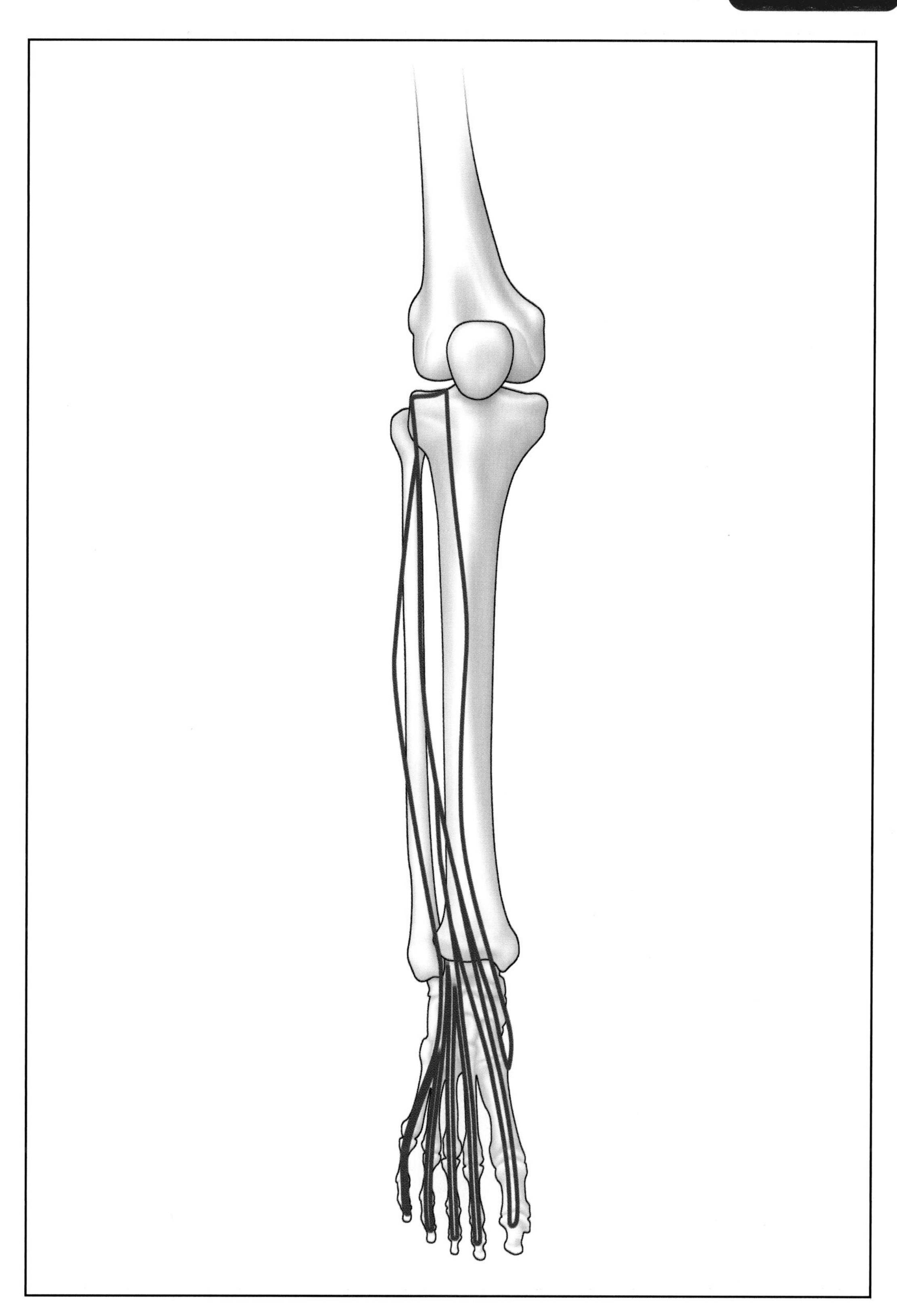

Plate 31
Draw the anterior muscles of the leg.

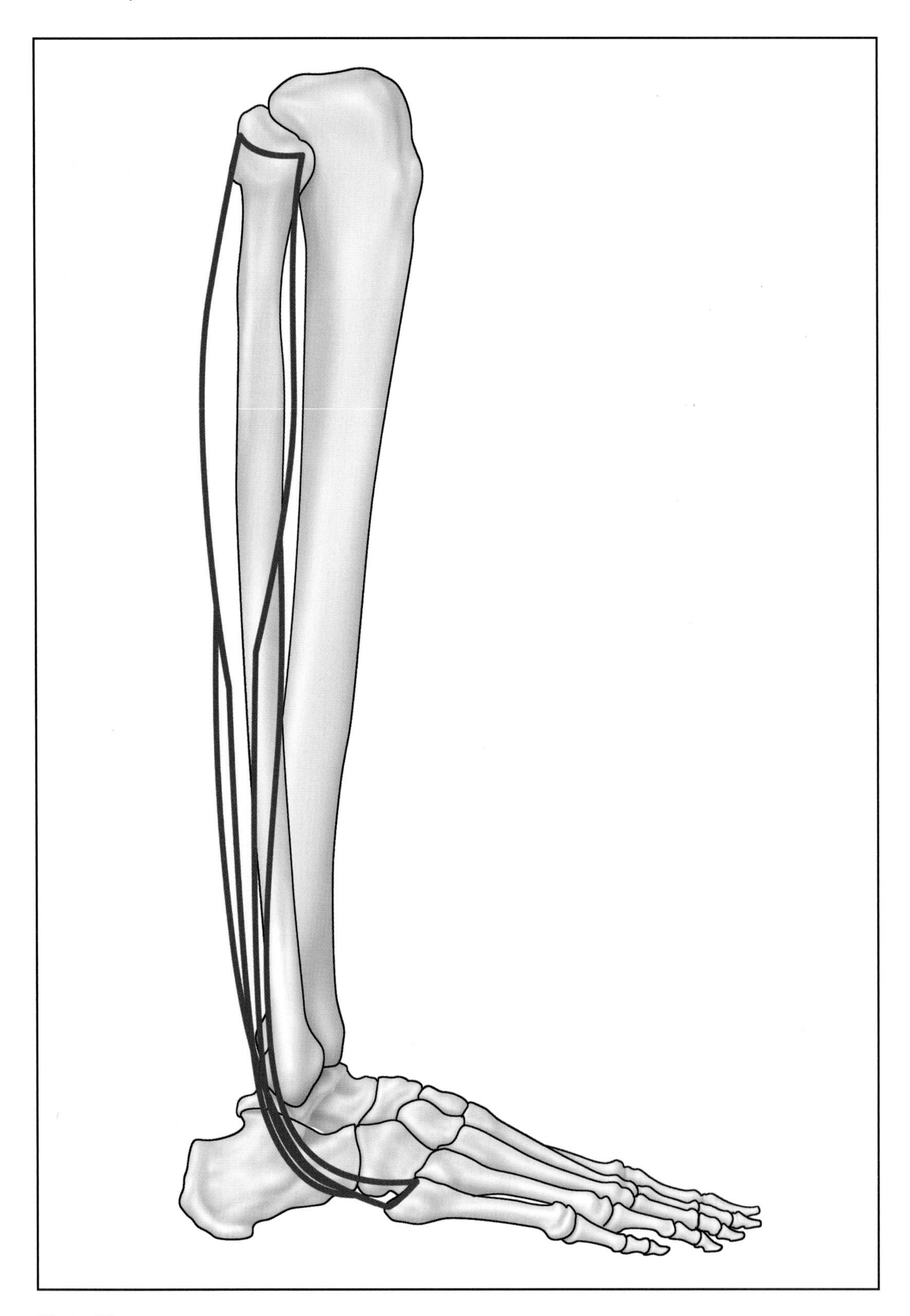

Plate 32
Draw the peroneal muscles. Note that these evert the ankle.

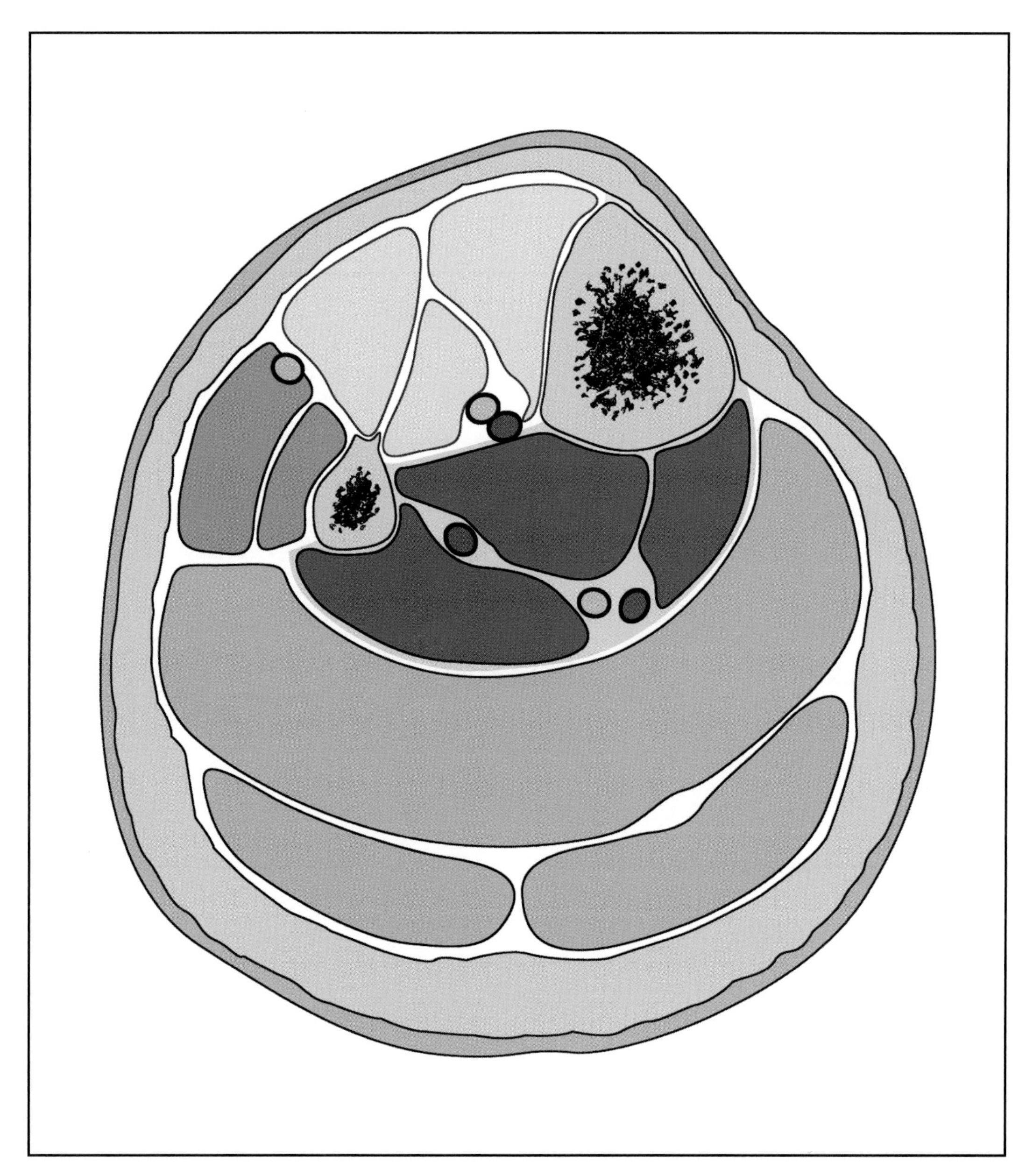

Plate 33

Draw a cross section of the leg illustrating the four compartments. Note that the anterior and deep posterior compartments contain both nerves and arteries, whereas the lateral compartment does not contain an artery and the superficial posterior compartment contains neither arteries nor nerves.

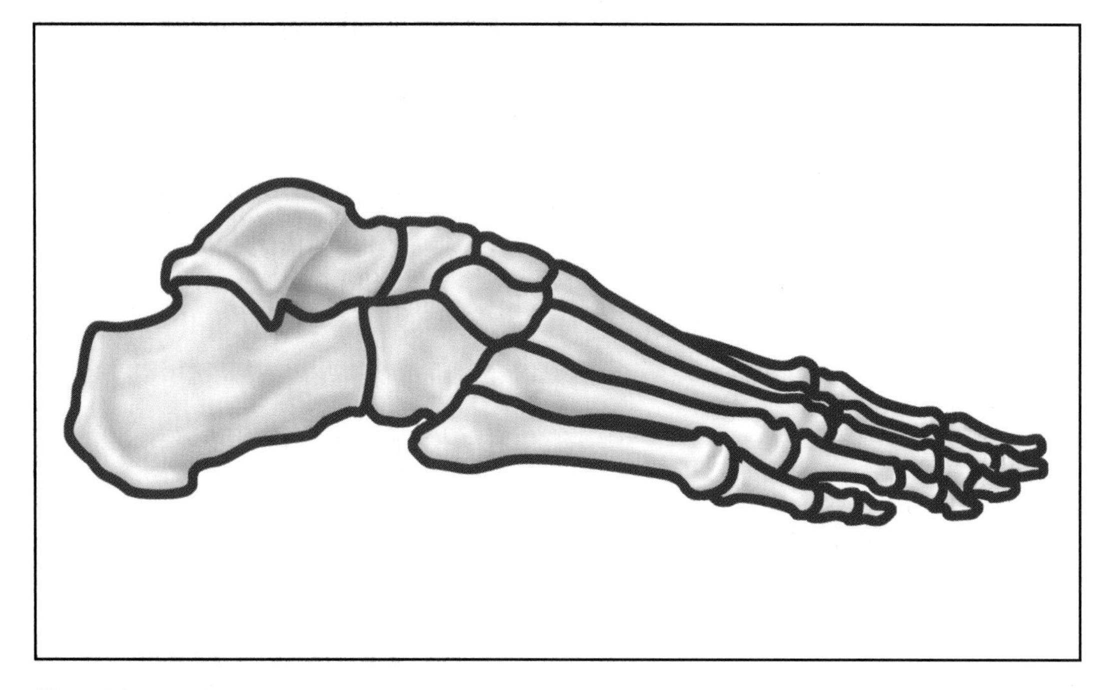

Plate 34
Draw a profile view of the foot.

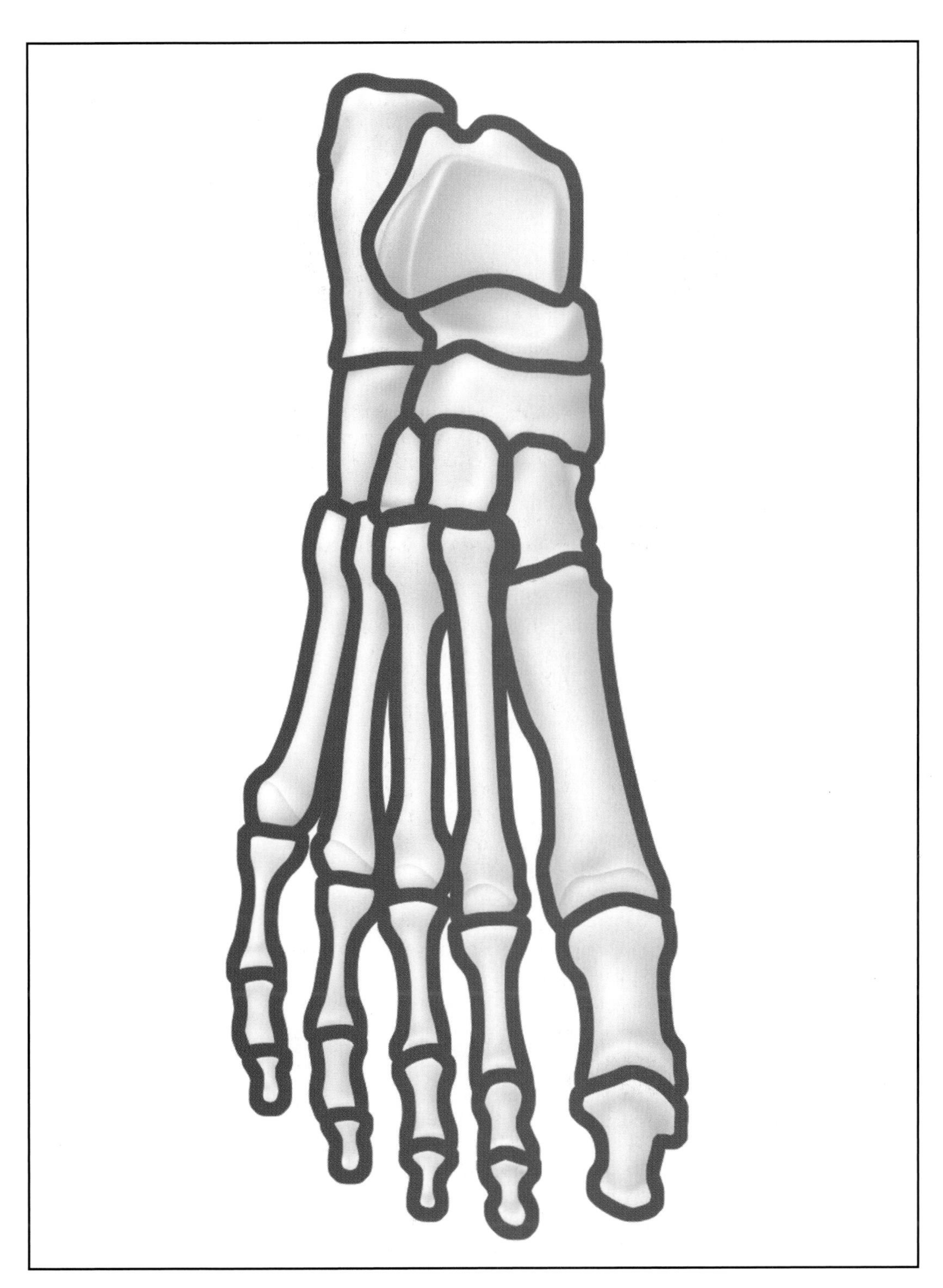

Plate 35
Draw the bones of the foot as seen from above.

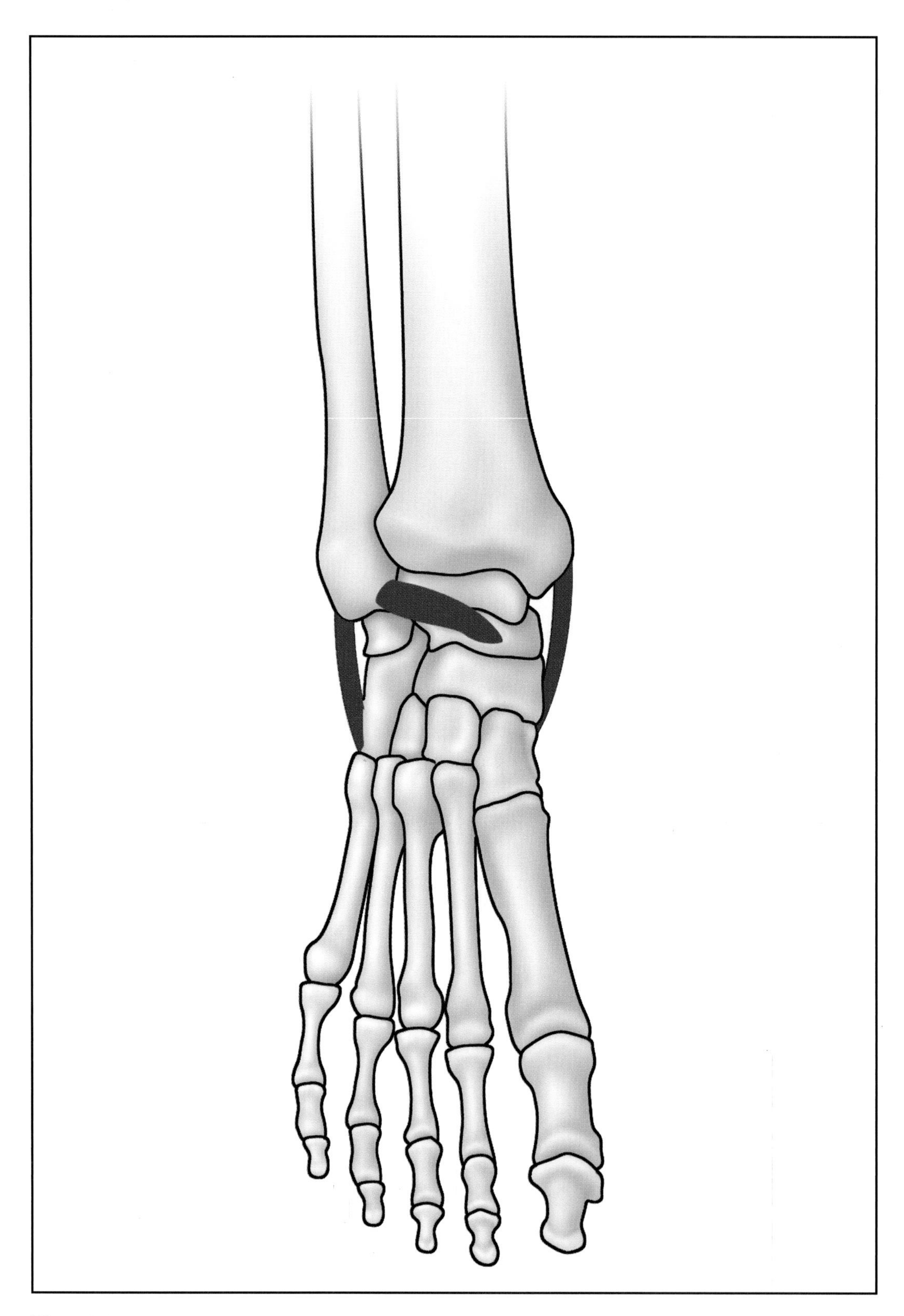

Plate 36
Draw the ankle ligaments as seen from the front.

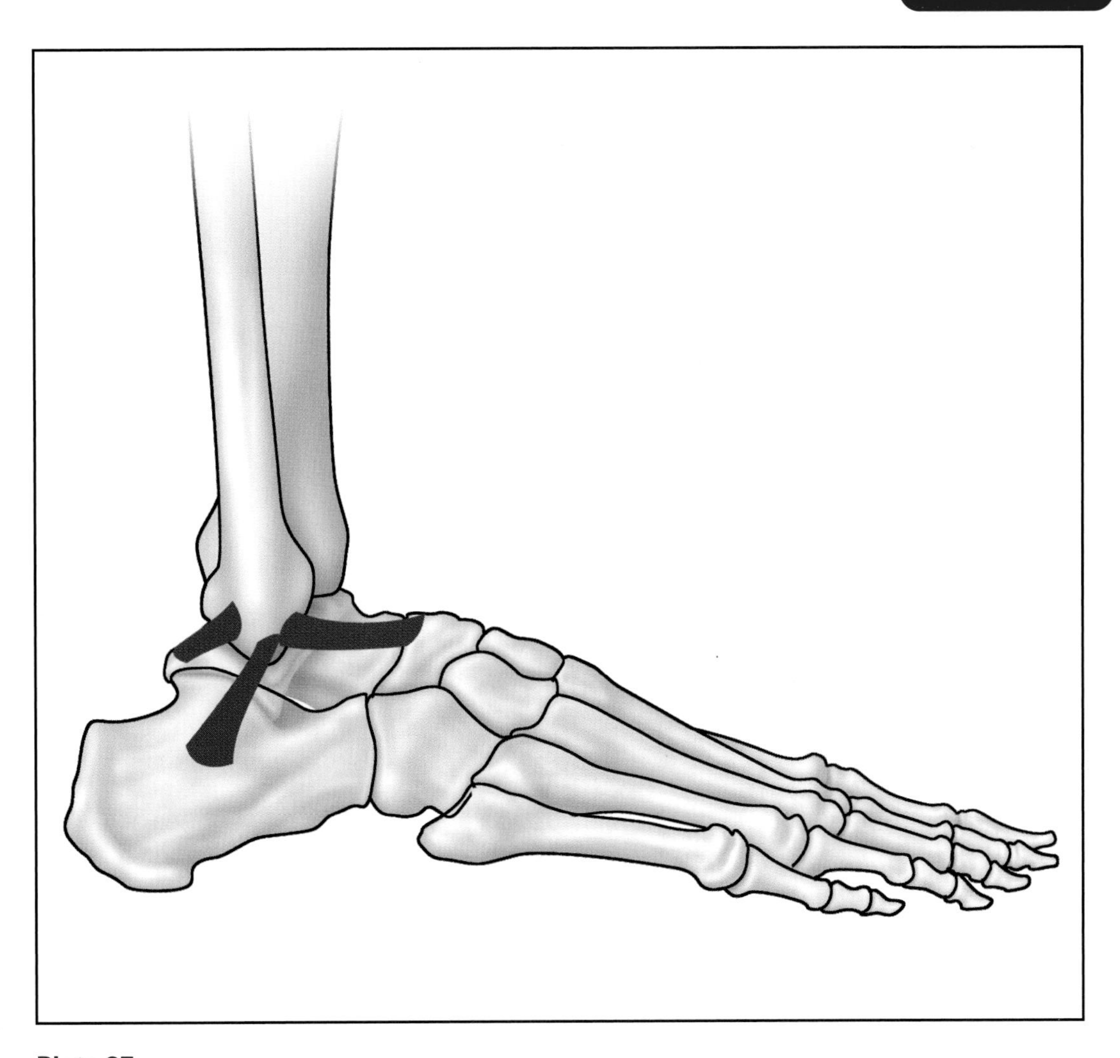

Plate 37
Draw a lateral view of the ankle showing the talofibular and calcaneofibular ligaments.

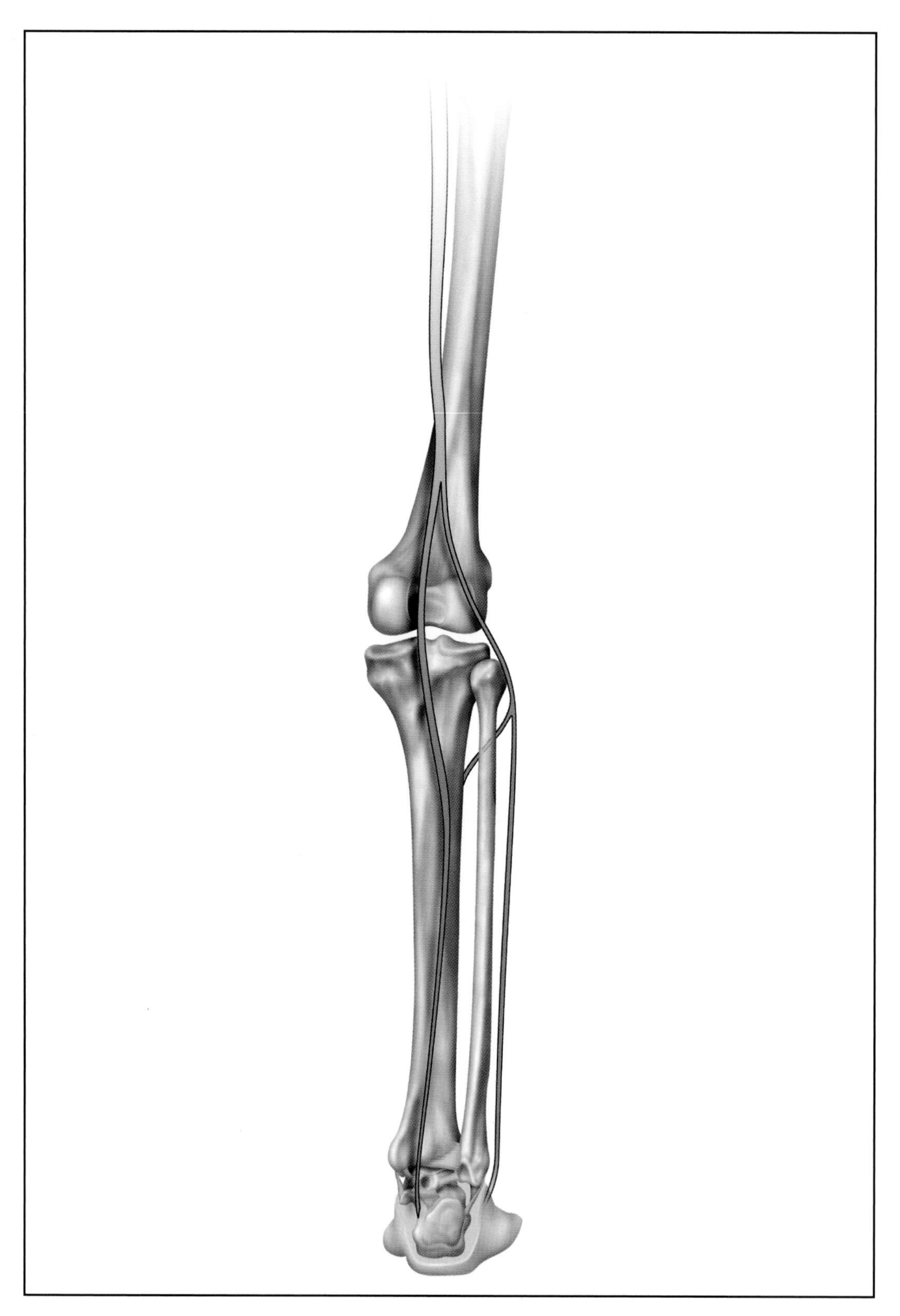

Plate 38
Draw the course of the tibial and peroneal nerves.

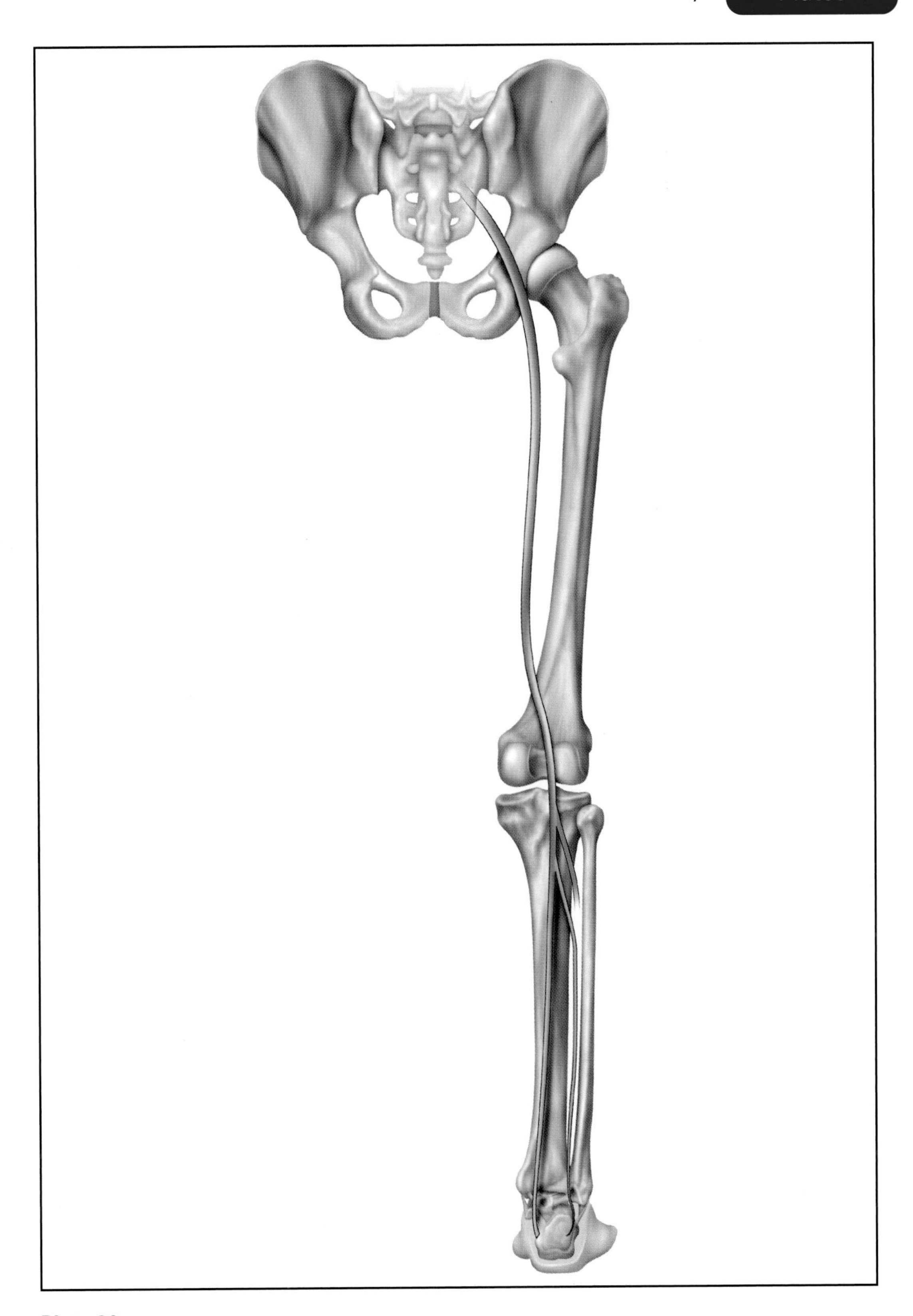

Plate 39
Draw the blood vessels of the leg. Note that the anterior tibial artery breaks from the popliteal artery. The popliteal artery continues as the posterior tibial artery and gives off the peroneal artery.

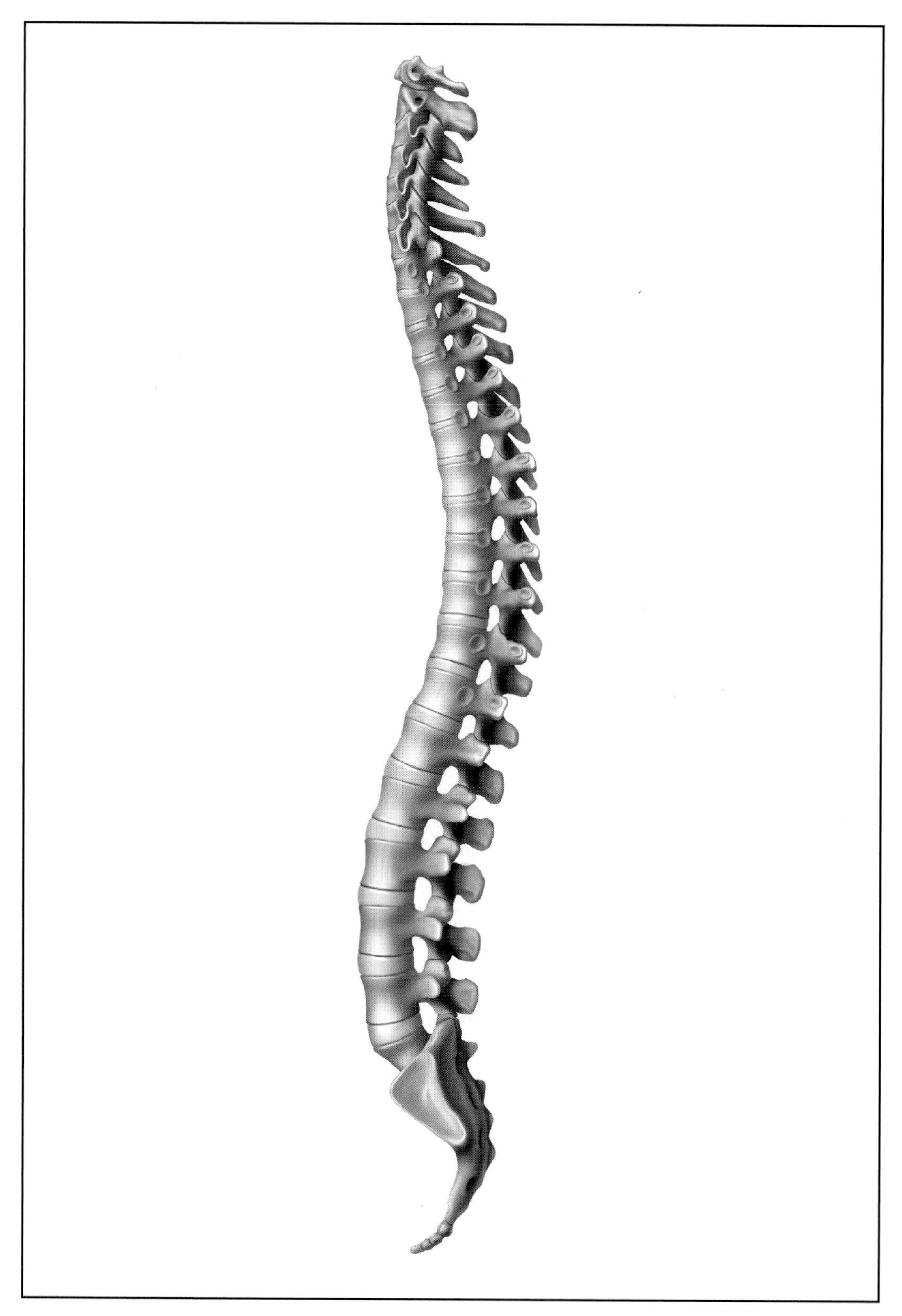

Plate 40

Draw the cervical, thoracic, lumbar, and sacral regions of the spine. Note the directions of the curves (kyphosis versus lordosis).

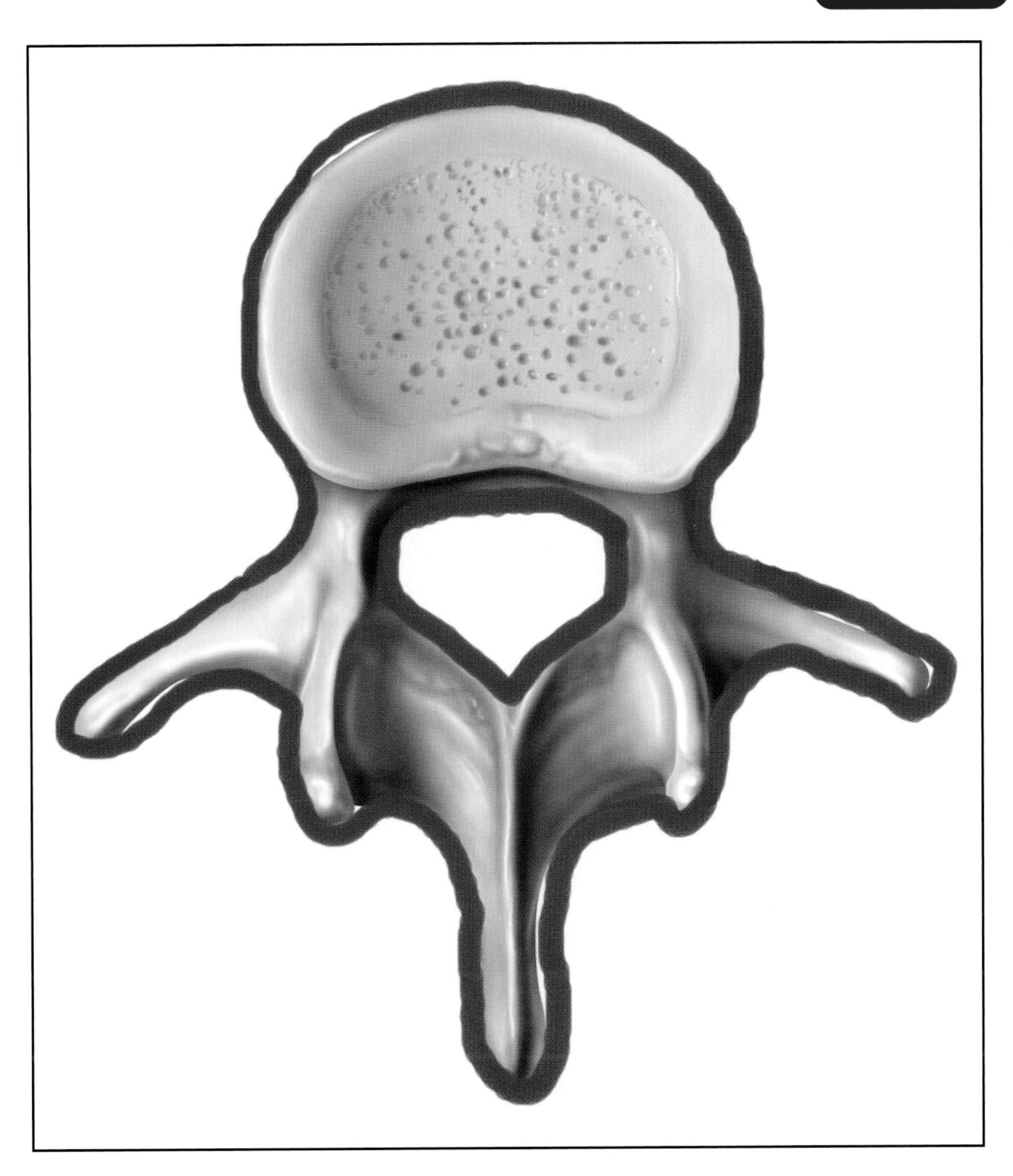

Plate 41
Draw an axial view of a lumbar vertebral body. Note that the space for the spinal cord itself (the vertebral canal) has the shape of a home plate formed by the body in front, the pedicles on the sides, and the lamina in the back.

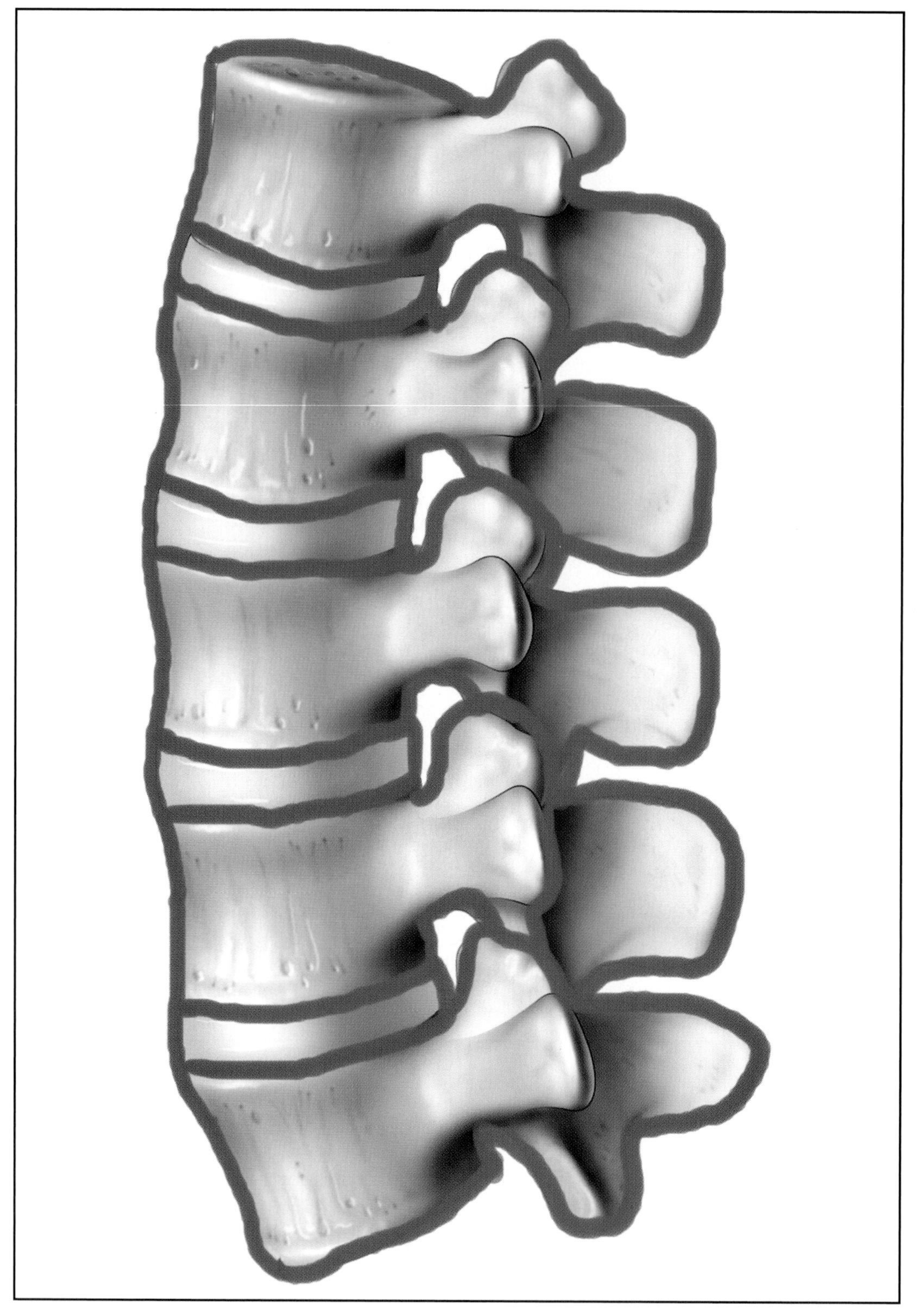

Plate 42
Draw a lateral view of the lumbar spine. Note how the inferior and superior articular processes articulate and how the pedicles do not touch, thereby forming the foramen. Note also how the vertebral body and disk interact.

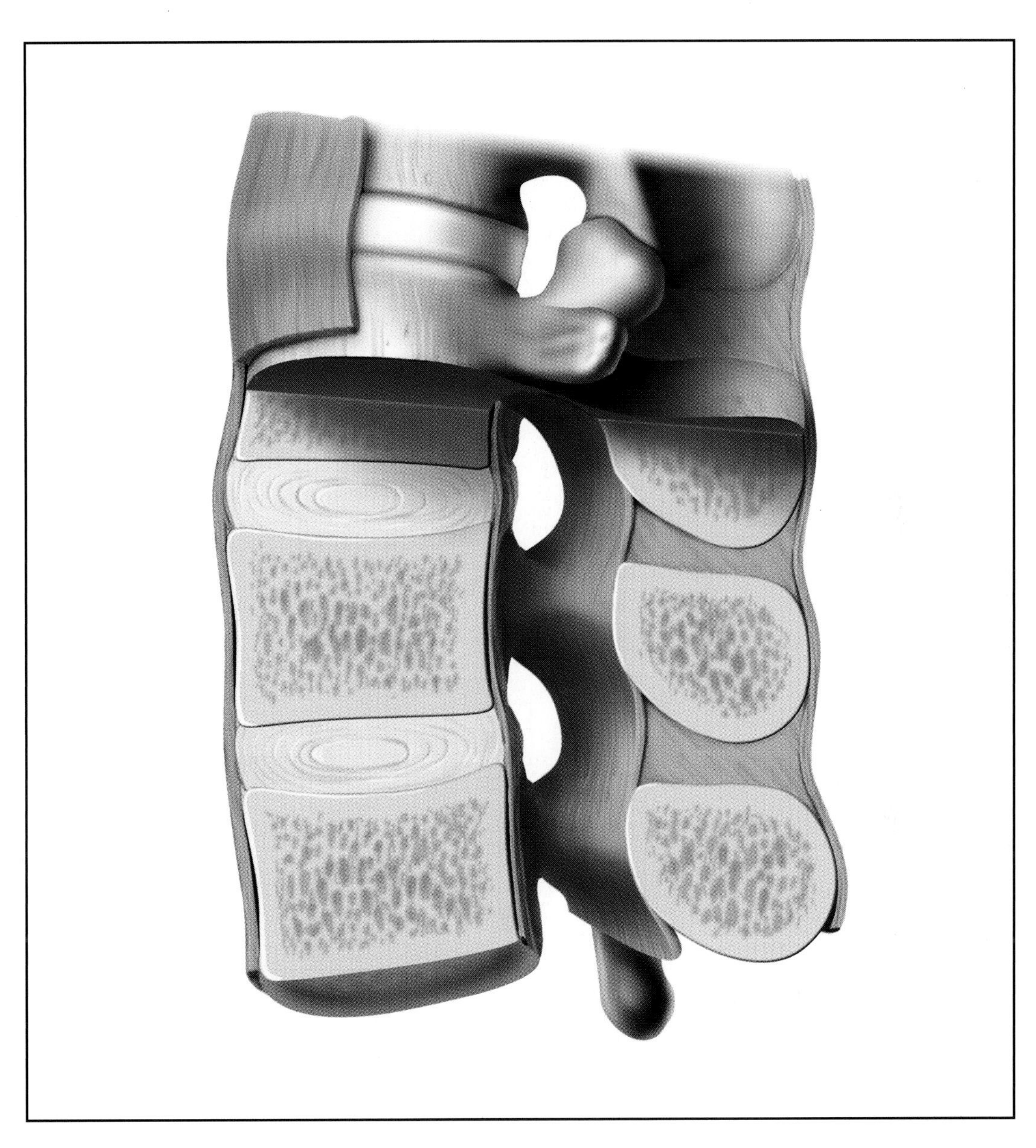

Plate 43

Draw a cut-away profile view of the lumbar spine to demonstrate the following ligaments, all shown here in red: the anterior longitudinal, posterior longitudinal ligaments (adjacent to the vertebral body, front and back), the ligamentum flavum (adjacent to the lamina, in front of the spinous process), and the interspinous and supraspinous ligaments (between and behind the spinous process).

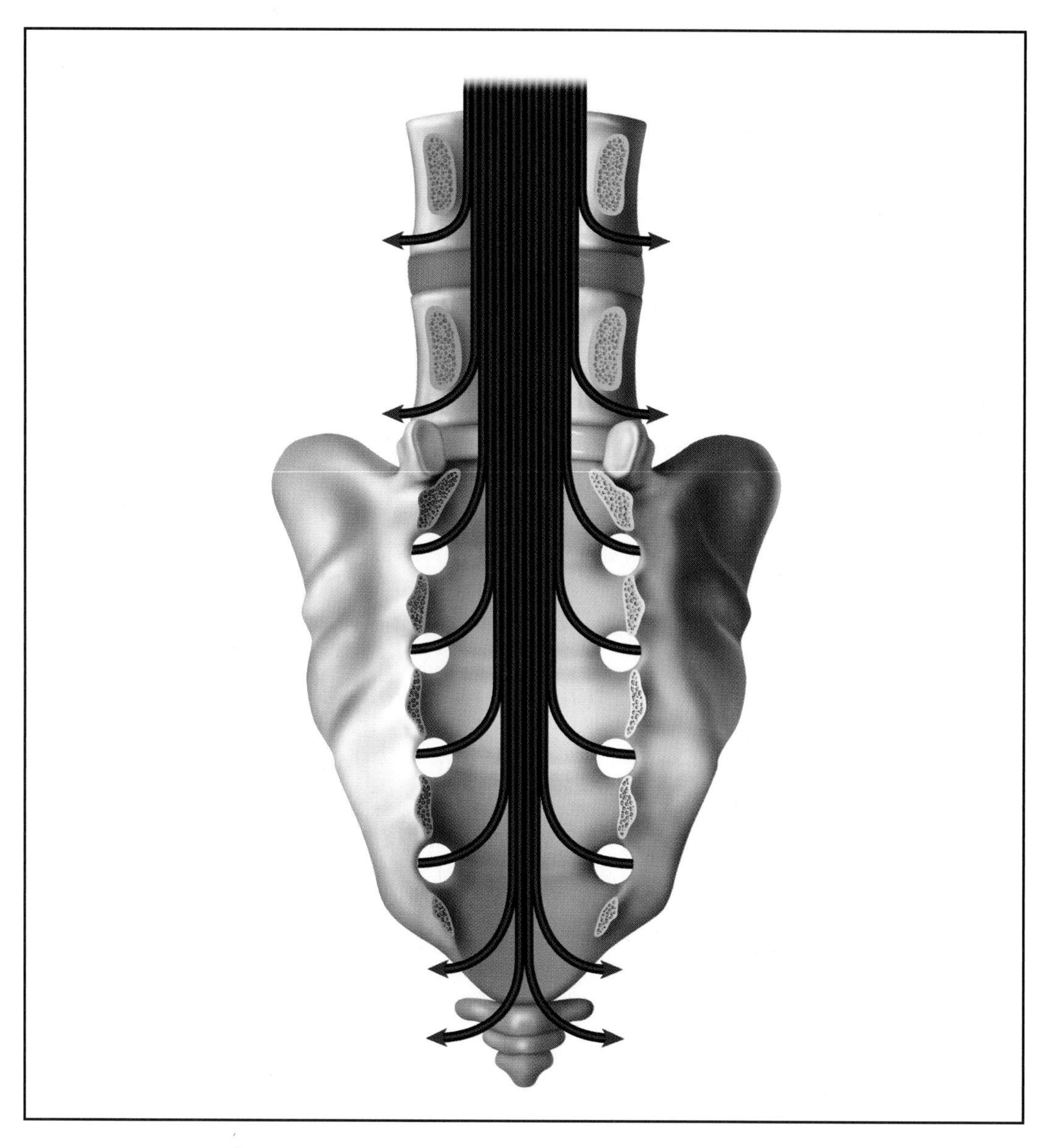

Plate 44

Draw a posterior view of the lumbar spine with the posterior elements cut away to show the nerve roots existing under the pedicles. Note that at the L4-5 level (shown here in red) both the L4 and L5 nerve roots are present, but the root above is always more lateral. Therefore, a disk herniation at L4-5 may hit the L4 root if the disk herniation is lateral and L5 if it is central.

Section Editors
Susan A. Scherl, MD
Kimberly Templeton, MD

Section 3

Disorders

Contributing Authors

Donald S. Bae, MD
Holly J. Benjamin, MD
Joseph Bernstein, MD, MS
Douglas C. Burton, MD
Peter Foster Cronholm, MD
John L. Esterhai, Jr, MD
Worth W. Everett, MD
Meral Gunay-Aygun, MD
Michael H. Huo, MD
Mark R. Hutchinson, MD
Kerwyn C. Jones, MS, MD
Rosemary D. Laird, MD, MHSA
Arabella I. Leet, MD
Carol B. Lindsley, MD
Herbert B. Lindsley, MD
Robert Lohman, MD
Catherine E. Lucasey, RN, MA
Barbara P. Lukert, MD, FACP
Jason J. Mickels, MD
Giuseppe Militello, MD
Craig S. Phillips, MD
Sally K. Rigler, MD, MPH
Haynes Robinson, MD
Jonathan Scherl, MD
Susan A. Scherl, MD
Kimberly Templeton, MD
Brian C. Toolan, MD
Carole S. Vetter, MD
Dennis S. Weiner, MD
Brent B. Wiesel, MD
A.J. Yates, Jr, MD

Disorders

The study of disease is the study of patterns. Making a diagnosis is a process of matching a patient's presentation to a set of signs, symptoms, and findings that represent a defined biologic process. This matching process allows you to offer interventions and make prognoses on the basis of your observations.

Students beginning their study of disease would do well to create a mental template—an outline—for each disease to make recognizing its pattern easier. You begin to understand a given disease once you are able to define it and describe its prevalence, pathology, presentation, differential diagnosis, and associated methods of treatment and prevention. With this information in hand, the novice will be appropriately knowledgeable and ideally receptive to new information as it becomes known.

Definition For each disease, construct a one-paragraph description of the problem using terms and concepts understandable to an educated layperson. If you must use medical terms (ischemia, for example), be certain that you clearly understand the definition; don't define a buzzword with another buzzword. Also, remember that the precise medical meaning of some words differs from the vernacular. Inflammation, for example, connotes a specific biologic process. Do not use inflammation as a synonym for irritation or swelling.

Prevalence It may seem that the prevalence of disease is interesting only to public health officials and trivia buffs, but that is not so. Prevalence is germane to clinical decision making for individual patients because most diagnostic tests do not establish or exclude the existence of a disease; rather, they simply refine the estimate of its likelihood. The chance that a disease is present after a test result is known depends not only on the power of the test but also on the probability of disease before the test. The prevalence of disease is often the best first estimate of that pretest probability. For example, you may wonder whether a mass that appears on an MRI scan is a ganglion cyst or a synovial cell sarcoma. Without scrutinizing the images, you will soon know that a ganglion is much more likely because it is much more common. The pearl to recall is "When you hear hoofbeats in Texas, think horses, not zebras."

Pathology From the student's perspective, pathology is the most essential information: a description of processes gone wrong. The key to understanding how things can go wrong, of course, is to have a strong sense of how things can go right. For example, to understand that a parathyroid adenoma may cause hypercalcemia, you must first understand the basics of calcium homeostasis. When learning pathology, try to correlate the disturbances in normal anatomy and physiology with the clinical manifestation of disease. That, unfortunately, may not always be easy. For instance, do not simply assume that arthritis is painful because two rough joint surfaces rub together. Although that may make intuitive sense, we know that many patients with rough joint surfaces are asymptomatic.

Presentation Try to create a mental image of the clinical presentation of each condition. Naturally, it will be easier to do this once you have actually seen patients with the disease. For musculoskeletal conditions, I use the mnemonic SOIL to summarize the presentation: Subjective data (the patient's symptoms), Objective signs and manifestations of the disease, Imaging (most often radiographs), and Laboratory data. Of course, making diagnoses can be more complicated than simply recognizing a specific SOIL pattern. You must have the wisdom to order the proper tests or perform the right maneuver on physical examination. Obviously, you will not diagnose Paget's disease by detecting an elevated serum alkaline phosphatase level if you fail to order that test.

Differential Diagnosis To make a correct diagnosis, you must first contemplate all of the possibilities. If you do not consider a condition, it is highly unlikely that you will recognize it. To do so, you can refer to preprinted lists—all conditions that cause knee pain in middle-aged people, for example. However, it is more effective (not to mention educationally useful) to use lists of your own creation. Therefore, when you study a disease, learn its differential diagnosis—that is, the list of conditions that may occur with similar symptoms. This will not only help you when you are compelled to make a diagnosis, but also as you learn about diseases: you will focus on those features that make it unique.

Treatment and Prevention As a medical student, your studies do not focus specifically on treatment and prevention. Nonetheless, the study of therapeutic modalities may help reinforce concepts of pathology. For example, if you observe that sciatic pain is relieved by removal of the herniated disks that impinge on the nerve root, you can infer that the disk material causes pain. Likewise, if you observe that surgery does not always eliminate pain, it is hoped that you will be motivated to question that dogma too. Studying therapeutics will help you to think like a physician, but don't get bogged down with the details of treatment at this point. There is plenty of time for that later in your career.

Miscellaneous Facts Collecting seemingly random facts about diseases can help you remember important information. For example, you may know that English sailors have been called "limeys." The reason for this is that all English sailors since 1795 were issued a daily ration of lime juice (or other citrus), which is rich in vitamin C. This seemingly random fact is helpful in recalling essential data about collagen biology. It turns out that vitamin C is needed to convert proline and lysine into hydroxyproline and hydroxylysine. Without this hydroxylation, collagen cross-links break down, weakening connective tissue. This bad collagen causes many problems, including gum disease. The image of a toothless sailor sucking on a lime, thus, serves as a hook into a trove of biologic information.

The outlines suggested in this section are necessarily simple, but the subject matter of each chapter is complex. Entire courses of study are devoted to each of these topics. In preparing this textbook, our challenge, as Einstein put it, was to "make things as simple as possible—but no simpler," a tricky balance we did not always maintain. Still, by carefully reading these chapters and constructing your own disease outlines, it is hoped that you will not only be able to acquire information rapidly, but that you will also develop a more sophisticated understanding of disease patterns as you gain additional experience.

—Joseph Bernstein, MD, MS

Chapter 9

Osteoarthritis

Osteoarthritis is a family of degenerative joint diseases characterized by chronic pain, deformity, and progressive physical and psychological disability. Osteoarthritis without an identified cause is called primary osteoarthritis. Secondary osteoarthritis results from known precipitants such as bone ischemia, trauma, and neuropathy, among others. The common terminal state for all forms of osteoarthritis is the destruction of the joint. The diagnosis is made when a patient reports joint pain; radiographs demonstrate cartilage loss, hardening of the bone below the cartilage, and bone spur formation; and no other disease is responsible. It is essential to remember that not all patients with radiographic evidence of osteoarthritis are in pain. Some who report pain are troubled to an extent far less than suggested by the radiographs. Thus, the common dictum in musculoskeletal medicine "treat the patient, not the x-rays" is especially germane to osteoarthritis.

Primary Osteoarthritis

Primary osteoarthritis is characterized by progressive loss of articular cartilage and reactive changes in the bone, leading to the destruction and painful malfunction of the joint. It is the most common arthropathy among adults—particularly the elderly. Osteoarthritis can involve all components of the joint: synovium, articular cartilage, subchondral bone, capsule, ligaments, and periarticular muscles. The principal mechanism of the disease is the destruction of articular cartilage, but osteoarthritis is abetted by the bone's attempt to make new bone. Specifically, it is known that bone reacts to stress concentration by remodeling. In osteoarthritis, the loss of cartilage leads to increased stress in the bone. This, in turn, promotes new bone formation, development of osteophytes, and sclerosis of the underlying bone. Because sclerotic bone is hard and less yielding, sclerosis decreases the joint's ability to absorb and dissipate force and makes it prone to further damage.

Epidemiology and Risk Factors

Osteoarthritis is very common. More than 20 million active cases of osteoarthritis are thought to be found in the United States alone. At least 80% of individuals age 65 years and older will have radiographic signs of the disease, but not all will have pain.[1] Among those with symptoms, functional limitations will be reported by approximately 10% to 20%.[2] Pain and difficulty with ambulation attributed to osteoarthritis may account for up to 30% of physician office visits. As the baby boomer generation ages, osteoarthritis will become an even greater public health problem.

The most important determinant of risk for osteoarthritis is age. With aging, articular cartilage loses proteoglycans and collagen. This process leads to softening of the articular cartilage, with eventual fissuring and fibrillation. Genetic transmission of osteoarthritis appears possible as well. Obesity is also considered a risk factor of osteoarthritis, but this risk is not likely the result of increased load bearing alone. The joint at the base of the thumb, which is a non–weight-bearing joint, of course, is one at highest risk for osteoarthritis in obese patients.

Ligamentous laxity and congenital bone deformity also can cause osteoarthritis, probably because of altered biomechanics and abnormal peak stress concentration. That is, joints function best when the two opposing surfaces come into near-perfect contact; however, deformity and laxity may prevent that. This is the same reason that car tires wear out prematurely if the lug nuts securing the wheel to the car are loose (laxity) or if the tire is mounted imperfectly (deformity).

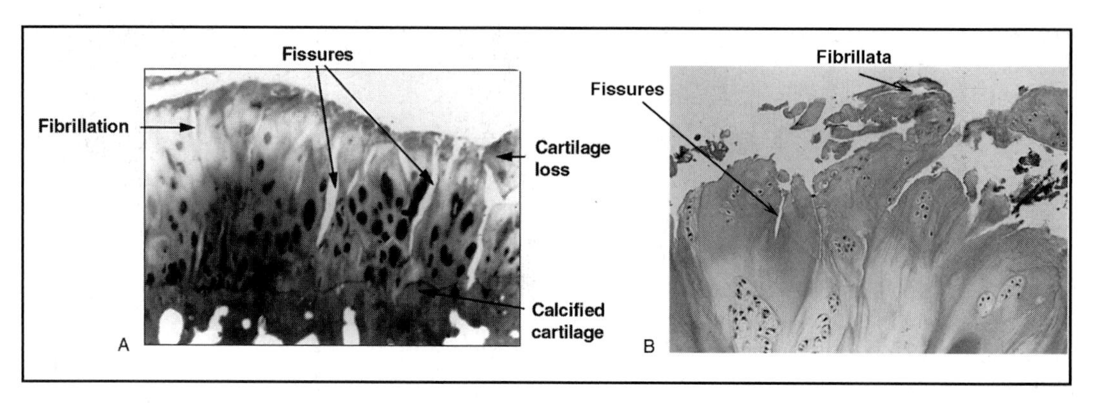

Figure 1
A, Low-power magnification of a section of a glenohumeral head of osteoarthritic cartilage removed during total shoulder arthroplasty. Note the significant fibrillation and fissures. **B**, High-power magnification of the articular surface of a humeral head with osteoarthritis. There are clefts in the cartilage and large dead areas devoid of cells.

(Reproduced from Mankin HJ, Mow VC, Buckwalter JA: Articular cartilage repair and osteoarthritis, in Buckwalter JA, Einhorn TA, Simon SR (eds): ***Orthopaedic Basic Science: Biology and Biomechanics of the Musculoskeletal System****, ed 2. Rosemont, IL, American Academy of Orthopaedic Surgeons, 2000, p 478.)*

Repetitive normal loading of a joint may also abet the development of clinically significant osteoarthritis, but the alternative, inactivity, may have worse consequences.

Pathophysiology

In primary osteoarthritis, the homeostatic mechanisms that maintain the cartilage no longer sustain the balance between normal breakdown and synthesis (regeneration). In healthy cartilage, chondrocytes continually replace the collagen and proteoglycans that compose the matrix. Collagen provides tensile strength, and the proteoglycans—and the water bound to them—provide the cushioning. When the chondrocytes cannot synthesize enough matrix, the tissue will fail mechanically. This loss of cartilage then leads, ultimately, to exposure of the underlying bone. Weight-bearing forces on the exposed bone stimulate new bone formation. As noted, this bony sclerosis makes the bone less compliant and more apt to break than bend with load. A vicious cycle is thus established (Fig. 1).

Osteoarthritis is distinct from normal aging. First, not all elderly people demonstrate the changes associated with the disease. Second, and perhaps more telling, is that in early osteoarthritis, the water content of the cartilage actually increases, whereas in aging, cartilage loses water. Water is held in the cartilage by its polar interactions with the matrix, suggesting that the primary lesion in osteoarthritis is one of matrix abnormality—either abnormal synthesis or increased production of matrix-destroying enzymes.

The pathologic findings in osteoarthritis depend on the stage. Early on, the cartilage is merely soft. As the disease progresses, there is abrasion of the cartilage, which may lead to full-thickness loss and exposure of subchondral bone. Once the protective layer of the cartilage (the lamina splendens) is lost, mechanical wear accelerates through the remaining cartilage. Also, a single small defect in the cartilage can spread geographically, as pieces of adjacent cartilage can be pulled off the bone when the edge of the defect catches on the opposing surface. This process not only enlarges the area of damage but also creates loose bodies that may inflict harm elsewhere in the joint.

Reactive changes in the bone cause it to harden, making it more likely to crack with load rather than bend (as a healthy subchondral bone would) leading to further joint destruction. Cracks in the bone create channels through which synovial fluid can enter the bone. This synovial fluid creates the subchondral cysts seen on radiographs. The bone repair process also includes new bone formation in the joint itself or at its margins in the form of bone spurs known as osteophytes.

Clinical Presentation

Physical Examination

The chief complaint of osteoarthritis is deep, aching pain, often of insidious onset. Typically, patients report pain in the interphalangeal joints and the first carpometacarpal joint of the hand, the knees and hips, the first metatarsophalangeal joint in the foot, and the spine. Patients generally report that the pain gradually becomes worse as the day goes on because mechanical load accumulates over time. Pain that is worse at the end of the day stands in contrast to the complaints of inflammatory conditions, which tend to improve as the day progresses. Other common signs and symptoms are stiffness, limited motion, and swelling. The typical patient is also at least somewhat overweight.

Obtaining a detailed medical history is important to identify the risk factors and associated problems that may affect clinical management. Most patients will have tried some form of treatment on their own, so obtaining a detailed medical history, including questions about conventional and alternative medicine therapies, is critical.

Early in the disease, there may be no objective findings. Later, the joint may be enlarged with distinct bony prominences. Joints may be slightly red, warm, and tender, reflecting the synovitis and inflammation caused by the cartilaginous debris. Osteoarthritis of the knee may produce an effusion. Range of motion should be documented. While testing range of motion, crepitus may be felt. The patient's gait and posture should be observed for signs of muscle weakness and neurologic imbalance. Muscle testing, reflexes, and neurologic examination of the peripheral nerves should be performed, and joint stability should be assessed as well.

The key feature of the examination is to exclude other causes of pain. In the lower extremity, for example, spinal stenosis, vascular disease, or bursitis may cause pain. In the presence of osteoarthritis detected on radiographs, it may be easy to dismiss these causes and falsely attribute all of the pain to the arthritis.

Laboratory Studies

There are no specific laboratory tests for osteoarthritis, although many tests may be performed in the course of excluding other causes of the patient's complaints. For example, a white blood cell count, erythrocyte sedimentation rate, and C-reactive protein level are obtained to exclude inflammatory arthritis.

Imaging Studies

Osteoarthritis is characterized by four cardinal features on plain radiographs: asymmetric joint space narrowing, subchondral sclerosis, osteophyte formation, and subchondral cysts (Figs. 2 and 3). Radiographs should be obtained while the patient is weight bearing to allow the opposing surfaces of the joint to come maximally close and thus outline the remaining cartilage (Fig. 4). MRI is typically not needed, unless it is used to exclude other diagnoses, such as osteonecrosis. Any patient with osteoarthritis that is detected on radiographs will also have cartilage abnormalities; therefore, ordering MRI to identify a meniscal tear in the knee in a patient with advanced osteoarthritis is often a poor diagnostic strategy.

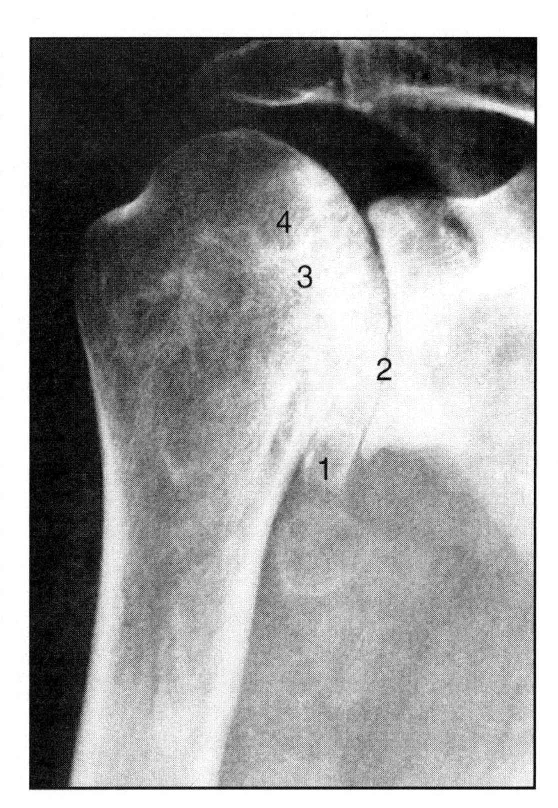

Figure 2

Severe osteoarthritis of the glenohumeral joint, with large inferior osteophyte (1), loss of joint space (2), sclerotic bone formation (3), and periarticular cyst formation (4).

*(Reproduced from Crosby LA (ed): **Total Shoulder Arthroplasty**. Rosemont, IL, American Academy of Orthopaedic Surgeons, 2000, p 18.)*

Differential Diagnosis

The key to diagnosis is to first establish that the patient's complaints are caused by disease in the joint itself and not referred from other areas. Once the diagnosis of arthritis (of any sort) is made, inflammatory causes, including rheumatoid arthritis, gout, and infection, must be considered. If these causes

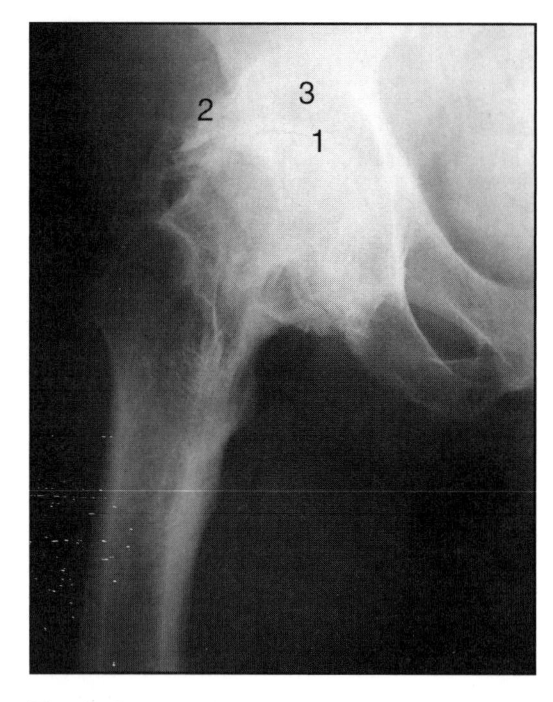

Figure 3
Osteoarthritis of the hip with loss of joint space (1) and complete degeneration of the articular surfaces of the femoral head and the acetabulum. There is also marginal osteophyte formation (2) as well as a subchondral cysts in the acetabulum (3).

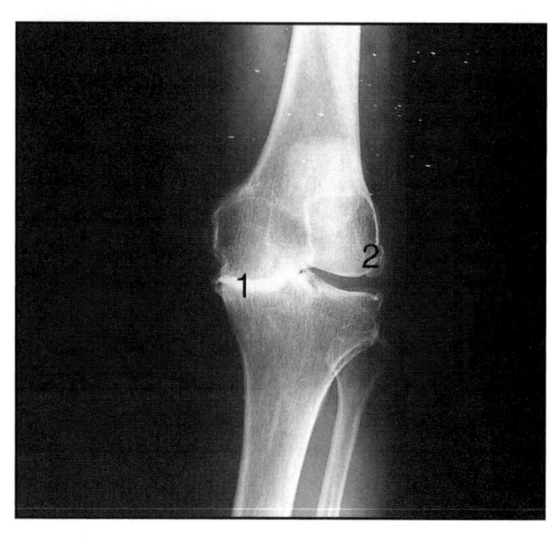

Figure 4
Osteoarthritis of the knee with varus alignment. There is destruction of the medial compartment with bone-on-bone contact and subchondral sclerosis (1) as well as osteophyte formation (2).

are excluded, a diagnosis of primary or secondary osteoarthritis can be made.

Prevention and Treatment

Medical Management

There is no known cure for osteoarthritis. The mainstay of medical treatment includes pain management, maintaining mobility and function, and preventing disease progression. Nonpharmacologic options include patient education, weight reduction, physical and occupational therapy, and a general cardiovascular fitness exercise program.[1] Weight reduction is beneficial in decreasing the mechanical load on painful joints. An exercise program should be followed to strengthen periarticular muscles, reduce weight, and promote general well-being. Although a strengthening regimen can be recommended on empirical grounds—fitness for its own sake—there is good reason to believe that it will mitigate the painful effects of osteoarthritis.

Normal muscle function protects the joints. Coordinated muscular contraction limits joint instability. Muscle contraction also can absorb some of the stresses in load bearing. Thus, strong and responsive muscles can replace, to some extent, dysfunctional ligaments and cartilage.

Bracing can occasionally be used to unload painful joints, especially if there is significant malalignment or associated ligamentous instability. Physical modalities such as heat or cold therapy, ultrasound treatment, and iontophoresis can be used for pain relief.

The principal pharmacologic approach to osteoarthritis has been the use of nonsteroidal anti-inflammatory drugs (NSAIDs), particularly aspirin. NSAIDs offer both pain relief and control of inflammation, but they have significant adverse effects, including impaired renal function and gastrointestinal bleeding. Therefore, long-term use demands careful monitoring. Recent guidelines have recommended the use of acetaminophen as the first-line pharmacologic treatment of pain from osteoarthritis. Published studies have demonstrated that the efficacy of acetaminophen equals that of ibuprofen for the management of mild to moderate osteoarthritis of the knee but offers a better safety profile.[3] If acetaminophen alone is inadequate, an NSAID can be prescribed in combination.

Other Nonsurgical Options

Joint injection with steroids or hyaluronate may offer symptom relief. The potential adverse effects of steroid therapy, such as decreased resistance to local infection, limit the

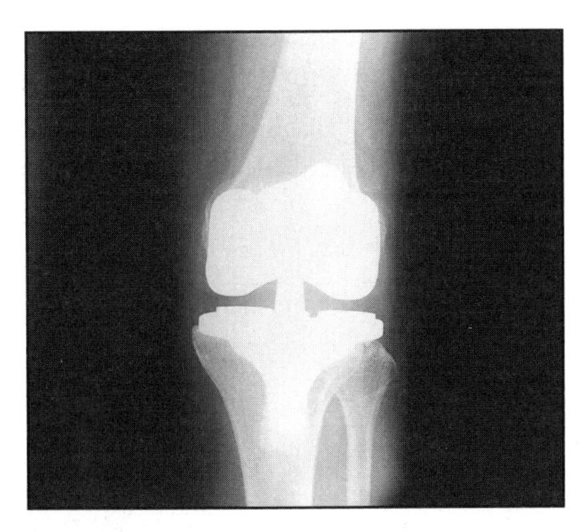

Figure 5
The same patient as in Figure 4 after total knee arthroplasty. The articular surface and adjacent bone was removed and replaced with metal and plastic components.

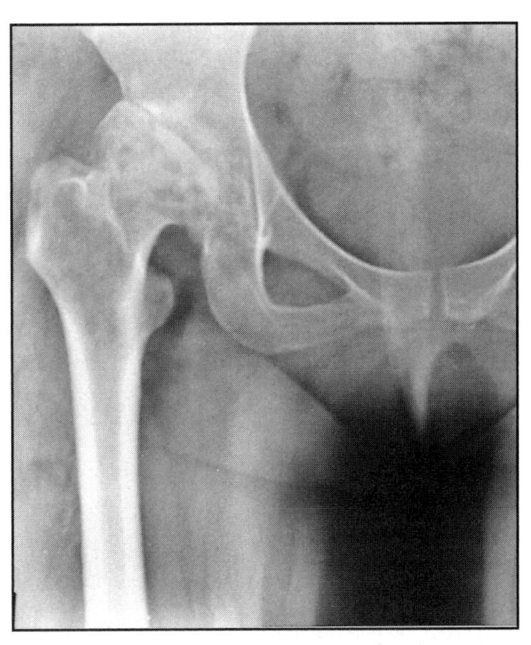

Figure 6
A patient with osteonecrosis of the femoral head from a hemoglobinopathy. There has been significant collapse of the femoral head. In time, the irregularities of the femur will destroy the acetabulum.

frequency with which this approach can be applied. Alternative medications (eg, nutritional supplements, such as glucosamine and chondroitin sulfate) are commonly used, but their effectiveness in treating osteoarthritis remains unproven.

Surgical Management
Many patients with osteoarthritis of the hip and knee eventually seek surgical treatment, particularly for pain relief and functional improvement. Joint replacement (arthroplasty) is highly effective in relieving pain (Fig. 5). The success rate, if measured by pain-relieving effects, is greater than 90%.[4] However, the success rate for functional improvement is not as high. The major limitation of joint arthroplasty is the lack of long-term durability of the prostheses. Additionally, medical complications, although rare, can be devastating. Arthroplasty is a good operation for patients with end-stage joint destruction with pain and loss of function. It is less than ideal for patients who can play 17 holes of golf but want to play 18: the biologic costs outweigh the benefits.

Osteonecrosis

Pathophysiology

Osteonecrosis is a common cause of secondary osteoarthritis. It represents the final pathologic process of a number of conditions that kill the bone from ischemia.[5] Other terms used to describe this phenomenon are avascular necrosis, aseptic necrosis, and ischemic bone necrosis. Osteonecrosis results from a reduction of blood flow to bone such that its metabolic demands are not met. Once the bone dies, it cannot repair cumulative damage and is prone to collapse (Fig. 6). The collapse of bone under the cartilage produces in an irregular surface and, ultimately, destruction of the other side of the joint as well. Osteonecrosis occurs in the setting of intravascular obstruction, vascular compression, or physical disruption of vessels that supply the affected bones.[6] Conditions associated with osteonecrosis include systemic lupus erythematosus, alcoholism, pancreatitis, hemoglobinopathies, Gaucher's disease, coagulopathies, Legg-Calvé-Perthes disease, corticosteroid use, displaced fractures and joint dislocations, organ transplantation, hemophilia, obesity, and even pregnancy.[6,7] Osteonecrosis can develop in some patients without a recognized cause.

Ischemia may result from processes originating within or outside of the vascular system. Intravascular obstruction by emboli or thrombosis directly leads to regions of ischemia. Infiltrative processes can indirectly compress the osseous sinusoids and lead to venous congestion and decreased blood flow. Bone cannot expand under pressure. Thus, venous congestion creates a region of increased pressure analogous to a compartment syndrome of the soft tissues. Displaced fractures and dislocations may disrupt vessels near the joint. Such disruption will cause osteonecrosis when the remaining vessels cannot provide enough blood flow to meet metabolic demands.

Certain regions of the skeleton are especially prone to osteonecrosis, specifically the femoral head and condyles, the humeral head, the carpal bones (scaphoid and lunate), and the talus.[6]

Clinical Presentation

Physical Examination

Osteonecrosis may be painful, even before secondary osteoarthritis develops. Patients typically report that the pain develops slowly over a long period and is generally associated with load-bearing activities. With disease progression, pain may be constant. Obtaining a comprehensive medical history is essential for identifying risk factors and conditions associated with osteonecrosis. Identifying and treating these conditions may prevent osteonecrosis from occurring in other bones.

Clinical findings for osteonecrosis are similar to those for osteoarthritis. The opposite extremity should be examined carefully because the incidence of bilateral involvement of osteonecrosis can approach 50%. Care should be taken to assess for possible referred pain from adjacent joints or the spine.

Imaging Studies

Plain radiography can confirm the diagnosis when changes in the bone, such as subchondral lucency or patchy sclerosis, are present. Although bone scanning can detect changes in the affected bones before they become apparent on plain radiographs, it is not highly specific for osteonecrosis. MRI is now the modality of choice for confirming the diagnosis. It is especially useful for those patients who are at risk but remain asymptomatic. It can also estimate the size of the necrotic segment, which may help plan treatment and estimate prognosis.

Prevention and Treatment

Medical Management

The most effective treatment is prevention, which is achieved by the proper identification of risk factors and the modulation of those factors. This is done, for example, by minimizing the dose and duration of corticosteroid and cytotoxic medications. There is no specific medical treatment for osteonecrosis.

Surgical Management

Surgical treatment has been shown in retrospective reviews to be effective; however, without randomized trials, it is hard to prove that surgery alters the natural history of the disease. Surgical options include core decompression with or without bone grafting. Core decompression is a drilling procedure aimed to relieve the increased intraosseous pressure within the bone and to provide a channel for new blood vessels to develop in the necrotic segment. This approach is useful only in early osteonecrosis before significant bone collapse and joint degeneration occurs.

Posttraumatic Arthritis

Pathophysiology

Posttraumatic arthritis is a form of secondary osteoarthritis caused by a loss of joint congruence and normal joint biomechanics. The healthy joint is congruent, and its articular facets align perfectly, allowing smooth motion and even distribution of forces across these surfaces. Without such congruence and smooth flow, the joint surfaces will be abraded by constant impact—rattling like dice in a cup—and the forces of weight bearing will be concentrated on small areas.

The course of joint degeneration is much more rapid in posttraumatic arthritis than primary osteoarthritis. Causes of posttraumatic osteoarthritis include intra-articular fractures, damage to the joint surface, periarticular fractures that produce malalignment of the extremity or altered biomechanical loading of the joint, and ligament injuries that upset the normal biomechanics and motion of the joint. A special form of posttraumatic arthritis of the knee can occur after meniscal injury. The loss of the shock-absorbing effect of the normal meniscus places the joint at risk for early and progressive destruction. The findings of secondary osteoarthritis after meniscal loss are known as Fairbank's changes (Fig. 7).

Clinical Presentation

Physical Examination

Patients with posttraumatic osteoarthritis typically present with pain and limited joint mobility. Symptoms are similar in all ways to those of primary osteoarthritis, but patients with posttraumatic osteoarthritis are typical-

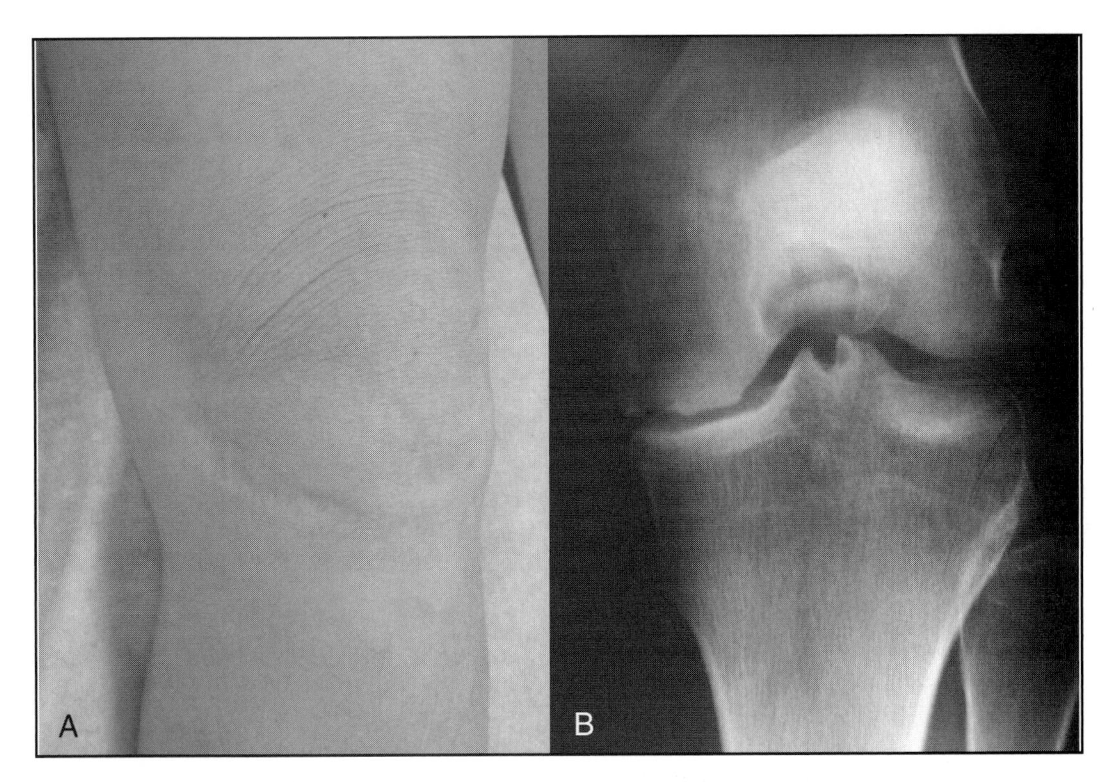

Figure 7
A, Clinical photograph of a 44-year-old patient 15 years after open medial menisectomy. **B**, Standing radiograph of the same patient demonstrates medial compartment degeneration (Fairbank's changes).

ly younger and have a history of trauma or injury of the affected joint. Examination should focus on identifying any deformity because the joint may yet be saved if these deformities are corrected.

Imaging Studies
Plain radiography is usually adequate for diagnostic purposes. CT and MRI can be useful adjuncts; however, once the diagnosis of posttraumatic arthritis is made, they typically provide little additional information; as such, they should be used primarily for preoperative planning or when an additional diagnosis such as a loose body is being considered.

Prevention and Treatment
The most effective measures to prevent posttraumatic arthritis are to identify the injuries, restore proper joint surface anatomy and alignment, provide soft-tissue support, and initiate rehabilitation. Treatment of intra-articular fractures focuses on reestablishing joint congruency and normal joint mechanics. For patients with established posttraumatic arthritis, treatment is similar to that for primary osteoarthritis.

Neuropathic Arthritis
Pathophysiology
A neuropathic joint is one that has been destroyed by an underlying neurologic dysfunction.[8] Other terms used to describe neuropathic arthritis are Charcot arthropathy and neuropathic arthropathy.[7] Some of the conditions associated with the development of neuropathic joints include the peripheral neuropathy of diabetes mellitus, syringomyelia, head or spinal trauma, meningomyelocele, multiple sclerosis, congenital insensitivity to pain, alcoholism, and familial dysautonomia.[7,9] Often, the destroyed joint is not painful.

Diabetes mellitus is most commonly associated with neuropathic joints. In approximately 5% of diabetic patients with sensory neuropathy, neuropathic arthritis eventually develops.[9]

Two principal theories have been proposed concerning the pathophysiology of neuropathic arthritis. The first cites cumulative trauma as the initiating mechanism. Repetitive loading of a joint, left unchecked, leads to fracture of the joint surface. Ordinarily, normal sensory feedback mechanisms will signal the individual to stop the inciting

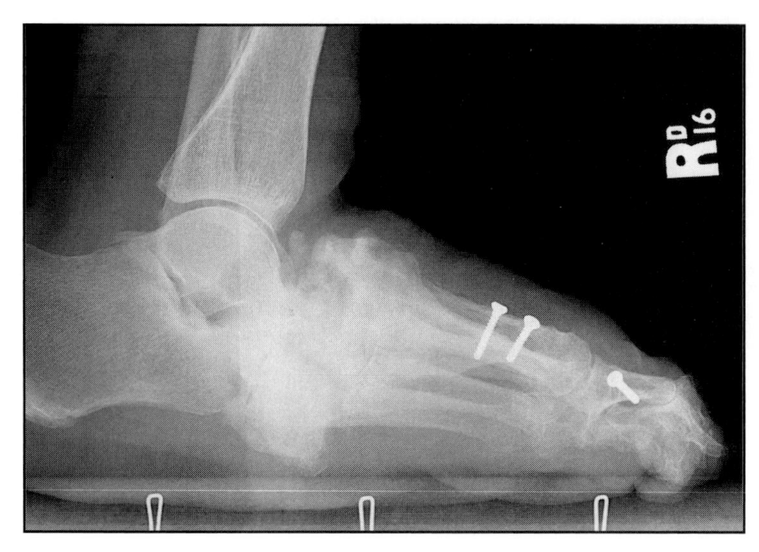

Figure 8
Characteristic radiographic features of the neuropathic joint demonstrating the five "Ds": joint distention, bone debris, dislocation, disorganization, and increased bone density.

activity and thus protect the joint. In neuropathy, this does not happen, and the person leaves his hand in the fire, so to speak. The second mechanism is associated with dysfunction of the sympathetic fibers innervating the smooth muscles of arterioles, which results in enhanced blood flow and hyperemia. Although patients with diabetes and peripheral vascular disease typically have decreased microcirculation, the foot itself may have increased blood flow, which can lead to increased osteoclast activity and bone resorption. This weakens the bone and predisposes it to fracture. It has yet to be explained why arthropathy develops in some patients but not in others with apparently similar patterns of neuropathy.[10]

The gross pathology of neuropathic arthritis is similar to that observed in severe osteoarthritis, with destruction of cartilage, sclerosis and fragmentation of bone, osteophyte formation, and loose bodies within the joint.[7] The joints most commonly affected are those in the foot and ankle. In these joints, erosive synovitis develops first, possibly followed by instability and subluxation. Ultimately, the joint is completely destroyed[11](Fig. 8). In the foot, collapse of the midtarsal joints can lead to a "rocker-bottom deformity." The shape of the foot and ankle can be dramatically altered, changing the distribution of pressure on the plantar surface. This, coupled with peripheral neuropathy, produces skin ulcerations that can progress to cellulitis, osteomyelitis, and gangrene and lead to amputation in some instances.[11]

Clinical Presentation

Physical Examination

Patients with neuropathic joints often present with joint swelling. They also typically describe a sudden onset of symptoms, occasionally associated with an audible "pop" after a minor injury that will not resolve. Pain is often much less than that expected for the amount of swelling or deformity. Patients also often have a long-standing history diabetes with significant sensory neuropathy. Examination reveals swelling with signs of inflammation and variable degrees of bony deformity and joint laxity.

Imaging Studies

Radiographs often show findings suggestive of osteomyelitis with aggressive bony destruction. In cases of slow progression, the radiographic features may appear similar to those of osteoarthritis.

Prevention and Treatment

Medical Management

As with most forms of secondary osteoarthritis, the most effective treatment is prevention. Patient education regarding proper foot care and footwear and regular follow-up by health care providers is required. Once neuropathic arthritis has been diagnosed, treatment centers on protecting the joint to allow for healing and preventing further damage to the joint and soft tissues. Total contact casting protects the limb and allows for healing. However, casts need to be changed regularly to accommodate the reduction in swelling and to assess the potential for pressure sores. In some studies, bisphosphonates have been effective in hastening bone healing in the acute phase.[8] Another important goal is to protect the unaffected extremity from overload and possible joint damage. If there is residual deformity, treatment should be focused on preventing the development of pressure ulcers. Orthotic devices can be molded to protect prominences around the foot and ankle.

Surgical Management

Surgery may be unavoidable as salvage for patients with end-stage conditions. Options include fusion (arthrodesis) of unstable joints, excision of bony prominences to reduce the risk of ulceration, correction of joint malalignment to reduce the risk of further joint destruction, and amputation.

Key Terms

Lamina splendens The protective top layer of articular cartilage

Neuropathic arthritis The chronic, progressive destruction of a joint that is caused by the loss of sensation from an underlying neurologic dysfunction; also known as Charcot arthropathy

Osteonecrosis The death of bone, often as a result of obstruction of its blood supply

Posttraumatic arthritis A form of secondary osteoarthritis caused by a loss of joint congruence and normal joint biomechanics

Primary osteoarthritis Osteoarthritis without an identified cause; characterized by progressive loss of articular cartilage and reactive changes in the bone, leading to the destruction and painful malfunction of the joint

Secondary osteoarthritis Osteoarthritis resulting from known precipitants such as bone ischemia, trauma, and neuropathy

References

1. Brandt KD: The role of analgesics in the management of osteoarthritis pain. *Am J Ther* 2000;7:75-90.
2. Brandt KD: (ed): *Diagnosis and Nonsurgical Management of Osteoarthritis*, ed 2. Caddo, OK, Professional Communications, Inc, 2000.
3. Shamoon M, Hochberg MC: The role of acetaminophen in the management of patients with osteoarthritis. *Am J Med* 2001;110(suppl 3A): 46S-49S.
4. Katz JN: Preferences, quality, and the (under) utilization of total joint arthroplasty. *Med Care* 2001;39:203-205.
5. Pavelka K: Osteonecrosis. *Baillieres Best Pract Res Clin Rheumatol* 2000;14:399-414.
6. Mirzai R, Chang C, Greenspan A, et al: Avascular necrosis. *Compr Ther* 1998;24:251-255.
7. Klippel JH, Weyand CM, Wortmann RL: (eds): *Primer on the Rheumatic Diseases*, ed 11. Atlanta, GA, Arthritis Foundation, 1997, pp 216-221, 322-324, 378-381.
8. Klenerman L: The Charcot joint in diabetes. *Diabet Med* 1996;13(suppl 1):S52-S54.
9. O'Connor BL, Brandt KD: Neurogenic factors in the etiopathogenesis of osteoarthritis. *Rheum Dis Clin North Am* 1993;19:581-605.
10. Guyton GP, Saltzman CL: The diabetic foot: Basic mechanisms of disease. *J Bone Joint Surg Am* 2001;83:1084-1096.
11. Wilson M: Charcot foot osteoarthropathy in diabetes mellitus. *Mil Med* 1991;156:563-569.

Inflammatory Arthritis

Adult Rheumatoid Arthritis

Rheumatoid arthritis (RA) is a chronic inflammatory disease that is probably triggered by an antigen and presents as an inflammatory reaction against the synovium in the joint (Fig. 1). This immune process leads to a proliferative synovitis (pannus formation), which proceeds to destroy the joint. Clinically, RA follows a course of waxing and waning, but it results in progressive disability over time. Patients with RA may be disabled from the disease beyond its effect in the joints. For example, RA can cause systemic manifestations, such as fatigue, weight loss, or anemia. In addition, RA is often treated with potent drugs; thus, patients are potentially exposed to complications from medications.

The initial effects of RA in the joint are manifest as stiffness, and the disease may progress to complete destruction of the articular surfaces. The damage is inflicted by the proliferative synovium itself, which tends to physically invade the joint space, as well as by enzymes that the synovium and reactive white blood cells release. RA also affects the ligaments of the joint; therefore, instability and deformity are characteristic (Fig. 2).

RA represents the intersection between musculoskeletal medicine and the basic science of immunology and inflammation. It is therefore a rich and, at times, recondite subject. This chapter perforce is but an overview.

Epidemiology

RA affects approximately 1% of the population in the United States. Approximately two of every three patients with RA are female.[1] RA typically occurs in patients between the ages of 40 and 60 years, although there is a juvenile form that can be found in those younger than 16 years. There appears to be a genetic susceptibility to this condition, as a monozygotic twin of a patient with RA is more likely to have the condition than a dizygotic twin. However, genes are not the only factor affecting susceptibility because even a monozygotic twin has only approximately a 15% chance of having RA when the twin has it.[2,3]

Pathophysiology

The exact etiology of RA is not known. It is thought that exposure to an antigen in a genetically susceptible host leads to an antibody production directed against the synovium. In this model, the antigen activates T cells, which produce cytokines. These me-

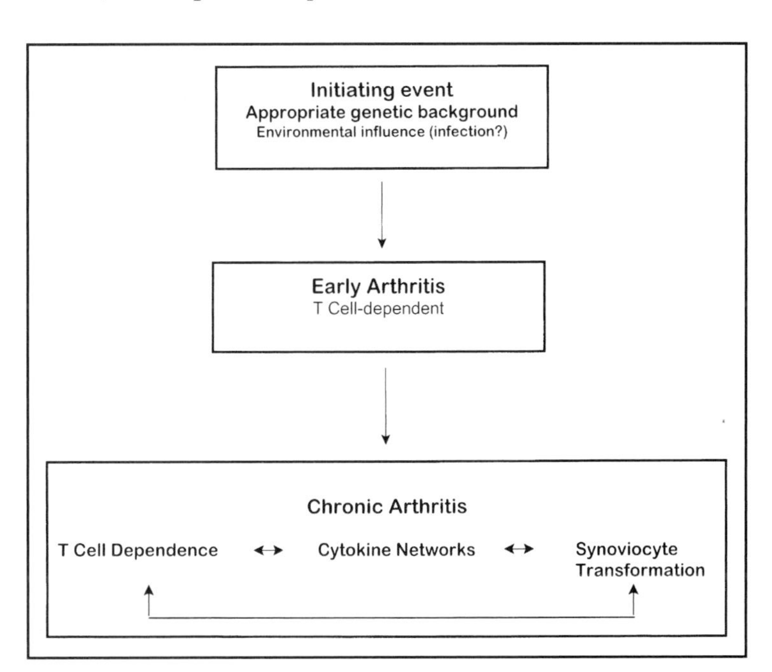

Figure 1
Pathogenesis of RA is complex, involving immune processes lead by T cells, cytokine-driven events (responsible for most of the erosive events), and fibroblast-like synoviocyte transformation, resulting in pannus formation, hypercellularity, and hyperplasia.

(Reproduced from Recklies AD, Poole AR, Banerjee S, et al: Pathophysiologic aspects of inflammation in diarthrodial joints, in Buckwalter JA, Einhorn TA, Simon SR (eds): ***Orthopaedic Basic Science: Biology and Biomechanics of the Musculoskeletal System****, ed 2. Rosemont, IL, American Academy of Orthopaedic Surgeons, 2000, pp 489-530.)*

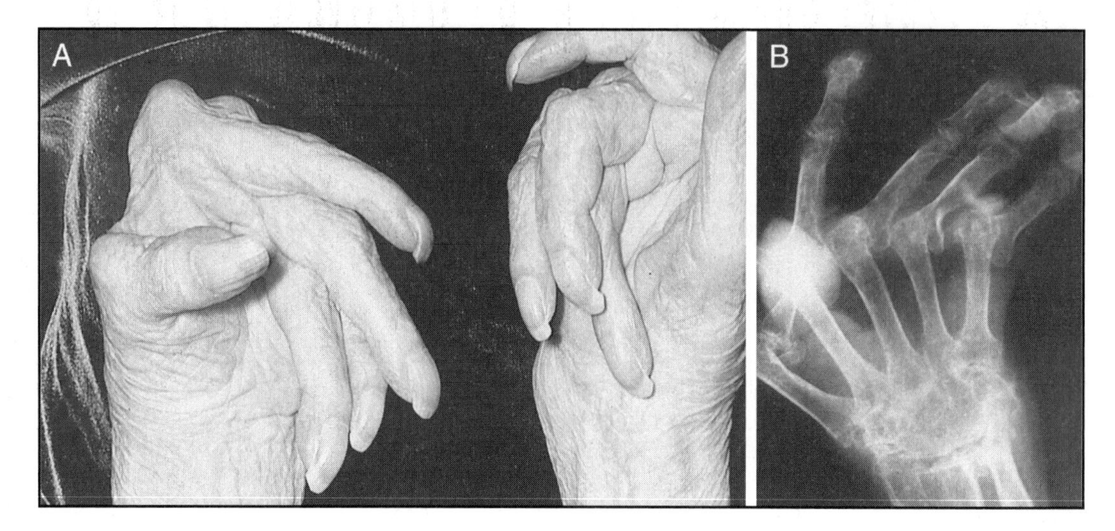

Figure 2
Advanced RA. Note that the fingers are deviated toward the ulnar side of the hand because synovitis destroys the tendon sheath. **A**, Clinical appearance. **B**, Radiographic appearance.

*(Reproduced from Greene WB: **Essentials of Musculoskeletal Care**, ed 2. Rosemont, IL, American Academy of Orthopaedic Surgeons, 2001, p 213.)*

diators stimulate antibody production by B cells.[4-6] Among these antibodies is the classic rheumatoid factor that binds to IgG. These antibodies are then deposited as immune complexes in the synovium and in cartilage. Activation of other white blood cells promotes a chronic inflammatory state (Fig. 3).

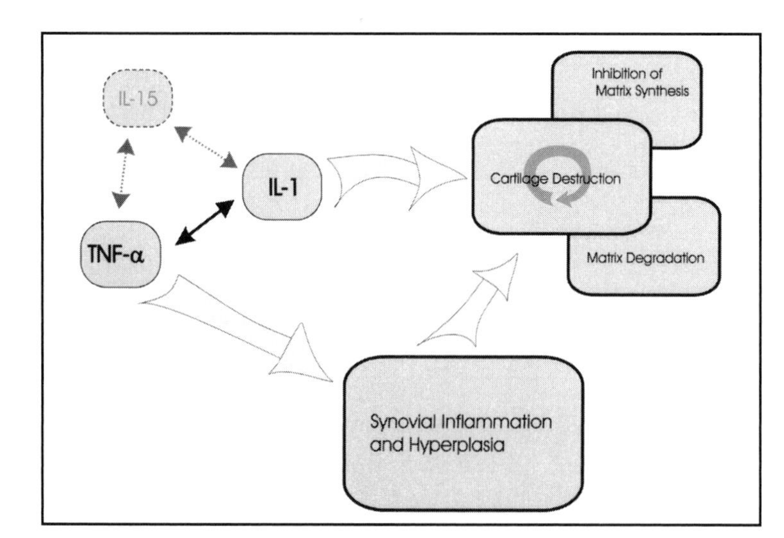

Figure 3
Cytokines found in the joint initiate and perpetuate many different inflammatory processes. TNF-α is thought to control many of the processes attributed to inflammatory cells in the joint fluid, and IL-1 is thought to act within the articular cartilage itself. The role of IL-15 in this hierarchy is currently hypothetical, but existing data suggest that IL-15 may act to initiate the T-cell driven autoimmune process.

*(Reproduced from Recklies AD, Poole AR, Banerjee S, et al: Pathophysiologic aspects of inflammation in diarthrodial joints, in Buckwalter JA, Einhorn TA, Simon SR (eds): **Orthopaedic Basic Science: Biology and Biomechanics of the Musculoskeletal System**, ed 2. Rosemont, IL, American Academy of Orthopaedic Surgeons, 2000, pp 489-530.)*

Synovial proliferation follows this immune activation, a process called pannus formation. The pannus is hyperplastic synovium with fibroblasts, blood vessels, macrophages, and lymphocytes. The synovial cavity of the joint becomes filled with fluid (an inflammatory effusion). The pannus itself, synovial fluid inflammatory cells, and bone-based osteoclasts attack cartilage and bone, resulting in joint destruction.[7-9] This destruction produces two characteristic radiographic features of RA: marginal erosions and symmetric narrowing of the joint space medially and laterally. Osteoarthritis, by contrast, tends to produce asymmetric joint space narrowing, beginning on the point of maximal force bearing or previous trauma.

Extra-articular manifestations of RA can occur, the most common of which is a subcutaneous nodule found near the joint. Rheumatoid nodules contain granulation tissue and inflammatory white blood cells. The synovitis of RA often extends to the tendon sheath as well, leading to its destruction. In patients with RA of the hand, loss of the tendon sheath causes ulnar deviation of the fingers because the tendons no longer remain tethered to the digit. In some patients with RA, inflammation in the heart, lung, or blood vessels may also occur.

The destruction of the joint in patients with RA is modulated by various factors. These include various cytokines as well as specific matrix-destroying enzymes.[10] These factors are elaborated from the synovial macrophages and invading T cells. The activation of these cells is thought to be responsible for the extra-articular manifestations of the disease.

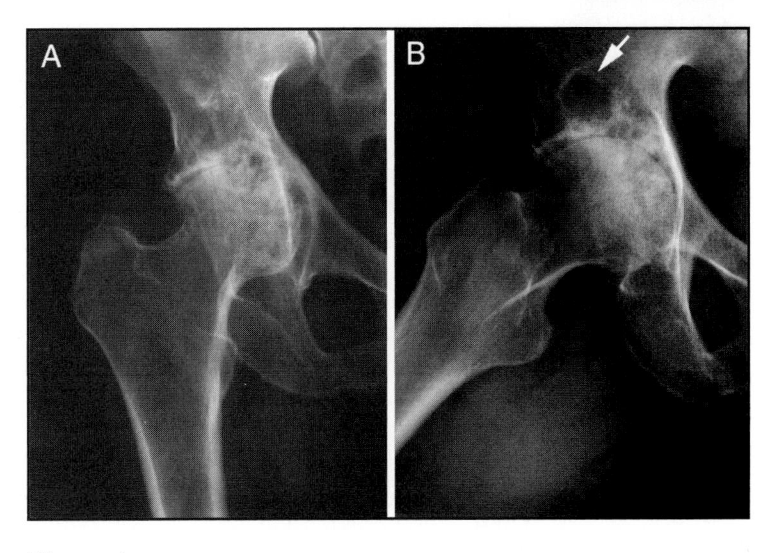

Figure 4
AP (**A**) and lateral (**B**) radiographs of the hip of a 56-year-old man with RA show moderate diffuse periarticular osteopenia, loss of joint space, and extensive acetabular cyst formation (arrow).

(Reproduced from Lachiewicz PF: Rheumatoid arthritis of the hip. ***J Am Acad Orthop Surg*** *1997;5:332-338.)*

Presentation

The presentation of RA depends on the stage of the disease. Moreover, expression of the disease varies not only from individual to individual but also within the same individual over time. RA often begins as an acute period of nonspecific complaints, such as malaise and joint pain. Characteristically, patients have diffuse joint tenderness and swelling, accompanied by morning stiffness lasting for at least 1 hour. The commonly accepted criterion is that these findings must be present for at least 6 weeks to make a diagnosis of RA (transient inflammation may have other, more benign causes). Patients with RA will demonstrate tender, warm, and swollen joints and have pain with motion. When the disease is long-standing, there will be characteristic objective findings, such as ulnar deviation of the fingers and radial deviation of the wrist caused by erosion of the extensor tendon sheath. Flexion or extension deformities of the fingers are also common.

Diagnostic Imaging

Imaging for RA in its early stages may not be diagnostic. Because RA begins by attacking soft tissue before bone, in its early stages, signs are not readily apparent on plain radiographs. As the disease progresses, osteopenia caused by increased blood flow and osteoclast activity can be detected. Marginal erosions of the bone are seen as the pannus begins to destroy bone. As the disease progresses, symmetric joint space narrowing occurs, culminating in total destruction of the articular surface (Fig. 4). RA can cause instability of the cervical spine when tenosynovitis near the C1-C2 junction destroys the transverse ligament holding the odontoid.[11,12] A cervical spine radiograph is therefore mandatory prior to elective endotracheal intubation of patients with RA.

Laboratory Studies

Laboratory studies may be helpful in the diagnosis of RA, but RA is not conclusively diagnosed on the basis of any single laboratory test result. The most useful laboratory test is probably the aspiration and analysis of the joint fluid, which is performed not so much to make the diagnosis of RA, but to exclude two other conditions that may mimic it: infectious arthritis and gout. When infectious arthritis or gout is present, bacteria or urate crystals usually are apparent on analysis of aspirated joint fluid. The fluid in RA is typically characterized by increased neutrophil concentration but only to moderate levels; cell counts are usually in the 5,000 to 50,000 range. A complete blood cell count of patients with RA may show an elevated white blood cell count in the peripheral circulation and anemia as well. Two useful tests for evaluating systemic inflammation involve determining the serum C-reactive protein concentration and the erythrocyte sedimentation rate.

Circulating rheumatoid factor also can be measured. The sensitivity of rheumatic factor for the diagnosis of RA is approximately 80%; therefore, it is a poor screening test. However, its relatively high specificity (approximately 95%) allows it to be used as a confirmatory test. Serial monitoring of rheumatoid factor levels does not offer valuable clinical information. In general, patients with RA who do not have circulating rheumatoid factor have a more benign course.

A number of clinical features have been associated with an unfavorable prognosis. These include persistent polyarticular synovitis for more than 2 years, the presence of articular erosions, and the presence of extra-articular manifestations. Extra-articular manifestations may be general and include fatigue, low-grade fever, weight loss, or anorexia; or they may be organ-specific and include subcutaneous nodules, interstitial lung disease, pleural effusions, or neuropathy. The presence of extra-articular manifestations usually mandates aggressive antirheumatic therapy.

Treatment and Prevention

There is no known prevention for RA. Nonetheless, aggressive treatment of early forms of the disease can be viewed as a preventive measure in that damage is prevented—that is, the disease process is arrested before it is able to destroy joints.

The management of RA is centered on drug therapies, although patient education, supportive counseling, and exercise are also essential to attain a good outcome. The pharmacologic therapy of RA is directed to relieve joint pain, limit inflammation, and retard joint destruction. Because the inflammation of RA is mediated by cytokines, drugs that block them are most effective. Nonsteroidal anti-inflammatory drugs (NSAIDs) are nonspecific inhibitors of prostaglandin synthesis. Accordingly, they can control joint pain, but they do not retard joint destruction. Corticosteroids are also used to limit inflammation. Both NSAIDs and corticosteroids expose the patient to the risk of complications. NSAIDs increase the risk of gastrointestinal bleeding and renal dysfunction, whereas the long-term use of corticosteroids frequently produces osteoporosis and other serious complications.

Another category of drugs used to treat RA is called disease-modifying antirheumatic drugs (DMARDs). The goal of using DMARDs is to prevent joint destruction. The most frequently used DMARD is methotrexate, which can slow the progression of joint destruction in patients with RA. Methotrexate is a known inhibitor of folate metabolism; however, its precise effect on RA is not clear. Other DMARDs include immunosuppressive medications, such as sulfasalazine or cyclosporine, and newer biologic agents directed at inhibiting the actions of tumor necrosis factor (TNF)-α or interleukin-1 (IL-1).

The surgical treatment of patients with RA includes joint arthroplasty for damaged joints, fusion of unstable joints, and tenosynovectomy to prevent tendon rupture.

Differential Diagnosis

RA can be confused with other forms of inflammatory arthritis, such as systemic lupus erythematosus and viral arthritis. It may also be confused with septic arthritis and gout because both are inflammatory conditions that produce fluid in the joints and pain. Appropriate laboratory tests, however, should be able to differentiate these conditions from RA. RA can also be confused with seronegative spondyloarthropathies, inflammatory joint diseases that occur in the absence of rheumatoid factor (hence, the term seronegative). These conditions affect the sacroiliac joint and frequently the spine and include ankylosing spondylitis and psoriatic arthritis. Although clinically important and scientifically interesting, an in-depth discussion of seronegative spondyloarthropathies is beyond the scope of this book.

Juvenile Rheumatoid Arthritis

Juvenile rheumatoid arthritis (JRA) may present as a systemic disease or as one affecting only the joints. Approximately 5% to 10% of children with JRA present with systemic signs, including fever, rash, pericarditis, or hepatosplenomegaly. These patients may have significant laboratory abnormalities, including leukocytosis and anemia. Although this form of disease can occur at any age, most cases are seen in patients younger than 10 years. The degree of joint involvement in this form of JRA can range from mild arthalgias to florid swelling of all joints. Children with this form of the disease appear acutely and gravely ill and may require hospitalization.

Epidemiology

JRA is classified as pauciarticular when it affects four or fewer joints over the first 6 months of the disease and as polyarticular when more than four joints are affected. Both types present with joint swelling over a period of weeks to months. Girls with pauciarticular disease are at high risk for eye involvement (chronic uveitis); boys with this subtype may develop a spondyloarthropathy (spinal arthritis). In children with polyarticular disease, the presence of rheumatoid factor in the blood (found in about 20% of patients) is a risk factor for a more persistent and destructive arthritis.

Pathophysiology

Factors similar to those implicated in RA are thought to affect JRA as well. Specifically, genetic predisposition (HLA-DR4 in seropositive JRA, HLA-DR4 in pauciarticular type I, and HLA-B27 in pauciarticular type II), unknown environmental triggers, and immune reactivity are all thought to be possible causes. The inflammatory synovitis that is the hallmark of JRA does not differ among subtypes and is generally more vascular than that seen in adults.

Diagnosis

Diagnosis is made by the demonstration of persistent arthritis in one or more joints for a minimum of 6 weeks, the exclusion of other diagnoses and, in the case of systemic disease, the presence of either fever or rash.

Treatment

Once the diagnosis of JRA is established, a therapeutic plan is devised based on the subtype and severity of disease. NSAIDs are standard initial therapy. In systemic disease, second-line agents such as hydroxychloroquine or methotrexate may be added early in the disease course; corticosteroids may be required for pericarditis or disease that is unresponsive to other therapies. Most children with pauciarticular disease will respond to an appropriate NSAID. Intra-articular corticosteroids may be needed, and occasionally second-line agents such as sulfasalazine are added. Patients with persistent, severe systemic or polyarticular disease who do not adequately respond to methotrexate therapy are candidates for treatment with anti-TNF drugs. Low-dose, short-term corticosteroid therapy may be needed. Long-term corticosteroid use in children is avoided because steroids may cause growth retardation, iatrogenic Cushing's disease, osteoporosis, fractures, and hypertension.

In addition to drug therapy, children with JRA must be carefully monitored for growth abnormalities, nutritional problems, and school and social dysfunction. Psychological and emotional health may be difficult to maintain in the presence of chronic disease. Therapeutic exercise programs can maximize joint motion and minimize muscle atrophy.

Other Inflammatory Arthritides

Gout is an inflammatory arthritis associated with elevated serum levels of urate (hyperuricemia). Hyperuricemia results from increased production or impaired excretion of uric acid. Acute inflammatory arthritis results when crystals of uric acid are deposited in joints. The synovium subsequently becomes hyperemic and swollen, but pannus formation does not occur in acute cases. Neutrophils ingest urate crystals and release mediators of inflammation. With treatment, these crystals can be removed, and the joint can be returned to normal. Gout has a chronic form in which tophi (chalky deposits of urate surrounded by inflammatory cells) are formed adjacent to the joint. In chronic gout, the synovium reacts to the formation of these tophi by becoming hyperplastic, and pannus is formed that leads to erosion of bone. Tophi typically arise close to the joints, but they may also appear in the soft tissue at some distance from the joint or even in other tissues located far from joints (eg, the ear).

Epidemiology

Gout is less common than RA. Unlike RA, it is more prevalent in men by a factor of 9:1. Elevated levels of urate are not synonymous with the presence of gouty arthritis.[13] Many people with transiently elevated urate levels will never have arthritis; in those who do, however, the urate levels are often elevated for years before the disease is manifest in the joints.

Pathophysiology

The most common form of gout occurs because of decreased excretion of urate by the kidney. Less common causes of this condition include enzymatic defects or diseases with increased nucleotide production. Diet tends to have a direct effect on gout. Excessive organ meat consumption can lead to increased urate levels, and drinking alcohol tends to block the excretion of urate. Taken together, a diet rich in alcohol and organ meats is likely to aggravate gout.

Presentation

Most people with hyperuricemia are asymptomatic. Some will have an acute gouty flare and experience increasing pain over a short number of hours, which typically occurs in only one joint—most often the metatarsophalangeal joint of the great toe. People who experience these types of episodes may report recent alcohol consumption or have a known history of asymptomatic hyperuricemia. Objective findings include marked tenderness, erythema, warmth, swelling, and severe limitation of motion. Patients with chronic forms of the disease may also have palpable tophi.[14]

Diagnostic Imaging

In its initial stage, gout is comparable on radiographs to RA because only soft-tissue swelling is seen; however, as the disease progresses, erosions and ultimately joint destruction may be apparent. The erosions that occur with gout are characteristic and may be easily discerned from those that occur with RA. In patients with gout, new bone formation partially surrounding a tophus creates a characteristic overhanging margin (Fig. 5).

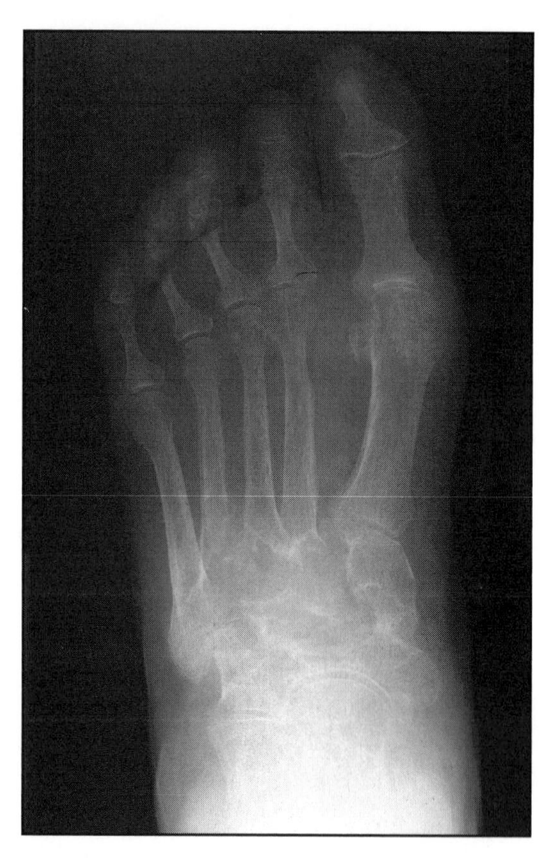

Figure 5
AP view of the foot shows erosions with an overhanging margin in the distal medial metatarsal of the metatarsophalangeal joint of the great toe. Note that the central joint space is not narrowed or compromised. A lateral view will further delineate the size and location of the erosion.

Laboratory Studies

The measurement of serum urate levels is not particularly helpful in diagnosing gout because gout will not develop in many patients with elevated urate levels. Some patients with active gout may have a normal serum urate level at the time of presentation. Other circumstances may result in hyperuricemia, such as renal insufficiency or ingestion of drugs (eg, thiazide diuretics).

Gout is diagnosed by laboratory testing of the synovial fluid. This fluid will show high levels of white blood cells and, more importantly, characteristic needle-shaped monosodium urate crystals. Because monosodium urate crystals are negatively birefringent (doubly refractive), the precise identification of crystals is made with polarized light. Calcium pyrophosphate may also deposit as crystals in the joint (a condition called pseudogout); these crystals are positively birefringent.

Treatment and Prevention

Gouty arthritis can be prevented by reducing elevated urate levels with diet modification or medication. Dietary recommendations include decreased ingestion of purines and decreased alcohol consumption. Drugs that lower serum urate levels include probenecid, which promotes the excretion of urate in the urine, and allopurinol, a xanthine oxidase inhibitor that decreases the synthesis of uric acid.

Colchicine can be used to treat acute episodes of gout because it blocks neutrophil phagocytosis of urate crystals. NSAIDs, of course, can be used to relieve pain and acute inflammation. Historically, indomethacin has been the agent of choice, but other NSAIDs can be equally effective.[15] For patients with chronic renal insufficiency, corticosteroids are the drug of choice.

Key Terms

Gout An inflammatory arthritis associated with deposition of urate crystals in the joint

Hyperuricemia Elevated serum levels of urate

Juvenile rheumatoid arthritis A chronic inflammatory disease in children that is characterized by pain, swelling, and tenderness in one or more joints and may result in impaired growth and development

Rheumatoid arthritis A chronic inflammatory disease that is probably triggered by an antigen-mediated inflammatory reaction against the synovium in the joint

Tophi Chalky deposits of urate surrounded by inflammatory cells

References

1. Lee DM, Weinblatt ME: Rheumatoid arthritis. *Lancet* 2001;358:903-911.
2. Aho K, Koskenvuo M, Tuominen J, et al: Occurrence of rheumatoid arthritis in a nationwide series of twins. *J Rheumatol* 1986;13:899-902.
3. Silman AJ, MacGregor AJ, Thomson W, et al: Twin concordance rates for rheumatoid arthritis: results from a nationwide study. *Br J Rheumatol* 1993;32: 903-907.
4. Jenkins RN, Nikaein A, Zimmermann A, et al: T cell receptor V beta gene bias in rheumatoid arthritis. *J Clin Invest* 1993;92:2688-2701.
5. Uematsu Y, Wege H, Straus A, et al: The T-cell-receptor repertoire in the synovial fluid of a patient with rheumatoid arthritis is polyclonal. *Proc Natl Acad Sci USA* 1991;88:8534-8538.
6. Wagner UG, Koetz K, Weyand CM, et al: Perturbation of the T cell repertoire in rheumatoid arthritis. *Proc Natl Acad Sci USA* 1998;95:14447-14452.
7. Shiozawa S, Shiozawa K, Fujita T: Morphologic observations in the early phase of the cartilage-pannus junction: Light and electron microscopic studies of active cellular pannus. *Arthritis Rheum* 1983;26:472-478.
8. McCachren SS, Haynes BF, Niedel JE: Localization of collagenase mRNA in rheumatoid arthritis synovium by in situ hybridization histochemistry. *J Clin Immunol* 1990;10:19-27.
9. Gravallese EM, Darling JM, Ladd AL, et al: In situ hybridization studies of stromelysin and collagenase messenger RNA expression in rheumatoid synovium. *Arthritis Rheum* 1991;34:1076-1084.
10. Martel-Pelletier J, McCollum R, Fujimoto N, Obata K, Cloutier JM, Pelletier JP: Excess of metalloproteases over tissue inhibitor of metalloprotease may contribute to cartilage degradation in osteoarthritis and rheumatoid arthritis. *Lab Invest* 1994;70:807-815.
11. Bland JH: Rheumatoid arthritis of the cervical spine. *J Rheumatol* 1974;1:319-342.
12. Kauppi M, Sakaguchi M, Konttinen YT, et al: Pathogenetic mechanism and prevalence of the stable atlantoaxial subluxation in rheumatoid arthritis. *J Rheumatol* 1996;23:831-834.
13. McCarty DJ: Gout without hyperuricemia. *JAMA* 1994;271:302-303.
14. Lawry GV II, Fan PT, Bluestone R: Polyarticular versus monoarticular gout: A prospective, comparative analysis of clinical features. *Medicine (Baltimore)* 1988;67:335-343.
15. Emmerson BT: The management of gout. *N Engl J Med* 1996;334:445-451.

Back Pain and Sciatica

Acute low back pain will affect most adults at least once in their lives, yet nearly all will recover spontaneously, often without a precise diagnosis ever being made. Most back pain is idiopathic, meaning that its exact cause is unknown. Many patients with back pain have abnormalities that appear on imaging studies, such as osteoarthritis or disk disease; however, many asymptomatic patients demonstrate radiographic findings. Therefore, it is incorrect to always assume a clear cause-and-effect relationship between symptoms and findings on imaging studies.

Low back pain can occur with or without leg pain. Pain that radiates down the back of the legs is known as radiculopathy or sciatica because the pain is thought to be caused by irritation of the roots of the sciatic nerve. Moreover, patients with back pain may have disk disease. Over time, aging and cumulative trauma may rob intervertebral disks of much of their ability to absorb shock. Spinal stenosis, characterized by bony or soft-tissue overgrowth in the spinal canal, may be the source of pain, especially in patients older than 60 years. Other sources of back pain include degeneration of the facet joints between the vertebrae, inflammation of the surrounding soft tissues, or diseases of any of the nearby organs, such as the aorta or kidney. Although most back pain is benign and self-limiting, in some cases it is caused by diseases that require more aggressive treatment, such as tumors, infections, fractures, and instabilities.

Epidemiology

Back pain is an extremely common medical complaint, with a lifetime prevalence as high as 70%.[1] It is estimated that at any given time 1% of the entire work force is disabled because of back pain. Given its multiple causes, back pain has many associated risk factors. Heavy lifting at work has been cited by many, although many people who do not lift routinely have both back pain and disk damage. Weakness of the abdominal musculature may also increase the risk of back pain. Obesity, smoking, and depression are all associated with an increased risk of back pain.[1]

Pathophysiology

Disk Disease

Disk herniation is thought to be a major cause of back pain and sciatica. Intervertebral disks consist of an outer ring and a central core. The outer ring, the anulus fibrosus (translated literally as a "fibrous ring") can tear as a result of an acute event or attrition. This tear permits the central core of the disk, the nucleus pulposus, to bulge (herniate) from its normal location. The injured disk can compress the nerve roots (Figs. 1 and 2). The nucleus pulposus consists of a gel that contains negatively charged proteins that are able to bind water. Normally, the water will be extruded with load bearing and then imbibed again when the load is lifted, allowing the disk to act as a cushion between adjacent vertebral bodies. The disk therefore acts as a hydraulic shock absorber. When there is a break in the anulus fibrosus, however, some of the nucleus pulposus may escape its central location, resulting in a bulge or herniation. Protruding pieces of nucleus pulposus can then compress a spinal nerve or cause local irritation.

The anulus fibrosus can tear at any point, but the direct anterior and posterior positions are reinforced by the longitudinal ligaments of the spine; thus, the path of least resistance is posterolaterally, toward the neural foramen. A large herniation in the lumbar region can compress the cauda equina, the terminal nerve roots of the spinal cord that have not yet left via their respective foramina. Bilateral sciatic pain or saddle anesthesia (decreased sensation around the perineum) and loss of bowel or bladder control suggests such compression. This presenta-

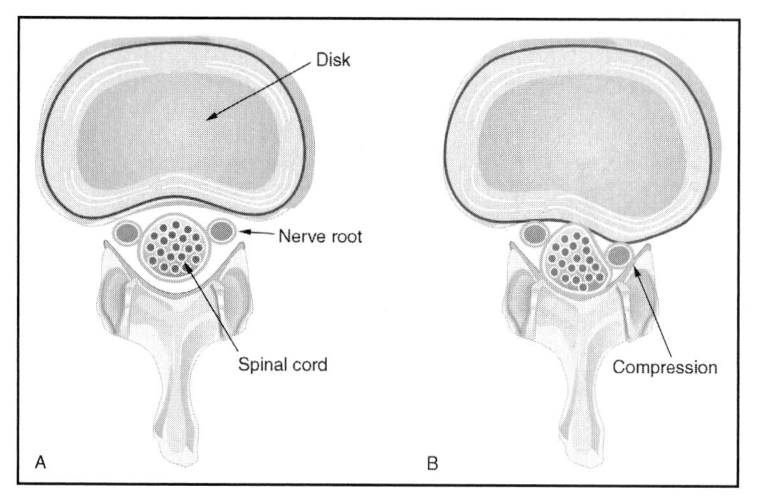

Figure 1
Simplified axial view of the lumbar spine at the level of the disk space. **A**, Normal anatomy. **B**, The same showing a left-sided, posterolateral herniated disk.

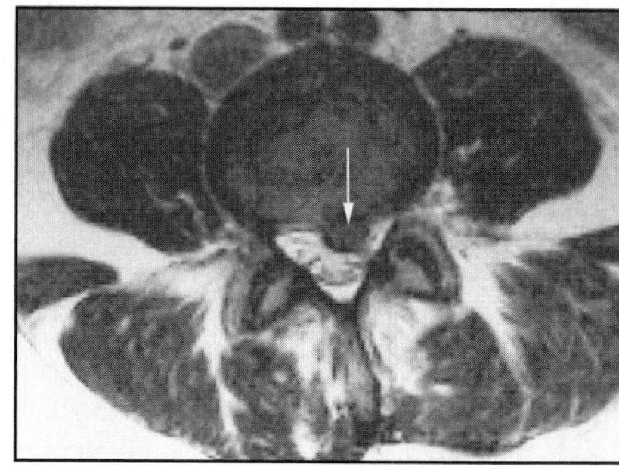

Figure 2
Preoperative axial MRI scan showing a herniated lumbar disk (arrow).

*(Reproduced from Greene WE (ed): **Essentials of Musculoskeletal Care**, ed 2. Rosemont, IL, American Academy of Orthopaedic Surgeons, 2001, p 561.)*

tion demands urgent MRI and possibly emergency surgery.

Degeneration of the disk without herniation can occur after trauma but is also a function of aging. As people age, the disk dehydrates and loses its resilience. Thus, even without overt trauma, it is likely that the disk space of an elderly patient will be radiographically abnormal, although not necessarily symptomatic.

Osteoarthritis and Spinal Stenosis

Osteoarthritis is often a continuation of lumbar disk disease. As the disk is no longer able to cushion the joint, abnormal forces and motions are applied to the posterior facet joints and will tend to erode the normal architecture (Fig. 3). This process is similar to the changes in the knee joint seen after the meniscus is severely damaged or completely removed.

Degenerative disk disease and arthritis can cause spinal stenosis, a condition characterized by narrowing of the canal housing the spinal cord (Fig. 4). Spinal stenosis is commonly caused by overgrowth of the bone osteophytes or by enlargement of the ligamentum flavum. Such overgrowth can put pressure on the neural elements and produce pain in the legs, especially after activity.

Rheumatologic Conditions

The facet joints of the spine are true synovial joints; thus, they are susceptible to any of the inflammatory conditions affecting the synovium. Rheumatoid arthritis can affect the low back, but it is more common in the cervical spine. Ankylosing spondylitis is an inflammatory disorder that affects the low back and pelvis, almost always involving the sacroiliac joint. After a period of years, it may lead to autofusion of vertebral bodies. The characteristic finding of most rheumatologic conditions is morning stiffness and discomfort that actually improve as the day progresses.

Spondylolysis and Spondylolisthesis

The term spondylolysis refers to a defect of the bone between the superior and inferior articular processes of the vertebra, or pars interarticularis (Fig. 5, *A*). This is a common condition, occurring in 5% of the population.[2] It is also seen in athletes, such as gymnasts or football linemen, who routinely hyperextend their spines. This bony defect can progress to the point at which the vertebral body above can slip forward onto the one below, causing a condition called spondylolisthesis (Fig. 5, *B*). This progression can be avoided by early detection with bone scanning or MRI and activity modification.

Tumors and Infections

Tumors and infections can be described together because both are typically caused by hematogenous (bloodborne) seeding. Tumor masses and epidural abscesses can cause a cauda equina syndrome by direct impingement on the neural elements. They also can cause bone destruction and pathologic frac-

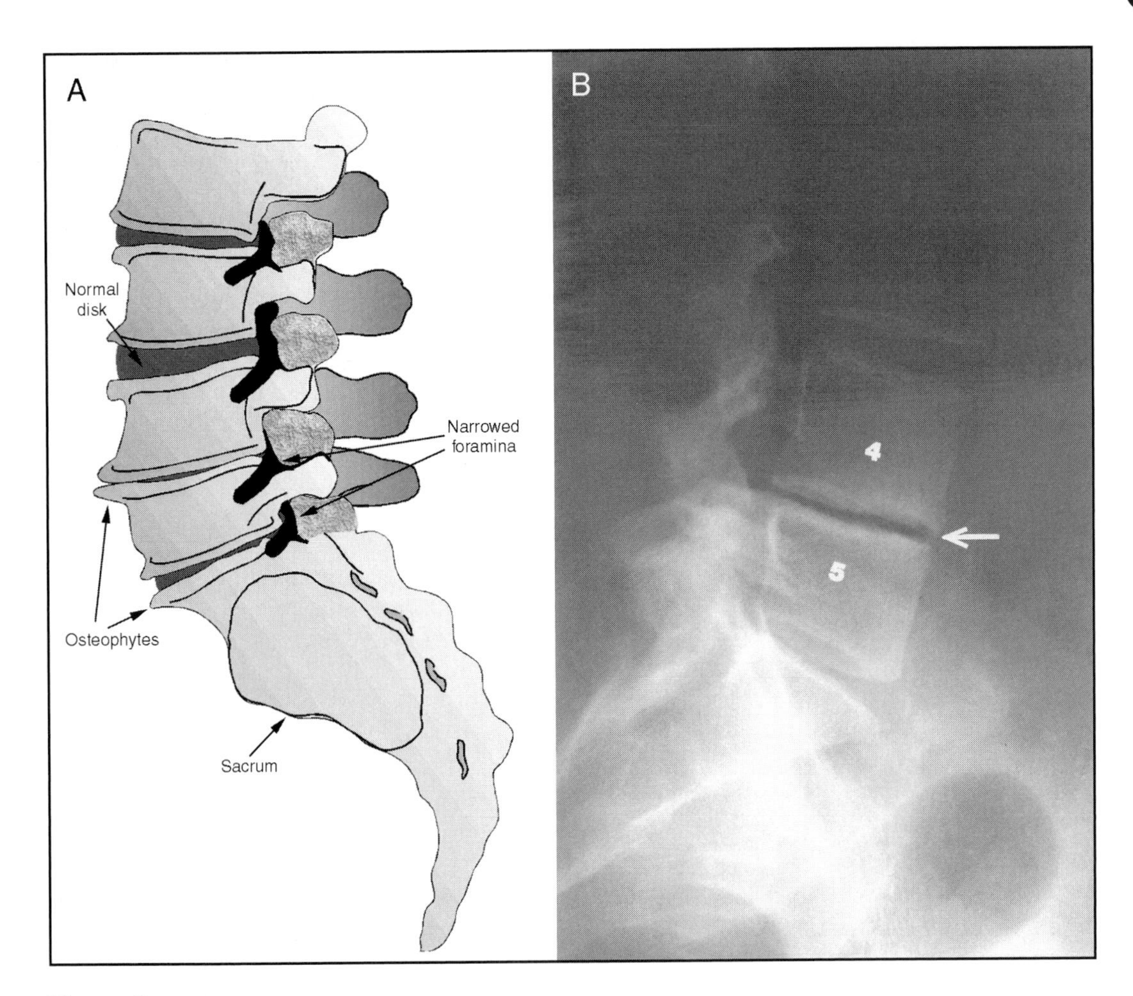

Figure 3
A, Schematic representation of degenerative changes of the spine. Notice the advanced degeneration of the L4-L5 and L5-S1 intervertebral disks, the formation of osteophytes, and the narrowing of the intervertebral foramina (seen here as black space).

(Reproduced from Buckwalter JA, Boden SD, Eyre DR, Mow VC, Weidenbaum M: Intervertebral disk aging, degeneration, and herniation, in Buckwalter JA, Einhorn TA, Simon SR (eds): ***Orthopaedic Basic Science: Biology and Biomechanics of the Musculoskeletal System****, ed 2. Rosemont, IL, American Academy of Orthopaedic Surgeons, 2000, p 561.)*

B, Lateral radiograph showing marked degenerative changes affecting the disk between L4 and L5.

(Reproduced from Greene WE (ed): ***Essentials of Musculoskeletal Care****, ed 2. Rosemont, IL, American Academy of Orthopaedic Surgeons, 2001, p 557.)*

tures (Fig. 6). Almost all spine tumors are either metastatic or blood-cell malignancies. One exception is the osteoid osteoma, a benign tumor found typically in teenagers. This tumor can cause back pain severe enough to warrant excision, but it may resolve without treatment.

Clinical Presentation

History and Physical Examination

Patients with back pain can be said to have either mechanical or nonmechanical pain. The key distinction is whether there is pain with activity. Many patients with sciatica report a sharp pain in the back and a burning pain that radiates down the back of the leg, sometimes as far as the toe, but often terminating behind the knee. Numbness and tingling (paresthesias) are also reported. Postural change may affect the quality of the pain; pain with movement implies disease at the joints of the spine, and pain worsened by sitting implicates the disks. The physical examination of a patient with low back pain should include observing gait, assessing spinal flexibility, testing motor strength (hip to toes), and assessing knee and ankle reflexes.

Patients with spinal stenosis may have neurogenic claudication, or leg pain during activity. This pain is superficially similar to that

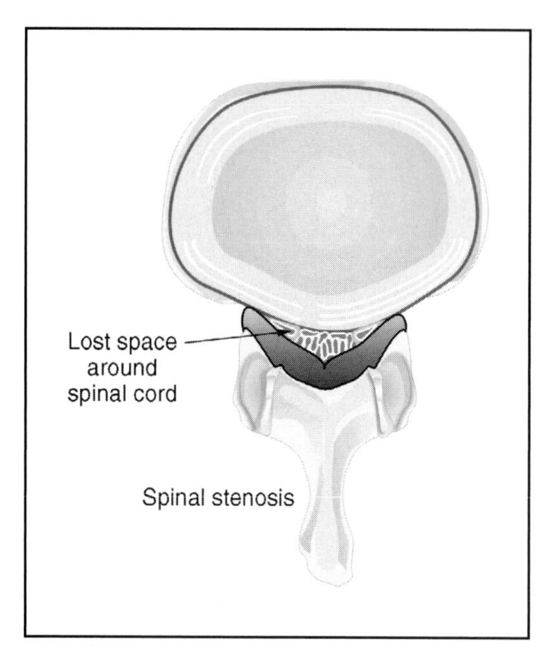

Figure 4

The lumbar spine at the level of the disk space showing spinal stenosis. The shaded region represents bone overgrowth and enlargement of the ligamentum flavum.

of vascular claudication. One key distinction between them is that forward bending, which distracts the spine and enlarges the spinal canal, improves the symptoms of neurogenic claudication only. Therefore, a patient with neurogenic claudication may have no pain while riding a bicycle, whereas a patient with vascular claudication will. Another distinction is that vascular claudication appears invariably with activity and is relieved consistently and promptly with rest, neither of which occurs with neurogenic claudication.

Other signs associated with back pain and sciatica include decreased spinal mobility; a positive straight leg raising test; altered sensory, motor, or reflex examination findings; and increased pain with increasing intra-abdominal pressure (Valsalva maneuver). Acute pain to palpation suggests a nonorganic cause.

Imaging Studies

MRI is the diagnostic imaging modality of choice for disk disease, although the incidence of disk abnormalities in patients who have no pain is probably 30% or higher.[3] MRI, therefore, is not clinically specific. However, it is sensitive. A sagittal view of the

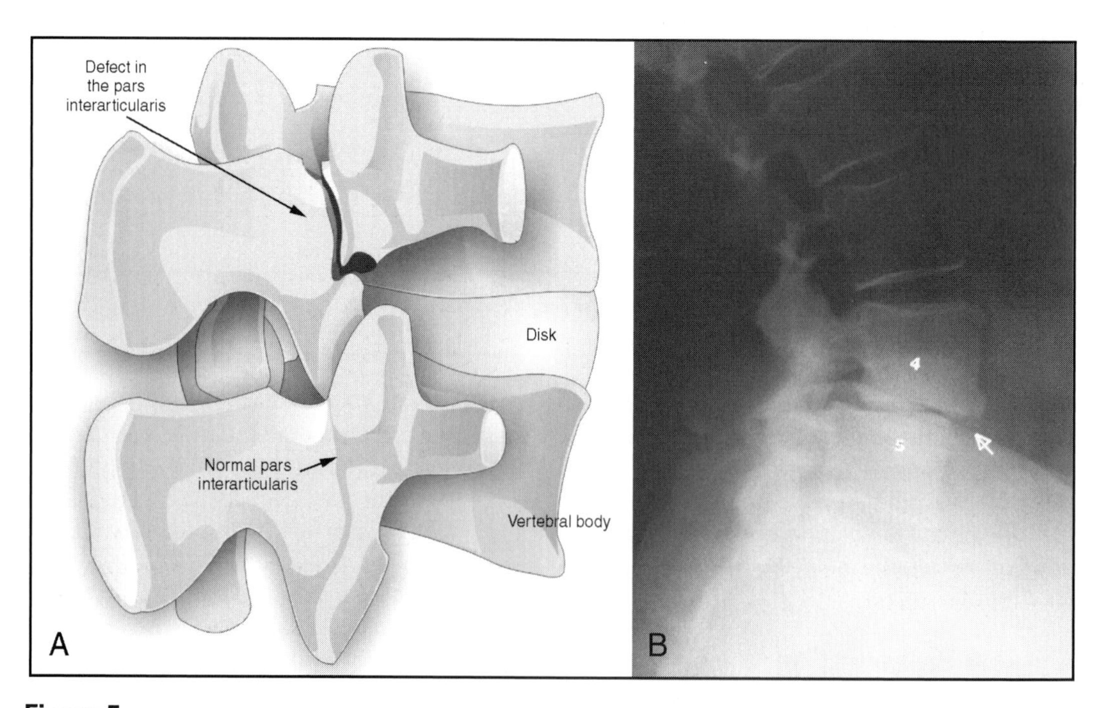

Figure 5

A, A lumbar spine segment showing a defect in the right pars interarticularis of the upper vertebra. **B**, Lateral radiograph of a patient with degenerative spondylolisthesis at L4 on L5 (arrow).

(Reproduced from Greene WE (ed): ***Essentials of Musculoskeletal Care****, ed 2. Rosemont, IL, American Academy of Orthopaedic Surgeons, 2001, p 564.)*

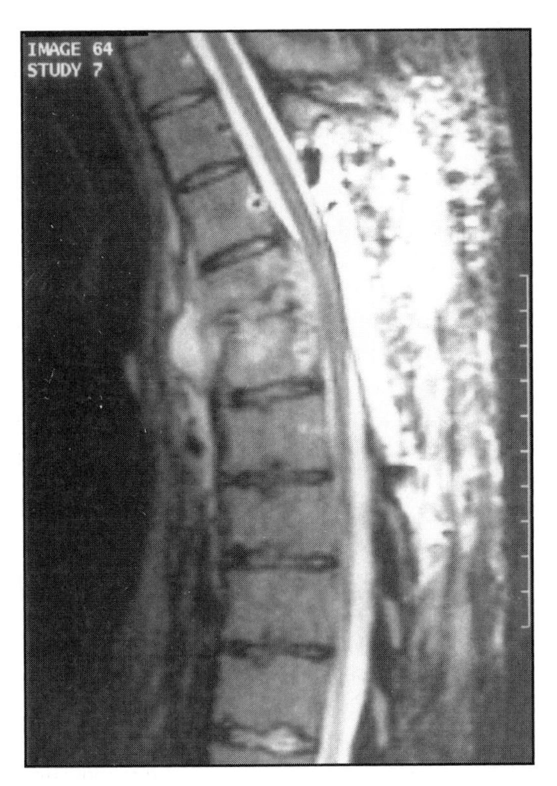

Figure 6
Sagittal T2-weighted MRI scan of a 35-year-old man in whom spinal osteomyelitis is causing bone destruction in the vertebral body.

(Reproduced from ***Orthopaedic In-Training Examination, 2000.*** *Rosemont, IL, American Academy of Orthopaedic Surgeons, 2000, p 46.)*

entire lumbar spine will clearly show bulges into the spinal canal. Axial views can show encroachment on the nerve root by highlighting disk material or osteophytes.

Plain radiographs may show degenerative joint disease of the spine, which is manifest as decreased space between the disks and osteophyte formation. Bone scanning is useful to exclude infection, occult fracture, or tumors; however, multiple myeloma, a blood cell malignancy commonly found in the spine, is not apparent on bone scans. The presence of vertebral compression fractures on radiographs suggests that an osteoporosis workup is indicated.

Laboratory Studies
Back pain is usually not associated with disease processes that can be identified with routine blood screening. If infection, tumor, or an inflammatory cause is suspected, a complete blood cell count and an erythrocyte sedimentation rate or C-reactive protein level can be obtained for general screening.

Differential Diagnosis
The first step of evaluation is to ensure that the back pain is not a harbinger of a serious medical condition. Thus tumor, infection, fracture, and instability should be excluded. A patient's medical history often provides enough information to determine whether radiographs are needed on the first visit. When a patient has had back pain for 6 weeks or longer, imaging studies are usually required.

It is unlikely that an adult younger than 50 years with no history of a prior malignancy has a tumor; the odds are even lower when a patient has no constitutional complaints, such as fever, weight loss, or night pain. Infection, likewise, is rare in an otherwise healthy patient. The medical history should include questions about possible immune system compromise (eg, human immunodeficiency virus or use of immune-suppressing drugs, such as steroids) or intravenous drug use. Patients with diabetes mellitus or sickle cell anemia are also prone to bone infections.

Fractures and instabilities are rare unless there is a history of trauma. The exception is a patient with osteoporosis. Osteoporosis may be a typical postmenopausal condition, but it also should be considered as a secondary condition in patients with a history of long-term steroid use, a metabolic abnormality, or recent lactation.

Low back pain may originate from the many discrete anatomic structures in the spine itself, such as the disk, the vertebral body, the spinal nerve roots, the facet joints and their ligaments, and the paraspinal muscles. There also may be a visceral cause for the pain, such as an abdominal aortic aneurysm or kidney infection. The leg pain of sciatica may mimic peripheral nerve diseases or vascular insufficiency. Worrisome findings that should prompt diligent investigation include night pain, pain at rest, fever, unintentional weight loss, acute motor weakness, and unremitting pain of increased severity.

Treatment and Prevention
Back pain resolves spontaneously in most patients; therefore, the principal treatment for most mechanical back pain is to simply "do no harm" as natural recovery takes place. Should diagnostic imaging studies reveal disk

degeneration without compression of the spinal nerves, physical therapy with an emphasis on lumbar strengthening and flexibility is indicated. Patients with sciatica and MRI evidence of nerve root compression caused by a disk herniation may benefit from epidural steroid injections to reduce the inflammation around the nerve roots.

Nonsurgical Treatment

The use of anti-inflammatory and mild analgesic medications may aid symptom relief, although some patients will require short courses of narcotic pain relievers. Bed rest is considered inappropriate because it induces atrophy and weakness, although activity modification, such as the avoidance of lifting, may be useful in hastening recovery and preventing recurrence. In the acute phase of low back pain, physical therapy modalities, such as heat, ice, and ultrasound may provide relief; however, there is no evidence-based literature to support these treatments.[4] Therapy should also include "back school" to educate the patient about proper posture and lifting techniques.

Surgical Treatment

Surgery for sciatica is reserved for patients who do not improve after a few months and are willing to assume the risks of surgery. Diskectomy (surgical removal of the herniated disk material) offers patients the chance for a more rapid initial recovery, especially for radicular symptoms in the legs. However, after a period of years, patients who have had surgery are functionally indistinguishable from patients who have not had surgery.[5] Surgical fusion for back pain remains a controversial topic.

Prevention

Preventing back pain is also an area of active investigation. Some believe that a program of spinal and abdominal muscle strengthening exercises may reduce the incidence of low back pain. Furthermore, patients are less debilitated by pain if they are fit and exercise regularly. Using education about spinal biomechanics as a preventive strategy has produced mixed results. Decreasing risk factors, such as smoking and obesity, has the force of logic, but there is no evidence that attempting to modulate these risk factors is effective in preventing back pain.

Key Terms

Ankylosing spondylitis An inflammatory disorder that affects the low back and pelvis and produces stiffness and pain

Anulus fibrosus The outer ring of fibrous material surrounding the nucleus pulposus of the intervertebral disks

Cauda equina The terminal nerve roots of the spinal cord located within the vertebral canal; so named because they resemble the tail of a horse

Neurogenic claudication Leg pain that occurs as a result of compression of nerves within the spinal canal; often associated with spinal stenosis

Nucleus pulposus The central core of gelatinous material within intervertebral disks

Osteophytes Overgrowth of bone, common in osteoarthritis and spinal stenosis

Saddle anesthesia Decreased sensation around the perineum

Spinal stenosis Narrowing of the canal housing the spinal cord; commonly caused by encroachment of bone

Spondylolysis A defect or fracture of the pars interarticularis, the portion of bone located between the superior and inferior articular processes of the vertebrae

Spondylolisthesis Anterior displacement of a vertebral body relative to the adjacent vertebral body below

Vascular claudication Leg pain that is caused by ischemia secondary to vascular disease; the pain uniformly occurs with activity and is relieved with rest

References

1. Andersson GBJ: The epidemiology of spinal disorders, in Frymoyer JW (ed): *The Adult Spine: Principles and Practice*, ed 2. New York, NY, Raven Press, 1997, pp 93-141.
2. Hensinger RN: Spondylolysis and spondylolisthesis in children. *Instr Course Lect* 1983;32:132-151.
3. Boden SD, Davis DO, Dina TS, et al: Abnormal magnetic-resonance scans of the lumbar spine in asymptomatic subjects: A prospective investigation. *J Bone Joint Surg Am* 1990;72:403-408.
4. van Tulder MW, Goossens M, Waddell G, et al: Chapter 12: Conservative treatment of chronic low back pain, in Nachemson AL, Jonsson E (eds): *Neck and Back Pain: The Scientific Evidence of Causes, Diagnosis, and Treatment*. Philadelphia, PA, Lippincott, Williams & Wilkins, 2000, pp 271-304.
5. Weber H: Lumbar disc herniation: A controlled, prospective study with ten years of observation. *Spine* 1983;8:131-140.

Chapter 12

Infection

Infections of the bones and joints, although less common than infections in other parts of the body, are nonetheless significant clinical problems. These infections generally fall into three broad categories: (1) infection of the joint (septic arthritis), (2) infection of the bone (osteomyelitis), and (3) soft-tissue or bone infection adjacent to a surgical implant or joint prosthesis. The presentation, pathophysiology, and treatment principles of each type of infection are distinct; thus, they will be described separately. Some rarer but still important types of musculoskeletal infection are described as well—namely, the musculoskeletal effects of human immunodeficiency virus (HIV) infection and other immunocompromised states, tuberculosis infections of the bone and joints, Lyme disease, and diabetic foot infections.

Bone and joint infections differ from infections in other parts of the body in three fundamental ways. First, the blood supply to bone and joint tissue is not as rich as that to most other tissues in the body. Although decreased blood flow makes it more difficult for bacteria to reach skeletal tissues and cause infection, it also makes the elimination of established infections much more troublesome. Second, the mechanical function of many joints is predicated on the precise geometry of the adjacent bones and cartilage surfaces. When this geometry is disturbed by infection, the function of a joint or even a system of joints may be compromised. For example, destruction of cartilage in even a small area of the knee as a result of septic arthritis can upset the mechanics of the knee, significantly interfere with walking, and possibly cause problems in other joints, such as the hip, ankle, or foot. Thus, treatment of musculoskeletal infections is imperative before tissue destruction can occur.

Finally, the frequent use of surgical implants increases the risk of infection following many orthopaedic procedures. These materials disrupt the normal biologic environment. Even when the material is totally inert, the presence of foreign bodies allows bacteria to establish a nidus of infection.

Septic Arthritis

Septic arthritis is infection and inflammation of a joint caused by bacterial, fungal, or viral invasion of the synovium. Bacterial septic arthritis involves a single joint in 90% of cases. The knee is the most commonly involved joint, followed by the hip, shoulder, ankle, and wrist. Patients with septic arthritis typically have acute swelling and warmth around the joint, effusion, tenderness to palpation, and extreme pain with minimal range of motion.

Bacteria can reach a joint in four ways: (1) from the blood, that is, via hematogenous seeding during bacteremia;(2) from the outside environment through direct inoculation of organisms following penetrating trauma, joint aspiration, or surgery;(3) from the localized spread of a nearby soft-tissue infection, such as cellulitis or bursitis; or (4) from spread of a bone infection near the joint (periarticular osteomyelitis).[1] Hematogenous seeding is the most common cause of infectious arthritis.

Epidemiology

Septic arthritis is rare in the normal adult population; fewer than 5% of cases of acute arthritis have an infectious cause. Bacterial septic arthritis is more common in infants, the elderly, and patients with impaired immunity. Risk factors include a history of rheumatoid arthritis (RA) or intravenous drug use. Patients with RA are more likely to have multiple joint involvement. Intravenous drug users often have infections in atypical joints, such as the sternoclavicular, sacroiliac, and manubriosternal joints.[2,3]

Information about patient age, medical history, and risk factors must be obtained because these factors not only identify the potential for infection but also the possible infecting organism. With this information, the physician can initiate empiric antibiotic treatment once the diagnosis is established but before identification of a specific bacterium. *Staphylococcus aureus* is the most common cause of all joint infections; therefore, empiric antibiotics must cover it, even if there is a specific bacteria associated with a particular demographic. For example, patients with a history of illicit drug use are known to have an increased risk of Pseudomonal infection, but even among these patients staphylococcal infection is common as well. Thus, empiric treatment of infection in these patients must cover that organism as well.

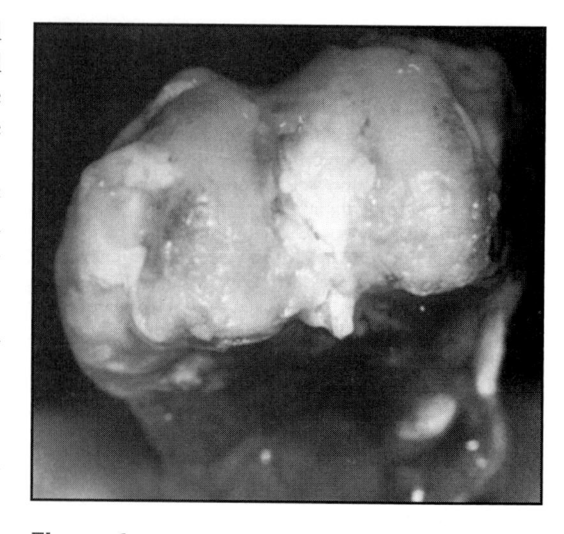

Figure 1
Photograph of the distal femur of a rabbit that was infected with *S aureus* but did not receive antibiotics. Notice the severe cartilage destruction.

(Reproduced with permission from Goodman SB, Chou LB, Schurman DJ: Management of pyarthrosis, in Chapman MW (ed): ***Operative Orthopaedics.*** *Philadelphia, PA, JB Lippincott, 1988, pp 847-858.)*

Pathophysiology

Hematogenous septic arthritis occurs when bacteria are able to seed the joint after escaping from the synovial capillaries, which do not have a basement membrane. In normal joints, host defense mechanisms are able to destroy these bacteria; however, when these mechanisms are compromised, the organisms can survive and cause infection.

While open trauma to a joint is an obvious risk factor for infection, closed trauma is also a risk factor for reasons that are not entirely clear. Possible explanations include a reactive hyperemia increasing exposure to bacteria, disruption of the local anatomy allowing bacteria easier access to the joint and interfering with defense mechanisms, and formation of a hematoma providing an ideal culture medium for bacterial growth.[1]

Many bacteria can bind to articular cartilage. *S aureus* has an increased ability to bind to articular cartilage compared with other bacteria, which may explain why it is the most common pathogen in bacterial septic arthritis.[1] Within hours after entering the synovium, bacteria induce a neutrophilic infiltration of the synovium. Cartilage destruction begins within 48 hours as a result of the release of proteases and cytokines from inflammatory cells and increased intra-articular pressure (Fig. 1). While neutrophils are one of the primary host defenses against joint infection, they also are a major cause of joint destruction in septic arthritis.

The cytokine interleukin-1 (IL-1), released by macrophages and tissue monocytes in response to bacteria, plays an important role in the destruction of cartilage. Exposure of the joint surface to IL-1 leads to proteoglycan release, increased secretion of collagenase, release of metalloproteases, activation of latent collagenases, and inhibition of glycosaminoglycan synthesis.[1]

If the septic process is terminated early, the proteoglycan losses can be reversed and normal function restored. If too much matrix is lost, the chondrocytes are exposed to increased mechanical stress and die, which leads to further matrix loss. The remaining chondrocytes are exposed to more stress and are more apt to die, perpetuating a vicious cycle that can lead to complete joint destruction.[4]

Bacteriology

In adults, *S aureus* accounts for 60% of the cases of bacterial septic arthritis.[5] Other common organisms include streptococci and gram-negative organisms. Although *S aureus* is still the most common infectious organism in patients who are intravenous drug users, gram-negative infections, especially Pseudomonal infections, can also be transmitted via intravenous drug use. [2]

S aureus is a pathogen that may be found in all children, but when infection occurs, several other organisms need to be considered, based on the age of the patient. In infants from birth to 6 weeks of age, the most likely pathogens are group A and group B streptococci, *Streptococcus pneumoniae*, and *Escherichia coli*.[6] *Neisseria gonorrheae* infection acquired via maternal transmission at birth also needs to be considered in this age group.

In children younger than 5 years, an increasing number of cases of septic arthritis are caused by *Kingella kingae*. This organism is thought to colonize the nasopharynx and then spread to the joints via invasion of the bloodstream. *Haemophilus influenzae* type B (HIB) was a common pathogen in children of this age in the past; however, with the advent of widespread vaccination, HIB infection has become extremely rare. *S pneumoniae* is also an important organism in this age group.[5]

In children older than 5 years, *S aureus*, group A streptococci, and *S pneumoniae* are the most common pathogens. As children get older, the incidence of staphylococcal infections increases, and the incidence of streptococcal infections decreases.

Disseminated gonococcal infection, classically associated with a syndrome of fever, chills, rash, and migratory arthritis of the large joints, always precedes gonococcal septic arthritis, normally involving a single joint. Although gonococcal septic arthritis is always preceded by systemic gonococcal infection, this stage may go unnoticed in up to 30% of patients.[7] Gonococcal infection is the most common type of infectious arthritis in healthy adults. It generally affects patients younger than 40 years, and women are two to three times more likely to be affected than men.

Clinical Presentation

A typical medical history for septic arthritis would be an 80-year-old female nursing home resident being treated with corticosteroids and methotrexate for RA. She has had increasingly worse knee pain for the past 2 days, with perhaps a history of minor trauma. Upon presentation, her temperature is 100.3°F, and she has a large warm effusion on her knee that is diffusely tender to palpation. She also has extreme pain with minimal range of motion. Although she was able to walk 3 days ago, she now finds weight bearing painful. Her skin appears normal, although cellulitis, a skin infection, is certainly common in such patients.

Radiographs of the knee joint typically show soft-tissue swelling and periarticular osteopenia consistent with long-term RA. In this setting, there is no particular need for special imaging studies. When hip infection is suspected, ultrasound or MRI may be needed because these studies can detect an effusion with great sensitivity.

When evaluating a patient for infection, it is helpful to obtain a white blood cell count, erythrocyte sedimentation rate, and C-reactive protein level. The essential laboratory test, however, is the analysis of fluid obtained from joint aspiration. This fluid must be examined microscopically (Gram stain) and then sent to the laboratory for a white blood cell count and culture. Aspirates with more than 50,000 white blood cells per milliliter are thought to represent infection. Some patients with inflammatory arthritis but no infection may have more than 50,000 white blood cells per milliliter, whereas some patients with infection and immune suppression (which limits white cell production) may have less. The best criterion to use depends on the utility (value) of each possible outcome.

The differential diagnosis of septic arthritis depends on the clinical scenario. Conditions to consider include inflammatory arthritis, reactive arthritis (a noninfectious joint inflammation following infection elsewhere in the body), trauma, superficial infection or abscess near but not in the joint, and collagen vascular disorders.

Treatment and Prevention

The goals of treatment include sterilization of the joint, removal of inflammatory cells and their enzymes, elimination of the destructive synovial pannus, and restoration of function.[1] Prompt administration of antibiotics and drainage of the involved joint can prevent cartilage destruction, minimizing the risk of secondary arthritis, joint instability, deformity, and loss of function. Empiric antibiotics should be started as soon as cultures of the blood and synovial fluid are obtained (ie, before interpretation).

The initial choice of antibiotic should be based on the morphology of organisms visualized on Gram stain, if any, and the likely pathogens based on the patient's age and risk factors. Antibiotics should be given intravenously to achieve rapid peak serum concentrations. Direct injection of the joint with antibiotics is not advised because it can cause a chemical synovitis.

In order to limit cartilage destruction and

decrease intra-articular pressure, early drainage of pus and necrotic material is required. Three methods are available: (1) repeated needle aspiration, often at the bedside; (2) arthroscopic lavage; and (3) arthrotomy (open surgery). For easily accessible joints, such as the knee, repeated needle aspirations can be used; however, patient discomfort and the difficulty draining thick collections of pus often make arthroscopy a better choice. For less accessible joints, such as the hip, arthrotomy is the best means for irrigation and débridement. Following arthrotomy, the joint should be closed over suction drainage, which can be removed after several days as the volume of drainage decreases and the patient's condition improves. If repeated aspirations are attempted and systemic antibiotics are administered, failure to improve within 48 hours is an indication for surgical drainage.

Postoperatively, the joint should be splinted in the functional position with assisted range-of-motion exercises initiated once the inflammatory response has decreased. Prolonged immobilization should be avoided because it promotes the formation of adhesions, atrophy, and contractures. For joints in the lower extremity, weight bearing should be protected until the inflammation has subsided and range of motion and strength are almost normal.[3]

Gonococcal septic arthritis is unique in that surgical drainage may not be needed, even with large effusions. This form of septic arthritis is less destructive and rarely requires surgical decompression. Instead, intravenous ceftriaxone and aspiration may be sufficient. Once there is clear improvement, oral antibiotics can be given for a total of 14 days.

Osteomyelitis

Osteomyelitis, technically defined as inflammation of the bone and marrow, signifies infection of the bone. Microorganisms can enter the bone by hematogenous spread, by extension from adjacent soft-tissue infections, or direct inoculation following trauma or surgical procedures. In adults, most cases of osteomyelitis are caused by direct inoculation, especially in the setting of open fractures. The most common location of osteomyelitis is the distal tibia. This location is at risk because the bone lacks a good envelope of muscle and robust blood supply.[8]

Acute hematogenous osteomyelitis occurs primarily in children and typically involves the metaphysis of a single long bone, especially the tibia, femur, or humerus. Most children with acute hematogenous osteomyelitis are systemically ill, with localized pain and limited use of the affected extremity. In neonates with acute hematogenous osteomyelitis, multiple sites of infection are more common. Presenting symptoms may include listlessness, poor feeding, and pseudoparalysis (unwillingness to move) of the involved limb.[9] In children, acute hematogenous osteomyelitis can be considered part of the spectrum of disease that includes septic arthritis. Unlike the disease in adults, pediatric bone infection is often found in otherwise healthy, normal hosts.

Epidemiology

Osteomyelitis in adults without risk factors is extremely rare. Among the common risk factors to consider are a history of prior open fracture, immune compromise, intravenous drug use, and blood disorders such as hemophilia and sickle cell disease. Osteomyelitis may also be found in the feet of patients with diabetes mellitus, vascular insufficiency, and prior puncture wounds through shoes.

Pathophysiology

In children, acute hematogenous osteomyelitis occurs in the metaphyses of the long bones because these areas are well perfused, yet blood flow is slow in the network of venous sinusoids (Fig. 2). Bacteria are able to escape from the blood and adhere to the collagen of the hypertrophic zone of physis. This process may be facilitated by minor trauma that can further decelerate blood flow by disrupting blood vessels and forming intraosseous hematomas, which, in turn, can serve as culture medium for bacterial growth.[7]

Once the metaphyseal region is seeded, there is a proliferation of neutrophils and the formation of pus. This pus expands under pressure toward the subperiosteal region at the surface of the bone. As the pus collects there, it can elevate the periosteum from the underlying bone or rupture through the periosteum, allowing the infection to spread into the adjacent soft tissue. Bone infarction (ie, ischemic death) occurs because the endosteal blood supply is blocked by thrombosis and the superficial blood supply is diverted as the periosteum is elevated off the bone. The area of

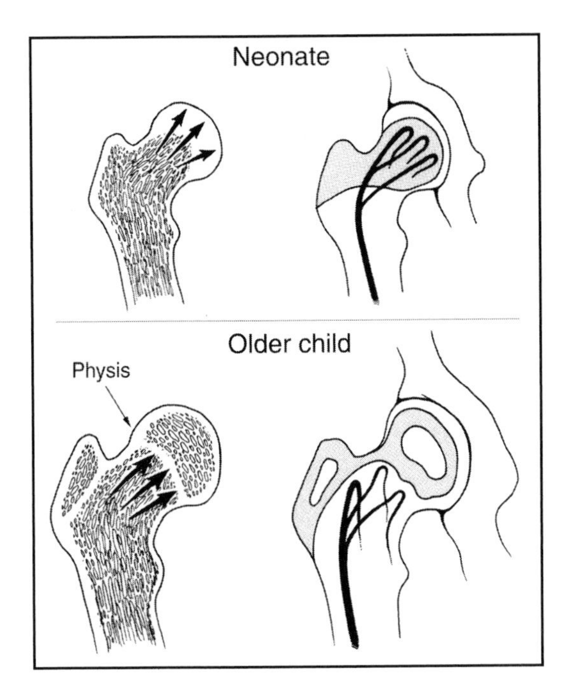

Figure 2
Relation of blood supply to the proximal femur and spread of infection (arrows). **Top,** In the neonate, the metaphyseal vessels penetrate directly into the chondroepiphysis, allowing an infection in the metaphysis to readily invade and destroy the chondroepiphysis and subsequently invade the joint. **Bottom,** In the older child, the physis serves as a mechanical barrier to the spread of infection.

(Reproduced from Dormans JP, Drummond DS: Pediatric hematogenous osteomyelitis: New trends in presentation, diagnosis, and treatment. ***J Am Acad Orthop Surg*** *1994;2:333-341.)*

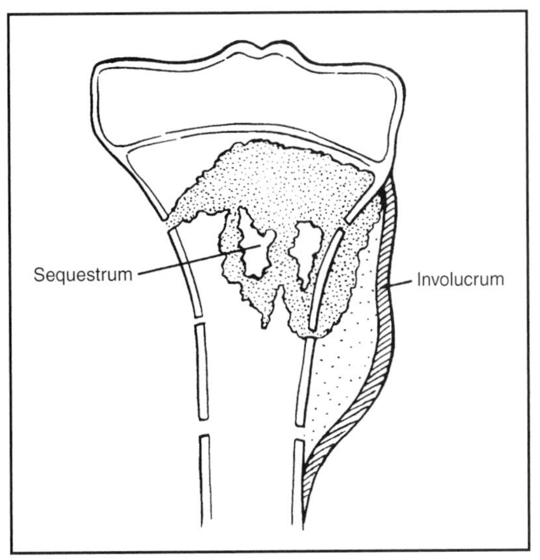

Figure 3
In osteomyelitis, the area of necrotic bone is called the sequestrum. Periosteal elevation stimulates the production of new bone, which is referred to as the involucrum.

(Adapted with permission from Wiesel SW, Delahay JN: ***Essentials of Orthopaedic Surgery****, ed 2. Philadelphia, PA, WB Saunders, 1997, p 87.)*

necrotic bone is called the sequestrum. The periosteum responds by producing new bone over the sequestrum. This new bone is called the involucrum[10] (Fig. 3).

In children, septic arthritis can develop from osteomyelitis in joints in which the metaphysis is within the joint capsule, such as the hip, shoulder, elbow (the proximal radius), and ankle (distal fibula).[7] In these joints, spontaneous decompression of pus through the surface at the metaphysis can seed the inside of the joint.

Bacteriology

The bacteriology of acute hematogenous osteomyelitis in children is similar to that of septic arthritis, with *S aureus* being the most common pathogen in all age categories. In neonates, group B streptococci and gram-negative bacteria should also be considered. Patients with sickle cell disease are at increased risk for osteomyelitis caused by *Salmonella*, which can often occur in the diaphysis rather than the metaphysis of the bone.[11]

Posttraumatic osteomyelitis occurs because the combination of clotted blood, dead space, and injured soft tissue provides an ideal medium for bacterial growth. Patients with open fractures can have bacterial contamination from the outside environment at the time of their injury. Fractures that do not involve breaks in the skin can nonetheless become infected when they are treated with surgical fixation because the bone may be exposed to bacteria during surgery.

If infection develops, there is a proliferation of inflammatory cells as in acute hematogenous osteomyelitis. The infection may become chronic as the increased intraosseous pressure (from the pus) impedes blood flow, causing bone necrosis. In adults, periosteal elevation by the inflammatory infiltrate is less common. Instead, the infiltrate breaks through the cortex and periosteum, resulting in soft-tissue abscesses and sinus tracts to the skin.[12]

S aureus is the most common organism responsible for posttraumatic osteomyelitis because it has receptors for many host proteins that are exposed following trauma, including collagen and fibronectin.[10] As a re-

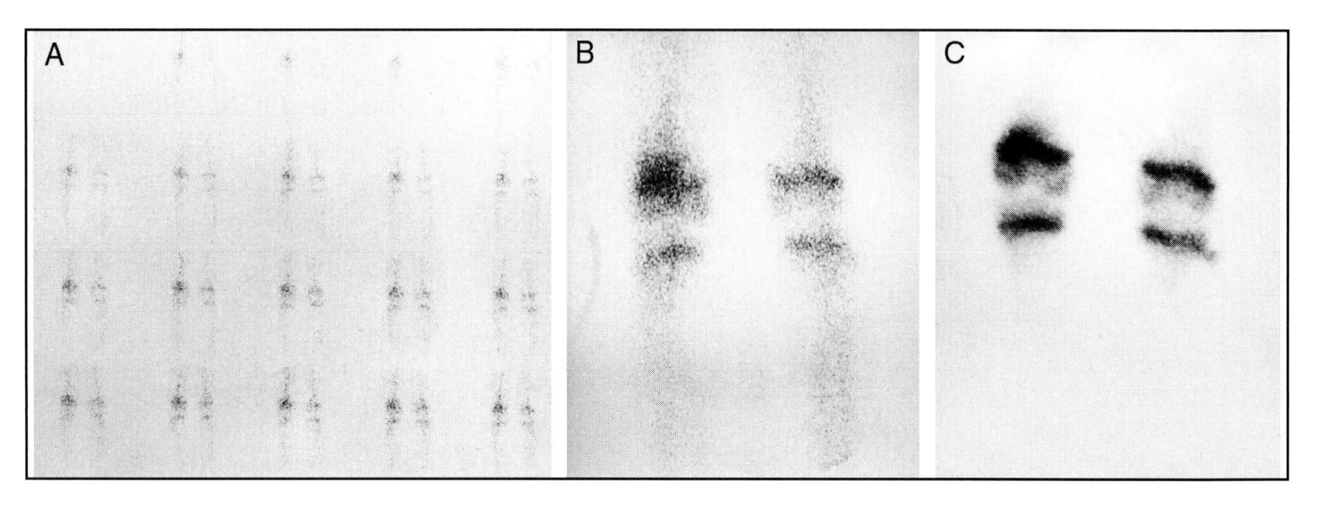

Figure 4
Three-phase bone scan of the knees of a child shows increased vascularity on the dynamic flow (**A**) and blood pool (**B**) scans and increased uptake on the delayed static images (**C**) in the metaphyseal region of the right femur as the result of osteomyelitis.

(Reproduced from Schneider R, Rapuano B: Radioisotopes in orthopaedics, in Buckwalter JA, Einhorn TA, Simon SR (eds): ***Orthopaedic Basic Science: Biology and Biomechanics of the Musculoskeletal System****, ed 2. Rosemont, IL, American Academy of Orthopaedic Surgeons, 2000, p 291.)*

sult, this bacteria adheres to the surface of the injured tissue.

Clinical Presentation

The onset of osteomyelitis often is missed initially because the signs of acute infection, although typically at hand, are often attributed to the initial injury or to a soft-tissue infection. Accordingly, the infection may not be recognized until there is surgical wound breakdown or failed healing of the fracture.[13] At times, infection is not noticed until it forms a draining sinus tract to the skin surface.

A typical presentation for pediatric osteomyelitis would include a short history of pain in the affected area. Most young children will not seem very sick; however, infants with bone infections may have a toxic appearance. Parents may report a history of minor trauma 1 week prior to the onset of pain. Usually, the child refuses to bear weight on the extremity. The child may have a fever, and on physical examination there will be tenderness to palpation over the bone with or without erythema or effusion. The child may report pain with movement of the joint but can typically still demonstrate a full range of motion; thus, the presence of a full range of motion does not exclude the diagnosis.

Plain radiographs can show soft-tissue swelling, but bone changes are typically absent acutely. A technetium 99m bone scan will show increased uptake in the bone (Fig. 4). MRI can show signal changes that are analogous to those seen on a bone scan, but greater sensitivity and anatomic localization is also provided (Fig. 5).

As is the case with septic arthritis, blood test results in patients with osteomyelitis may be abnormal: the white blood cell count, erythrocyte sedimentation rate, and C-reactive protein level can all be elevated, albeit only mildly in some cases. In children, the bone can be aspirated, and Gram stain and culture can help guide treatment.

A typical patient with osteomyelitis would be a 33-year-old man who was referred to a specialist because the closed fracture of the right tibia he sustained in a motorcycle accident 2 years ago has not healed. This fracture was treated with a surgical implant. On presentation, the patient denies any fevers or chills. Physical examination reveals tenderness over the site of the fracture but no erythema, skin breakdown, or sinus tracts. The white blood cell count and erythrocyte sedimentation rate are normal; the C-reactive protein level is slightly elevated. Radiographs show a nonunion of the fracture. In this case, special nuclear medicine imaging, such as indium-111–labeled white blood cell scanning, can be used. (The metal surgical implant may interfere with the production of a signal on MRI.) This test shows intense signal in areas where white blood cells collect (Fig. 6).

Because infections tend to destroy the bone, they can often be visually similar to malignancies; hence, the dictum "biopsy what you culture and culture what you biopsy." In other words, if an infection is suspected, then the correct diagnosis may be tumor and vice versa.

Treatment and Prevention

Historically, the mainstay of treatment for acute hematogenous osteomyelitis has been intravenous antibiotic therapy. The current trend, however, is to use oral antibiotics after a shorter course of intravenous antibiotic therapy.[13] Empiric treatment is based on the common pathogens given the patient's age. Children can typically be switched to oral antibiotic therapy after 5 to 10 days of intravenous treatment if signs of active infection have resolved. A normalization of C-reactive protein levels can be used as an objective indicator of improvement. Current consensus favors a minimum of 3 weeks of antibiotic treatment.[13]

Surgical débridement for acute hematogenous osteomyelitis is necessary in three situations: (1) when pus is obtained on initial aspiration, (2) when there is radiographic evidence of a metaphyseal sequestrum, and (3) when there is no clinical improvement within 24 to 48 hours of antibiotic therapy.

Prompt treatment of osteomyelitis in children is especially important because continued infection can have devastating effects on the developing skeleton. One of the most serious consequences of infection is physeal destruction with subsequent growth disturbance. Children with combined osteomyelitis and septic arthritis, especially of the hip joint, have a slower recovery and are at a higher risk for complications.[9]

Treatment for posttraumatic osteomyelitis includes adequate irrigation and débridement of all infected tissues followed by 4 to 6 weeks of organism-specific antibiotic therapy.[14] Once the infected tissue has been excised, muscle flaps and skin grafts may be needed to cover soft-tissue defects. Bone defects can be treated with a variety of techniques, including bone grafts or vascularized bone transfer. Alternatively, the bone may be allowed to heal in a shortened position and then lengthened thereafter if needed.[8]

For infections involving fracture fixation devices, treatment should include débridement and suppressive antibiotics until the fracture heals. Thereafter, the implant can be removed and definitive treatment of the infection pursued. If the fracture has healed at the time of initial presentation, all hardware should be removed and antibiotics should be given.[15]

Before considering surgical treatment, the risks and benefits must be carefully weighed. In some patients, especially in the elderly and the chronically ill, the effects of the multiple procedures required to eliminate infection and restore skeletal integrity

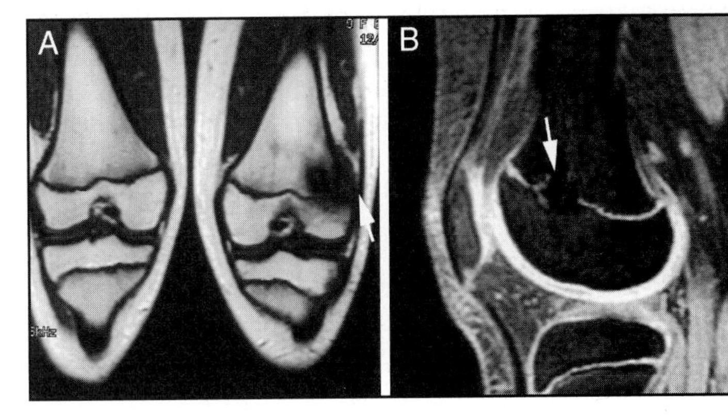

Figure 5
A, T1-weighted MRI scan of the knee shows infection in the metaphysis (arrow). **B,** A gradient-echo MRI scan demonstrates destruction of the distal femoral physis by the infection (arrow).

(Reproduced from Song KM, Sloboda JF: Acute hematogenous osteomyelitis in children. ***J Am Acad Orthop Surg*** *2001;9:166-175.)*

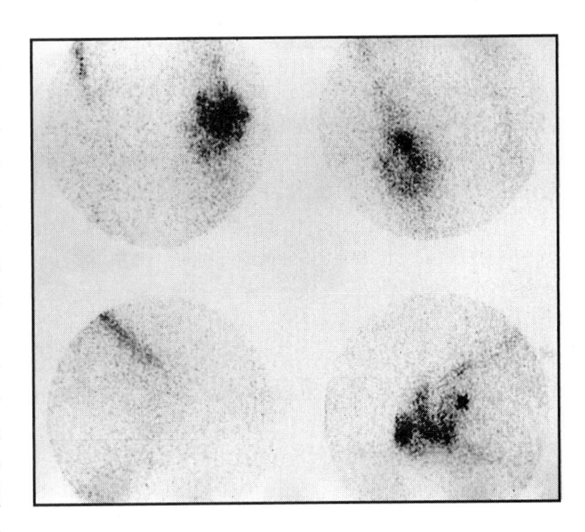

Figure 6
An indium-111 white blood cell scan shows high uptake around the prosthesis in a patient with an infected left total knee prosthesis (seen as intense black signal).

(Reproduced from Schneider R, Rapuano B: Radioisotopes in orthopaedics, in Buckwalter JA, Einhorn TA, Simon SR (eds): ***Orthopaedic Basic Science: Biology and Biomechanics of the Musculoskeletal System****, ed 2. Rosemont, IL, American Academy of Orthopaedic Surgeons, 2000, p 301.)*

may be more detrimental than the disease itself. If this is the case, intermittent courses of antibiotics can be used to suppress exacerbations as needed.

Periprosthetic Infection

Infections of total joint prostheses are a combination of septic arthritis and osteomyelitis; the infection involves not only the joint but also the adjacent bones in which the prostheses have been implanted. At least 1% of joint replacements performed in the United States are complicated by infection.[8] While infections can occur in the immediate postoperative period, they can also occur months to years after implantation. Diagnosing these late infections can be difficult because patients may report symptoms similar to those attributable to simple mechanical dislodging of the prosthesis, a process called aseptic loosening.

Pathophysiology

Prosthetic implants can become infected by direct inoculation of organisms at the time of implantation or through hematogenous spread of bacteria any time thereafter. A key factor in the development of these infections is the adherence of bacteria to the foreign material. The most common pathogens in periprosthetic infections are *S aureus* and *Staphylococcus epidermidis* because these two bacteria have surface proteins that enhance adherence to foreign material such as metal.[12]

Treatment and Prevention

The best treatment for prosthetic infections is prevention. With few exceptions, prophylactic antibiotics should be used before all orthopaedic surgical procedures, especially those that involve the implantation of foreign material. The need for prophylactic antibiotics in patients with total joint arthroplasties who are undergoing procedures that are associated with transient bacteremia remains controversial. The current recommendations call for all patients with joint prostheses to be given oral prophylactic antibiotics prior to dental procedures for the first 2 years postoperatively. After that period, only those patients at high risk for hematogenous infection, such as those with RA or type 1 diabetes mellitus, should be given prophylactic antibiotics.[16]

The treatment of infected joint arthroplasties depends on the time that has elapsed since implantation. For infection detected within the first month following surgery, the prosthesis can be retained. Treatment consists of irrigation and débridement followed by 4 weeks of antibiotic therapy.[12] If infection is detected more than 1 month after implantation, the prosthesis must be removed and the patient treated with intravenous antibiotics. Two weeks after the patient has completed antibiotic therapy, the joint should be reaspirated; if sterile, reimplantation of a new prosthesis can be considered. Positive cultures require further débridement and antibiotic therapy.[17]

Special Situations

Immunocompromised Hosts and HIV Infection

Immunocompromised hosts are at risk for septic arthritis and osteomyelitis caused by atypical mycobacteria and fungi. These infections tend to be more indolent than bacterial infections and often have a chronic course. Common mycobacterial infections in this patient population include *Mycobacterium kansasii* and *Mycobacterium avium-intracellulare*. Fungal infections may be caused by several *Candida* species, *Cryptococcus,* or *Aspergillus*.[18] Treatment involves thorough irrigation and débridement of the infected tissue followed by an extended course of systemic antimycobacterial or antifungal medications. Immunocompromised patients are also at risk for ordinary bacterial infections, but these infections may present in an atypical manner.

HIV infection can cause an acquired immunodeficiency syndrome associated arthropathy. This syndrome consists of a subacute arthritis developing in a few joints simultaneously over a period of a few weeks. It primarily involves the knees and ankles. This syndrome can be the first manifestation of HIV infection.[4] Nonsteroidal anti-inflammatory drugs (NSAIDs) and intra-articular steroid injections may help relieve associated pain.

Tuberculosis

Tuberculosis is a chronic granulomatous infection caused by the bacteria *Mycobacterium tuberculosis*. Tuberculosis of the bones and joints is a locally destructive disease that spreads hematogenously from a primary focus of infection, typically the lungs (Fig. 7). Prior to the advent of antibiotics, tuberculosis of the musculoskeletal system was a major cause of morbidity. It is now uncommon

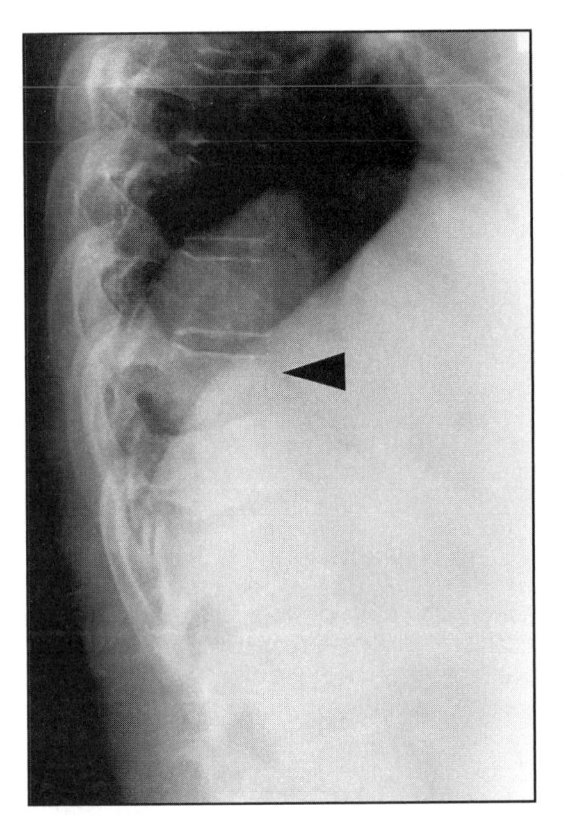

Figure 7
Radiograph of the spine of a 35-year-old man who had persistent pain and fever despite medical treatment for tuberculosis. Arrowhead indicates kyphosis with the apex at T9. Note that T9 through T11 have been eroded by the tuberculosis.

(Reproduced from Garvin KL, Luck Jr JV, Rupp ME, Fey PD: Infections in orthopaedics, in Buckwalter JA, Einhorn TA, Simon SR (eds): ***Orthopaedic Basic Science: Biology and Biomechanics of the Musculoskeletal System****, ed 2. Rosemont, IL, American Academy of Orthopaedic Surgeons, 2000, p 251.)*

in the developed world,[2] but it may become more common as the number of patients with HIV and other causes of immunosuppression continues to increase.

Skeletal tuberculosis accounts for 2% of all tuberculosis cases.[19] Although skeletal tuberculosis can affect almost any bone or joint, it most commonly involves large weight-bearing joints and the spine. Spinal tuberculosis is referred to as Pott's disease. Although most patients with tuberculosis typically have a positive tubercular skin test result, definitive diagnosis requires biopsy of the infected tissue. Treatment is similar to that for pulmonary tuberculosis, requiring the use of multiple agents for 6 to 12 months. Necrotic bone and soft tissue must be surgically débrided. Surgery is also used to prevent progression of spinal deformities.[9]

Lyme Disease

Lyme disease is caused by the spirochete *Borrelia burgdorferi*, which can be transmitted to humans via the bite of deer ticks from the *Ixodes ricinus* complex.[20] The disease course is divided into three stages, with musculoskeletal involvement occurring in the second and third stages. In the first stage, localized lesions (called erythema migrans) appear on the skin. In stage two, disseminated infection occurs, following the first stage by days to weeks. The musculoskeletal manifestation of stage two consists of migratory pain without swelling of the articular and periarticular surfaces. In stage three, late or persistent infection occurs. This can involve an intermittent arthritis of the large joints, primarily the knee, but at times a few joints at once. Approximately 60% of untreated patients will progress to stage three within 6 months after initial infection.[4] Joints affected by Lyme arthritis tend to be more swollen than painful. Initial episodes often present with significant but brief effusions. If the disease is untreated, the duration of the arthritis attacks tends to lengthen. Approximately one of five untreated patients progresses to chronic Lyme arthritis, which is characterized by an episode of continuous joint involvement lasting longer than 1 year.[21] Although initially less destructive than bacterial septic arthritis, continued episodes of Lyme arthritis can lead to joint destruction.

In the United States, the diagnosis of Lyme disease is based primarily on a history of exposure to an area with deer (including many suburban areas) and an antibody response to *B burgdorferi*. This antibody test may be done by enzyme-linked immunosorbent assay or using the Western blot test. Analysis of joint fluid also has been shown to be effective in the detection of *B burgdorferi* in patients with Lyme arthritis.[16]

Treatment of Lyme disease is based on the stage of involvement. For patients with stage one and stage two disease, 14 to 21 days of oral doxycycline is effective. For patients with stage three Lyme disease, a 30-day course of oral doxycycline or intravenous ceftriaxone is effective. Given its decreased cost and easier administration, oral antibiotic therapy is recommended as initial treatment. Intravenous antibiotic therapy is generally reserved for those who do not respond to oral therapy. Although most patients with Lyme arthritis respond well to antibiotic treatment, approximately 10% of patients have arthritis that persists for months or years despite appropriate therapy. The results of analyses of the synovium and joint fluid of these patients tend to be negative for *B burgdorferi* DNA. The current consensus is that the chronic arthritis is an autoimmune response caused by molecular mimicry rather than continued infection.[16] Chronic arthritis may be treated with NSAIDs, intra-articular steroid injections, or arthroscopic synovectomy. For patients living in epidemic areas with frequent exposure to deer ticks, a vaccine is available to protect against Lyme disease. In phase three trials, the vaccine was 76% effective in the prevention of Lyme disease after three injections.[16]

Diabetic Foot Infections

Approximately 25% of all hospitalizations associated with diabetes mellitus occur for the treatment of foot infections.[8] People with diabetes mellitus are at increased risk for foot infections for several reasons. First, they are more likely to injure their feet because they often have decreased sensation from peripheral neuropathy. Second, diabetes mellitus also causes peripheral vascular disease, which leads to poor wound healing. Third, people with diabetes mellitus often have impaired vision, which increases the risk of injury and prevents full appreciation of the extent of injury. Once infected, the patient's impaired systemic immunity and compromised local immunity caused by poor blood flow make eradication of foot infections extremely difficult.[22]

Foot infections in those with diabetes mellitus tend to be polymicrobial, involving both aerobic and anaerobic organisms, including *Pseudomonas*. The ability to reach the bone with a blunt probe when examining these patients indicates osteomyelitis in 80% of patients.[18] There are often no systemic markers of infection, although loss of glucose control may be suggestive.

Before planning treatment for diabetic foot infection, the vascular status of the leg should be evaluated with duplex ultrasonography. The ankle-brachial index can be normal in patients with diabetic vascular disease because arterial calcification prevents compression of the vessel with the blood pressure cuff; toe pressures, therefore, are a better indicator of healing potential. Intravenous antibiotics should be administered after cultures are obtained. Ticarcillin with clavulanate or other broad-spectrum agents should be used. Necrotic tissue and bone should be surgically removed. For severe infections with systemic involvement or for patients whose vascular status precludes adequate wound healing, amputation may be needed.

Key Terms

Bone infarction Bone death that occurs as the result of ischemia

Erythema migrans Localized skin lesions that appear during the first stage of Lyme disease

Hematogenous seeding The dissemination of bacteria or cancerous cells via the blood

Lyme disease A recurrent, multisystem infection caused by the spirochete *Borrelia burgdorferi*, which can be transmitted to humans via the bite of deer ticks from the *Ixodes ricinus* complex

Tuberculosis A chronic granulomatous infection caused by the bacterium *Mycobacterium tuberculosis*

References

1. Esterhai JL Jr., Ruggiero V: Adult septic arthritis, in Esterhai JL Jr, Gristina AG, Poss R (eds): *Musculoskeletal Infection,* Park Ridge, IL, American Academy of Orthopaedic Surgeons, 1992, pp 409-419.
2. Mikhail IS, Alarcon GS: Nongonococcal bacterial arthritis. *Rheum Dis Clin North Am* 1993;19: 311-331.
3. Friedland GH, Selwyn PA: Infections (excluding AIDS) in injection drug users, in Fauci AS, Braunwald E, Isselbacher KJ, et al (eds): *Harrison's Principles of Internal Medicine,* ed 14. New York, NY, McGraw-Hill Health Professions Division, 1998, pp 831-835.
4. Goodman SB, Chou LB, Schurman DJ: Management of pyarthrosis, in Chapman MW (ed): *Chapman's Orthopaedic Surgery,* ed 3. Philadelphia, PA, Lippincott Williams & Wilkins, 2001, pp 3561-3575.
5. Baker DG, Schumacher HR Jr.: Acute monoarthritis. *N Engl J Med* 1993;329:1013-1020.
6. Morrissy RT: Bone and joint sepsis, in Morrissy RT, Weinstein SL (eds): *Lovell and Winter's Pediatric Orthopaedics,* ed 5. Philadelphia, PA, Lippincott, Williams & Wilkins, 2001, pp 459-505.
7. Thaler SJ, Maguire JH: Infectious arthritis, in Fauci AS, Braunwald E, Isselbacher KJ, et al (eds): *Harrison's Principles of Internal Medicine,* ed 14. New York, NY, McGraw-Hill Health Professions Division, 1998, pp 1944-1949.
8. Mader JT, Cripps MW, Calhoun JH: Adult posttraumatic osteomyelitis of the tibia. *Clin Orthop* 1999;360:14-21.
9. Cooper PS, Delahay JN: Musculoskeletal infections, in Wiesel SW, Delahay JN (eds): *Principles of Orthopaedic Medicine and Surgery.* Philadelphia, PA, WB Saunders Company, 2001, pp 139-158.
10. Song KM, Sloboda JF: Acute hematogenous osteomyelitis in children. *J Am Acad Orthop Surg* 2001;9:166-175.
11. Griffin PP: Bone and joint infections in children, in Chapman MW (ed): *Chapman's Orthopaedic Surgery,* ed 3. Philadelphia, PA, Lippincott Williams & Wilkins, 2001, pp 4469-4484.
12. Tsukayama DT: Pathophysiology of posttraumatic osteomyelitis. *Clin Orthop* 1999;360:22-29.
13. Maguire JH: Osteomyelitis, in Fauci AS, Braunwald E, Isselbacher KJ, et al (eds): *Harrison's Principles of Internal Medicine,* ed 14. New York, NY, McGraw-Hill, 1998, pp 824-827.
14. Laughlin RT, Wright DG, Mader JT, et al: Osteomyelitis. *Curr Opin Rheumatol* 1995;7:315-321.
15. Garvin KL, Mormino MA, McKillip TM: Management of infected implants, in Chapman MW (ed): *Chapman's Orthopaedic Surgery,* ed 3. Philadelphia, PA, Lippincott Williams & Wilkins, 2001, pp 3577-3593.
16. Hanssen AD, Osmon DR: The use of prophylactic antimicrobial agents during and after hip arthroplasty. *Clin Orthop* 1999;369:124-138.
17. Aaron AD: Orthopaedic Infections in Wiesel SW, Delahay JN (eds): *Essentials of Orthopaedic Surgery,* ed 2. Philadelphia, PA, WB Saunders, 1997, pp 83-95.
18. Meier JL, Beekmann SE: Mycobacterial and fungal infections of bone and joints. *Curr Opin Rheumatol* 1995;7:329-336.
19. Gristina AG, Naylor PT, Myrvik QN: Mechanisms of musculoskeletal sepsis. *Orthop Clin North Am* 1991;22:363-371.
20. Steere AC: Lyme Disease. *N Engl J Med* 2001;345:115-125.
21. Tetsworth KD: Infection, in Beaty JH (ed): *Orthopaedic Knowledge Update 6: Home Study Syllabus.* Rosemont, IL, American Academy of Orthopaedic Surgeons, 1999, pp 191-203.
22. Saltzman CL, Pedowitz WJ: Diabetic foot infections. *Instr Course Lect* 1994;48:317-320.

Chapter 13

Metabolic Bone Disease

Metabolic bone diseases are those in which abnormal formation or maintenance of bone leads to clinical manifestations such as fracture and deformity. Metabolic bone diseases are characterized by failed skeletal homeostasis. This failure may be caused by an abnormality in mineral homeostasis, such as calcium wasting in kidney disease; however, not all metabolic bone diseases are caused by defects in mineralization. The most common and perhaps most important metabolic bone disease, osteoporosis, is one in which mineralization is normal. Other significant metabolic bone diseases include osteomalacia (and its juvenile form, rickets) and Paget's disease of bone. Metabolic bone diseases merit entire textbooks of their own. Therefore, this chapter serves only as an introduction to these complex conditions.

Osteoporosis

Osteoporosis is a systemic skeletal disease characterized by low bone mass, increased bone fragility, and susceptibility to fracture. Among patients with osteoporosis, fractures are most commonly seen in the wrist, vertebral column, and hip. Osteoporosis is a disease of abnormal bone remodeling. Specifically, bone resorption is uncoupled from bone formation, and over time, osteoclasts remove more bone than osteoblasts create. This net imbalance leads to decreased bone density and decreased mechanical strength. Because spicules of osteoporotic bone appear normal on microscopic examination, it has been suggested that osteoporotic bone is otherwise normal but merely less dense than healthy bone. However, the lack of sufficient normal tissue renders the gross structure of osteoporotic bone distinctly abnormal.

Osteoporosis is categorized as either primary or secondary. Primary osteoporosis is further divided into postmenopausal (type 1) or senile (type 2). In senile osteoporosis, an age-related decline in renal production of active vitamin D is the probable cause of bone loss. Secondary osteoporosis is characterized by conditions in which bone is lost because of the presence of another disease, such as hormonal imbalances, malignancies, or gastrointestinal disorders, or because of corticosteroid use. Although osteoporosis occurs primarily in women, senile and secondary osteoporosis are not uncommon in men.

Epidemiology and Risk Factors

The most prevalent form of primary osteoporosis is postmenopausal osteoporosis. This condition is a major threat to public health in the United States. It is estimated that 20% to 30% of postmenopausal Caucasian women in the United States have sustained an osteoporosis-related fracture and most of those without fractures nonetheless have low bone density measured in the spine, hip, or wrist. The National Osteoporosis Foundation estimates that more than 8 million women have osteoporosis and more than 14 million have low bone mass. It is further estimated that 40% to 50% of women age 50 years or older are at risk for an osteoporosis-related fracture during their lifetime.[1] In 1995, there were 3.4 million outpatient and emergency department visits for fractures attributable to osteoporosis. Although the incidence of osteoporosis in men is about half that in women, the lifetime risk for future fractures in men age 60 years is nearly 30%.

Hip fracture from osteoporosis can be

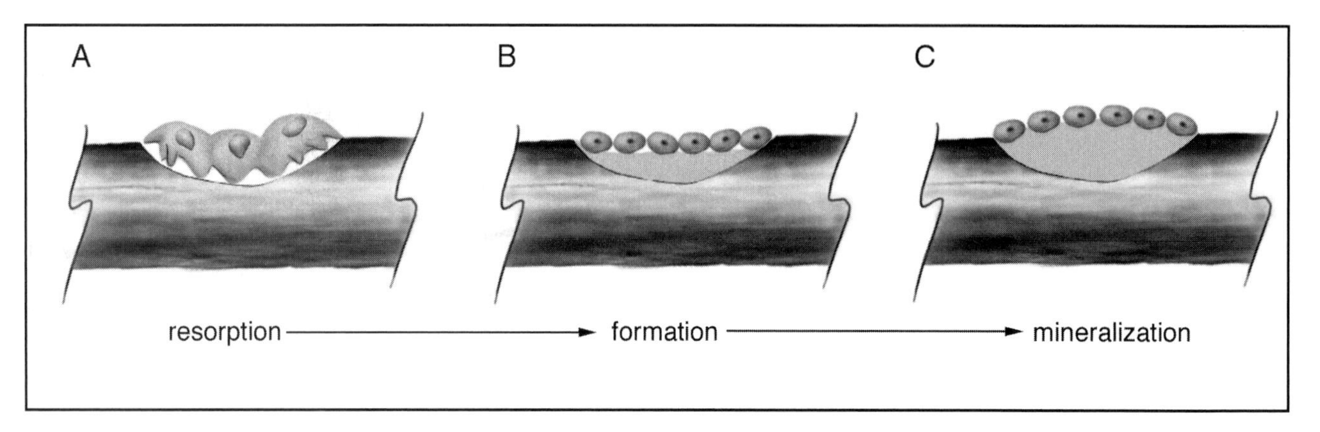

Figure 1
The bone remodeling cycle begins with osteoclastic bone resorption (**A**), which continues for approximately 3 weeks. This is followed by the recruitment of osteoblasts (**B**) that produce a collagen-rich protein matrix, osteoid, to fill the resorption cavity. This matrix must then become mineralized (**C**) to produce sturdy bone. The completion of this cycle requires 3 to 4 months.

medically devastating. One of the most serious consequences of hip fractures is the loss of functional independence. Fewer than half of the patients with a hip fracture return to their preinjury level of mobility within 6 months after fracture, and further improvement is rare. About 25% return to their prefracture functional status, and 25% will require nursing home care for at least a limited period. Between 3% and 4% of patients older than 50 years with hip fractures die during the hospital admission for the fracture, and the risk of mortality in the year after hip fracture increases by 20% to 25%. Compression fractures of vertebrae have received less attention, but they are also medically devastating. Patients with compression fractures may have severe back pain for 2 to 3 months after the fracture and 70% continue to have chronic back pain that limits their activities. These seemingly innocuous fractures are also associated with a 15% increase in mortality for the first 5 years after fracture.[1]

Risk factors for osteoporosis include alcoholism, smoking, low body mass, sedentary lifestyle, low calcium and vitamin D intake, treatment with glucocorticoids, and illnesses that cause accelerated bone loss. Caucasians, particularly fair-skinned people from northern Europe, are at increased risk for acquiring osteoporosis as well.

Pathology

Bone is a metabolically active tissue; remodeling (resorption and formation) occurs throughout life (Fig. 1). Because the rate of bone formation by osteoblasts exceeds the rate of bone resorption by osteoclasts in normal, young adults, bone mass accumulates until the late 20s or early 30s. At about age 30 years, bone resorption begins to exceed formation, and a subtle loss of bone occurs (Fig. 2). With the onset of menopause and falling levels of estrogen, osteoclastic bone resorption increases without an equal increase in osteoblastic bone formation. As a result, bone loss accelerates (Fig. 3). Trabecular bone, the principal type of bone in vertebrae and in the femoral neck, is particularly susceptible to osteoporosis because it has a higher surface-to-volume ratio than cortical bone: remodeling occurs strictly on the surface of bone. As a result of this bone loss and the deformation of the microscopic architecture, structural strength is lost (Fig. 4). Osteoporosis is called a "silent" disease because this progressive loss of bone goes on without signs or symptoms until a fracture occurs.

Presentation

The first sign of osteoporosis is often fracture, typically in the distal radius or the thoracic spine (compression fracture) among women in the first decade following menopause. Among older people, hip fractures are more common. Ordinarily, the mechanism of injury leading to an osteoporosis-related hip fracture is insufficient to fracture healthy bone.

A typical patient is an otherwise healthy, petite 58-year-old woman who experiences severe pain between her shoulder blades while lifting a turkey out of the oven. Because the pain is so severe, she goes to the

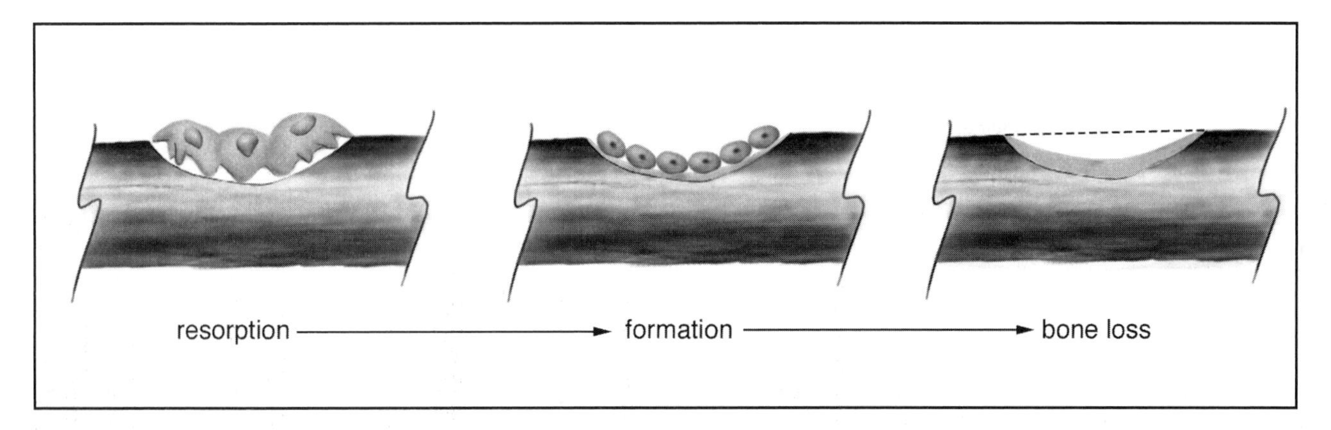

Figure 2
By about 30 years of age, bone formation is no longer able to keep up with resorption, and resorption cavities are left partially unfilled. This imbalance between resorption and formation leads to a gradual decrease in bone mass.

emergency department, where physical examination reveals that she has acute pain, a heart rate of 92 beats/min, and blood pressure of 155/90 mm Hg. She is unable to move without pain and is exquisitely tender to percussion of the spine just below the shoulder blades. Radiographs show a compression fracture of the 10th thoracic vertebra.

Because this is the woman's first fracture, her risk factors for osteoporosis must be assessed. Based on her height and weight, she has a low body mass index. She states that she never drinks milk and does not particularly like other dairy products. She has been relatively sedentary, and smoked one pack of cigarettes a day from age 18 to 45 years. She stopped menstruating at age 48 years and has never taken hormone replacement therapy. She drinks socially, her diet is erratic, and she frequently eats fast foods. Her mother is also petite and fractured her hip last year at the age of 78 years.

This patient demonstrates several of the risk factors for osteoporosis listed in Table 1. She may have had a low peak bone mass because she is petite and is genetically predisposed to osteoporosis because her mother has the disease. She likely experienced accelerated bone loss at menopause when estrogen levels fell. This bone loss was probably accentuated by her low calcium intake and lack of exercise. As is usually the case, her bone loss was asymptomatic until she experienced a fracture. The fracture was typical in that it occurred without significant trauma and involved the spine, which is composed primarily of trabecular bone as opposed to the cortical bone comprising the long bones.

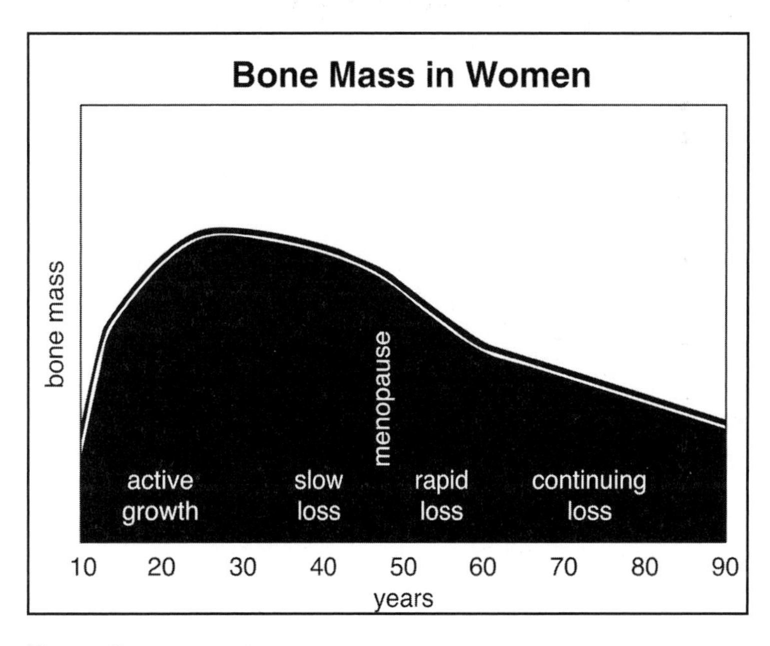

Figure 3
Bone mass as a function of age. Bone loss begins in young adulthood, but at menopause, the rate of bone loss accelerates as the result of an increase in osteoclastic bone resorption without a compensatory increase in formation.

For many patients, a fracture of the spine or distal radius is an event heralding a potentially lethal hip fracture.

Imaging Studies

Radiography

Plain radiographs can demonstrate osteoporosis, but to do so the bone loss must be severe: radiographs are not sensitive for the detection of bone loss of less than 30%. Accordingly, radiography is not a good screening modality for osteoporosis because there will be many false-negative results.

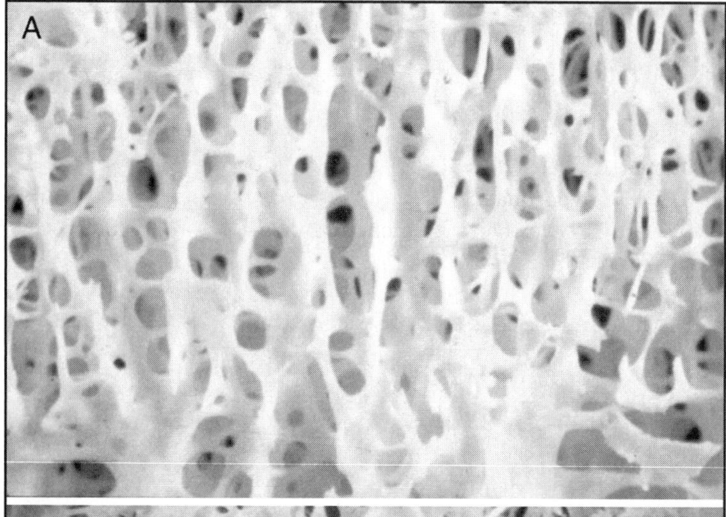

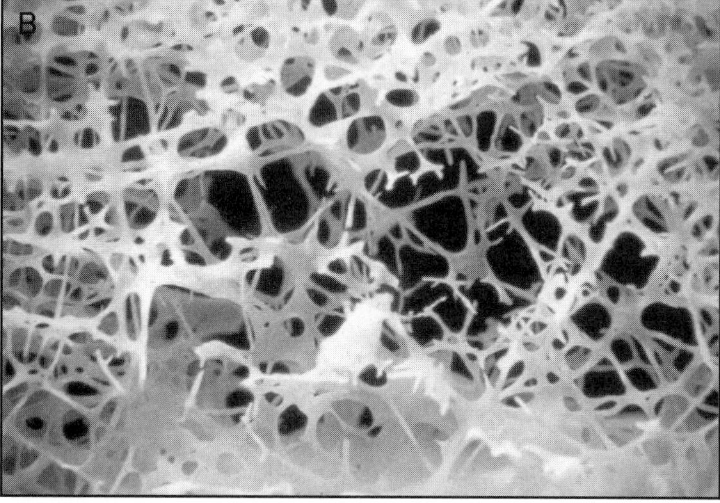

Figure 4
A, Normal trabecular bone from a vertebral body. **B,** Osteoporotic trabecular bone from a vertebral body. As bone resorption continues, trabecular bone is lost, which leads to increased fragility.

(Reproduced from Einhorn TA: The structural properties of normal and osteoporotic bone. Instr Course Lect 2003;52:533-539.)

Table 1
Risk Factors for Osteoporosis

Risk Factors for Osteoporosis
Female sex
Caucasian or Hispanic race
Estrogen deficiency
Low calcium intake
Alcoholism
Inadequate physical activity
Low body mass index
History of smoking
Family or personal history of fractures
Use of medications associated with accelerated bone loss

Dual-Energy X-ray Absorptiometry (DEXA)
Individuals at risk for osteoporosis-related fractures can be identified before their first fracture occurs with the use of **dual-energy x-ray absorptiometry (DEXA)**. The bone density results are expressed as grams of mineral per square centimeter of bone. This test measures the density of the spine and hip with good precision and minimal radiation exposure. The guidelines of the National Osteoporosis Foundation for identifying those who should be tested are shown in Table 2. The World Health Organization categorized bone density results based on a comparison of the subject's bone density and the peak bone density of a healthy, young adult population. The difference is expressed as a T-score.[2] A **T-score** is equivalent to one standard deviation above or below ideal bone mass. The definitions based on T-scores are as follows:

Normal 0 to −1
Osteopenia −1 to −2.5
Osteoporosis < −2.5

Peripheral Bone Densitometry
Measuring bone density in the forearm, hand, or heel rather than in the hip or spine requires less sophisticated equipment. With this method, however, interpretation may be incorrect because there is discordance of bone mineral density (BMD) at different sites; that is, the spine may be osteoporotic but the heel or forearm may be normal. This discordance is noted particularly in younger or perimenopausal women and produces false-negative results. As a result, the risk of fracture may be underestimated. Another disadvantage of screening for osteoporosis by measuring BMD at peripheral sites is that the technique is not precise enough to be used for long-term monitoring of response to therapy.

Laboratory Studies
There are no laboratory tests to diagnose primary osteoporosis; rather, the purpose of such testing is to determine if the bone loss that has already been identified (either by imaging studies or because of fractures) is caused by another process. Accordingly, it may be helpful to measure serum calcium, vitamin D, thyroid, and parathyroid hormone levels as well as concentrations of pituitary hormones. Markers of bone formation or resorption can be used to assess

whether there is accelerated bone turnover. Alkaline phosphatase is an enzyme that becomes elevated with increased bone formation and mineralization. Total serum alkaline phosphatase, however, is an imperfect marker because it represents the total concentration of four isoenzymes of which the bone isoform contributes less than half. Osteocalcin is a small protein made by osteoblasts and is a sensitive marker of bone formation. Hydroxyproline and hydroxylysine, products of collagen breakdown, can be measured in the urine; however, because collagen is found in many tissues, they are not specific markers of bone turnover. Among young men with osteoporosis, the assessment of serum testosterone levels may help identify a cause.

Differential Diagnosis

Once the diagnosis of osteoporosis is made, the next step is to determine whether the osteoporosis is primary (postmenopausal or senile) or secondary. If the diagnosis is suggested on the basis of a low-energy fracture, other possible causes of pathologic fracture, such as malignancy or infection, must be excluded.

Treatment and Prevention

Prevention of osteoporosis must begin in young adulthood with lifestyle and dietary modifications. Recommendations include maintaining normal body weight; adequate intake of calcium and vitamin D (1,000 to 1,500 mg/day and 800 IU/day, respectively); avoiding smoking and excessive alcohol intake; and engaging in weight-bearing, impact exercises, such as walking, for 30 minutes three to five times per week. Among older people with established low bone mass, measures to prevent falls, such as "fall proofing" the home by providing adequate lighting and removing obstacles, having frequent vision examinations, and engaging in strength-training exercises, will decrease the risk of fractures independent of bone mass.

Osteoporosis can also be addressed with pharmacologic agents. Conjugated estrogens (0.625 mg) and estradiol (0.5 mg) given orally or transdermally have been shown to increase bone density. Although there are no prospective, placebo-controlled trials proving that drugs decrease the risks of vertebral and hip fractures, several retrospective studies have shown lower rates of fracture in women who have been taking estrogen for more than 5 years.[3] The greatest risk reduction was observed in users of estrogen. If the patient has an intact uterus, a progestational agent must be given with the estrogen to prevent endometrial hyperplasia and endometrial carcinoma. A recent study demonstrated an increase in the risk for breast cancer, heart disease, and stroke in women taking conjugated estrogen and medroxyprogesterone for more than 5 years.[4] Women with a history of thromboembolic events are not candidates for estrogen therapy. When deciding whether to prescribe estrogen, the risk-benefit ratio must be carefully weighed for each patient.

Table 2
Indications for Bone Mineral Analysis

All postmenopausal women
All men older than 65 years
Men with hypogonadism
Women who have taken hormone replacement therapy for prolonged periods
Men and women with radiologic evidence of vertebral deformities
Men and women with primary hyperparathyroidism
Men and women requiring long-term glucocorticoid therapy
Anyone taking medication for osteoporosis (to monitor response)

Selective Estrogen Receptor Modulators (SERMs)

Selective estrogen receptor modulators (SERMs) are drugs that behave either as an agonist or antagonist of estrogen depending on the site. They have been shown to prevent bone loss and lower serum cholesterol levels in postmenopausal women. One such drug, raloxifene, has been approved by the US Food and Drug Administration (FDA) for both the prevention and treatment of osteoporosis. It has estrogen-like effects on bone and inhibitory effects on the breast and endometrium. It is hoped that raloxifene may retain the beneficial effect of estrogen and eliminate the associated cancer risk to the endometrium and breasts. Raloxifene causes a modest increase in BMD and reduces fracture risk for the vertebrae but not for the hip.[5] Adverse effects include hot flashes, peripheral edema, leg cramps, and venous thrombosis, but the incidence is low. Preliminary evidence suggests that the incidence of breast cancer may be reduced in women taking raloxifene, but additional studies are needed. Several other SERMs are under investigation for the treatment of osteoporosis.

Bisphosphonates

Bisphosphonates are structurally similar to naturally occurring pyrophosphates. Because they have a strong chemical affinity for hydroxyapatite, a major inorganic component of bone, they are potent inhibitors of bone resorption. They reduce the rate at which bone remodeling occurs and reduce the depth of resorption. Together this increases bone mass. Alendronate sodium and risedronate sodium are FDA-approved for the prevention and treatment of osteoporosis. Both agents increase bone density and reduce fracture risk for the hip and spine.[6-8] Alendronate has also been shown to be equally effective in men, increasing bone density and reducing fracture risk.

Calcitonin

Calcitonin is a natural polypeptide hormone that regulates serum calcium and bone metabolism. Administration of calcitonin decreases the rate of bone resorption by decreasing the number and activity of osteoclasts. A nasal spray form of the drug is commonly used. Calcitonin causes a modest increase in bone density, but studies have not shown conclusive evidence for a reduction in fracture risk.[9,10]

Drugs for Steroid-Induced Osteoporosis

Estrogen and bisphosphonates are both effective in treating steroid-induced osteoporosis. Both classes of drugs can increase bone density when given to patients taking glucocorticoids.[11,12] Calcitonin has also been reported to reduce bone loss, but whether it prevents fractures in these patients is less clear. A bisphosphonate should be given to all patients with osteopenia or osteoporosis who require treatment with glucocorticoids for more than 3 months.

Parathyroid Hormone

Parathyroid hormone (PTH) is a major regulator of calcium homeostasis. Although continuous administration stimulates bone resorption, intermittent, low doses of PTH stimulate bone formation. Such usage markedly increases bone density and reduces the incidence of vertebral and nonvertebral fractures in women with postmenopausal osteoporosis.[13] Intermittent, low-dose PTH has also been shown to increase bone density significantly in patients with glucocorticoid-induced osteoporosis.[14] It can be administered daily via subcutaneous injection. Its major adverse effects are hypercalciuria and hypercalcemia.

Other Anabolic Therapies Under Investigation

Bone morphogenic proteins, IGF-1 and IGFBP3 complex, IGF-binding protein, transforming growth factor-β, and PTH-related peptide analogs all have clinical potential as effective anabolic agents for bone.

Evaluating Response to Therapy

After instituting therapy for osteoporosis, it is important to assess the response and modify treatment accordingly. Response is monitored by measuring bone density and biochemical markers of bone remodeling. Bone density should be measured 1 to 2 years after patients begin taking an antiresorptive drug. A change in bone density is considered significant when the change is greater than twice the coefficient of variation of the method being used to measure bone density. For DEXA instruments, the coefficient of variation is usually < 2%; therefore, changes of 4% to 5% or greater are considered significant.

The markers of bone remodeling most frequently used are serum osteocalcin and bone-specific alkaline phosphatase to assess formation and the N-telopeptide of type I collagen (NTX) to assess resorption. NTX levels are the most helpful for assessing response to an antiresorptive agent. A decrease in serum or urine NTX levels of 50% or greater is considered a good response.[14]

Osteomalacia and Rickets

Osteomalacia is a metabolic bone disease characterized by abnormal mineralization of normal osteoid. The most common cause of osteomalacia is vitamin D deficiency, although rarer forms may be seen, such as resistance to adequate levels of vitamin D, renal disease, or excess aluminum intake. Osteomalacia in childhood is called rickets. This is typically caused by inadequate vitamin D intake, but gastrointestinal disease, renal disease, or inadequate exposure to sunlight may also be contributing factors.

Epidemiology and Risk Factors

Dietary osteomalacia is less common in the United States because nutritional awareness has increased and many foods are now fortified with vitamin D. Osteomalacia, however, is common among patients with renal disease and must be considered in the diagnostic evaluation of most patients with metabolic bone disorders.

Pathology

Patients with rickets demonstrate a failure of bone mineralization (deposition of calcium in the bones), specifically at the zone of provisional calcification in the growth plate. The key histopathologic findings, therefore, are a preponderance of unmineralized osteoid within the bone and a gross widening of the physis, which represents uncalcified cartilage. This lack of mineralization weakens the bone, and gross bowing deformities may occur. In the adult with osteomalacia, the gross contours of the bone are not deformed, but the weakened bones are subject to pathologic fracture.

Vitamin D deficiency leads to poor mineralization because low levels of vitamin D result in lowered levels of serum calcium. Since calcium is needed for the nervous system and other functions, the body responds to low calcium levels by increasing synthesis of PTH. This hormone mobilizes calcium from the skeleton and encourages the kidney to spill phosphorus in the urine, which produces low serum levels of phosphorus, or hypophosphatemia. The lack of phosphate and the mobilization of calcium out of the skeleton impairs mineralization and produces the characteristic osteopenia.

Presentation

The signs and symptoms of osteomalacia are nonspecific. Osteomalacia may be silent at first; it also may be associated with vague bone pain. In children with rickets, the presentation may range from failure to thrive to grossly visible deformity of the bone. Children with rickets are often breastfed exclusively (with no vitamin supplementation). Adult malnutrition also can lead to osteomalacia and, therefore, should be suspected in malnourished adults with bone pain, including those with anorexia nervosa, for example.

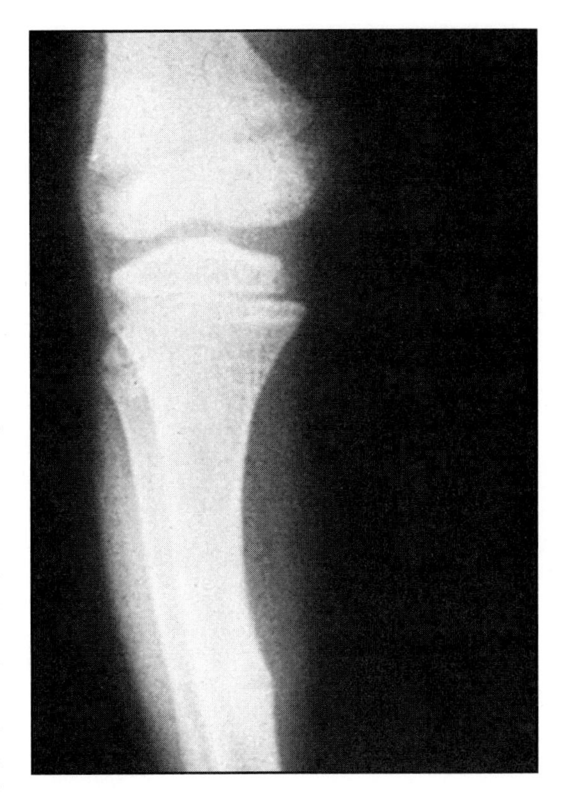

Figure 5
Radiograph showing deformed bone and widening of the physis in a patient with osteomalacia.

*(Reproduced from Bostrom MPG, Boskey A, Kaufman JK, Einhorn TA: Form and function of bone, in Buckwalter JA, Einhorn TA, Simon SR (eds): **Orthopaedic Basic Science**, ed 2. Rosemont, IL, American Academy of Orthopaedic Surgeons, 2000, p 362.)*

Imaging Studies

The radiographic diagnosis of rickets is made by the detection of widening of the physis, especially when the shaft of the bone is deformed and less radiographically dense. These findings suggest decreased mineralization (Fig. 5). Additional radiographic findings that suggest the presence of osteomalacia are Looser's transformation zones, also known as pseudofractures. These zones are lines of radiolucency that represent stress fractures with unmineralized osteoid.

Laboratory Studies

Laboratory test findings, including increased alkaline phosphatase levels and low or normal serum calcium, phosphorus, and vitamin D levels, help define abnormal mineralization. Definitive diagnosis is made by bone biopsy when widened osteoid seams are found. Tetracycline can be given as a marker because it is deposited in the bone. Tetracy-

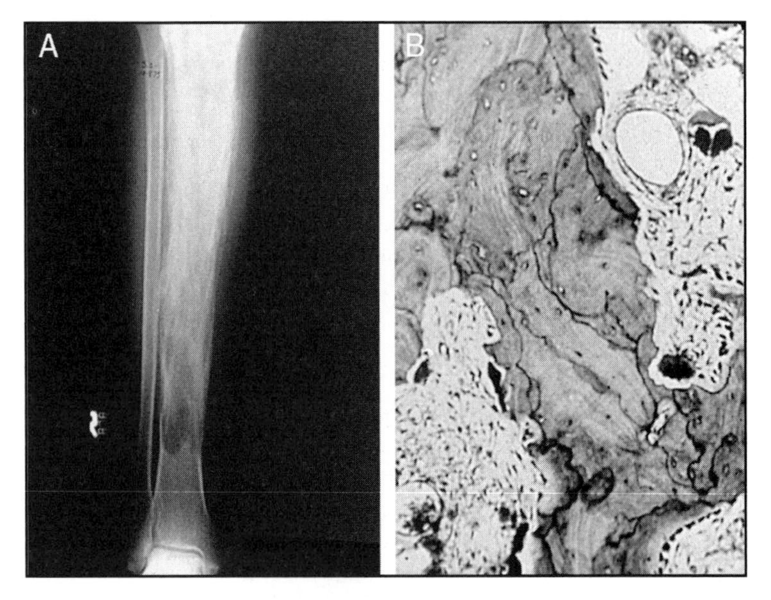

Figure 6
Paget's disease. **A,** Radiographic features include areas of lucency and cortical thickening. **B,** Histologic features include osteoclasis and disorganized bone.

(Reproduced from Bostrom MPG, Boskey A, Kaufman JK, Einhorn TA: Form and function of bone, in Buckwalter JA, Einhorn TA, Simon SR (eds): ***Orthopaedic Basic Science****, ed 2. Rosemont, IL, American Academy of Orthopaedic Surgeons, 2000, p 366.)*

cline lines can help measure the rate of mineralization. Since many cases of osteomalacia are associated with renal disease, laboratory assessment of serum creatinine levels is also performed.

Differential Diagnosis
The specific point of contrast between osteomalacia and osteoporosis is that mineralization in patients with osteoporosis is normal, whereas in those with osteomalacia it is abnormal. Therefore, in instances in which there may be confusion between the two diagnoses, a bone biopsy can provide definitive diagnosis.

Treatment and Prevention
When the cause of osteomalacia or rickets is decreased intake of vitamin D, the treatment naturally centers on administering vitamin D along with dietary calcium. Treatment of rickets associated with renal disease requires attention to that underlying disorder and to the hyperparathyroidism the renal disease produces. Specific attention to bone deformities caused by poor mineralization is often not required because when vitamin D levels are restored, the deformities can correct themselves. At times, however, splinting or even surgical realignment of the bone is needed.

Paget's Disease
Paget's disease (also called osteitis deformans) is a condition of abnormally increased and disorganized bone remodeling (Fig. 6). It is not a systemic disorder; rather, Paget's disease is localized within the bones, where foci of bone remodeling are formed for uncertain reasons. Current thinking holds that a viral infection of osteoclasts in genetically susceptible hosts may be responsible. Because of rapidly increased bone resorption, a rapid increase in bone formation also occurs, leading to a mosaic of disorganized, immature (woven) bone. This bone is at times grossly deformed and apt to fracture. It may be painful and, in rare instances, prone to malignant degeneration. Paget's disease near the articular surfaces can deform them and lead to secondary osteoarthritis. Paget's disease is found commonly in the femur, tibia, pelvis, and skull.

Epidemiology and Risk Factors
Paget's disease is most common among people of European descent (excluding Scandinavia). Approximately 3% of Americans older than 45 years will have radiographic evidence of Paget's disease. It affects men slightly more commonly than women.

Pathology
The first stage of Paget's disease is increased bone resorption. Inclusion bodies seen in osteoclasts suggest that this may occur as the result of a viral infection. Osteoclast activity during this lytic phase recruits osteoblasts to synthesize bone to replace that which was lost. Because so much bone must be synthesized at once, the structure formed during this blastic phase is disorganized (Fig. 7). Over time, the disease extinguishes itself, and the bone reaches a sclerotic phase in which neither bone formation nor resorption takes place. Within a given bone, there may at once be areas in which resorption, formation, or sclerosis dominate.

Presentation
Paget's disease is often clinically silent. The most common presentation is an incidental finding on radiographs. Symptoms may include bone pain at the site of remodeling, pain from arthritis, warmth from increased

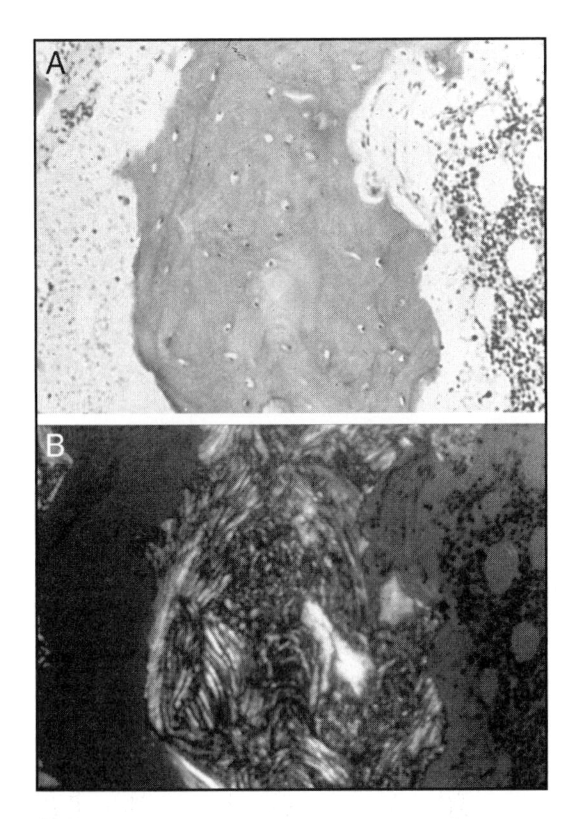

Figure 7
In Paget's disease, the osteoid appears ordinary under normal light (**A**), but polarized light demonstrates the disorganization (**B**).

(Reproduced from Bostrom MPG, Boskey A, Kaufman JK, Einhorn TA: Form and function of bone, in Buckwalter JA, Einhorn TA, Simon SR (eds): ***Orthopaedic Basic Science****, ed 2. Rosemont, IL, American Academy of Orthopaedic Surgeons, 2000, p 366.)*

vascularity in the area, or neurologic complaints from compression of nerves by deformed bone. Paget's disease of the vertebral bodies can produce signs similar to those of spinal stenosis; Paget's disease of the skull may induce hearing loss (from compression of the eighth cranial nerve).

Imaging Studies
In its early phases, Paget's disease shows discrete areas of bone lysis. Thereafter, regions of cortical thickening can be found, along with gross deformity, such as bowing. Paget's disease of the skull produces a thickening described as a "cotton wool" appearance.

Laboratory Studies
Elevated serum alkaline phosphatase levels reflect increased osteoblast activity, whereas high concentration of hydroxyproline and hydroxylysine in the urine reflect bone resorption. These findings are not specific to Paget's disease. The utility of these markers is primarily to measure the response to treatment and to follow the course of the disease over time.

Differential Diagnosis
Paget's disease, in its sclerotic phase, has a distinct radiographic appearance. In earlier phases, however, the radiographic appearance of Paget's disease may suggest a tumor (primary or metastatic) or an infection. A key question is whether pain in a patient with known Paget's disease has pain from the Paget's disease or some other process. A bone scan can help make the distinction because only "active" Paget's disease (ie, that which appears on a bone scan) is thought to cause pain. Sclerotic Paget's disease may cause secondary symptoms from deformation or compression.

Treatment and Prevention
The lytic phase of Paget's disease can be treated with a drug aimed to decrease osteoclast activity, such as calcitonin or a bisphosphonate. Because the disease is so prevalent, it is thought that only symptomatic patients require treatment. Bisphosphonates, drugs that inhibit bone resorption and also block bone mineralization, may be helpful during the lytic phase. Nevertheless, because patients with Paget's disease often have areas of lytic and blastic activity occurring simultaneously, continuous use of bisphosphonates may be harmful: although they block increased osteoclast activity in areas of resorption, they impede needed mineralization elsewhere. As such, the cyclic use of bisphosphonates is suggested; this allows new bone to become mineralized, while decreasing osteoclast activity.

Arthritis caused by Paget's disease is amenable to joint replacement, but this may be complicated by blood loss, heterotopic bone formation, or loosening of the implant.

Key Terms

Bisphosphonates Potent inhibitors of osteoclasts and bone resorption

Calcitonin A peptide produced by the thyroid parafollicular cells that inhibits bone resorption through a direct inhibition of osteoclasts

Dual-energy x-ray absorptiometry (DEXA) A diagnostic imaging technology that uses two different x-ray voltages to assess bone density

Osteomalacia A metabolic bone disease characterized by abnormal mineralization of normal osteoid

Paget's disease A condition of abnormally increased and disorganized bone remodeling

Parathyroid hormone A major regulator of calcium homeostasis; promotes increased levels of serum calcium

Postmenopausal osteoporosis The most prevalent form of primary osteoporosis

Pseudofractures Lines of radiolucency that represent stress fractures with unmineralized osteoid

Rickets The childhood form of osteomalacia

Secondary osteoporosis Osteoporosis caused by the presence of another disease, such as hormonal imbalances, malignancies, or gastrointestinal disorders, or because of corticosteroid use

Senile osteoporosis Osteoporosis in which an age-related decline in renal production of active vitamin D is the probable cause of bone loss

T-score A score used to express the results of bone density tests derived from the difference between the bone density of the patient being tested and the peak bone density of a healthy, young adult population; a T-score is equivalent to one standard deviation above or below ideal bone mass

References

1. Osteoporosis: Review of the evidence for prevention, diagnosis, and treatment and cost-effectiveness analysis: Introduction. *Osteoporos Int* 1998;8(suppl 4):S7-S80.
2. WHO Study Group: Assessment of fracture risk and its application to screening for postmenopausal osteoporosis: Report of a WHO Study Group. *World Health Organ Tech Rep Ser* 1994;843: 1-129.
3. Torgerson DJ, Bell-Syer SE: Hormone replacement therapy and prevention of nonvertebral fractures: A meta-analysis of randomized trials. *JAMA* 2001;285:2891-2897.
4. Rossouw JE, Anderson GL, Prentice RL, et al: Risks and benefits of estrogen plus progestin in healthy postmenopausal women: principal results from the Women's Health Initiative randomized controlled trial. *JAMA* 2002;288:321-333.
5. Ettinger B, Black DM, Mitlak BH, et al: Reduction of vertebral fracture risk in postmenopausal women with osteoporosis treated with raloxifene: Results from a 3-year randomized clinical trial: Multiple Outcomes of Raloxifene Evalation (MORE) Investigators. *JAMA* 1999;282:637-645.
6. Black DM, Cummings SR, Karpf DB, et al: Randomised trial of effect of alendronate on risk of fracture in women with existing vertebral fractures: Fracture Intervention Trial Research Group. *Lancet* 1996;348:1535-1541.
7. Harris ST, Watts NB, Genant HK, et al: Effects of risendronate treatment on vertebral and nonvertebral fractures in women with postmenopausal osteoporosis: A randomized controlled trial: Vertebral Efficacy With Risedronate Therapy (VERT) Study Group. *JAMA* 1999;282:1344-1352.
8. McClung MR, Geusens P, Miller PD, et al: Effect of risedronate on the risk of hip fracture in elderly women: Hip Intervention Program Study Group. *N Engl J Med* 2001;344:333-340.
9. Overgaard K, Hansen MA, Jensen SB, et al: Effect of salcatonin given intranasally on bone mass and fracture rates in established osteoporosis: A dose-response study. *BMJ* 1992;305:556-561.
10. Downs RW Jr., Bell NH, Ettinger MP, et al: Comparison of alendronate and intranasal calcitonin for treatment of osteoporosis in postmenopausal women. *J Clin Endrocrinol Metab* 2000;85:1783-1788.
11. Lukert BP, Johnson BE, Robinson RG: Estrogen and progesterone therapy reduces glucocorticoid-induced bone loss. *J Bone Miner Res* 1992;7:1063-1069.
12. Saag KG, Emkey R, Schnitzer TJ, et al: Alendronate for the prevention and treatment of glucocorticoid-induced osteoporosis: Glucocorticoid-Induced Osteoporosis Intervention Study Group. *N Engl J Med* 1998;339:292-299.
13. Reid DM, Hughes RA, Laan RF, et al: Efficacy and safety of daily risedronate in the treatment of corticosteroid-induced osteoporosis in men and women: A randomized trial: European Corticosteroid-Induced Osteoporosis Treatment Study. *J Bone Miner Res* 2000;15:1006-1013.
14. Lane NE, Sanchez S, Modin GW, et al: Parathyroid hormone treatment can reverse corticosteroid-induced osteoporosis: Results of a randomized controlled clinical trial. *J Clin Invest* 1998;102:1627-1633.

Chapter 14

Musculoskeletal Aspects of Trauma

Trauma resuscitation begins with securing and maintaining the ABCs: airway, breathing, and circulation. Evaluation of the musculoskeletal system must be deferred to these priorities because when a patient cannot get oxygen into the lungs or cannot circulate oxygenated blood to the brain, ischemic damage occurs within 4 minutes. Muscle and bone are not as sensitive. This central fact of physiology, therefore, establishes the priorities of patient care. All injuries to the extremities—even major injuries—must be ignored until the airway is secured, breathing established, and circulation ensured. That said, musculoskeletal medicine is still an important part of trauma care because bones and soft tissues and the neurovascular systems they surround are often affected by high-energy trauma. In fact, injuries to the bones can impede both breathing (as in the case of multiple rib fractures) and circulation (as in the case of pelvic and femoral fractures, which can cause blood loss to the point of shock). This chapter will explore some of the musculoskeletal aspects of trauma medicine. It is not intended to be a definitive summary. The interested reader is referred to the Advanced Trauma Life Support (ATLS) manual published by the American College of Surgeons for a more complete overview.[1]

Epidemiology

Trauma, which includes unintentional injury, homicide, and suicide, is the leading cause of death for people between the ages of 1 and 44 years, and it is eighth among the leading causes of death at any age.[2] However, fewer than 1% of patients who sustain trauma actually die. For example, in 1995, approximately 150,000 deaths occurred as a result of traumatic injury; however, that same year, 2.6 million people were hospitalized and 37 million required emergency department care for trauma-related injuries.[3] In fact, almost 40% of all emergency department visits are the result of patients seeking care for trauma-related injuries.

In the United States, motor vehicles accidents account for most accidental deaths and are the second most common cause of nonfatal accidental injury at all ages.[4] The most important factors that influence the risk of a motor vehicle accident are speed and alcohol or drug intoxication. Speed increases the risk of an accident by reducing driver reaction time and vehicle stability; it also increases the risk that a given accident will lead to a fatality.

Falls are the leading cause of nonfatal injury requiring hospital admission, but they do not often result in death. The most important factor that influences both the risk and consequences of a fall is age. For patients younger than 5 years, one third of all emergency department visits for injury are the result of falls, but 97% of these patients are not admitted to the hospital. In older children, injuries from falls are associated with play; in adults, injuries from falls are associated with working on ladders, roofs, etc. For patients age 65 years or older, falls are the leading cause of accidental death. In this age group, nonfatal falls often result in musculoskeletal trauma, most commonly fractures of the hip and distal radius. Hip fractures will significantly impair the quality of life even if death does not result. Both environmental and host factors increase the risk of accidental falls in people older than 65 years. Environmental factors include inadequate lighting and unstable or slippery floor surfaces. Host factors include impaired vision, altered mental status, and muscle weakness.[3,5,6]

Presentation

The goal of trauma care is to maximize patient survival and functional outcome after injury. Reaching this goal requires an accurate, organized appraisal of the injuries followed by prompt resuscitation. The methods of evaluation and resuscitation promoted by ATLS are widely used in the United States and call for a sequence of care that includes a primary survey to identify immediately life-threatening injuries, resuscitation to stabilize the patient's vital signs, reevaluation of the vital signs, and then a more detailed secondary survey, with specific radiographic and laboratory tests performed as indicated.

The primary survey is meant to identify and treat immediately life-threatening injuries. In addition to the ABCs of trauma care, emergency medical personnel evaluate disability resulting from central nervous system injury and make sure the patient is fully exposed by removing all clothing to rule out occult life-threatening injuries.[1] Using the mnemonic ABCDE (airway, breathing, circulation, disability, exposure), these five steps are considered from the perspective of the musculoskeletal medicine physician as follows:

Airway

An intact or patent airway implies that the patient is capable of moving air between the mouth and nose and the lungs. Control of the airway by endotracheal intubation is indicated for airway obstruction or pending airway obstruction, as might be seen with neck swelling. Endotracheal intubation should be undertaken with particular care in unconscious patients or in those who have documented or suspected spinal cord injuries (ie, those who report neck pain or have sustained significant trauma). The patient's head should be immobilized with rolled towels, foam padding, or a combination of sandbags, a plastic cervical spinal collar, and tape (Fig. 1).

Breathing

Inadequate ventilation is usually caused by hemothorax, pneumothorax, or musculoskeletal trauma to the chest wall resulting in a flail chest. A flail chest occurs after multiple rib fractures and results when a segment of the chest wall is no longer stable. The loose segment floats freely and moves in during inspiration and out during exhalation, exactly the opposite of normal chest wall dynamics. Air may enter the pleural space at the site of injury during inspiration, resulting in a sucking chest wound. Pulmonary contusion and limited respiratory effort as a result of pain (splinting) are associated with chest trauma. Treatment includes evacuation of the pneumothorax, application of an occlusive dressing on sucking chest wounds, and stabilization of the floating segment of the chest wall.

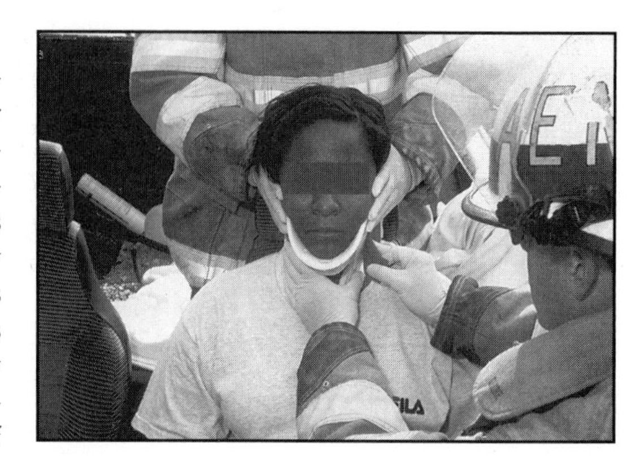

Figure 1
Immobilization of the cervical spine is essential to prevent further injury.

(Reproduced with permission from Browner BD, Pollak AN, Gupton CL (eds): **American Academy of Orthopaedic Surgeons** *Emergency: Care and Transportation of the Sick and Injured, ed 8. Sudbury, MA, Jones and Bartlett Publishers, 2002, p 702.)*

Circulation

Shock occurs when oxygen-rich blood is not delivered in adequate amounts to the peripheral tissues. Clinical signs of shock include rapid, faintly palpable, peripheral pulses; pallor; reduced skin temperature; and sweating. Initially, patients may seem anxious, but, as shock progresses, lethargy and decreased level of consciousness are seen. Tachycardia is the most sensitive sign of shock in the injured patient; a heart rate that is persistently faster than 120 beats/min in adults or 160 beats/min in children younger than 6 years should be treated as shock.

The most common cause of shock resulting from trauma is bleeding. A source of external bleeding should be identified and stopped during the primary survey. Common sources of bleeding are extremity, facial, and scalp wounds. Bleeding from extremity wounds is best controlled by direct pressure. The use of hemostats and suture ligatures to control bleeding in the emergency department is discouraged because it may lead to the inadvertent damage of nerves and tendons near the source of bleeding. Life-

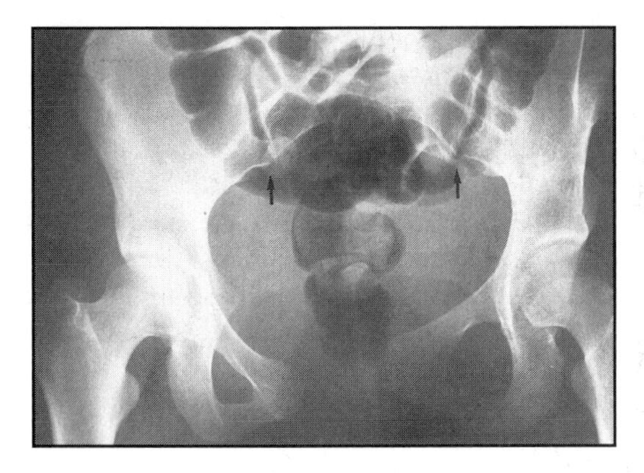

Figure 2
Pelvic fracture can be associated with significant blood loss. Closure of the "open-book" pelvis, by manually compressing the right and left ilium to bring the pubic bones together, can often stop ongoing blood loss. In this radiograph, in addition to the obvious widening of the symphysis, there is subtle widening of the sacroliliac joints posteriorly (arrows).

(Reproduced with permission from Tile M (ed): ***Fractures of the Pelvis and Acetabulum****, ed 2, Baltimore, MD, Williams & Wilkins, 1995.)*

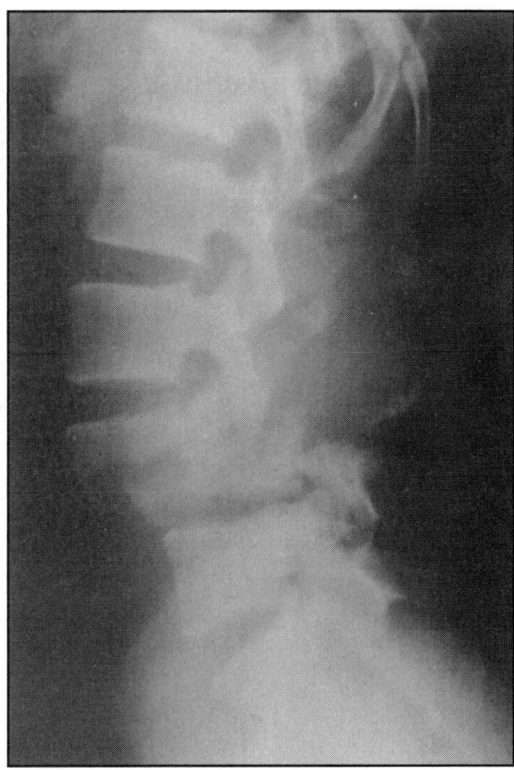

Figure 3
Lateral radiograph showing a cervical spine fracture with displacement, which poses a high risk of neurologic injury.

(Reproduced with from Spivak JM, Vaccaro AR, Cotler JM: Thoracolumbar spine trauma: I: Evaluation and classification. ***J Am Acad Orthop Surg*** *1995;3:345-352.)*

threatening bleeding from an extremity wound usually indicates a partial laceration of an artery. Complete division of an artery produces spasm and clot formation that will usually stop the bleeding. Following partial transection of an artery, spasm occurs, but it will not completely occlude the vessel lumen and bleeding will continue. Because the venous system maintains lower pressure, life-threatening bleeding from extremity veins is unusual.

If no external source of bleeding is identified during the primary survey, other potential causes of shock must be investigated, such as internal bleeding from pelvic, abdominal, or thoracic injuries. Even closed extremity or pelvic fractures (ie, fractures that have not penetrated the skin) can be responsible for significant blood loss (Fig. 2). Note that not all shock is a result of blood loss. Although there is adequate blood volume, the heart may be incapable of circulating it (cardiogenic shock). For example, cardiac tamponade from bleeding into the pericardium should be suspected in the setting of a penetrating trauma near the heart. Injury to the spinal cord, resulting in loss of vasomotor tone and peripheral vasodilation, can also produce shock. This type of shock, called neurogenic shock, should be suspected in the setting of a spinal injury (Fig. 3). Unlike hemorrhagic and cardiogenic shock, neurogenic shock is characterized by hypotension without tachycardia. Initial treatment is with crystalloid administration, but vasopressors also may be required. Before vasopressors are administered, the potential for concurrent hemorrhagic shock should be evaluated and then treated if necessary.

Neurologic Disability
A brief neurologic examination is indicated to identify patients with potential intracranial injuries and/or intoxication. A detailed neurologic examination is not part of the primary survey. However, it is important to document both the initial neurologic status and any changes in status occurring in the emergency department.

Exposure
The final step of the primary survey is to remove all the patient's clothing so that a complete inspection for injuries can be performed.

Care is taken to maintain body temperature by using radiant warmers, administering warm fluids, and covering the patient with warm blankets after the primary survey.

Treatment and Prevention

Resuscitation

The initial steps of resuscitation proceed simultaneously with the primary survey. Once the airway is secure, ventilation with 100% oxygen is initiated. At least two large-bore (16-gauge or larger) intravenous catheters are inserted in almost all trauma patients. Insertion sites in the extremities are preferred. If there is significant musculoskeletal or vascular injury, a site remote from the trauma should be selected. In children, fluids can be administered directly into the medullary canal of the tibia or femur, but intravenous catheterization is preferred because fluid reaches the circulating volume faster.

For patients with hypotension and tachycardia, 2 L of lactated Ringer's solution (20 mL/kg for children) is rapidly infused. If the vital signs are corrected after 2 L of crystalloid is administered, maintenance fluids are initiated. If the vital signs are not stabilized, an additional 2-L bolus is given, and transfusion with blood products is indicated. Blood that is not crossmatched (type O negative) is used if exsanguination is imminent; otherwise, type-specific, crossmatched (type-specific) blood is used.

Secondary Survey

The secondary survey is a more complete evaluation that is performed once the ABCs have been addressed. The physical examination is performed in a systematic fashion, moving from head to toe. The neck is inspected for penetrating wounds, swelling, hematoma, and other abnormalities. Any patient who is not conscious, has been involved in a high-energy accident, or has evidence of direct trauma to the head and neck requires radiographs to rule out cervical spinal injury. If a complete series of plain radiographs is not adequate to rule out injury, CT is indicated because it is often difficult to visualize the lower cervical spine on radiographs. Until cervical spinal injury can be ruled out, the patient should remain immobilized on a spine board and in a rigid collar. Abnormalities in the reflexes and deficits in sensory and motor function must be documented at this point.

Next, the anterior, posterior, and flanks of both the abdomen and chest must be inspected. This requires logrolling the patient from side to side with care to maintain the neck in a neutral position. A rectal examination is performed at this time and should always precede insertion of an indwelling urinary catheter in case of urethral injury. Significant findings during the rectal examination include gross blood in the rectal vault, indicating anorectal injury; loss of rectal sphincter tone, indicating spinal cord injury; and a high-riding prostate, indicating urethral injury. Stability of the pelvis is assessed by rocking the iliac crests from side to side. Instability implies a pelvic fracture.

Each extremity, including the hands and feet, are inspected for deformity and penetrating wounds. Evidence of fractures includes abnormal position, swelling, and instability of the extremity. Potentially fractured bones are initially splinted until a definitive evaluation is performed. Sensation and motor function of the extremities are assessed. Potential vascular injuries are evaluated by inspection for penetrating wounds in proximity to major vessels and associated hematoma. The peripheral pulses are palpated and compared with the contralateral side. Diminished or asymmetric pulses, isolated extremity ischemia, and expanding hematoma are indications for surgical exploration of the involved vessel.

Complications

Extremity trauma may be complicated by compartment syndrome, which occurs if the interstitial pressure within a fascial compartment rises above capillary closing pressure. This prevents oxygen delivery to tissue within the involved compartment. If untreated, compartment syndrome can lead to irreversible tissue injury and necrosis. Compartment syndrome is traditionally associated with injuries such as open and closed fractures but may also occur after soft-tissue crush injury and vascular trauma. Compartment syndrome may also appear following repair of a vascular injury. The first sign is pain out of proportion to the apparent injury, especially with passive motion. Nerve dysfunction, including paralysis and paresthesia, and loss of arterial pulses are late signs and cannot be used to make the diagnosis. By the time these appear, irreversible damage may have already occurred.

If compartment syndrome is suspected be-

Table 1
Mangled Extremity Severity Score (MESS)

Category	Examples	Score
Energy of Trauma		
Low	Stab wound	1
Medium	Open fracture	2
High	Gunshot or high-speed motor vehicle accident	3
Massive crush	Industrial injury	4
Shock		
Normal	Stable vital signs in the field and hospital	0
Brief hypotension	Shock that responds to fluid resuscitation	1
Prolonged hypotension	Systolic pressure < 90 mm Hg; resuscitated only in operating room	2
Limb Ischemia		
None		0
Mild	Diminished pulses but good capillary refill	1 (2 if ischemic 6 hours)
Moderate	No pulse, paresthesia, and weak motor response	2 (4 if ischemic 6 hours)
Severe	No capillary refill, no pulse, insensate, and no motor response	3 (6 if ischemic 6 hours)
Age (years)		
< 30		0
30 to 50		1
> 50		2

cause of the mechanism of injury or physical findings, pressures can be directly measured by devices that can be inserted into the muscle compartment of concern. If compartment pressure is sufficiently elevated, fasciotomy is indicated. Fasciotomy allows the soft tissues to expand, thereby reducing compartment pressure and improving capillary function. Ideally, fasciotomy is performed in the operating room, but if transfer to the operating room will result in undue delay, fasciotomy can be done in the emergency department.

The mangled extremity severity score (MESS) can help physicians make decisions about when to attempt to save an extremity as opposed to performing primary amputation. Four characteristics of the injured extremity are scored, including the energy of the trauma, the presence of shock, the extent of limb ischemia, and the patient's age (Table 1). Primary amputation should be considered if the MESS is greater than six. It is unlikely, given such a score, that attempted limb salvage would be successful. Primary amputation spares the patient the complications of repeated and perhaps futile surgeries. Primary amputation of any extremity should be carried out when 6 hours or more of warm ischemia have transpired, which on the basis of ischemia and energy alone, would result in a MESS score of at least seven. In this circumstance, the muscle will not be viable even if perfusion could be restored. Also, for leg injuries associated with division of the tibial nerve, amputation should be considered. The reason is that even if the tibial nerve were repaired, a neurogenic pain syndrome would likely develop that is so disabling that most patients would be more functional with an amputation and rehabilitation with a prosthesis. Finally, complex extremity salvage operations are not appropriate when there are other life-threatening injuries that preclude a lengthy surgical procedure.[7]

Imaging

Patients with a history of significant trauma are presumed to have a cervical spine injury until proven otherwise. If the patient is awake and alert, the cervical spine can be evaluated clinically. If the patient has normal range of motion and no pain, imaging studies are not required. Otherwise, obtaining plain radiographs in the emergency department is the first step in evaluation. Three views, including lat-

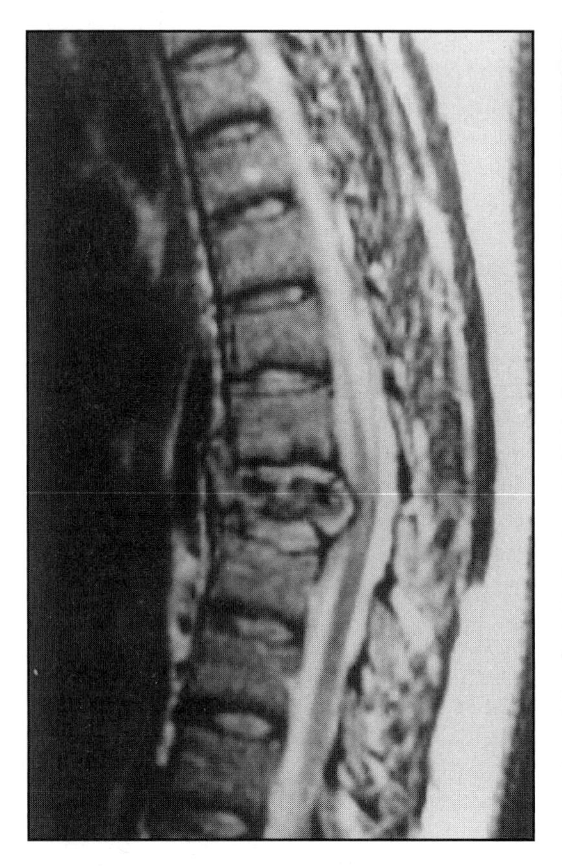

Figure 4
T2-weighted MRI scan of the thoracic spine showing canal encroachment by fracture fragments.

(Reproduced from Spivak JM, Vaccaro AR, Cotler JM: Thoracolumbar spine trauma: I. Evaluation and classification. ***J Am Acad Orthop Surg*** *1995;3:345-352.)*

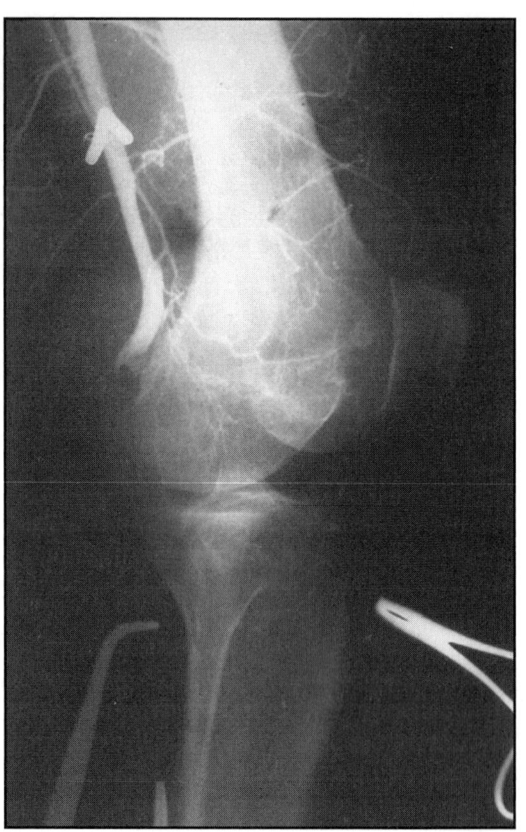

Figure 5
Intraoperative angiogram shows disruption of the popliteal artery above the knee after a dislocation.

(Reproduced from Good L, Johnson RJ: The dislocated knee. ***J Am Acad Orthop Surg*** *1995;3:284-292.)*

eral, AP, and odontoid projections, are obtained.[8] For patients with neurologic deficits, neurosurgical or orthopaedic consultation is indicated; plain radiographs, CT, or MRI is required (Fig. 4). When high-quality radiographs cannot be obtained or when the patient reports neck pain or is not able to cooperate with a physical examination, CT of the cervical spine is indicated. A protective cervical collar should not be removed until the patient is alert enough to participate in an examination. A pelvic radiograph is indicated to diagnose pelvic fractures, which can be a source of significant occult bleeding and lead to hemorrhagic shock. Rapid application of an external fixator may compress the fragments to reduce the bleeding.

Patients with diminished but not absent pulses are candidates for vascular imaging studies. Angiography has a sensitivity of up to 100% and a specificity of up to 98% for diagnosing arterial injuries (Fig. 5). However, peripheral angiography after trauma carries a risk of minor complications of up to 4% and a risk of iatrogenic injury that requires surgical repair in nearly 1% of patients.[9] In addition, angiography is expensive and time consuming. Unlike angiography, duplex ultrasound scanning is noninvasive and poses essentially no risk. Duplex studies have a sensitivity that approaches 100% and a specificity of up to 97% for diagnosing vascular injuries when interpreted by expert radiologists.[10] However, most surgeons find ultrasound images substantially more difficult to interpret than angiograms, and their usefulness is accordingly limited.

Bleeding from bony fragments of the pelvis or from the rich plexus of veins surrounding the pelvis usually stops once the fragments are reduced and compressed. This can be rapidly accomplished with external fixation. Pelvic fractures often disrupt branches of the internal iliac arteries, which can result in brisk bleeding and hemorrhagic shock.

Less commonly, bleeding occurs from branches of the external iliac arteries. Embolization is a very effective way to treat this type of bleeding. If there is persistent shock and no specific site of bleeding is identified, embolization of one or both internal iliac arteries should be performed. This maneuver is both well-tolerated and effective. Surgical exploration of an acutely bleeding and fractured pelvis is difficult and carries a higher risk of complications.[11]

Prevention

The term "accident" implies a random, uncontrollable event. However, scientific study of accidents has found that they are highly complex but reproducible events. Therefore, it is reasonable to consider the role of prevention in the management of trauma. This includes the identification of potential hazards and their elimination. Prevention also implies taking steps to decrease the consequences of accidents that do occur, including minimizing the chances that a given accident will lead to injury, by using seat belts, for instance; minimizing the chances that a given injury will lead to disability or death, which is what emergency medical services strive to do; and minimizing the chances that a given disability will lead to impairment, by employing high-quality health care and rehabilitative services.[12]

The responsibility for preventing motor vehicle accident trauma lies with government (designing safe roads and intersections) and industry (making safe cars), but ultimately it is the individual who must take appropriate steps to ensure his or her own safety. This includes driving at reasonable speeds and not driving under the effects of alcohol.

Key Terms

Compartment syndrome Ischemia of the nerves and muscles within a fascial compartment caused by elevated pressure within the compartment; frequently seen in association with tibial fractures

Flail chest A condition in which three or more ribs are fractured in two or more places, or in association with a fracture of the sternum, so that a segment of chest wall is effectively detached from the rest of the thoracic cage.

Mangled extremity severity score (MESS) A rating scale for classifying lower extremity trauma based on age, shock, limb ischemia, and skeletal and/or soft-tissue damage

Primary survey An emergency examination during which a patient is assessed and treatment priorities are established based on the patient's injuries, stability of vital signs, and injury mechanism

Secondary survey A comprehensive evaluation of a patient that is performed after resuscitation to detect injuries not noted on the primary survey

References

1. American College of Surgeons Committee on Trauma: *Advanced Trauma Life Support Program for Doctors: ATLS*, ed 6. Chicago, IL, American College of Surgeons, 1997.
2. National Center for Injury Prevention and Control: *Ten Leading Causes of Death, 1995.* Atlanta, GA, Centers for Disease Control and Prevention, 1998.
3. Health: *United States: Injury Chartbook: 1996-1997.* Hyattsville, MD, National Center for Health Statistics, 1997.
4. Annest JL, Conn JM, James SP: (eds): *Inventory of Federal Data Systems in the United States for Injury Surveillance, Research and Prevention Activities.* Atlanta, GA, National Center for Injury Prevention and Control, 1996.
5. Rivara FP, Alexander B, Johnston B, et al: Population-based study of fall injuries in children and adolescents resulting in hospitalization or death. *Pediatrics* 1993;92:61-63.
6. Tinetti ME, Speechley M, Ginter SF: Risk factors for falls among elderly persons living in the community. *N Engl J Med* 1988;319:1701-1707.
7. Johansen K, Daines M, Howey T, et al: Objective criteria accurately predict amputation following lower extremity trauma. *J Trauma* 1990;30:568-573.
8. Mirvis SE, Hastings G, Scalea TM: Diagnostic imaging, angiography, and interventional radiology, in Mattox KL, Feliciano DV, Moore EE (eds): *Trauma*, ed 4. New York, NY, McGraw-Hill Health Professions Division, 2000, pp 285-288.
9. Reid JD, Weigelt JA, Thal ER, et al: Assessment of proximity of a wound to major vascular structures as an indication for arteriography. *Arch Surg* 1988;123:942-946.
10. Shackford SR, Rich NH: Peripheral vascular injury, in Mattox KL, Feliciano DV, Moore EE (eds): *Trauma*, ed 4. New York, NY, McGraw-Hill Health Professions Division, 2000, pp 1011-1046.
11. Agolini SF, Shah K, Jaffe J, et al: Arterial embolization is a rapid and effective technique for controlling pelvic fracture hemorrhage. *J Trauma* 1997;43:395-399.
12. Haddon W Jr.: The changing approach to the epidemiology, prevention, and amelioration of trauma: The transition to approaches etiologically rather than descriptively based. *Am J Public Health Nations Health* 1968;58:1431-1438.

Fracture

Bones break when they are subjected to forces greater than their mechanical tolerance. Whether a bone will fracture under a given load depends on both the inherent strength of the bone and the magnitude of the force. The direction of the force and rate at which it is applied are also important considerations. Bone has a remarkable capacity to regenerate; it and the liver are the only two organs in the body that repair themselves without scar and form new, intact tissue.

Fractures are clinically important for a variety of reasons. They can be painful, debilitating, and in some settings even life threatening. They may cause (or be related to) medical complications, such as compartment syndrome or thrombosis. They also may reflect underlying medical diseases, such as endocrinopathies or malignancies. Above all, fractures are common.

Describing fractures entails mastering a common vocabulary. Although physicians may name a fracture by its eponym (a fracture of the distal radius, for instance, is known as a Colles' fracture) or a classification number (a method perhaps limited to orthopaedic surgeons), this practice is not suggested for students or practitioners because it is easy to use the wrong eponym or classification number without realizing it or to confuse others who may not be familiar with these systems. Therefore, fractures should be described in such a way that the description matches the radiographic findings.[1] To do this, the following questions must be answered: (1) Which bone is broken? (2) Which region or segment of the bone is broken? (3) What is the pattern of the break? and (4) Is the skin broken?

The first question is straightforward but requires some knowledge of skeletal anatomy. Although it may be difficult to recall all the names of the small bones of the hand and feet, the distinctions among them are critical. For example, a nondisplaced fracture of the scaphoid is a much more serious injury than a similar break of the metacarpal; the scaphoid fracture has a much higher risk of nonunion. (Thus, when you are confronted with a radiograph of a fracture but are uncertain of the name of the bone, look it up.)

The segment of bone that is broken also must be described. In the long bones, the diaphysis, metaphysis, and epiphysis are clearly separate (Fig. 1). It is also important to note whether the fracture violates the articular surface. Slight displacement of the cartilage that lines the joint can impair the smooth mechanics of the joint. As such, displacement is less well tolerated (and more important to detect) in fractures near the joint compared with, for example, fractures in the midshaft of the bone. Angulation of fractures near the joint line also may interfere with joint function or dislocate the joint (Fig. 2); thus, an assessment of the alignment of the adjacent joint is required.

Certain bones have distinct areas that are subject to fracture, and breaks there are described with their own nomenclature. For example, an oblique fracture of the proximal femur between the greater and lesser trochanter is known as an intertrochanteric fracture. A diagonal fracture of the distal radius is called a radial styloid fracture. Also, the regions of the flat bones, such as the pelvis, are named distinctly because long-bone terminology (which uses the physis as a point of reference) does not apply to flat bones.

The pattern of the fracture is described in terms of the geometry of the fracture line or lines (Table 1 and Fig. 3). First, is there one line or more than one? Breaks with more than one fracture line are called comminuted fractures. Comminution usually implies a high-energy mechanism of injury but also can result from a low-energy injury to a bone already weakened by disease. A transverse

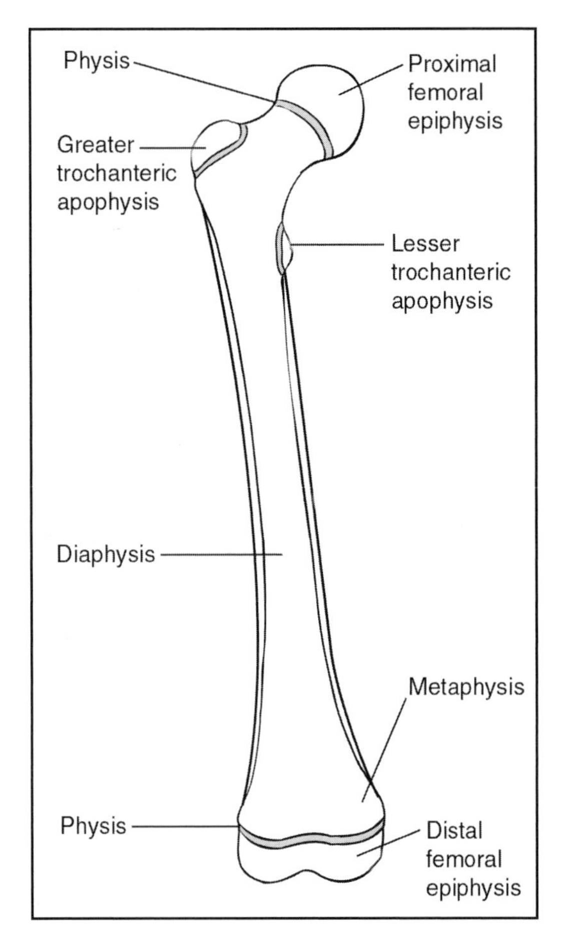

Figure 1
The regions of a long bone, shown here in the femur.

(Reproduced from Sullivan JA: Introduction to the musculoskeletal system, in Sullivan JA, Anderson SJ (eds): ***Care of the Young Athlete****. Rosemont, IL, American Academy of Orthopaedic Surgeons, 2000, pp 243-258.)*

fracture is a fracture that occurs at an angle perpendicular to the shaft of the bone. An oblique fracture crosses the bone diagonally from one cortex to the other. A spiral fracture, as the name implies, wraps around the bone like a coil. The pattern of a fracture also often gives clues as to how the bone was broken and therefore may suggest the presence of other injuries. For instance, an oblique fracture of the medial portion of distal tibia at the ankle (the medial malleolus) implies that the injury was caused by inversion—the talus impacts the tibia and breaks it. This positioning increases the possibility that there is a tension injury on the lateral side, most likely a ligament rupture. Thinking mechanistically is important in this case because the ligament rupture will not appear on radiographs.

Finally, the condition of the soft tissues must be assessed. Fractures associated with breaks of the skin are called open fractures. These are true medical emergencies because if open fractures are not cleaned out surgically, there is an increased risk of bone infection. Therefore, it is critical to detect them. All fractures associated with blood on the skin, even without an obvious break in the skin, must be considered open unless and until another explanation for the blood is found. One method to determine whether a fracture is open is to probe the laceration for the presence of bone. Another indication that an open fracture may be present is when bleeding does not stop despite attempts at holding pressure over the skin laceration.

With these questions answered, almost all fractures can be described completely. Additional information including the age of the patient, the mechanism of injury, and the quality of the bone stock completes the picture and allows the physician to derive an appropriate treatment plan.

How Bones Break

The fact that bone is alive influences how it responds to injury and even how it prevents injury—active remodeling removes areas of small damage and prevents fatigue fractures.[2] Nonetheless, these processes, typical of biologic responses, are slow and progress over days to weeks. Thus, when a force is applied to a bone instantaneously, the fate of the bone—whether it will break or not—depends not on its living biologic properties but on its inert mechanical properties.

A fracture represents the failure of a material (bone) to withstand a force. Failure occurs as a function of the inherent properties of the material as well as at least three additional factors: (1) the magnitude of the force applied to it, (2) the direction of that force, and (3) the rate at which force is applied. The importance of the inherent properties is intuitively clear: stronger material can withstand greater force. In clinical terms, the mechanism of injury (specifically, how much energy is imparted to the bone) will influence whether a fracture occurs. For example, a fall onto the buttocks from a standing height typically does not fracture the pelvis, whereas a similar landing after a fall from a three-story roof almost always does. The quality of bone and the ability of the individual to use his or her muscles to protect bone from absorbing energy are also key factors in determining whether a fracture will occur.

The duration and rate of the force application are critical factors in determining whether an injury to bone will occur. Materials such as bone whose mechanical properties are dependent on the loading rate of an applied force are said to be viscoelastic. The viscoelasticity of bone is important clinically because when force is applied at low speeds, bone is weaker than ligament; when applied at higher speeds bone is stronger. Accordingly, a rapidly applied anterior force to the tibia will rupture the anterior cruciate ligament. By contrast, if the force were applied at a lower rate, the ligament would remain intact, but the bone would break. This low-speed mechanism of injury would produce an avulsion fracture of the piece of bone to which the cruciate ligament is attached. Likewise, an inversion force to the ankle at lower speeds might produce a transverse fracture of the fibula, whereas that same force applied at a greater speed would lead to a rupture of the lateral ankle ligaments.

Even with a given amount of energy applied to the bone at a given rate of application, it is still not certain that the bone will fracture because the direction of the force is also critical. The forces applied to bone can be categorized by their direction as compressive (push), tensile (pull), or torsional (twist). Materials such as bone whose mechanical properties are dependent on the direction of loading are said to be anisotropic.[3] (A material such as steel, which has the same mechanical properties regardless of direction, is isotropic.) The anisotropic properties of bone make sense from the perspective of engineering efficiency. Most of the forces applied to bone in ordinary use are compressive (ie, resulting from loading of the skeleton during walking or running); thus, the mineral component of bone is aligned along the axis of the bone to resist such forces. A greater force is required to fracture the bone in axial compression. If a bending force is applied, the bone experiences tension forces on one cortex and compression forces on the opposite cortex. The failure of the bone will begin on the side under tension and move across to the compression side (Fig. 4).

As a material, adult bone is brittle and tends to deform only slightly before it breaks. Comparatively, younger bone is able to deform a great deal more before it fails. The property of greater plastic deformation before failure is called ductility. Because of the ductile nature of pediatric bone, children can develop fracture patterns not seen in adults. A torus (buckle) fracture occurs when the bone under load bends but does not break. Another pattern unique to children is the greenstick fracture, which occurs when forces applied to young bone break it through only one cortex but not completely across. This fracture pattern gets its name from a similar fracture pattern that occurs when one bends a small, living tree branch.

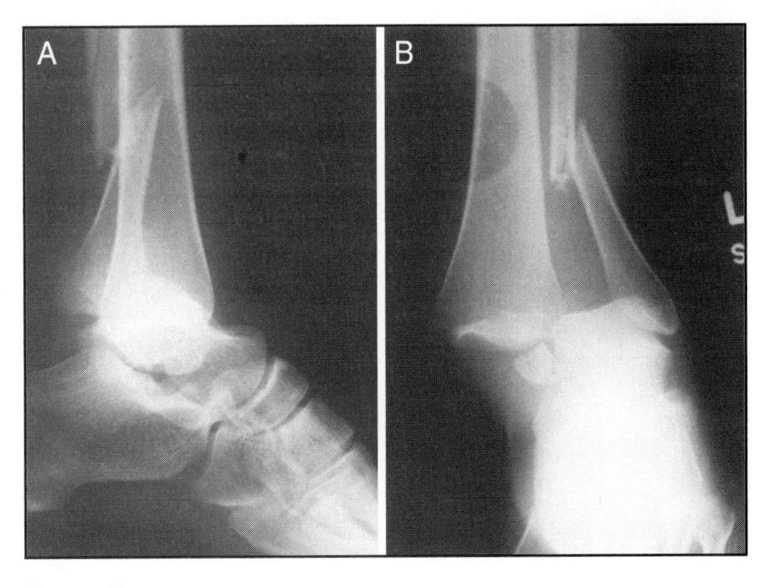

Figure 2
A and **B**, In this ankle fracture, the joint between the tibia and talus, referred to as the mortise, is disrupted, producing a fracture-dislocation.

(Reproduced from Stephen DJG: Ankle and foot injuries, in Kellam JF, Fischer TJ, Tornetta P III, Bosse MJ, Harris MB (eds): ***Orthopaedic Knowledge Update: Trauma 2****. Rosemont, IL, American Academy of Orthopaedic Surgeons, 2000, pp 203-225.)*

Of course, the major difference between adult and pediatric bone is that young bone contains a physis (growth plate). This region of bone is not only mechanically weaker than bone and ligament, but it is also critically important for the growth of the bone. Injury to the physis may cause growth disturbance or arrest.

How Bones Heal

Bone healing is divided into two types: primary and secondary. Primary bone healing, requiring precise reapproximation of the fracture, is rare in nature. Primary healing requires rigid immobilization of the fracture, such as that which is seen with surgical plating and compression of the cortices of the bone together. Because rigidity is required,

Table 1
Fracture Descriptions

Location in the bone	Description
Epiphyseal	The end of the bone, forming part of the adjacent joint
Metaphyseal	The flared portion of the bone at the ends of the shaft
Diaphyseal	The shaft of a long bone
Orientation/extent of the fracture line(s)	**Description**
Transverse	A fracture that is perpendicular to the shaft of the bone
Oblique	An angulated fracture line
Spiral	A multiplanar and complex fracture line
Comminuted	More than two fracture fragments
Segmental	A completely separate segment of bone bordered by fracture lines
Intra-articular	The fracture line crosses the articular cartilage and enters the joint
Torus	A buckle fracture of one cortex, often seen in children
Compression	Impaction of bone, such as in the vertebrae or proximal tibia
Greenstick	An incomplete fracture with angular deformity, seen in children
Pathologic	A fracture through bone weakened by disease or tumor
Amount of displacement of the fracture fragments	**Description**
Nondisplaced	A fracture in which the fragments are in anatomic alignment
Displaced	A fracture in which the fragments are no longer in their usual alignment
Angulated	A fracture in which the fragments are malaligned
Bayonetted	A fracture in which the distal fragment longitudinally overlaps the proximal fragment
Distracted	A fracture in which the distal fragment is separated from the proximal fragment by a gap
Integrity of the skin and soft-tissue envelope around the fracture	**Description**
Closed	The skin over and near the fracture is intact
Open	The skin over and near the fracture is lacerated or abraded by the injury

*(Reproduced from Greene WB (ed): **Essentials of Musculoskeletal Care**, ed 2. Rosemont, IL, American Academy of Orthopaedic Surgeons, 2001, p 37.)*

this form of healing almost never occurs without surgical intervention. Primary bone healing may be thought of as simply the deposition of new bone across the fracture by osteoblasts. This new bone integrates into the two opposing sides through tunnels created by osteoclasts called cutting cones.[4] In the cutting cone, there is local bone resorption and eventual recreation of normal bone structure.

In secondary bone healing, bone first produces a mass of cartilage scar. This mass ossifies and then remodels to form normal bone. The secondary response of bone is best understood by recalling that the bone is a highly vascularized structure: breaking a bone tears some blood vessels, and a fracture always creates some degree of hematoma (blood collection). The hematoma is the source of much of the initial healing re-

Open fracture

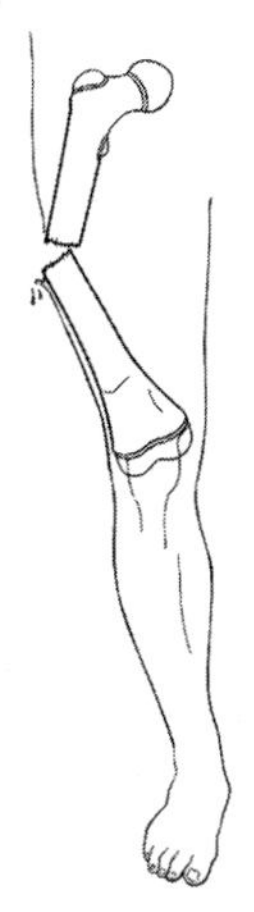

A fracture in which the skin is broken, exposing the fracture site to the environment

Closed fracture

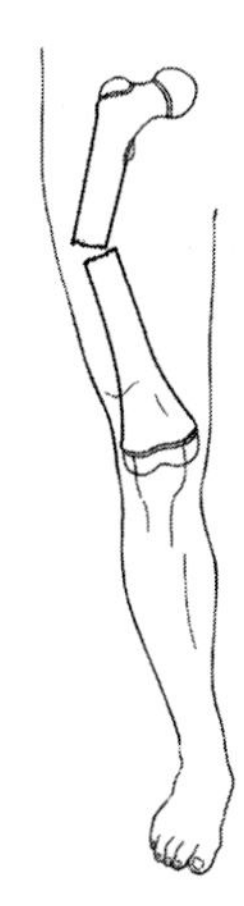

A fracture in which the skin is not broken

Comminuted fracture

A fracture that results in three or more bone fragments

Spiral fracture

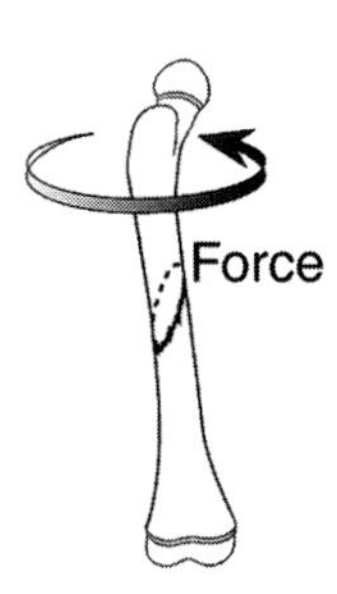

A fracture caused by a twisting force that results in a spiral-shaped fracture line about the bone

Oblique fracture

A fracture in which the fracture line is angled with respect to the long axis of the bone

Transverse fracture

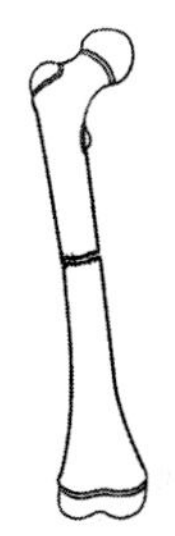

A fracture in which the fracture line is at a right angle to the long axis of the bone

Torus (buckle) fracture

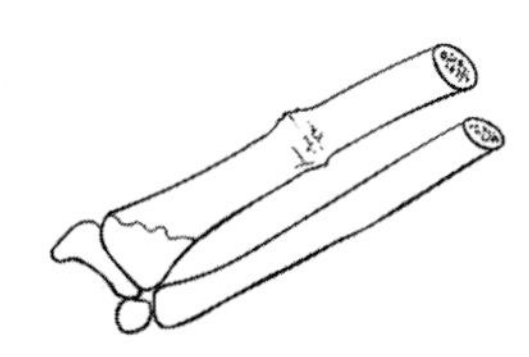

A fracture that disrupts, but does not completely break the cortex (appears wrinkled or buckled)

Greenstick fracture

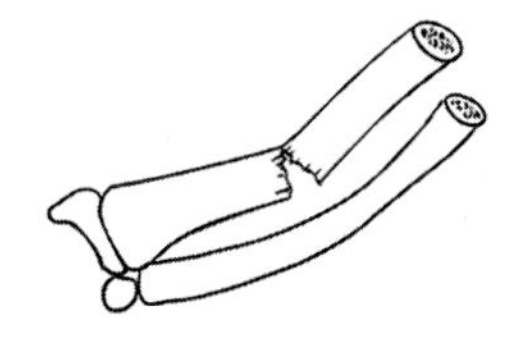

A fracture that breaks one side of the cortex and causes plastic deformation of the other side of the same bone

Figure 3
Patterns of fractures.

*(Reproduced from Sullivan JA: Introduction to the musculoskeletal system, in Sullivan JA, Anderson SJ (eds): **Care of the Young Athlete**. Rosemont, IL, American Academy of Orthopaedic Surgeons, 2000, pp 243-258.)*

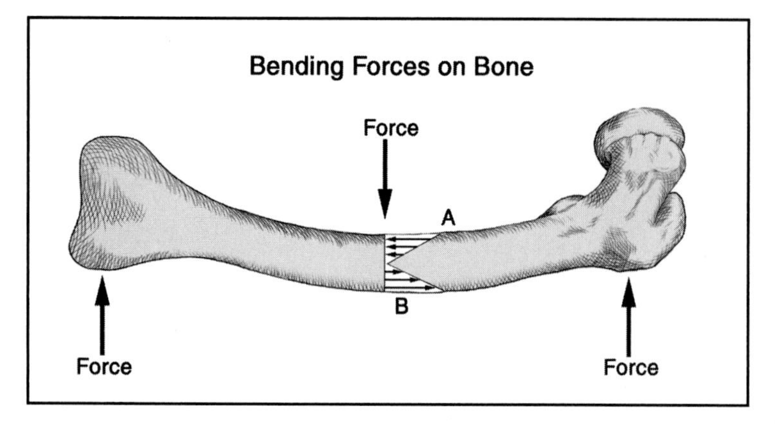

Figure 4
Bending the bone (with forces shown as arrows) stresses the shaft. The concave side (A) is compressed, whereas the convex side (B) is under tension. Thus, the fracture will begin on the convex side because bone fails under tension before compression.

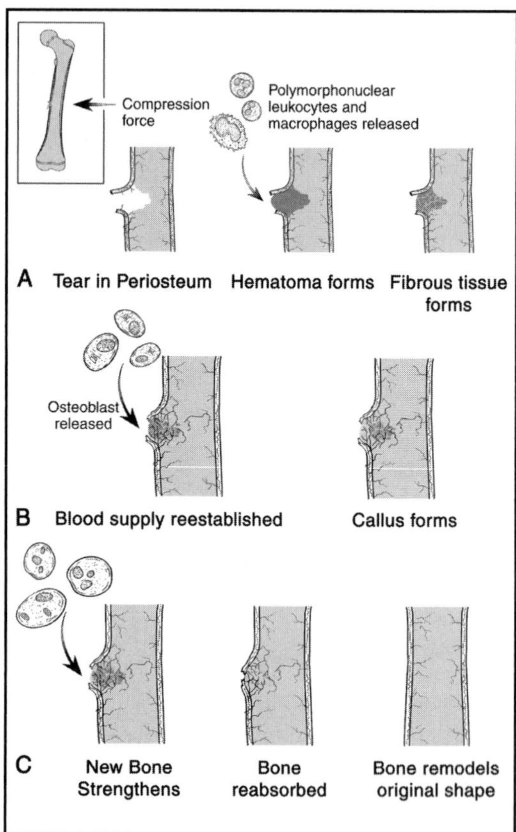

Figure 5
Phases of secondary fracture healing. **A**, Early inflammatory stage. **B**, Callus formation occurs in the second phase. **C**, In the third phase, new bone formation strengthens the bone, and the bone remodels to restore its normal shape.

(Reproduced from Sullivan JA: Introduction to the musculoskeletal system, in Sullivan JA, Anderson SJ (eds): ***Care of the Young Athlete****. Rosemont, IL, American Academy of Orthopaedic Surgeons, 2000, pp 243-258.)*

sponse—being both the mechanical scaffolding upon which healing takes place and a depot of biologic factors that initiate and sustain the development of new bone.

The process of secondary healing can be described as follows: the bone breaks, and a hematoma is formed. Through the process of inflammation and angiogenesis, the hematoma becomes organized. Cells proliferate, and granulation tissue (primitive scar) is formed. This cartilaginous tissue is then calcified. Blood vessels subsequently invade, bringing the cellular machinery of bone remodeling, and the calcified cartilage of the callus is thereby converted to normal bone (Fig. 5). Secondary healing, in a sense, repeats the process of endochondral ossification and bone remodeling.

The periosteum is the main source of cells during the healing response. From the periosteum (and to some extent the surrounding soft tissues) progenitor mesenchymal stem cells are recruited to differentiate into a bone-producing lineage. Although the biochemistry and molecular biology of fracture healing is beyond the scope of this chapter, these processes will assume greater importance as new therapeutic methods are developed to manipulate them.[5,6]

Currently, there is great interest in studying growth factors that signal bone cells to proliferate or differentiate. Substances such as bone morphogenic proteins are being studied intensely and have been demonstrated to increase new bone formation.[7-9] The challenge of using growth factors effectively will be controlling them so that the bone is formed only in the affected area to the extent desired. Using precursor cells such as bone marrow stromal cells on carriers may help to promote bone formation locally and holds promise.[10]

Other research directions in fracture healing involve studying the actual genetic expressions of bone healing and formation. The Human Genome Project has resulted in the formation of genetic libraries. A technique called microarrays on bone filters will allow researchers to identify which genes and proteins are recruited for fracture healing and which genes are downregulated. The idea is first to understand which individual genes are involved in fracture healing to develop techniques to manipulate gene expression in the future.

Table 2
Methods for Immobilizing Fractures

Technique	Advantages	Disadvantages	Common Example
Casting	Relatively noninvasive Easy to apply	Skin breakdown/maceration Loss of reduction if cast becomes loose Pressure on nerves and blood vessels	Minimally displaced distal radius fracture
Open reduction and internal fixation (ie, plates and screws)	Allows perfect alignment of bone ends at the fracture site Holds bone in compression (promotes primary healing)	Soft-tissue stripping inevitable and biologic environment disturbed Stress shielding at the site Possible infection from opening fracture	Tibial plateau fracture
Intramedullary rod	Smaller incisions than open reduction and internal fixation Early weight bearing	Disruption of endosteal blood supply Reaming the medullary canal may trigger fat embolism syndrome	Midshaft femur or tibia fracture
External fixation	Allows access to soft tissue if wound is open Easy to adjust angulation of bone during treatment	Pin tract infections Cumbersome Fracture through empty pin sites	Open tibia fracture
Traction	Minimal surgical time Does not violate fracture site or growth plates	Prolonged bed rest causes medical complications	While awaiting surgery for hip fracture Definitive treatment for pediatric femur fracture

Clinical Evaluation of Fractures

Physical Examination

The single most important factor to consider when evaluating a patient with a suspected fracture is that attention to musculoskeletal injuries must be deferred to those that are more immediately life threatening.[11] Thus, the ABCs (airway, breathing, circulation) of trauma must first be considered.

Once these critical factors are evaluated and stabilized, suspected fractures should be immobilized (Table 2). Although fractures typically hurt, not all injured patients can reliably report their symptoms. Examine any body part that is painful, deformed, swollen, or otherwise abnormal. The physical examination assesses skeletal integrity; this includes not only palpating the bones to identify deformity, tenderness, or crepitus but evaluating the surrounding joints for motion as well.[12] The status of the nerves and blood vessels distal to the injury should also be assessed. Many physicians seek certification in Advanced Trauma Life Support (ATLS), and training manuals for certification are excellent primers on the evaluation of the injured patient.[13] Nonetheless, the best way to learn these skills is with hands-on experience.

When evaluating fractures, particular attention must be given to the soft tissues because an acute fracture is almost always accompanied by soft-tissue trauma. The forces that fracture bone can also crush, tear, avulse, or otherwise damage the surrounding tissues. Thus, the skin should be inspected for abrasions, bruises, and swelling. If a fractured bone is stripped of its soft tissue, the potential for osseous healing is significantly impaired. Thus, careful treatment of the soft tissues is critical to a successful outcome in fracture management.

Open fractures are often high-energy injuries associated with extensive soft-tissue damage and therefore need special attention. A fractured bone exposed to the environment through a traumatic wound is frequently contaminated with bacteria. Open fractures require urgent surgery to excise necrotic tissue and irrigate the wound to reduce the risk of subsequent infection.[14,15] Repeated débridement is needed in some cases to ensure that only viable, uninfected tissue remains in the wound. Because of the loss of the initial hematoma, bone healing occurs more slowly in open fractures. A system of classifying the soft-tissue injury associated with open fractures has been devised for clinical use (Table 3).

Imaging Studies

The radiographic assessment of fractures begins with plain radiographs, typically AP and lateral views. The joints above and below a suspected fracture must also be included in the study because some of the energy of the injury may have been absorbed at a site distal from the injury. A classic example of this is a twisting injury to the ankle in which the tibia is fractured at the medial malleolus (ankle) but the fibula is fractured proximally at its neck near the knee.

Certain injuries are known to occur in patterns. For example, a history of fracture of both calcanei (heel bones) sustained when a patient jumps out of a window requires that radiographs of the spine be obtained because energy may have been transmitted up the skeleton.[16] Patients who sustain high-energy trauma of any sort may benefit if radiographs of the cervical spine, chest, and pelvis are obtained routinely, even in the absence of specific signs or symptoms.

Table 3
Classification of Soft-Tissue Injury in Open Fractures

Type	Description
I	< 1 cm of skin laceration
II	> 1 cm laceration
IIIA	Extensive soft-tissue damage but closure of skin without a flap is possible
IIIB	Open fracture that needs a muscle flap for closure
IIIC	Extensive soft-tissue damage with vascular injury needing repair

*(Adapted with permission from Gustilo RB, Anderson JT: Prevention of infection in the treatment of 1025 open fractures of long bones. **J Bone Joint Surg Am** 1976; 58:453-458.)*

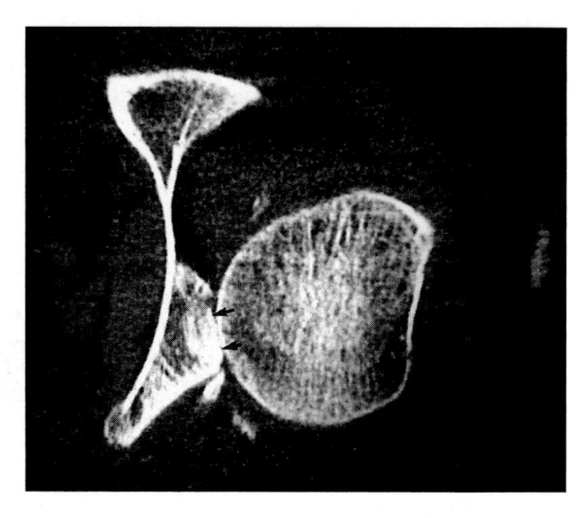

Figure 6
CT section of a nondisplaced transverse and posterior wall fracture of the acetabulum demonstrates that the femoral head is subluxated and identifies the area of impaction (arrows). No articular surface of the posterior wall remains intact.

*(Reproduced from Tornetta P III: Displaced acetabular fractures: Indications for operative and nonoperative management. **J Am Acad Orthop Surg** 2001;9:18-28.)*

In some instances, plain radiographs will not conclusively exclude the possibility of fracture. When a prompt diagnosis is essential, CT or MRI can be used. CT is used to diagnose fractures of the spine and pelvis (Fig. 6). MRI is particularly useful for detecting occult hip fractures[17] (Fig. 7).

Radiographic assessment should include some qualitative analysis of the bone adja-

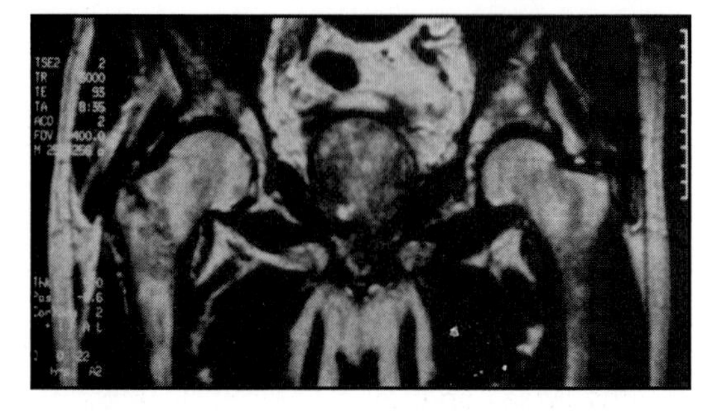

Figure 7
This fracture of the right hip was not apparent on radiographs but can be seen clearly on an MRI scan.

*(Reproduced from Koval KJ, Zuckerman JD: Hip fractures: I. Overview and evaluation and treatment of femoral-neck fractures. **J Am Acad Orthop Surg** 1994;2:141-149.)*

cent to the fracture. The term pathologic fracture is used to describe fractures that occur in the setting of abnormal bone. The presence of a malignancy or infection will weaken the bone and make it more susceptible to fracture, but detecting these conditions goes beyond academic interest: their presence obviously influences treatment.

Treatment

Soft-Tissue Care

The primary goal of treating soft-tissue injuries associated with fractures is to halt the continuing trauma to the tissues. This begins by realigning displaced fractures (fracture reduction) and dislocated joints and immobilizing the site of injury. The reduction of edema (and surgical release of fascial compartments under pressure, if needed) increases perfusion in the injured tissues and thus augments healing. Although diffuse cell death is common after trauma, the surviving tissue becomes highly active metabolically as it initiates repair. The delivery of oxygen and essential nutrients is impaired in swollen tissue; therefore, soft-tissue care can be thought of as the first step in promoting bone healing.

Surgical Options

If the quality of the clinical outcome from treatment depended only on the quality of the fracture reduction, the ideal treatment plan would be to surgically repair all fractures directly using open reduction and internal fixation (ORIF). Yet surgery is not done in many cases because the quality of the reduction does not solely determine outcome.

The main reason that ORIF is not used to treat every fracture is that it interferes with the biology of healing. Understanding the adverse effects of ORIF may better explain the biology of fracture healing and as such serve as a useful mnemonic. In this case, what seem to be details of treatment are actually cues for understanding the basic science.

ORIF has two adverse properties: (1) it strips the soft tissue (periosteum), and (2) it creates stress shielding. The fact that soft-tissue stripping occurs is fairly intuitive—to secure a plate along a fracture, a surgeon must surgically expose the bone. This process disrupts the soft-tissue envelope—specifically, the periosteum—and increases the risk that the bone will not heal and that infection will develop. Recall that the periosteum is the source of many of the cells needed in the healing response.

Rigid fixation of the bone also prevents the fracture line from experiencing and responding to loading stresses (stress shielding). Total rigidity, in this case, can be thought of as too much of a good thing. Immobilization is good, but too much immobilization may be bad. Wolff's law states that bone grows in response to mechanical stress. Thus, some (but not too much) loading at the fracture site is optimal for bone regeneration.[18] ORIF blocks that needed mechanical load.

Recognizing the limitations of ORIF helps in the understanding of the first principle of fracture treatment: it is essential to create the correct biologic and mechanical environment for repair and remodeling. From this first principle stem the following rules of fracture care:

1. Ensure that the fracture is surrounded by a healthy soft-tissue envelope (ie, débride open wounds and maintain soft-tissue coverage, but do not cut the skin unnecessarily).
2. Ensure that there is good perfusion (ie, release compartments with elevated pressure, loosen tight casts).
3. Align the bone (which may also help with perfusion).
4. Ensure that just the right amount of load is applied (ie, enough to stimulate bone growth but not so much that gross motion leading to nonunion is allowed).
5. Be mindful of the growth plate, if present; it is weaker than the surrounding bone, and injuries to it may lead to growth disturbances.

Casting

Most fractures are treated with casting, although this treatment may result in adverse effects as well. A cast that is too tight may compromise blood supply or put pressure directly on the skin or the nerves. In some cases casting is obviously impractical; for example, casting a femur fracture requires a body cast, which would not allow an adult patient to get out of bed. Enforced bed rest subjects adults to risks associated with prolonged immobilization, such as clot formation and pneumonia.

Immobilization of the joint adjacent to the fracture site, which is typically required to fully immobilize it, can produce joint stiff-

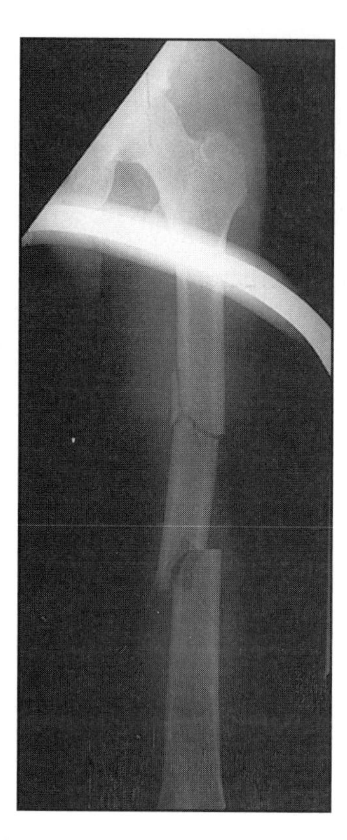

Figure 8
Radiograph of a segmental diaphyseal fracture.

(Reproduced with permission from Moed BR, Watson JT: Retrograde nailing of fractures of the femoral shaft. ***Orthop Traumatol*** *1998;6:193-204.)*

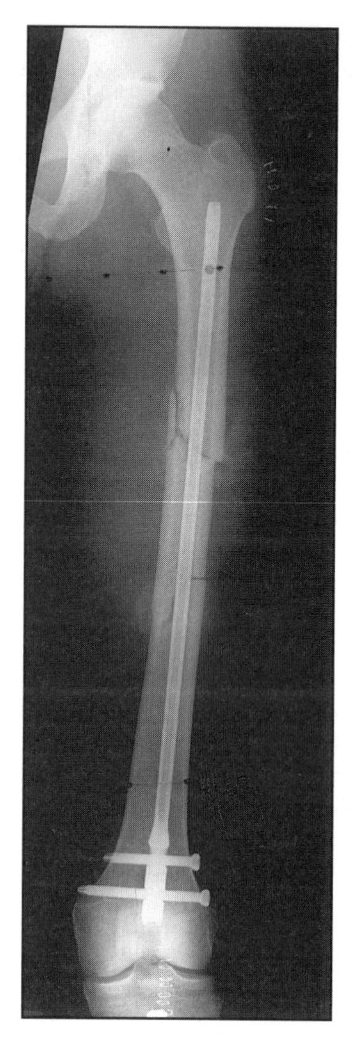

Figure 9
Radiograph of an intramedullary rod used to repair a fractured femur.

(Reproduced with permission from Moed BR, Watson JT: Retrograde nailing of fractures of the femoral shaft. ***Orthop Traumatol*** *1998;6:193-204.)*

ness. Immobilization may also produce muscle atrophy, which is problematic itself but also because it can cause the cast to loosen. Unfortunately, a cast is apt to become loose at exactly the time when the callus is least stable. Continued monitoring of the fracture by physical examination and serial radiographs is essential. At times, new casts must be made to account for atrophy.

Once radiographs show evidence of bone healing, typically 6 weeks after injury in an adult, the clinical strategy changes. At this point, since the callus holds the reduction provisionally, a cast is used more for protection than for immobilization. The initial cast can be converted to something smaller. For instance, a long leg cast extending over the knee to prevent rotation of the tibia at the fracture site can be converted to a short leg cast or a short cast may be exchanged for a removable cast boot.

Traction

Traction was frequently used prior to the advent of modern surgical techniques and is still used in children who do not seem to be at comparable risk for the adverse effects of prolonged bed rest observed in adults.

Case Presentation

A typical example is a 49-year-old man who sustains a femur fracture as a result of a rollover motor vehicle accident. He is awake and alert at the scene and has an obvious deformity and swelling in the left thigh. He is transported to the emergency department in a splint. Following initial evaluation of the ABCs, examination in the hospital reveals that the skin on the leg appears intact, the leg is well perfused, and the nerves to the foot and ankle region are intact. After examining the injured leg, the remaining extremities and spine are examined for any evidence of injury.

AP and lateral radiographs of the left femur, including the hip joint and the knee joint, are ordered to fully assess angulation and displacement. The radiographs of the femur show two fracture lines located in the middle of the bone (Fig. 8); thus, the fracture is classified as a closed, comminuted diaphyseal fracture. Because of the high-energy mechanism of injury, additional radiographs of the pelvis and the spine should be obtained as well.

Treatment consists of intramedullary fixation, an operation in which a rod is placed inside the bone (Fig 9). With this type of fixation, knee motion will be possible soon after surgery. Postoperatively, a second musculoskeletal survey is required because some injuries may be initially masked by the severe pain caused by the broken bone.

Complications

Complications can occur as a result of delays in treatment or sometimes as a result of the treatment itself. Any patient with a lower extremity fracture is at risk for venous blood clotting (thrombosis) as a consequence of either mediators released by the injury or decreased mobilization afterwards. These clots can be asymptomatic or produce a local phlebitis. Some clots (whether symptomatic or not) may travel to the lungs—a pulmonary embolism. For that reason, prophylaxis against thrombosis is offered to some patients with lower extremity fractures.

A fat embolism syndrome can develop as fatty marrow exposed at the fracture site is carried in the bloodstream and circulates to the lungs. The signs of fat embolism include respiratory compromise, change in mental status, and a petechial rash. The treatment is supportive ventilatory care. Delayed treatment of long bone fractures has been associated with a higher incidence of pulmonary complications.[19]

With femur fractures, compartment syndromes resulting from pressure on the nerves and blood vessels are rare but possible. A similar fracture of the tibia would be at significantly higher risk. Patients who report pain that is out of proportion to the injury or pain with passive stretching of the toes require close monitoring.[20]

Although the risk of compartment syndromes and thrombosis can be reduced with the tools of modern medicine, physicians cannot err on the side of too much caution. The methods of diagnosis and treatment are themselves not without risk. For example, the risk of thrombosis can be reduced with anticoagulant therapy, but this increases the risk of hemorrhage. It is therefore essential to consider the unique attributes of each clinical situation.

Finally, recall that a "simple" broken bone can have major psychosocial implications for the patient. Unlike elective surgery, fractures are obviously unplanned, and patients will find their lives are disrupted suddenly for a lengthy period. Returning to work may not be immediately possible, and the concerns regarding lost income and lost function can trigger a reactive depression. That fact (coupled with the reality that many injuries that occur as a result of motor vehicle accidents are associated with substance abuse) signals that complete care of the musculoskeletal injury may need to include psychological treatment as well.

Key Terms

Anisotropic Having unlike mechanical properties in different directions; that is, the mechanical properties depend on the orientation of the applied force

Comminuted fractures A fracture with more than two fracture fragments

Ductility The property of greater plastic deformation prior to material failure

Fat embolism syndrome Respiratory distress and cerebral dysfunction caused by droplets of marrow fat released at the fracture site and deposited in the lungs or brain

Fracture reduction The realignment of fracture fragments to restore normal anatomy of the bone

Greenstick fracture An incomplete fracture of one cortex only, seen typically in children whose bones are more flexible

Isotropic A material that has the same mechanical properties regardless of direction of loading

Mechanism of injury A representation of the patterns of energy that cause traumatic injuries

Oblique fracture A fracture in which the fracture line crosses the bone diagonally

Open fractures A fracture in which the skin is broken, exposing the fracture site to the external environment

Pathologic fracture A fracture caused by a normal load on abnormal bone, which is often weakened by tumor, infection, or metabolic bone disease

Physis The horizontal growth plate located at the ends of immature long bones; a site of endochondral ossification

Primary bone healing The end-to-end repair process that occurs when the bone ends are anatomically opposed and held together rigidly; no callus forms

Pulmonary embolism Migration of a thrombus from a large vein (often in the leg) to the lung, causing obstruction of blood flow, respiratory distress, or even death

Secondary bone healing The repair process that is characterized by the formation of fracture callus, which then remodels to form new bone

Spiral fracture A fracture caused by a twisting force that results in a helical fracture line

Torus (buckle) fracture A fracture that warps but does not completely break the cortex

Transverse fracture A fracture in which the fracture line is perpendicular to the shaft of the bone

Viscoelastic Having mechanical properties that depend on the loading rate of an applied force

References

1. Martin RE: Initial assessment and management of common fractures. *Prim Care* 1996;23:405-409.
2. Verborgt O, Gibson GJ, Schaffler MB: Loss of osteocyte integrity in association with microdamage and bone remodeling after fatigue in vivo. *J Bone Miner Res* 2000;15:60-67.
3. Cowin SC: Wolff's law of trabecular architecture at remodeling equilibrium. *J Biomech Eng* 1986;108: 83-88.
4. Greenbaum MA, Kanat IO: Current concepts in bone healing: Review of the literature. *J Am Podiatr Med Assoc* 1993;83:123-129.
5. Lieberman JR, Daluiski A, Einhorn TA: The role of growth factors in the repair of bone: Biology and clinical applications. *J Bone Joint Surg Am* 2002;84:1032-1044.
6. Hannouche D, Petite H, Sedel L: Current trends in the enhancement of fracture healing. *J Bone Joint Surg Br* 2001;83:157-164.
7. David SM, Gruber HE, Meyer RA Jr., et al: Lumbar spinal fusion using recombinant human bone morphogenetic protein in the canine: A comparison of three dosages and two carriers. *Spine* 1999;24:1973-1979.
8. Yasko AW, Lane JM, Fellinger EJ, et al: The healing of segmental bone defects, induced by recombinant human bone morphogenetic protein (rhBMP-2): A radiographic, histological, and biomechanical study in rats. *J Bone Joint Surg Am* 1992;74:659-670.
9. Sumner DR, Turner TM, Purchio AF, et al: Enhancement of bone ingrowth by transforming growth factor-beta. *J Bone Joint Surg Am* 1995;77:1135-1147.
10. Krebsbach PH, Kuznetsov SA, Bianco P, et al: Bone marrow stromal cells: Characterization and clinical application. *Crit Rev Oral Biol Med* 1999;10:165-181.
11. Melio FR: Priorities in the multiple trauma patient. *Emerg Med Clin North Am* 1998;16:29-43.
12. Rudman N, McIlmail D: Emergency department evaluation and treatment of hip and thigh injuries. *Emerg Med Clin North Am* 2000;18:29-66.
13. Browner BD, Jacobs LM, Pollak AN: (eds): American Academy of Orthopaedic Surgeons. *Emergency Care and Transportation of the Sick and Injured*, ed 7. Sudbury, MA, Jones and Bartlett Publishers, 1999.
14. Alonso JE, Lee J, Burgess AR, et al: The management of complex orthopedic injuries. *Surg Clin North Am* 1996;76:879-903.
15. Gustilo RB, Anderson JT: Prevention of infection in the treatment of one thousand and twenty-five open fractures of long bones: Retrospective and prospective analyses. *J Bone Joint Surg Am* 1976;58:453-458.
16. Barei DP, Bellabarba C, Sangeorzan BJ, et al: Fractures of the calcaneus. *Orthop Clin North Am* 2002;33:263-285.
17. Eustace S, Adams J, Assaf A: Emergency MR imaging of orthopedic trauma: Current and future directions. *Radiol Clin North Am* 1999;37:975-994.
18. O'Sullivan ME, Chao EY, Kelly PJ: The effects of fixation on fracture-healing. *J Bone Joint Surg Am* 1989;71:306-310.
19. Beck JP, Colins JA: Theoretical and clinical aspects of posttraumatic fat embolism syndrome. *Instr Course Lect* 1973;22:38-87.
20. Blick SS, Brumback RJ, Poka A, et al: Compartment syndrome in open tibial fractures. *J Bone Joint Surg Am* 1986;68:1348-1353.

Musculoskeletal Emergencies

A musculoskeletal emergency is any injury or disease of the bones, joints, muscles, and adjacent neurovascular structures that, if not treated immediately, can result in significant impairment. Conditions that must be recognized promptly include spinal injuries, crush injuries, compartment syndrome, fractures (specifically open fractures, pelvic fractures, and long bone fractures), infections, dislocations, deep venous thrombosis (DVT), and soft-tissue injuries.

Pain is the most common reason for patients to seek medical care. However, physicians must look beyond this presenting complaint because it may mask a second and more serious condition. For example, a patient involved in a motor vehicle accident who reports shoulder pain may have a clavicle or rib fracture. Although this is not an emergency, per se, there may be a more serious associated injury, such as a myocardial contusion. Additional information about damage to the car or the patient's inability to walk should raise the index of suspicion for other occult injuries.

Common injury complexes can be identified when a reasonable history of events is obtained. For example, a fall from a height is commonly associated with lumbar, calcaneal, and pelvic fractures. Thus, a patient who fractures the calcaneus jumping out of a window requires a detailed examination of the entire axial skeleton, even if heel pain is the only presenting complaint (Fig. 1). Of course, if there is any question about the patient's mental status (and ability to report symptoms), then reliance on a focused examination alone is not appropriate.

Patients with chronic musculoskeletal pain pose an additional dilemma: differentiating new symptoms and possibly an unrelated new diagnosis from preexisting symptoms. Neurologic deficits, absent or diminished pulses, or signs of infection should prompt additional investigation, including imaging studies, laboratory studies, and appropriate specialty con-

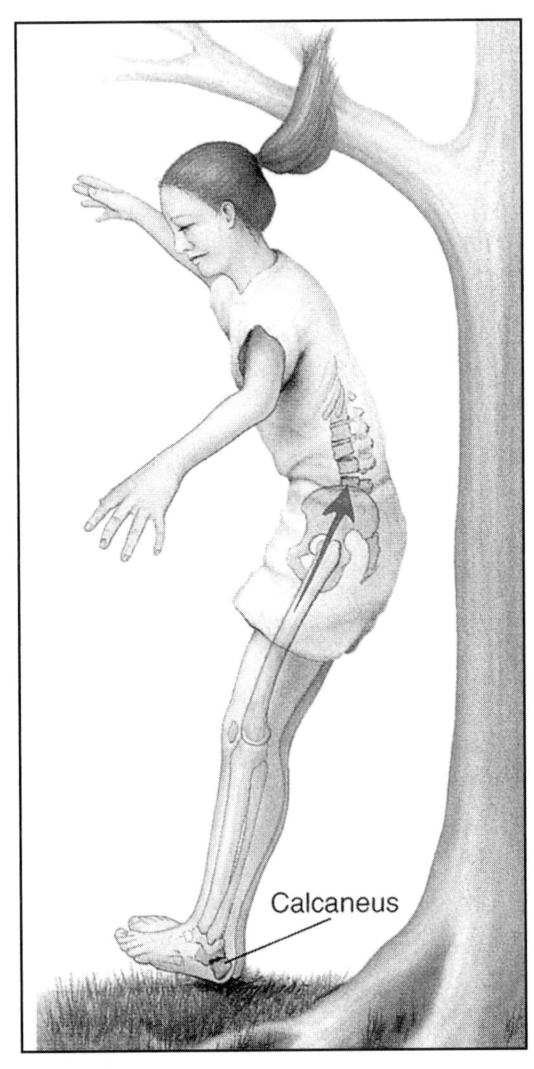

Figure 1
The force of a calcaneal injury is transmitted up the legs to the spine, often resulting in a fracture of the lumbar spine.

sultation. This process may be tedious, and often nothing will be found. Nonetheless, this systematic approach to trauma is essential.

This chapter presents common musculoskeletal emergencies or conditions that require immediate identification based on the potential for complications. Each section describes an anatomic and physiologic basis for consideration as a musculoskeletal emergency. Discussions of diagnosis, treatment, and potential complications are also presented.

Spinal Injuries

Damage to the vertebrae can be associated with varying degrees of injury—from simple fractures without neurologic impairment to spinal cord injuries resulting in paraplegia, quadriplegia, or even death. Forces that damage the vertebrae can injure the spinal cord through stretching, laceration, ischemia, or compression.

Spinal cord injuries are most commonly the result of motor vehicle accidents, falls from heights, or gunshot wounds, with young men affected most often (>80%).[1] The cervical spine is most commonly injured, followed by the lumbar spine. The thoracic spine is injured less often because the ribs provide additional support. Patients who report pain anywhere along the spine, loss of sensory or motor function, or incontinence following a traumatic event should be considered at high risk for having a spinal fracture or possible spinal cord injury.

Pertinent medical history may reveal clues suggesting other causes of spinal pain or neurologic deficit. Cancer can metastasize to the spine resulting in bone destruction and spinal cord compression. Intravenous drug use is a risk factor for epidural abscesses. Rheumatic diseases, chronic steroid use, and osteoporosis can result in compression fractures from seemingly minor or low-energy injuries. In addition, prior spine surgery and the presence of spinal hardware make the spine stiff and less capable of absorbing the energy of forceful blows without injury.

Treatment of life-threatening injuries should have priority, but immobilization of the entire spinal column should be initiated at once and maintained until a thorough neurologic evaluation is completed. Spinal immobilization consists of applying a rigid cervical collar and using a backboard. Emergency care personnel typically initiate these measures before the patient arrives at the hospital. Initial assessment should include evaluating level of consciousness, testing motor and sensory function, and assessing anal sphincter tone. Sensory deficits can be mapped based on the dermatomal pattern of innervation. The level of spinal cord injury can be identified by testing the motor function of the sequential nerve roots.

Radiographic evaluation for suspected cervical spine injuries includes AP, lateral, and odontoid views. The lateral view identifies most cervical spine fractures but will fail to detect about 15% of cases.[2] Thoracic spine and lumbosacral views should be obtained for low back pain or with neurologic deficits that correspond to those levels. CT is required for patients with identified fractures, neurologic deficits without identifiable fracture, pain out of proportion to the injury, or equivocal findings on plain radiographs. Urgent surgical consultation is required in all cases of spine fracture with neurologic injury.

Crush Injuries

Prolonged, continuous pressure on an extremity can result in a crush injury. Many of the first descriptions of crush injuries came from military records, which describe the association between prolonged partial burials and renal failure. More recently crush injuries have been reported following earthquakes, collapse of buildings and tunnels, and extended periods of extremity compression after poisonings or drug overdoses.[3] Without prompt diagnosis and intervention, metabolic derangements, compartment syndrome, and multiple system complications can occur.

Continuous pressure on muscle results in cellular ischemia and loss of cellular integrity. This loss of integrity causes a massive spillage of potassium and myoglobin out of the cells. There is also an influx of sodium, chloride, calcium, and water into the cells. The influx of calcium leads to irreversible cellular damage.[4] Fluid shifts can be so great that hypovolemic shock results. Vascular damage further increases tissue pressure, contributing to swelling and the disordered flow of ions and water.

Release of the cellular muscle components can cause hyperkalemia, myoglobinemia, hypocalcemia, hyperphosphatemia, metabolic acidosis, and hyperuricemia.[4] Normal functioning kidneys can handle small amounts of potassium, phosphorus, and myoglobin released into the bloodstream from minor muscle damage. In high concentrations, myoglobin can precipitate in the distal renal tubules, causing an obstructive nephropathy and acute renal failure. Microvascular blood clots, hypovolemia (from massive fluid shifts), and acidosis (from release of intracellular acids) are factors that increase the likelihood of renal failure.[5] Acute renal failure with severe crush injuries is associated with 20% to 40% mortality.[4] Cardiac arrhythmias occur because of hyperkalemia and hypocalcemia. Hypoperfusion and hypovolemia also depress cardiac function.

Patients with crush injuries may also have a range of neurologic, cardiovascular, and respiratory problems. Therefore, attention to the trauma ABCs (airway, breathing, and circulation) takes priority. Assessment may reveal obvious injuries to the soft tissues and bones. Examination may also reveal flaccid paralysis with patchy loss of sensation that mimics a spinal cord injury. The distinction is usually made by the presence of normal anal sphincter tone and bladder function, as well as asymmetry of the deficit. Severe, tense, and painful swelling of the extremity may be present. The presence of palpable pulses is not a reliable indicator of normal compartment pressures. If there is a question of compartment syndrome, pressure measurements are mandatory.

Laboratory studies will reveal elevated levels of creatine kinase, potassium, and myoglobin. Urine may be pinkish to dark brown, and dipstick analysis suggests high levels of blood. This finding may be misleading because dipstick testing does not differentiate between hemoglobin and myoglobin. The absence of red blood cells by microscopy suggests myoglobinuria. Cardiac monitoring may show peaked T waves, widened QRS complexes, heart blocks, and other signs of hyperkalemia.

Early and aggressive therapy for crush injuries is aimed at preventing renal failure and minimizing metabolic fluctuations, specifically myoglobinuria, hyperkalemia, hyperphosphatemia, hyperuricemia, and metabolic acidosis. Fluid resuscitation with normal saline solution should begin as soon as intravenous access is established in the field. The incidence of acute renal failure approaches 100% if fluid therapy is delayed longer than 12 hours.[4] Large volumes of saline solution are given (1.5 L/h initially, averaging 12 L/day.) Sodium bicarbonate is added to the fluids to alkalinize the urine (pH >6.5); this prevents renal myoglobin precipitation and facilitates myoglobin excretion.

Compartment Syndrome

Many sites in the body have muscle groups that are separated by fascial sheaths that are relatively nondistensible. Increased pressure within these closed myofascial spaces causes decreased perfusion and oxygen deprivation. Anoxia damages cells in the muscles, nerves, blood vessels, and the supporting tissue matrix. The damage that results from elevated tissue pressure is known as compartment syndrome.[6] The most common site of compartment syndrome is the leg, which has four muscle compartments (Fig. 2). The next most common site is the forearm. The foot, the hand, and the thigh can also be affected. If compartment syndrome is untreated or the diagnosis is missed, permanent tissue damage may occur. This includes death of the tissues in the compartment or loss of function in the muscles distal to the compartment but supplied by nerves that course through it.

Factors responsible for increased compartment pressures are classified as either external or internal. External factors are those that reduce the size of muscle compartments and include tight casts and splints, various types of occlusive dressings, and the eschar of burns. Internal factors are those that increase compartment volume and include bleeding (especially from fractures), tissue swelling, and iatrogenic fluid infusion into the soft tissue. Fractures of the tibia, fibula, and forearm are responsible for most cases of compartment syndrome.

Because of the devastating consequences of compartment syndrome, early recognition is essential. A high index of suspicion must be maintained in patients with fractures, crush injuries, or other injuries involving extremity pain. Physical findings vary, but the earliest and most important findings are pain out of proportion to the injury and pain with passive stretching of the muscle. Other findings may include sensory deficits, paresthesias, weakness, and pallor. Palpable pulses do not exclude the possibility of compartment syndrome, but the absence of palpable pulses should raise concern about arterial in-

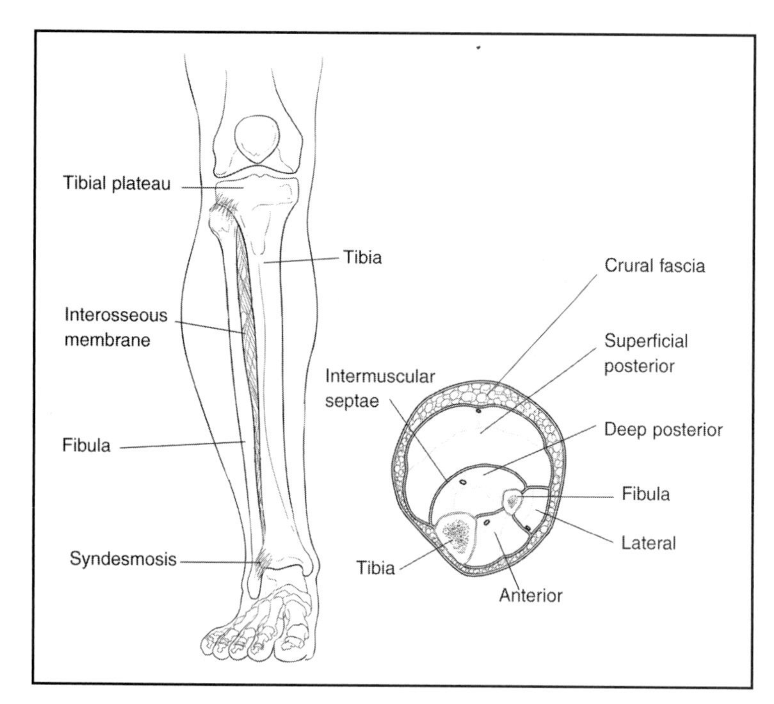

Figure 2
The muscular compartments of the leg are enveloped by strong septae. Swelling within the compartments will increase the pressure on the nerves and compromise the perfusion of the muscles.

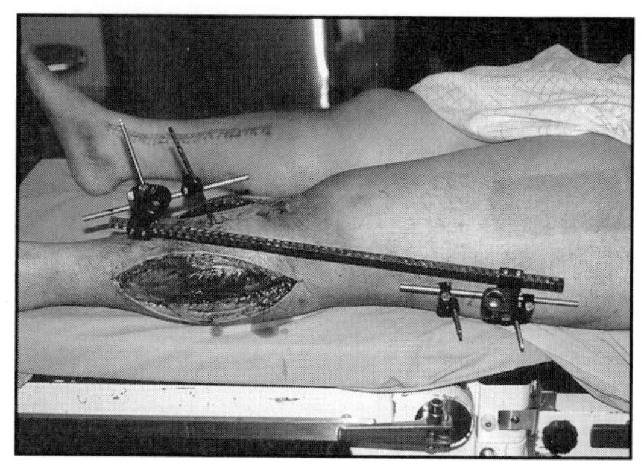

Figure 3
Treatment of a compartment syndrome involves surgical release of the compartments (fasciotomy). In this photo, a patient with a dislocated knee is shown with an external fixator used to stabilize the knee; the open wound shows the surgical release of the compartments.

(Reproduced from Schenck RC, Hunter RE, Ostrum RF, Perry CR: Knee dislocations. ***Instr Course Lect*** *1999;48:519.)*

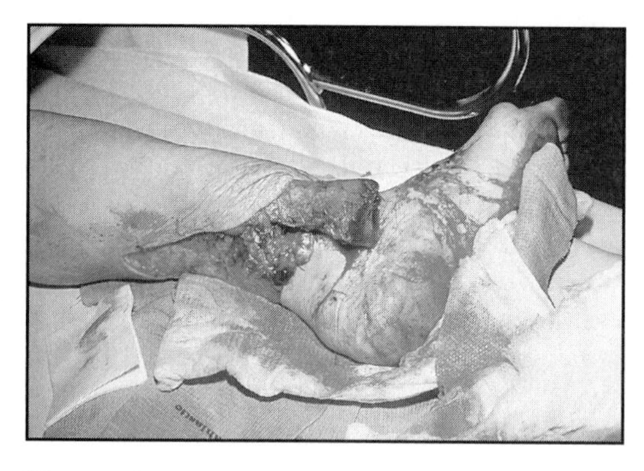

Figure 4
An open fracture of the tibia. Urgent treatment is needed for reduction and wound management.

jury. The compartment may be visibly swollen with taut skin and exquisite tenderness to palpation. There are no physical findings that exclude the diagnosis. If the diagnosis is considered, pressures must be measured.

Compartment syndrome is confirmed by the measurement of an elevated compartment pressure. Although the exact pressure at which tissue viability is compromised cannot be cited with certainty, in general, compartment pressures greater than 40 mm Hg should be considered dangerous. A lower criterion should be applied when the patient is hypotensive; in this instance, there is less force driving perfusion of the extremity. Commercial handheld instruments to measure compartment pressure are now widely available in emergency departments and may be used when the diagnosis is considered.

Definitive treatment is surgical decompression of the compartment (Fig. 3). Delays of longer than 8 hours are associated with high levels of permanent myoneural damage.[7] While awaiting the definitive treatment, tight dressings should be removed and casts or splints cut off.

Open Fractures

Any fracture with a puncture wound or laceration of the overlying soft tissue or frank exposure of bone is considered an open fracture (formerly referred to as compound fracture). Exposure of open bone makes the diagnosis straightforward (Fig. 4). However, a puncture wound can easily be dismissed as an isolated soft-tissue injury. Open fractures are considered musculoskeletal emergencies and require aggressive treatment to prevent infection. Surgical consultation is necessary in every instance because many of these fractures require surgical irrigation and débridement, along with definitive alignment and stabilization. Moreover, early recognition helps to prevent infection, promote normal healing, and restore normal function.

Table 1
Grade of Open Fracture and Recommended Duration of Antibiotic Therapy

Grade	Severity of Injury	Antibiotic Therapy/Duration
I	< 1 cm long wound size Minimal contamination Low-energy mechanism of injury	First/second cephalosporin for 3 days
II	1 to 10 cm long wound size Moderate contamination Moderate mechanism of injury	First/second cephalosporin plus aminoglycoside for 3 days
III	> 10cm long wound size High-energy mechanism of injury Comminuted fracture Extensive tissue damage Extensive contamination	First/second cephalosporin plus aminoglycoside for 5 days

Potential complications include infection of the soft tissues and bone, nonunion of the fracture, limb shortening, and impaired lymph drainage. In extreme cases, complications may lead to amputation.

Up to 30% of patients with open fractures have multiple system injuries.[8] The presence of concurrent life-threatening injuries may delay diagnoses of fractures, but securing a patent airway and hemodynamic stabilization should always be the priorities in these patients.

Open fractures can be classified by the degree of soft-tissue damage overlying the fracture site.[9] This classification system is imperfect, but it provides a common lexicon with which to communicate information to other physicians and can help with prognostic and management issues (Table 1).

Emergency treatment of an open fracture includes removing obvious debris, covering the wound, giving tetanus prophylaxis and antibiotics, and temporarily stabilizing the limb. Covering an open wound early provides a warm environment that prevents tissue desiccation, promotes healing, and reduces continued exposure to potential pathogens. Infection rates are lower in wounds covered early and left covered until débridement.[10]

By definition, all open fractures are contaminated wounds. Thus, early treatment with antimicrobial agents is necessary for all open fractures, preferably within the first 3 hours after injury. Antibiotics should cover gram-positive (*Staphylococcus aureus*), gram-negative (*Pseudomonas aeruginosa* and *Escherichia coli*), and anaerobic organisms. Duration of antibiotic coverage correlates with the severity of the injury. With few exceptions, open fractures require high-volume irrigation in the operating room. The goals of irrigation are to remove debris and bacteria and to provide a clean field to better identify the extent of tissue and bone damage. Tissue integrity is improved and wound sepsis decreased by thorough irrigation and débridement.

Pelvic Fractures

A pelvic fracture can be one of the most severe types of fractures; fortunately, it is one of the least common, accounting for only approximately 3% of all fractures. Although pubic rami fractures can occur with minimal trauma—and are often clinically benign—most pelvic injuries occur as a result of high-impact forces, such as motor vehicle, motorcycle, and pedestrian collisions. Because of the large forces necessary for pelvic injury, its presence suggests the likelihood of other injuries. One feared consequence of a pelvic fracture is death from hemorrhagic shock. The pelvic compartment is spacious and can accommodate a large volume of blood. Consequently, a patient with a pelvic fracture can exsanguinate from venous or osseous bleeding.

The pelvis contains the iliac vessels, urogenital organs, the distal portion of the gastrointestinal tract, and a multitude of neural plexi, including the pudendal and sciatic nerves. Surgical repair of arterial injuries or embolization of bleeding vessels by interventional radiologists is sometimes required to control hemorrhage.

Immediate treatment begins with assessment of the trauma ABCs. The physical examination should focus on evaluation for signs of pelvic injury, such as ecchymosis (bruising or discoloration) or hematoma in the perineal and scrotal areas, focal pelvic bony tenderness, crepitus, or pelvic instability. Pelvic instability, assessed by pain or laxity when the iliac spines are compressed laterally or anteroposteriorly suggests a pelvic fracture until proven otherwise. Gross blood at the urethral opening, a high-riding or nonpalpable prostate on rectal examination, or vaginal bleeding indicates urologic injury that requires radiologic workup and urologic consultation. A complete neurologic examination, including assessment of anal sphincter tone, is necessary to exclude neurologic injury. Intra-abdominal injury also should be ruled out in patients with multiple system injuries and those found to have pelvic fractures.

Angiography should be considered for hemodynamically unstable patients believed to have massive bleeding in the pelvis. This invasive procedure identifies which vessels are bleeding so that immediate embolization can be performed. Angiographic intervention has a high success rate, but if embolization fails to control the bleeding, surgical treatment is difficult and the mortality rate is high.

Long Bone Fractures

Fractures of the long bones—the femur, humerus, tibia, and fibula—are rarely missed or overlooked. The mechanism of injury is most commonly blunt trauma, such as a motor vehicle accident or a fall. However, the incidence of penetrating trauma, such as that from gunshot wounds, is steadily increasing. Significant energy is necessary to break large bones; therefore, a thorough examination is needed to identify associated injuries. Long bone fractures can result in significant blood loss, but shock from these injuries alone is rare. Although complications still can occur as a result of long bone fractures, they are less common now because of advances in equipment, surgical techniques, and rehabilitation programs. Early complications include blood loss, fat embolism syndrome, and infection. Fat embolism syndrome is characterized by progressive respiratory decline, fever, change in mental status, and thrombocytopenia (low platelet count). These findings are a consequence of the release of fat globules into the circulation. Long-term complications of fracture include nonunion, limb shortening, and posttraumatic arthritis.

The diagnosis of a long bone fracture is often straightforward; the extremity appears swollen, painful, and may be malaligned because of the deforming forces of the attached muscle groups. Attention to the neurovascular status of the affected extremity is necessary because arterial and nerve injuries may occur. AP and lateral radiographs typically confirm the diagnosis.

Upon confirmation of a long bone fracture, three steps must be taken immediately. First, a splint must be applied to the extremity to stabilize the fracture. Repositioning the fracture in anatomic alignment decreases bleeding and pain. Second, adequate pain medication should be given. Third, immediate surgical consultation is required. Definitive management will depend on the type of fracture, the site of the fracture, and the type of associated injuries. Early treatment is associated with shorter hospital stays, quicker time to mobilization, and fewer complications.[11]

Long bone fractures can be associated with significant blood loss (femur, 1.0 to 1.5 L; humerus, 0.2 to 0.5 L; and tibia/fibula, 0.4 to 0.8 L).[12,13] However, hemodynamic instability should not be attributed to a long bone fracture until the other injuries are excluded.

Bite Wounds

Bite wounds account for more than 3 million hospital and office visits each year. Bite wounds become a true musculoskeletal emergency because patients often delay seeking medical attention. Therefore, urgent treatment is needed at the time of presentation.

Dog bites account for most of all bites (>80%) and are easily treated without major complications.[14] Cat bites are the second most common of all bites. Human, rodent, and wild and exotic animals bites occur but are less common. Special consideration should be given to human bites, particularly closed-fist injuries. These occur when a closed fist hits another individual's mouth and directly inoculates the wound with oral secretions. The skin overlying the metacarpal head is the typical injury site. That which appears to be a simple abrasion or laceration on the hand can quickly develop into an extensive closed-space infection.

A variety of mechanical injuries to the soft-tissue structures and bone occur with

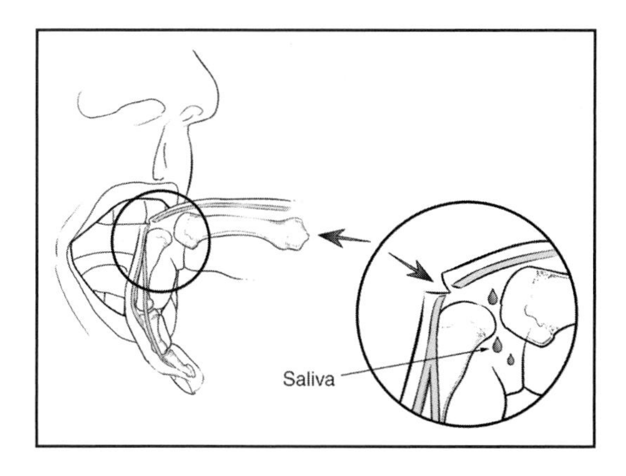

Figure 5
Mechanism of tendon laceration in closed-fist human bite injury.

(Adapted with permission from Carter PR: ***Common Hand Injuries and Infections: A Practical Approach to Early Treatment.*** *Philadelphia, PA, WB Saunders, 1983.)*

bite wounds. Dog and human bites tend to cause crush injuries, while cat bites are predominately puncture wounds and abrasions. Scratches rarely result in serious infection, whereas puncture wounds and closed-fist injuries have high rates of infection. The types of infection include cellulitis, necrotizing fasciitis, tenosynovitis, septic arthritis, and osteomyelitis. Most infections from bites involve both aerobic and anaerobic bacteria. *S aureus* and streptococci are the most common bacteria in all bite wounds. *Pasteurella multocida* is common in cat bites.

The patient's medical history should include information about the type of animal, time of the injury, the circumstances of the bite, and the patient's tetanus inoculation status. Information about any history of immune system depression, diabetes, splenectomy, and peripheral vascular disease should be noted as well because these factors place the patient at a higher risk for infection. Wounds treated within 8 hours of the incident usually result in a lower rate of infection. Unprovoked animal attacks should raise the suspicion of rabies.

The physical examination should focus on the skin, tendons, joints, bones, and neurovascular status. Signs of fluctuance, drainage, and erythema indicate infection. Closed-fist injuries should be examined closely for joint involvement and extensor tendon injury (Fig. 5).

Radiographs are useful for all but superficial bites to rule out foreign bodies (teeth), presence of air or gas (signifying gangrene), and involvement of the underlying bone. All bite wounds should be irrigated. Necrotic tissue should be débrided and any foreign bodies removed. Bites to the extremities may benefit from a period of elevation and immobilization. If rabies is suspected (eg, with raccoon, bat, skunk, fox, and unvaccinated dog bites), both active and passive rabies vaccinations should be administered and state or local public health officials notified. Tetanus prophylaxis should be updated as well. Table 2 outlines the general treatment options based on the bite wound characteristics.

Effusions

Abnormal fluid accumulation in the joint is called an effusion. The presence of an effusion is not always considered an emergency; fluid may accumulate in many chronic conditions. Nonetheless, the presence of two types of fluid, pus and blood, signify an acute condition. Establishing the cause requires a complete history of preceding events and associated symptoms, examination, imaging, and analysis of the joint fluid. The goals of timely examination are threefold: (1) to make a diagnosis; (2) to initiate treatment with antibiotics and anti-inflammatory medications or surgical intervention; and (3) to prevent persistent pain and joint damage that could result in decreased mobility.

Most joint swelling represents isolated inflammation resulting from soft-tissue or bony injury. While any joint can be involved, the knee, hip, shoulder, wrist, ankle, and elbow are most commonly affected. In the absence of a history of local or overuse injury, effusions may indicate an infection, especially in a susceptible host. Infections require immediate treatment not only to prevent the spread of bacteria but to prevent damage to the articular surfaces by the body's own immune response. Inflammation of multiple joints often suggests a systemic condition such as gout, pseudogout, rheumatoid arthritis, septic arthritis, or osteoarthritis. A history of anticoagulant use or hemophilia may account for the presence of blood in the joint (hemarthrosis) in the absence of trauma.

Needle aspiration of joint fluid—a procedure called an arthrocentesis—may yield important information about the cause of the joint swelling and may also relieve symptoms by decompressing the joint.[15] Joint fluid should be analyzed for cell count, Gram stain, presence of crystals, and culture. A

Table 2
Treatment Guidelines for Bite Wounds

Bite Wound Characteristics	Antibiotics	Treatment
< 8 hours old Not on hand or face Clean with no soft-tissue injury	Controversial: none versus 3 to 5 days of oral amoxicillin/clavulanic acid or doxycycline therapy versus ceftriaxone administered intramuscularly	Primary closure
Superficial hand or facial bites Limited cellulitis	14 days of amoxicillin or doxycycline	Delayed primary closure
Puncture or closed-fist injury Deep infection	Broad-spectrum antibiotic therapy (intramuscular or oral) with close follow-up or intravenous administration	Extensive débridement Secondary closure
Involvement of tendon, nerve, joint, or bone Extensive cellulitis Necrotizing fasciitis Persistence of the wound despite oral antibiotic therapy	Intravenous antibiotic therapy	Extensive débridement Possible amputation

culture of the fluid is the best way to exclude or confirm septic arthritis and will also identify the organism. The presence of fat droplets on microscopic analysis of the joint fluid is evidence of a joint fracture. A clinical decision rule, such as "consider a joint infected if the white blood cell count is greater than 50,000/mm^3," can be used as a proxy until the culture results are known. A lower white blood cell count threshold will increase the sensitivity of the rule at the expense of specificity.

Fractures involving the articular surfaces with subsequent development of effusions require surgical consultation. Although not all fractures require surgical treatment, timely consultation with a surgeon will help ensure that those with healing potential are treated before displacement or damage to the fragment occurs. When the culture findings are consistent with infection, antibiotic therapy should be initiated. The most common bacterial sources of septic arthritis are *Neisseria gonorrhoeae* and *S aureus*. Purulent effusions require drainage, but debate continues as to whether surgical lavage or serial needle aspiration is the more effective treatment.

Dislocations

A dislocation occurs when the two articular surfaces of a joint are no longer in contact. Subluxation occurs when there is only partial contact between articular surfaces, resulting in disruption of normal joint alignment. Dislocations are considered musculoskeletal emergencies because of possible associated neurovascular stretch injuries, the impact of which can be diminished by immediate reduction of the joint. Missed dislocations can lead to ischemic bone death, permanent neurologic injury, and in rare cases amputation. The potential for eventual degenerative joint disease and functional impairment can be reduced with appropriate early treatment.

The structures surrounding a joint can become impinged, stretched, or lacerated when the bones are suddenly moved out of alignment, as can occur as the result of a motor vehicle accident or a fall (Fig. 6). Vascular injury may cause bony damage from ischemia, resulting in osteonecrosis (bone death). Nerve injury from impingement can cause neurapraxia, a temporary loss of neural function, and paresthesias, which are tingling, burning, or prickly sensations. Laceration or severe nerve impingement may cause permanent injury. Although the peripheral nervous system has some regenerative capabilities, preinjury neural function is rarely restored completely. To avoid osteonecrosis and neurologic injury, immediate reduction should be attempted to relieve pressure on the arteries or nerves. Direct damage to the articular surfaces may cause early arthritis, even if the joint is reduced in a timely fashion.

The shoulder is the most commonly dislocated joint. Although the dislocation can occur in any direction, anterior dislocations are the most common (>95%).[16] Posterior dislocations can result from seizures or noncontact mechanisms, such as electrocutions or lightning strikes. These mechanisms result in a forceful contraction of the muscles, and power imbalances can displace the humeral head posteriorly. Posterior dislocations are associated with fractures of the lesser tuberosity. Because the shoulder joint has a large humeral head in contact with a shallow socket, the ligamentous capsule and rotator cuff muscles play a major role in shoulder stability. These restraints are often torn during a dislocation. Therefore, chronic instability may result from a single traumatic event. The axillary nerve lies close to the shoulder joint and may be injured in an acute shoulder dislocation. Vascular compromise is less of a concern in shoulder dislocations than in the hip; however, if a fracture is present, ischemia of the humeral head can occur, resulting in osteonecrosis.

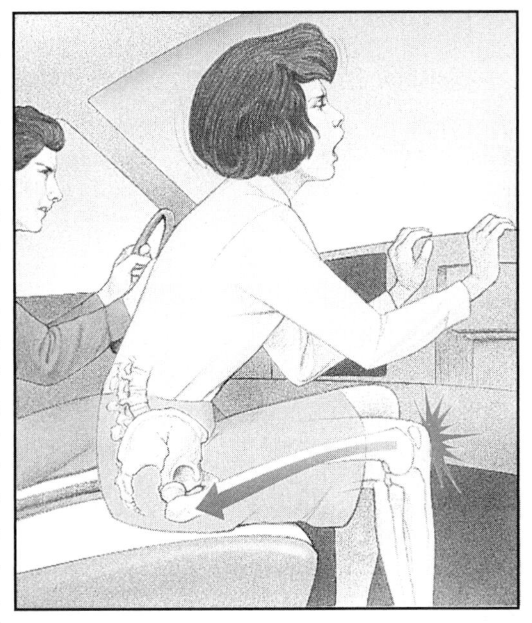

Figure 6
Significant force is required to cause hip dislocations.

The hip, by contrast, is a remarkably stable structure. Hip dislocations typically occur as a result of high-energy injuries, such as motor vehicle accidents. The hip joint is reinforced anteriorly by the iliofemoral ligament and posteriorly by the ischiofemoral ligament. The ligaments are stronger anteriorly; therefore, most hip dislocations (>85%) occur posteriorly.[17] A posterior dislocation occurs when the hip and knee are flexed and the extremity experiences an anterior blow to the knee, as in the case of a car passenger's knee hitting the dashboard during a sudden stop or collision. Associated fractures of the acetabulum and pelvis are not uncommon. Hip dislocations may result in damage to the arterial network supplying the femoral head. Closed reduction is usually successful in restoring blood flow to the femoral head; long delays to reduction increase the risk of osteonecrosis.

Dislocations around the knee often involve the patellofemoral joint (the joint between the patella and femur). Given the strong ligaments that support it, the joint between the tibia and femur is rarely dislocated; however, when a dislocation does occur, significant damage to the knee joint results. The popliteal artery lies behind the knee and is highly susceptible to traction injury if the knee dislocates. Reducing a knee dislocation may not resolve an arterial injury because there may be damage to the inner lining of the vessel (also called an intimal tear) that promotes thrombosis. Loss of blood flow may require amputation if the arterial injury is not detected. Other potentially serious problems associated with knee dislocations are compartment syndrome and damage to the common peroneal nerve, which is tethered to the fibula and thus cannot tolerate much stretch.

Table 3 lists the features of shoulder, hip, and knee dislocation. For any dislocation, serial neurovascular examinations should be documented. Absence of pulses or nerve deficits should prompt immediate reduction. Postreduction neurovascular assessment is essential as well. Plain radiographs of the joint should be obtained before and after reduction to assess for fractures and misalignment.

Deep Venous Thrombosis

Deep venous thrombosis (DVT) is a dangerous complication in that it may lead to development of pulmonary embolism (PE) and thus result in sudden death. Elective hip and knee surgery and lower extremity fractures are significant risk factors for DVT.[18] Therefore,

Table 3
Management of Dislocation

Joint	Physical Findings	Imaging Studies	Initial Management
Shoulder	Squared appearance Loss of acromial fullness Possible axillary nerve damage	AP, lateral, and axillary views	Closed reduction Sling immobilization
Hip	Posterior dislocation (shortened, adducted, flexed, internally rotated) Anterior dislocation (abducted, externally rotated)	AP and lateral views, CT	Immediate closed reduction Open reduction if fracture fragments are in the joint
Knee	Gross instability Possible popliteal artery injury Possible compartment syndrome findings	AP and lateral views, angiography, MRI	Closed reduction Serial vascular examinations Surgery for vascular compromise Fasciotomy for compartment syndrome

maintaining a high index of suspicion forboth DVT and PE and providing prophylacticmedications to patients scheduled to undergo surgery may decrease morbidity and mortality.

DVT is caused by at least one of three factors first described by Virchow in 1856: venous stasis, endothelial injury, and hypercoagulability. In musculoskeletal medicine, all three may play a role. Patients are normally supine when undergoing surgery; this position decreases venous return to the heart, resulting in stasis. Casts and splints also contribute to venous stasis. Endothelial injury and the postoperative release of tissue factors increase the risk of hypercoagulability. Other risk factors include prior DVT, congestive heart failure, malignancy, pregnancy, use of oral contraceptives, certain genetic traits, and a history of long-term immobilization.

Most thrombi form in the lower extremities. Thrombus formation can cause acute thrombophlebitis, an inflammation of the vein, manifested as erythema and pain. The clot will usually reorganize within the vessel lumen over time. However, an unstable clot may embolize, travel proximally to the heart, and lodge in the pulmonary arteries causing a PE. Acute PE causes hypoxia by shunting blood through regions of low perfusion—a condition known as a ventilation-perfusion mismatch. If one or both of the major pulmonary arteries are totally occluded (called a saddle embolus), cardiac failure and immediate death may occur.

Physical examination of a patient with DVT typically reveals pain and asymmetric swelling and erythema of the involved extremity. The degree of pain, swelling, and redness does not correlate with the size of the thrombus, however. The physical findings are not perfectly sensitive; hence, imaging studies are needed to make the diagnosis if treatment is being considered. Doppler ultrasound has greater than 90% sensitivity and specificity for detecting thrombi above the knee[19] (Fig.7). Impedance plethysmography, which measures electrical resistance caused by changes in venous flow, has been largely replaced by Doppler ultrasound but is still used in some centers. The

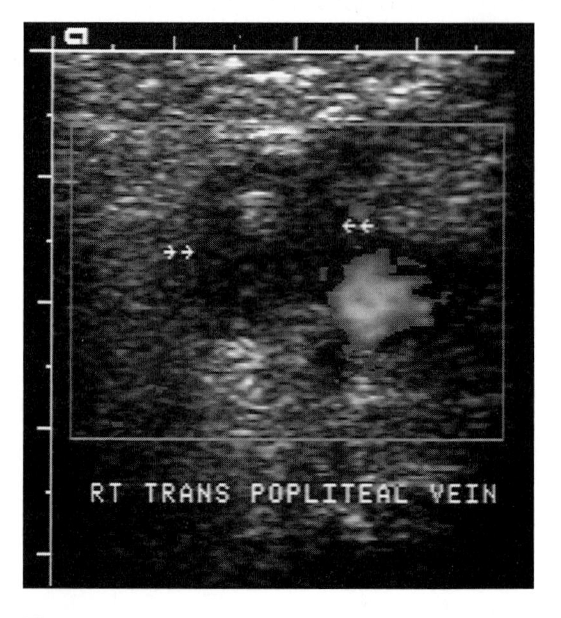

Figure 7
Doppler ultrasound image of a patient with swelling in the leg shows an area of occlusion in the popliteal vein (arrows).

gold standard is contrast venography, but it is an invasive procedure and contraindicated in patients with allergies to the contrast medium or those who have renal insufficiency. MRI has a high sensitivity and specificity but is not yet commonly used. It may, however, become one of the first-line tests used in the future, especially for diagnosing pelvic thrombi.

Signs and symptoms of PE include dyspnea, tachypnea, hypoxia, pleuritic chest pain, and hemoptysis in the context of possible risk factors discussed above. The ventilation-perfusion scan has become the first-line test for detecting a PE because it compares the pulmonary distribution of an inhaled, inert, radioactive gas with that of an intravenous contrast. Mismatch defects are highly suspicious for the presence of a PE. An adjunctive test for diagnosing PE is the lower extremity Doppler ultrasound. The gold standard, however, is contrast pulmonary angiography, which must be used with caution because it is an invasive procedure for which contrast material is required.

Anticoagulation medication is used for both prophylaxis and treatment. Treatment options include unfractionated heparin, low-molecular-weight heparin, selective thrombin inhibitors, and warfarin. Each medication acts through different mechanisms on the clotting cascade. Each form of anticoagulation has associated risk and benefits, and the identification of the best medication continues to be the subject of intense debate and research. The main risk associated with use of anticoagulation therapy is bleeding. Therapy using unfractionated heparin and warfarin needs to be monitored to achieve and maintain a therapeutic range; otherwise, these agents can cause hemorrhagic stroke or massive internal bleeding. Venous valve destruction and chronic venous hypertension may result from the presence of chronic DVT.

Occult Fractures

Not all fractures are obvious on plain radiographs. For example, fractures of the scaphoid (also called the carpal navicular) and certain hip and growth plate fractures can elude radiographic detection. For some of these fractures, delayed diagnosis may be harmful.

A complex of ligaments supports the articulations between the carpal bones and the distal radius and ulna. A feature unique to the scaphoid is that its blood supply is derived from the distal aspect of the palmar arterial arch, which enters from the distal aspect of the bone and proceeds proximally. Thus, an injury to the distal portion of the bone, especially with displacement, may interrupt the tenuous blood supply to the more proximal parts of the scaphoid. A scaphoid fracture is considered an emergency in that failure to detect a nondisplaced fracture may allow it to displace, resulting in osteonecrosis. Emergency treatment of a suspected scaphoid fracture, even if radiographs do not identify it, includes protective immobilization.

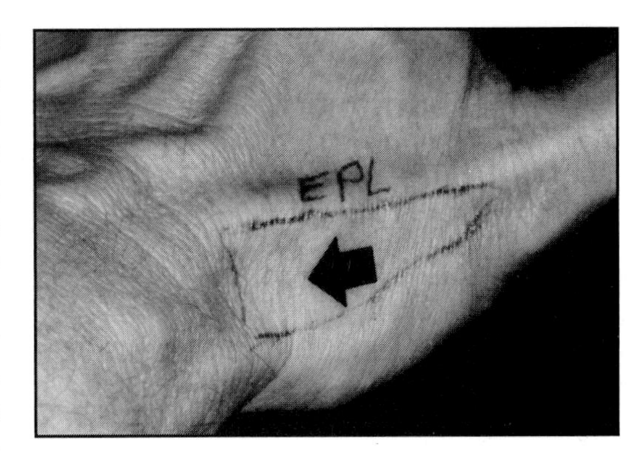

Figure 8
Palpation of the anatomic snuffbox can help detect an occult scaphoid fracture.

Patients with scaphoid fractures usually have pain and swelling of the hand, wrist, or forearm. Moreover, three clinical findings are classically associated with scaphoid fractures: (1) snuffbox tenderness; (2) tenderness to palpation of the scaphoid tubercle; and (3) pain on axial loading of the first metacarpal.[20] The anatomic snuffbox is the depression formed by the tendons of the extensor pollicis longus (on the ulnar side), the extensor pollicis brevis, and abductor pollicis longus (on the radial side) (Fig. 8). Pain is elicited by applying digital pressure on the floor of the depression. The scaphoid tubercle, which is located at the distal radial aspect of the flexor crease, is best felt with the wrist in radial deviation. Tenderness with digital pressure suggests a scaphoid fracture. Axial loading is performed by gripping the first metacarpophalangeal joint, which is extended and slightly abducted, and compressing proximally. This pressure directs a compressive force on the scaphoid. If pain is elicited by any of the maneuvers, then a scaphoid injury should be suspected.

Up to 20% of scaphoid fractures do not appear on radiographs. Thus, a careful history and clinical examination are critical. Bone scanning and CT are sometimes used several days after an injury if the diagnosis is still in question, but these imaging studies are not appropriate in the acute setting.

Immobilizing suspected fractures in a neutral position prevents further movement and damage. Immobilization can be discontinued at the time of follow-up if no fracture is present.

Key Terms

Arthrocentesis A procedure in which a needle is used to aspirate joint fluid

Compartment syndrome Ischemia of the nerves and muscles within a fascial compartment caused by elevated pressure within the compartment; frequently seen in association with tibial fractures

Crush injury An injury produced as a result of continuous pressure applied to a part of the body, usually an extremity

Deep venous thrombosis Venous clot formation caused by immobilization, hypercoagulability, obstructed venous flow, or endothelial injury, among others

Dislocation Complete displacement of a bone from its normal position in the joint, resulting in a complete loss of contact between articular surfaces; usually implies ligament damage or preexisting laxity

Ecchymosis Bruising or discoloration associated with bleeding within or under the skin

Effusion The presence of fluid within a joint

Ischemic Lacking oxygen, usually as the result of partial or complete blockage of blood flow

Neurapraxia A temporary loss of neural function

Open fracture A fracture in which the skin is broken, exposing the fracture site to the environment

Osteonecrosis The death of bone, often as a result of obstruction of its blood supply

Paresthesias Abnormal sensations such as tingling, burning, or prickling

Pulmonary embolism Migration of a thrombus from a large vein (often in the leg) to the lung, causing obstruction of blood flow, respiratory distress, or even death

Saddle embolus A condition in which one or both of the major pulmonary arteries are totally occluded

Subluxation Partial or incomplete dissociation of joint surfaces.

References

1. Frohna WJ: Emergency department evaluation and treatment of the neck and cervical spine injuries. *Emerg Med Clin North Am* 1999;17:739-791.
2. Ruoff B, West OC: The cervical spine, in Schwartz DT, Reisdorff EJ(eds): *Emergency Radiology*. New York, NY, McGraw-Hill Health Professions Division, 2000, pp 269-318.
3. von Schroeder HP, Botte MJ: Crush syndrome of the upper extremity. *Hand Clin* 1998;14:451-456.
4. Gans L, Kennedy T: Management of unique clinical entities in disaster medicine. *Emerg Med Clin North Am* 1996;14:301-326.
5. Better OS, Rubinstein I, Winaver J: Recent insights into the pathogenesis and early management of the crush syndrome. *Semin Nephrol* 1992;12:217-222.
6. Mabee JR, Bostwick TL: Pathophysiology and mechanisms of compartment syndrome. *Orthop Rev* 1993;22:175-181.
7. Lagerstrom CF, Reed RL II, Rowlands BJ, et al: Early fasciotomy for acute clinically evident posttraumatic compartment syndrome. *Am J Surg* 1989;158:36-39.
8. Gustilo RB, Merkow RL, Templeman D: The management of open fractures. *J Bone Joint Surg Am* 1990;72:299-304.
9. O'Meara PM: Management of open fractures. *Orthop Rev* 1992;21:1177-1185.
10. Alonso JE, Lee J, Burgess AR, et al: The management of complex orthopedic injuries. *Surg Clin North Am* 1996;76:879-903.
11. Dunham CM, Bosse MJ, Clancy TV, et al: Practice management guidelines for the optimal timing of long-bone fracture stabilization in polytrauma patients: The EAST Practice Management Guidelines Work Group. *J Trauma* 2001;50:958-967.

12. Simon RR, Koenigsknect SJ: Fracture principles, in Simon RR, Koenigsknect SJ(eds): *Emergency Orthopedics: The Extremities*, ed 4. New York, NY, McGraw-Hill, 2001, pp 3-23.
13. Rudman N, McIlmail D: Emergency department evaluation and treatment of hip and thigh injuries. *Emerg Med Clin North Am* 2000;18:29-66.
14. Callahan ML: Bites and injuries inflicted by mammals, in Auerbach PS (ed): *Wilderness Medicine: Management of Wilderness and Environmental Emergencies*, ed 3. St. Louis, MO, Mosby Year Book, 1995, pp 927-973.
15. Johnson MW: Acute knee effusions: A systematic approach to diagnosis. *Am Fam Physician* 2000;61:2391-2400.
16. Daya M: Shoulder, in Rosen P (ed): *Emergency Medicine: Concepts and Clinical Practice*, ed 3. St. Louis, MO, Mosby Year Book, 1992, pp 626-658.
17. Phillips AM, Konchwalla A: The pathologic features and mechanism of traumatic dislocation of the hip. *Clin Orthop* 2000;377:7-10.
18. Heit JA: Prevention of venous thromboembolism. *Clin Geriatr Med* 2001;17:71-92.
19. Kearon C, Julian JA, Math M, et al: Noninvasive diagnosis of deep venous thrombosis: McMaster Diagnostic Imaging Practice Guidelines Initiative. *Ann Intern Med* 1998;128:663-677.
20. Parvizi J, Wayman J, Kelly P, et al: Combining the clinical signs improves diagnosis of scaphoid fractures: A prospective study with follow-up. *J Hand Surg Br* 1998;23:324-327.

Chapter 17

Overuse Injuries

Overuse injuries are caused by the repetitive application of forces, none of which individually is great enough to damage the tissue. Rather, it is the repeated application of force that results in an accumulation of microscopic damage and clinical evidence of injury. Overuse injuries can affect many tissue types including bone, cartilage, tendon, fascia, and bursa. Overuse injuries can be categorized broadly into the following three types: (1) excessive loads on normal tissue, (2) normal loads abnormally applied (because of poor mechanics or altered anatomic alignment, for example), and (3) normal loads on abnormal tissue.

The universal theme in all overuse injuries is relative overload: more force is applied than the tissue is able to bear. An example of an overuse injury caused by excessive loads on normal tissue would be the classic march fracture. A march fracture is a stress fracture of the foot initially described among military recruits who are required to march for long distances at the outset of their training.

An overuse injury from abnormal application of a normal force may be seen in patients who have altered alignment of the patella. With lateral tilt of the patella, there is focal concentration of the normal loads of the patellofemoral joint on the lateral side. Since these forces are not applied uniformly across the entire normal area of contact, the lateral side experiences overuse, even though the total load is normal. Because of this overuse, pain and cartilage breakdown may result.

Overuse injuries in the setting of abnormal tissue include Achilles tendon ruptures in middle-aged patients who have preexisting degenerative tendinopathy and stress fractures in the weakened bone of patients with poor bone remodeling (eg, women with the female athlete triad) or those who have been immobilized in casts. In these latter patients, the lack of weight bearing leads to a lack of bone deposition—a "use it or lose it" phenomenon.

One of the most thoroughly characterized overuse injuries in musculoskeletal medicine is the stress fracture, which is discussed as a representative example in this chapter. A stress fracture is, in engineering terms, an instance of fatigue failure—tissue damage from the cyclic application of a series of loads, all of which are below the threshold for breakage. Not every musculoskeletal overuse injury is necessarily a fatigue failure, but in biologic terms, many overuse syndromes involve the process of tissue breakdown overwhelming the process of tissue repair.

Stress Fractures

A stress fracture in bone appears after relative overuse, which implies that there is no predetermined amount of force or number of loading cycles that injures the bone; rather, what matters is the magnitude and frequency of load relative to the tolerance of the tissue. Bone, as Wolff's law states, adapts to load. If the increase in load is introduced over a sufficiently long period, the bone will grow and make itself able to withstand this new demand. However, increasing the load too rapidly can lead to mechanical failure, even if the bone was initially healthy.

A stress fracture may seem, to lay people at least, like a misnomer because the bone is not overtly broken. The structural failure is present only at the microscopic level, but it is real: although the bone may look grossly normal, it is weakened and may break overtly with only slight additional force. Consequently, a stress fracture is similar to the damage that occurs when a paper clip is subjected to repetitive bending. No single bend is enough to break the paper clip, but the sum of the force applied cyclically weakens the structure such that one additional bend may break it.

Before that analogy is taken too far, there are two distinctions between bones and paper clips that must be recalled:

1. Bones heal. Thus, a stress fracture is an injury in which the damage outpaces the healing.[1] It is not simply a matter of wear and tear.
2. Bones have responsibilities beyond skeletal homeostasis and therefore can be weakened for reasons other than overuse. For instance, in response to metabolic calcium demands, osteoclasts may resorb structurally significant bone tissue. Some use the term "insufficiency fracture" to describe a stress fracture in abnormal bone.

Epidemiology

No general population statistics are available, but the experience of military physicians and sports medicine doctors indicates that stress fractures represent perhaps 1% to 3% of all athletic injuries and 15% of runner's injuries. The female-to-male ratio ranges from 2:1 to more than 10:1.[2-4] African Americans have greater bone density and are therefore less prone to stress fractures.[5] But these epidemiologic data do not yield hard and fast rules. For example, an African American man with focal anterior medial shin pain after doubling the distance he runs weekly overnight can have a stress fracture, even though he is in a low-risk group.

In one military study, stress fractures typically occurred during the third week of basic training, and simply allowing some rest to all soldiers at that time reduced the overall incidence. This latter finding implies that stress fractures are caused by not only increasing activity, but also by trying to play through the pain—ie, continuing to run despite the signal to stop.

Runners represent the majority of patients with stress fractures. These injuries can be traced to changes in gross mileage and also to changes in intensity, method of training (ie, more hill work), surface, and footwear. Changes in footwear can occur by simply doing nothing because with each mile run, the shock-absorbing capacity of footwear decreases. Thus, failure to replace running shoes frequently may cause a stress fracture.

Although the lower extremity is most commonly affected, stress fractures can be seen in non-weight-bearing bones, such as the ribs.[6] Stress fractures of the ribs can develop in rowers as a result of repetitive muscle tension acting directly on bone.

Pathology

Grossly, bones with stress fractures appear normal until the bone actually breaks. Microscopically, a bone with a stress fracture may not appear to be different than normal bone because there is always some damage and some remodeling occurring in normal bone—a bone with a stress fracture just has more remodeling occurring. In a stress fracture, there is more osteoclastic activity, larger cavities in the osteons, and over time, more osteoblastic bone deposition. Of course, biopsy specimens of bones with stress fractures are not routinely obtained.

Stress fractures in bone can be categorized into three stages: (1) crack initiation, (2) crack propagation, and (3) gross failure of the bone. The crack initiation creates a stress riser—a point of weakness that makes it easy for the damage to spread. Think of a small tear in a piece of folded paper; that cut and crease is a stress riser that allows the paper to be torn more easily by simply pulling on the edges.

Pathophysiology

Bone is constantly being damaged and renewed—that is why our skeleton lasts longer than the average automobile. Stress fractures occur when damage outpaces the repair. This imbalance causes pain and increases the risk of overt fracture.

According to Wolff's law, bone remodels in direct response to the forces applied. Mechanical loads (and the damage they cause) are actually stimulants for bone repair. But there is a question of balance. Under normal circumstances, bone is able to keep up with necessary repairs and thereby avoid injury. However, there appears to be a physiologic limit for bone remodeling. When the bone's reparative capacity is overwhelmed, microscopic damage results.[1]

Abnormally weak bone can tolerate less loading. Thus, any condition that weakens the bone—and especially those conditions that weaken the bone by interfering with normal remodeling—is a risk for stress fracture. The classic example of this is a female athlete who exercises to the point of complete fat depletion. This depletion of fat decreases the stores of estrogen (which is made from fat). The low estrogen level subsequently tilts the balance of bone remodeling to favor resorption over formation, yielding osteopenia.[7] Osteopenic bone has a lower tolerance for repetitive load and is vulnerable to stress fracture damage.

Differential Diagnosis

The presenting symptom of a stress fracture is bone pain in the setting of relative overuse. Typically, the shin is affected. In the shin, the differential diagnosis includes periostitis (inflammation in the periosteum) and chronic exertional compartment syndrome. A crescendo-like pain localized to the bone that becomes more severe with increased running distance and does not abate within minutes of stopping is highly suspicious for stress fracture. In time, there may even be pain at rest. Common characteristics of stress fracture are shown in Table 1. Note that none of these findings is completely sensitive or specific. Therefore, diagnostic imaging studies are often required.

Table 1
Characteristics of Stress Fracture

Factors	Findings
History	Increased activity within 4 weeks prior to onset of symptoms Pain with activity that has become progressively worse since onset and subsides hours or days after activity stops
Physical examination	Pain with pressure on affected area or percussion of bone distal to affected area Occasionally local swelling Normal range of motion, muscle strength, and tone
Nutritional	Decreased body fat (with possible amenorrhea and osteopenia) Calcium deficiency (with possible osteoporosis) Vitamin D deficiency (with possible osteomalacia)

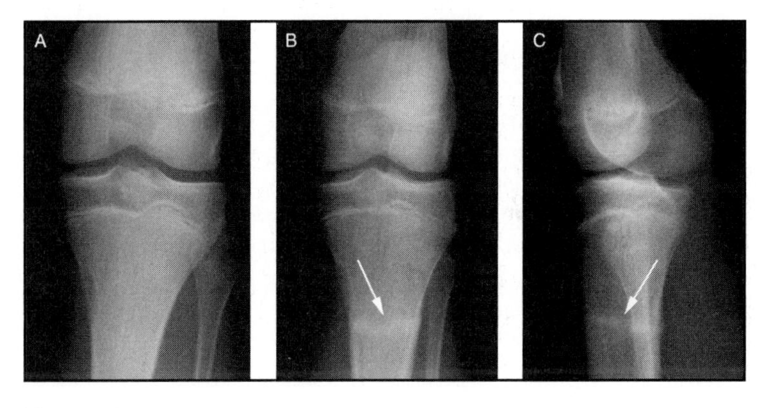

Figure 1
Stress fracture of the tibia. **A**, AP radiograph shows a fibrous cortical defect medially. Follow-up AP (**B**) and lateral (**C**) radiographs taken 3 weeks later show a sclerotic line across the tibia (arrows).

(Reproduced from Sullivan JA, Anderson SJ (eds): ***Care of the Young Athlete****. Rosemont, IL, American Academy of Orthopaedic Surgeons and the American Academy of Pediatrics, 2000, p 282.)*

Imaging Studies

Plain radiographs appear normal early in the course of a stress fracture because the damage is too small and too subtle to be detected with this mode of diagnostic imaging.[8] Except in bones with a high concentration of trabecular regions, such as the calcaneus, plain radiography is not sensitive in the early stages—precisely when the information would be clinically useful.

Endosteal or periosteal callus formation, representing healing, can be seen in established stress fractures. Occasionally, in an established stress fracture that is not given adequate rest to heal, radiographs will show evidence of a sclerotic line that is consistent with a stress fracture (Fig. 1).

Bone scanning detects metabolic activity, and bone scans will be positive within 1 or 2 days after the onset of symptoms (Fig. 2). Bone scanning is extremely sensitive. If the bone scan shows no evidence of focal uptake, the diagnosis of stress fracture is quite unlikely. Because bone scanning is very sensitive, the test occasionally remains positive (with focal tracer uptake at the site of fracture healing) for weeks to months after the patient becomes asymptomatic.[9] Thus, this test is not clinically specific. Also, the nature of nuclear medicine scans are such that fine anatomic resolution is not achieved.

The temporal pattern of uptake during three-phase bone scanning often is useful in distinguishing between periostitis (in the soft tissue only) and acute tibial stress fractures. Both entities typically demonstrate uptake on the first and second phases; however, only a stress fracture results in focal uptake on the third phase.[8] Note that other processes besides stress fracture, including osteomyelitis and tumor, can have a similar appearance on

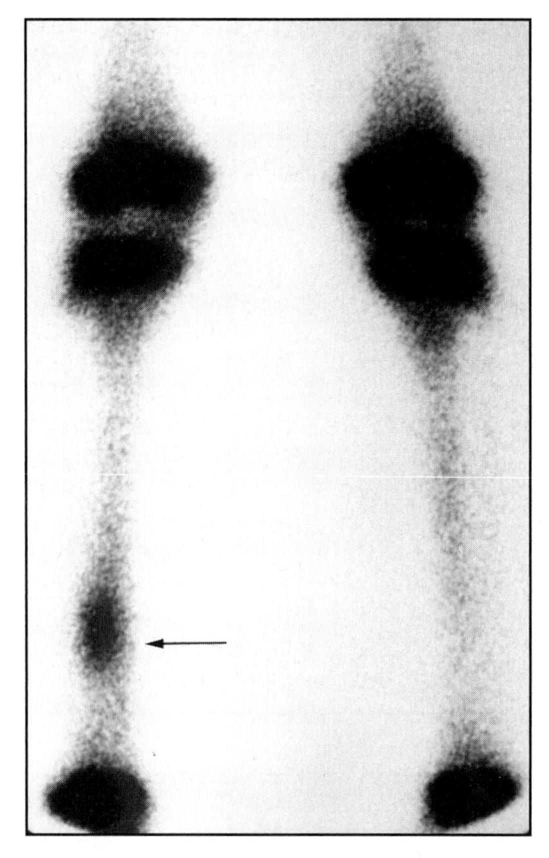

Figure 2
Bone scan of a stress fracture with focal uptake along the shaft of the tibia (arrow). The other regions of intense uptake, near the knee and ankle, represent growth plate activity.

(Reproduced from Sullivan JA, Anderson SJ (eds): ***Care of the Young Athlete****. Rosemont, IL, American Academy of Orthopaedic Surgeons and the American Academy of Pediatrics, 2000, p 408.)*

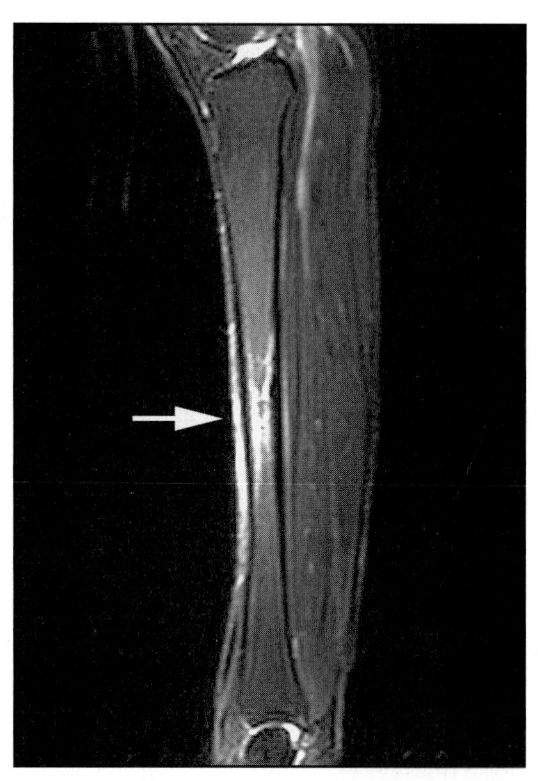

Figure 3
MRI scan of the tibia of a female high-school basketball player who reported tibial pain 2 weeks into the season. Note the edema and microfracture lines in the midportion of the tibia (arrow).

a three-phase bone scan. A positive scan indicates only that there is an area of heightened metabolic activity (as seen in the growth plates in Figure 2). The physician must be able to interpret the meaning of that activity.

MRI can also be used to diagnose stress fractures. It is likely to become the diagnostic method of choice because it spares the patient an injection of radioactive material, which is needed for bone scanning; it provides more anatomic detail; and its temporal response more closely matches the clinical situation[10] (Fig. 3).

Treatment

The management of stress fractures centers on a basic principle: the treatment of an overuse injury is underuse—that is, relative rest is needed. It is not necessary, and in many cases not desirable, to recommend total rest because total rest can lead to atrophy. Rather, what is needed is sufficient abatement of the load to allow the healing process to catch up.

Patients should decrease or discontinue the inciting activities for 4 to 12 weeks and maintain other activities to prevent disuse osteoporosis. For athletes, cross-training (ie, engaging in a different activity, such as swimming to replace running) is helpful to maintain aerobic fitness. Ice and analgesics help relieve pain, but they do not speed healing.

Occasionally, crutches are used to unload affected areas completely for a short time; casting to immobilize affected areas is also used occasionally. Gradual reintroduction of activity is essential.

Surgical treatment for stress fractures includes fixation and/or bone grafting for fractures that will not heal. Surgeons recommend treating stress fractures of the superior femoral neck (Fig. 4) with screw fixation to prevent displacement. Displacement of a femoral neck fracture can lead to osteonecrosis of the femoral head.

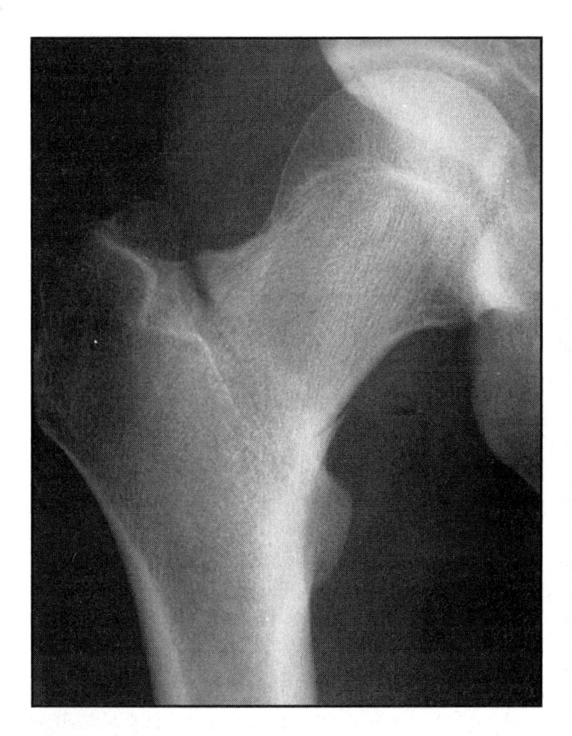

Figure 4
AP radiograph of the proximal femur revealing a femoral-neck fatigue fracture. There is cortical disruption on the superior surface with propagation of the fracture across the femoral neck. This fracture must be stabilized surgically to prevent displacement and subsequent disruption of the blood supply to the femoral head.

(Reproduced from Shin AY, Gillingham BL: Fatigue fractures of the femoral neck in athletes. ***J Am Acad Orthop Surg*** *1997;5: 293-302.)*

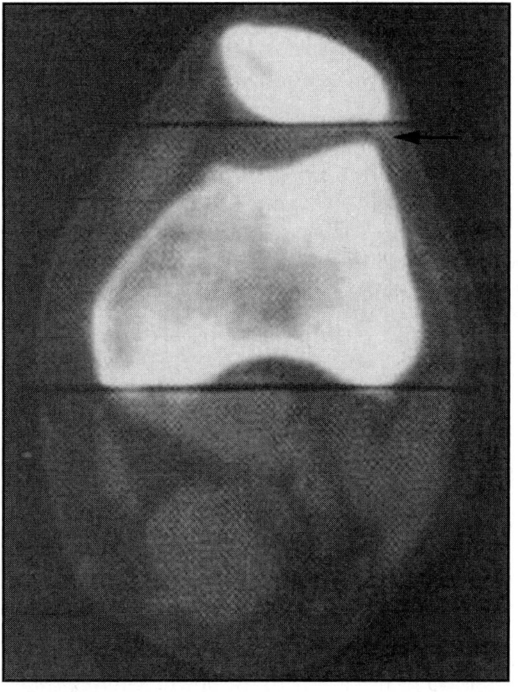

Figure 5
CT scan of a knee in 15° of flexion; the patella is tilted laterally, as seen by the asymmetric joint space (arrow).

(Reproduced from Fulkerson JP, Kalenak A, Rosenberg TD, Cox JS: Patellofemoral pain. ***Inst Course Lect*** *1992;41:57-71.)*

Treatment should also include at least an assessment of the psychosocial impact of the condition. It has been observed that athletes who cannot play can get depressed; that young people who experience any medical condition for the first time, which introduces them to mortality, can get depressed; and that patients who begin vigorous exercise in an attempt to lose weight may lose motivation if they sustain an overuse injury.

Women and young female athletes who are diagnosed with a stress fracture should be evaluated for the presence of the female athlete triad.[11] This triad consists of the following three components: (1) disordered eating, (2) abnormal or absent menses, and (3) osteoporosis. The evaluation of female athletes with stress fractures should include a complete medical history with information on nutritional status, menstrual cycle, age, height, weight, and orthopaedic history as well as screening for excessive exercise. The best treatment is awareness and prevention of the disease. In general, exercise should be encouraged in all women to promote general health and improve bone density.[12]

Women with irregular menstrual cycles or amenorrhea may benefit from hormone therapy to regulate their cycles. Women with eating disorders, nutritional deficiencies, or metabolic disturbances will need treatment.[11]

Patellofemoral Pain

Anterior knee pain is one of the most common complaints of active adolescents and adults, and often no clear cause is found. Occasionally, the pain is related to lateral tracking of the patella in the femoral trochlea (the femoral groove for the patella). Pressure on the lateral side may be caused by lateral tilt of the patella (Fig. 5) or by excessive lateral pull by the quadriceps. If the patella does not sit symmetrically in the trochlea, the lateral facet and corresponding surface of the femur will sustain overload. That is because the forces once distributed over a wide area are now borne on focal points, leading to stress concentration.[13]

The term "chondromalacia patella" is often used to describe anterior knee pain, but it is best reserved for cases when it is literally applicable. The Greek word "chondromalacia" means "soft cartilage"; thus, it should be used only after arthroscopic confirmation of that finding.

A common case of anterior knee pain would be a young active woman who reports a chronic, dull, aching pain that is usually localized to the patella. The pain is often worse with squatting, climbing stairs, or sitting for long periods with the knee flexed. On physical examination, the lateral patellar facet may be tender to palpation, especially on its undersurface. The patient may also report pain with compression of the patella against the femur. Quadriceps weakness or atrophy can be seen in long-standing cases, as the pain reflex causes inhibition of the muscle. Of note, signs such as swelling or locking of the knee are not indicative of overuse and should prompt investigation for another explanation.

Tendon Injury

Applied mechanical loads of high intensity or high frequency can cause injury to tendons. This clinical condition goes by the name tendinitis, although a better name may be the more general tendinosis (a term denoting a condition of the tendon, not necessarily an inflammatory one). Patients are at risk for tendinosis if they use improper technique while performing repetitive tasks or simply from too much activity. (This is why baseball pitchers throw only every fourth or fifth day, for example.) There is also an issue of age-related degeneration of tendons and diminution of their blood supply, a process that makes older people more predisposed to injury.[14] Tendinosis occurs most commonly in areas where tendons attach to bone, such as the rotator cuff in the shoulder, the patellar tendon in the knee, and the origin of the common wrist extensors in the elbow. The latter is the source of pain in lateral epicondylitis. The midsubstance of the tendon is often spared. In overuse injuries of the Achilles tendon, however, midsubstance failure is the norm.

A common example of tendinosis is seen in the shoulders of those who perform repetitive overhead activity—tennis players, for example. This excess can cause cumulative trauma to the rotator cuff tendon and overlying bursa, producing a condition known as impingement syndrome. Impingement implies that there is abnormal contact between the rotator cuff and the overlying coracoacromial arch (made up of the acromion, the coracoacromial ligament, and the acromioclavicular joint). Because there is a putative association between the presence of a hooked acromion and rotator cuff pathology, some believe that the damage to the tendon is inflicted by pressure from the bone. It is also plausible that inherent damage in the tendon stimulates an inflammatory process that also induces changes on the acromial surface.[15]

Patients with impingement syndrome are typically in their 30s, 40s, or 50s. Patients who are younger than this but appear to have an overuse tendinosis are more likely to have instability; they have cuff impingement because their shoulder slides out of the joint and abuts the acromion. Patients with impingement syndrome initially report activity-related pain localized to the anterior and lateral shoulder. In time, they may report difficulty lying on the affected shoulder, pain with any overhead activity, and weakness lifting the arm above shoulder level. On physical examination, they note tenderness to palpation over the humeral head and demonstrate a mild loss of active elevation of the shoulder and pain with passive forced forward elevation or resisted active elevation.

Impingement is one form of overuse injury that is treated with active exercise. The goal of treatment is to strengthen the parts of the rotator cuff that are not injured (the internal and external rotators, typically), in the hope that added strength there can enable the irritated portion of the cuff to rest.

Osteochondritis Dissecans

Osteochondritis dissecans is a localized abnormality of a focal portion of the subchondral bone, which can result in loss of support for the overlying articular cartilage. This loss of support can cause breakdown and fragmentation of the cartilage and underlying bone. There is currently no single universally accepted cause for osteochondritis dissecans, but one theory is that it is the result of overuse. Cumulative microtrauma to the subchondral bone, it is thought, leads to stress fracture and ultimately to collapse.[16] Another popular theory is that osteochondritis dissecans is a form of osteonecrosis of the subchondral bone caused by idiopathic is-

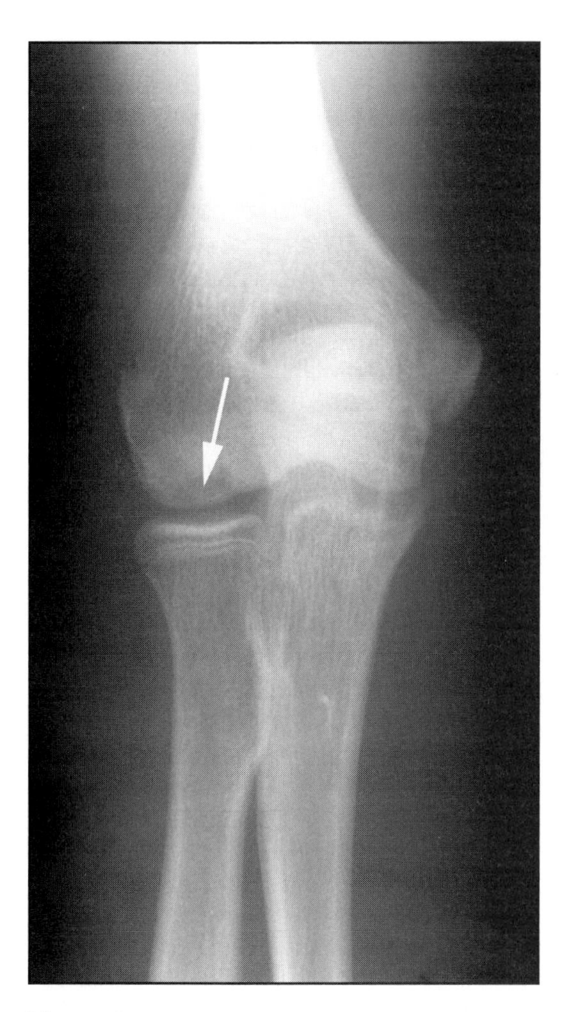

Figure 6
A lucent area in the capitellum of the humerus (arrow) associated with restriction of elbow motion and pain is characteristic of osteochondritis dissecans.

(Reproduced from Sullivan JA, Anderson SJ (eds): ***Care of the Young Athlete****. Rosemont, IL, American Academy of Orthopaedic Surgeons and the American Academy of Pediatrics, 2000, p 319.)*

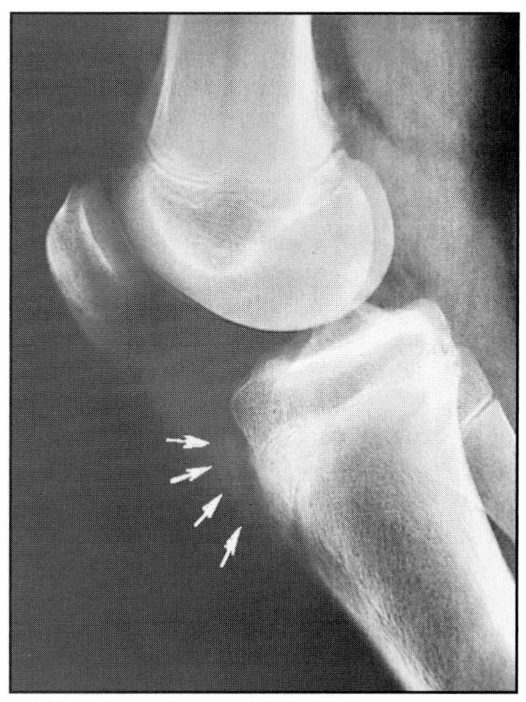

Figure 7
Osgood-Schlatter's disease. An overuse injury to the growth center (apophysis) to which the patellar tendon attaches may lead to separation of the apophysis and formation of an ossicle (arrows).

(Reproduced from Greene WB (ed): ***Essentials of Musculoskeletal Care,*** *ed 2. Rosemont, IL, American Academy of Orthopaedic Surgeons, 2001, p 588.)*

chemia. The acronym OCD should not be used to describe this condition, as this is the acronym for osteochondral defect, which is a far more general term for damage to the cartilage and underlying bone.

Osteochondritis dissecans of the elbow occurs in the capitellum of the distal humerus. It is seen in younger athletes who throw or bear weight on the arm, such as baseball players and gymnasts. Intense throwing activities apply abnormal forces on the otherwise normal elbow, resulting in compression and stress concentration between the radial head and the capitellum of the humerus. A similar process occurs in gymnastics, in which the elbow is often used as a weight-bearing joint.[17]

The typical patient with osteochondritis dissecans of the elbow is an adolescent baseball pitcher between the ages of 11 and 15 years.[18,19] He will report insidious onset of elbow pain that is relieved by rest, mild swelling, inability to fully extend the elbow, and pain on palpation of the lateral elbow. Catching and locking of the elbow are late symptoms that are indicative of articular cartilage fragmentation and loose body formation. The diagnosis of osteochondritis dissecans is made radiographically with identification of focal changes in the subchondral bone of the capitellum (Fig. 6). Early detection and evaluation of lesion size and fragment detachment is made with MRI.

Treatment of the lesion is based on whether the lesion is intact, partially attached, or completely detached. Intact lesions are treated with complete cessation of the offending activity and occasional immobilization (no more than 3 weeks) to allow acute symptoms to resolve. Large, partially detached lesions can be reduced and fixed with or without bone grafting. Loose bodies are treated with excision and curettage of the base to stimulate a healing response. Significant fragmentation leads to bony arthritic changes and ongoing symptoms in about half of the patients. Most athletes are discouraged from continuing provocative activities, such as pitching and gymnastics.

Apophysitis

Overuse injuries in growing children or adolescents can cause injury to the growth centers, especially to the apophysis. An apophysis is a cartilaginous growth plate at the insertion of major muscle groups. This area, like the open growth plate between the metaphysis and epiphysis, has cartilage that is weaker than the surrounding bone and tendons. Apophysitis (injury to these growth centers) occurs when repetitive traction is applied to the muscle group attaching to the bone.[20] Common sites of apophysitis include the tibial tubercle (where the patellar tendon inserts) and the medial epicondyle (where the wrist flexors originate). Apophysitis at the tibial tubercle is known as Osgood-Schlatter's disease.

Osgood-Schlatter's disease is often seen in adolescents during rapid growth (age 13 years, typically) who are active in sports. Osgood-Schlatter's disease is frequently associated with sports that require repetitive jumping, such as basketball, track, and gymnastics. It is more common in boys than girls. Typical symptoms include pain and swelling over the tibial tubercle. Radiographs may show a discreet separate ossicle (Fig. 7).

Medial epicondyle apophysitis is seen in adolescents involved in intense throwing activities, such as baseball pitching. Overhead throwing applies a valgus strain across the elbow with traction to the medial side. Clinical examination reveals pain, tenderness, and swelling over the medial epicondyle with a subtle flexion contracture of the elbow.

Apophysitis is generally a self-limiting problem that is treated with rest from the aggravating activity. Symptoms of Osgood-Schlatter's disease resolve completely by skeletal maturity when the tibial tubercle fuses to the main body of the tibia, usually by age 16 years in boys and 15 years in girls. Medial epicondyle apophysitis is treated with rest from throwing, stretching of the anterior elbow capsule, strengthening (especially the triceps), and instructing the child in proper throwing technique.

Key Terms

Apophysitis Injury from repetitive traction to the cartilaginous growth plate near the origin or insertion of muscle

Impingement syndrome Shoulder pain caused by tendinosis of the rotator cuff tendon or irritation of the subacromial bursa

Osteochondritis dissecans A localized abnormality of a focal portion of the subchondral bone, which can result in loss of support for the overlying articular cartilage

Osgood-Schlatter's disease Apophysitis of the tibial tubercle

Stress fracture An overuse injury in which the body cannot repair microscopic damage to the bone as quickly as it is induced, leading to painful, weakened bone

Tendinosis Injury to the tendon or musculotendinous unit caused by the application of mechanical loads of high intensity or high frequency; not truly an inflammatory condition; also called tendinitis

References

1. Sallis RE, Jones K: Stress fractures in athletes: How to spot this underdiagnosed injury. *Postgrad Med* 1991;89:185-192.
2. Bennell KL, Brukner PD: Epidemiology and site specificity of stress fractures. *Clin Sports Med* 1997;16:179-196.
3. Jones BH, Bovee MW, Harris JM III, et al: Intrinsic risk factors for exercise-related injuries among male and female army trainees. *Am J Sports Med* 1993;21:705-710.
4. Reinker KA, Ozburne S: A comparison of male and female orthopaedic pathology in basic training. *Mil Med* 1979;144:532-536.
5. Brudvig TJ, Gudger TD, Obermeyer L: Stress fractures in 295 trainees: A one-year study of incidence as related to age, sex, and race. *Mil Med* 1983;148:666-667.
6. Karlson KA: Rib stress fractures in elite rowers: A case series and proposed mechanism. *Am J Sports Med* 1998;26:516-519.
7. Lindberg JS, Fears WB, Hunt MM, et al: Exercise-induced amenorrhea and bone density. *Ann Intern Med* 1984;101:647-648.
8. Spitz DJ, Newberg AH: Imaging of stress fractures in the athlete. *Radiol Clin North Am* 2002;40:313-331.
9. Anderson MW, Greenspan A: Stress fractures. *Radiology* 1996;199:1-12.
10. Arendt EA, Griffiths HJ: The use of MR imaging in the assessment and clinical management of stress reactions of bone in high-performance athletes. *Clin Sports Med* 1997;16:291-306.
11. Callahan LR: Stress fractures in women. *Clin Sports Med* 2000;19:303-314.
12. Yeager KK, Agostini R, Nattiv A, et al: The female athlete triad: Disordered eating, amenorrhea, osteoporosis. *Med Sci Sports Exerc* 1993;25:775-777.
13. Koh TJ, Grabiner MD, De Swart RJ: In vivo tracking of the human patella. *J Biomech* 1992;25:637-643.
14. Fukuda H, Hamada K, Yamanaka K: Pathology and pathogenesis of bursal-side rotator cuff tears viewed from en bloc histologic sections. *Clin Orthop* 1990;254:75-80.
15. Almekinders LC: Impingement syndrome. *Clin Sports Med* 2001;20:491-504.
16. Singer KM, Roy SP: Osteochondrosis of the humeral capitellum. *Am J Sports Med* 1984;12:351-360.
17. Jackson DW, Silvino N, Reiman P: Osteochondritis in the female gymnast's elbow. *Arthroscopy* 1989;5:129-136.
18. Stubbs MJ, Field LD, Savoie FH III: Osteochondritis dissecans of the elbow. *Clin Sports Med* 2001;20:1-9.
19. Baumgarten TE: Osteochondritis dissecans of the capitellum. *Sports Med Arthros Rev* 1995;3:219-223.
20. Micheli LJ, Fehlandt AF Jr: Overuse injuries to tendons and apophyses in children and adolescents. *Clin Sports Med* 1992;11:713-726.

Chapter 18

Bone and Soft-Tissue Tumors

Bone and soft-tissue tumors arise from undifferentiated mesenchymal tissue. When these tumors develop, they express many tissue types, including fat, bone, cartilage, blood vessels, lymphatics, and fibrous tissue (Tables 1 and 2). Although specific chromosomal alterations have been noted in some tumors, the etiology of most bone and soft-tissue tumors is unknown.

As in other organ systems, a musculoskeletal tumor is considered malignant if it has the potential to spread to distant locations. The process of tumor spread, metastasis, is usually hematogenous (via the bloodstream). A malignant tumor of the musculoskeletal system is called a sarcoma, which is derived from a Greek word for "fleshy tumor."

Musculoskeletal tumors that do not spread are considered benign. Although not life threatening, benign tumors can be clinically significant on the basis of damage incurred at the primary site. Benign tumors can be biologically aggressive, destroying bone and impinging on nearby structures, such as nerves and blood vessels. Consequently, some of these lesions demand aggressive treatment similar to treatment for malignant tumors.

Many lesions that appear to be primary bone tumors on radiographs are subsequently found to be caused by other disease processes. These lesions may be metastases from nonskeletal malignancies, sites of infection (osteomyelitis), or sites of secondary bone damage from adjacent arthritis. As such, the detection of a suspected tumor on radiographs is the beginning of the diagnostic process, not the end.

Table 1
Classification of Benign Bone Tumors

Bone-forming tumors
Osteoma
Osteoid osteoma
Osteoblastoma
Cartilage-forming tumors
Chrondroma
Osteochondroma
Chondroblastoma
Chondromyxoid fibroma
Giant cell tumor
Benign vascular tumors
Hemangioma
Lymphangioma
Glomus tumor
Other benign connective tissue tumors
Desmoplastic fibroma
Fibrous histiocytoma
Lipoma
Neurilemmoma
Neurofibroma
Tumor-like conditions
Solitary bone cyst
Aneurysmal bone cyst
Metaphyseal fibrous defect
Eosinophilic granuloma
Fibrous dysplasia
Osteofibrous dysplasia
Myositis ossificans
Brown tumor of hyperparathyroidism
Intraosseous epidermoid cyst
Giant cell (reparative) granuloma

*(Reproduced with permission from Simon MA, Springfield D: **Surgery for Bone and Soft-Tissue Tumors**. Philadelphia, PA, Lippincott-Raven Publishers, 1998, p 120.)*

Table 2
Classification of Malignant Bone Tumors

Cartilaginous, bony, and fibrous tissue origin
- Chondrosarcoma
- Fibrosarcoma
- Osteosarcoma
- Secondary osteosarcoma

Marrow origin
- Ewing's sarcoma
- Leukemia
- Multiple myeloma or plasmacytoma

Epidemiology

The incidence of benign bone and soft-tissue tumors is not well defined because these tumors are frequently found only incidentally.[1,2] The incidence of malignant tumors, however, is known more precisely. Approximately 2,000 primary sarcomas of bone and 5,000 primary sarcomas of soft tissue are diagnosed in the United States each year.[3] The most common types of primary sarcomas of bone are chondrosarcoma, Ewing's sarcoma, and osteosarcoma. These three tumor types account for approximately 75% of primary bone sarcomas. The incidence rates are shown in Table 3.

Table 3
Age-Adjusted Incident Rates* for Selected Neoplasms by Race and Gender: United States 1995

Tumor Type	Incidence (Rate per 1 million persons) White Male	White Female	Black Male	Black Female
Bone and joint	11.8	7.7	11.0	5.9
Osteosarcoma	4.1	2.7	7.0	0.7
Chondrosarcoma	3.5	2.1	0.0	3.2
Ewing's sarcoma	1.6	1.7	0.0	0.0
Soft tissue	29.7	20.9	28.8	20.8
Multiple myeloma	49.0	31.0	104.0	72.0

*Adjusted to the 1970 US standard population

Source: National Cancer Institute, Surveillance, Epidemiology and End Results Program, CD-ROM, 1973-1995

(Reproduced from Praemer A, Furner S, Rice DP: ***Musculoskeletal Conditions in the United States****. Rosemont, IL, American Academy of Orthopaedic Surgeons, 1999, p 49.)*

Chondrosarcoma is found primarily in adults, with increasing incidence noted during and after the sixth decade. In contrast, Ewing's sarcoma is found primarily in children, with most patients between the ages of 5 and 25 years at the time of diagnosis. Osteosarcoma is also found primarily in children and adolescents between the ages of 10 and 25 years, with the peak age of incidence in the second and third decades. However, osteosarcoma also demonstrates a second incidence peak, with a significant number of patients diagnosed during the sixth and seventh decades of life. These secondary tumors arise within other lesions such as those associated with Paget's disease of bone, osteonecrosis, or irradiated bone.

Most malignancies in bone are not bone tumors per se; they are the result of metastasis from somatic primary tumors located elsewhere in the body. Of the estimated 1.2 million people in the United States who are newly diagnosed with cancers of all types each year, about half will have a type of cancer that is known to readily metastasize to bone (ie, breast, lung, thyroid, kidney, or prostate cancer). Blood cells are another source of tumors in bone. Multiple myeloma, a malignancy of the B lymphocytes, is the most common type of cancer arising in bone itself. This bone-based malignancy, however, is much less common than metastatic disease. Accordingly, the tumors most commonly seen in bone, metastatic lesions, are those that do not form in the bone; and the most common tumor that does form within the bone is not a malignancy of bone cells but rather a blood-cell disease. Bone tumors arising from bone cells represent a minority of tumors found in bone.

Clinical Presentation

History and Physical Examination

The diagnosis of a musculoskeletal tumor may be delayed at the outset because the patient's symptoms (eg, pain and swelling) may be attributed to more common bone or joint conditions, such as sprains or bursitis. For both bone and soft tissue, benign tumors are more frequent than sarcomas. Knowledge of the distinguishing clinical and radiographic findings can help identify aggressive lesions early in their course.

When examining a patient with a suspected or known musculoskeletal tumor, symptoms that are directly attributable to the lesion, such as pain or pressure at the site, should be identified first. This information may help identify the biologic characteristics of the lesion, such as rate of growth and aggressiveness. It is also important to identify whether the lesion was noted incidentally, whether the patient has localized symptoms, and the duration and circumstances (time of day, relationship to activity, position of the extremity, etc) of the symptoms.

A bone with a tumor may be painful because of edema and consequent stimulation of periosteal nerves. The bone can also be painful when the lesion compromises the strength of the remaining bone, resulting in a pathologic fracture (a fracture through abnormal bone under normal load). Soft-tissue tumors are frequently less painful than tumors in the bone. Pain is manifested with compression of surrounding nerves, erosion into adjacent bone, or when a hemorrhage into the lesion causes a rapid enlargement of the mass.

Functional causes of bone and joint pain, such as arthritis or muscle strains, are primarily present during periods of activity and improve with rest. However, pain caused by musculoskeletal tumors is typically present at rest as well as during activity. Patients with tumors (and infections) may also report night pain. Although patients may relate a history of injury to the involved extremity, the intensity and duration of pain is usually out of proportion to the injury.

Systemic symptoms such as fever, anorexia, weight loss, or fatigue should be documented, as should any medical and family history of musculoskeletal lesions or cancer.

Once the history is complete, the physical examination should include assessments for tenderness, swelling, or a firm mass. The skin should be examined closely. Skin changes (eg, café au lait spots), distant areas of tenderness or swelling, lymphadenopathy, and organomegaly can signal systemic involvement or provide other clues to etiology. The range of motion in adjacent joints should be tested carefully when the patient is in pain or when there is any question of impending pathologic fracture.

Imaging Studies

Imaging studies should begin with radiography for both suspected bone and soft-tissue lesions. Because there is a characteristic pattern to many types of tumors, a precise radiographic description coupled with the patient history and clinical presentation may be sufficient in many cases to make a diagnosis without a biopsy or additional imaging studies.

The first characteristic to describe is the precise area of bone involved. For example, a tumor of the knee should be described as a tumor of the distal femur or proximal tibia. Patellar tumors are far less common. The next characteristic is the region of the bone involved. Certain tumors have a propensity to develop in particular regions of the bone. Tumors are most frequently found in the metaphyses, but epiphyseal or diaphyseal lesions are also possible tumor sites. Tumors may also be found in the bodies of the vertebrae. It is important to identify the type of bone as well, specifically whether cancellous bone (eg, marrow space) is involved or whether the lesion has developed within or adjacent to the cortex. Cancellous bone is usually involved, but tumors may also be found within or on the surface of the cortex.

Once the location of the lesion is identified, three questions must be answered: (1)What is the tumor doing to the bone? (2) What is the bone doing to the tumor? (3) Is there any type of mineralization noted within the lesion?

What Is the Tumor Doing to the Bone?

Inactive tumors typically involve little of the surrounding bone whereas active tumors can erode into adjacent bone. Erosion caused by intramedullary lesions is called endosteal scalloping (Fig. 1). The more aggressive the lesion, the greater the loss of adjacent bone. With aggressive lesions, an adjacent soft-tissue mass may also be noted.

What Is the Bone Doing to the Tumor?

The normal bone surrounding a tumor makes an attempt to wall it off with additional bone. Inactive or slow-growing tumors will have a sclerotic border. In such tumors, the transition between tumor and normal bone is sharp, producing a geographic margin (Fig. 2). Tumors that grow faster afford less of an opportunity for the surrounding bone to produce this sclerotic border. Thus, as tumors become more active, the margin between tumor and normal bone becomes less distinct. Such tumors are said to have a permeative margin. On occasion, the tumor grows diffusely through large areas of bone,

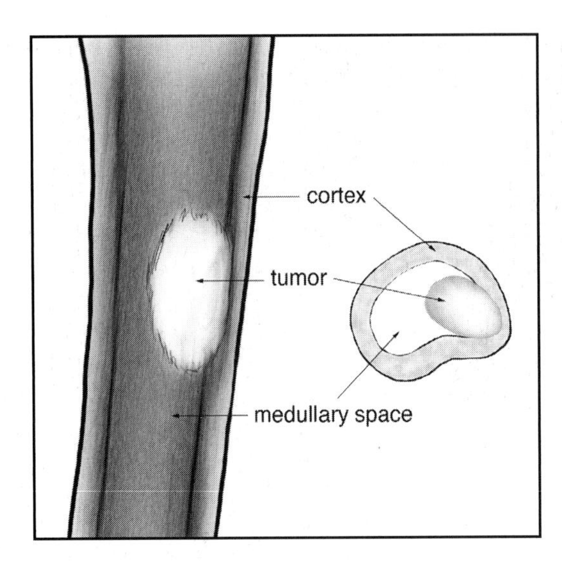

Figure 1
Longitudinal and axial views of endosteal scalloping, showing erosion of the cortex by intramedullary lesions.

yielding a moth-eaten margin, with no border of the tumor being distinct.

Is There Any Type of Mineralization Noted Within the Lesion?
Aggressive tumors also can begin a reaction within the periosteum as they erode the bone surface. This reaction can appear in a variety of ways, all of which result from mineralization of periosteum as it is lifted from the bone surface by tumor extension. Tumors producing cartilage or bone have specific patterns of internal mineralization. Tumors producing fluid, such as a simple cyst, or fibrous tissue demonstrate no mineralization.

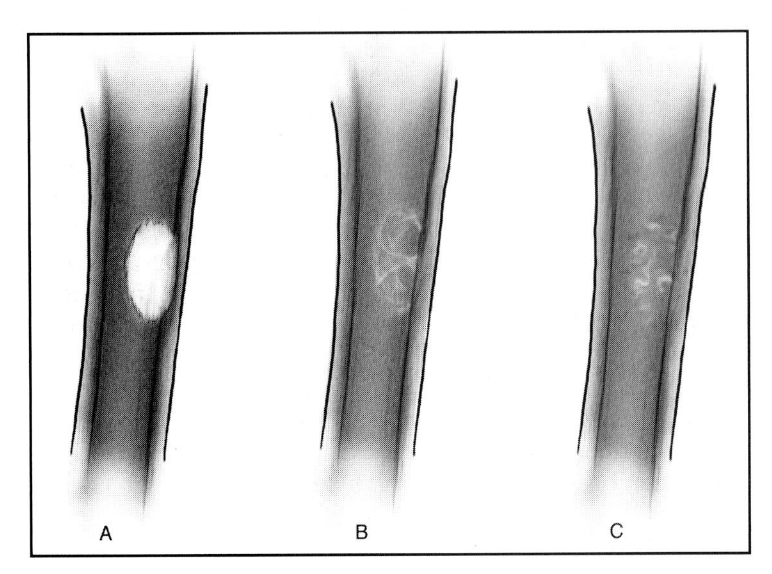

Figure 2
Tumor margins. **A**, Geographic. **B**, Permeative. **C**, Moth-eaten.

Advanced Imaging Studies
Additional imaging modalities include technetium 99m bone scanning, CT, and MRI. Bone scanning, performed by injecting the patient with a radioisotope prior to obtaining the scan, indicates areas of active bone formation because the isotope is incorporated into the bone as it is produced by osteoblasts. Bone scanning is used to evaluate the biologic activity of primary lesions and to assess other areas for potential metastatic involvement. CT is used primarily to evaluate the geometry of the lesion and to identify areas of cortical destruction. CT of the chest and abdomen is also used to locate visceral metastases. MRI allows for precise localization of the tumor in multiple planes and may, on occasion, suggest the type of tissue produced by the lesion. MRI is particularly useful for evaluating the soft tissues. An algorithm for the radiographic evaluation of bone tumors is presented in Figure 3.[4]

Laboratory Tests
The role of laboratory tests in the evaluation of musculoskeletal tumors is somewhat limited. Laboratory tests are used primarily to evaluate adult patients with lesions that appear to be aggressive. These lesions are most frequently either metastases from other, more common cancers (such as breast, lung, thyroid, kidney, or prostate cancer) or myelomas (tumors arising from lymphocytes in the bone marrow). Some of these cancers can be detected by specific blood tests. Although breast and prostate cancers can express specific markers that allow for early detection, no such markers for bone or soft-tissue malignancies have been identified.

When evaluating bone lesions that appear aggressive on radiographs, osteomyelitis should always be in the differential diagnosis. Blood tests that measure the erythrocyte sedimentation rate and C-reactive protein level may help in this evaluation because infection normally elevates these values. However, the results of these tests may be abnormal in some patients with malignant bone tumors.

Treatment
Benign Tumors
Not all tumors of bone and soft tissue need to be removed. If a tumor appears to be benign

based on its history and radiographic presentation and is asymptomatic, usually it can be ignored. If a tumor is painful yet appears benign on radiographs, it can be removed, both to ensure a correct diagnosis as well as to relieve symptoms. Removing these tumors typically involves excising them from surrounding tissue, after which they rarely recur. Any tumor that grows rapidly must be suspected of harboring aggressive potential. Therefore, if the patient or family thinks that the tumor is growing or if the diagnosis remains unclear despite radiographs, the tumor should be biopsied or removed to confirm the diagnosis.

Tumors for which the history and radiographic appearance suggest malignancy warrant a biopsy, either with a needle or via an open biopsy. Histologic characteristics of malignancy include the presence of hyperchromatic nuclei and nuclear pleomorphism. Tumor grade is assigned to reflect the degree of activity seen histologically.[5]

Malignant Tumors

Malignant tumors differ from benign tumors in many ways, especially the propensity to spread to adjacent areas and, via the bloodstream, to distant areas such as the lungs and other bones. This characteristic gives malignant tumors a much higher rate of recurrence. Metastases are detected through the use of CT and bone scanning. The results of these imaging studies and the histologic grade of the tumor can be combined to yield the stage of the tumor. Identifying a tumor by its stage is a shorthand method of assessing prognosis. While the specifics of staging tumors is beyond the scope of this text, generally the higher the number the more serious the tumor. Bone tumors that are known to have a high risk of metastasis are also treated with chemotherapy. If chemotherapy is given after surgery, it is called adjuvant chemotherapy; if given before removing the tumor, it is called neoadjuvant chemotherapy. Neoadjuvant chemotherapy is often given because it allows the treating physician to assess the amount of necrosis in the biopsy specimen and assess the tumor's response. In addition to helping determine the prognosis, neoadjuvant chemotherapy may also help shrink a lesion that is going to be removed.

Chemotherapy has been found to have substantial beneficial effects on the long-term control of primary cancers in bone, but this effect is less well substantiated in the treatment of soft-tissue malignancies. This may be the result of a relative insensitivity of these tumors to chemotherapy or because patients with soft-tissue malignancies tend to be older than those with primary bone tumors and may not tolerate the full dosage of chemotherapeutic agents.

The treatment of malignant tumors involves surgery to remove the main tumor mass and a cuff of normal tissue around it (called wide resection) to remove the satellite tumor cells. Before the advent of MRI, malignancies of the extremities were frequently treated with amputation to ensure the removal of the entire lesion. Because MRI now provides a more precise way to localize the margins of tumors in the extremities, less aggressive surgical treatment options are now available. Currently, malignant bone and soft-tissue tumors of the extremities are treated with limb salvage, which involves removing the tumor and maintaining the extremity. In the case of soft-tissue malignancies, radiation therapy may also be used to attack any residual tumor cells, which further decreases the chance of recurrence.

After resection of malignant tumors of bone, the area is reconstructed either with bone from organ donors (allograft) or with a prosthetic device. The decision regarding how to reconstruct the area depends on the amount and location of bone removed and the age and activity level of the patient.[1,6] If a wide resection requires the removal of so much tissue that the extremity is not viable or functional, amputation is the preferred method of treatment. Adequate tumor removal should never be sacrificed in order to improve function. However, because a proximal amputation is so functionally debilitating, extensive surgical techniques, such as vascular bypass grafts, may be used to save the distal limb when the tumor is located in the pelvis or shoulder.

The role of surgery in the treatment of metastatic tumors in bone is more limited. Biopsies may be needed to confirm that the lesion has metastasized from a known visceral primary tumor. Surgery may also be required if the lesion has caused or is expected to cause a fracture. Surgery in this instance is intended to relieve pain and improve function but does not affect the course of the primary disease process.[7]

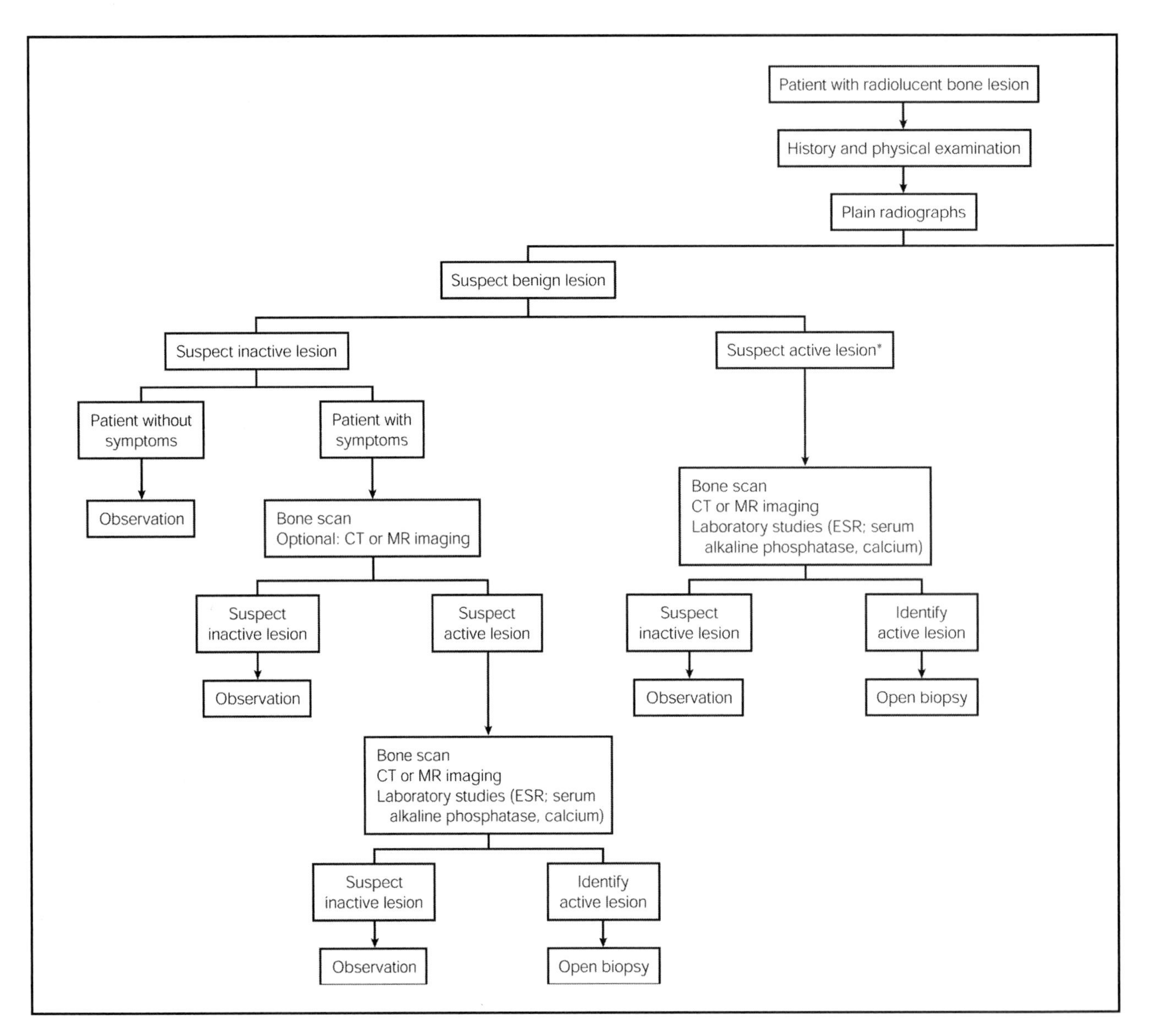

Figure 3
Algorithm for the evaluation of a patient with a radiolucent lesion of an extremity. Asterisk indicates a point at which referral is indicated if the evaluating physician is not prepared to treat the patient. ESR = erythrocyte sedimentation rate, IEP = individualized education plan, CBC = complete blood cell count, LDH = lactate dehydrogenase, ORIF = open reduction and internal fixation, PSA = prostate-specific antigen.

(Reproduced from Springfield DS: Radiolucent lesions of the extremities. ***J Am Acad Orthop Surg*** *1994;2:306-316.)*

Benign Bone Tumors

The etiology of most benign bone tumors is unknown. In some syndromes, benign lesions occur in multiple bones. Based on their clinical and radiographic appearance, benign bone tumors are classified into the following stages of increasing severity: latent, active, and aggressive.

Latent benign tumors are frequently noted incidentally on radiographs. Symptoms, if present, are typically mild and long standing. The onset of pain attributable to these lesions either indicates that the lesion was more active than anticipated initially or that surrounding structures have been irritated. Patients with active benign tumors will report more consistent, progressive pain that is not related to activity. Occasionally, these lesions can produce a palpable soft-tissue mass. With aggressive but benign tumors, the difference between benign and malignant becomes less distinct, except that ag-

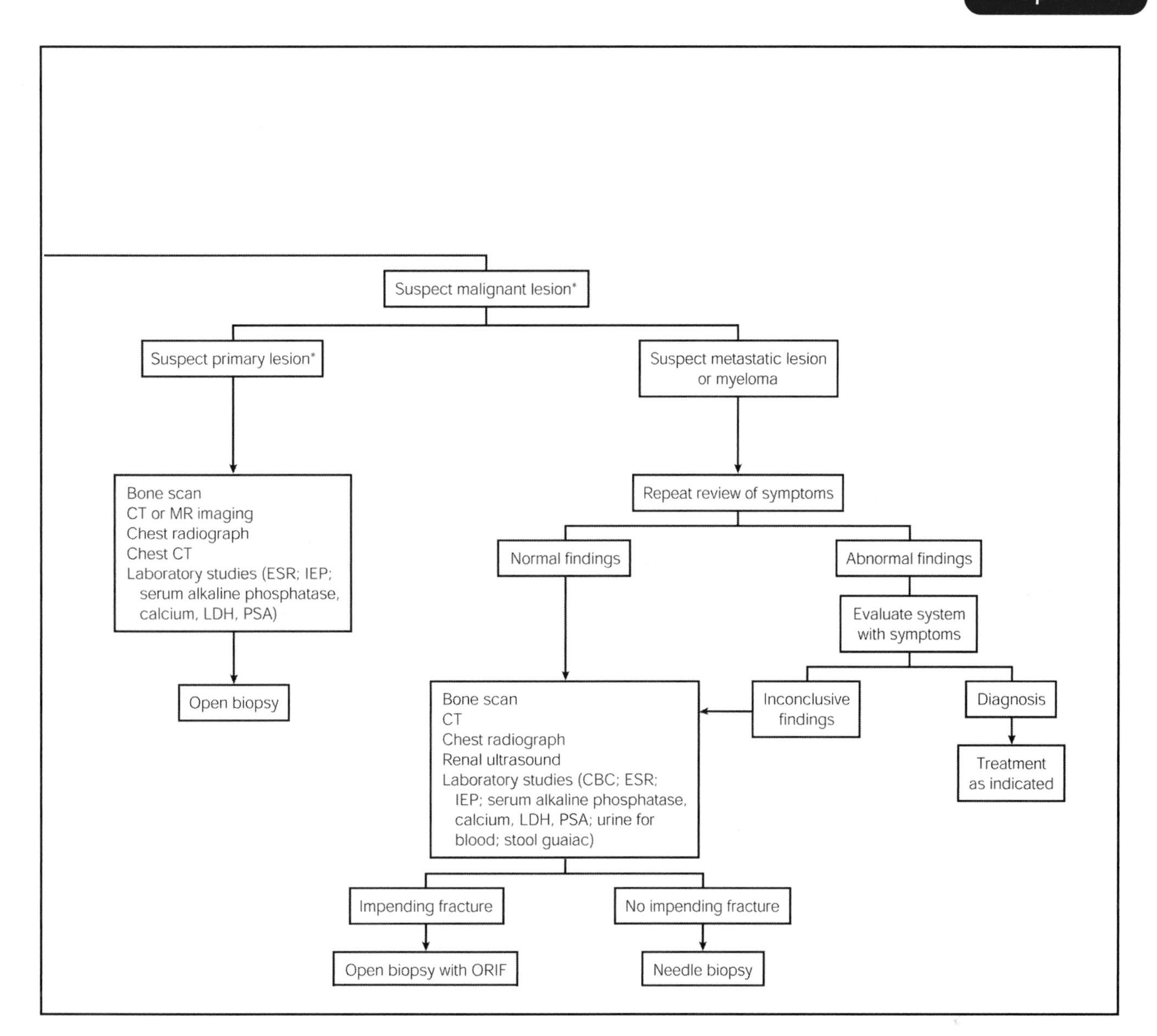

gressive but benign tumors are not life threatening.

Nonossifying Fibroma

Nonossifying fibromas are benign asymptomatic tumors that commonly occur in children. Microscopically, a nonossifying fibroma is composed of spindle (fibrous) cells. Approximately 20% of all children have this lesion, most frequently in the posterior distal femur. As a child matures, the lesions tend to disappear. In rare instances, however, they enlarge and become more central in location, placing the bone at risk for pathologic fracture.

A typical case presentation would be a 12-year-old boy who injured his ankle playing soccer a week ago, at which time radiographs were obtained in the emergency department. The patient and his mother report that he had no pain prior to the injury. He did not play soccer in the week following the accident, and his pain is now resolving. Physical examination demonstrates no area of palpable tenderness. In the history, the lack of antecedent pain and the rapid resolution of symptoms indicate that the tumor is not the source of the pain.

Radiographs show a distal tibia metaphyseal lesion with minimal endosteal scalloping, which is not uncommon in latent benign tumors (Fig. 4). Rather, it reflects a previous episode of growth. The margin between the lesion and the surrounding bone is distinct. The well-defined sclerotic rim indicates that the tumor is now minimally active, if at all. The lack of internal mineralization indicates that the lesion is composed either of fluid or fibrous tissue. Therefore, this type of lesion

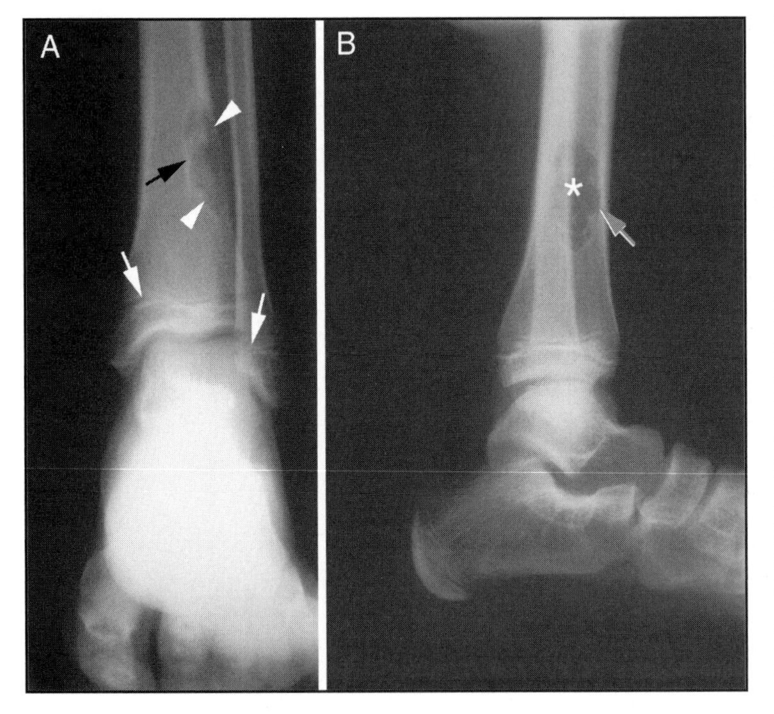

Figure 4
Nonossifying fibroma. **A**, AP view of the distal tibia and fibula. The open growth plates (white arrows) are seen in both bones. The tumor (arrowheads) is eccentric within the bone, involving both the cortex and medullary space. The bone has produced a sclerotic border (black arrow) around the lesion. **B**, Lateral view shows some endosteal scalloping of the cortex by the tumor (gray arrow). The interior of the lesion demonstrates no mineral formation (ie, it is osteolytic) (asterisk), indicating that it is composed of something other than bone or cartilage (eg, fluid or fibrous tissue). In this patient, it was benign fibrous tissue.

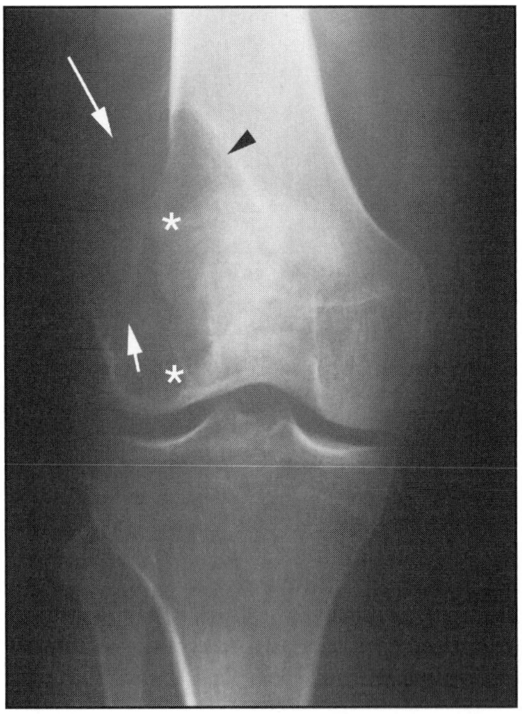

Figure 5
AP view of the knee in a patient with a giant cell tumor of bone. The growth plates are closed, indicating that this patient has reached skeletal maturity. The tumor has eroded through the lateral cortex of the distal femur (short arrow), suggesting an adjacent soft-tissue mass (long arrow). The bone is producing a sclerotic border around the remainder of the lesion (arrowhead). The lesion itself is not producing bone (ie, it is osteolytic). The small areas of mineralization within the tumor (asterisks) represent residual bone trapped by the tumor. This appearance indicates that the tumor is biologically active.

does not require surgical intervention or further radiographic workup.[8]

Giant Cell Tumor of Bone

Unlike a nonossifying fibroma, a giant cell tumor of bone is an aggressive lesion. It typically affects adults in the third to fifth decades of life. Histologically, it is composed of multinucleated giant cells. It is one of the few tumors that involve the epiphysis.

A typical case presentation would be a 40-year-old woman who has a 6-month history of progressive knee pain without any previous injury. The pain was initially noted with walking but now is present at rest and at night. She limited her activities, but this modification did not resolve the pain, and she is no longer able to bear weight on the affected leg. Her distal femur is diffusely tender to palpation, and a soft-tissue mass is noted. The aggressive nature of the tumor is manifest in the severe pain at rest, even after an attempt at limiting activity.

Radiographs show a lesion of the distal femur, including the epiphysis, with a less distinct margin in the surrounding bone (Fig. 5). There is no evidence of bone production adjacent to the mass or mineralization produced by the tumor. The lateral cortex is eroded, and there is the suggestion of an adjacent soft-tissue mass. These findings also indicate that this tumor is extremely active. To better assess the amount of bone involved and size of the soft-tissue mass, MRI should be ordered.

Untreated giant cell tumors will continue to enlarge, resulting in further bone destruction and potential destruction of the adjacent joint. Therefore, this lesion requires surgical treatment, consisting of curettage (scraping the tumor out of the bone). The more aggressive the lesion, the more thorough the curettage needs to be to prevent local recurrence. On occasion, the segment of involved bone needs to be removed because of the extent of bone destruction.

Osteoid Osteoma

Osteoid osteoma is a benign tumor that typically occurs in patients younger than 25 years, with a male to female ratio of 2:1. The lesions are usually intracortical and found either in a long bone or in the posterior elements of the spine. Histologically, the interior of the lesion shows areas of fibrous and vascular tissue. The pain associated with osteoid osteoma comes from the lesion itself, not from any weakening of the surrounding bone. These lesions are believed to cause pain because they synthesize prostaglandins, a process that results in edema in the surrounding bone and bone marrow. Nonsteroidal anti-inflammatory drugs provide pain relief in this instance because they block prostaglandin production.

A typical case presentation would be a 5-year-old boy who has a 3-month history of thigh pain that occurs primarily at night but no history of significant trauma. His parents report that the night pain is relieved with ibuprofen. On examination, the pain occurs with attempts at hip motion, but there is no palpable mass.

Radiographs demonstrate a lesion within the cortex of the anterior proximal femur (Fig. 6). The margin of the lesion is distinct, with an area of dense bone surrounding the lesion. The interior of the lesion demonstrates no evidence of mineralization. The lesion shows intense uptake on a bone scan (Fig. 7). The bone surrounding this particular lesion did not originate from the lesion itself. The tumor stimulates the production of a significant amount of new bone at the periphery of the lesion, which accounts for the findings on the radiograph and bone scan.

Surgical intervention is reserved for patients in whom the diagnosis is in doubt or for those who cannot control their pain with medication. In addition to surgical resection, ablation with microwave radiation may be attempted.

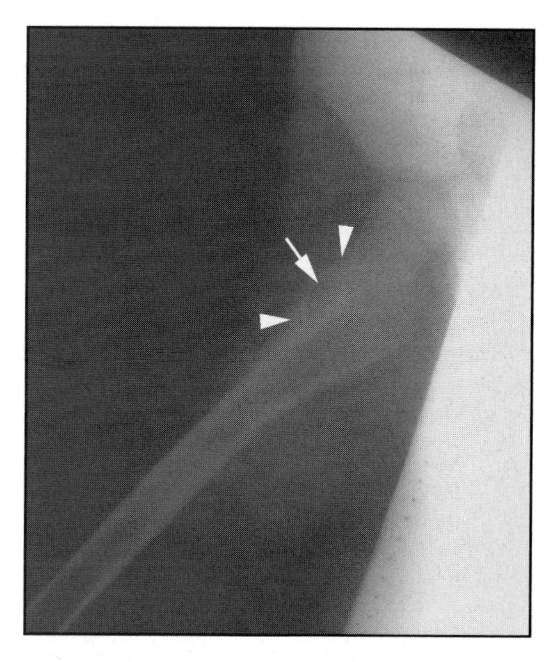

Figure 6
Frog-lateral view of a patient with an osteoid osteoma shows the tumor (the lytic area) arising within the anterior cortex of the femur (arrow). The arrowheads point to bone that is denser than that typically produced by bone surrounding a tumor and is a response to the prostaglandins produced by the osteoid osteoma.

Malignant Bone Tumors

Patients with malignant tumors in bone report pain and occasionally the presence of a soft-tissue mass. The duration of symptoms varies. Patients may also have systemic constitutional symptoms, such as malaise, anorexia, or weight loss. In children, bone tumors are usually sarcomas arising primarily from bone. In adults, most malignancies found in bone do not arise from bone cells but represent metastases from visceral neoplasms. In both children and adults, hematologic malignancies can be found in bone.

Osteosarcoma

Osteosarcoma is the most common malignant lesion to arise primarily from bone cells. It is found in the metaphyseal regions of long bones, especially around the knee. As in the case of most malignancies, pain at rest and night is indicative of an aggressive lesion. The diagnosis must be confirmed with a biopsy because the treatment is intensive. An osteosarcoma is composed of malignant cells that produce immature bone (osteoid). Because this tumor can spread hematogenously, the potential for metastases must be considered. Thus, CT of the chest should be ordered to identify pulmonary metastases, and bone scanning is needed to look for bone metastases. This additional testing is called a staging workup.

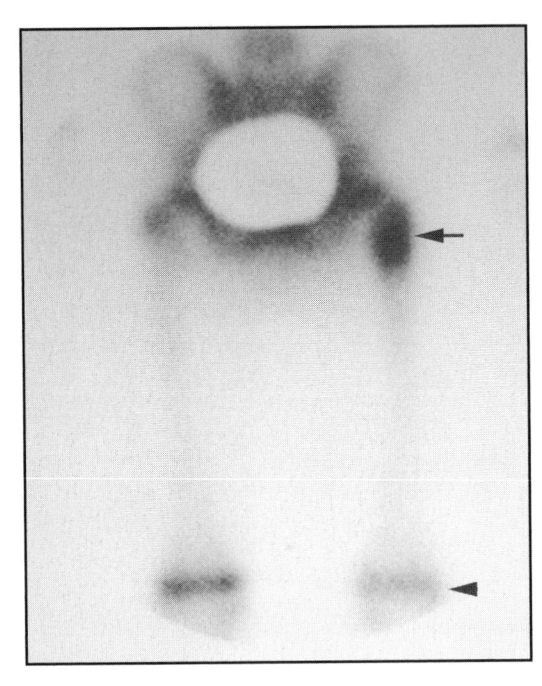

Figure 7
Bone scan of a patient with osteoid osteoma shows increased uptake (arrow), suggesting bone formation around the tumor. Note that the uptake is greater than that seen in the growth plates (arrowhead).

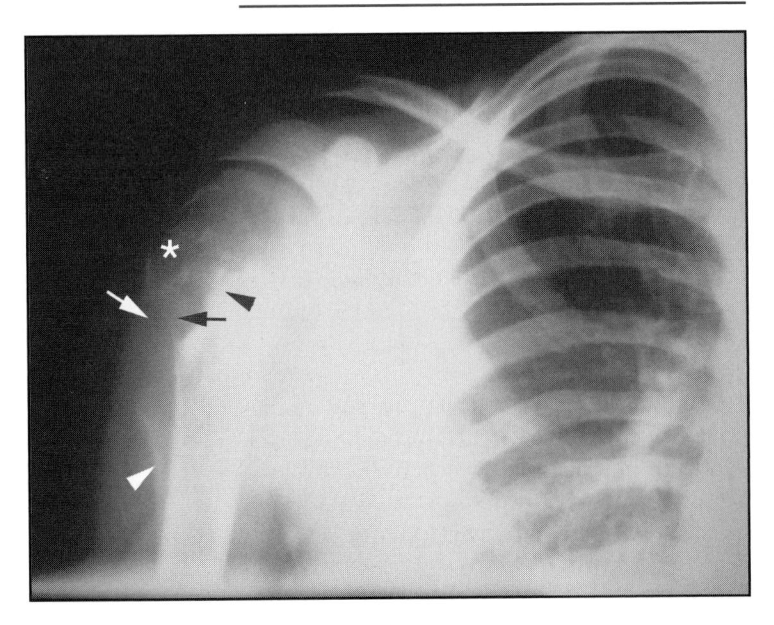

Figure 8
AP view of the shoulder in a patient with osteosarcoma shows that the tumor has eroded the lateral cortex of the humerus (black arrow) and extended into the soft tissues (white arrow). The tumor induces periosteal reaction, forming a Codman's triangle (white arrowhead). The border with the medullary bone is indistinct (permeative), and there is no evidence of a surrounding sclerotic reaction. The interior of the tumor is mixed, with lytic areas (asterisk) and osteoblastic (bone-forming) areas (black arrowhead). These findings are diagnostic of an aggressive, bone-forming tumor (ie, osteosarcoma).

A typical case presentation would be a 17-year-old boy who has had pain and swelling in the shoulder for the past 8 months that was initially attributed to an injury sustained while playing basketball. However, concern arose when his symptoms progressed to pain at rest and at night, despite ceasing all athletic activities. Physical examination demonstrates pain in the shoulder and an adjacent soft-tissue mass.

Radiographs show a lesion in the metaphyseal region of the humerus with internal mineralization (Fig. 8). The border with the remainder of the humeral metaphysis is indistinct. The mineralized lesion extends outside of the bone. Adjacent periosteal reaction is noted, which produces an elevation of the periosteum that, along with the shaft of the bone, forms a Codman's triangle (Fig. 9). This radiographic appearance, including indistinct margins, periosteal reaction, and appearance of mineralization intermingled with lytic areas confirms the diagnosis. In other cases, the mineralization may be more dominant, creating a "sunburst" appearance (Fig 10).

Treatment of the primary lesion has changed in recent years, primarily as a result of improvement in diagnostic imaging modalities. More precise identification of the borders of the tumor allows for removal of the tumor and limb salvage. The survival rate for limb salvage is the same as amputation, but the functional results are better.[9]

Survival rates associated with osteosarcoma have been poor in the past, even with amputation. Patients with osteosarcoma often succumb to metastatic disease. However, the use of both neoadjuvant and adjuvant chemotherapy has improved survival rates dramatically.

Metastatic Carcinoma

Although almost all somatic malignancies can spread to bone, those arising in the breast, lung, thyroid, kidney, and prostate represent a vast majority of cases. The reason these tumors have a predilection for skeletal metastasis is unclear. It may relate to the ease with which these tumors enter the bloodstream or the presence of receptors to allow attachment to bone. Because skeletal metastases are far more common than primary malignancies, an adult with a malignant tumor in bone should be evaluated concurrently for the presence of a primary malignancy elsewhere. This can be accom-

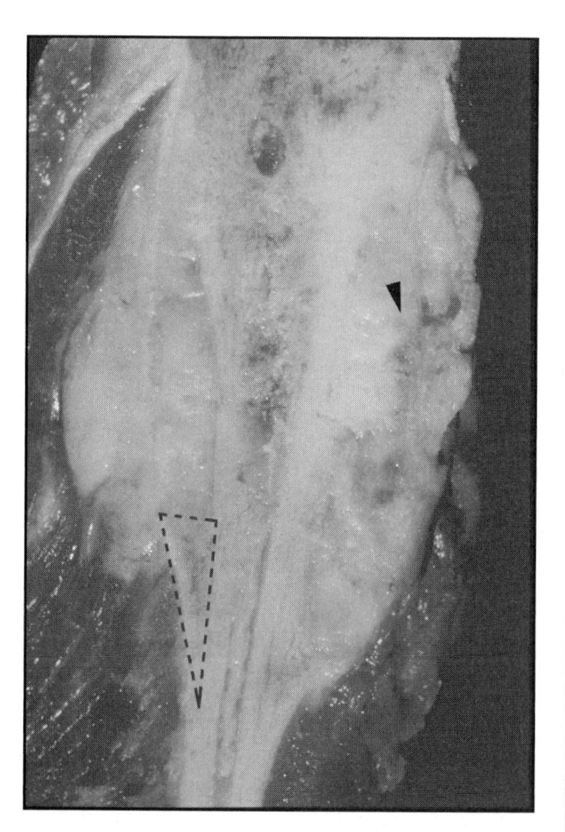

Figure 9
Photograph of a gross specimen after removal of an osteosarcoma. The tumor is seen within the bone, extending out into the soft tissues (arrowhead). The periosteal reaction, forming a Codman's triangle, is also seen (dashed lines). In this type of periosteal reaction, the tumor erodes through the bone, elevating the periosteum. In turn, the periosteum lays down reactive bone.

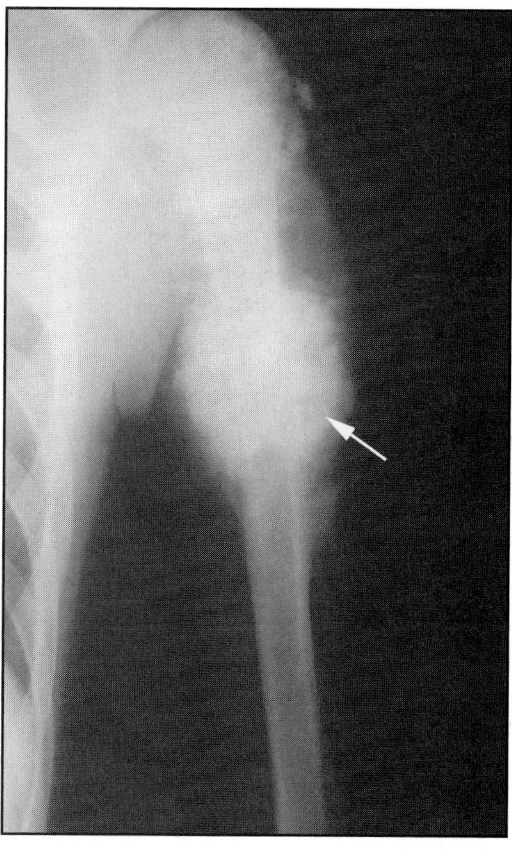

Figure 10
AP view of the humerus in a patient with osteosarcoma shows that the tumor is primarily osteoblastic and extends out into the soft tissues, with some areas appearing to arise perpendicular to the long axis of the bone, creating a "sunburst" appearance (arrow).

plished with CT and blood testing. Bone scanning is also indicated to detect distal bone lesions.

A typical case presentation would be a 60-year-old woman with a 6-week history of progressive thigh pain who reports increased fatigue, an unexplained 10-lb weight loss, and pain with any attempt to move the knee. Radiographs show a permeative lesion in the diaphysis of the femur with a surrounding periosteal reaction (Fig. 11). No mineralization is detected within the lesion. During a subsequent radiographic workup, a primary lesion was found in the lung. Metastatic lesions tend to enlarge quickly, which accounts for the rapid progression of the patient's symptoms. The metabolic demands of this aggressive lesion are responsible for the fatigue and weight loss.

Metastatic disease in bone is usually treated with radiation to control pain and halt bone destruction by the tumor. However, if lesions are large enough at the time of presentation to create the risk of pathologic fracture, prophylactic fixation (that is, surgically stabilizing the bone before it breaks) should be performed. Prophylactic fixation is also indicated if pain does not improve despite radiation. Pain after radiation suggests that the remaining bone is under stress and that the structural integrity of the bone is at risk. It is often easier and medically safer to place a rod in a long bone before it breaks; it is also far more comfortable for the patient.

Benign Soft-Tissue Tumors

Benign soft-tissue tumors are common, especially in areas with a greater volume of soft tissue (eg, the thigh and upper arm). Even though these tumors are not tumors of the bone itself, they are frequently evaluated with radiographs because imaging studies can initially confirm that the lesion is in the soft tis-

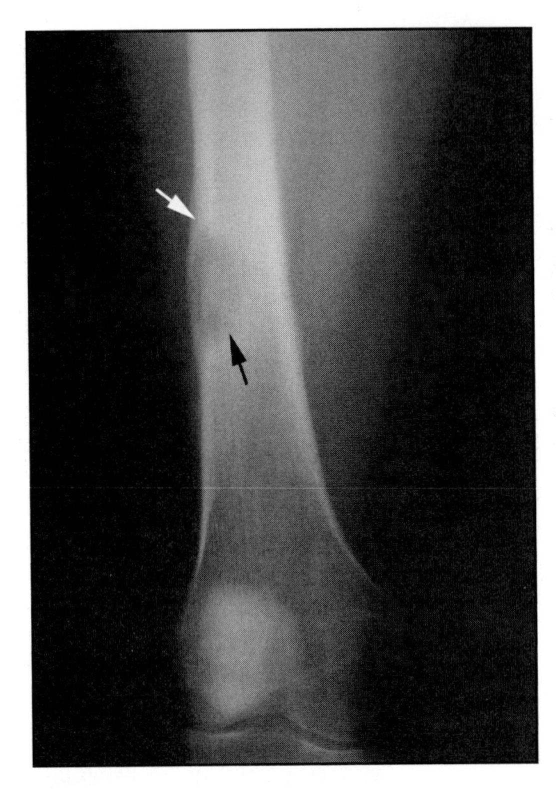

Figure 11
AP view of a femur in a patient with metastatic carcinoma shows a tumor in the diaphysis that involves both cortical and medullary bone. The tumor has an indistinct (permeative) border, with scant surrounding bone formation (black arrow). There is periosteal reaction (white arrow), indicating that the tumor has eroded through the outer portion of the cortex and into the periosteum. Both features indicate that the lesion is biologically active. Because there is no evidence of mineral formation within the lesion, osteosarcoma or chondrosarcoma is not considered in the differential diagnosis.

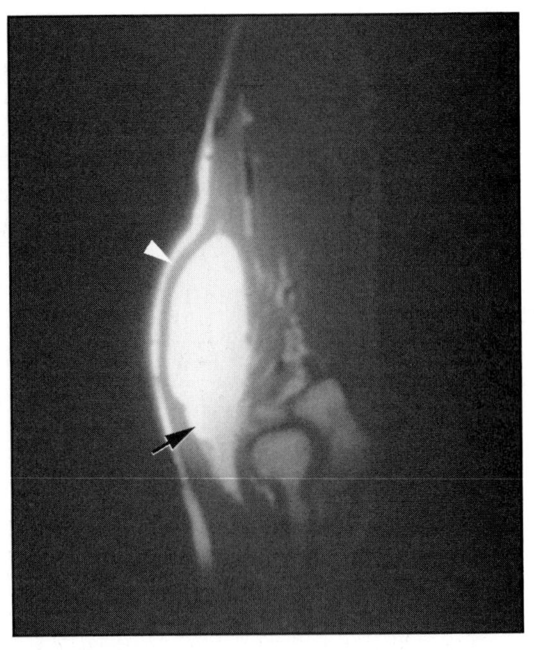

Figure 12
Sagittal T1-weighted MRI scan shows a lipoma of the elbow. The lesion located anterior to the elbow (arrow) has the same high signal (bright) intensity as subcutaneous fat (arrowhead).

sue and not originating from the surface of the bone. Radiographs may also demonstrate mineralization within the lesion, which can help to narrow the differential diagnosis. If further evaluation of the tumor is necessary, MRI is the modality of choice because it can localize the lesion within the extremity and, in certain instances, indicate tissue type.[10]

Benign soft-tissue tumors are frequently found incidentally. They are usually asymptomatic or, at worst, produce a cosmetic deformity. If they enlarge at all, they typically do so slowly. However, enlargement is often difficult to detect as the surrounding soft tissue may obscure the tumor. Benign soft-tissue lesions tend to be smaller and more superficial than malignant lesions.

A lipoma is one of the most common types of soft-tissue tumor. Most frequently seen as a subcutaneous lesion in older patients, it affects both sexes equally and presents spontaneously without pain. Benign lipomas may appear in multiple locations (lipomatosis). They may grow fast at first but then stop growing and never turn malignant. Histologically, these tumors are almost identical to fat in the subcutaneous tissue (Fig. 12).

A typical case presentation would be a 29-year-old man with a history of benign soft-tissue masses who now has a mass on his elbow. This lesion has not changed in several years and is not symptomatic. Physical examination demonstrates a 3-cm, soft, mobile, superficial lesion on the anterior aspect of the elbow. Similar lesions are noted on the opposite arm and back.

The mobility of a lipoma indicates that it is most likely superficial; such lesions do not demand radiographic studies. Characteristics that should prompt further evaluation include enlargement of the mass, large size at the time of presentation, and immobility of the mass when palpated.

Malignant Soft-Tissue Tumors

Malignant soft-tissue tumors usually develop in middle-aged and older adults. These lesions tend to enlarge more quickly than benign lesions. On occasion, a quiescent benign lesion may rapidly expand, indicating malignant degeneration. Patients with soft-tissue sarcomas may present with a history of minor trauma and resultant swelling, which calls attention to the mass. However, unlike typical posttraumatic swelling, it does not improve with time and rest and may actually worsen. Swelling may be caused by injury to the immature blood vessels at the periphery of the tumor, which tend to bleed readily.

The most common malignant soft-tissue tumor in adults is malignant fibrous histiocytoma, which is seen most often in men between the ages of 50 and 70 years. The tumor mass usually occurs deep within the larger muscles about the hip, shoulder, thigh, and retroperitoneum.

A typical case presentation would be a 70-year-old man with a 2-month history of an enlarging mass on his hip. He denies any weight loss but does note increased fatigue and limited hip motion. Physical examination reveals a firm, fixed mass adjacent to the greater trochanter but no other palpable masses. Radiographs suggest the presence of a soft-tissue mass, with no internal mineralization or underlying bone erosion. An MRI scan shows a 20-cm heterogeneous tumor within the soft tissue (Fig. 13).

The enlarging nature of this patient's mass, as well as its deep-seated location, most likely indicates a malignancy. The overall size and appearance on the MRI scan support this diagnosis. Heterogeneity within a mass on MRI usually signals necrosis, an indicator that the tumor is growing so fast it outpaces its blood supply. The next step in the evaluation of this tumor is a biopsy. Histologically, malignant fibrous histiocytoma has both fibrous elements and histiocytes (large, multinucleated cells). Soft-tissue sarcomas may also spread hematogenously to the lungs. Therefore, the workup should also include a CT scan of the chest.

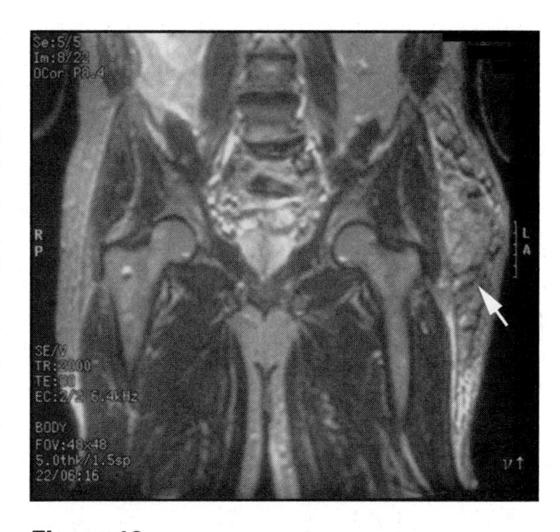

Figure 13
Coronal MRI scan of a patient with malignant fibrous histiocytoma shows a tumor arising within the subcutaneous tissue surrounding the left hip. Note that the interior of this lesion is heterogeneous (arrow), indicating that at least some of this tumor is necrotic. This typically reflects the rapid growth of the tumor.

Historically, amputation has been the standard treatment for soft-tissue sarcomas. However, if the lesion can be removed and a useful extremity maintained, limb salvage surgery is performed, especially for more proximal lesions. MRI can delineate the extent of the tissue that needs to be removed. However, soft-tissue sarcomas can disperse satellite cells from their edges into adjacent tissue. If only the grossly involved tissue is removed, the risk of local recurrence is high. Therefore, adjuvant radiation therapy is used to eradicate these satellite cells, and it decreases the risk of local recurrence to below 10%. The role of chemotherapy in the treatment of soft-tissue sarcomas is less clear than it is for bone sarcomas. Currently, studies do not clearly indicate whether chemotherapy affects survival.

Key Terms

Café au lait spots Congenital pigmented skin marks; the color of coffee with milk

Chondrosarcoma A primary sarcoma formed from cartilage cells or their precursors but without direct osteoid formation

Codman's triangle A radiographically visible triangular-shaped area formed by the elevation of the periosteum by a bone tumor and the adjacent cortex of normal bone

Curettage The removal of growths from within cavity walls; in the treatment of musculoskeletal tumors, the scraping of tumor out of bone

Endosteal scalloping The erosion of bone in the medullary canal caused by active tumors

Ewing's sarcoma A primary sarcoma of the bone that usually arises in the diaphyses of long bones, ribs, and flat bones of children and adolescents.

Giant cell tumor of bone A tumor composed of multinucleated giant cells resembling osteoclasts

Limb salvage Surgical removal of a tumor without amputation of the affected extremity

Lipoma A benign, soft, rubbery tumor usually composed of mature fat cells

Malignant fibrous histiocytoma A soft-tissue sarcoma histologically characterized by the presence of fibrous elements and large, multinucleated histiocytes

Metastasis The transfer of disease from one part of the body to another; tumor metastasis usually occurs via the bloodstream

Multiple myeloma A disseminated malignancy of the B lymphocytes that can result in widespread osteolytic lesions

Nonossifying fibromas Osteolytic and sometimes painful proliferative lesions composed of spindle-shaped (fibrous) cells

Osteoid osteoma A small, benign but painful tumor usually found in the long bones or the posterior elements of the spine

Osteosarcoma A primary sarcoma of the bone that is characterized by the direct formation of bone or osteoid tissue by the tumor cells

Pathologic fracture A fracture caused by a normal load on abnormal bone (often weakened by tumor, infection, or metabolic bone disease)

Prophylactic fixation Surgical stabilization of a bone before it breaks; often performed to stabilize bones weakened by tumors

Staging workup Evaluation using CT of the chest and nuclear bone scanning to search for metastases

References

1. Simon MA, Springfield DS, Conrad EU: (eds): *Surgery for Bone and Soft-Tissue Tumors.* Philadelphia, PA, Lippincott-Raven, 1998.
2. Unni KK, Dahlin DC: (eds): *Dahlin's Bone Tumors: General Aspects and Data on 11,087 Cases,* ed 5. Philadelphia, PA, Lippincott-Raven, 1996.
3. Praemer A, Furner S, Rice DP: (eds): *Musculoskeletal Conditions in the United States.* Rosemont, IL, American Academy of Orthopaedic Surgeons, 1999.
4. Springfield DS: Radiolucent lesions of the extremities. *J Am Acad Orthop Surg* 1994;2:306-316.
5. Huvos AG: (ed): *Bone Tumors, Diagnosis, Treatment, and Prognosis,* ed 2. Philadelphia, PA, WB Saunders, 1991.
6. Malawer MM, Sugarbaker PH: (eds): *Musculoskeletal Cancer Surgery: Treatment of Sarcomas and Allied Diseases.* Dordrect, Netherlands, Kluwer Academic Publishers, 2001.
7. Swanson KC, Pritchard DJ, Sim FH: Surgical treatment of metastatic disease of the femur. *J Am Acad Orthop Surg* 2000;8:56-65.
8. Aboulafia AJ, Kennon RE, Jelinek JS: Benign bone tumors of childhood. *J Am Acad Orthop Surg* 1999;7:377-388.
9. Heinrich SD, Scarborough MT: (eds): *The Orthopedic Clinics of North America: Pediatric Orthopedic Oncology.* Philadelphia, PA, WB Saunders, 1996, vol 27.
10. Sim FH, Frassica FJ, Frassica DA: Soft-tissue tumors: Diagnosis, evaluation, and management. *J Am Acad Orthop Surg* 1994;2:202-211.

Chapter 19

Disorders of Musculoskeletal Growth and Development

Disorders of musculoskeletal growth and development are relatively uncommon, with an incidence of approximately 1 in 5,000 births.[1] Nonetheless, they provide critical insights into the normal physiology of growth and development. Historically, physicians have often been more concerned with clinical diagnosis and surgical treatment than the molecular biology and genetics of these rare diseases. A more complete understanding of the skeletal dysplasias has become increasingly important, however, because it enables health care providers to offer genetic counseling and education to patients, determine prognoses, and provide more effective treatment.

Because the general objective of this chapter is to provide a framework upon which to approach disorders of growth and development, the specific goals of this chapter are as follows: (1) to define the terms used to describe skeletal dysplasias; (2) to review certain aspects of musculoskeletal development; (3) to outline a general approach to patient evaluation; and (4) to provide a brief clinical description of several representative disorders of musculoskeletal development, emphasizing the basic science concepts underlying each disease.

Definitions

Dysplasia is a broad term that describes a condition affecting growth or development in which the primary defect is intrinsic to affected tissue. As such, nearly all musculoskeletal dysplasias arise from genetic abnormalities (Table 1). Dysostosis refers to conditions of abnormal cartilage ossification or bony remodeling. Often in these disorders, a single bone or group of bones is affected. Dystrophy is technically defined as a condition resulting from defective or faulty nutrition, although nutrition here broadly refers to nourishment of tissue by all essential substances, not simply dietary needs.[1] In skeletal dystrophies, normal fetal cartilage is affected by extrinsic factors, such as hormonal abnormalities or metabolic disturbances, resulting in abnormal skeletal development. In cases of muscular dystrophies, a lack of an essential factor leads to progressive degeneration of initially normal muscle, often resulting in significant weakness and functional compromise.

Discussions of disorders of growth and development often include descriptive terms that express characteristic clinical or phenotypic manifestations (Table 2). Most skeletal dysplasias result in short stature, commonly called dwarfism. Dwarfism is more specifically characterized by an adult height of less than 58 inches in males or a standing height below the third percentile for age. Dwarfism is typically divided into two general categories, based on the degree to which the trunk and limbs are affected. Proportionate dwarfism refers to short stature in which both the trunk and extremities are equally affected,

Table 1
Genetic Abnormalities With Musculoskeletal Manifestations

Disease	Subtype	Inheritance Pattern	Chromosomal Mapping	Affected Gene/Gene Product
Skeletal dysplasias				
Achondroplasia		AD	4p16.3	FGRC-3
Apert syndrome		AD	10q26	FGRC-2
Chondrodysplasia punctata		XLD	Xq28	Unknown
Cleidocranial		AD	6p21	Unknown
Diastrophic dysplasia		AR	5q31-34	Diastrophic dysplasia sulfate transporter
Hypochondroplasia		AD	4p16.3	FGRC-3
Kniest syndrome		AD	12q13.11-q13.2	Type II collagen
Metaphyseal chondrodysplasia				
	Jansen type	AD	3p22-21.1	Parathyroid hormone related peptide receptor
	McKusick type	AR	9p13-q11	Unknown
	Schmid type	AD	6q21-q22.3	Type X collagen
McCune-Albright syndrome		Unknown	20q13.2	Guanine-nucleotide binding protein alpha
Mucopolysaccharidoses				
	Type I (Hurler)	AR	4p16.3	Alpha-L-iduronidase
	Type II (Hunter)	XLD	Xq28	Sulfoiduronate sulfatase
	Type IA (Morquio)	AR	16q24.3; 3p21.33	Galactosamine-6-sulfate-sulfatase, beta-galactosidase
Multiple epiphyseal dysplasia				
	Type I	AD	19p13.1	Cartilage oligomeric matrix protein
	Type II	AD	1p32	Type IX collagen
Nail-patella syndrome		AD	9q34	Unknown
Osteopetrosis		AR	1p21-13	Macrophage colony stimulating factor
Pseudo-achondroplasia		AD	19p13.1	Cartilage oligomeric matrix protein
Strickler syndrome		AD	12q13.11-q13.2	Type II collagen
Spondyloepiphyseal dysplasia				
	Congenital	AD	12q13.11-q13.2	Type II collagen
	Tarda	AR	12q13.11-q13.2	Type II collagen
	X-linked	XLD	Xp22-22.1	Unknown
Neuromuscular disorders				
Angelman syndrome		AR	15q11-13	Unknown
Dystrophinopathies				
	Duchenne muscular dystrophy	XLR	Xp21.2	Dystrophin
	Becker muscular dystrophy	XLR	Xp21.3	Dystrophin
Charcot-Marie-Tooth disease				
	Type IA	AD	17p11.2	Peripheral myelin protein-22
	Type IB	AD	1q22	Myelin protein zero
	Type IIA	AD	1p36-35	Unknown
	Type IVA	AR	8q13-21.1	Unknown
	X-linked	XL	Xp22.1	Connexin-32

Table 1 (cont.)
Genetic Abnormalities With Musculoskeletal Manifestations

Disease	Subtype	Inheritance Pattern	Chromosomal Mapping	Affected Gene/Gene Product
Neuromuscular disorders (cont.)				
Friedreich ataxia		AR	9q13-q21.1	Frataxin
Myotonic dystrophy		AD	19q13.2-13.3	Myotonin-protein kinase
Myotonia congenita		AD	7q35	Muscle chloride channel-1
Prader-Willi syndrome		AR	15q11-13	Unknown
Spinocerebellar ataxia				
	Type I	AD	6p23	Ataxin-1
	Type II	AD	14q24.3-32	MJD/SCA1
Spinal muscular atrophy		AR	5q12.2-13.3	Survival motor neuron
Connective tissue disorders				
Ehlers-Danlos syndrome				
	Type IVA	AD	2q31	Type III collagen
	Type VI	AR	1p36.3-36.2	Lysine hydroxylase
	Type X	AR	2q34-36	Fibronectin-1
Marfan syndrome		AD	15q21.1	Fibrillin-1
Osteogenesis imperfecta				
	Type I	AD	17q21.31-22.05; 7q22.1	Type I collagen (COL1A1, COL1A2)
	Type II	AR		Type I collagen (COL1A1, COL1A2)
	Type III	AR		Type I collagen (COL1A1, COL1A2)
	Type IVA	AD		Type I collagen (COL1A1, COL1A2)

AD = Autosomal dominant, XLD = X-linked dominant, AR = autosomal recessive, XLR = X-linked recessive.
A portion of this table was adapted with permission from Dietz FR, Matthews KD: Update on the genetic bases of disorders with orthopaedic manifestations. *J Bone Joint Surg Am* 1996;78:1583-1598.

whereas disproportionate dwarfism refers to the condition in which the extremities are relatively more (or less) affected than the trunk. Dysmorphisms are morphologic variations of musculoskeletal appearance. Often affecting the limbs, face, and cranium, dysmorphisms are usually characteristic of a specific disorder and may provide insight into diagnosis.

Table 2
Descriptive Terms of Dwarfism

Type	Musculoskeletal Characteristics
Disproportionate	Relative shortening differs in trunk and extremities.
Proportionate	Relative shortening affects trunk and extremities equally.
Rhizomelia	Proximal segments are disproportionately shorter than middle and distal segments.
Mesomelia	Middle segments are disproportionately shorter than proximal and distal segments.
Acromelia	Distal segments are disproportionately shorter than proximal segments.
Micromelia	Shortness of a limb.

Normal Skeletal and Muscle Biology

The basic biology of musculoskeletal development is described elsewhere in this text. However, as these principles provide a critical foundation for the study and treatment of developmental disorders, a review of certain aspects of skeletal growth and muscle formation is warranted.

Endochondral ossification is the process by which longitudinal growth of long bones occurs, typically at the physis. In this type of bone formation, chondrocytes secrete a cartilaginous extracellular matrix (ECM), which is subsequently mineralized by osteoprogenitor cells. Osteoclasts then serve to resorb calcified cartilage, and osteoblasts concurrently form mature bone. In this model, therefore, bone replaces cartilage. Consequently, individuals with defects in endochondral ossification will have abnormalities in the growth of long bones.

The physis, or horizontal growth plate located at the ends of immature long bones, is a primary site of endochondral ossification. The physis is divided into several histologic zones, each with its own characteristic function. In the reserve zone, closest to the long bone epiphysis, cells synthesize and store proteoglycans, glycogen, lipids, and other matrix proteins to be used during bony growth. The juxtaposed proliferative zone is the site of cellular proliferation and histologic organization of chondrocytes into columns. The hypertrophic zone lies next to it and is subdivided into the zones of maturation, degeneration, and provisional calcification.

Intramembranous ossification is characterized by the aggregation of undifferentiated mesenchymal cells, which differentiate into osteoblasts. These cells form bone without a cartilage model. Sites of intramembranous ossification include the skull, clavicle, and pelvis. Defective intramembranous ossification is seen in patients with cleidocranial dysplasia, for example. Predictably, these patients typically have craniofacial abnormalities and often complete absence of the clavicles.

The functional unit of skeletal muscle is the myofibril, which is a collection of sarcomeres contained within a myocyte. Each sarcomere is made up of an arrangement of thin and thick filaments, primarily actin and myosin proteins, respectively. Muscle contraction is stimulated by the activation of acetylcholine receptors on the myocyte cell surface at the neuromuscular junction. The resulting cell membrane depolarization triggers a release of calcium contained within the sarcoplasmic reticulum. Intracellular calcium subsequently binds to troponin molecules along the thin filaments, exposing actin and allowing for the formation of cross-bridging between actin and myosin proteins. The thin and thick filaments then slide past one another, resulting in muscular contraction. Clearly, since contraction is mediated by ion flux, the integrity of the muscle cell membrane plays a critical role in skeletal muscle physiology. Disruption of muscle cell membrane physiology (eg, Duchenne muscular dystrophy) will result in abnormal muscle function.

Collagen synthesis is fundamental to the biology of all connective tissues, including bone, cartilage, and muscle. Type I collagen is the predominant organic ECM component of bone, tendon, and ligament and accounts for much of the mechanical properties of these tissues. At a molecular level, collagen is comprised of a triple helix. This molecule undergoes many modifications after the initial transcription and translation of the collagen gene(s), including hydroxylation, glycosylation, cross-linking, extracellular transport, and cleavage of the procollagen protein. Defects in this elaborate mechanism of collagen synthesis (eg, osteogenesis imperfecta) may result in significant disorders of bony, tendinous, and ligamentous structure and function.

Clinical Approach

Skeletal dysplasias cause abnormal growth and usually manifest as short stature; dysmorphic features of the face, hands, and feet; or characteristic deformities of the axial skeleton and extremities. Short stature is a common presentation, and in males an adult height of less than 58 inches should raise suspicion of an underlying disorder of musculoskeletal growth and development. Short stature may have a wide variety of etiologies, including malnutrition, endocrinopathy, skeletal dysplasia, and constitutional short stature. Therefore, a thorough history and physical examination is imperative.

Careful examination of proportionality is equally important because skeletal dysplasias typically produce disproportionate dwarfism. Furthermore, the relative lengths of the segments of the upper and lower extremities should be noted. These characteristics may be objectively recorded via the use of arm span-to-height, humerus-to-radius, or femur-to-tibia ratios.

Dysmorphisms, or morphologic variations of bony and soft-tissue structures, also provide helpful information and insight into skeletal dysplasias. Several conditions are characterized by very distinctive, almost pathognomonic, dysmorphisms; typically, these involve the face, hands, and feet.

Because many skeletal dysplasias are caused by genetic mutations transmitted by Mendelian patterns of inheritance, a complete family history should be obtained in the evaluation of any patient with a suspected disorder of growth and development. A family history of musculoskeletal, endocrine, and metabolic diseases should be obtained as should the height of the patient's parents and siblings, if any. Many of these conditions develop de novo from spontaneous mutations despite no family history of skeletal dysplasias.

Plain radiographs provide additional diagnostic information and guide treatment. Lateral skull, AP pelvis and hip, lateral lumbosacral spine, AP and lateral hand and wrist, and AP and lateral knee radiographs often suffice. These are the anatomic locations that require orthopaedic care, and initial studies are useful to establish a baseline for future comparison. When assessing radiographs, not only should the joint or body part affected be identified but also the specific bony and/or cartilaginous abnormality. The epiphysis, physis, metaphysis, and diaphysis must be independently evaluated to make a precise and accurate diagnosis.

Genetic analysis has become increasingly important in the diagnosis and treatment of skeletal dysplasias because the molecular or genetic bases for many of the skeletal dysplasias have been characterized. In some disease entities, the specific mutation or gene product has not been identified, but gene mapping and chromosomal localization data exist that may aid in diagnosis, treatment, and genetic counseling.

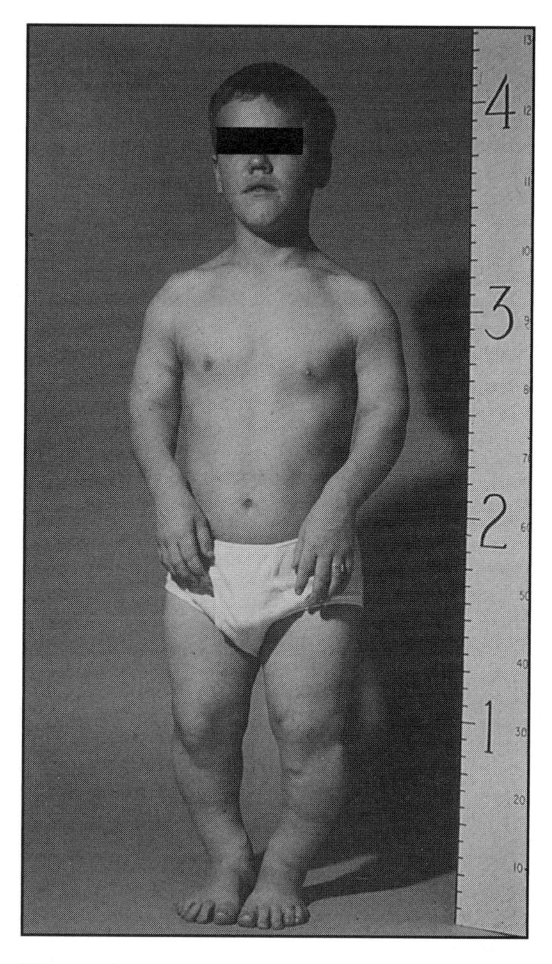

Figure 1
Achondroplasia is characterized by rhizomelic short-limbed short stature, abnormal facies, short iliac wings, horizontal acetabula without early hip arthritis, and an adult height of approximately 48 inches.

(Reproduced from Beals RK, Horton W: Skeletal dysplasias: An approach to diagnosis. ***J Am Acad Orthop Surg*** *1995;3:174-181.)*

Skeletal Dysplasias: Achondroplasia

Epidemiology

Achondroplasia is the most common form of dwarfism, occurring in approximately 1.5 per 10,000 live births. Although it can be transmitted via an autosomal dominant inheritance pattern with near complete penetrance, several population studies have demonstrated that most patients with achondroplasia have no family history of the disorder.[2,3] New mutations are thought to account for most affected individuals. In these cases, advanced paternal age at the time of conception is thought to be a factor.[4]

Etiology

The etiology of achondroplasia has been well established.[5,6] Although the term "achondroplasia," first coined by Parrot in 1878, would suggest an absence of cartilage formation, a genetic defect has been identified—namely, a point mutation in the gene coding for fibroblast growth factor receptor-3 (FGFR-3).[7] FGFR-3 is expressed in all cartilaginous tissue and functions to inhibit chondrocyte proliferation within the proliferative zone of the physis. It is thought that this function regulates bony growth by limiting endochondral ossification in growing bones. Patients with achondroplasia are thought to have constitutively activated FGFR-3 as a result of their genetic mutation. This excess activity of FGFR-3 results in profound inhibition of endochondral ossification and thus short stature.

Clinical Presentation

Classically, patients with achondroplasia are disproportionate rhizomelic dwarfs, with an average adult height of 48 inches (Fig. 1). As the defect in FGFR-3 affects endochondral and not intramembranous ossification, patients with achondroplasia have short ex-

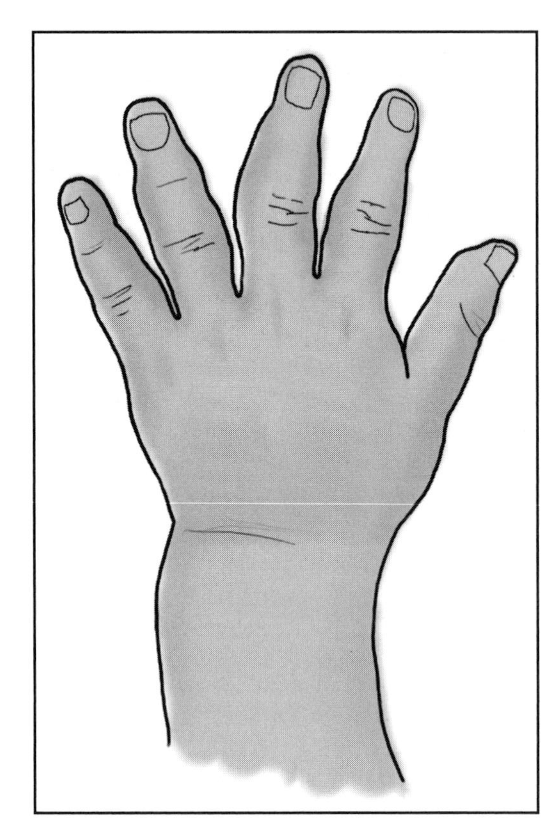

Figure 2
Trident hand characteristic of achondroplasia.

tremities but normal trunk lengths. Furthermore, as appositional growth is similarly unaffected, affected patients will have normal growth of the iliac wings and clavicles.[8]

In addition to shortening of the upper limbs, the shoulders appear broad by virtue of normal clavicular growth. Typically, loss of full elbow extension as a result of developmental abnormalities of the joint is present, as are trident hands, which are so called because in full extension the bases of the digits can touch but the distal tips cannot (Fig. 2). These patients typically are unable to touch the top of their heads or reach below their hips, resulting in difficulties with personal hygiene. In the lower extremities, genu varum, or bowlegs, is common. Loss of full hip extension is also frequently seen. Consequently, most patients with achondroplasia have a circumducting (waddling) gait.

In addition to rhizomelic limbs, patients with achondroplasia demonstrate characteristic dysmorphic features, including a large head, midface hypoplasia and frontal bossing, a depressed nasal bridge, and a prominent lower jaw. Perhaps the most significant deformities in achondroplasia, however, occur in the cervical spine. The lower lumbar spinal canal is often developmentally narrowed, as a result of a shortened interpedicular distance and short, thickened pedicles. Spinal canal narrowing is often compounded by hyperlordosis. Spinal stenosis, when advanced, may result in low back pain, lower extremity paresthesias, weakness, and neurogenic claudication.

Radiographic Findings

Consistent with their clinical appearance, radiographs of the long bones in patients with achondroplasia appear short and thick with metaphyseal flaring. Sites of muscular insertions appear prominent, as appositional bone growth is unaffected. The distal femoral physis exhibits a characteristic inverted-V appearance during childhood. The fibula appears relatively long compared with the tibia.

The pelvis is broad, and the iliac wing takes on a short, square appearance. The acetabulum is horizontal. Lateral radiographs of the skull will exhibit shortening of the base of the skull (that part of the skull formed via endochondral ossification) and narrowing of the foramen magnum. Spine radiographs will demonstrate short pedicles with narrowing of the interpedicular distance. The thoracolumbar kyphosis of infancy will give way to lumbar lordosis with forward inclination of the pelvis.

Treatment

The spinal manifestations of achondroplasia are among the most common developmental deformities treated by orthopaedic surgeons. The spinal stenosis caused by shortened pedicles, narrowed interpedicular distances, lumbar hyperlordosis, and degenerative disk disease may result in low back pain, leg pain (neurogenic claudication), sensory deficits, and weakness. These symptoms affect approximately 50% of adults with achondroplasia.[9] In addition to weight reduction and physical therapy, surgical spinal decompression with laminectomy with or without spinal fusion is often necessary for symptomatic relief. Care must be taken, however, during spinal fusion because the narrowed spine is more prone to neurologic complications during instrumentation of the spinal canal.

Genu varum is another skeletal manifestation of achondroplasia that commonly requires surgical treatment. Symptoms of knee pain accompanied by progressive varus deformity are usually refractory to physical therapy or bracing. When the severity of symptoms warrants surgical treatment, either bony realignment procedures (osteotomies) or growth plate arrest (epiphyseodeses) may result in improved lower extremity alignment and relief of symptoms.

Limb lengthening is advocated by some for achondroplasia and other diseases resulting in short stature.[10,11] While lengthenings in excess of 30 cm have been reported, these procedures have a high complication rate and therefore are not universally accepted.[12]

Muscular Dystrophies: Duchenne Muscular Dystrophy

Epidemiology

Duchenne muscular dystrophy is an X-linked recessive disease characterized by progressive loss of muscle function and replacement of normal muscle by fibrous and fatty tissue. The incidence of Duchenne muscular dystrophy is thought to be 2 in 10,000 live male births. Seventy percent to 80% of cases are inherited, while the remaining 20% to 30% of cases arise from spontaneous genetic mutations.[13]

Etiology

Duchenne muscular dystrophy is caused by an abnormality in the gene that encodes for dystrophin, which is located on the small arm of the X chromosome.[14] Dystrophin is a cell membrane protein that stabilizes the muscle cell membrane and links the cytoskeleton to the ECM.[15] In patients with Duchenne muscular dystrophy, there is a deletion of nucleotides within the dystrophin gene, resulting in a frame-shift mutation and a nonfunctional gene product. The absence of the dystrophin protein results in cell membrane fragility, leakage of intracellular contents into the ECM that is clinically manifest by elevated serum creatine kinase levels, and decreased synthesis of dystrophin-associated proteins, all of which result in functional deficits and muscle cell death.

Clinical Presentation

Muscular dystrophies are disorders of progressive muscle deterioration. Unlike myopathies, which are characterized by functional but nonprogressive abnormalities, patients with muscular dystrophies continue to deteriorate. Unfortunately, significant physical impairment and premature death, usually from respiratory failure, is the norm.

Patients with Duchenne muscular dystrophy have normal skeletal muscle at birth; however, over time, this tissue is replaced with fibrofatty tissue. As a result, patients will typically present well after birth. Parents will report delayed walking, abnormal gait, and clumsiness. Any boy who is unable to ambulate by the age of 18 months should be screened for Duchenne muscular dystrophy by measuring serum creatine kinase levels. At later ages, affected patients will walk with a wide-based gait, will be unable to run, and will not be able to maneuver stairs in a reciprocal fashion.

By the age of 3 to 4 years, patients with Duchenne muscular dystrophy will demonstrate the characteristic pseudohypertrophy of the calves, which is caused by the replacement of muscle with fibrofatty tissue. Lumbar lordosis can also be quite pronounced. Perhaps the most characteristic clinical feature is the way in which these patients arise from the floor to a standing position—the so-called Gowers' maneuver (Fig. 3). As a result of decreased muscle capacity, these patients will roll prone, rise on all four limbs, extend the lower extremities, and walk their hands along the floor and up their legs until they are standing upright. Diagnosis is suggested by elevated serum creatine kinase levels and confirmed by genetic analysis and muscle biopsy, with absence of dystrophin on immunohistochemical analysis.[14,16]

Treatment

Corticosteroid treatment can improve muscle strength and reduce loss of strength over time.[17] Although the mechanism of action is unclear, corticosteroid therapy may provide a beneficial effect by reducing the local inflammatory response to the leakage of intracellular contents into the ECM. Preliminary investigations have suggested that long-term corticosteroid treatment may prolong ambulatory potential, delay pulmonary failure, and reduce the need for surgical treatment.

Musculoskeletal care for patients with Duchenne muscular dystrophy focuses primarily on the prevention of deformity and preservation of function. During childhood, plantar contractures of the ankle develop, in part, to compensate for quadriceps weakness during gait.[18] To prevent excessive contractures, heel cord stretching, ankle-foot

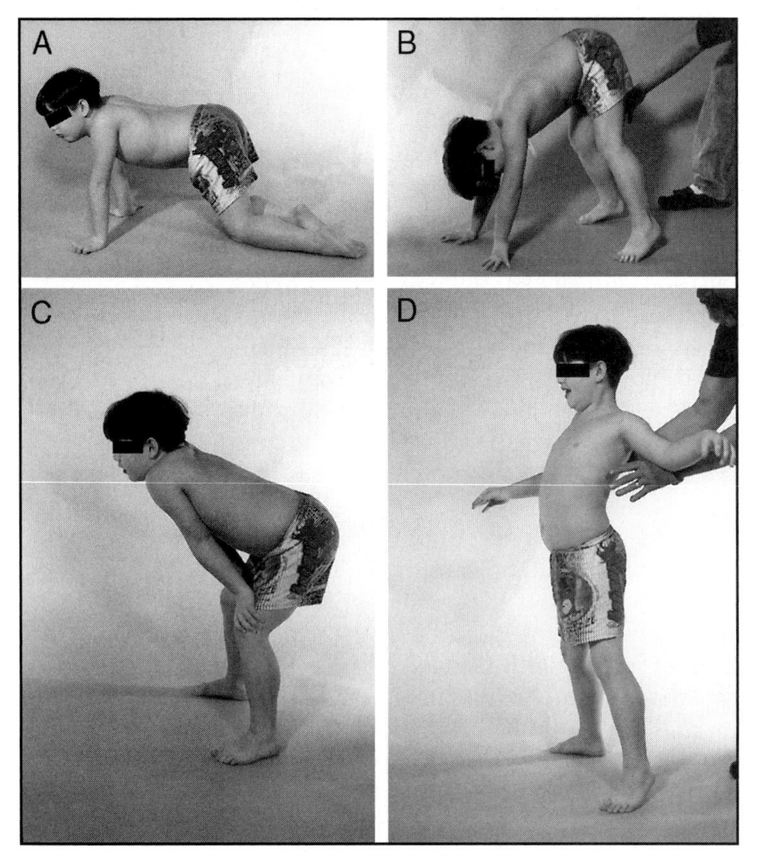

Figure 3
Patient with Duchenne muscular dystrophy demonstrating Gowers' maneuver. **A**, The prone position. **B**, Walking the hands along the floor. **C**, Moving the hands up the thighs to help upright the trunk and augment knee extension. **D**, The upright position.

(Reproduced from Sussman M: Duchenne muscular dystrophy. J Am Acad Orthop Surg 2002;10:138-151.)

orthoses, and even surgical lengthening of the Achilles tendon may be considered. Excessive lengthening should be avoided, however, as the equinus positioning is an adaptive response; overcorrection may compromise the ability to walk.

Over time, the proximal muscles of the lower extremity will progressively weaken. Quadriceps weakness resulting in a knee flexion contracture is particularly debilitating because patients with Duchenne muscular dystrophy usually lock their knees in extension to walk. A contracture will compromise their ability to ambulate. For this reason, an aggressive regimen of stretching, strengthening, and bracing therapy is often recommended. At present, however, there are no scientific data suggesting that these measures prolong the ability to ambulate.

Eventually, patients will become too weak to ambulate independently. At this stage of disease, typically in the second decade of life, a powered wheelchair and adaptive devices should be obtained in the hope of preserving as much independence as possible. Some have advocated surgical treatment of ankle contractures and/or hip and knee flexion contractures; however, because these surgical procedures do not prevent recurrence or improve function, they are of dubious value. Nearly all patients with Duchenne muscular dystrophy will eventually have progressive scoliosis from muscle imbalance, typically between the ages of 11 and 13 years. Significant scoliosis may affect posture, wheelchair fitting, personal hygiene, and ultimately pulmonary function. For this reason, spinal fusion is indicated for any patient with Duchenne muscular dystrophy and a spinal curvature of greater than 20°.[19,20] Delaying surgery until the deformity is more severe may only result in higher complication rates, primarily as a result of diminished pulmonary function.[21]

Connective Tissue Disorders: Osteogenesis Imperfecta

Epidemiology

Osteogenesis imperfecta (OI) is a hereditary disorder of connective tissue caused by mutations in the gene for type I collagen. OI is actually a family of disorders with genetic phenotypic heterogeneity. The incidence of OI is 1 to 5 per 100,000 live births for the nonlethal forms of the disorder; the incidence may be as high as 1 per 50,000 births for the lethal forms.[22] OI is transmitted via Mendelian patterns of inheritance, but new genetic mutations account for a significant number of cases. As would be expected, there is significant ethnogeographic distribution of cases.

Etiology

OI results from mutations in the genes encoding type I collagen, COL1A1 and COL1A2.[23,24] Unlike achondroplasia, in which the genetic mutation is identical among affected individuals, OI is characterized by well over 100 different genetic mutations. In part, this phenomenon may be explained by the shear size of the collagen genes, which are made up of over 50 exons and thus may be affected by a multitude of base-pair substitution, deletion, or insertion mutations. This variation likely accounts for the phenotypic heterogeneity characteristic of OI.

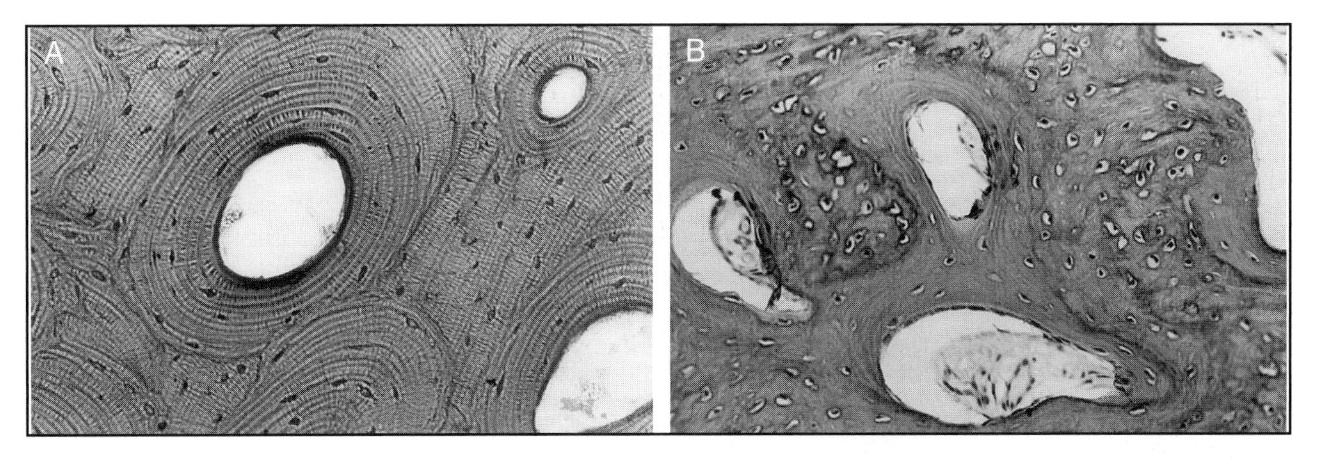

Figure 4
Histologic appearance of a femoral cortex biopsy specimen (hematoxylin-eosin; original magnification, x 65). **A,** Type I OI demonstrates a near-normal lamellar structure. **B,** Type III OI demonstrates a mixed pattern of woven and lamellar bone.

(Reproduced from Kocher MS, Shapiro F: Osteogenesis imperfecta. ***J Am Acad Orthop Surg*** *1998;6:225-236.)*

The COL1A1 and COL1A2 genes encode for the pro-α1(I) and pro-α2(I) chains of the procollagen triple helix. Normally, this promolecule is transported to the extracellular environment, where it is enzymatically processed into the mature collagen molecule. In patients with OI, however, the quantity or quality of the procollagen molecules is severely diminished, resulting in insufficient or defective type I collagen. In general, mutations that affect the quantity of collagen production have milder phenotypic manifestations, whereas those that affect the quality of the collagen molecules have more severe clinical abnormalities.

Clinical Presentation

As might be expected in a disease affecting type I collagen, patients with OI demonstrate abnormalities in connective tissues, including bone (Fig. 4), ligament, dentin, and sclerae. Skeletal manifestations include short stature, pectus excavatum, trefoil-shaped pelvis, and bowing of the long bones. The skull is often deformed, resulting in a broad forehead and triangular-shaped face. Most patients with OI will have spinal deformities, usually resulting from compression fractures and ligamentous laxity. Thoracic scoliosis is the most common deformity.[25] The hallmark feature of OI, however, is bone fragility. Early and frequent fractures of the long bones are seen in many patients. Although these fractures heal, they do so with abnormal tissue, often resulting in malunion and subsequent deformity.

Sites of extraskeletal disease include ligaments, which may result in lax and often hypermobile joints. Frequent dislocations, flat feet, and tendon ruptures may occur. Abnormal collagenous tissue of the eye leads to the blue sclerae characteristic of OI as the uveal pigments and underlying vessels are seen. Teeth are often soft and brown, signifying abnormalities of dentinogenesis. As a result, dental caries are common, and the teeth are often short and weak. Finally, the skin is usually thin, translucent, and easily stretched because of the absence of normal collagen. The vascular and auditory systems may also be affected, manifest as aortic root dilation and hearing loss, respectively.

There is significant phenotypic heterogeneity of OI, depending on the nature of the genetic mutation. This has prompted a number of classification systems for OI, the most commonly used being that of Sillence.[26] Sillence type I OI is the most common form and has the mildest clinical manifestations. Inheritance is autosomal dominant. This group is often subdivided into types A and B, denoting the absence or presence of dentin abnormalities, respectively. Patients with type I OI have blue sclerae, exhibit long bone fractures after walking age, and have normal stature and life expectancy.

Sillence type II OI is the lethal form. Affected individuals are stillborn or die within the first weeks of life, usually because of respiratory failure or intracranial hemorrhage. Multiple rib and long bone fractures are seen and sclerae are blue. This form of OI is autosomal recessive.

Sillence Type III OI is a severe, nonlethal form (Fig. 5). Sclerae are of normal color, but

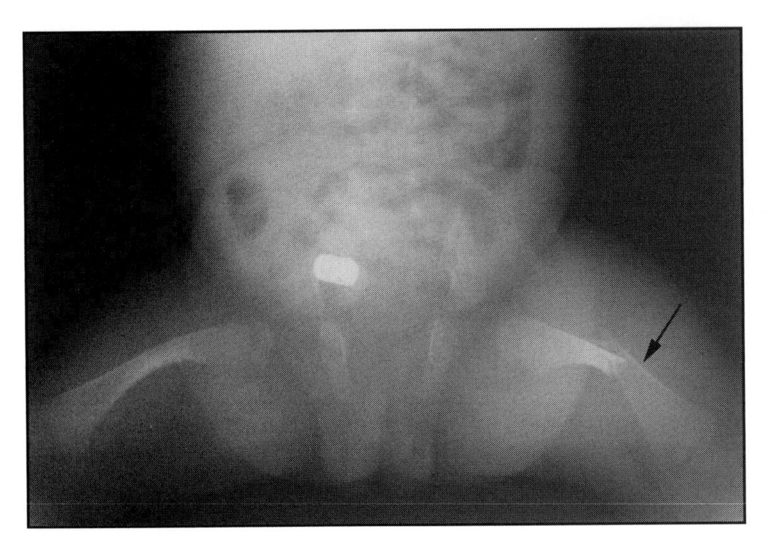

Figure 5
Radiograph of a newborn with type III OI. Note the osteopenia and a fracture of the left femur (arrow).

(Reproduced from Kocher MS, Shapiro F: Osteogenesis imperfecta. J Am Acad Orthop Surg 1998;6:225-236.)

multiple fractures are seen at birth. Stature is short and life expectancy is reduced, with respiratory failure, intracranial hemorrhage, and basilar invagination accounting for premature mortality. Most cases are autosomal dominant.

Sillence type IV OI, the intermediate form, is the most rare. Sclerae are normal, adult height is variably affected, and abnormal dentition and bony fractures are common. Life expectancy varies depending on the severity of disease.

Treatment

Ultimately, treatment of OI will involve correcting the underlying genetic defect via gene therapy. Currently, there have been no human studies of gene-based therapies for OI. Nonetheless, there have been promising in vitro and animal studies, the results of which may lead to successful strategies in the future. In one such study, transgenic mice were created that expressed an anti-sense RNA for the pro-α1 (I) collagen molecule.[24] When these mice were crossed with those that expressed a phenotype similar to OI, mortality was reduced and bony strength improved. Further investigation may eventually result in treatment modalities that correct the fundamental defect of OI.

Until such gene-based therapies are available, treatment will remain primarily supportive, with an emphasis on injury prevention and maximizing function. Fracture treatment should be as minimally debilitating as possible. Most fractures may be managed through nonsurgical means because patients with OI retain the capacity to heal bony injuries. However, prolonged immobilization and general deconditioning should be avoided as much as possible. The use of braces, supportive devices, and physical therapy is encouraged during recovery, both for bony healing and further injury prevention. Medical therapies, ranging from fluoride and calcitonin to bisphosphonates and growth hormones, have been proposed. To date, however, there are no data suggesting that such medical treatments are effective in patients with OI.

Surgical treatment of fractures in OI is typically reserved for patients older than 5 years with recurrent fractures or symptomatic bony deformity. Long-term use of orthoses is common to ensure bony healing and prevent recurrence of deformity. Scoliosis is common, and surgical treatment with arthrodesis should be considered for progressive spinal curves or curves greater than 40°.[27]

Key Terms

Achondroplasia The most common form of congenital dwarfism; characterized by misshapen epiphyses and deformed long bones

Disproportionate dwarfism Short stature in which the extremities are relatively more (or less) affected than the trunk

Duchenne muscular dystrophy An X-linked recessive disease characterized by progressive loss of muscle function and replacement of normal muscle by fibrous and fatty tissue

Dwarfism Short stature characterized by an adult height of less than 58 inches in males or a standing height below the third percentile for age

Dysmorphisms Morphologic variations of musculoskeletal appearance

Dysostosis Conditions of abnormal cartilage ossification or bony remodeling

Dysplasia A broad term that describes a condition affecting growth or development in which the primary defect is intrinsic to affected tissue

Dystrophy A condition resulting from defective or faulty nutrition, broadly construed to include nourishment of tissue by all essential substances, including those normally manufactured by the body itself

Intramembranous ossification Bone formation in the absence of a cartilage model characterized by the aggregation of undifferentiated mesenchymal cells, which differentiate into osteoblasts

Osteogenesis imperfecta A hereditary disorder of connective tissue caused by mutations in the gene for type I collagen

Physis The horizontal growth plate located at the ends of immature long bones; a site of endochondral ossification

Proportionate dwarfism Short stature in which both the trunk and extremities are equally affected

References

1. Andersen PE Jr, Hauge M: Congenital generalised bone dysplasias: A clinical, radiological, and epidemiological survey. *J Med Genet* 1989;26:37-44.
2. Murdoch JL, Walker BA, Hall JG, et al: Achondroplasia: A genetic and statistical survey. *Ann Hum Genet* 1970;33:227-244.
3. Oberklaid F, Danks DM, Jensen F, et al: Achondroplasia and hypochondroplasia: Comments on frequency, mutation rate, and radiological features in skull and spine. *J Med Genet* 1979;16:140-146.
4. Wynne-Davies R, Walsh WK, Gormley J: Achondroplasia and hypochondroplasia: Clinical variation and spinal stenosis. *J Bone Joint Surg Br* 1981;63:508-515.
5. Muenke M, Schell U: Fibroblast-growth-factor receptor mutations in human skeletal disorders. *Trends Genet* 1995;11:308-313.
6. Shiang R, Thompson LM, Zhu YZ, et al: Mutations in the transmembrane domain of FGFR3 cause the most common genetic form of dwarfism, achondroplasia. *Cell* 1994;78:335-342.
7. Parrot J: Sur la malformation achondroplasique et le Dieu Phtah. *Bull Soc d'Antrop de Par* 1878;1: 296-308.
8. Ponseti IV: Skeletal growth in achondroplasia. *J Bone Joint Surg Am* 1970;52:701-716.
9. Kahanovitz N, Rimoin DL, Sillence DO: The clinical spectrum of lumbar spine disease in achondroplasia. *Spine* 1982;7:137-140.
10. Aldegheri R, Dall'Oca C: Limb lengthening in short stature patients. *J Pediatr Orthop B* 2001;10:238-247.
11. Saleh M, Burton M: Leg lengthening: Patient selection and management in achondroplasia. *Orthop Clin North Am* 1991;22:589-599.
12. Correll J: Surgical correction of short stature in skeletal dysplasias. *Acta Paediatr Scand Suppl* 1991;377:143-148.
13. Hoffman EP: Muscular dystrophy: Identification and use of genes for diagnostics and therapeutics. *Arch Pathol Lab Med* 1999;123:1050-1052.
14. Hoffman EP, Fischbeck KH, Brown RH, et al: Characterization of dystrophin in muscle-biopsy specimens from patients with Duchenne's or Becker's muscular dystrophy. *N Engl J Med* 1988;318:1363-1368.
15. Matsumura K, Lee CC, Caskey CT, et al: Restoration of dystrophin-associated proteins in skeletal muscle of mdx mice transgenic for dystrophin gene. *FEBS Lett* 1993;320:276-280.
16. Pasternak C, Wong S, Elson EL: Mechanical function of dystrophin in muscle cells. *J Cell Biol* 1995;128:355-361.
17. Griggs RC, Moxley RT III, Mendell JR, et al: Duchenne dystrophy: Randomized, controlled trial of prednisone (18 months) and azathioprine (12 months). *Neurology* 1993;43:520-527.
18. Khodadadeh S, McClelland MR, Patrick JH, et al: Knee moments in Duchenne muscular dystrophy. *Lancet* 1986;2:544-545.
19. Bridwell KH, Baldus C, Iffrig TM, Lenke LG, Blanke K: Process measures and patient/parent evaluation of surgical management of spinal deformities in patients with progressive flaccid neuromuscular scoliosis (Duchenne's muscular dystrophy and spinal muscular atrophy). *Spine* 1999;24:1300-1309.

20. Sussman MD: Advantage of early spinal stabilization and fusion in patients with Duchenne muscular dystrophy. *J Pediatr Orthop* 1984;4:532-537.
21. Miller F, Moseley CF, Koreska J, Levison H: Pulmonary function and scoliosis in Duchenne dystrophy. *J Pediatr Orthop* 1988;8:133-137.
22. Byers PH, Steiner RD: Osteogenesis imperfecta. *Annu Rev Med* 1992;43:269-282.
23. Kivirikko KI: Collagens and their abnormalities in a wide spectrum of diseases. *Ann Med* 1993;25: 113-126.
24. Prockop DJ, Kuivaniemi H, Tromp G: Molecular basis of osteogenesis imperfecta and related disorders of bone. *Clin Plast Surg* 1994;21:407-413.
25. Benson DR, Newman DC: The spine and surgical treatment in osteogenesis imperfecta. *Clin Orthop* 1981;159:147-153.
26. Sillence DO, Senn A, Danks DM: Genetic heterogeneity in osteogenesis imperfecta. *J Med Genet* 1979;16:101-116.
27. Hanscom DA, Bloom BA: The spine in osteogenesis imperfecta. *Orthop Clin North Am* 1988;19:449-458.

Chapter 20

Chronic Pain Syndromes

Pain is one of the most common complaints in clinical medicine. Pain, however, has a purpose. Acute pain alerts us to the many hazards in our environment and motivates us to withdraw from danger. People who are not able to feel pain are prone to injury. This is seen, for example, in the joints of patients with neuropathy caused by diabetes mellitus, in which the lack of protective sensation leads to rapid joint destruction. Pain is a basic part of normal life and serves as an important factor in our cognitive development. Our experiences, culture, personality, and a host of psychosocial factors influence our perception of pain. Pain is a source of information and should be appreciated for its purpose.

Problems arise when pain is no longer protective and assumes a persistent pathologic state. Because of the subjective nature of pain, definitions of chronic pain syndromes may be vague. In its most general form, chronic pain refers to pain that endures after the time normally required to heal the underlying injury. Chronic pain has been defined as lasting anywhere from 6 weeks to 6 months or more. Although pain may be an indication of a serious underlying condition, chronic pain itself can become a disease state.[1]

Chronic pain has been redefined over the past 20 years, with more attention being given to states in which pain is the predominant symptom but for which no underlying organic condition can be identified. Patients with chronic pain syndromes can be undertreated because physicians may attribute the pain to abnormal mental processes. Chronic pain often has a significant impact on a patient's relationship with family, friends, and employers. The disruption in their lives may leave patients angry and afraid that they are no longer able to continue their role as partner, friend, or employee. A significant proportion of patients with chronic pain syndromes have an associated psychiatric diagnosis; however, it is difficult to determine whether psychologic symptoms are effects or causes of pain syndromes. After all, living with chronic pain is apt to induce mental anguish.

Chronic pain syndromes are often classified as diffuse or regional pain syndromes. This chapter describes the most common of each of these types of chronic pain: fibromyalgia and complex regional pain syndrome, respectively. Each condition includes a descriptive example. When treating patients with chronic pain, physicians are professionally obligated to determine whether there is an underlying disorder. Thus, patients with chronic pain are best approached in a systematic manner. Questions concerning how the pain began, how the pain is described, the intensity of the pain, known aggravators and alleviators of the pain, and how symptoms impact the patient's life can help to identify the cause. The descriptions of these conditions illustrate many of the features common to most chronic pain syndromes.

When examining a patient whose sole complaint is pain, two caveats are in order. The first is that there may be no structural disease present. If this is the case, then the patient may be malingering (feigning illness). Physicians treating musculoskeletal disease must be aware of the possibility of malingering, although the patient should be given the benefit of the doubt, as there is no objective way to verify symptoms. The second caveat is that patients with chronic pain can have objective but unrelated musculoskeletal ab-

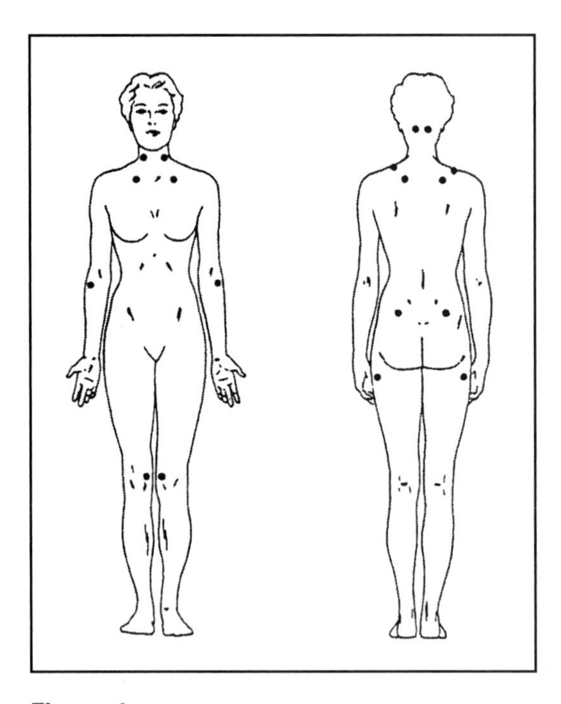

Figure 1
Anterior and posterior tender points associated with fibromyalgia.

(Reproduced with permission from the Arthritis Foundation, Atlanta, GA.)

normalities. Herniated spinal disks and tears of the menisci and rotator cuff are common MRI findings, but these may not be the source of the patient's complaints. Careful correlation of data obtained from the history, physical examination, and imagining studies is essential, particularly in patients with chronic pain syndromes.

Fibromyalgia

Fibromyalgia is a syndrome of diffuse musculoskeletal pain, sleep disturbances, and exhaustion. Fibromyalgia is now recognized as the most common cause of generalized musculoskeletal pain in women 20 to 55 years of age, yet because it defies objective identification, it remains a controversial diagnosis. The physical examination findings of patients with fibromyalgia are typically normal beyond disproportionate tenderness to palpation. Because results of laboratory tests and imaging studies are also usually normal, an organic cause of fibromyalgia has been questioned. Whether the physical symptoms found in fibromyalgia are a manifestation of psychological processes or whether the organic abnormalities that give rise to this condition are too subtle to be detected continues to be debated.

Definition

Before the term fibromyalgia was coined, similar syndromes were described as fibrositis. Despite the connotation of a term ending in "itis," histologic evidence of inflammation of either muscle or connective tissue has never been reproducibly reported. In the late 1980s, the Fibromyalgia Multicenter Criteria Committee was formed under the sponsorship of the American College of Rheumatology (ACR) to create a standardized approach to the clinical diagnosis and treatment of fibromyalgia. In 1990, the ACR published diagnostic criteria for fibromyalgia, foremost among them being a history of chronic widespread pain.[2] Widespread pain is defined as pain on both sides of the body, above and below the waist, as well as pain in the cervical spine, anterior chest, thoracic spine, and/or lower back. According to the ACR criteria, fibromyalgia can be diagnosed only if patients also report pain on palpation in 11 of 18 tender point sites[2] (Fig. 1). Other examiners use control points, such as the midforearm and over the thumbnail, as points of reference that should not be as tender as the diagnostic points.

The ACR classification criteria for fibromyalgia and the reproducibility of symptoms by different observers have been validated in several international studies.[2] Many symptoms are known to be associated with fibromyalgia, including fatigue (81%), stiffness (77%), sleep disturbance (75%), fluctuating symptoms with changes in the weather (67%), paresthesias (63%), headaches (53%), and anxiety (48%). Moreover, ACR classification criteria have been shown to be 88% sensitive and 81% specific in distinguishing fibromyalgia from other chronic pain syndromes.[2]

It is essential to note that the ACR criteria for diagnosing fibromyalgia were developed in order to define a reproducible population of patients to be studied. That is, the ACR did not prove that fibromyalgia exists; rather, it provided classification guidelines that allow physicians to communicate. This is the rationale of the *Diagnostic and Statistical Manual* used by psychiatrists, for example, and it is an accepted approach to the description of disease. Without criteria to define a disease, it is difficult, if not impossible, to conduct the research needed to validate it.

Epidemiology

Fibromyalgia consists of two cardinal features: diffuse musculoskeletal pain and a heightened sense of tenderness in certain areas of the body. There are no known objective markers of the disease. Several other features are associated with this condition, including sleep disruption, stiffness, and fatigue. For reasons not understood, this condition affects women 10 times more often than men. The estimated prevalence is 3.4% among women and 0.5% among men.[3] Although fibromyalgia most often affects women who are 20 to 55 years of age, it has been reported in patients up to 79 years of age as well. No genetic predisposition has ever been proven. Over 75% of patients with this diagnosis were examined by a physician in the preceding 6 months, and 66% were taking pain medications for their symptoms. Almost 20% had applied for disability benefits, and approximately 7% had actually received such benefits.[4]

Clinical Presentation

A typical patient with fibromyalgia is a 20- to 55-year-old woman who reports chronic aching pain and stiffness. These pains frequently involve the entire body, but most of the pain occurs around the neck, shoulders, low back, and hips. The intensity of the pain may vary from day to day. The patients may have difficulty distinguishing joint pain from muscle pain and may report a sense of swelling in the affected areas. Poor sleep, numbness, and a persistent headache also are common complaints. Patient may associate the onset of their symptoms with a traumatic event or flulike illness. They may have changed employers or stopped work altogether as a result of their symptoms. Other than excessive pain to palpation in the characteristic distributions, physical examination findings and laboratory and radiograph test results are essentially normal.

Several conditions are commonly confused with fibromyalgia. Some of these include depression, irritable bowel syndrome, migraine, and chronic fatigue syndrome; a psychiatric diagnosis was found in 16.9% of patients with chronic widespread pain.[5,6] Most patients with chronic fatigue syndrome meet the tender point criteria for fibromyalgia, and 70% of patients with fibromyalgia meet the criteria for chronic fatigue syndrome.[7] Fibromyalgia has clinical similarities to other, better-defined, rheumatologic conditions, such as rheumatoid arthritis, Sjögren's syndrome, and systemic lupus erythematosus, with 22% of patients with systemic lupus erythematosus meeting the ACR criteria for fibromyalgia.[8]

Potential Etiologic Mechanisms

As many as 50% of patients with fibromyalgia can cite an event that triggered the onset of symptoms. Because of this pattern, some have suggested that fibromyalgia may occur as a result of trauma, surgery, or a medical illness. Although the etiology remains unknown, several pathophysiologic models have been proposed and are supported by studies used to explain this condition. The proposed mechanisms can be categorized as follows: sleep, muscle, neuroendocrine, neurotransmitter, and cerebral blood flow.

More than 75% of patients with fibromyalgia report a nonrestorative sleep pattern. Sleep abnormalities have been investigated as both a primary cause and a secondary finding for the past 20 years. It has been shown in normal subjects that disruption of non–rapid-eye-movement (REM) sleep can result in symptoms similar to fibromyalgia.[9] Alpha-wave intrusion into deeper, non-REM stages of sleep have been demonstrated in patients with fibromyalgia. However, these findings lack both sensitivity and specificity.

Many studies have suggested histologic and histochemical markers for fibromyalgia. Some have reported abnormalities in muscle fiber specimens obtained from biopsies of tender point areas in patients with fibromyalgia. Others reported evidence of local hypoxia and decreased high-energy phosphate levels in the muscle tissue of patients with fibromyalgia; however, these findings have not been reproduced by researchers looking for biochemical and histologic markers of fibromyalgia. It has been suggested that biopsy specimen abnormalities may be the result of deconditioned muscles, which is common in patients with fibromyalgia because abnormal pain leads to inactivity.

Although there is no known etiologic role for some of the viruses associated with chronic fatigue syndrome and other chronic conditions, such as Epstein-Barr virus and parvovirus, some infections have been associated with triggering fibromyalgia. One study found 8% of patients with Lyme disease later had fibromyalgia that lasted months to years despite adequate treatment of the infection.[10]

One subjective finding that has been reproducibly demonstrated is a qualitatively altered sense of nociception (pain sensing). Patients with fibromyalgia are usually noted to be overly sensitive to both pain and auditory stimuli. Whether this process is a central or peripheral abnormality is unknown, but studies have suggested that the increased pain noted on tender points examination may result as a manifestation of a central nervous system alteration in the processing of nociceptive stimuli.

Depression is more prevalent in patients with fibromyalgia than in the general population. Although it is often difficult to distinguish depression as a comorbidity versus a secondary complication of living with chronic pain, there is little disagreement that the psychologic state of the patient plays a large role in both the development and treatment of fibromyalgia. One study demonstrated increased rates of lifetime sexual abuse (17% versus 6%), physical abuse (18% versus 4%), and drug abuse (16% versus 3%) in women with fibromyalgia who were compared with controls.[11] The authors of this study suggest that these factors may affect the expression and persistence of fibromyalgia in adult life. Patients who attribute their pain to a traumatic or inciting event are less likely to show clinical improvement. This may be a consequence of the belief that people who are in pain need to find meaning for their suffering or the observation that people who are in pain may better recall traumatic events.

Treatment and Prevention

General Approaches

Treating fibromyalgia can be difficult and unsatisfying for both the patient and the physician, and no known methods of preventing fibromyalgia exist. One study reported that less than 50% of patients with fibromyalgia reported adequate relief of their symptoms after treatment.[12] One 3-year study found that only 3% of patients reported complete remission of all pain symptoms after treatment.[13] As a general rule, when a given medical condition is treated in many different ways, it is likely that no single treatment is highly effective. This is the case with fibromyalgia; many different therapies have been tried and none have demonstrated a dramatic clinical effect.

A supportive attitude is essential in treating patients with fibromyalgia. Physicians should emphasize the benign nature of the disorder, stressing that fibromyalgia is neither life threatening nor a deforming disease. Educating the patient about the natural history of the disorder and the current understanding of possible pathologic mechanisms can help patients understand and develop reasonable expectations about their treatment and prognosis. As when treating any disease, relating possible mechanisms to the proposed treatments may put patients more at ease. Physical therapy and exercise, for example, can be supported by discussions concerning evidence of decreased blood flow and changes in muscles that are found with deconditioning.

Pharmacologic Approaches

Although no drugs are currently indicated specifically for the treatment of fibromyalgia, antidepressants and analgesics are commonly prescribed. Studies report that about 25% of patients with fibromyalgia demonstrated significant improvement in their symptoms with use of tricyclic antidepressants (TCAs), such as amitriptyline.[14] Amitriptyline is usually taken a few hours before the patient goes to bed in order to take advantage of its sedating properties. Physicians should prescribe the lowest possible dose to avoid adverse effects.

It is unclear whether selective serotonin reuptake inhibitors (SSRIs) provide any clinical benefit. Some studies have found little benefit, whereas others have found the effects of SSRIs to be similar to those of TCAs. SSRIs have also been shown to be effective when used in conjunction with a TCA taken at night, but their effectiveness may be limited to the treatment of the comorbid depression that is so common in fibromyalgia.

Analgesics have been used in the treatment of fibromyalgia for many years. Although nonsteroidal anti-inflammatory drugs are prescribed by physicians for almost all patients with fibromyalgia, they are probably no more effective than a placebo. Acetaminophen and tramadol hydrochloride, however, have been shown to be of some benefit. There are no clinical data available concerning the effectiveness of narcotics for the treatment of fibromyalgia, but consensus suggests that prolonged use of narcotics should be avoided. Injections of lidocaine into tender points has been shown to improve symptoms; however, because this improvement can also be demonstrated with an injection of saline solution, it is thought that the efficacy of this therapy may be the result of a placebo effect or from a reaction that takes place after inserting a needle into the affected area.

Nonpharmacologic Approaches

Several nonpharmacologic approaches have been studied in the treatment of fibromyalgia. Randomized, clinical trials have shown that cardiovascular fitness training, regional sympathetic blocks, electromyographic biofeedback, hypnotherapy, and electroacupuncture provide some benefit. Lifestyle and behavior modification is also commonly used to improve outcomes.[15] Since prognosis has been shown to improve with the development of favorable coping strategies, physical therapy and exercise are often used to treat fibromyalgia, as are cognitive behavioral therapy and stress management programs. Aerobic exercise performed three times weekly has been shown to reduce symptoms at tender points. Two thirds of patients have tried at least one form of alternative medicine in their search for an effective treatment. In one study, 42% of patients receiving acupuncture noted no significant benefit; 39% had a satisfactory benefit; and 19% had a greater benefit than expected.[16] There are no trials available that have compared alternative therapies with more traditional approaches.

Complex Regional Pain Syndrome

Complex regional pain syndrome, occasionally still referred to as reflex sympathetic dystrophy or RSD, is a disorder of the extremities characterized by severe and continuous pain, decreased range of motion, and demineralization of adjacent bony structures. Vasomotor instability also occurs, resulting in swelling and changes in skin color and temperature. The symptoms are typically not confined to the distribution of a single peripheral nerve or nerve root.

The presentation of complex regional pain syndrome changes over time. Warmth of the skin characterizes the early stages. Thereafter, a relative coolness and, as more trophic changes begin to occur, decreased mobility, contractures, and chronic pain develop. The affected area typically enlarges after onset, beginning focally but often spreading to the entire distal limb.

Historically, complex regional pain syndrome has had many synonyms, including algodystrophy, causalgia, shoulder-hand syndrome, Sudeck's atrophy, and reflex sympathetic dystrophy. In 1994, the International Association for the Study of Pain proposed diagnostic criteria for this disorder, and the World Health Organization renamed it complex regional pain syndrome. It was subdivided into two types, depending on whether a definable nerve lesion was present. As a result, complex regional pain syndrome type 2 is diagnosed when symptoms are attributable to a nerve lesion. Complex regional pain syndrome type 1 is diagnosed when comparable symptoms are present but there is no nerve lesion present.

Even with this attempt to precisely characterize complex regional pain syndrome, no unequivocal diagnostic criteria have been established. As with fibromyalgia, no specific pathologic mechanism has been identified as the source. The role of the autonomic nervous system in the development of this characteristic pain remains unproven. Moreover, psychiatric disease is also found in patients with complex regional pain syndrome.

Complex regional pain syndrome is difficult to diagnose because of its many variations and its evolution. In one study, the median time for diagnosis after presentation was 12 weeks, with 15% of patients waiting as long as 12 months before a diagnosis was made. Even with increased attention accorded to complex regional pain syndrome, this condition is not often recognized or diagnosed, especially among children. Children are less likely to provide a history of trauma associated with the onset of this condition, and their functional outcomes are better than those of the adult population.[17]

The scarcity of objective findings in some patients with complex regional pain syndrome makes the clinical history even more important. Characteristic descriptions of the pain and its distribution can provide helpful clues. Laboratory test results of patients with complex regional pain syndrome are usually normal. Autonomic and radiographic test results and response to therapies thought to be specific to this condition all reinforce the diagnosis.

Epidemiology

Mitchell and associates[18] first described a form of complex regional pain syndrome during the Civil War. As the disorder began to be recognized, it was noted that the symptoms tended to develop out of proportion to the magnitude of the inciting trauma. It has been shown to occur in about 1 of every 2,000 instances of fracture, blunt trauma, or surgery involving an extremity.[19] Complex regional pain syndrome has also been recognized in patients who have had myocardial infarction or a cerebrovascular accident or stroke. It may affect both sexes, but it usually affects women more frequently than men.

Clinical Presentation

A typical patient with complex regional pain syndrome is a 20- to 55-year-old woman who seeks medical attention after sustaining a wrist fracture in a fall. There are no other detectable injuries and no evidence of a peripheral nerve injury. Approximately 1 week later, the patient may report a change in symptoms, describing relief of the acute pains of the fracture but onset of a constant, diffuse, burning pain. The patient may note swelling and a sensation of increased warmth in the affected arm, along with changes in her fingernails and increased hair growth on the affected side. She may report that pressure applied to the affected extremity aggravates the pain, whereas resting and cooling reduces the symptoms. The fingers of the affected extremity may feel weak and be difficult to move. The affected extremity is warmer than the opposite extremity, and some swelling is apparent. Results of the physical examination will not correspond with those of a patient with a single nerve lesion. Complex regional pain syndrome can eventually be diagnosed after combining information from this patient's clinical presentation with the results of bone scanning that demonstrate diffuse uptake of tracer around the distal joints on the affected side (Fig. 2).

Complex regional pain syndrome may affect either the upper or lower extremities; however, a patient in whom both arms and legs are affected would be unusual. Three stages of this disease have been described, but not all patients progress through all three stages (Fig. 3). Pain in the limb, edema, and vasomotor changes characterize the early (acute) stage, which may last for several months. Some patients describe the pain as a burning or throbbing sensation, others describe more of a chronic aching feeling. The extremity may be sensitive to touch or cold during this period. It also may differ in color or temperature compared with the unaffected side. Some patients report a difference in sweating as well as increased nail and hair growth on the affected side. Radiographs of the extremity are usually normal during this stage.

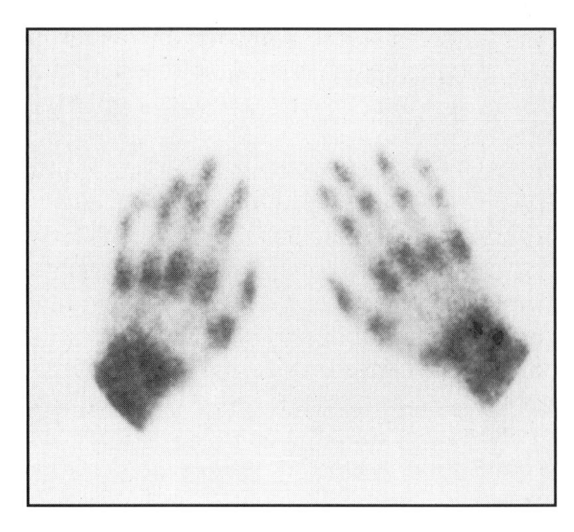

Figure 2
A positive delayed-phase bone scan of a patient with bilateral complex regional pain syndrome. Note the uptake of tracer near the joints.

*(Reproduced from Gellman H, Nichols D: Reflex sympathetic dystrophy in the upper extremity. **J Am Acad Orthop Surg** 1997;5: 313-322.)*

The middle (dystrophic) stage of complex regional pain syndrome usually develops 3 to 6 months after the onset of symptoms. This stage is characterized by soft-tissue edema, which likely results from increased regional sympathetic activity. Burning pain typically persists during this period. Changes in skin temperature in the affected extremity may be noted. Unusual nail and hair growth ceases, and hair loss may occur during this period. A thickening of the skin and articular soft tissues may also begin at this time, resulting in the limited movement of involved joints. Some patients with this syndrome have described the skin as undergoing brawny changes. Early atrophy of the surrounding muscles has also been described during this period.

The late (atrophic) stage of complex regional pain syndrome is the most severe. Continued changes in hair growth and nail formation may worsen as hair falls out and nails may become rigid and brittle. Some patients report improvements in their symptoms of pain, but they begin to note significant limitations in their ability to move the extremity as fibrosis and contractures develop. Capsular retractions have been described that may leave a joint immobilized and dysfunctional. These contractures may involve the fingers as well. Radiographs may show severe demineralization.

Diagnostic Imaging

Radiographs can assist in making the diagnosis of complex regional pain syndrome. A characteristic demineralization takes place during the later stages. A patchy demineralization has been described in the early stages but is not always observed. Later radiographs may reveal severe osteoporosis in the bones of the affected limb. Radiographs may also demonstrate destruction of joints and adjacent bones, as well as subluxation and evidence of new bone formation. Because no diagnostic gold standard exists, sensitivity and specificity of imaging studies cannot be stated.

MRI may be useful in all stages of complex regional pain syndrome, but especially in the acute and dystrophic stages. Soft-tissue edema, contrast enhancement of the involved tissues, and muscle atrophy may be seen. CT may show focal osteopenia.[20] Bone scanning can provide additional useful information. After the onset of symptoms, it may reveal decreased perfusion in the affected areas. Bone scanning is thought to be most useful in the early stages of this disorder.

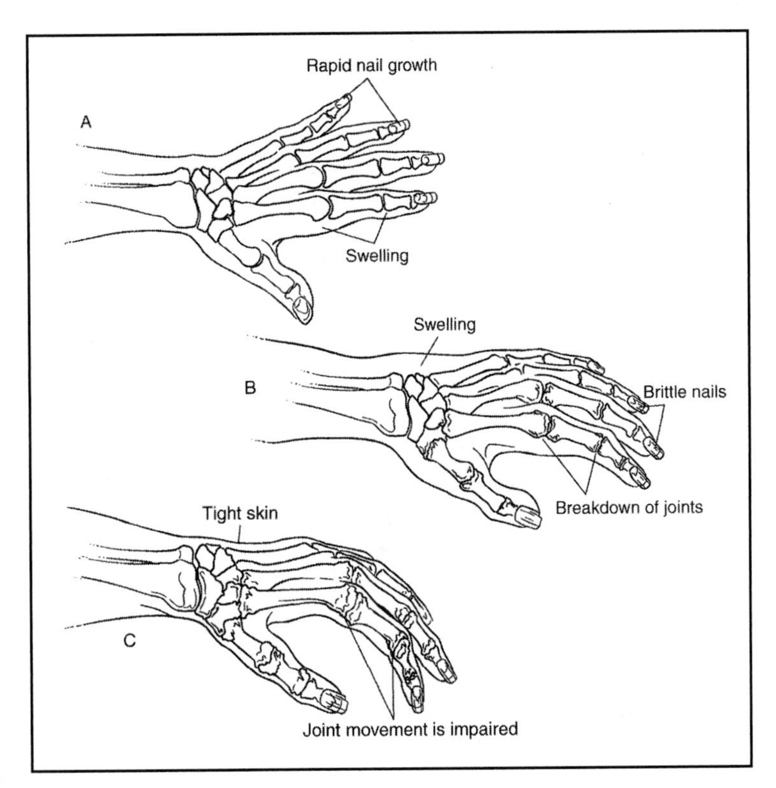

Figure 3
Clinical presentation of complex regional pain syndrome. **A**, Early (acute) stage 1. **B**, Middle (dystrophic) stage 2. **C**, Late (atrophic) stage 3.

(Reproduced from Greene WB (ed): ***Essentials of Musculoskeletal Care****, ed 2. Rosemont, IL, American Academy of Orthopaedic Surgeons, 2001, p 66.)*

Other Testing

Autonomic testing has been shown to be useful in the diagnosis of complex regional pain syndrome and may help identify patients who will respond to some therapies directed at the autonomic nervous system. Examples of autonomic findings that can be measured are the resting sweat output and the quantitative sudomotor axon reflex test.

A final consideration that many physicians incorporate into their diagnosis is how the patient responds to therapy. If the role of the autonomic nervous system in the pathophysiology of complex regional pain syndrome is accepted, then it follows that treatments directed at blocking sympathetic nerve transmission should improve symptoms. As a result, regional sympathetic nerve blocks can be used both to alleviate symptoms and support the diagnosis of complex regional pain syndrome.

Potential Etiologic Mechanisms

The pathophysiology of complex regional pain syndrome is not completely understood. Most episodes occur after some provoking event, although some studies suggest that as many as 35% of patients have no such event identified.[21] Some common precipitating events include soft-tissue injuries (40%), fractures (25%) (Fig. 4), myocardial infarction (12%), and cerebrovascular accident or stroke (3%).[22] The relationship of myocardial infarction and stroke with complex regional pain syndrome has been further illuminated by the decreased rate of complex regional pain syndrome among patients with these conditions who receive early ambulation and passive arm stretching during prolonged treatment with intravenous medications.

The role of the autonomic nervous system in the pathophysiology of complex regional pain syndrome continues to be debated. Many of the clinical findings, such as temperature changes of an extremity, cyanosis, mottling, hyperhydrosis, and abnormal hair growth, can be caused by derangements of the autonomic nervous system. The skin changes that occur may result from an increase in the sensitivity of skin microvessels to circulating catecholamines, a mechanism that has been supported by studies of pa-

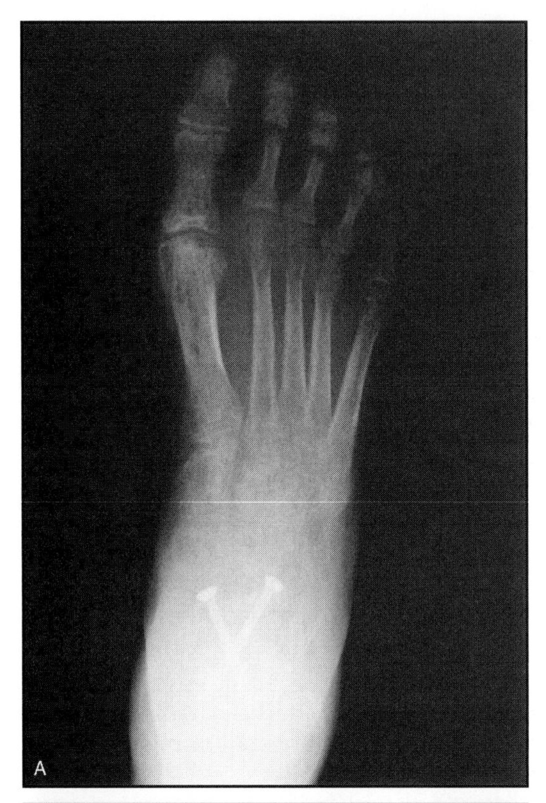

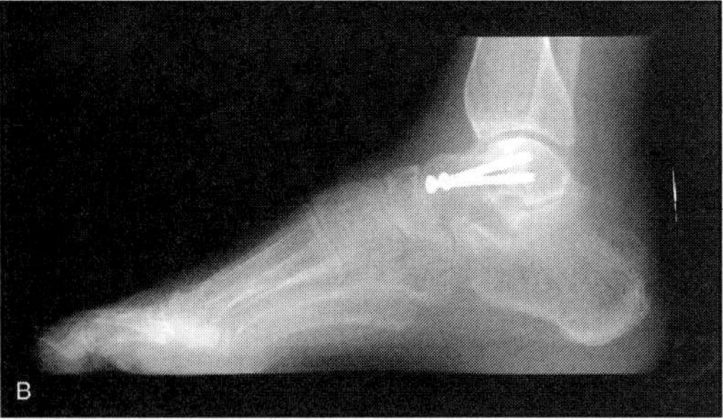

Figure 4
AP (**A**) and lateral (**B**) radiographs of the right foot of a middle-aged woman 3 months after open reduction and fixation for a talar neck fracture. She had pain in the entire foot and stiffness of the ankle and all joints. She could not tolerate bearing weight or wearing shoes. The amount of osteopenia seen is more than can be expected from disuse, a finding typical of the middle (dystrophic) stage of complex regional pain syndrome.

(Reproduced from Hogan CJ, Hurwitz SR: Treatment of complex regional pain syndrome of the lower extremity. ***J Am Acad Orthop Surg*** *2002;10:281-289.)*

tients with complex regional pain syndrome who have patterns of microcirculatory blood flow that differ from controls.[23]

It is important to note that not all of the regional vascular changes found in patients with complex regional pain syndrome can be explained by a simple disruption of sympathetic nerve fibers nor the denervation of blood vessels. Additionally, some experts believe that the muscle weakness found in patients with complex regional pain syndrome is better characterized clinically as a disuse syndrome. Some clinicians have suggested that the inciting trauma may result in injury to the nocioceptive nerves of the surrounding area, thereby increasing their sensitivity. Another suggested mechanism is that a reflex arc is established after the inciting event. This arc follows the route of the sympathetic nervous system and is modulated by areas within the cerebral cortex. The presence of this reflex arc has been supported by the findings of abnormal somatosensory evoked potentials in patients with complex regional pain syndrome.[24]

Treatment and Prevention

General Approaches

Although there may be no practical means of preventing most injuries, it may be possible to decrease the incidence of complex regional pain syndrome after trauma by initiating early mobilization of injured extremities. Even if this syndrome is not preventable, this approach appears to be an important factor for achieving a good outcome after musculoskeletal injury in general.

Pharmacologic Approaches

The most commonly used, but nonspecific, therapies are nonsteroidal anti-inflammatory drugs and corticosteroids. Patients whose bone scans suggest active inflammation are more apt to respond to corticosteroid treatments. Daily use of prednisone (30 to 60 mg) has been shown to be effective in treating patients with complex regional pain syndrome.[25] In general, efforts should be directed at tapering steroids as soon as the patient shows a clinical response because these medications can cause many adverse effects; however, some patients have shown benefit from a prolonged course of low-dose corticosteroids.

Regional nerve blocks are the most commonly used specific therapy. The rationale of this approach is that overactivity of the sympathetic nervous system contributes to the development of complex regional pain syndrome. Injections near neural ganglia or regional intravenous infusions of anesthetics (Bier blocks) are the most commonly used methods. It has been estimated that two thirds of patients with this syndrome will respond to regional sympathetic blocks.[26] Blocks can be repeated every few days for six to 12 treatments; they are usually stopped if there is no significant response following the first or second treatment. Complete surgical or chemical sympathectomies have been performed in patients who have responded to regional nerve blocks but for whom complex regional pain syndrome progressed despite treatments.

Nonpharmacologic Approaches

The cornerstone of nonpharmacologic treatment is aggressive physical therapy. It has been shown that passive shoulder movement in patients receiving prolonged intravenous treatment decreases the risk of complex regional pain syndrome. In addition, early mobilization of patients after injury or myocardial infarction also decreased the risk of complex regional pain syndrome.[27] It is best to start therapy as soon as the patient can tolerate it because the best results occur when therapy begins before radiographic changes are seen or movement disorders become manifest.

Innovative Therapies

Spinal cord stimulation and intrathecal baclofen are two new therapies that have been considered in the treatment of complex regional pain syndrome. In spinal cord stimulation, an electrode is positioned on the dorsal aspect of the spinal cord in the epidural space at the level of the nerve roots that correlate with the patient's symptoms. The goal is to produce paresthesias that interrupt the sensations of pain. The safety and efficacy of spinal cord stimulation has been documented in the treatment of other chronic pain syndromes. Baclofen is a specific gamma-aminobutyric acid-receptor agonist that inhibits sensory input to the neurons of the spinal cord. Intrathecal baclofen has been shown to improve symptoms of dystonia in patients with complex regional pain syndrome.[28]

Key Terms

Autonomic nervous system The set of nerves not under conscious control that regulates key bodily functions, including activity of the heart, smooth muscles (especially within blood vessels), and the gut

Nociception Pain sensing

Tender points A set of 18 characteristic locations on the surface of the body at which patients with fibromyalgia report discomfort when palpated

Vasomotor instability Abnormal constriction or dilation of blood vessels owing to dysfunction of the autonomic nervous system

References

1. National Institutes of Health: *Chronic Pain: Hope Through Research.* Bethesda, MD, US Dept Health Human Services, 1982, DHHS publication 82-2406.
2. Wolfe F, Smythe HA, Yunus MB, et al: The American College of Rheumatology 1990 criteria for the classification of fibromyalgia: Report of the Multicenter Criteria Committee. *Arthritis Rheum* 1990;33:160-172.
3. Wolfe F, Ross K, Anderson J, et al: The prevalence and characteristics of fibromyalgia in the general population. *Arthritis Rheum* 1995;38:19-28.
4. White KP, Speechley M, Harth M, et al: Fibromyalgia in rheumatology practice: A survey of Canadian rheumatologists. *J Rheumatol* 1995;22:722-726.
5. Benjamin S, Morris S, McBeth J, et al: The association between chronic widespread pain and mental disorder: A population-based study. *Arthritis Rheum* 2000;43:561-567.
6. Hudson JI, Goldenberg DL, Pope HG Jr., et al: Comorbidity of fibromyalgia with medical and psychiatric disorders. *Am J Med* 1992;92:363-367.
7. Buchwald D, Garrity D: Comparison of patients with chronic fatigue syndrome, fibromyalgia, and multiple chemical sensitivities. *Arch Intern Med* 1994;154:2049-2053.
8. Middleton GD, McFarlin JE, Lipsky PE: The prevalence and clinical impact of fibromyalgia in systemic lupus erythematosus. *Arthritis Rheum* 1994;37:1181-1188.
9. Moldofsky H, Scarisbrick P, England R, et al: Musculoskeletal symptoms and non-REM sleep disturbance in patients with "fibrositis syndrome" and healthy subjects. *Psychosom Med* 1975;37:341-351.
10. Dinerman H, Steere AC: Lyme disease associated with fibromyalgia. *Ann Intern Med* 1992;117: 281-285.
11. Boisset-Pioro MH, Esdaile JM, Fitzcharles MA: Sexual and physical abuse in women with fibromyalgia syndrome. *Arthritis Rheum* 1995;38:235-241.
12. Goldenberg DL: Treatment of fibromyalgia syndrome. *Rheum Dis Clin North Am* 1989;15:61-71.
13. Felson DT, Goldenberg DL: The natural history of fibromyalgia. *Arthritis Rheum* 1986;29:1522-1526.
14. Carette S, Bell MJ, Reynolds WJ, et al: Comparison of amitriptyline, cyclobenzaprine, and placebo in the treatment of fibromyalgia: A randomized, double-blind clinical trial. *Arthritis Rheum* 1994;37:32-40.
15. Godfrey RG: A guide to the understanding and use of tricyclic antidepressants in the overall management of fibromyalgia and other chronic pain syndromes. *Arch Intern Med* 1996;156:1047-1052.
16. Deluze C, Bosia L, Zirbs A, et al: Electroacupuncture in fibromyalgia: Results of a controlled trial. *BMJ* 1992;305:1249-1252.
17. Murray CS, Cohen A, Perkins T, et al: Morbidity in reflex sympathetic dystrophy. *Arch Dis Child* 2000;82:231-233.
18. Mitchell SW, Morehouse GR, Keen WW: (eds): *Gunshot Wounds and Other Injuries of Nerves.* Philadelphia, PA, JB Lippincott, 1864.
19. Plewes LW: Sudeck's atrophy in the hand. *JBone Joint Surg Br* 1956;38:195-203.
20. Schweitzer ME, Mandel S, Schwartzman RJ, et al: Reflex sympathetic dystrophy revisited: MR imaging findings before and after infusion of contrast material. *Radiology* 1995;195:211-214.
21. van Laere M, Claessens M: The treatment of reflex sympathetic dystrophy syndrome: Current concepts. *Acta Orthop Belg* 1992;58(suppl 1):259-261.
22. Pak TJ, Martin GM, Magness JL, et al: Reflex sympathetic dystrophy: Review of 140 cases. *Minn Med* 1970;53:507-512.
23. Schwartzman RJ: Editorial: Explaining reflex sympathetic dystrophy. *Arch Neurol* 1999;56:521-522.
24. Ritchlin C, Chabot R, Kates S, et al: Abstract: Cortical abnormalities in patients with reflex sympathetic dystrophy. *Arthritis Rheum* 1994;37(suppl): S275.
25. Kozin F, McCarty DJ, Sims J, et al: The reflex sympathetic dystrophy syndrome: I. Clinical and histologic studies: Evidence of bilaterality, response to corticosteroids and articular involvement. *Am J Med* 1976;60:321-331.
26. Chelimsky TC, Low PA, Naessens JM, et al: Value of autonomic testing in reflex sympathetic dystrophy. *Mayo Clin Proc* 1995;70:1029-1040.
27. Sheon RP: Reflex sympathetic dystrophy (complex regional pain syndrome). UpToDate Web site. Available at: http://www.uptodate.com/.
28. van Hilten BJ, van de Beek WJ, Hoff JI, et al: Intrathecal baclofen for the treatment of dystonia in patients with reflex sympathetic dystrophy. *N Engl J Med* 2000;343:625-630.

Compression Neuropathies

Compression neuropathies are clinical conditions in which pressure on a peripheral nerve produces dysfunction in the nerve. These may be manifest as sensory, motor, or autonomic changes in the involved nerve distribution. A peripheral nerve arises from the spinal root as it exits the neural foramen and terminates at the end organ. Irritation of the peripheral nerve causes a characteristic molecular response, resulting in typical symptoms and signs.

A basic understanding of the neuroanatomy and nerve function, placed in a context of a thorough history and physical examination, is often enough to allow accurate diagnosis of the various compression neuropathies. At times, however, special testing of the nerve by electrodiagnostic studies is necessary for further clarification.

Compression neuropathies in the upper extremity typically occur at or around the joints; for example, carpal tunnel syndrome (CTS) affects the median nerve at the wrist, and cubital tunnel syndrome affects the ulnar nerve at the elbow. Thus, the evaluation should focus on the anatomic path of the affected nerve and the structures that may compress it. The diagnosis of most upper extremity compression neuropathies is clinical, with confirmation provided by electrodiagnostic studies.

Each peripheral nerve is comprised of three distinct layers: the epineurium, perineurium, and endoneurium (Fig. 1). The epineurium, or outer covering, cushions the nerve against external pressure and contains concentric layers of dense collagenous connective tissue. The perineurium, or intermediate layer, surrounds the fascicle, each containing a group of axons. Each axon is surrounded by a collagen shell called the endoneurium. The perineurium is an analogue of the blood-brain barrier; it controls the intraneural environment by limiting diffusion, blocking the entry of foreign matter (especially bacteria), and by maintaining a slightly positive intrafascicular pressure. Segmental nutrient vessels supply longitudinally oriented vascular networks in the epineurium, which feed small plexi within the perineurium. From the perineurial plexi, vessels arise that penetrate the endoneurium to feed the axon.

The intricacy of the vascular system is responsible, in part, for the development of compression neuropathies. Compression of the nerve produces characteristic signs and symptoms, most likely on the basis of ischemia. This mechanism is inferred from the observation that there is a dramatic reversal of symptoms often seen following surgical decompression of a nerve that has not yet

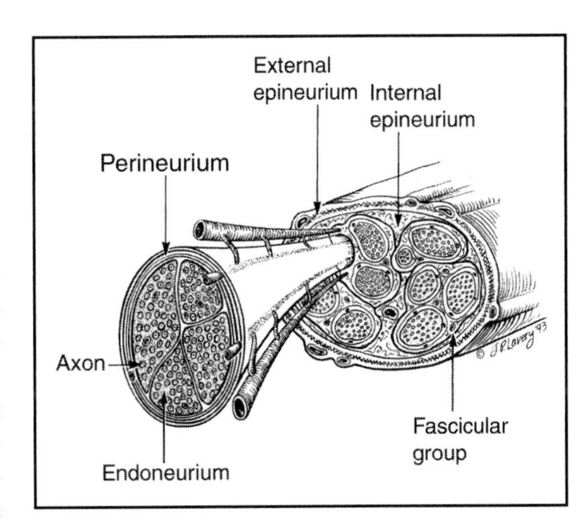

Figure 1
Functional anatomy of the peripheral nerve, including the epineurium around the nerve, perineurium around the fascicle, and endoneurium around the individual axons.

(Copyright ©1993 JP Lavery)

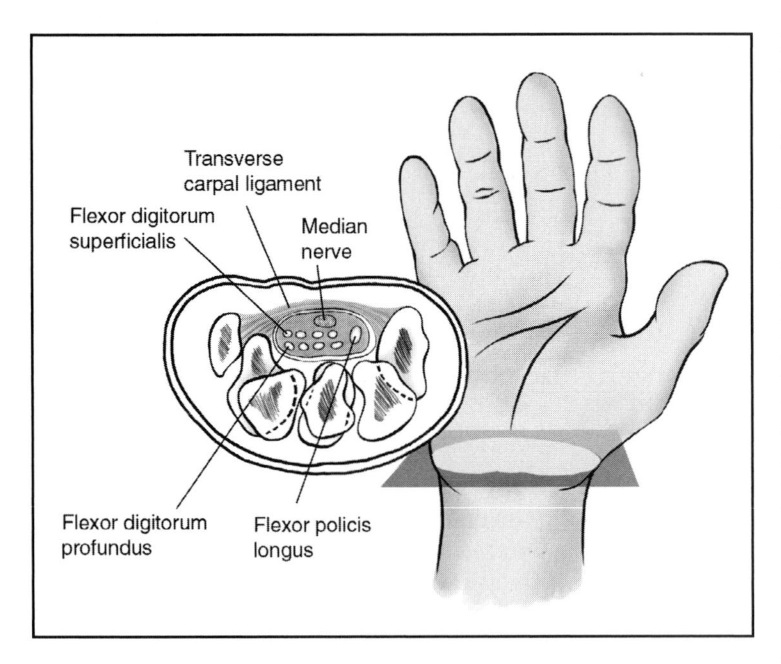

Figure 2
Cross-sectional schematic representation of the constraints and contents of the carpal canal.

been permanently damaged. Compression of a peripheral nerve leads to impaired venous return, intraneural edema, and an altered ionic milieu. Diminished axoplasmic transportation and decreased efficiency of the sodium pump promote membrane instability. All of these factors impede signal conduction along the path of the nerve. Neurophysiologic changes with symptoms of paresthesia, for example, occur with 30 to 40 mm Hg of pressure on the median nerve.[1,2] Complete intraneural ischemia occurs with pressures of 60 mm Hg or greater, resulting in total sensory block. Prolonged compression may lead to epineurial fibrosis, which exacerbates intraneural edema, decreases signal transmission, and leads to permanent nerve dysfunction.[2]

Compression causes damage not only by direct pressure but also by tethering nerves to the surrounding tissue. This limits the physiologic motion of the nerves, which would otherwise occur in response to normal joint motion. This tethering restricts mobility and places the nerve at risk for traction injuries in response to repetitive joint motion.

Carpal Tunnel Syndrome

Carpal tunnel syndrome, characterized by pain, numbness, and weakness in the median nerve distribution of the hand, is caused by pressure on the median nerve at the wrist. It is perhaps the most common compression neuropathy, and studying it as a paradigm offers insights that can be applied to all such neuropathies.

CTS is a collection of signs and symptoms arising from median nerve compression at the wrist. The median nerve exits the forearm and enters the wrist through the carpal canal. This rigid canal is formed by the transverse carpal ligament on the palmar surface and the carpal bones on the dorsal surface (Fig. 2). Joining the median nerve in this closed compartment are nine flexor tendons (two to each finger and one to the thumb). Within the carpal canal, the median nerve lies superficially, hugging the undersurface of the transverse carpal ligament. The carpal canal extends from approximately the area of the radial styloid to a point distal to the carpometacarpal joint in the palm. Upon exiting the canal, the median nerve divides into branches to supply the thumb, index, and long fingers and the radial half of the ring finger.

Epidemiology

CTS is the most common of the compression neuropathies, and affects women more often than men. Its prevalence in the general population is approximately 2%.[3,4] CTS was thought to affect individuals of about 50 years of age, but younger people, especially industrial workers, are now being diagnosed with CTS with increasing frequency. CTS may be closely related to both health status and industrial exposure. Risk factors for CTS beyond gender and occupation include rheumatoid arthritis (in which inflammation of the tendon sheath may compress the nerve); endocrine disorders, such as diabetes mellitus and hypothyroidism; trauma; hormonal changes, such as those that occur with pregnancy and menopause; and the presence of masses, such as lipomas and ganglions.

Clinical Presentation

Early symptoms include vague wrist pain, with numbness or tingling in the thumb, index and long fingers, and the radial half of the ring finger, all of which are innervated by the median nerve. With persistent compression, symptoms progress to include pain in the distal forearm or palm and in the hand and fingers at night. Discomfort or numbness may be provoked by activities that place the wrist in a prolonged flexed or extended position, such as typing or bicycling. With long-standing CTS, symptoms include a sense of grip weakness caused by denervation of the thenar muscles (abductor pollicis brevis and opponens pollicis) and difficulty with fine motor activities, such as buttoning a shirt. This loss of fine motor control probably occurs because of decreased proprioception. People with CTS often awake at night with pain because the wrist assumes a flexed position during sleep; this position increases pressure in the carpal tunnel.

The only objective physical signs associated with CTS are atrophy and loss of sweating in the fingers that are innervated by the median nerve. Two-point discrimination (ie, the ability to discern correctly between contact with one pin and contact with two pins on the fingertip placed a few millimeters apart) will be decreased in patients with CTS. This loss is quantified by identifying the minimum distance between pins that is recognized as two distinct points of contact. Weakness may be noted objectively but can reflect poor effort or inhibition caused by pain. Physical examination must always include evaluation for proximal sources of compression, especially in the cervical spine, and for masses in the carpal canal causing compression of the median nerve.

Clinical Tests

Tinel's test, which is performed by percussing the median nerve at the wrist, demonstrates nerve irritability by reproducing symptoms of CTS. A median nerve compression test, performed by continued manual compression of the median nerve at the entrance to the carpal canal, is sensitive and specific for the diagnosis of CTS.[5] In addition, Phalen's test shows that patients with CTS are able to reproduce their symptoms fairly rapidly with sustained symmetric wrist flexion. Each of these provocative tests is considered positive if symptoms of CTS are reproduced. Finally, weakness of the thenar muscles, which are innervated by the median nerve, results in difficulty with pinching.[6] This finding represents advanced disease.

Diagnostic Imaging

Diagnostic imaging tests include studies of the cervical spine (to rule out a herniated disk and osteophytic compression of the neural foramina) and of the wrist itself. Radiographs of the wrist are especially important after fracture of the distal radius or in the presence of thumb carpometacarpal arthritis. Additional testing, such as ultrasound or MRI, may be indicated to identify a soft-tissue mass that is either involving the median nerve or compressing it.

Laboratory Tests

There are no laboratory tests for CTS, but tests can be used to assess whether CTS is caused by an underlying medical condition, such as diabetes mellitus, rheumatoid arthritis, or thyroid disease.

Nerve Conduction Testing

Although CTS is a clinical diagnosis, nerve conduction velocity studies are frequently used. Nerve conduction velocities and latencies can be measured in both the sensory and motor components of the median nerve.

Differential Diagnosis

Compression of the median nerve at the elbow (pronator syndrome) usually can be distinguished from CTS based on the involvement of muscles supplied by the median nerve proximal to the wrist. For example, involvement of the flexor digitorum superficialis to all fingers and the flexor digitorum profundus of the index and long fingers indicates more proximal compression. Cervical radiculopathy involving the C6 or C7 nerves may produce symptoms in regions overlapping with CTS. Radiculopathy commonly affects patients with cervical disk disease or arthritis. Cervical instability, as seen in patients with rheumatoid arthritis or trauma, may also produce a radiculopathy similar to CTS.

Symptoms in the hand can also be produced by ulnar nerve compression at the wrist in Guyon's canal (Fig. 3) or at the elbow in the cubital tunnel. Patients with ulnar nerve compromise at Guyon's canal report numbness in the regions innervated by the ulnar nerve: the little finger and the ulnar half of the ring finger. Motor dysfunction is limited to the intrinsic muscles of the hand. Percussion over Guyon's canal usually provokes symptoms.[7] Because compression of the ulnar nerve at the wrist is fairly uncommon, nerve conduction velocity studies should be obtained to support the diagnosis. Ulnar nerve compression is more commonly found near the elbow where the nerve is superficial. Compression of the ulnar nerve

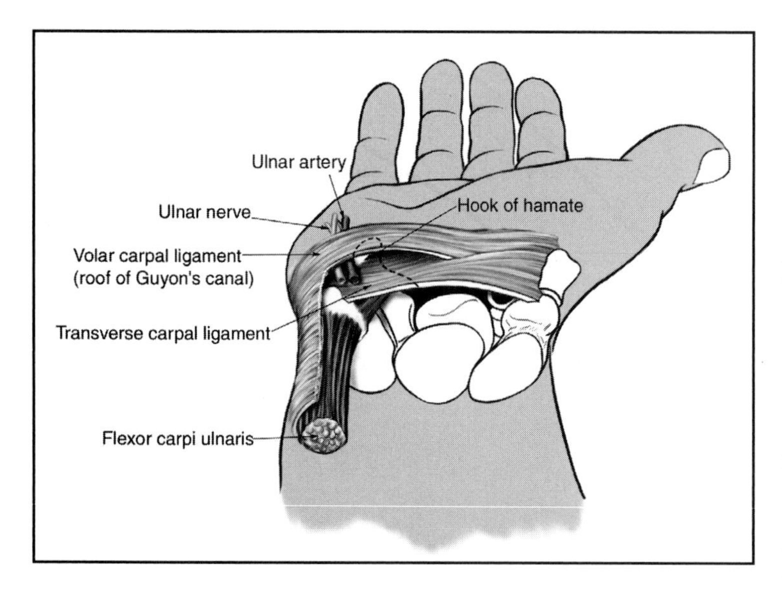

Figure 3
Guyon's canal with the radial and ulnar constraints. Note that Guyon's canal is superficial to the carpal canal.

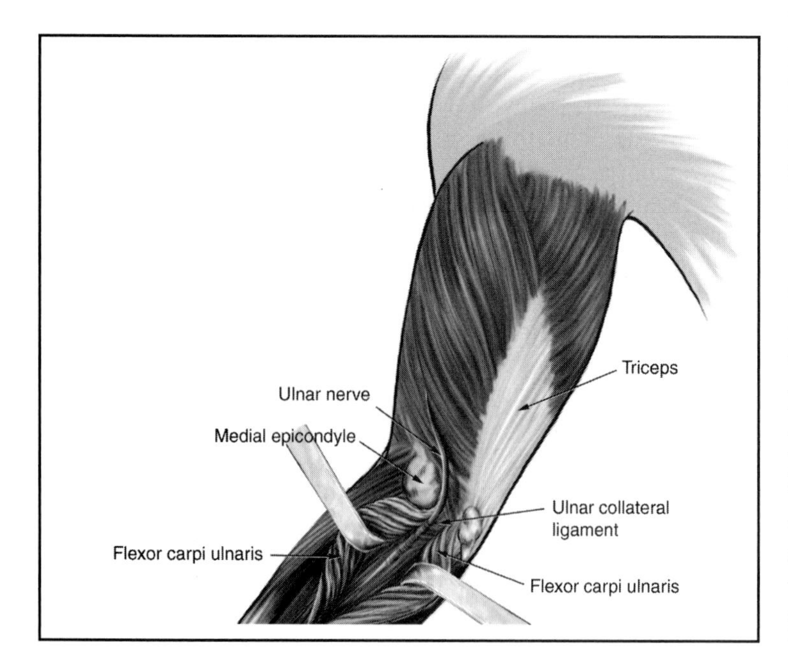

Figure 4
The path of the ulnar nerve in the cubital tunnel. Also depicted are the various structures that may compress or tether the ulnar nerve.

at the elbow is called cubital tunnel syndrome (Fig. 4). Patients with cubital tunnel syndrome report numbness along the little finger and the ulnar half of the ring finger, and they often note grip weakness, especially when using tools. Patients also commonly report a dull ache along the medial forearm and admit to resting their elbows on hard surfaces during daily activities. The physical examination for cubital tunnel syndrome centers on percussion along the course of the ulnar nerve posterior to the medial epicondyle of the elbow, which can reproduce symptoms in the fingers innervated by the ulnar nerve. Weakness and atrophy may be present in the intrinsic muscles of the hand that are innervated by the ulnar nerve.

Whereas CTS affects the radial side of the hand, thoracic outlet syndrome (TOS) typically affects the ulnar side of the hand. TOS is a constellation of symptoms arising from compression of the vascular or neural components of the brachial plexus in the thoracic outlet (the space between the clavicle, the first rib, and the subclavius and anterior scalene muscles). The brachial plexus and the subclavian artery and vein cross this space; however, in almost all cases, there is no vascular compression. Symptoms usually begin promptly after starting an inciting activity (usually shoulder abduction and external rotation) and resolve within minutes after cessation of the activity. Numbness usually occurs along the ulnar half of the hand and forearm. A number of provocative tests have been used to help diagnose TOS. Unfortunately, these tests lack clinical specificity.

Treatment

The treatment of early-onset nontraumatic CTS is nonsurgical. Night splints that maintain the wrist in a neutral position and thereby decrease pressure on the median nerve are effective.[1,8] Additionally, cortisone injected into the carpal canal may decrease edema in the nerve or tendon sheaths, thereby reducing the volume of the carpal canal. Injections are a helpful diagnostic tool as well as an effective therapeutic modality. The response to injection can confirm the diagnosis of CTS.[8] It is generally believed that nonsurgical treatment will be ineffective if symptoms have persisted for more than 1 year or if symptoms develop soon after a fracture of the radius.[9,10] When this is the case, surgical decompression of the carpal tunnel is recommended.[11] This treatment is extremely effective in decreasing symptoms and halting the progression of CTS. Treatment approaches for ulnar nerve compression are similar to those for CTS. The treatment of TOS is more controversial because surgical decompression of the thoracic outlet is major surgery and the diagnosis and prognosis are often less certain.

Key Terms

Carpal tunnel syndrome Median nerve compression at the wrist that is characterized by pain, numbness, and weakness in the median nerve distribution of the hand

Cubital tunnel syndrome Compression of the ulnar nerve at the elbow, producing symptoms in the forearm and hand

Endoneurium The sheath that surrounds each axon in the nerve

Epineurium The outer covering of the nerve that cushions the nerve against external pressure and contains concentric layers of dense collagenous connective tissue

Perineurium The intermediate layer of the nerve that surrounds each fascicle and contains a group of axons

Thoracic outlet syndrome A constellation of symptoms arising from compression of the vascular or neural components of the brachial plexus in the thoracic outlet

Tinel's test Percussion of the median nerve at the wrist to demonstrate the degree of nerve irritability by reproducing symptoms of carpal tunnel syndrome

References

1. Gelberman RH, Hergenroeder PT, Hargens AR, et al: The carpal tunnel syndrome: A study of carpal canal pressures. *J Bone Joint Surg Am* 1981;63:380-383.
2. Gelberman RH, Szabo RM, Williamson RV, et al: Sensibility testing in peripheral-nerve compression syndromes: An experimental study in humans. *J Bone Joint Surg Am* 1983;65:632-638.
3. Papanicolaou GD, McCabe SJ, Firrell J: The prevalence and characteristics of nerve compression symptoms in the general population. *J Hand Surg Am* 2001;26:460-466.
4. Tanaka S, Wild DK, Seligman PJ, et al: The US prevalence of self-reported carpal tunnel syndrome: 1988 National Health Interview Survey Data. *Am J Public Health* 1994;84:1846-1848.
5. Durkan JA: A new diagnostic test for carpal tunnel syndrome. *J Bone Joint Surg Am* 1991;73:535-538.
6. Levine DW, Simmons BP, Koris MJ, et al: A self-administered questionnaire for the assessment of severity of symptoms and functional status in carpal tunnel syndrome. *J Bone Joint Surg Am* 1993;75:1585-1592.
7. Cobb TK, Carmichael SW, Cooney WP: Guyon's canal revisited: An anatomic study of the carpal ulnar neurovascular space. *J Hand Surg Am* 1996;21:861-869.
8. Gelberman RH, Aronson D, Weisman MH: Carpal-tunnel syndrome: Results of a prospective trial of steroid injection and splinting. *J Bone Joint Surg Am* 1980;62:1181-1184.
9. DeStefano F, Nordstrom DL, Vierkant RA: Long-term symptom outcomes of carpal tunnel syndrome and its treatment. *J Hand Surg Am* 1997;22:200-210.
10. Rosen B, Lundborg G, Abrahamsson SO, et al: Sensory function after median nerve decompression in carpal tunnel syndrome: Preoperative vs. postoperative findings. *J Hand Surg Am* 1997;22:602-606.
11. Atroshi I, Breidenbach WC, McCabe SJ: Assessment of the carpal tunnel outcome instrument in patients with nerve-compression symptoms. *J Hand Surg Am* 1997;22:222-227.

Chapter 22

Soft-Tissue Trauma

Muscles and tendons move joints, and ligaments stabilize them. Skin and its surrounding connective tissue provide a durable cover to the skeleton and protect vital neurovascular structures. Unfortunately, injury to these soft tissues is common. The clinical implication of a soft-tissue injury depends on its unique characteristics, including its type, severity, and location. This chapter offers a broad overview of injury to soft tissues, including ligament, tendon, skin, and muscle.

Ligament Injuries

Sprains

Ligaments are elastic structures that stabilize the joints. A sprain occurs when a tensile (stretching) force elongates a ligament beyond its elastic limit.[1,2] A ligament can be elongated a small percentage of its overall length without structural injury. When stretched within this limit, the ligament can recoil to its normal length. Stretched beyond its limit of deformation, however, a ligament will tear—either in its midsubstance or at the attachment to bone. When this latter injury pulls off a sliver of the bone, it is called an avulsion fracture, although this injury is conceptually equivalent to a sprain and is treated similarly.

A sprain can be described by the extent of the injury to the ligament. The terms "degree" and "grade" are often used interchangeably; a second-degree sprain may also be called a grade 2 sprain.

A first-degree sprain denotes only slight stretching of the ligament and microscopic damage limited to the collagen fibrils. A first-degree sprain is associated with no discernable instability to the joint. A second-degree sprain represents a partial tearing of the ligament. Although the gross continuity of the ligament is preserved, abnormal laxity of the joint is produced when the joint is stressed. A third-degree sprain represents a complete tear. This injury, paradoxically, may cause less discomfort than a second-degree sprain because there is no remaining intact ligament to be stretched, and it is the stretch that produces pain. At times, gross instability is initially absent in third-degree sprains because muscle spasm may temporarily hold the joint in place.

Pathophysiology

Sprains are extremely common. Literally thousands of ankle sprains occur each day in the United States.[3] The joints of the fingers, knee, and shoulder are also frequently sprained. Although sprains may occur during adolescence, a growth plate fracture masquerading as a sprain must be considered.

With mild sprains, injury is limited to the ligament itself. With complete tears, however, the joint may become unstable, leading to damage to the joint surfaces. Instability of the joint is termed dislocation when the articular surfaces completely lose contact with each other. If the joint surfaces begin to dissociate but do not completely lose contact with each other, it is called subluxation. Instability and dislocations are described in terms of the direction of the pathologic motion. By convention, the abnormal motion is named according to the direction that the distal portion of the limb moved in relation to the proximal part. For example, in an anterior knee dislocation, the tibia and fibula are anterior to the femur.

The damage from sprains can be grouped into four categories: (1) injury of the ligament itself, causing local pain, swelling, and acute dysfunction of the joint; (2) residual instability of the joint; (3) disruption of the articular surface from impact at the time of injury (or from continued abnormal joint mechanics because of ligament laxity); and (4) injury to the surrounding arteries and nerves that were stretched at the time of the ligament injury.

Table 1
Classification of Ankle Sprains

Severity	Physical Examination Findings	Impairment	Pathophysiology	Typical Treatment*
Grade 1	Minimal tenderness and swelling	Minimal	Microscopic tearing of collagen fibers	Weight bearing as tolerated No splinting/casting Isometric exercises Full range-of-motion and stretching/strengthening exercises as tolerated
Grade 2	Moderate tenderness and swelling Decreased range of motion Possible instability	Moderate	Complete tears of some but not all collagen fibers in the ligament	Immobilization with air splint Physical therapy with range-of-motion and stretching/strengthening exercises
Grade 3	Significant swelling and tenderness Instability	Severe	Complete tear/rupture of ligament	Immobilization Physical therapy similar to that for grade 2 sprains but over a longer period Possible surgical reconstruction

*Patients must receive treatment that is tailored to their individual needs. This table outlines common treatment protocols.

Injury within the ligament producing swelling and compromised function of the joint is characteristic of mild ankle sprains. With such an injury, no instability may be detected. However, the patient may have significant disability. This may be because the proprioceptive nerve fibers in the ligament are damaged or because there is swelling that constrains normal motion.

Damage to the articular surface of the joint is frequently seen with anterior shoulder dislocations. When the humeral head is forced forward out of the glenoid socket, its posterior surface can collide with the front of the glenoid and cause an impaction injury. This injury is called a Hill-Sachs lesion and is definitive proof that a dislocation occurred, even if no instability is seen on examination.

Residual joint instability may be found with sprains of the anterior cruciate ligament (ACL) of the knee. A chronically torn ACL does not hurt, but without a functional ACL stabilizing the knee, the normal mechanics of the knee are impeded.

Trauma to the vascular bundle is characteristic of hip dislocations. When the hip is pulled out of joint, the blood vessels that are tightly bound to the femur may be stretched beyond their limit and tear. As a result, hip dislocation can produce osteonecrosis of the femoral head because of the disruption of the blood supply.

Ankle Sprain

The ankle ligaments stabilize the ankle primarily against twisting injuries (Fig.1). The deltoid ligament on the medial side resists eversion, and the lateral ankle ligaments resist inversion. The syndesmotic ligament complex between the tibia and the fibula resists separation (diastasis).

The ankle ligaments are commonly injured when a person turns or rolls the ankle. The severity of the sprain can vary widely (Table 1). After sustaining a sprain, a patient may or may not be able to bear weight on that ankle. Immediate pain and swelling are frequently noted. As swelling increases, the joint becomes stiff and painful to move. Physical examination will reveal tenderness to palpation over the injured ligaments. Manipulation of the injured joint may show excessive or abnormal motion. Palpation of the proximal fibula is also necessary because the energy of a twisting injury may be transmitted up the syndesmosis or leg, producing a fracture of the fibula near knee. Radiographs are usually normal; nonetheless, they are usually obtained because a bony injury may also be present. With severe sprains, abnormal alignment of a partially or completely dislocated joint may be seen.

Prompt treatment of ankle sprains improves outcome. The general treatment approach is referred to as RICE, which stands for rest, ice, compression, and elevation. The goal of treatment is to reduce the swelling that contributes to pain and loss of motion. When the acute symptoms subside, the patient may benefit from rehabilitation to restore strength and range of motion. The proper mechanics of gait may also need to be retaught. Because the tendons crossing the ankle provide stability, even complete ligament tears may not necessarily lead to functional instability. Surgical reconstruction is reserved for injuries that fail to respond to nonsurgical treatment.

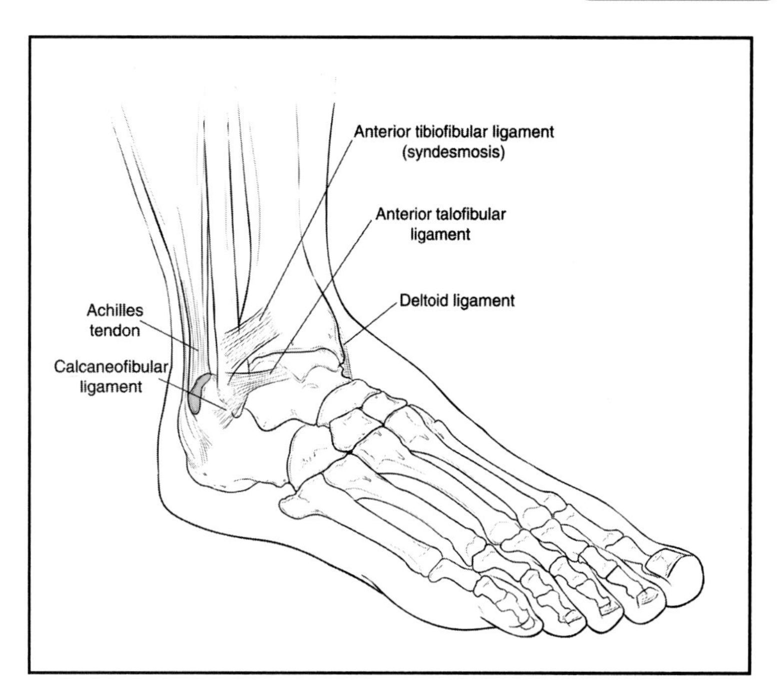

Figure 1
Ligaments of the ankle. The Achilles tendon in also shown.

(Adapted with permission from ***Briner WW Jr, Carr DE, Lavery KM: Anterior tibiofibular ligament injury: Not just another ankle sprain. Phys Sportsmed*** **1989;17:63-69.)**

Shoulder Dislocation and Instability

No single lesion is responsible for all cases of glenohumeral instability because the shoulder is stabilized by different anatomic structures, depending on the position of the shoulder in space and the direction of the distracting force. The shoulder is stabilized statically by capsular ligaments: the superior, middle, and inferior glenohumeral ligaments. The inferior glenohumeral ligament is the main stabilizer against anterior translation, especially with the arm abducted. The glenoid labrum, a fibrous structure attached to the circumference of the glenoid, deepens the saucer-like glenoid socket to lend additional stability. The rotator cuff, biceps, and scapular rotator muscles all provide dynamic stability to the shoulder as well.

In the late fifth century BC, Hippocrates classified shoulder dislocations as traumatic or atraumatic—a system that still applies today. Neer and Welsh[4] added a third category —acquired— to describe shoulder dislocations that result from repeated minor injuries.

Atraumatic instability is characterized by subluxations and dislocations of the joint in the absence of specific trauma. Generalized joint laxity is almost always seen in these patients. For example, Marfan syndrome is a possible cause of atraumatic instability. This syndrome is characterized by excessive height, long fingers and toes, and laxity of connective tissue, including the connective tissue in the heart valves, aorta, and eyes. Many instances of atraumatic instability, however, have no identifiable cause.

Traumatic dislocation typically occurs as a result of a fall. Ninety-eight percent of shoulder dislocations are anterior, with the head of the humerus lying anterior to the glenoid[5] (Fig. 2). In most cases of an anterior dislocation, the anterior labrum and capsule are torn away from the glenoid; this tear is called a Bankart lesion[6] (Fig. 3). Many shoulder dislocations are associated with an impression fracture of the humeral head on its posterolateral surface, the point where the head contacts the glenoid when it is out of place (Hill-Sachs lesion)[7] (Fig. 4).

The patient with a shoulder dislocation usually has an obvious deformity. It is important to consider the age of the patient at the time of the initial dislocation because the younger the patient is at the time of initial dislocation, the more likely that dislocation will recur.[8,9] Older patients are less likely to have a recurrence, but dislocations in older patients are associated with rotator cuff

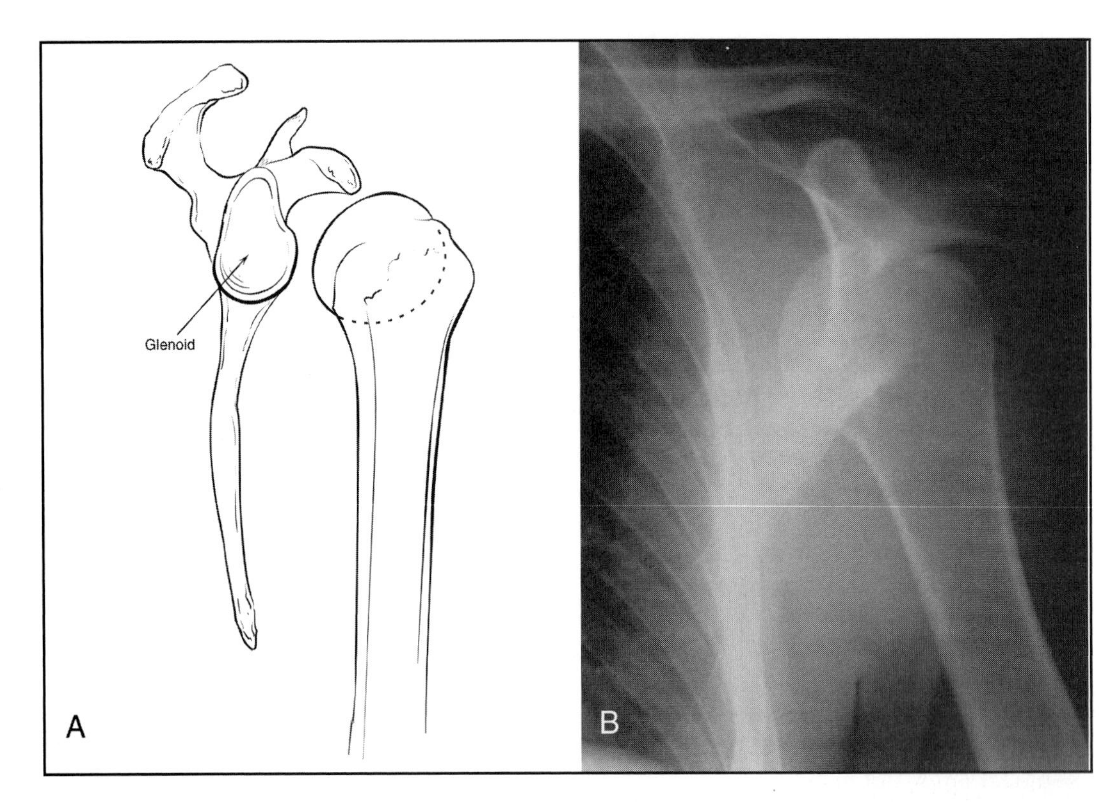

Figure 2
Anterior shoulder dislocation. **A**, Schematic representation of the humeral head dislocated anteriorly. **B**, AP radiograph of same.

*(Figure **2A** is reproduced from Greene WB (ed): **Essentials of Musculoskeletal Care**, ed 2. Rosemont, IL, American Academy of Orthopaedic Surgeons, 2001, p 147.)*

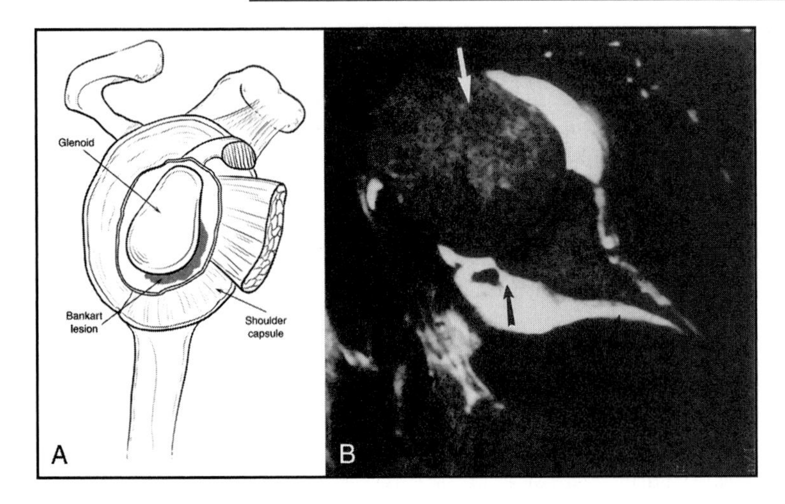

Figure 3
Bankart lesion. **A**, Schematic representation of a true lateral view with the humerus removed showing detachment of the labrum and capsule from the glenoid. **B**, An axial T2-weighted MRI obtained after an acute primary dislocation. Black arrows indicate an anterior labral tear with capsular stripping along the glenoid neck. White arrow indicates marrow edema (Hill-Sachs lesion).

*(Figure **3A** is reproduced from Greene WB (ed): **Essentials of Musculoskeletal Care**, ed 2. Rosemont, IL, American Academy of Orthopaedic Surgeons, 2001, p 148. Figure **3B** is reproduced from Lintner SA, Speer KP: Traumatic anterior glenohumeral instability: The role of arthroscopy. **J Am Acad Orthop Surg** 1997;5:233-239.)*

tears. A neurovascular assessment with special attention to the axillary nerve is needed because a traction injury may occur during dislocation.

Radiographs should include AP transscapular lateral (Y view) and axillary views. The Hill-Sachs lesion is seen well on the axillary view. It is easy to miss a dislocation when only an AP view is obtained because a single two-dimensional representation cannot identify where the head lies in relation to the glenoid.

Traditionally, treatment for a first-time traumatic shoulder dislocation consists of closed reduction followed by a period of immobilization and then a program of rehabilitation. Recurrent dislocation is treated with surgery, preferably restoration of normal anatomy by repairing the detached labrum and tightening the capsule.

ACL Injuries

Stability of the knee joint depends on both the ligaments and the surrounding muscles. The ACL stabilizes the knee against anterior

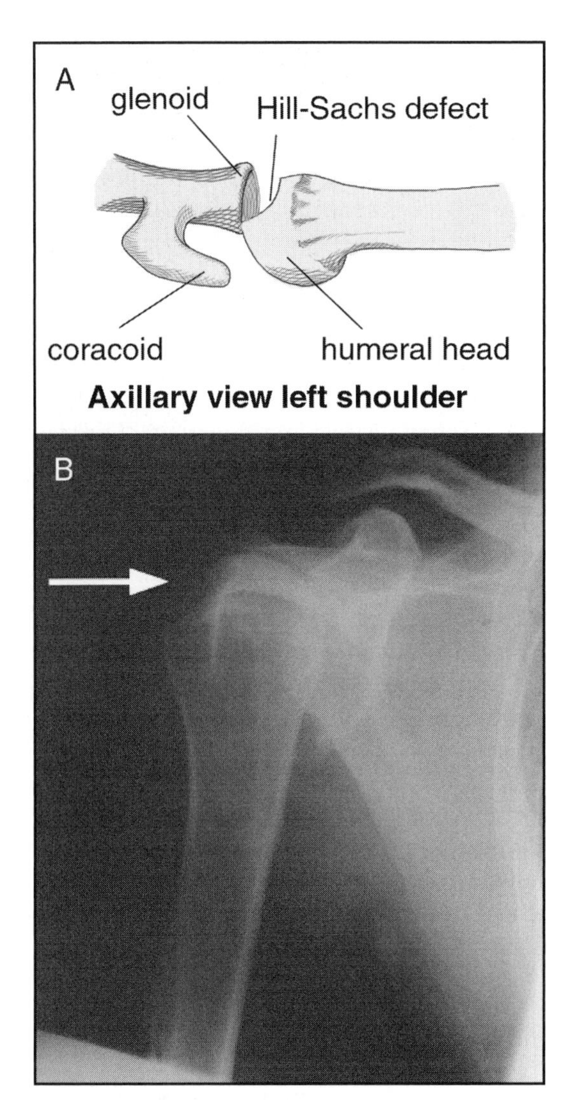

Figure 4
Hill-Sachs lesion. **A**, An impression fracture occurs on the humeral head posterolaterally when the humeral head impacts on the glenoid. **B**, Radiograph of a Hill-Sachs lesion (arrow).

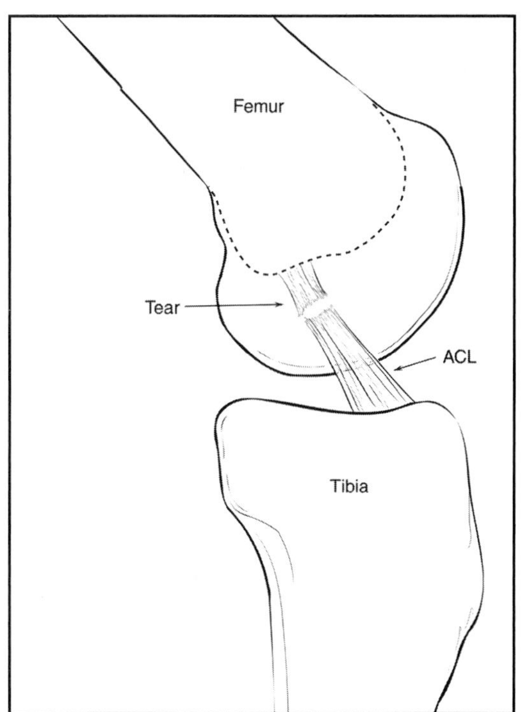

Figure 5
Complete ACL tear. The deforming force is generated as the tibia moves anteriorly.

*(Reproduced from Greene WB (ed): **Essentials of Musculoskeletal Care**, ed 2. Rosemont, IL, American Academy of Orthopaedic Surgeons, 2001, p 360.)*

subluxation of the tibia, and its action is supplemented by the hamstrings and the gastrocnemius. When the tibia is forced forward, which typically occurs when an individual is attempting to change directions at a high rate of speed, the ACL may rupture (Fig. 5). This injury commonly occurs in skiing when the ski and boot tether the leg to the ground as the body begins to move in another direction.

Abnormal anterior translation of the tibia can also damage the articular surfaces (as the posterior tibia collides with the femur) or tear the meniscus. Both are serious injuries and may cause posttraumatic arthritis. Even without these secondary injuries, however, ACL tears can lead to functional instability. The patient simply does not "trust" his or her knee and feels unstable when pivoting. This instability may be subtle and perceived by only the patient, or it may be grossly apparent on examination. The development of symptomatic knee instability after an ACL injury is variable. A patient can usually stabilize the knee at low speeds and when moving in a straight direction but will experience instability when pivoting at high speeds. Given this variability, not all tears are surgically reconstructed; a reasonable therapeutic approach for some patients is to simply avoid activities that produce instability.

Treatment of the symptomatic knee instability is surgical replacement of the ruptured ACL; a torn ACL rarely heals. Those that do heal usually are only partially torn. The torn ACL is replaced with biologic tissue; either an autograft (from the patient) or allograft (from a cadaver) is used. Tissues used include the patellar tendon, hamstrings, and Achilles tendon (allograft only).

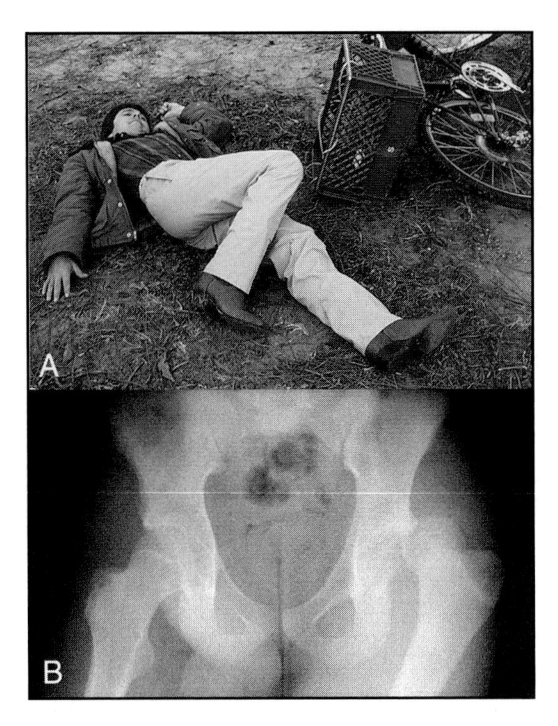

Figure 6
A, Clinical appearance of a posterior dislocation of the right hip. **B**, AP radiograph of a posteriorly dislocated hip.

(Figure ***6A*** *is reproduced from Heckman JD (ed):* ***Emergency Care and Transportation of the Sick and Injured****, ed 4. Park Ridge, IL, American Academy of Orthopaedic Surgeons, 1987, p 208.)*

Hip Dislocation

Traumatic hip dislocations differ from developmental hip dislocations (discussed in chapter 24) in many ways. Chief among these differences is that traumatic dislocation is characterized by sudden and violent displacement of the femoral head from the acetabulum. This can damage the blood vessels that travel along the femoral neck to supply the head of the femur. In developmental dislocation, however, gradual displacement occurs over time, and the blood vessels can adapt and grow along with the deformity.

The hip joint is inherently stable because the pelvis has a deep socket (the acetabulum). This deep socket is further enhanced by a rim of surrounding connective tissue that includes the labrum, capsule, and iliofemoral ligament. Given the strength of the hip joint, dislocations require high forces applied in precisely the right direction. One common mechanism for a posterior dislocation is the knee striking the dashboard during a motor vehicle collision, transmitting forces along the femur to the flexed hip joint.

The typical presentation of a traumatic posterior hip dislocation includes a history of high-energy trauma, severe pain, and classic positioning of the hip: flexion, internal rotation, and adduction (Fig. 6). Radiographs may identify an acetabular fracture as well. Incongruence of the joint on radiographs following reduction suggests the interposition of fragments of bone or soft tissue, and CT evaluation is then needed.

Hip dislocation is an emergency; the longer the hip is out of place, the greater the risk to the blood vessels and the greater the risk of osteonecrosis. The damage from osteonecrosis results from a failure of bone remodeling and thus does not appear immediately. Even though hip dislocations require urgent treatment, attention to the trauma ABCs takes precedence. Many, if not all, hip dislocations occur in context of high-energy trauma; therefore, initial management must be directed toward life-saving measures.

Hip reduction requires in-line traction. Since the joint is inherently stable, it may be difficult to move the femoral head back in place; therefore, surgical reduction may be needed. Fortunately, the inherent stability provided by the bone anatomy results in a low recurrence rate.

Tendon Injuries

Direct, penetrating trauma from sharp objects can lacerate tendons.[10,11] Lacerations commonly occur in the hand and foot, where the tendons lie near the surface of the skin. Tendons are also subject to rupture. The typical mechanism of injury in a tendon rupture is an eccentric contraction of the musculotendinous unit—that is, a contraction of the muscle as it is pulled in the opposite direction by an external force. An eccentric contraction of the quadriceps occurs, for example, when landing from a jump; the force of landing tends to flex the knee, and the muscle tightens in an attempt to decelerate that motion. Rupture of the quadriceps tendon will occur when the tensile forces within the tendon exceed its strength.

Flexor tendon lacerations in the hand will threaten essential features, such as power and precision grasp. Without intact flexor tendons, joints cannot be moved to place the hand, wrist, or fingers in the proper position for use. In contrast to lacerations, ruptures tend to occur in tendons that move the large joints of the lower extremity. The quadriceps tendon, the patellar tendon, and the Achilles tendon are all subject to rupture, especially in middle-aged individuals who participate in recreational sports. In some cases, a prodrome of pain and inflammation of the tendon precedes a rupture.

Lacerations and ruptures can result in either partial or complete disruption of the tendon. Lacerations tend to transect the tendon perpendicular to its long axis and leave two relatively clean edges for repair. The lacerating object can also cut nearby structures, such as nerves or blood vessels. Ruptures usually damage a segment of tendon adjacent to the bony insertion, although occasionally a piece of bone is avulsed. After rupture, the ends of the tendon are ragged and disorganized, giving them a "mop-end" appearance. With rupture, injury to surrounding nerves and blood vessels is rare.

The diagnosis of a tendon injury is not always obvious. When a laceration occurs, the depth and path of the penetrating object cannot always be surmised from the skin wound. Also, a partial laceration (or a complete injury to a tendon whose function can be subsumed by another—the superficial flexor of the finger, for example) may not produce deficits that are immediately obvious to the patient. A tendon rupture will usually produce enough pain that the patient will understand that he or she has been injured; however, the patient may not realize that a rupture has occurred if there are no apparent functional deficits. For example, a rupture of the Achilles tendon prevents powerful plantar flexion of the ankle, but some flexion is possible by action of the flexor digitorum longus.

Treatment of ruptured tendons focuses on positioning the torn ends to allow healing, which often requires surgery. However, in some cases surgery is not needed because appropriate positioning of the extremity may reapproximate the tendon ends well enough to allow healing. For example, plantar flexion of the ankle (and casting in that position) will bring the edges of the Achilles tendon close enough to allow healing. Even with surgical repair, suturing is needed only to bring the edges together because the ultimate tensile strength comes from healing.

Laceration of the flexor tendons of the hand represents a therapeutic challenge because these tendons pass through sheaths and pulleys to work effectively. Thus, the challenge is not only to get the tendon to heal but also to ensure that it is neither too bulky nor too sticky, lest it not slide within its soft-tissue sleeve. Bulk is minimized with the use of small sutures and meticulous surgical technique. Prevention of adhesion is attempted by a postoperative protocol that allows early motion, which itself can be a source of problems—not only because of the obvious risk of rupture from motion that is too vigorous but also because some of the healing comes from the surrounding sheath. Even in the best of cases, some loss of power and excursion of the tendon often results from these injuries.

Tendons may also rupture in a setting of chronic attrition, as commonly occurs with rheumatoid arthritis. With these ruptures, extensive degeneration may preclude repair, making a tendon graft or transfer necessary. Additionally, chronic tendon ruptures may be associated with deformation of the joint as well; thus, fusion may be required. For example, a chronic rupture of the posterior tibial tendon may produce a loss of the arch of the foot. Over time, this collapse changes the shape of the joint such that fusion, not tendon repair, is needed to restore normal geometry.

Skin Injuries

Approximately 45,000 burns severe enough to require hospitalization occur annually in the United States.[12] Thermal burns resulting from contact with fire or hot objects are the most common; chemical burns and burns from ionizing radiation occur much less frequently. An electrical burn is essentially a thermal burn from the heat generated by current conducting through the body. These burns are especially harmful to the nerves because higher electrical resistance in other soft tissues promotes travel of the current within the neural tissue. Nevertheless, the deep soft tissue, such as muscle, can be injured by electrical burns because the skin has high resistance to conducting electricity and essentially traps the heat in the body.

Burns are classified by the depth of injury (Fig. 7). First-degree (superficial) burns involve the epidermis only. Patients with first-degree burns present with pain, edema, and erythema of the skin. Second-degree (partial-thickness) burns extend down to the dermis. The hallmark of this injury is painful blistering of the skin. Third-degree (full-

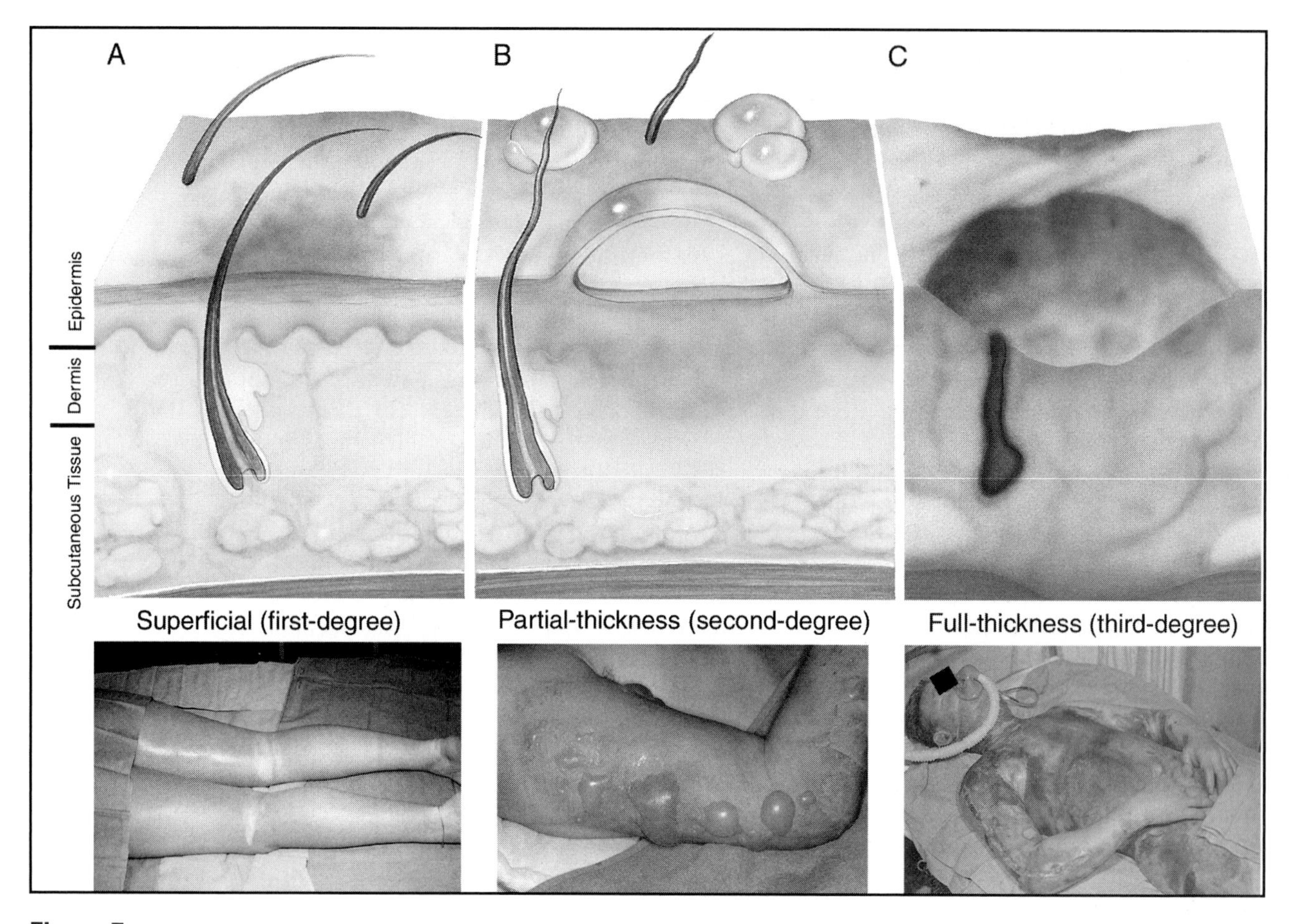

Figure 7

Classification of burns. **A**, Superficial or first-degree burns involve only the epidermis. The skin turns red but does not blister or actually burn through. **B**, Partial-thickness or second-degree burns involve some of the dermis, but they do not destroy the entire thickness of the skin. The skin is mottled, white to red, and is often blistered. **C**, Full-thickness or third-degree burns extend through all layers of the skin and may involve subcutaneous tissue and muscle. The skin is dry, leathery, and often either white or charred.

(Reproduced with permission from Browner BD, Pollak AN, Gupton CL (eds): **American Academy of Orthopaedic Surgeons** *Emergency: Care and Transportation of the Sick and Injured, ed 8. Sudbury, MA, Jones and Bartlett Publishers, 2002, p 576.)*

thickness) burns appear waxy and dry. Full-thickness burns can also penetrate deep into muscle and bone. There is often charring of the involved tissue, and the associated nerve death results in injuries that are not particularly painful. Full-thickness injuries require vigilant and aggressive management, including multiple surgical débridements and tissue coverage procedures.

A large percentage of all burns involve the hand.[13] The thin skin on the back of the hand is more susceptible to injury than the thicker skin on the palmar surface. Full-thickness burns to the dorsal skin can result in contractures that render the hand useless. With extensive burns, the hand assumes a "clawed" position. Once these contractures occur, they are difficult to correct; thus, prevention is paramount to preserving the function of the burned hand. Nonadherent dressings with topical antibiotic cream should be applied, and the hand should be splinted. Splinting the wrist in extension with the fingers slightly flexed allows the hand to remain functional, even if the joints eventually become stiff in that position. The hand must be kept elevated to overcome the significant edema that develops in burned tissue; the dressings must also be changed frequently to reduce the risk of infection. Range-of-motion exercises should begin early in the course of treatment to prevent joint contractures.

Burns can cause circulatory problems. In the extremity, a circumferential burn can create a constricting band of tissue (eschar) that impairs arterial blood flow and prevents the drainage of edema fluid. A circumferential eschar must be released surgically to eliminate this restriction to circulation. Full-thickness burns may cause the underlying muscles to swell within their fascial compartments. In severe cases, the swelling greatly increases the interstitial pressure and compromises blood flow to the muscle, creating a compartment syndrome that must be released surgically.

Muscle Injuries

Myositis ossificans is the abnormal production of bone within muscle.[14] Heterotopic ossification refers to the formation of bone in any nonosseous tissue. Although the exact etiology is unknown, both conditions can develop as a result of trauma or as a secondary manifestation of a systemic disease or a genetic disorder. A deep contusion of the quadriceps muscle, fracture of the leg or arm, or dislocation of the elbow is associated with heterotopic bone formation. Myositis ossificans is characterized by the formation of bone between muscle fibers. The patient often reports pain with motion of the involved muscle. Tenderness to palpation and swelling are noted on physical examination. A mass may become palpable as bone formation continues over time.

Heterotopic ossification arises from the proliferation of precursor cells and their maturation into osteoblasts. The cellular mechanisms that effect this transformation have not been identified, but the observation that heterotopic ossification at times develops in the extremities of patients with closed head injuries implicates a circulating mediator. Heterotopic ossification is also observed in patients with pelvic and elbow fractures, perhaps as a result of excessive bone healing.

Microscopically, heterotopic ossification and myositis ossificans are similar. Three distinct zones are seen. Each zone appears to represent a specific stage in the process of the disease. The central zone consists of mitotically active, undifferentiated mesenchymal cells. This is surrounded by a layer of osteoblasts within organic bone matrix (osteoid) and cartilage. The outer zone contains mature, calcified cancellous bone.

The formation of heterotopic ossification and myositis ossificans may be prevented by oral administration of bisphosphonates or nonsteroidal anti-inflammatory drugs. Radiation is another method used to prevent ectopic bone formation after fractures, but it may impede healing. All methods must be initiated soon after the injury to be effective.

Key Terms

Allograft Biologic tissue from a cadaver that is used to surgically replace damaged tissue

Autograft Biologic tissue from the patient's own body that is used to surgically replace damaged tissue

Avulsion fracture A fracture that occurs when a ligament or tendon pulls off a sliver of the bone

Dislocation Complete displacement of a bone from its normal position in the joint, resulting in a complete loss of contact between articular surfaces

First-degree burns Superficial burns that involve the epidermis only; characterized by pain, edema, and erythema of the skin

Heterotopic ossification The formation of bone in any nonosseous tissue

Hill-Sachs lesion The indentation of the posterolateral humeral head that is formed when it collides with the front of the glenoid in an anterior shoulder dislocation

Myositis ossificans Abnormal production of bone within muscle

Second-degree burns Partial-thickness burns that extend down to the dermis; characterized by painful blistering of the skin

Sprain The injury that occurs when a tensile (stretching) force elongates a ligament beyond its elastic limit

Subluxation Partial or incomplete dissociation of joint surfaces

Third-degree burns Full-thickness burns that can penetrate deep into muscle and bone; often characterized by charring of the involved tissue and associated nerve death

References

1. Clanton TO: Athletic injuries to the soft tissues of the foot and ankle, in Coughlin MJ, Mann RA (eds): *Surgery of the Foot and Ankle*, ed 7. St. Louis, MO, Mosby, 1999, vol 2, pp 1090-1209.
2. Toolan BC, Zapson DS: Ankle, in Spivak JM, Di Cesare PE, Feldman DS, Koval KJ, Rokito AS, Zuckerman JD (eds): *Orthopaedics: A Study Guide*. New York, NY, McGraw-Hill Health Professions Division, 1999, pp 633-636.
3. Praemer A, Furner S, Rice DP: (eds): *Musculoskeletal Conditions in the United States*, ed 2. Rosemont, IL, American Academy of Orthopaedic Surgeons, 1999, p 9.
4. Neer CS II, Welsh RP: The shoulder in sports. *Orthop Clin North Am* 1977;8:583-591.
5. Rowe CR: Prognosis in dislocations of the shoulder. *J Bone Joint Surg Am* 1956;38:957-977.
6. Lintner SA, Speer KP: Traumatic anterior glenohumeral instability: The role of arthroscopy. *J Am Acad Orthop Surg* 1997;5:233-239.
7. Hintermann B, Gachter A: Arthroscopic findings after shoulder dislocation. *Am J Sports Med* 1995;23:545-551.
8. Simonet WT, Cofield RH: Prognosis in anterior shoulder dislocation. *Am J Sports Med* 1984;12: 19-24.
9. Hovelius L: Anterior dislocation of the shoulder in teen-agers and young adults: Five-year prognosis. *J Bone Joint Surg Am* 1987;69:393-399.
10. Strickland JW: Flexor tendons: Acute injuries, in Green DP, Hotchkiss RN, Pederson WC (eds): *Green's Operative Hand Surgery*, ed 4. New York, NY, Churchill Livingstone, 1999, vol 2, pp 1851-1897.
11. Coughlin MJ: Disorders of tendons, in Coughlin MJ, Mann RA (eds): *Surgery of the Foot and Ankle*, ed 7. St. Louis, MO, Mosby, 1999, vol 2, pp 786-861.
12. American Burn Association: Burn Incidence and Treatment in the US: 2000 Fact Sheet. Available at: http://www.ameriburn.org/pub/Burn%20Incidence%20Fact%20Sheet.htm. Accessed October 30, 2002.
13. Achauer BM: The burned hand, in Green DP, Hotchkiss, RN, Pederson WC (eds): *Green's Operative Hand Surgery*, ed 4. New York, NY, Churchill Livingstone, 1999, vol 2, pp 2045-2060.
14. Albright JP: Musculotendinous problems about the knee, in Evarts CM (ed): *Surgery of the Musculoskeletal System*, ed 2. New York, NY, Churchill Livingstone, 1990, vol 4, pp 3499-3537.

Chapter 23

Special Concerns of the Athlete

Sports medicine addresses many medical concerns of the athlete, including concerns about physiology, nutrition, and psychology; however, the most common health concern of athletes is musculoskeletal injury. The first section of this chapter is a review of some of the musculoskeletal conditions that commonly affect athletes, grouped by anatomic region. The second section addresses common complaints associated with specific sports. The final three sections briefly address the special concerns of the female athlete, athletes at both extremes of the age spectrum, and athletes with disabilities.

Common Musculoskeletal Conditions Affecting Athletes

Shoulder and Arm

The athlete's shoulder is subject to certain characteristic problems, especially those resulting from throwing—specifically, the cocking and acceleration phases that stress the shoulder capsule and ligaments. Throwing can stretch these restraints and produce instability. Athletes rarely dislocate the shoulder completely by throwing; rather, the joints usually slip out of place only slightly, a condition called subluxation.[1] This subluxation and the sense of instability it produces can impede high-level athletic performance. Moreover, when the shoulder joint slips out of place even to only a small degree, it can produce traction on the brachial plexus, with a resultant "dead-arm" phenomenon in which there is a sense of profound arm weakness.[2] Traction injuries resulting from repetitive use can occur within the shoulder joint itself because the biceps can be pulled from its anchor at the top of the glenoid. This can involve the surrounding labrum as well (Fig. 1).

Repetitive overhead athletes such as throwers, swimmers, and tennis players are subject to rotator cuff disorders.[3] In middle-aged adults, irritation of the rotator cuff is usually caused by intrinsic pathology of the tendons or extrinsic impingement from spurring of adjacent structures. However, in young athletes, irritation of the rotator cuff is likely to be caused by glenohumeral instability, which is often induced by overuse. Such instability permits proximal and anterior migration of

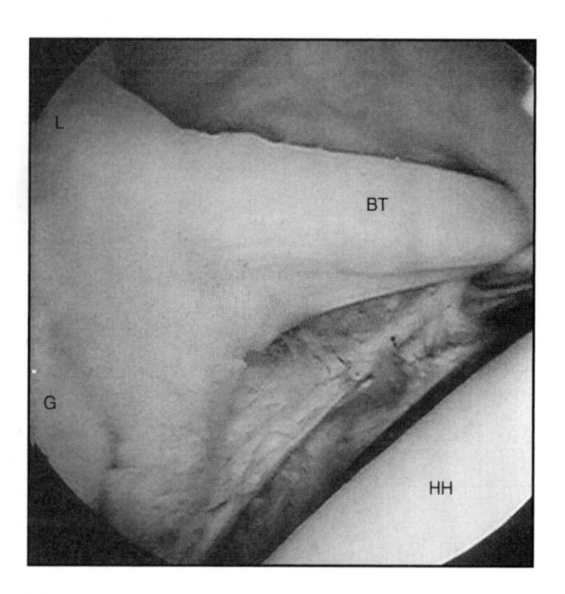

Figure 1

The origin of the long head of the biceps is an intra-articular structure of the shoulder. In this arthroscopic view, the biceps tendon (BT) is seen at its normal attachment to the glenoid labrum. This connection is subject to injury in throwers. G = glenoid, HH = humeral head, L = labrum.

(Reproduced from Eakin CL, Faber KJ, Hawkins RJ, Hovis WD: Biceps tendon disorders in athletes. J Am Acad Orthop *1999;7: 300-310.)*

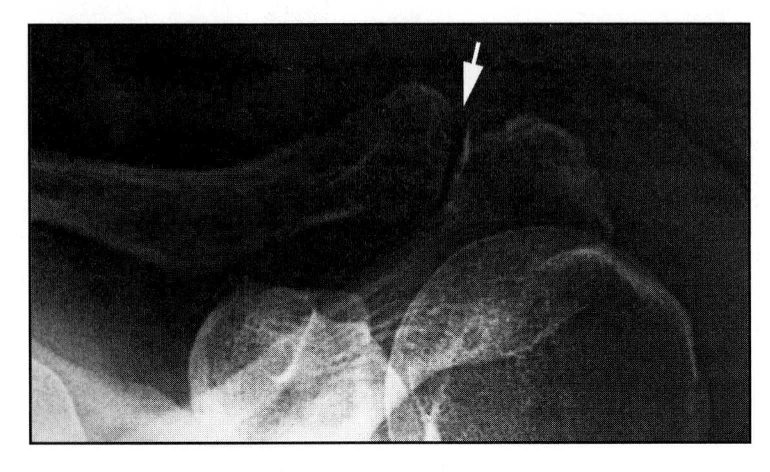

Figure 2
Acromioclavicular arthrosis. Radiograph of the acromioclavicular joint (arrow) demonstrates the typical findings of arthritis, including joint space irregularity, sclerosis, subchondral cyst formation, and the presence of an osteophyte on the acromial facet.

(Reproduced from Shaffer BS: Painful conditions of the acromioclavicular joint. **J Am Acad Orthop Surg** *1999;7:176-188.)*

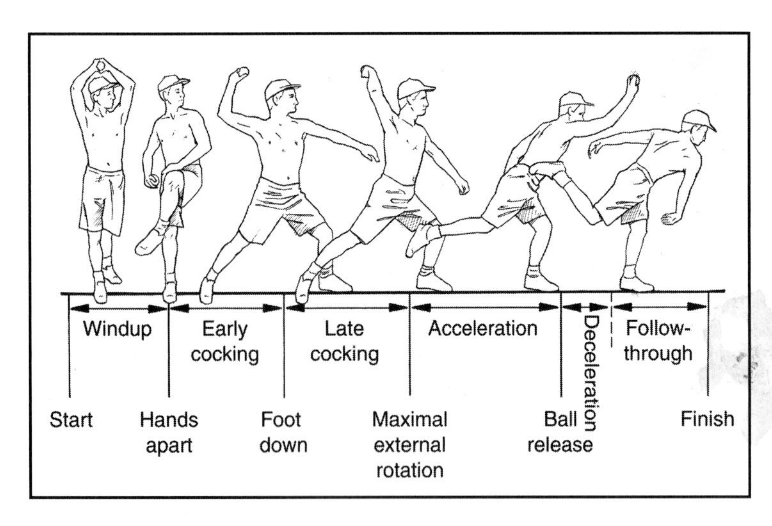

Figure 3
The main stages of the overhead throwing motion.

(Adapted with permission from DiGiovine NM, Jobe FW, Pink M, Perry J: An electromyographic analysis of the upper extremity in pitching. **J Shoulder Elbow Surg** *1992;1:15-25.)*

the humeral head against the acromion. This abutment of tendon on bone creates an extrinsic impingement on the cuff tendons—much the way an acromial bone spur may scrape the tendon surface. In young athletes, however, the appropriate treatment does not involve removal of the acromion (as might be done in cases of primary impingement from a bone spur) but instead involves eliminating the instability. Although this is often achieved via physical therapy, surgery (capsular tightening) is occasionally required.

Acromioclavicular joint injuries (shoulder separations) occur among athletes who participate in sports in which there is a risk of impact against the point of the shoulder. Athletes at risk commonly participate in ice hockey (in which participants get checked into the boards), wrestling, football, and martial arts.[4] Treatment is nonsurgical for all but the most severe injuries. The acromioclavicular joint can also be subject to wear and tear (Fig. 2). In extreme cases of overuse, osteolysis (dissolution) of the distal clavicle occurs—a condition seen most notably among power lifters. Although osteolysis of the distal clavicle may mimic the appearance of an aggressive lesion on radiographs, it typically follows a benign course.

With the exception of the clavicle, bones of the shoulder rarely fracture in athletes. Breaks of the midshaft of the clavicle usually can be treated with simple sling immobilization and avoidance of contact until healing is complete.[5] Glenoid fractures are rare, but they can occur with shoulder dislocations in the athlete. Collision sports can impart significant energy that may result in fractures of the body of the scapula or proximal humerus. Midshaft fractures of the humerus in noncollision sports should raise suspicion of an underlying bone pathology or tumor.

Elbow and Forearm

The elbow is subjected to high stress during throwing (Fig. 3). Accordingly, the throwing athlete may have irritation of the medial collateral ligament (MCL), which is under tension during throwing, or articular cartilage abnormalities on the lateral side, which is subjected to compressive forces. Over time, persistent traction along the MCL can stretch it to the extent that valgus instability occurs and surgical reconstruction is required.[6] Tension injuries may place traction on the ulnar nerve, which produces weakness or paresthesias in the hand. Compression injuries involve the articular surfaces of the radiohumeral joint. Arthropathy here often presents insidiously, but it can progress to a full flexion contracture and the creation of loose bodies.

Another common elbow problem experienced by athletes is lateral epicondylitis, or tennis elbow. Lateral epicondylitis is an irritation or partial tear of the extensor tendons of the wrist, which originate from the lateral epicondyle of the humerus. The wrist exten-

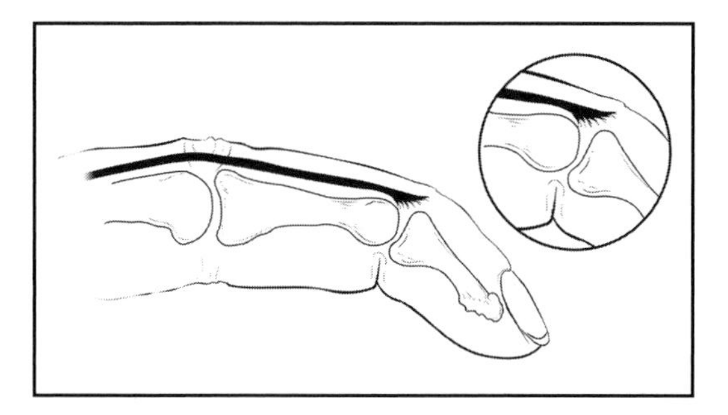

Figure 4
Mallet finger deformity from tendon rupture.

*(Reproduced from Sullivan JA, Anderson SJ (eds): **Care of the Young Athlete**. Rosemont, IL, American Academy of Orthopaedic Surgeons and American Academy of Pediatrics, 2000, p 356.)*

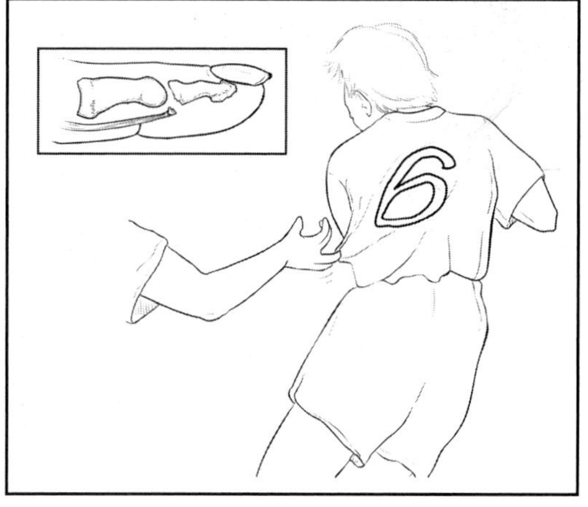

Figure 5
Mechanism of injury in rupture of the profundus tendon.

*(Reproduced with permission from Carter PR: **Common Hand Injuries and Infections: A Practical Approach to Early Treatment**. Philadelphia, PA, WB Saunders, 1983.)*

sors are active not only when the wrist is moving in dorsiflexion (extension), but also when the wrist is completely motionless and resisting forced flexion. A classic example of this action is when a tennis player hits a tennis ball with a backhand stroke. The wrist is motionless as the hand is held rigid against the force of the ball, which, left unopposed, will force the wrist into flexion. This force may create a traction injury on the extensor tendons at the elbow. Initially, only participation in sports activities is affected; however, over time, even simple grasping may become painful. Treatment for lateral epicondylitis includes analgesics, rest, and stretching exercises. Injections of steroids into the painful area have also been used with some success.[7]

Wrist and Hand

Hand injuries in sports are common. Direct impact from catching a ball or colliding with an opponent may produce sprains and fractures. If the finger is forced into flexion against resistance, the extensor tendons attaching to the distal phalanx may rupture (Fig. 4). The resulting injury, known as mallet finger, is often caused by direct contact from a ball.[8] Splinting the finger in a slightly hyperextended position for 6 weeks is usually sufficient treatment. With such treatment, complete healing and restoration of function is the norm.

Flexor tendons may also rupture during sports activities, for example, when a football player attempts to tackle an opponent by grabbing his jersey with the fingers. As the opponent tries to escape, the flexor digitorum profundus of the tackling player avulses off the distal phalanx, resulting in a jersey finger injury (Fig. 5). This injury is analogous to the mallet finger injury seen on the extensor side of the digit. The injuries are similar in that both are separations caused by traction. The key distinction, however, is that the flexor tendons are apt to retract much further—at times, they may even retract all the way into the palm. Accordingly, a splint is inadequate treatment, and surgery is required. The flexor tendons of the wrist, originating from the medial epicondyle, can also be injured as the result of athletic activity. Known as golfer's elbow, this is an overuse injury. Treatment is similar to that for lateral epicondylitis.

Ligaments of the fingers can be injured as well. One of the more significant injuries is rupture of the ulnar collateral ligament of the thumb. Skiers are especially prone to damaging this ligament when they fall while grasping their poles. Mild injuries are treated with protective splints and taping. Complete ruptures require surgical repair.

Hip and Thigh

Pain in the hip joint can be caused by ligament injuries, stress fractures, referred pain from the spine, bursitis, and muscle or tendon injuries. Colloquially, muscle strains are referred to as muscle pulls, but the damage usually occurs at the musculotendinous junction or within the tendon itself. The two common mechanisms of injury are overuse (weakening of the tendon) and a violent force

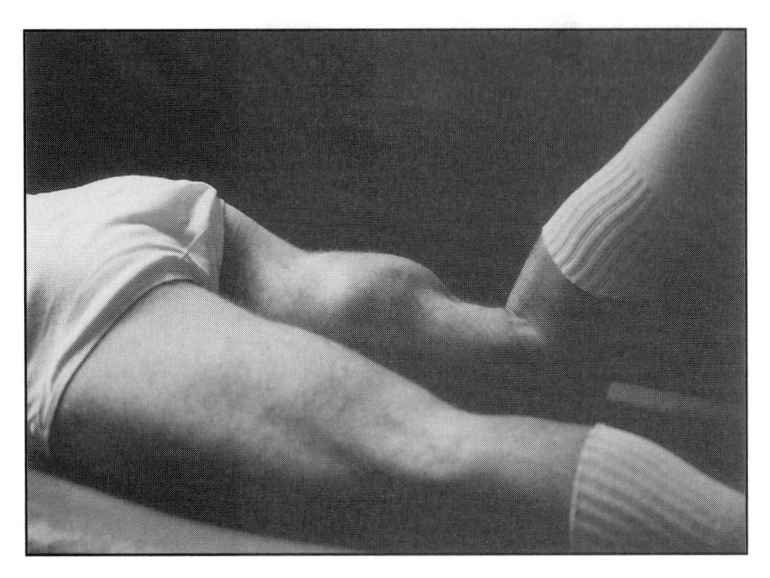

Figure 6
Clinical photograph of an athlete with a large tear in the right hamstring muscle group causing a large bulge.

(Reproduced from Clanton TO, Coupe KJ: Hamstring strains in athletes: Diagnosis and treatment. J Am Acad Orthop Surg 1998;6:237-248.)

against a contracted muscle. Avulsion injury can occur at the area of insertion of tendon into bone, especially at secondary centers of ossification. Typical areas include the ischial tuberosity or the anterior-superior iliac spine, where the hamstring and the sartorius respectively originate. Muscle belly injuries can occur in the hamstrings or quadriceps (Fig. 6).

The bones of the hip joint are rarely fractured in sports, but one such bony injury, a femoral neck stress fracture, can be devastating if not promptly diagnosed. Specifically, a femoral neck stress fracture on the superior surface of the neck (the tension side of bone) is a serious injury because if the fracture propagates and displaces, the displacement can injure the arteries feeding the femoral head. Without its blood supply, the femoral head will die, and the hip joint will collapse.[9] Management of this stress fracture demands immediate cessation of activity, protected weight bearing, and consideration of prophylactic surgical fixation. A stress fracture on the inferior surface of the neck (compression side of bone) is less likely to displace and more likely to heal with nonsurgical treatment.

The thigh is a common site for contusions among athletes, especially those who participate in contact sports. The thigh has a large muscle mass; consequently, when it is hit with sufficient force, there may be damage to the muscle and bleeding within it. This is not usually a serious problem, but the hematoma that forms can organize and ultimately calcify, a condition called myositis ossificans. This is a form of heterotopic ossification in which bone forms within the bruised muscle itself. Contusions, periosteal injuries, and bursitis can occur at any site of bony prominence or where the bone is relatively subcutaneous. About the hip, common sites include the iliac crest, the greater trochanter, and the ischium. Treatment includes ice, stretching, and padding of the prominent bone.

Knee and Leg

Knee

The knee is one of the most frequently injured joints in sports, second only to the ankle. There are two types of athletic knee injuries: repetitive motion (ie, overuse) injuries and acute traumatic injuries. Overuse injuries include tendinopathy of the quadriceps tendon and the patellar tendon, bursitis of any of the bursae surrounding the knee, and patellofemoral pain. Traumatic injuries include patellar dislocations, sprains of the collateral and cruciate ligaments, and tears of the medial and lateral menisci.

Anterior knee pain associated with overuse can be a diagnostic challenge and even more difficult to treat (as the vigorous athlete may not want to rest as needed). Chondromalacia patella, which literally means "softening of the cartilage of the patella," is commonly used to describe many cases of nonspecific anterior knee pain, but this term should be avoided unless true softening has been documented. Patellofemoral syndrome is more accurate and includes issues of malalignment, instability, early arthritis, and abnormal pressure and tilt. Treatment of anterior knee pain centers on minimizing abnormal forces.[10] Overuse is kept to a minimum, and sitting for long periods with the knee flexed is discouraged. Some patients may find it helpful to wear a centralizing sleeve that keeps the patella within its groove on the femur (the trochlea), thereby distributing the patellofemoral forces over the widest possible area.

Sprains of the MCL are commonly caused by a clipping mechanism in which an athlete is struck on the outside of the knee while the foot is planted, creating tension on the MCL. By contrast, anterior cruciate ligament (ACL) tears are usually noncontact injuries. When an athlete changes direction at high speed, forces that are ordinarily resisted by the ACL tend to subluxate the tibia on the femur. When these forces exceed the tensile strength of the ACL, the ligament ruptures.[11] This injury can cause persistent symptomatic instability. This instability is particularly problematic during high-speed athletic activity and twisting and cutting sports. When instability is present with low-speed activities, suspicion should be raised for associated injuries of the meniscus or other articular surfaces. When ACL rupture produces symptomatic instability, surgical reconstruction is indicated.

Athletic activities also place the menisci at risk. Pivoting activities, especially with the knee hyperflexed, place sheer forces across the posterior horn of the meniscus. Tears of the meniscus are often seen in concert with an ACL injury as well. Meniscal tears can cause acute pain and, when displaced, blocked motion.[11] The treatment of symptomatic meniscal tears is surgical and involves removing torn fragments that lack healing potential. However, every effort should be made to repair and not remove the meniscus because it is an important shock absorber of the knee: without it, posttraumatic arthritis is inevitable.

Leg

Three distinct clinical entities often cause leg pain in the athlete. Pain in a focal area on the anteromedial aspect of the shin, where the tibia resides, suggests a problem in the bone, such as a stress fracture. Pain absent at rest but exacerbated by activity may be chronic exertional compartment syndrome. Pain is most commonly located in the anterior and lateral muscle compartments but may be deep and posterior. Pain in the muscle belly or at the musculotendinous junction is consistent with muscle strains. Shin-splints is a nonspecific term that should be avoided.

Stress fracture is an overuse injury in which the body cannot repair microscopic damage as quickly as it is induced, and pain is generated to signal the need for rest and healing. Exertional compartment syndrome differs from posttraumatic compartment syndrome in that pressures are not elevated to the point at which any damage occurs. Rather, the increased pressure causes decreased perfusion and therefore a painful buildup of lactic acid. This prompts athletes to stop what they are doing. Rest, in turn, leads to dissipation of the pain. Most athletes with an exertional compartment syndrome are pain free at rest and begin to have leg pain and tightness only after at least several minutes of exertion. The exertion leads to relative muscle swelling within a restricted fascial compartment. With no room to expand, the blood vessels collapse, which reduces circulation and leads to muscle ischemia. This, in turn, favors anaerobic metabolism, which produces lactic acid. The diagnosis of exertional compartment syndrome is confirmed by a documented increase in compartmental pressure with activity. At times, surgical release of the tight fascia is indicated.

Foot and Ankle

Ankle

Ankle sprains are perhaps the most common of all athletic injuries. Because the bony configuration of the ankle joint is not inherently stable, the ankle relies on its ligaments for support. These ligaments are prone to injury as the foot experiences great forces during landing. Also, the leg is often at a distance from the body's center of gravity, creating a lever effect and exposing the ankle ligaments to forces many times the body's weight.

Ligaments may tear partially or completely. Most partial ankle ligament injuries need only a period of protection to heal.[12] Early motion may prevent stiffness and will also promote restoration of the ligament fibers along appropriate stress lines. Motion may also help restore proprioception. A complete ligament tear can heal without surgical repair if appropriately immobilized, and an ankle with a chronically torn ligament can be highly functional (with the overlying tendons offering stability). These facts argue against early surgical intervention: even with a severe tear, functional recovery may be attained with nonsurgical treatment. Nonetheless, persistent instability may benefit from surgical reconstruction.

Rupture of the Achilles tendon is usually the result of an athletic event and most commonly seen among middle-aged adults who participate in sports only intermittently (Fig. 7). The tendon will rupture as the gastrocnemius-soleus complex contracts eccentrically, as when landing. Injured athletes will often report feeling as though they had been kicked from behind.[12] Even with a com-

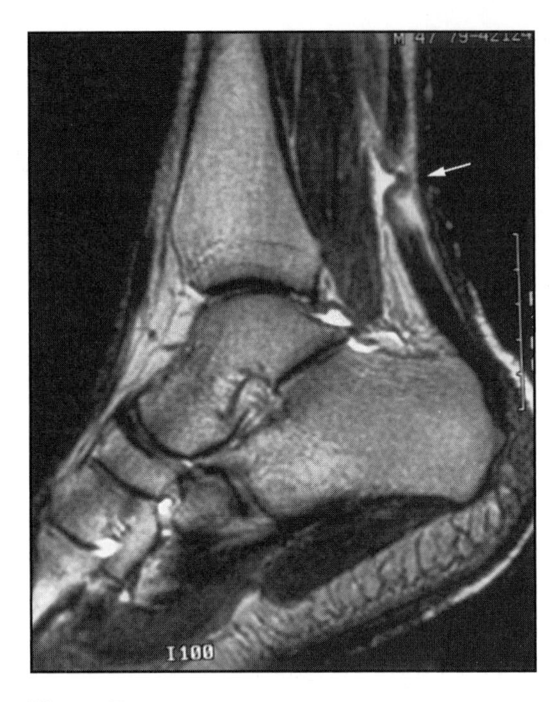

Figure 7
T2-weighted sagittal MRI scan of an acute tear of the Achilles tendon (arrow).

(Reproduced from Saltzman CL, Tearse DS: Achilles tendon injuries. ***J Am Acad Orthop Surg*** *1998;6:316-325.)*

plete rupture of the Achilles tendon, the patient may be able to plantar flex the ankle. This flexion is accomplished by actions of the toe flexors, which course behind the medial malleolus. An accurate diagnosis of Achilles tendon rupture is made by using the Thompson test.

Foot

The foot is not immune to athletic injury. Plantar fasciitis and stress fractures of the calcaneus can cause heel pain. Pain in the midfoot is often caused by stress fractures. Even the toes may be injured in sports. For example, the sesamoids under the great toe can become inflamed in runners, and football players can hyperextend the first metatarsophalangeal joint, thereby damaging the joint capsule.

Spine

Cervical spine injuries may occur during athletic activities, especially when tackling is performed incorrectly. Therefore, tackling with the head first (spearing) is prohibited. After this rule was introduced, the incidence of quadriplegia from football injuries decreased significantly. Instead of dozens of such injuries per year, fewer than 10 per year typically occur. The mechanism of injury to the spine during spearing is axial loading. Sports activities can also cause hyperextension injuries. These occur, for example, when diving into shallow water. Although cervical spine injuries are rare in sports, team physicians caring for athletes at risk must be familiar with the protocols for evaluating players with suspected neck injuries. For example, the physician should not attempt to move a football player with a suspected neck injury without complete spine immobilization.[13] The helmet and pads should remain in place during transport to the hospital.

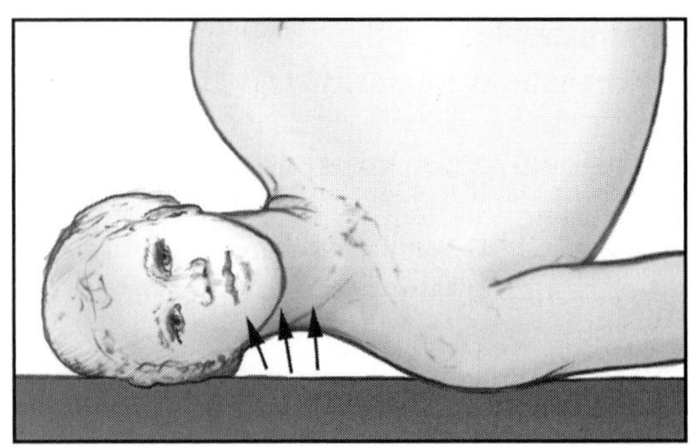

Figure 8
Nerve roots can be injured when stretched during a fall, causing the shoulder to be pulled in the opposite direction of the head.

(Reproduced from Pfeffer GB, Jimenez RL, Sarwark JF, Yurko Griffin L (eds): ***The 2003 Body Almanac: Your Personal Guide to Bone and Joint Health At Any Age****. Rosemont, IL, American Academy of Orthopaedic Surgeons, 2003, p 254.)*

Participation in contact sports may also cause compression or traction injuries to the cervical nerve roots. These injuries, called stingers or burners, are associated with painful sensations radiating down the arm and possibly numbness or weakness.[14] Stingers are common in football and are caused by shoulder depression or lateral bending of the neck during tackling (Fig. 8). Although these are typically minor injuries, the athlete must be evaluated and not allowed to return to play unless full strength and painless range of neck motion are restored.

Low back pain is common in a variety of sports, but it is especially associated with gymnastics and football. A stress fracture of the pars interarticularis, known as spondylosis, is often caused by hyperextension and occurs among gymnasts, football lineman, and other athletes who place axial loads on the spine while in full extension. Oblique radiographs may demonstrate the lesion. The diagnosis is

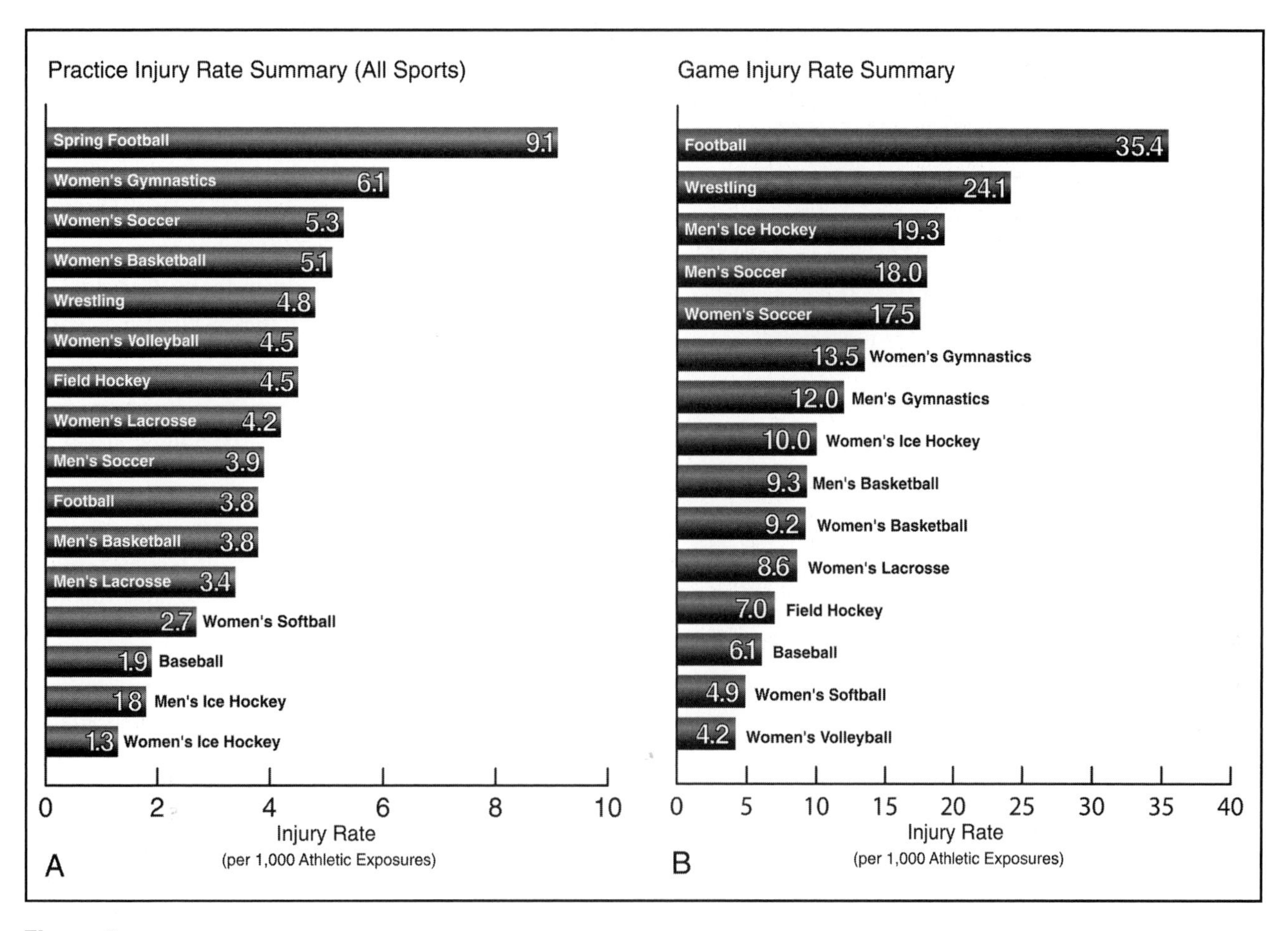

Figure 9
A comparison of practice (**A**) and game-related (**B**) injury rates of various collegiate sports.

(Reproduced with permission from the National Collegiate Athletic Association: ***Sports Medicine Handbook****. Indianapolis, IN, National Collegiate Athletic Association, 2001, pp 82-83.)*

confirmed with bone scanning or MRI. The treatment of this overuse injury is rest and bracing. Progressive instability and anterior displacement of one vertebral body on the one below, a condition called spondylolisthesis, is a rare complication in athletes.

Sport-Specific Injuries

Because such a high correlation exists between a given sport and the injuries it produces, diagnosis can often be made on knowledge of the athlete's sport and presenting complaint alone.[15] The National Collegiate Athletic Association Surveillance System studies many collegiate sports, and its rigid data collection criteria, consistent definition of injury, and control of the study population size provide an excellent method of assessing injury risk for each particular sport (Fig. 9). The National Institutes of Health also has evaluated injury patterns in young athletes (Fig. 10).

Baseball

Baseball is a sport in which throwing-related injuries are common, especially among pitchers. Frequently encountered conditions include shoulder instability with secondary impingement of the rotator cuff and valgus overload syndromes of the elbow, especially in the skeletally immature thrower. In young throwers, osteochondritis dissecans of the elbow and proximal humeral physis injuries must be considered. Sliding into bases may cause injuries to the fingers, hands, and ankles. All ball sports, especially baseball, have been associated with mallet finger injuries. Batting has also been associated with injuries to the hook of the hamate bone in the wrist.

Lacrosse

Lacrosse is a high-energy, physically brutal running sport in which participants use netted sticks to pass a ball. The running and twisting de-

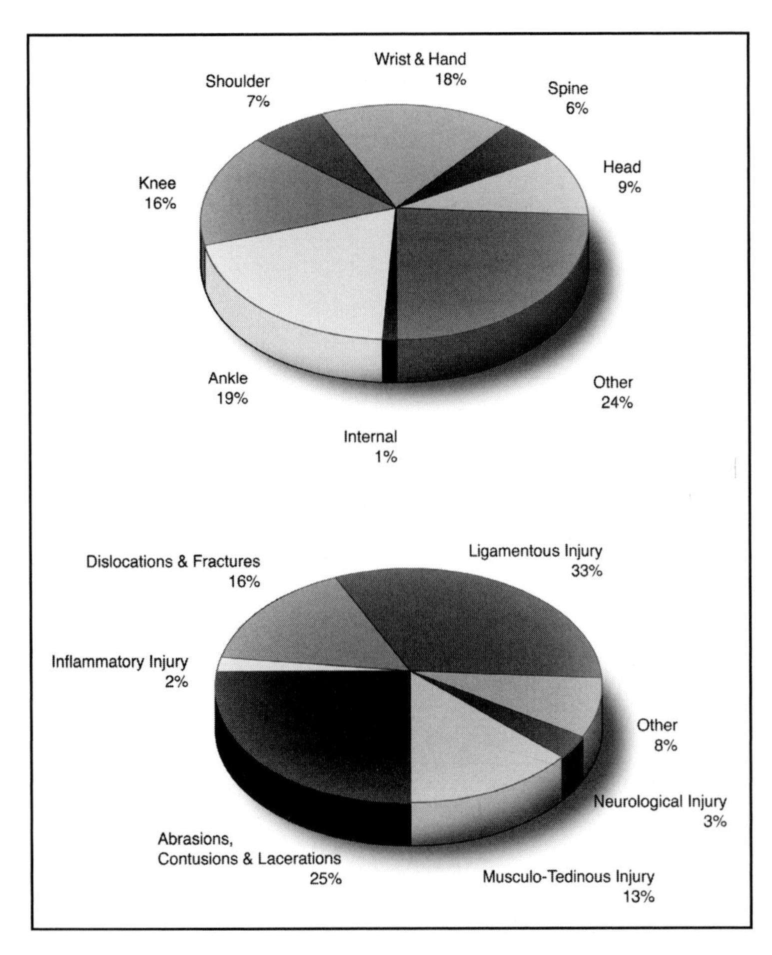

Figure 10
Distribution of athletic injury by anatomic location (**A**) and types (**B**) of injuries sustained by youth across all sports.

(Reproduced with permission from ***Sports Injuries in Youth Surveillance Strategies****. Bethesda, MD, National Institutes of Health, 1992, p 43. NIH Publication No. 93-3444.)*

mand of such activity increases the risk of knee and ankle injuries. In addition, a netted stick will occasionally hit an athlete, which can cause concussions and forearm, facial, and acromioclavicular joint injuries.

Gymnastics
Gymnastics demands flexibility, strength, and balance. It is unique in the weight-bearing demands it places on the upper extremity, which lead to a relative increased risk of soft-tissue damage of the wrist and impingement of the shoulder. Twisting and cutting increases the possibility of ACL and meniscus damage at the knee. Landing awkwardly also exposes the ankle to damage. Repetitive hyperextension of the lumbar spine places gymnasts at increased risk of spondylolysis. Gymnasts may have an increased likelihood of having eating disorders or the female athlete triad (ie, eating disorders, abnormal or absent menses, and osteoporosis/stress fractures). As noted in Figure 9, gymnastics has the highest injury rate among noncontact sports.

Football
American football is a classic collision sport that has been described as being an excellent laboratory for the study of trauma. Contusions are nearly universal among football players. Fractures and dislocations occur less commonly but are more common than in noncontact sports. Football linemen who crouch into a semisquatted stance are at increased risk of spondylolysis because of the repetitive hyperextension of the spine. Forced contact to the head, neck, and arms increases the risk of injuries to the cervical roots and brachial plexus (stingers). Turf toe, a condition that is not unique to football but was initially described as occurring among football players, is a hyperextension injury of the first metatarsophalangeal joint associated with athletic activity on hard surfaces.

Tennis
Tennis and other racquet sports can result in injury patterns that in many ways mimic those caused by throwing sports. Shoulder instability and secondary impingement are frequently seen in tennis players. Low back pain has been associated with certain types of tennis or racquet sport serves, specifically the American twist, which demands a component of hyperextension. Lateral epicondylitis or tennis elbow is a chronic overuse injury involving tendinopathy of the wrist extensors. It is more common among older players and is often associated with a single-handed, backhand tennis stroke. Because tennis is a sport that can be played occasionally without extensive conditioning or training, players may not have an appropriate sense of their aging or deconditioning, resulting in injuries among middle-aged tennis players that are associated with sudden bursts of activity, such as rupture of the Achilles tendon or rotator cuff.

Basketball
Basketball is a ball sport as well as a jumping sport. Most ball sports pose an increased risk of little finger injuries from catching the thrown ball; in basketball, participants also attempt to swat the ball back at the passer or shooter. Thus, finger injuries are common.

Jumping sports pose an increased risk of knee injuries, particularly injuries of the extensor mechanism. Jumper's knee is a chronic tendinosis of the patellar tendon that is usually located at the distal pole of the patella. Rehabilitation focuses on hamstring stretching, quadriceps strengthening, and treatment with analgesic medications. The knee and ankle are the most common sites of injury in basketball players, and injuries at these sites include knee strains, meniscus injuries, ACL injuries, and repetitive ankle sprains.

Cheerleading

Cheerleading is not yet one of the sports that is routinely evaluated by the National Collegiate Athletic Association, but it certainly demands a high level of athletic activity. Cheerleaders, as might be expected, sustain injuries similar to those seen in gymnastics, although at a lower frequency.[9] The most common injuries are related to overuse and include muscle strains and ligament sprains. Low back pain is commonly caused by activities that involve repetitive hyperextension. Cheerleaders who serve as the base for pyramids have increased risk of shoulder complaints. Relative to the number of participants, cheerleaders also appear to have an inordinate number of catastrophic injuries (cervical spine or head injuries). This is likely related to the distance from the ground achieved with certain stunts, such as basket tosses and pyramids.

Soccer

The lower extremities are used extensively in soccer. Therefore, ankle sprains and knee injuries are more common among soccer players. Osteitis pubis is an injury that is somewhat unique to soccer players. It is caused by repetitive stress on the symphysis pubis from high kicking. In addition, repetitive headers (shooting or passing the ball by hitting it with the head) have been implicated as possibly being related to chronic brain injury in young athletes who play soccer.

Hockey

Ice hockey has the potential to cause a broad range of traumatic injuries. The speed at which ice hockey is played is in itself enough to make it physically risky; however, the use of sharp skates and rigid sticks makes it all the more dangerous. Skate blades can cause lacerations, and high-sticking contact can cause clavicle fractures. The clavicle and the acromioclavicular and sternoclavicular joints are at particular risk of injury when players are checked (ie, body blocked) vigorously into the boards, and catastrophic cervical spine injuries have occurred as a result of being checked from behind.

Martial Arts

The specific demands of martial arts vary. All martial arts that include throwing the opponent, such as judo, pose an increased risk of injury to the clavicle and the acromioclavicular and sternoclavicular joints. This can occur when the thrown athlete lands on the point of the shoulder. Open- and closed-hand contact can lead to fractures and dislocations of the fingers and hands. Kicking can cause injuries in the foot and ankle.

Volleyball

Volleyball involves jumping, diving, and repetitive overhead use of the arms for blocks, serves, and spikes. The repetitive impact load can lead to stress fractures and jumper's knee, and twisting and cutting activities increase the risk of ACL and meniscal injury. In addition, abrasions and contusions are associated with digs and dives. Shoulder instability can develop from repetitive overuse shoulder activity. Contact with the volleyball net may cause fingertip injuries as well.

Swimming

Swimming has been correlated with a high incidence of shoulder complaints, which are probably caused by acquired instability. There may be injuries related to specific strokes. For example, the whip kick used with the breaststroke may irritate the medial side of the knee or the MCL, resulting in a condition known as breaststroker's knee.

Diving

Divers may present with injury patterns similar to those of swimmers, but diving also introduces some unique injury risks. Certain dives require significant hyperextension in the lumbar spine; therefore, diving may be related to an increased risk of spondylolysis. Platform diving from extreme heights has been associated with inferior shoulder instability secondary to the forces transferred down the arms when flat hands make contact with the water. If divers over-rotate, their shoulders can be forced into extremes of flexion and dislocate.

Running

Cross-country and long-distance runners have predominantly lower extremity problems, as might be expected. Stress fractures, especially of the tibia and metatarsals, are frequently seen in these runners, especially among those who adhere to aggressive training schedules. Chronic exertional compartment syndrome is also more common in cross-country and long-distance runners.

The Female Athlete

Female athletes have unique concerns compared with their male counterparts. First, women are not only smaller than men on average, but they have different anatomic proportions.[16] The relative differences in anatomy may be advantageous for some sports. For instance, a lower center of gravity is advantageous in certain gymnastic activities. Nevertheless, these differences may also be a source of problems. For example, the wider pelvis of the female athlete creates a higher valgus angle at the knee (ie, women are more knock-kneed than men). Accordingly, problems associated with lateral patellar tracking tend to occur more frequently among female athletes than among male athletes. Additionally, women tend to have a smaller notch between their femoral condyles than men, and this lack of space has been cited as the cause of the higher incidence of ACL tears among women.[17]

The second unique concern of female athletes is that women have monthly fluctuations of their estrogen levels. The menstrual cycle impacts the musculoskeletal system because estrogen couples osteoclastic bone resorption to osteoblastic bone formation. Thus, osteoporosis and poor repair of stress fractures may result when estrogen levels are low. The constellation of abnormal or absent menses, eating disorders, and osteoporosis/stress fractures has been termed the female athlete triad.[18] Female athletes who participate in aesthetic sports, such as figure skating, dance, or gymnastics, may be at increased risk for the female athletic triad. It should also be noted that a formal eating disorder does not require avoidance of eating: one may purge a large meal by overexercising.

The Aging Athlete

Although this category of athlete defies precise definition, the aging athlete can be characterized as one in whom degenerative processes contribute to sports-related musculoskeletal problems. Because aging athletes tend to be less flexible, they are prone to sprains and strains. Because they have lower bone density, they are at increased risk of fracture. It is also likely that because some aging athletes may be less aware of their physical limitations, they are prone to overuse injuries as well.

Given that many older people participate in athletic activities, physicians may need to consider what once may have been atypical diagnoses. For example, leg pain in a younger athlete is most likely caused by stress fractures, muscle strains, or exertional compartment syndrome. Among older athletes, vascular claudication (peripheral vascular disease and ischemia) and neurogenic claudication (spinal stenosis) must be considered as well. In both of these conditions, calf muscle pain occurs with activity. However, the ischemic pain associated with vascular claudication occurs invariably with activity and resolves with rest; pain associated with neurogenic claudication is more intermittent and dissipates more slowly with rest. Additionally, spinal flexion—a position that is assumed while bicycling, for instance—reduces or obliterates the symptoms of neurogenic claudication, a finding not seen with vascular claudication.

Aging athletes are apt to be more fit and healthier than their inactive counterparts. For this reason, older patients should be encouraged to participate in sports at whatever level is possible. Like other athletes, they should be reminded to warm up before any activity.

The Pediatric Athlete

Children are not simply small adults. Specifically, their open physes (growth plates) introduce the risk of a unique group of injuries, especially in early adolescence. The single most important factor to consider when evaluating pediatric athletes is that, under tension, the physis tends to fail before the ligament fails. Thus, all sprains in this group should be considered growth plate fractures until proven otherwise. Nondisplaced physeal injuries carry an increased risk of displacement if not adequately treated.

Certain symptoms in adults usually suggest a diagnosis of tendinosis, but in growing children the attachment site of tendons tends

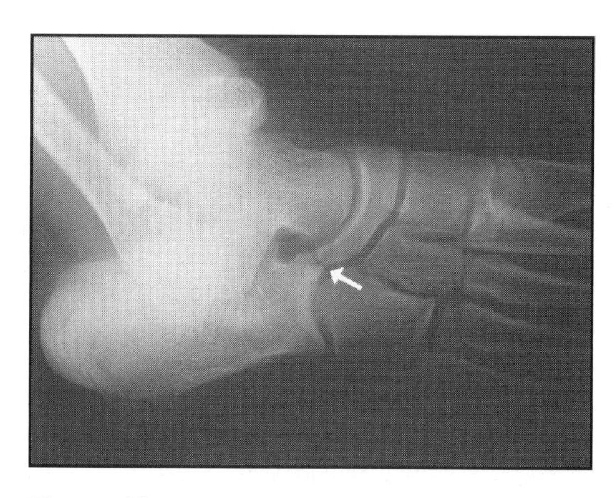

Figure 11
Oblique radiograph of a foot with a cartilaginous calcaneonavicular coalition (arrow).

*(Reproduced from Greene WB (ed): **Essentials of Musculoskeletal Care**, ed 2. Rosemont, IL, American Academy of Orthopaedic Surgeons, 2001, p 715.)*

to be irritated rather than the tendons themselves. Over time, the bony attachment may become irritated (a condition called apophysitis) or pull off completely. A classic location for such an avulsion is the tibial tubercle (a condition called Osgood-Schlatter's disease).[19] Muscle strains are unlikely in this age group because the muscle is stronger than the bony attachment. Instead of a hamstring strains a pediatric athlete will usually avulse part of the ischial tuberosity at the hip.

Of special consideration in young athletes is a condition called osteochondritis dissecans. This is a lesion of the articular cartilage that is probably induced from repetitive overload.[19] It is seen, for example, in the elbows of young pitchers. This lesion may heal if the inciting activity is stopped, but it may also require surgical treatment.

Finally, occult congenital defects should be considered—minor problems that are not manifest with normal activity but appear when children become more active than usual. For example, a child may have been born with a tarsal coalition, a congenital abnormal attachment of the bones of the feet (Fig. 11). This child could be unaware of this condition until athletic activities involving running are attempted. When athletes who are 11 to 15 years of age develop recurrent ankle sprains, a tarsal coalition should be considered.

Athletes with Disabilities

The Special Olympics has been a great motivator of athletes with disabilities who might not otherwise have an opportunity to participate in athletic competition. Nonetheless, because they are at higher risk for injury, athletes with disabilities should participate in competitive sports only after appropriate medical screening. Some risk factors may be obvious. For example, a wheelchair racer is at increased risk for rotator cuff injury, and an amputee sprinter may have chafing of the tibial stump from the prosthesis. Other risk factors are less apparent. For example, athletes with Down syndrome are at increased risk of odontoid dysplasia and atlantoaxial (C1-C2) instability. Preparticipation screening in these athletes should include lateral flexion-extension radiographs of the cervical spine if participation in athletic activities involving possible collisions is being considered.

Key Terms

Exertional compartment syndrome A condition in which exertion leads to relative muscle swelling within a restricted fascial compartment, reduced circulation, muscle ischemia, and the painful buildup of lactic acid

Female athlete triad The constellation of abnormal or absent menses, eating disorders, and osteoporosis/stress fractures

Jersey finger Traumatic rupture of the deep flexor tendon of the ring finger, often as a result of grabbing an opponent's jersey

Jumper's knee Chronic tendinosis of the patellar tendon that is usually located at the distal pole of the patella

Lateral epicondylitis An irritation or partial tear of the extensor tendons of the wrist near their origin at the elbow; also called tennis elbow

Mallet finger An injury often caused by direct contact with a ball in which the finger is forced into flexion against resistance and the extensor tendons attaching to the distal phalanx may rupture

Myositis ossificans Abnormal production of bone within muscle

Osteitis pubis An inflammatory condition of the pubic bones caused by repetitive stress on the symphysis pubis

Osteochondritis dissecans A localized abnormality of a focal portion of the subchondral bone, which can result in loss of support for the overlying articular cartilage

Osteolysis Dissolution of bone, particularly as resulting from excessive resorption

Patellofemoral syndrome Patellar and peripatellar pain resulting from patellar malalignment, instability, abnormal pressure and tilt, or early arthritis

Stingers (also known as burners) Injuries to the cervical nerve roots caused by compression or traction that can occur with shoulder depression or lateral bending of the neck

Stress fracture An overuse injury in which the body cannot repair microscopic damage to the bone as quickly as it is induced, leading to painful, weakened bone

Subluxation Partial or incomplete dissociation of joint surfaces

Turf toe A hyperextension injury of the first metatarsophalangeal joint associated with athletic activity on hard surfaces

References

1. Rowe CR, Zarins B: Recurrent transient subluxation of the shoulder. *J Bone Joint Surg Am* 1981;63:863-872.
2. Burkhart SS, Morgan CD, Kibler WB: Shoulder injuries in overhead athletes: The "dead arm" revisited. *Clin Sports Med* 2000;19:125-158.
3. Perry J: Anatomy and biomechanics of the shoulder in throwing, swimming, gymnastics, and tennis. *Clin Sports Med* 1983;2:247-270.
4. Burra G, Andrews JR: Acute shoulder and elbow dislocations in the athlete. *Orthop Clin North Am* 2002;33:479-495.
5. Brunelli MP, Gill TJ: Fractures and tendon injuries of the athletic shoulder. *Orthop Clin North Am* 2002;33:497-508.
6. Hyman J, Breazeale NM, Altchek DW: Valgus instability of the elbow in athletes. *Clin Sports Med* 2001;20:25-45.
7. Ciccotti MG, Charlton WP: Epicondylitis in the athlete. *Clin Sports Med* 2001;20:77-93.
8. Lee SJ, Montgomery K: Athletic hand injuries. *Orthop Clin North Am* 2002;33:547-554.
9. Scopp JM, Moorman CT III: Acute athletic trauma to the hip and pelvis. *Orthop Clin North Am* 2002;33:555-563.
10. Panni AS, Biedert RM, Maffulli N, et al: Overuse injuries of the extensor mechanism in athletes. *Clin Sports Med* 2002;21:483-498.
11. Iobst CA, Stanitski CL: Acute knee injuries. *Clin Sports Med* 2000;19:621-635.
12. Title CI, Katchis SD: Traumatic foot and ankle injuries in the athlete. *Orthop Clin North Am* 2002;33:587-598.
13. McAlindon RJ: On field evaluation and management of head and neck injured athletes. *Clin Sports Med* 2002;21:1-14.
14. Shannon B, Klimkiewicz JJ: Cervical burners in the athlete. *Clin Sports Med* 2002;21:29-35.
15. Luckstead EF Sr., Satran AL, Patel DR: Sport injury profiles, training and rehabilitation issues in American sports. *Pediatr Clin North Am* 2002;49:753-767.
16. Greydanus DE, Patel DR: The female athlete: Before and beyond puberty. *Pediatr Clin North Am* 2002;49:553-580.
17. Ireland ML: The female ACL: Why is it more prone to injury? *Orthop Clin North Am* 2002;33:637-651.
18. Hobart JA, Smucker DR: The female athlete triad. *Am Fam Physician* 2000;61:3357-3367.
19. Patel DR, Nelson TL: Sports injuries in adolescents. *Med Clin North Am* 2000;84:983-1007.

Chapter 24

Special Concerns of the Pediatric Patient

Children are not simply smaller versions of adults. Their vast growth potential allows children at times to correct deformities that would lead to poor outcomes among adults. Consideration of long-term outcome is of critical importance in children because their bones, joints, and muscles must remain functional and pain-free not just during childhood but throughout the rest of their lives. Finally, the social and psychological toll that illness and disability take on a child and his or her family must be considered.

This chapter describes a wide range of topics that have been grouped together rather broadly. The disorders unite around the common theme of occurrence in childhood; the effects of these disorders, however, may not be limited to a person's younger years.

Rotational Deformities

Intoeing and outtoeing are probably the most common nontraumatic musculoskeletal disorders that prompt parents to consult a physician. Because intoeing is more common than outtoeing, the chief complaint is generally that the child is "pigeon-toed" or that his or her feet turn in when he or she stands, walks, or runs. Parents will often express concern that the intoeing leads to falls, and they may state that they needed braces or special shoes for a similar problem when they were children. It is sometimes helpful to reassure the family that most rotational deformities are variations of normal development and that they correct spontaneously without the need for surgical intervention.

Intoeing has three possible locations of origin: in the foot itself; in the leg, between the knee and the ankle; or in the thigh, between the hip and the knee. When the origin is the foot itself, the condition is known as metatarsus adductus, meaning that the forefoot is angled toward the midline of the body (Fig. 1). Metatarsus adductus is a packaging defect, meaning that it is a deformity that is a direct result of the positioning of the baby inside the mother's womb.

Metatarsus adductus can also be classified according to flexibility, and this classification is useful in determining the necessary treatment. Metatarsus adductus is considered actively correctable if the infant straightens the foot in response to being tickled. Passively correctable metatarsus adductus does not correct when the foot is tickled but can correct with gentle laterally directed pressure on the first metatarsal head. Rigid metatarsus adductus does not correct even with stretching.

For actively correctable metatarsus adductus, no specific treatment is necessary; spontaneous correction is the rule. For passively correctable metatarsus adductus,

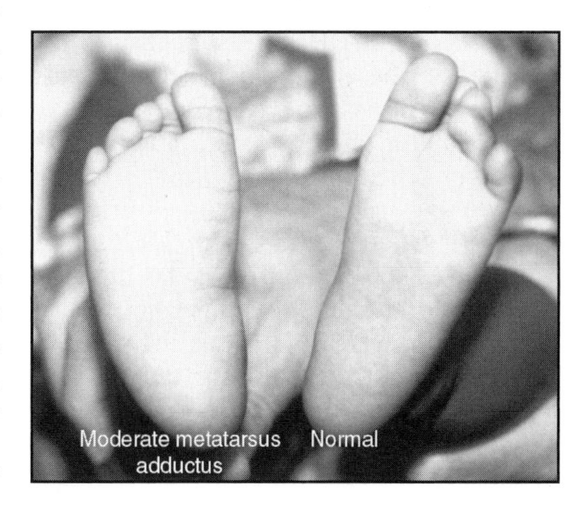

Figure 1
Metatarsus adductus.

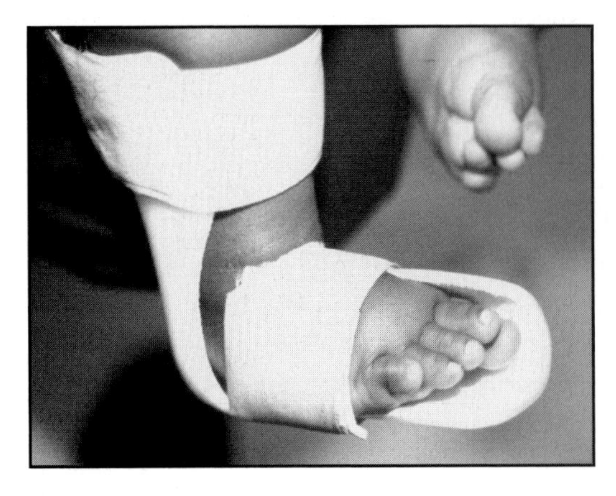

Figure 2
Wheaton brace for treatment of metatarsus adductus.

stretching exercises, in which the deformity is corrected as described above, are done five times per foot with each diaper change, holding the correction for 10 seconds each time. For rigid feet, either stretch casting, in which a series of casts is used to straighten out the deformity gradually, or the Wheaton brace, a removable, plastic version of a stretch cast, are treatment options (Fig. 2). It is important to differentiate between isolated metatarsus adductus and clubfoot.

Clubfoot is a more complex foot disorder that includes three separate deformities, of which metatarsus adductus is one. The hallmark of a clubfoot is a plantar flexion deformity called equinus, ("like a horse," which indeed walks on its toes). Plantar flexion is not a feature of metatarsus adductus. Finally, there may be an association between metatarsus adductus and developmental dysplasia of the hip (DDH).[1,2] Although it is not certain that these two conditions are truly associated, it is especially important to screen children with metatarsus adductus for DDH.

Angular Deformities

Common lower extremity angular deformities in children include genu varum (Fig. 3) and genu valgum (Fig. 4), colloquially known as bowlegs and knock-knees, respectively. Like the rotational deformities, these are usually a variation of normal, and a typical progression has been described.[3] All infants are born bowlegged, although parents usually notice this only when children start to pull themselves up to stand. Between the ages of 2 and 3 years, the bowlegging gradually decreases, and by the age of 3 years, children are maximally knock-kneed. The knees straighten somewhat over the next several years, and by the age of 7 years, most children will have reached the typical adult configuration, which is slightly valgus (knock-kneed). An explanation and reassurance to the parents are thus often the only intervention necessary for these common lower extremity angular deformities.

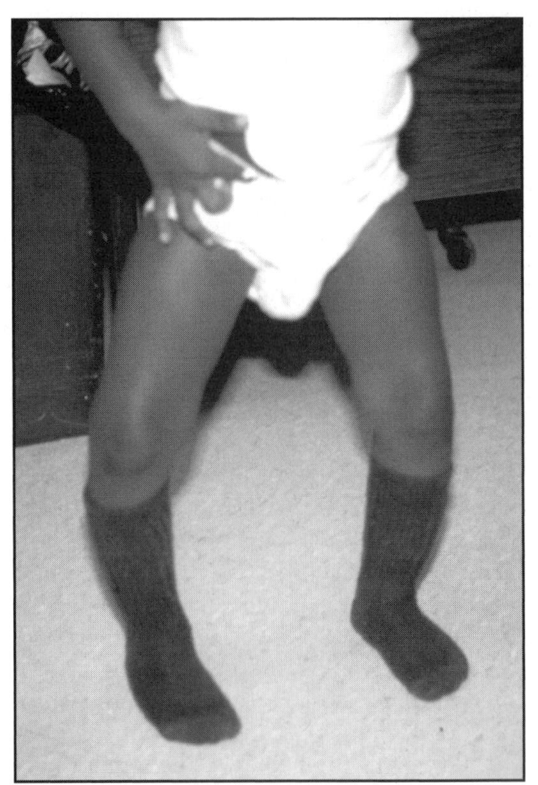

Figure 3
Genu varum.

An angular deformity is considered pathologic when it is asymmetric, unilateral, or painful or when the trend of progression is different than expected. The differential diagnosis includes entities such as Blount's disease, tumor, and infection, all of which can cause damage to the growth plate and result in angular deformity. Systemic conditions such as rickets, renal disease, and various dysplasias (dwarfisms) are possible as well.

Blount's disease is a condition of unknown etiology in which the medial proximal tibial physis ceases to function appropriately, leading to relative overgrowth of the lateral side of the tibia. This results in curving of the tibia and genu varum (Fig. 5). The disease is particularly common in African-

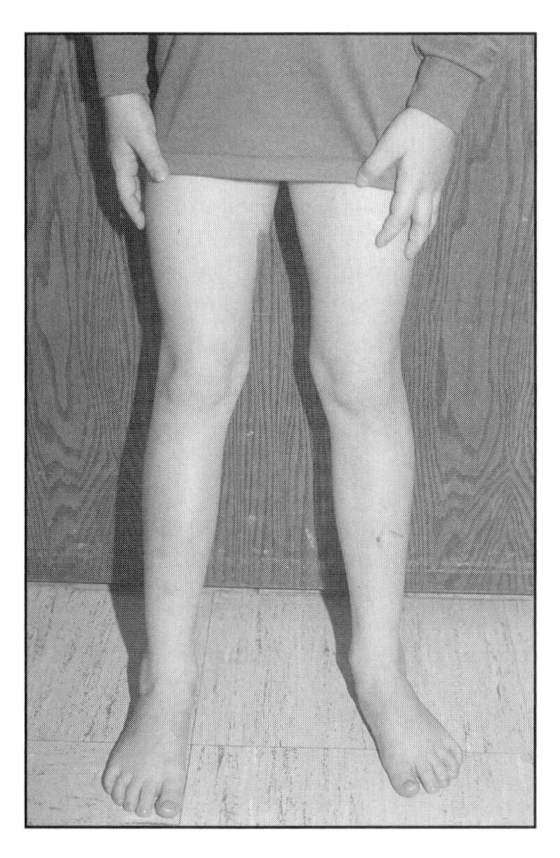

Figure 4
Genu valgum.

*(Reproduced from Sullivan JA, Anderson SJ (eds): **Care of the Young Athlete**. Rosemont, IL, American Academy of Orthopaedic Surgeons and American Academy of Pediatrics, 2000, p 380.)*

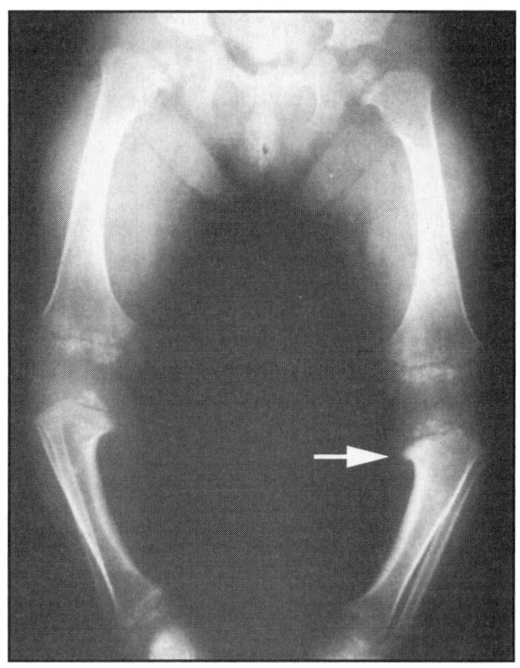

Figure 5
Radiograph of a patient with Blount's disease, showing medial collapse (arrow).

American children, girls, children who are large for their age, and children who were "early walkers" (ie, walked when they were younger than 11 months). Although bracing has been tried in the past, the treatment for infantile Blount's disease today is surgical realignment of the tibia and restoration of the proper mechanical axis of the leg.

Foot Deformities

Clubfoot

Clubfoot (talipes equinovarus) is a relatively common congenital deformity that occurs in 1 of every 1,000 live births, with about half of those cases being bilateral. The male-to-female ratio for affected children is 2.5:1. Inheritance is multifactorial; however, increased risk does seem to run in families in which one member is already affected. Clubfoot can also be a component of a dysmorphic syndrome or neuromuscular disorder.

The clubfoot deformity consists of three separate components: metatarsus adductus, equinus, and heel varus (Fig. 6). However, unlike isolated metatarsus adductus, clubfoot is not a packaging defect. The etiology of clubfoot is somewhat controversial, but it probably results from a primary germ-plasm

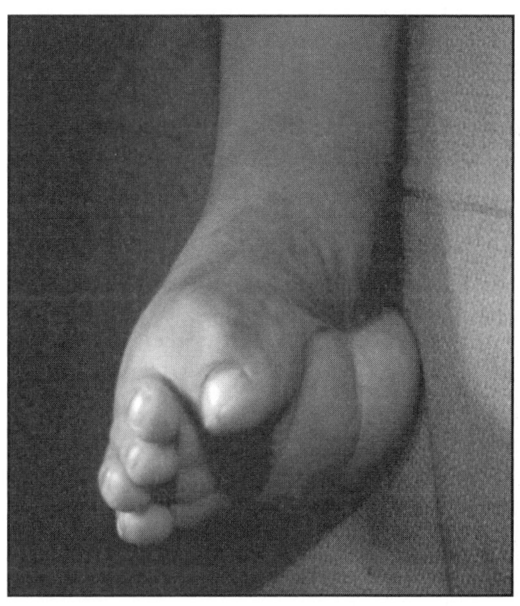

Figure 6
Recurrent clubfoot. This child has previously undergone surgery for clubfoot repair; the surgical scar is visible on the medial aspect of the foot.

defect or from an early intrauterine vascular event.[4] In an affected foot, not only are the tarsal bones misshapen, small, and misaligned, but the musculature in the posterior compartment of the leg is also reduced in size. Therefore, in unilateral cases, it is important to explain to the parents that the affected foot and calf will always be smaller than that of the opposite normal side.

As discussed earlier, clubfoot and DDH are likely to be associated in some way. Therefore, children with an obvious clubfoot must be screened very carefully for DDH, with both serial physical examinations, and appropriate imaging studies.

Treatment for a clubfoot consists of manipulation of the foot (usually with serially applied casts), surgery, or a combination of the two. Generally, casting is attempted for 3 months, and, if unsuccessful, surgery is planned. Traditionally, the success rate for casting alone has been reported to be about 25%, with the remainder of children ultimately requiring surgery.

Flat Feet

Flexible flat feet (pes planus), in which the arch seems to be absent but reappears when the patient stands on the toes, is an autosomal dominant condition associated with generalized ligamentous laxity. The chief complaint is often that the child's ankles "cave in." This condition was once thought to contribute to back and knee problems later in life, but there is no evidence to support this.[5] Most physicians prescribe shoe inserts only in cases in which the patient reports pain. Inserts, however, will not permanently form an arch; they only give the appearance of an arch as long as they are being worn. Finding a shoe with a built-in arch support will often achieve the same result, and wearing high-top shoes will help support the ankles. Reassuring the family that this is not a serious or dangerous condition is often a major component of intervention.

Rigid flat feet, in which the arch does not reform when the patient stands on the toes and in which the subtalar joint has limited motion on examination, is less common and more often symptomatic. In children, the etiology is usually a tarsal coalition, in which some of the tarsal bones (ie, calcaneus, talus, and navicular) are fused. The lack of motion in the fused area puts more stress on the neighboring joints that are initially normal and causes pain. Tarsal coalitions often become symptomatic in early adolescence; before that, the fused joints are cartilaginous and still relatively flexible. Although some tarsal coalitions can be seen on plain radiographs, CT is often the best diagnostic imaging study for this condition. When tarsal coalition is symptomatic, early treatment usually involves immobilization to decrease pain. However, since tarsal coalitions are essentially mechanical problems, they respond best to surgical treatment to remove the abnormal bony connection and restore motion throughout the affected joint.

Hip Disorders

Developmental Dysplasia of the Hip

Developmental dysplasia of the hip (DDH) is a spectrum of abnormalities of the developing hip joint that can include shallowness of the acetabulum (hip socket), capsular laxity and instability, or frank dislocation. DDH was previously known as congenital dislocation of the hip, but it is now understood that the condition is not purely genetic in origin and may arise during development. For this reason, it is critical that children be examined for DDH not only in the immediate newborn period but also periodically until walking age.

DDH is relatively common, occurring in 1 of every 1,000 live births. Risk factors include being female, being firstborn, having been carried or delivered in the breech position, and having a family history of hip dysplasia or ligamentous laxity. The screening examination for DDH consists of looking for asymmetries in the number of skin folds in the thigh, in the height of the affected knee (the Allis or Galeazzi sign), and in the range of abduction (Fig. 7). There are also two provocative tests, the Ortolani and Barlow maneuvers (Fig. 8) designed to elicit clunks when the dislocated femoral head moves in and out of the acetabulum.

Both maneuvers are performed with the infant lying supine on a relatively firm surface with his or her diaper removed. The Ortolani maneuver relocates a dislocated hip and is performed by the physician placing the fingers on the greater trochanters and the thumbs on the knees, abducting the legs and pulling up on the trochanters. As the femoral head reduces into the acetabulum, a clunk will be heard and felt. The Barlow maneuver dislocates the hip and is performed by placing the hands in the

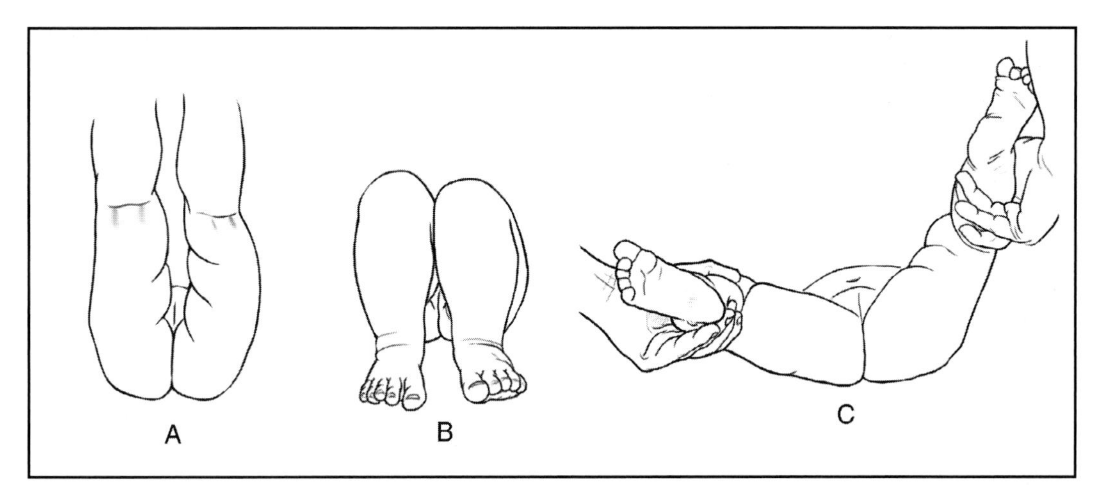

Figure 7
Positive signs for developmental dysplasia of the left hip include increased number of thigh folds (**A**), decreased knee height (**B**), and decreased abduction (**C**).

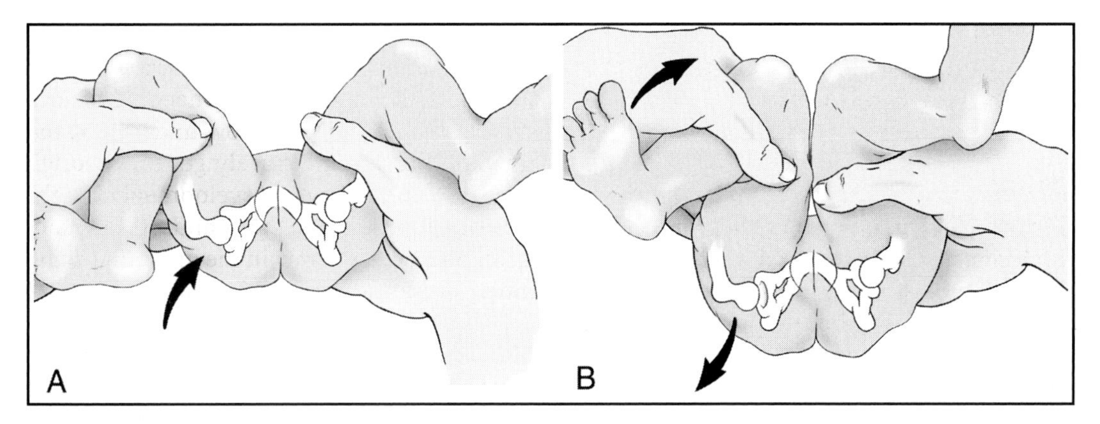

Figure 8
A, The Ortolani maneuver. **B**, The Barlow maneuver.

same position as for the Ortolani maneuver but adducting the legs and pushing down on the knees. A similar clunk will be felt as the femoral head dislocates. Most of the screening examination for DDH is based on detecting asymmetries between the affected and nonaffected sides; thus, bilateral DDH can be particularly difficult to detect.

Because a large portion of the pelvis and hips is not ossified at birth, plain radiographs are not sensitive for the diagnosis of DDH until a child is 4 to 6 months of age. Until that time, the diagnostic imaging study of choice is an ultrasound. This is performed when the child is at least 2 weeks old; if performed before that time, retained maternal ligament relaxing hormones in the infant's circulation may give a false-positive test result.

DDH can be difficult to diagnose, and up to 5% of cases are missed by even experienced examiners. It is therefore important not only to examine infants repeatedly during the first year of life but also to have a low threshold for using ultrasound. Note that in the absence of other disabilities, DDH does not cause significant functional impairment in children, even when the diagnosis is missed or delayed. Children with isolated DDH reach their developmental milestones on time, and they ambulate without difficulty. However, when left untreated, DDH leads to severe early osteoarthritis of the hip. Therefore, the goal of early definitive treatment is the prevention of future degenerative changes.

Slipped Capital Femoral Epiphysis

Slipped capital femoral epiphysis (SCFE) is a displacement (or slipping) of part of the femoral head (the epiphysis) through the growth plate. It occurs when the growth plate is

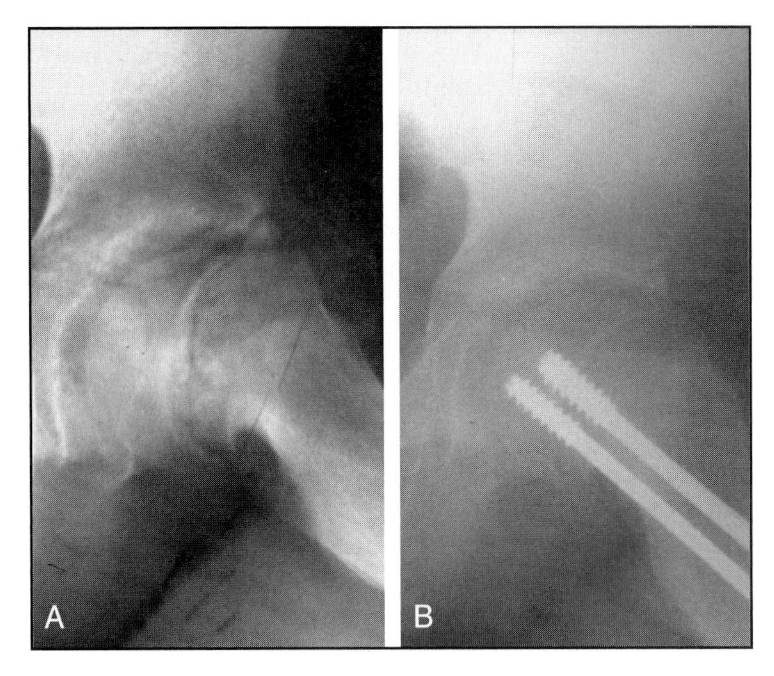

Figure 9
A, Radiograph of a slipped capital femoral epiphysis. **B**, Radiograph of a slipped capital femoral epiphysis after pinning.

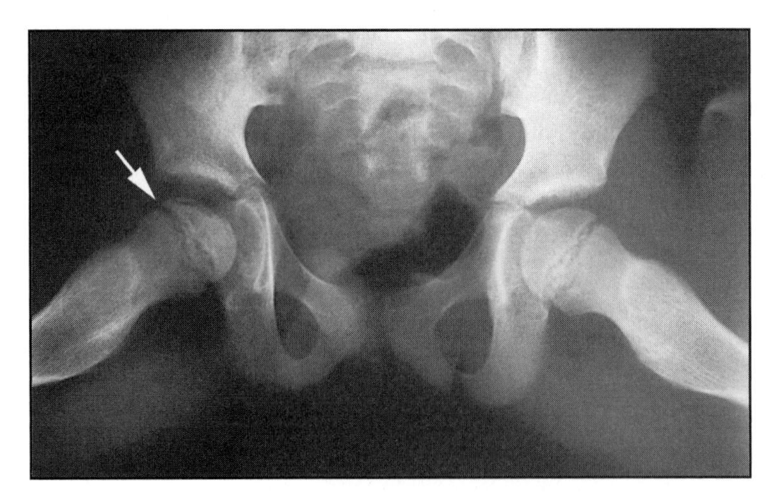

Figure 10
Radiograph showing Legg-Calvé-Perthes disease in a 7-year-old boy with a 2-month history of right hip pain. Note the subchondral fracture (arrow) through the femoral head.

(Reproduced from Beaty JH: Legg-Calvé-Perthes disease: Diagnostic and prognostic techniques, in Barr JS (ed): ***Instructional Course Lectures 38****. Park Ridge, IL, American Academy of Orthopaedic Surgeons, 1989, pp 291-296.)*

highly active, that is, during a period of rapid growth during adolescence (Fig. 9). Although children with SCFE usually have a history of insidious pain or a limp, the onset can also be acute. The typical patient is seen in early adolescence, with the average age at diagnosis being 11 to 13 years for girls and 13 to 15 years for boys. SCFE can be related to hormonal disorders; therefore, children younger than 10 years at diagnosis or those who are in the lowest 10th percentile for height should undergo an endocrine workup.[6,7]

SCFE occurs more commonly among African-Americans than Caucasians; and although the typical patient body habitus of those with SCFE is obese, up to one third of patients are not. The child with SCFE will localize the pain to the groin in less than 50% of cases; of the remainder, most will report pain in the knee or thigh because a branch of the obturator nerve originating in the hip joint ends in the knee joint capsule. Accordingly, hip pathology should be ruled out in any child reporting knee or thigh pain.

The physical examination will reveal a limp, external rotation of the hip, and limited or painful internal hip rotation. The diagnosis of SCFE is generally made with bilateral AP and lateral radiographs of the pelvis. It is imperative to obtain bilateral views of both sides because in 20% of patients SCFE is found on both sides, although one side is often asymptomatic. An additional 30% of patients will present with a SCFE on the opposite side within 1 year of the initial presentation.

The treatment for SCFE is surgical.[8] Many surgeons advocate that the pinning is done in situ, meaning that no attempt is made to reduce the epiphysis back into its original position. The rationale is that such maneuvers may damage the blood supply to the femoral head, which can lead to osteonecrosis. Prophylactic pinning of the opposite side in patients with unilateral SCFE remains controversial.

Legg-Calvé-Perthes Disease

Legg-Calvé-Perthes (LCP) disease, which is also typically known as Perthes disease, is idiopathic osteonecrosis of the femoral head (Fig. 10). The etiology of LCP disease is probably multifactorial, with an underlying genetic or hormonal predisposition and an external catalytic (often traumatic) event.[9] The typical patient is a 4- to 8-year-old boy, who is somewhat small for his age, very active, and has had an insidious onset of pain and limp.

The findings on physical examination of the affected extremity include limited abduction and internal rotation of the hip. AP and lateral radiographs of the pelvis are usually diagnostic. Typically, the affected side shows

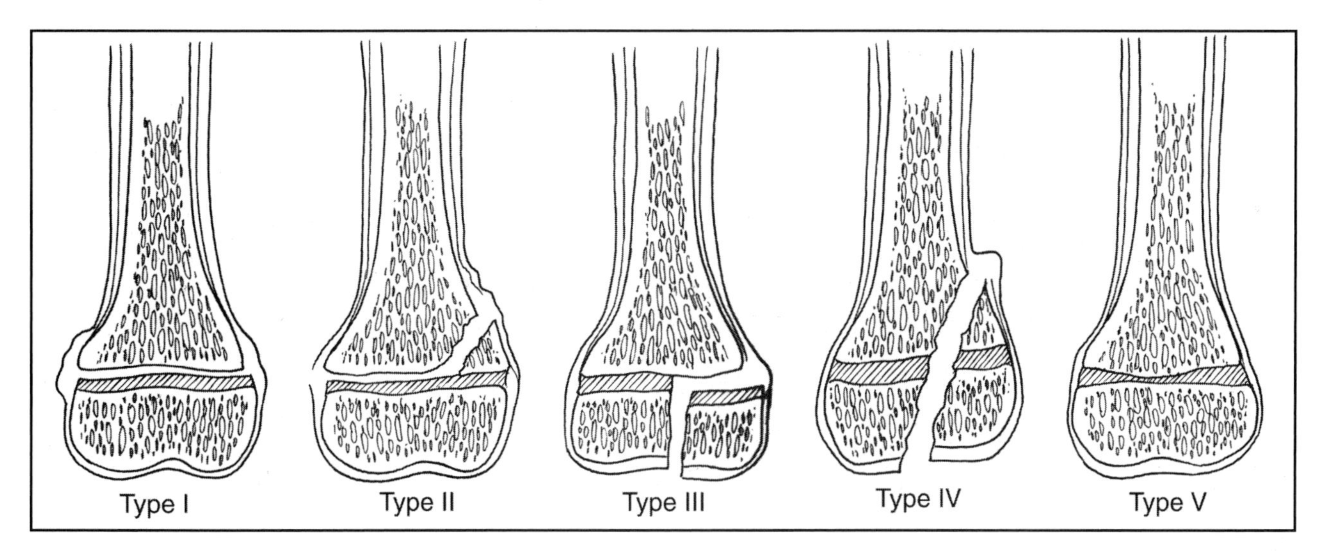

Figure 11
Salter-Harris classification of growth plate fractures.

(Reproduced from Kay RM, Matthys GA: Pediatric ankle fractures: Evaluation and treatment. ***J Am Acad Orthop Surg*** *2001;9:268-278.)*

sclerosis, flattening, and fragmentation of the femoral head. The disease runs a course of about 2 years from the time of diagnosis, during which time the femoral head fragments, subsides, and then slowly reforms.

Treatment includes physical therapy, bracing, and various types of surgery.[9] At present, the underlying disease process cannot be changed; all treatments are aimed at maintaining the femoral head position in the acetabulum, range of motion, and roundness of the head during the period of regrowth. Because reshaping and remodeling of the femoral head will continue throughout patient growth, the prognosis for LCP disease is better the younger the age of the patient at diagnosis.

LCP disease is rarely bilateral; and in such cases, the two hips are affected asymmetrically. Patients with a radiographic appearance of bilateral, symmetric osteonecrosis should be evaluated for possible hemoglobinopathy or skeletal dysplasia.

Growth Plate Fractures

Long bone fractures in children may involve the physis, or growth plate. Injury to the growth plate may impede future growth of the bone. Unlike adults, children rarely experience ligamentous sprains because the physis is weaker than the surrounding ligaments. When stressed, the physis will break before the ligament ruptures.

Salter and Harris[10] devised a classification system for fractures of the growth plate (Fig. 11). A Salter-Harris type I fracture goes directly through the physis. Because the physis is composed of cartilage and is radiolucent, patients with nondisplaced Salter-Harris type I fractures frequently have normal radiographs. In such a situation, if the clinical suspicion is high, the child is treated with immobilization as for a fracture, and a radiograph is rechecked 2 to 3 weeks later, at which time visible periosteal reaction will indicate fracture healing.

Salter-Harris type II fractures go through the physis and metaphysis, leaving a wedge-shaped metaphyseal fragment, connected to the epiphysis. Salter-Harris type II fractures are the most of the common physeal fractures and generally have a good prognosis.

Salter-Harris type III fractures go through the physis and the epiphysis and are therefore intra-articular. Because the contours of the joint must be lined up anatomically to prevent posttraumatic osteoarthritis, Salter-Harris type III fractures generally require surgical correction. They can also lead to growth arrest at the injured area of the physis, resulting in angular or longitudinal deformities of the bone.

Salter-Harris type IV fractures cross both the epiphysis and metaphysis. These fractures almost always must be treated surgically, and they have a high incidence of subsequent growth arrest. Finally, Salter-Harris type V fractures are compression injuries that crush the growth plate across its entire surface. These are quite rare and inevitably result in growth arrest. After acute treatment of a Salter-Harris fracture, patients are generally followed with radiographs for about 1 year to confirm that the physis is functioning properly.

Neuromuscular Disorders

Cerebral Palsy

Cerebral palsy (CP) is a nonprogressive neurologic condition caused by a brain lesion, with a continuum of involvement from mild to severe. The term static encephalopathy is also used to describe CP because the underlying brain injury does not change over time. The hallmark of CP is abnormal muscle control, but decreased sensory function or intellectual development is common as well.

The onset of CP usually occurs in the perinatal period, and there are a wide variety of possible etiologies, including brain malformation, vascular insult, trauma, toxins, fetal or maternal metabolic disease, and infection. CP is associated with prematurity, low birth weight, and perinatal hypoxia. The incidence of CP is about 3.5 in every 1,000 live births.[11]

The anatomic classification of CP describes the areas of the body affected, and the functional classification describes the manner in which the involvement is expressed. Anatomically, patients are described as quadriplegic, meaning that all four extremities are involved; diplegic, indicating that the lower extremities are involved more than the upper; or hemiplegic, in which one side of the body is more involved, typically with the upper extremity on the side most affected. Functionally, most patients exhibit spasticity, ie, increased muscle tone.

For patients with CP, the goals of treatment are to maintain a functional level for activities of daily living, mobility, and ambulation.[12] Treatment may include physical, occupational, and speech therapies; medications; bracing; and surgery. Patients with severe CP may have seizure disorders or nutritional problems. Therefore, a variety of modalities are needed to optimize functional outcome.[13]

Spina Bifida

Spina bifida describes a variety of neural tube defects. The clinical presentation depends on the severity of the defect and on the vertebral level at which it occurs. The incidence is 0.7 to 2 in every 1,000 live births, and the male-to-female ratio is 1:1.15.[14] The etiology is multifactorial; genetic, environmental, and nutritional factors all play a role.

Early in embryonic development, a dorsal thickening of the ectoderm forms. Within that thickening, the neural groove develops. The groove then deepens, and the lateral neural folds develop. By 21 days of gestation, the neural folds fuse to form the neural tube. Spina bifida appears either when there is a failure of fusion of the neural folds or, less commonly, when a rupture occurs after fusion. Clinically, neural tube defects may take the form of a meningocele, in which the vertebral arches are unfused and the meningeal sac is visible at birth; a myelomeningocele, in which neural elements are visible within the sac; or a rachischisis, in which the neural elements are exposed without a sac.

Because fusion of the neural tube occurs early in embryologic development, women must have adequate amounts of folate (400 to 800 µg/day) and avoid hot baths, saunas, and steam rooms, which have been associated with neural tube defects even before pregnancy is confirmed.[15,16]

Spina bifida can be diagnosed prenatally by 16 weeks of gestation. Diagnostic modalities include ultrasound, in which the vertebral defect and sac are usually readily visible; measurement of maternal serum levels of alpha fetoprotein, which are elevated in cases of spina bifida; and amniocentesis, which is used to confirm the diagnosis in patients with a persistently elevated serum alpha fetoprotein level.[14]

Initial treatment is immediate closure of the defect. This is a neurosurgical procedure that can often be scheduled electively in conjunction with a delivery via cesarean section in patients diagnosed prenatally. Immediate closure greatly decreases the incidence of perinatal central nervous system infection.

Clinically, patients with spina bifida manifest a wide range of central nervous system, urologic, anorectal, and musculoskeletal abnormalities. These vary depending on the vertebral level of the defect. Generally, the higher the level, the more severe the clinical problems. Patients with spina bifida often have low to midnormal intelligence and little or no sensory or motor function below the level of the neurosegmental defect.

From a musculoskeletal standpoint, patients with thoracic-level defects typically manifest spine and hip problems, whereas patients with lumbar- and sacral-level defects usually have more knee and foot problems. To ambulate effectively, patients must have quadriceps function, the loss of which corresponds to a defect at or below the L4 nerve root level.[17] Other factors affecting the patient's ability to ambulate include motivation, age, and size. Patients with spina bifida rely on upper body strength to propel themselves, which typically becomes more difficult as their weight increases. Because sensation is impaired, these patients are extremely prone to lower extremity wounds, skin breakdown, and fractures, and they must be carefully monitored for these problems.

Scoliosis

Scoliosis is a three-dimensional curvature of the spine (Fig. 12). Although most pronounced in the frontal plane and thus best viewed on an AP radiograph, scoliosis also includes a rotational component and is therefore three dimensional. There are three categories of scoliosis: idiopathic, congenital, and neuromuscular.

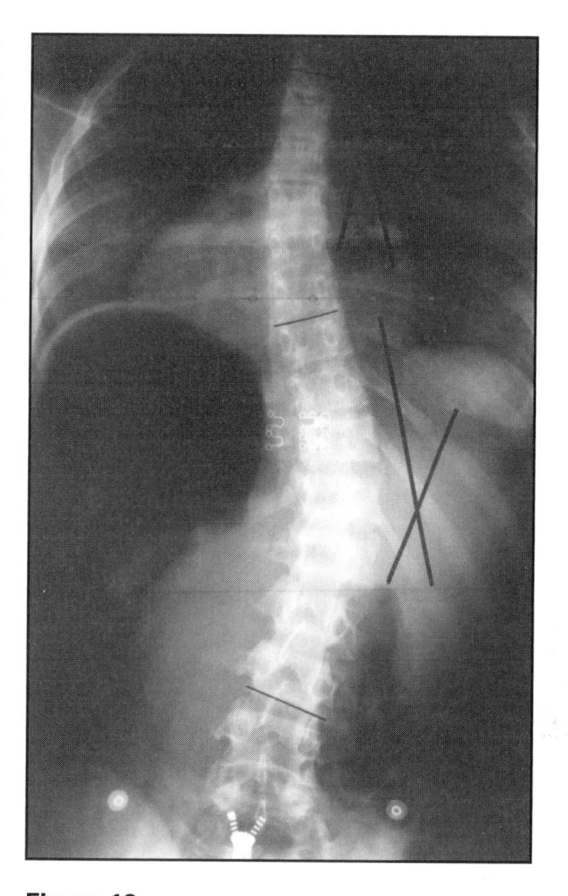

Figure 12
AP radiograph of the spine demonstrating a scoliotic deformity. There is a rotational component to scoliosis that is not necessarily apparent on the AP view.

(Reproduced from Sullivan JA, Anderson SJ (eds): ***Care of the Young Athlete****. Rosemont, IL, American Academy of Orthopaedic Surgeons and American Academy of Pediatrics, 2000, p 303.)*

Idiopathic Scoliosis

Idiopathic curvature of the spine can occur at any age but typically is detected between the ages of 10 and 12 years. Its etiology is uncertain, but there seems to be a genetic component. Idiopathic scoliosis is relatively common, with a prevalence in adolescents of about 2%. However, only 10% of these patients will have a curve severe enough to warrant treatment. Girls are more likely than boys to have a progressive deformity.

Idiopathic scoliosis is generally asymptomatic and is usually diagnosed through a screening examination administered at school or by a pediatrician, using the forward bend test (see chapter 26 for a description of this test). Because the patient's rib cage rotates along with the spine as it curves, the ribs will appear asymmetric when viewed from behind with the child bending forward. This rib hump is the clinical hallmark of scoliosis. An AP radiograph will confirm the diagnosis and allows for measurement of the degree of spinal curvature.

Generally, the curvature continues to progress as long as a child is growing; therefore, early-onset scoliosis carries a poorer prognosis than late-onset scoliosis. Treatment depends on the age of the patient and the degree of curvature. Experts believe that curves of less than 25° should be simply observed for further progression. Curves of 25° to 45° should be braced. The goal of brace treatment is to halt progression of the curve, not to reverse it. Curves greater than 45° are thought to benefit from surgery because they will continue to worsen even after the child stops growing. Surgical treatment involves inserting metal rods and then fusing the spine in position.[18]

Congenital Scoliosis

Congenital scoliosis is a curvature of the spine that occurs because of abnormalities in the vertebrae themselves. In about 50% of patients, congenital scoliosis is progressive and may require surgery. Congenital scoliosis may be a part of a syndrome with other congenital anomalies. Because the embryologic development of the kidneys and heart occurs at approximately the same time as that of the spine, these organs are affected in 15% to 20% of patients with congenital scoliosis. If such an abnormality is suspected, screening can be performed with ultrasound.[19]

Neuromuscular Scoliosis

Neuromuscular scoliosis occurs in patients with a wide variety of disorders, including cerebral palsy, spina bifida, muscular dystrophy, and spinal cord injuries. Scoliosis can develop in any patient whose trunk musculature is weak, unbalanced, or denervated. These curves tend to worsen with time. In nonambulatory patients, a severe curve can adversely affect pulmonary function and the ability to sit in a wheelchair. Bracing and custom-made adaptive seating can play a limited role in the management of these curves, but most curves ultimately require surgical fusion.

Child Abuse

Child abuse, the nonaccidental injuring of a child, is a common problem, but the diagnosis can be difficult. Moreover, there are serious adverse consequences to both underdiagnosis and overdiagnosis of child abuse, which may explain why physicians are uncomfortable with the topic.

The battered child syndrome was first described by Kempe and associates[20] in 1962. They estimated that 25% of fractures in children younger than 1 year and 10% to 15% of fractures in children younger than 3 years are the result of abuse. By 1967, all 50 states had adopted mandatory reporting laws for physicians. In 1968, Haggerty[21] estimated that an abused child who is returned to an unsafe home environment is at 50% risk for additional injury and 10% risk of death over 5 years. These figures are now widely quoted, although Haggerty's estimates were made on the basis of his experience with a sample of only 50 patients.

The National Committee on Prevention of Child Abuse stated in 1998 that there are 3.5 million reports of child abuse per year and that 33% of the reports were substantiated. These figures give a nationwide incidence of approximately 15 for every 1,000 children. According to data collected by the Child Welfare League in the same year, about 1,000 deaths annually result from child abuse. Child abuse is an affliction of young children: 33% are younger than 1 year and 50% are younger than 2 years. Of the victims, 10% to 70% sustain a skeletal injury and 30% to 50% require the care of an orthopaedist.[22]

Many types of abuse exist, including emotional, medical neglect (which typically is manifested as malnutrition or failure to thrive), sexual, and physical. Victims of physical abuse may have soft-tissue injuries, burns, head trauma, internal injuries, or fractures. Fifty percent of fractures secondary to child abuse occur in children younger than 1 year, and the incidence of nonaccidental fractures decreases with increasing patient age.

There is no particular fracture pattern, location, or morphology that is pathognomonic (ie, perfectly diagnostic) of child abuse. However, some fracture findings are more suggestive of abuse than others. Multiple fractures in various stages of healing, posterior rib fractures, bilateral acute long bone fractures, complex skull fractures, and long bone fractures in nonambulatory children are highly suggestive of abuse.[23] Spiral fractures of long bones are not pathognomonic of child abuse. Numerous studies of large series of fractures in abused children have shown that a single, transverse fracture is the most common long bone fracture pattern seen in cases of abuse, although it is not a specific finding.[22,24,25] The most frequently involved bone has been variously reported to be the humerus, tibia, or femur.

In cases of suspected abuse, a careful history and a full physical examination are necessary. Radiographs of areas of suspected fracture should be obtained as well. A skeletal survey (screening radiographs of the entire skeleton) is indicated in all children younger than 2 years with any evidence of physical abuse, all children younger than 1 year with evidence of medical neglect, and possibly all children younger than 5 years with a suspicious acute fracture. A variety of metabolic, genetic, and congenital disorders can mimic child abuse, and all must be considered in the differential diagnosis of a child with a suspicious fracture.

Infection

Musculoskeletal infections in children typically take the form of osteomyelitis, which is infection of the bone, and septic arthritis, which is infection of the joint. These infections are typically acute and caused by the spread of bacteria in the blood. They are also more common in children than in adults. Both osteomyelitis and septic arthritis can cause serious long-term damage to the bone or joint; thus, a timely diagnosis and initiation of appropriate treatment are crucial in their management.

Osteomyelitis

Pediatric osteomyelitis is relatively common, with a reported annual incidence of 1 in 250.[26] This figure represents a significant decrease since the advent of antibiotics. The incidence may now be rising slightly, however, because of increased antibiotic resistance and the increasing number of immunocompromised patients.

Osteomyelitis is generally spread hematogenously (via the dissemination of bacteria in the blood). Even healthy people become bacteremic several times a day; ordinarily, however, the bacteria are easily cleared out of the circulation. In children, the structure of the blood vessels of the metaphyseal regions of the long bones predisposes them to infection. In this region, there is a series of vascular loops with such sharp angles of curvature and small diameters that circulation through them is sluggish, allowing them to become clogged with bacteria. This thereby initiates infection. The role of trauma in this process is not fully understood. Although patients often report a recent injury and experimental models have shown that injury to the circulation may further predispose patients to osteomyelitis, trauma is probably not a necessary condition for the onset of infection.[27]

Because bone is relatively rigid, the pus that forms becomes rapidly pressurized and spreads down the paths of least resistance—down the medullary canal, out through the outer cortex of the bone, and into the subperiosteal space. The physis itself acts as a barrier to pus; therefore, only rarely (typically in neonates) does infection cross it and invade the epiphysis. Once pus has breached the cortex, the infection is described as a subperiosteal abscess. In joints such as the hip and shoulder, in which the metaphysis is essentially intra-articular, a subperiosteal metaphyseal abscess is, in effect, a septic joint.

Left untreated, acute osteomyelitis leads to bone necrosis and resorption. Radiolucent patches in bone and periosteal reaction are typically visible about 2 weeks after the onset of osteomyelitis. Chronic osteomyelitis usually represents either a delay or failure of diagnosis or treatment. In this case, necrotic infected bone is walled off by fibrotic tissue, forming a sequestrum. Because antibiotics typically cannot penetrate the involucrum surrounding the sequestrum, treatment of chronic osteomyelitis usually involves surgical débridement.

The bacteriology of osteomyelitis varies widely, depending on geographic location and patient age. In general, *Staphylococcus aureus* is the most common causative organism in all age groups. *Streptococcus* species are also a common cause of infection in children younger than 4 years, and osteomyelitis caused by enteric organisms (ie, those native to the gastrointestinal tract) commonly occur in neonates. *Haemophilus influenzae* has historically been a common osteomyelitic pathogen, but widespread use of a vaccine against it has greatly diminished its incidence.

Osteomyelitis is relatively common in patients with sickle cell disease. *Staphylococcus aureus* is still the most common pathogen in those with sickle cell disease; however, people with sickle cell disease are far more likely than any other group of patients to have a *Salmonella* infection. *Pseudomonas aeruginosa* may be the causative organism in osteomyelitis that results from stepping on a nail while wearing a shoe, especially a sneaker. The history of footwear is critical because the infectious organism is found in the shoe, not on the nail. These infections can be quite aggressive and may warrant combined antibiotic therapy or early surgical débridement.

The clinical features of acute osteomyelitis in a child include bone pain that restricts activity, tenderness to palpation, and possible swelling. Often, children show few systemic signs of osteomyelitis acutely, such as fever, malaise, or anorexia. Osteomyelitis in neonates may manifest only as generalized irritability or failure to thrive; therefore, it is easy to miss the diagnosis of infection in these patients.

Initial evaluation should include a history and physical examination, complete blood cell count, erythrocyte sedimentation rate (ESR) or C-reactive protein (CRP) level, blood cultures, radiographs, and possibly an aspiration of the bone. The white blood cell count and radiographs are usually normal early in the course of the illness, but they are useful to obtain as a baseline or if there is a question of the duration of illness. The ESR and CRP level are typically elevated, even early in the course of the infection. The CRP level rises and falls more rapidly in response to treatment than does the ESR; thus, it is useful for monitoring the efficacy of the treatment course. Blood cultures will identify the responsible organism in 50% of patients with infection, so it is important to obtain them prior to initiating treatment. Aspirating the bone in addition to obtaining blood cultures increases organism identification from 50% to 70%.

Imaging studies can be useful diagnostic aids when the clinical picture and laboratory study results are equivocal, the site of involvement is in question, the patient cannot cooperate with the examiner, or multifocal disease is suspected. Three-phase technetium 99m bone scanning is both sensitive and specific for osteomyelitis and is usually the initial study of choice. Ultrasound can be used for localizing a subperiosteal abscess. MRI is extremely sensitive for diagnosing and visualizing the extent of a bony infection, but it requires more patient cooperation and a more narrowly defined study area than bone scanning and is, therefore, sometimes less useful in children.

The recommended treatment for acute hematogenously spread osteomyelitis in children is 6 weeks of treatment with antibiotics, usually initially administered intravenously. Treatment begins with empiric coverage for *S aureus*, and antibiotic therapy is subsequently adjusted based on blood culture results and clinical response. Surgical drainage is rarely necessary for acute infections.

Septic Arthritis

Joint infections in children are usually caused by the spread of bacteria hematogenously, but they may also be caused by direct penetration of the joint or extension of an adjacent osteomyelitis, particularly in the hip or shoulder. Septic arthritis occurs when bacteria invade the joint synovium, which is a good medium for bacterial growth. The presence of pus in the joint causes damage to the hyaline cartilage on the joint surface. Changes to the cartilage start to occur between 8 and 12 hours after infection. Hyaline cartilage does not regenerate well; consequently, the damage that occurs is permanent. Septic arthritis is, therefore, a medical emergency, and immediate surgical drainage is often required.

The bacteriology of septic arthritis is similar to that of osteomyelitis, with *S aureus* being the most common pathogen. *Streptococcus* species are also frequently cultured in children with septic arthritis. The prevalence of *H influenzae* is decreasing with the widespread use of a vaccine against it. In sexually active adolescents, the possibility of *Neisseria gonorrhoeae* infection should be considered.

Clinically, children with joint sepsis tend to be slightly sicker than those with osteomyelitis. They are more likely to have a fever and to be lethargic and irritable. They are also more likely to have a history of recent illness, such as an ear or throat infection, or report a recent injury. The involved joint is often warm, tender, and swollen; however, the hallmark of septic arthritis is resistance to both active and passive motion. Because joint pain is mediated via stretch receptors in the joint capsule, the most comfortable position for the patient will be that which increases the volume of the joint. For most joints, this will be a position of flexion; for joints that rotate, it will be a position of flexion and external rotation.

As with children with suspected osteomyelitis, a complete blood cell count, ESR, CRP level, radiographs, and blood cultures should be obtained for children with a suspected septic joint. Aspiration of the suspicious joint is imperative because the gold standard for diagnosis is analysis of the synovial fluid. Aspiration can be performed blindly for accessible large joints like the knee, but it is best performed with ultrasound or fluoroscopy guidance for deeper, smaller joints, particularly the hip. The retrieved fluid should be sent for Gram stain, culture, and cell count. A white blood cell count of more than 50,000/mm^3 almost always indicates sepsis. Inflammatory processes, such as juvenile rheumatoid arthritis, can also present with a white blood cell count as high as 80,000/mm^3; thus, this clinical-decision rule threshold of 50,000/ mm^3 is not perfectly specific. But because the sequalae of missed septic arthritis can be severe, it is best to use a low threshold in equivocal cases.

Toxic synovitis, which is an acute nonbacterial joint inflammation, may appear clinically similar to bacterial sepsis. It can be differentiated, however, on the basis of a normal ESR or CRP level, lack of joint effusion, or low white blood cell count in the joint fluid.

Surgical decompression and drainage of the affected joint is the most important aspect of treatment of septic arthritis, particularly for infections of the hip. Surgical drainage should be performed as soon as possible to minimize damage to the joint surfaces. For accessible joints such as the knee, arthroscopic drainage may be used, but it may not clean the joint as completely as surgical drainage. Antibiotics should be administered after initial joint aspiration and should continue for 1 to 6 weeks after decompression.

Gonococcal septic arthritis is the one exception to the early drainage rule. For gonococcal infection, aspiration and administration of appropriate antibiotics to both the patient and recent partners is usually sufficient.[28,29]

Key Terms

Blount's disease A condition of unknown etiology in which the medial proximal tibial physis ceases to function well, leading to relative overgrowth of the lateral side of the tibia and genu varum

Clubfoot A complex foot disorder that includes three separate deformities: metatarsus adductus, ankle equinus, and heel varus

Developmental dysplasia of the hip (DDH) A spectrum of abnormalities of the developing hip joint that can include shallowness of the acetabulum, capsular laxity and instability, or frank dislocation

Equinus Plantar flexion deformity (so named because horses walk on their toes)

Genu valgum Angular deformity of the lower extremities known as knock-knees

Genu varum Angular deformity of the lower extremities known as bowlegs

Legg-Calvé-Perthes (LCP) disease An idiopathic osteonecrosis of the femoral head

Meningocele A neural tube defect in which the vertebral arches are unfused and the meningeal sac is visible at birth

Metatarsus adductus A condition in which the forefoot is angled toward the midline of the body

Myelomeningocele A neural tube defect in which neural elements are visible within the meningeal sac at birth

Packaging defect A deformity that is a direct result of the positioning of the baby inside the mother's womb

Rachischisis A neural tube defect in which the neural elements are exposed without a sac at birth

Scoliosis Three-dimensional curvature of the spine

Skeletal survey Screening radiographs of the entire skeleton

Slipped capital femoral epiphysis (SCFE) A displacement (or slipping) of part of the epiphysis through the growth plate at the head ("capital") of the femur

Spasticity The condition of having increased muscle tone that results in sudden or involuntary muscle contractions

Tarsal coalition Congenital fusion of the tarsal bones

References

1. Jacobs JE: Metatarsus varus and hip dysplasia. *Clin Orthop* 1960;16:203-213.
2. Kumar SJ, MacEwen GD: The incidence of hip dysplasia with metatarsus adductus. *Clin Orthop* 1982;164:234-235.
3. Salenius P, Vankka E: The development of the tibiofemoral angle in children. *J Bone Joint Surg Am* 1975;57:259-261.
4. Irani RN, Sherman MS: The pathological anatomy of idiopathic clubfoot. *Clin Orthop* 1972;84:14-20.
5. Staheli LT: Footwear for children. *Instr Course Lect* 1994;43:193-197.
6. Burrow SR, Alman B, Wright JG: Short stature as a screening test for endocrinopathy in slipped capital femoral epiphysis. *J Bone Joint Surg Br* 2001;83:263-268.
7. Loder RT, Wittenberg B, DeSilva G: Slipped capital femoral epiphysis associated with endocrine disorders. *J Pediatr Orthop* 1995;15:349-356.
8. Loder RT, Richards BS, Shapiro PS, et al: Acute slipped capital femoral epiphysis: The importance of physeal stability. *J Bone Joint Surg Am* 1993;75:1134-1140.
9. Skaggs DL, Tolo VT: Legg-Calvé-Perthes disease. *J Am Acad Orthop Surg* 1996;4:9-16.
10. Salter RB, Harris WR: Injuries involving the epiphyseal plate. *J Bone Joint Surg Am* 1963;45:587-622.
11. Ellenberg JH, Nelson KB: Birth weight and gestational age in children with cerebral palsy or seizure disorders. *Am J Dis Child* 1979;133:1044-1048.
12. Bleck EE: (ed): *Orthopaedic Management in Cerebral Palsy*. London, England, Mac Keith Press, 1987, pp 142-143.
13. Renshaw TS, Green NE, Griffin PP, et al: Cerebral palsy: Orthopaedic management. *J Bone Joint Surg Am* 1995;77:1590-1606.
14. Tachdjian MO: (ed): Myelomeningocele, in *Pediatric Orthopaedics*, ed 2. Philadelphia PA, WB Saunders, 1990, vol 2, pp 1774-1776.
15. Czeizel AE, Dudas I: Prevention of the first occurrence of neural-tube defects by periconceptional vitamin supplementation. *N Engl J Med* 1992;327:1832-1835.
16. Milunsky A, Ulcickas M, Rothman KJ, et al: Maternal heat exposure and neural tube defects. *JAMA* 1992;268:882-885.
17. Asher M, Olson J: Factors affecting the ambulatory status of patients with spina bifida cystica. *J Bone Joint Surg Am* 1983;65:350-356.
18. Pizzutillo PD: Idiopathic scoliosis and kyphosis. *Instr Course Lect* 1994;43:185-191.
19. Winter RB, Lonstein JE, Boachie-Adjei O: Congenital spinal deformity. *J Bone Joint Surg Am* 1996;8:300-311.
20. Kempe CH, Silverman FN, Steele BF, et al: The battered-child syndrome. *JAMA* 1962;181:17-24.
21. Haggerty RJ: Physically abused children, in Green M, Haggerty RJ(eds): *Ambulatory Pediatrics*. Philadelphia, PA, WB Saunders, 1968, pp 285-290.
22. Scherl SA, Miller L, Lively N, et al: Accidental and nonaccidental femur fractures in children. *Clin Orthop* 2000;376:96-105.
23. Cramer KE: Orthopaedic aspects of child abuse. *Pediatr Clin North Am* 1996;43:1035-1051.
24. Galleno H, Oppenheim WL: The battered child syndrome revisited. *Clin Orthop* 1982;162:11-19.
25. King J, Diefendorf D, Apthorp J, et al: Analysis of 429 fractures in 189 battered children. *J Pediatr Orthop* 1988;8:585-589.
26. Craigen MA, Watters J, Hackett JS: The changing epidemiology of osteomyelitis in children. *J Bone Joint Surg Br* 1992;74:541-545.
27. Morrissy RT, Haynes DW: Acute hematogenous osteomyelitis: A model with trauma as an etiology. *J Pediatr Orthop* 1989;9:447-456.
28. Green NE, Edwards K: Bone and joint infections in children. *Orthop Clin North Am* 1987;18:555-576.
29. Jackson MA, Nelson JD: Etiology and medical management of acute suppurative bone and joint infections in pediatric patients. *J Pediatr Orthop* 1982;2:313-323.

Chapter 25

Special Concerns of the Geriatric Patient

Older adults make up the most rapidly growing segment of the population in the United States. Although there is no clear definition for the term elderly, 65 years and older has generally been used to define this segment of the population, in large part because this is the traditional age of retirement and Medicare eligibility. The first baby boomers (those born in the population surge immediately following World War II) will be age 65 years and older in 2010; by 2030 this group is expected to grow from 13% to 20% of the US population.[1] Because older people experience relatively more disease and disability, the growth of this population segment is expected to have a significant and growing impact on health care delivery and financing systems in the future.[1] Therefore, a working knowledge of expected age-related physiologic changes and familiarity with common geriatric conditions is required of all physicians.[2] This is particularly true in the practice of musculoskeletal medicine because the incidence of degenerative disorders of the skeleton increases with age.[3]

Functional limitations occur more commonly with advanced age. While only one out of ten adults aged 65 to 69 years report limitations in daily activities, more than half of those 85 years and older have functional limitations.[1] These limitations are frequently related, at least in part, to musculoskeletal conditions. Osteoarthritis and osteoporosis are common examples of age-related musculoskeletal disorders, but other syndromes that affect older adults, including nutritional problems, polypharmacy, gait instability and falling, dementia, and elder abuse and neglect, can present as problems of the bones and joints. Arthritis, the most prevalent health condition among older adults, is a major source of functional impairment.[4] Osteoporosis and disorders of mobility also challenge the elderly. It is hoped that advances in health care will help meet that challenge.

Among these advances is the establishment of Acute Care of the Elderly units in hospitals specially designed for the frail elderly. These multidisciplinary care teams tailor their processes to avoid the complications and deconditioning that so often leads to nursing home placement.[1] Comprehensive programs of integrated outpatient and inpatient services for the elderly, who would otherwise qualify for nursing home placement, have also been piloted as Programs for All Inclusive Care of the Elderly.[1] Another advance in health care for the elderly involves advocating the use of hip protectors in individuals at high risk for falls, such as nursing home residents. These undergarments add special padding over the trochanteric area and may reduce the rates of hip fracture in individuals who are wearing them when they fall.[5] Further development is underway to make these undergarments easier to use in order to increase compliance.

Age-Related Changes

Physiologic Changes

The response to the physiologic stress of severe illness or injury, changes in temperature, electrolyte and water imbalances, and other factors is more sluggish as people age.[1] Physicians must recognize these physiologic limitations and the risk for a cascade of de-

Table 1
Key Physiologic Changes With Aging

Affected Areas of Body Systems	Physiologic Change
Body Size and Composition	Relative increase in body fat; relative decrease in muscle mass and body water
	Gradual decrease in height caused, in part, by loss of vertebral height
	Decreasing weight in men older than 50 years and in women older than 60 years
Skin	Thinning of the dermis; skin more susceptible to minor trauma
Sensory	Laxity of supporting tissues around the eyes; diminished tear production
	Pupillary responses more sluggish; lens changes; impaired dark adaptation
	Symmetric, bilateral hearing loss, particularly for high frequencies
	Declining olfactory function and taste
Oral	Dry mouth common and worsened by taking many medications; loss of dentition from disease
Endocrine	Peripheral insulin resistance
	Decreased growth hormone secretion
	Decreased vitamin D synthesis; increased parathyroid hormone level
	Increased norepinephrine level; increased sympathoadrenal activation
	Bone loss, accelerating particularly after menopause
Cardiac	Increased left ventricular mass; reduced compliance and diastolic dysfunction
	Resting cardiac function unchanged; diminished inotropic/chronotropic response
	Reduced compliance of large blood vessels; increased systolic blood pressure
Respiratory	Increased anteroposterior diameter of the chest wall
	Decreased chest wall compliance; reduced vital capacity and forced expiratory volumes
	Higher closing volume with ventilation-perfusion mismatching
	Less effective cough and mucociliary clearance mechanisms
Renal	Reduced renal mass, renal blood flow, and renal clearance
	Diminished tubular function; reduced maximal urinary concentration/dilution
Gastrointestinal/Hepatic	Reduced colonic function; frequent constipation
	Liver blood flow and liver mass decline with aging; metabolic capacity variable
Hematologic/Immune	Blood counts generally maintained with normal aging
	Changes in T-cell production, macrophage function, and cytokine secretion
	Diminished antibody response with immunization
	Reduced barrier functions of skin and mucous membranes
	Reduced acidity of urine and stagnation predispose to urinary tract infection
Musculoskeletal/Neurologic	Arthritis and osteoporosis very common
	Sarcopenia, or reduced muscle mass, associated with impaired mobility and frailty
	Decrease in walking speed; increasing balance problems
	Deep tendon reflexes in lower extremities may be modestly diminished
	Minimal cognitive changes, such as slower processing speeds, but most cognitive functions preserved in healthy aging
	Reduced sleep efficiency; less time spent in more restful sleep stages

cline during illness in which one illness triggers a series of other problems with resulting loss of functional ability.[6] For example, pneumonia may keep a patient in bed, and the extended bed rest may lead to muscle atrophy, which may never be restored.

Table 1 lists selected key physiologic changes that occur with aging.[1] Of specific importance to surgical patients, wound healing is impaired by age-related reductions in the blood flow to the skin and the effectiveness of immune responses. Incontinence can lead to maceration of wet skin, and pressure and shear forces are common during periods of bed rest, thereby increasing the risk for skin breakdown. Poor nutrition, common among older persons who are ill, also impairs wound healing.

Pharmacologic Changes

Numerous pharmacokinetic and pharmacodynamic changes of aging have direct clinical relevance.[7] In the absence of disease, normal aging does not significantly reduce drug absorption. In contrast, drug distribution and metabolism are significantly altered by the changing body composition that occurs with aging. Because total body water is relatively reduced and the fat compartment is relatively increased with age, lower doses of water-soluble drugs must be used to achieve the desired concentration. Fat-soluble medications remain in the body for relatively prolonged periods. When the serum albumin level is low, as it frequently is in older adults who are ill, the binding of protein-bound drugs decreases. This leads to increased concentration of the active unbound drug for a given dose. Therefore, measuring serum albumin levels in older patients can be extremely important for proper drug dosing.

Metabolic liver capacity varies among aging individuals. In general, oxidation, reduction, and hydrolysis of drugs declines as a result of decreasing hepatic blood flow. Primary conjugation is relatively less impacted by aging. Many older adults take numerous medications that use the cytochrome P450 system for breakdown, and drug interactions may inhibit the necessary metabolism of some agents.

Evaluating the Geriatric Patient

History and Physical Examination

The history and physical examination of older patients should be modified to account for their age.[8-10] In healthy older adults, the usual history and physical examination routines may not need to be altered, but in frail older adults extra time and assistance may be required in reaching the examination room, gowning, and moving about on the examination table. Medical histories of older persons tend to be more complex because of many medical conditions for which they take multiple medications.[4] History taking can be challenging with hearing-impaired patients but is greatly facilitated by the use of a voice amplifier. Communications should not be directed primarily to an accompanying family member or caregiver during an office visit.[11] However, supplemental information from family members or other caregivers is critically important when the patient is cognitively impaired.

Numerous age-related changes are typically found on physical examination.[8] For example, kyphosis resulting from osteoporosis of the spine is common and may be associated with atelectatic basilar lung crackles from impaired chest wall expansion. In severe cases of osteoporosis, the lower ribs may rest on the pelvic brim. The combination of osteoporosis and increasing laxity of abdominal musculature tends to result in flexed posture and a protuberant abdomen. Most neurologic functions remain essentially unchanged with healthy aging. However, deep tendon reflexes in the lower extremities may be more difficult to elicit in healthy older adults and gait speed slows somewhat.

Screening for Functional Impairment

Along with a thorough medical history and physical examination, assessment of the older patient should include screening for functional impairments typical of older adults.[12] Activities of daily living (ADLs) include mobility, bathing, dressing, grooming, eating, and toileting. Instrumental activities of daily living (IADLs) include activities such as getting out of the house, paying bills, answering the telephone, and preparing meals. Part of the screening should include determining whether these tasks are performed independently, with assistance, or not at all and identifying which factors contribute to any functional impairments. Special attention should be paid to review of all medication use, including prescription and over-the-counter medications and any vitamin and herbal supplements. Additional key elements of the screening assessment include evaluation for signs and symptoms of depression, confusion, weight loss, incontinence, instability and falling, sleep disturbances, constipation, dysphagia, dizziness, and sensory impairments.[1] Multidisciplinary care teams, including social workers, pharmacists, and pharmacy technicians, can contribute enormously to this assessment.[1,13]

Nutrition

Poor nutrition in this patient population usually arises from combinations of age-related physiologic changes, disease states or medications that impair appetite, deteriorating oral and dental health, functional impairments related to meal preparation and eating, social isolation, and economic hardship. Poor nutrition is often not identified until an advanced stage, at which time it can severely

Table 2
The Nutrition Screening Initiative's DETERMINE Checklist[12]

	YES
I have an illness/condition that made me change the kind and/or amount of food I eat.	2
I eat fewer than two meals per day.	3
I eat few fruits or vegetables or milk products.	2
I have three or more drinks of beer, liquor, or wine almost every day.	2
I have tooth or mouth problems that make it hard for me to eat.	2
I don't always have enough money to buy the food I need.	4
I eat alone most of the time.	1
I take three or more different prescribed or over-the-counter drugs a day.	1
Without wanting to, I have lost or gained 10 pounds in the last 6 months.	2
I am not always physically able to shop, cook, and/or feed myself.	2
TOTAL	

Patients who score 0 to 2 are doing well and should repeat the survey within 6 to 12 months. Patients scoring 3 to 5 are at moderate nutritional risk, their dietary consumption should be evaluated and revised for nutritional content, and they should repeat the survey in 3 months. Those who score 6 or higher are at high nutritional risk. These individuals need a full medical and dietary evaluation, including functional and emotional status, and likely will require dietary supplementation.

decrease the body's ability to withstand infection, heal fractures or wounds, and recover mobility and functioning after an illness. Older patients who are admitted to the hospital in a compromised nutritional state stay longer and have higher complications rates compared with well-nourished patients; additionally, malnourished older adults are at increased risk for morbidity and death in direct proportion to the severity of their nutritional deficits.[13]

Given the difficulties of reversing established nutritional deficits in the elderly, greater efforts should be made to prevent poor nutrition. The DETERMINE (an acronym for the following possible problems: disease, eating poorly, tooth loss/mouth pain, economic hardship, reduced social contact, multiple medications, involuntary weight loss/gain, needs assistance in self-care, elder years greater than 80) checklist, shown in Table 2, was developed by the Nutrition Screening Initiative as a screening tool for nutritional risk.[13]

Polypharmacy

Although the elderly currently constitute 12% of the US population, they receive 32% of all medication prescriptions.[14] Community-dwelling older adults use, on average, five prescription and over-the-counter medications daily.[15] Recent pharmaceutical advances have led to new and better-tolerated products for symptom relief and disease modification or prevention. In the elderly, however, medication use must be carefully scrutinized because age-related physiologic and pharmacologic changes increase susceptibility to adverse effects, as well as drug-drug and drug-disease interactions.[1]

The term polypharmacy means that many drugs are used, but the implication is that more medications are in use than is clinically warranted. Use of five to seven or more drugs per day is associated with adverse drug reactions, decreased medication compliance, poorer quality of life, a high rate of symptomatic adverse reactions, and potentially unnecessary drug expense.[15] Experts have identified medications that are generally inappropriate in most circumstances for use in older adults because of their adverse effects. General prescribing principles for older adults are shown in Table 3.

Gait Instability and Falling

Each year approximately 35% to 40% of community-dwelling adults 65 years and older experience a fall. Rates are higher with increasing age and among nursing home residents.[1] Falling is associated with considerable mortality, morbidity, reduced functioning, and premature nursing home admission.

Table 3
Prescribing Principles for Older Adults

Simplify drug regimens; use a single, daily dose when possible.
Use medication boxes to enhance compliance.
Limit the use of medications that are used as needed.
Consider new medications to be therapeutic trials.
Discontinue ineffective medications.
Monitor for adverse effects and discontinue or adjust dose, if necessary.
Attempt to choose medications that will treat more than one established condition.
Avoid using new medications to treat adverse effects from other medications.

There are many risk factors for falling, some of which are at least partially modifiable.[1] Such risks include intrinsic factors, such as lower extremity weakness, disturbed balance, cognitive impairment, and visual deficits. Extrinsic risk factors include the use of sedatives and environmental hazards.[1]

Fall prevention combines medical management with preventive measures in the environment. Table 4 outlines the appropriate steps to prevent recurrent falls, many of which can be instituted before the first fall. Patients who have experienced a fall-related fracture should be counseled about preventing subsequent falls. Restraints do not reduce the rate of fall-related injuries among nursing home residents and can lead to other complications, including skin breakdown, agitation, aspiration, injury, and even death by strangulation. Restraint use is thus highly discouraged and tightly regulated in these institutional settings.[16]

Table 4
Recommended Physician Interventions to Prevent a Fall

Conduct a medication review and modification, especially for psychoactive drugs.
Treat cardiovascular disorders, including symptomatic dysrhythmias.
Treat orthostatic hypotension, if present.
Evaluate gait and train patients with appropriate assistive devices.
Recommend an exercise program, including balance training.
Treat visual impairments.
Treat incontinence.
Treat foot disorders affecting gait mechanics.
Recommend that patients modify environmental hazards (eg, poor lighting and loose rugs or cords).
Recommend that patient equip bathroom with handrails, a raised toilet seat, and bath chair as needed.

Elder Neglect and Abuse

Neglect and abuse of older adults is an increasingly recognized public health problem in the United States. Nearly one in 25 people older than 65 years are victims of elder neglect and abuse each year.[1] The most common perpetrators are spouses or adult children, but they may also be other family members or paid or informal caregivers.

Abuse can occur as physical or psychological mistreatment, neglect, or financial exploitation of the elderly.[17] Physical abuse includes striking, shoving, restraining, and improper feeding, the results of which can be manifest as musculoskeletal injuries. Psychological abuse includes issuing threats (eg, threat of institutionalization), insults, harsh commands, or ignoring the older adult. Financial abuse is the exploitation or neglect of a person's possessions or funds. Neglect centers on the failure to provide food, medicine, personal care, or other necessities.

Abuse in older adults can be difficult to detect. Many signs are subtle and include clinical conditions already common among geriatric patients, such as unintentional weight loss, injuries, isolation, and dehydration. If cognitive impairment is known or suspected, complaints of abuse may be dismissed as the result of paranoia and confusion. In addition, the social isolation of a frail elderly adult is often difficult to detect. It is important to note, however, that not all abuse is intentional. Caregivers for elderly persons with chronic medical, emotional, cognitive, or functional needs might not realize their behaviors constitute abuse. For example, a caregiver may restrain a demented older adult in a well-intentioned but misguided attempt to prevent wandering. Therefore, the well-being of caregivers should be assessed, and they should be encouraged to take advantage of available local resources, such as adult day care and respite programs.

Reporting suspected or confirmed abuse is mandatory in all states when the abuse occurs in an institution and in most states when it occurs in a home. All states have laws protecting and providing services for vulnerable, incapacitated, or disabled adults. Documentation is critical, and the medical record should contain a complete report of the actual or suspected abuse. If the victim is able to make decisions, he or she should help determine his or her own fate. If the victim does not have decision-making capacity, a guardian or objective conservator should guide decision making. Collaboration with health professionals from social work, psychiatry, nursing, and the legal field can provide a comprehensive approach to these complex situations.

Osteoarthritis

Osteoarthritis is the most common degenerative joint disease and a leading cause of disability in persons older than 65 years.[1] In the elderly, osteoarthritis is characterized by constant joint pain that is usually accompanied by limited range of motion and joint deformity.

Comprehensive management of osteoarthritis involves cognitive, physical, pharmaceutical, and surgical measures. The goals of treatment are to minimize pain and to maximize function. The patient's functional deficits and preferences for treatment must be considered. Asymptomatic osteoarthritis, diagnosed by radiographic findings, does not require treatment.

Nonpharmacologic measures are important components of therapy and should be included in all management plans. Cognitive or behavioral therapy directed toward enhancing coping skills or increasing confidence in performing activities safely has been shown to improve function. A regimen of range-of-motion, strengthening, and endurance exercises is useful for pain relief and restoration of function.[18] Assistive aids (eg, braces, canes, and devices to increase hand function) may be indicated to restore function and improve independence. Weight loss is important for patients who are obese, and even modest weight loss can provide symptomatic relief.

Pharmacologic therapy should be used when nonpharmacologic measures provide insufficient pain relief and restoration of function. In the elderly, the most commonly used medications are acetaminophen and nonsteroidal anti-inflammatory drugs. Acetaminophen is the first choice for pain relief because it is much safer than nonsteroidal anti-inflammatory use and equally effective.[18] Total joint arthroplasty is highly effective for treating osteoarthritis of the hip and knee, and age alone is not a contraindication. However, treatment goals (pain relief and improved physical function) and the needs and rehabilitative capabilities of the patient must be clearly defined.[19]

Osteoporosis

Osteoporosis is a systemic skeletal disease characterized by low bone mineral density, bone fragility, and increased risk of fracture. The incidence of osteoporosis increases with age and most frequently affects postmenopausal women. The primary clinical manifestation of osteoporosis is fracture, which can also be the first indicator of the disease. Typical fracture sites include the vertebrae, hip, proximal humerus, wrist, and ribs. Insufficiency fractures can occur in the pelvis and the tibial plateau.[1]

The diagnosis of osteoporosis is often suggested by the medical history and physical examination. Patients with a history of fractures at the hip, wrist, or vertebral bodies; those with kyphosis on examination; and those in whom evidence of possible osteopenia is detected on radiographs should be presumed to have a metabolic bone disease and should be evaluated further. The World Health Organization has established diagnostic criteria for osteoporosis based on the measurement of bone mass and the optimal peak bone mass for adults. A T-score of −2.5 or lower on bone density testing confirms the presence of osteoporosis. Sequential bone mineral density tests may be helpful in assessing the rate of bone loss as well as a patient's response to pharmacologic therapy. Available therapies include calcium and hormone replacement therapy; bisphosphonates; and the calcitonin analogue, miacalcin, which is delivered as a nasal spray. Selective estrogen receptor modulators may also have a role in preventive therapy in women in whom hormone replacement therapy would not be advisable.

Key Terms

Activities of daily living (ADLs) The functions that are fundamental to maintaining independence and mobility, including ambulation, bathing, dressing, grooming, eating, and toileting; the assessment of the ability to perform these functions helps screen for functional impairments and other common syndromes that can affect older adults

Community-dwelling older adults Older adults who live independently within their communities

Instrumental activities of daily living (IADLs) The functions that require a higher level of function than ADLs, such as getting out of the house, paying bills, answering the telephone, and preparing meals; the assessment of the ability to perform these functions helps screen for functional impairments and other common syndromes that can affect older adults

Polypharmacy Literally "many drugs," but common usage of the term carries the implication that more medications are in use than is clinically warranted

References

1. Cobbs EL, Duthie EH, Murphy JB: (eds): *Geriatrics Review Syllabus: A Core Curriculum in Geriatric Medicine,* ed 4. Dubuque, IA, Kendall/Hunt Publishing Company, 1999.
2. Cassel CK: Why physicians need to know more about aging. *Hosp Pract* 2000;35:11-15.
3. Goldstein J, Zuckerman JD: Selected orthopedic problems in the elderly. *Rheum Dis Clin North Am* 2000;26:593-616.
4. Fried LP: Epidemiology of aging. *Epidemiol Rev* 2000;22:95-106.
5. Kannus P, Parkkari J, Niemi S, et al: Prevention of hip fracture in elderly people with use of a hip protector. *N Engl J Med* 2000;343:1506-1513.
6. Creditor MC: Hazards of hospitalization of the elderly. *Ann Intern Med* 1993;118:219-223.
7. Ugalino JA: Understanding the pharmacology of aging, in Dreger D, Krumm B (eds): *Hospital Physician Geriatric Medicine Board Review Manual.* Wayne, PA, Turner White Communications, 2001, pp 1-12.
8. Schneiderman H: Physical examination of the aged patient. *Conn Med* 1993;57:317-324.
9. Fields SD: History-taking in the elderly: Obtaining useful information. *Geriatrics* 1991;46:26-35.
10. Fields SD: Special considerations in the physical exam of older patients. *Geriatrics* 1991;46:39-44.
11. Adelman RD, Greene MG, Ory MG: Communication between older patients and their physicians. *Clin Geriatr Med* 2000;16:1-24.
12. Lachs MS, Feinstein AR, Cooney LM Jr., et al: A simple procedure for general screening for functional disability in elderly patients. *Ann Intern Med* 1990;112:699-706.
13. The Nutrition Screening Initiative: *Incorporating Nutrition Screening and Interventions into Medical Practice.* Washington, DC, The Nutrition Screening Initiative, 1994, p 73.
14. Stuck AE, Beers MH, Steiner A, et al: Inappropriate medication use in community-residing older persons. *Arch Intern Med* 1994;154:2195-2200.
15. Prince BS, Goetz CM, Rihn TL, et al: Drug-related emergency department visits and hospital admissions. *Am J Hosp Pharm* 1992;49:1696-1700.
16. Guttman R, Altman RD, Karlan MS: Report of the Council on Scientific Affairs: Use of restraints for patients in nursing homes: Council on Scientific Affairs, American Medical Association. *Arch Fam Med* 1999;8:101-105.
17. Swagerty DL Jr., Takahashi PY, Evans JM: Elder mistreatment. *Am Fam Physician* 1999;59:2804-2808.
18. American College of Rheumatology Subcommittee on Osteoarthritis Guidelines: Recommendations for the medical management of osteoarthritis of the hip and knee: 2000 update. *Arthritis Rheum* 2000;43:1905-1915.
19. Bagge E, Brooks P: Osteoarthritis in older patients: Optimum treatment. *Drugs Aging* 1995;7:176-183.

Section Editors
Kevin B. Freedman, MD
Robert A. Hart, MD

Section 4

Clinical Evaluation

Contributing Authors
Joseph Bernstein, MD, MS
Michael Brage, MD
Hans L. Carlson, MD
Nels L. Carlson, MD
Raymond M. Carroll, MD
Robert K. Eastlack, MD
Kevin B. Freedman, MD
Robert A. Hart, MD
Gregory N. Lervick, MD
William N. Levine, MD
Samir Mehta, MD
Donald Resnick, MD
Philip E. Rosen, MD,
Daphne J. Theodorou, MD
Stavroula J. Theodorou, MD
Robert H. Wilson, MD
Robert J. Winn, MD

Medical History

Much has been written about how to obtain the medical history, but there is a catch: these writings tend to make the most sense only to readers who are experienced enough to no longer need them. In light of this paradox, this book will not say much on the subject; there is no formal chapter on the medical history. Rather, what follows is only a simple list of guidelines. These instructions are an incomplete education, but they are a start.

Introduce yourself. By introducing yourself when you enter the examination room, you bring a necessary human dimension to the initial meeting. It matters little whether you introduce yourself as Jane Smith, Student-Doctor Smith, or some other variation on that theme. What matters is that you remind yourself and the patient that the encounter is personal. You should also repeat the patient's name, saying it when appropriate during your interview.

Wash your hands. When Semmelweiss introduced the idea that doctors must wash their hands before going from the anatomy laboratory to the obstetrics ward, he was ridiculed and shunned. Today, we understand that hand washing prevents infection. Hand washing also helps obtain a good history, I believe: washing your hands in the presence of the patient can help establish trust—you are showing that you respect the patient's physical space and therefore prepare yourself to enter it. Such a bond will be conducive to the sharing of intimate yet relevant detail. (For instance, you will learn that both gonorrhea and ligament sprains can cause effusions of the knee. A complete history is needed to make the correct diagnosis, and trust in the physician may be needed to obtain that history.) Assume patients *always* notice whether you do or do not wash your hands.

Sit down. Even if you are short on time, you will seem less hurried if you sit down. As Kurt Vonnegut noted, you become who you pretend to be. If you simply assume the mannerisms of a person who is not in a rush, you will not be rushed. Arguably, the act of sitting transforms you; it actually *makes* you less hurried.

After asking the patient to speak, listen without interruption for at least 90 seconds. The best way to ensure that the most important medical details are obtained is to allow the patient to tell his or her story. Do not interrupt. If you interrupt the patient, you will probably spend more time obtaining less information—and most patients will not even use the full 90 seconds you give them. This is not so much about good manners as it is about effectiveness. If you think of a pressing question and are afraid you will forget it, scribble a short note to yourself. (I confess that I have trouble with this.)

Find out what the patient wants. In musculoskeletal medicine particularly, your job is to discern mere incidental findings from those that are a source of distress or disability. You must also assess whether the patient is willing to assume the costs, risks, and hassles of treatment. Do not try to cure what is not bothering the patient. For example, many adults have herniated lumbar disks. Your job as diagnostician is not only to diagnose that disk herniation, but also to determine whether there are signs and symptoms attributable to it and whether they are severe enough to warrant treatment. (It is possible that a patient will seek medical attention for back pain only to be reassured that "it's not cancer.") Recommending aggressive treatment for asymptomatic and benign abnormalities is foolhardy.

The patient is always right—usually. In most cases, a patient's story alone will reveal the diagnosis to the perceptive listener. Still, remain aware of the fact that musculoskeletal conditions often produce referred pain, and the area in which the patient reports symptoms is not necessarily the area that harbors the pathology. Hip disease can cause knee area pain; similarly, a proximal nerve injury may be reported as pain in the distal area where that nerve terminates. And, of course, visceral organs can produce musculoskeletal symptoms—a patient who reports left shoulder pain may need a cardiologist, not an orthopaedist.

Collect pearls. Pearls are aphorisms that contain a kernel of medical wisdom. Until you have gained a lot of experience, these pearls may help you take advantage of the experiences of your predecessors—allowing you, in Newton's phrase, to see further by standing on the shoulders of giants. Pearls help you obtain a history by reminding you to ask the right question or by helping you understand the answers you hear. The best way to collect pearls is to see as many patients as you can. One pearl I still recall from medical school is "posterior shoulder dislocations are associated with seizures and isolated lesser tuberosity fractures of the humerus." I think I remember this not because I read it, but precisely because that rule helped reveal a subtle tuberosity fracture in a patient I saw.

Obtaining a perfect medical history may be beyond the ability of typical medical students—or even their teachers, for that matter. A reasonable medical history, however, can be obtained by following these rules, and time and experience will only make you better.

—Joseph Bernstein, MD, MS

Physical Examination

Medicine is an art, and like all art, it begins with fundamentals. In musculoskeletal medicine, the physical examination is the basis upon which much of clinical practice rests. Although there is certainly a place in musculoskeletal medicine for sophisticated diagnostic tests (or even not-so-sophisticated tests, such as plain radiography), to know when to use these tests and for their results to be meaningful, you must master the art of physical examination.

Certain elements in the patient's medical history often suggest the diagnosis, and the experienced clinician may be able to pounce on a few short physical examination maneuvers to confirm a diagnosis. The novice, while not as fortunate, has the advantage of not being fooled by too much knowledge. He or she is less likely to omit a seemingly irrelevant test, and in so doing may be able to make a diagnosis otherwise missed.

A beginner can expect to need more time for the examination, not only because each step is slower, but also because more steps will be taken. I urge you to embrace that. The only way to recognize that something is abnormal is to first recognize what is normal; and the best way to do that is to have collected the experience of examining many normal structures.

It may also be helpful to perform a screening musculoskeletal examination on all of your general medical and surgical patients, assessing strength, range of motion, and the absence or presence of pain or swelling in the major joints. The marginal utility of a screening test is fairly low—even if you were to detect an abnormality, it would not be clear from the screening examination alone that abnormality would require treatment. Yet without detecting the abnormality in the first place, a discussion of treatment cannot begin. Thus, fast and easy screening examinations are good for patients. They are also good for novice examiners. Such examinations will allow you to gain facility approaching and making contact with patients. You will also collect a mental database of normal values, which in turn will be invaluable when it comes time to recognizing the abnormal. The enclosed CD has a demonstration of the screening examination, which, as you will see, can be performed expertly in less than 5 minutes.

All physical examinations begin with inspection. When examining a patient, it is important that you have him or her undress. You will not detect muscle atrophy through a shirt. The examination of a man's shoulder is the rare instance when I will remain in the room and have the patient disrobe in my presence because important diagnostic information can be gained by watching the coordinated muscle action needed to remove a shirt. For women and all lower extremity examinations, I excuse myself while the patient puts on a gown.

After inspection, the next step is palpation—but before you palpate, make sure you wash your hands, preferably in the patient's presence. Palpation can be directed to the known surface landmarks, but it may be equally useful to note whether the patient is very tender in areas that do not correspond to discrete structures. Malingering is not uncommon, and tenderness reported in bizarre locations may reflect that.

Failing to adequately expose the area under examination is probably the most common mistake in physical diagnosis. The next most common mistake is to terminate the examination prematurely: by ending as soon as one positive finding is encountered, you may fail to detect others. Remember the wisdom of the veterinarians: a dog can have lice *and* fleas! Finding an abnormality is not a license to stop the examination. Indeed, it is known that some injuries and illnesses come in pairs: a sprained anterior cruciate ligament may be seen in association with a sprain of the medial collateral ligament, for example.

What follows is a collection of physical examination maneuvers. Although the photographs are probably worth a thousand words, viewing pictures of a physical examination is not as good as watching a video, and all forms of passive learning are dwarfed by the experience of examining live people. So please use the information that follows for grounding, but go see patients!

—Joseph Bernstein, MD, MS

Chapter 26

Physical Examination

Shoulder and Arm

Inspection

Look for erythema, ecchymosis, and swelling. Noting deformity and atrophy is particularly important. In the shoulder this may be difficult to appreciate because atrophy of the large muscles of the shoulder girdle must be severe to be apparent. Although infraspinatus and supraspinatus atrophy can be identified by inspection of their respective fossae posteriorly, the best way to note any asymmetry is by side to side comparison.

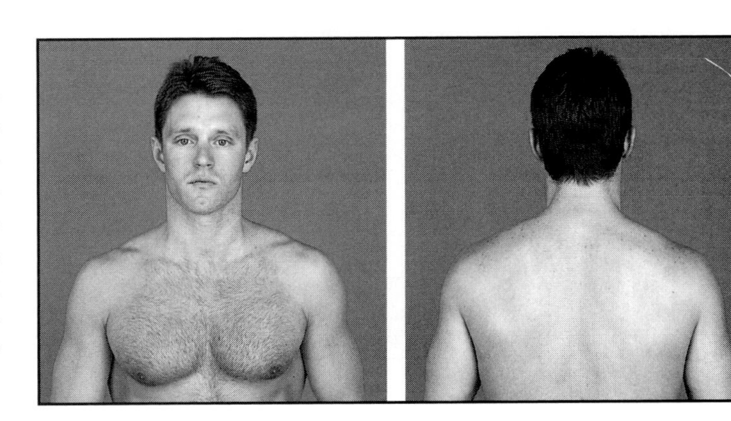

Palpation

Palpation Point	Significance
Sternoclavicular (SC) joint	Tenderness can indicate an SC joint sprain, arthritis, or dislocation.
Clavicle	Deformity without tenderness suggests an old (healed) fracture.
Acromioclavicular (AC) joint	Prominence suggests AC joint arthritis or separation.
Biceps tendon origin	Tenderness can indicate bicipital tendinitis, especially if the biceps is stressed.
Greater tuberosity	Pain suggests rotator cuff pathology.
Glenohumeral joint line	Tenderness can be caused by any intra-articular process but most likely arthritis.
Cervical spine	Tenderness can indicate a primary spine problem causing pain to radiate to the shoulder.

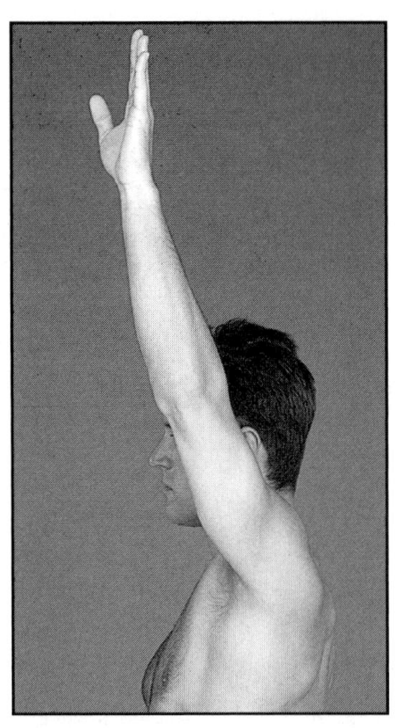

Forward flexion (elevation)

What to do: Ask the patient to elevate the arms above the head in the forward (sagittal) plane.

Normal: Ability to forward flex to 165°.

Interpretation: If the patient is unable to do this, passive motion must also be evaluated. Restricted active motion alone suggests a neurologic or muscular cause, such as a rotator cuff tear. Restricted passive motion suggests a mechanical block, most commonly frozen shoulder. The supraspinatus is typically needed to initiate elevation but not to hold it.

External rotation

What to do: Ask the patient to rotate the arm externally with the arm held against the chest wall and the elbow flexed to 90°.

Normal: 45° to 90°.

Interpretation: Lack of external rotation is often observed in patients with a frozen shoulder or glenohumeral arthritis. A radiograph can help distinguish between these two diagnoses.

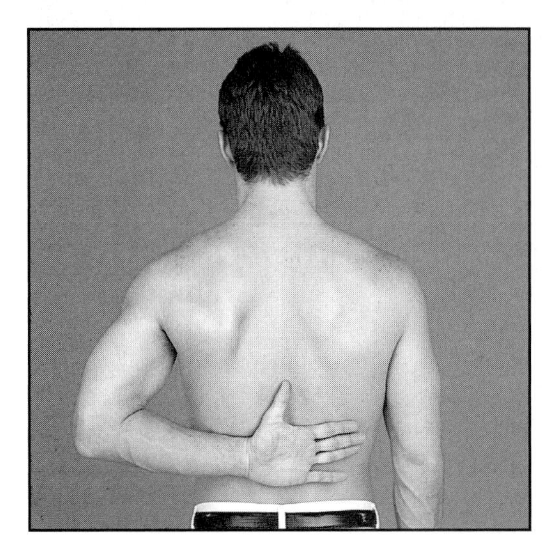

Internal rotation

What to do: Ask the patient to bring the arm behind the back. Record the vertebral level the patient can reach.

Normal: Internal rotation to the midthoracic spine.

Interpretation: Lack of internal rotation can be associated with a frozen shoulder or glenohumeral arthritis. In addition, patients with rotator cuff tears or contracture of the posterior capsule typically have restricted internal rotation on the basis of pain.

Motor Testing

Rhomboideus major and rhomboideus minor (dorsal scapular nerve, C5)

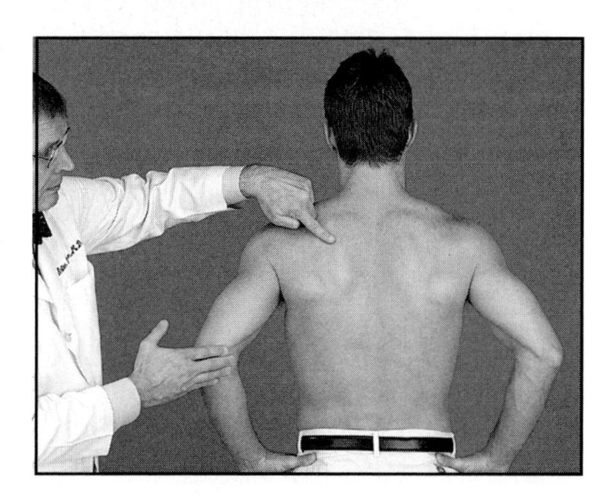

What to do: Ask the patient to place both hands on the side of the iliac crest. Stand beside the patient and push the patient's arm and elbow anteriorly with one hand and palpate the vertebral border of the scapula with the other hand.

Normal: Rhomboids palpable and scapula maintained against the chest wall.

Interpretation: Winging usually indicates an injury to the brachial plexus.

Supraspinatus (suprascapular nerve, C5-C6)

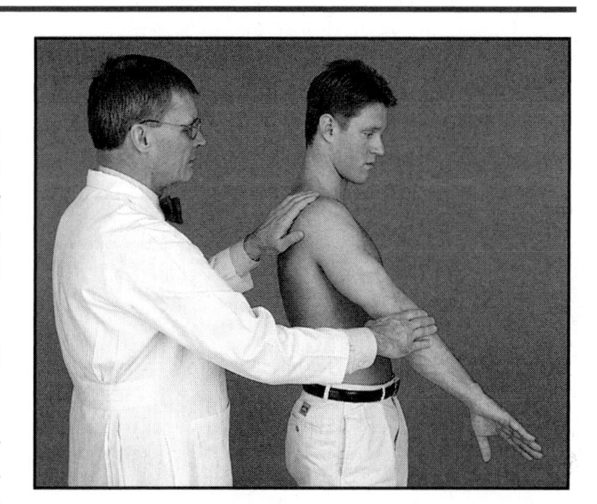

What to do: Ask the patient to abduct the arm in the scapular plane (ie, halfway between the front and the side of the body) and maximally rotate the arm internally so that the thumb points toward the floor. Then apply downward force to the patient's arm against resistance.

Normal: Good muscle strength maintained against resistance without pain.

Interpretation: Failure to maintain this position results in the "drop arm sign" and suggests a supraspinatus tear. Mild weakness or pain suggests rotator cuff tendinitis.

Infraspinatus (suprascapular nerve, C5-C6)

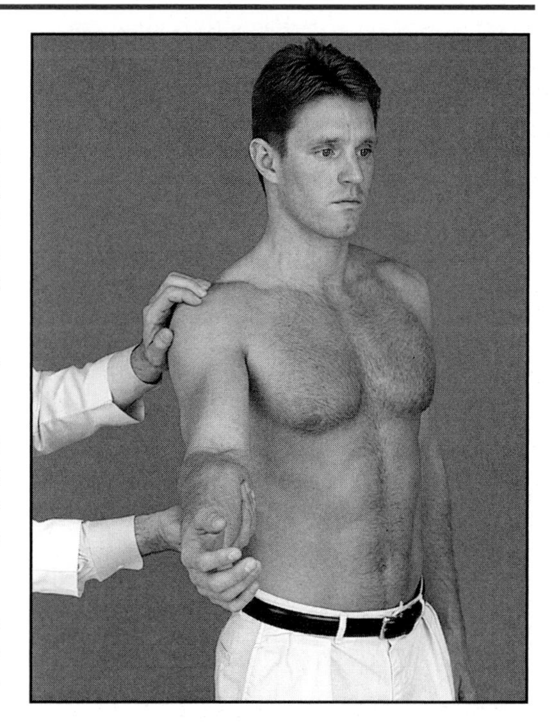

What to do: Ask the patient to place the arm comfortably at the side so that the elbow is positioned against the chest wall and flexed 90° with the forearm in neutral rotation. Then ask the patient to externally rotate the arm against internal rotation resistance applied at the wrist.

Normal: Adequate muscle strength maintained against resistance.

Interpretation: Inadequate muscle strength usually indicates a rotator cuff tear. Injury to the suprascapular nerve, including compression from a cyst, can also result in inadequate muscle strength.

Comment: If the arm is not held at the side, the posterior deltoid can externally rotate the arm and mask an injury to the posterior rotator cuff.

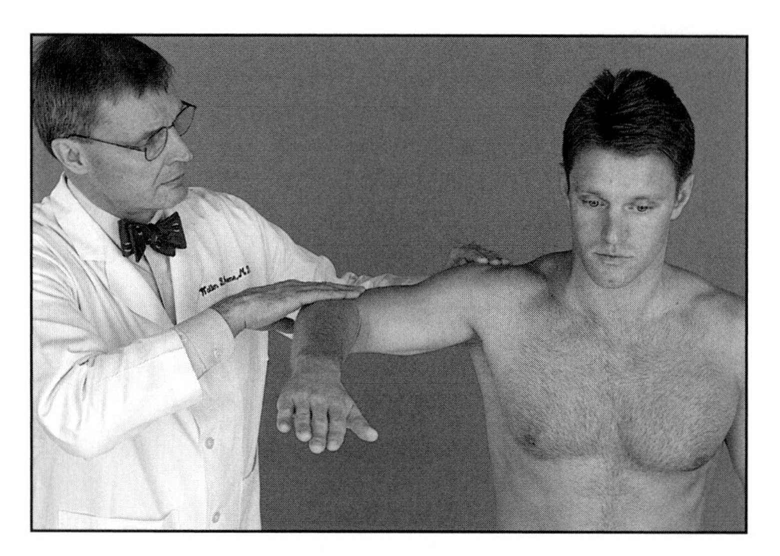

Deltoid (axillary nerve, C5-C6)

What to do: Ask the patient to abduct the arm to 90°. Then apply maximal downward force to the lateral aspect of the elbow with one hand and palpate all portions (anterior, middle, and posterior) of the deltoid muscle with the other hand.

Normal: Adequate muscle strength maintained; no defects felt in the muscle.

Interpretation: Deltoid weakness can be caused by injury to the axillary nerve, which commonly occurs with proximal humerus fracture or glenohumeral dislocation.

Comment: If the patient is in too much pain to move the arm (ie, after a fracture), test the sensory function of the axillary nerve by palpating the area where the deltoid inserts on the lateral humerus.

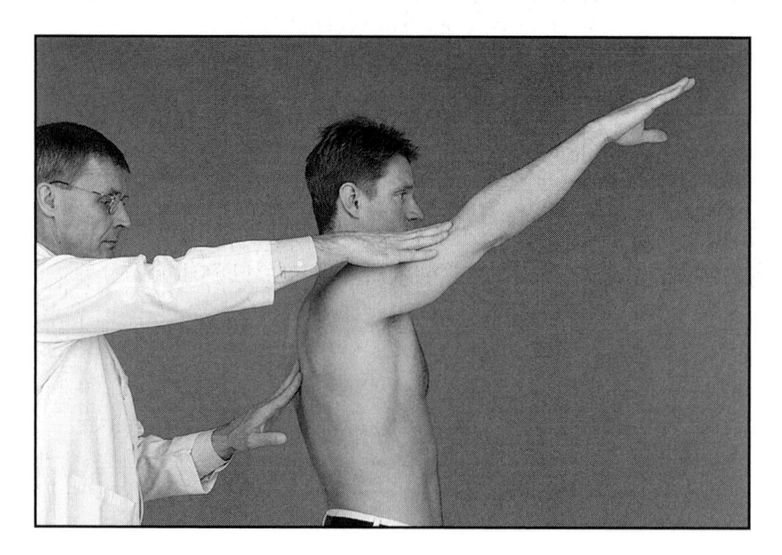

Serratus anterior (long thoracic nerve, C5-C6-C7)

What to do: Stand behind the patient and ask the patient to elevate the arm from the resting position at the side of the body. With one hand, apply steady downward force to the arm, resisting the patient, while using the other hand to palpate the scapula.

Normal: Scapula held against chest wall with arm elevation.

Interpretation: Inability to maintain the scapula in the normal position is called "scapular winging." This usually occurs from a stretch injury to the long thoracic nerve.

Comment: "Scapular winging" can also be observed by viewing the patient from behind as the patient performs a "wall push-up," ie, pushing the body away from a wall.

Biceps brachii (musculocutaneous nerve, C5-C6)

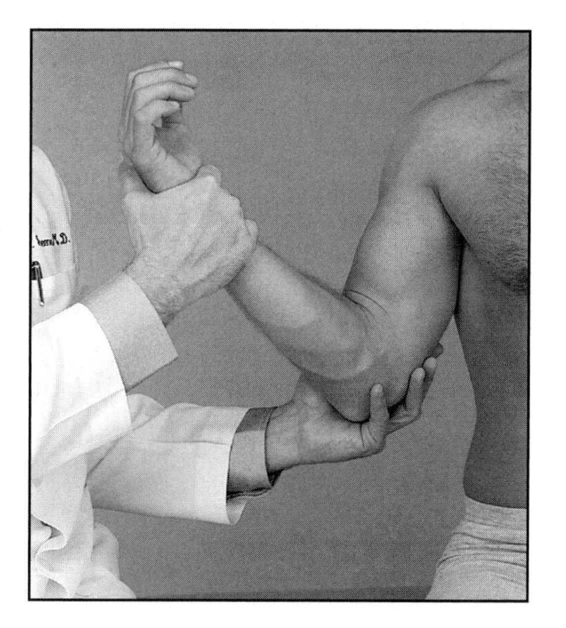

What to do: Stand in front of the patient and ask the patient to flex the elbow against your resistance.

Normal: Adequate muscle strength maintained against resistance.

Interpretation: A soft-tissue mass in the middle of the arm (Popeye sign) may be consistent with a rupture of the long head of the biceps, resulting in retraction of the muscle belly. Tendinitis causes pain near the origin of the tendon.

Comment: Rupture of the long head of the biceps may be seen as a component of subacromial impingement syndrome and rotator cuff tears.

Special Tests

Neer impingement sign

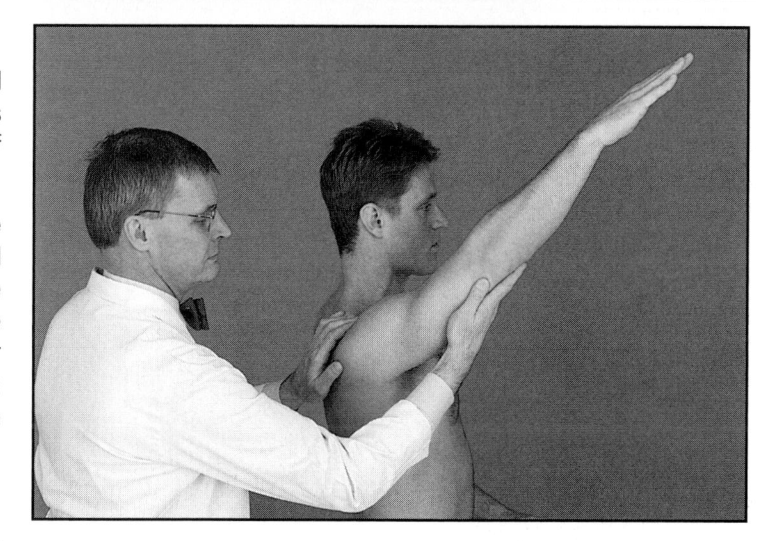

What it tests: The presence of subacromial impingement, a constellation of conditions that includes subacromial bursitis, rotator cuff tendinitis, and chronic rotator cuff tears.

What to do: Ask the patient to stand with the arms resting comfortably at the sides. Stand either behind or beside the patient. Grasp the patient's forearm with one hand and stabilize the shoulder girdle with the other hand over the trapezius, clavicle, and scapula. Then, with the arm in slight internal rotation, elevate the arm.

Normal: No pain.

Interpretation: If the patient has impingement, the affected structures are mechanically compressed (ie, impinged) beneath the acromion, producing pain as the arm is elevated above 90°. Limited passive elevation of the shoulder, as occurs with a frozen shoulder, invalidates the test because no mechanical impingement is produced.

Comment: To distinguish weakness caused by muscle deficiency and from that caused by pain, inject lidocaine into the subacromial space; weakness caused by pain is temporarily "cured" by this technique (the impingement test).

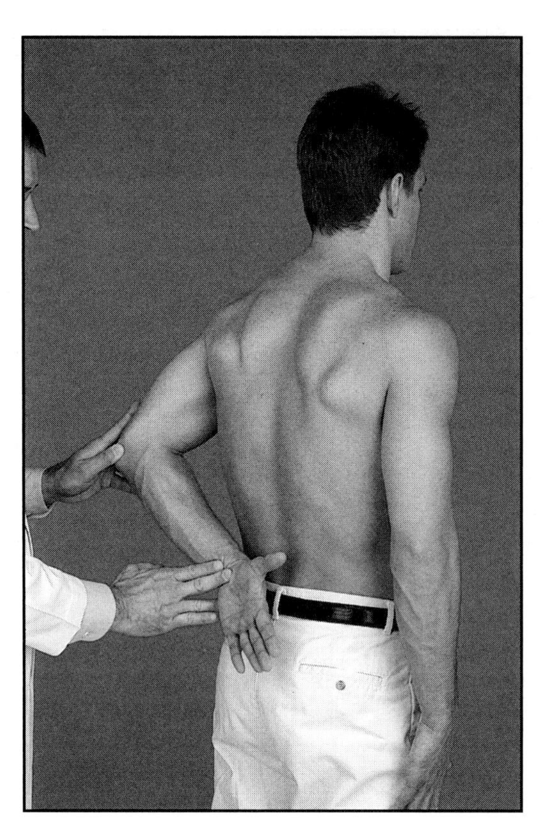

Lift-off test

What it tests: The integrity of the subscapularis musculotendinous unit. Because there are multiple internal rotators (ie, pectoralis major, latissimus dorsi), a tear of the subscapularis can be masked.

What to do: Ask the patient to maximally internally rotate the arm and then lift the hand off the back.

Normal: Patient can lift hand off of back.

Interpretation: With a rupture of the subscapularis musculotendinous unit, the patient will be unable to lift the hand off of the back.

Comment: This test may be difficult to perform in patients with severely limited motion or extreme pain. Massive tears of the subscapularis can be detected in such patients by demonstrating more passive external rotation on that side because the resistance of the subscapularis is lost. The belly press test may be used if the patient cannot internally rotate. To perform this test, the patient is asked to press the palm against the belly and keep the elbow out in front of the plane of the body. Inability to keep the elbow in front of the body is considered a positive test result.

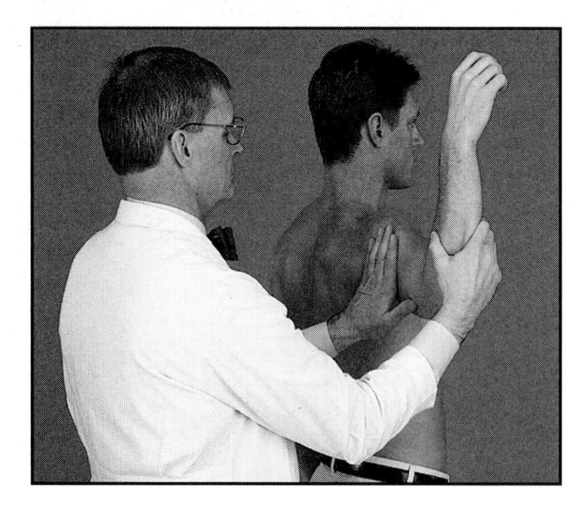

Apprehension test

What it tests: Anterior glenohumeral instability.

What to do: Position the patient's arm to 90° of abduction and maximal external rotation (typically to 90° or beyond—the position of an arm cocked to throw).

Normal: No sense of apprehension.

Interpretation: The test places the arm in a position in which the inferior glenohumeral capsulolabral complex normally tightens and resists anterior translation of the humeral head on the glenoid fossa. Injury to this complex allows subluxation or impending dislocation of the humeral head during the maneuver. When the patient demonstrates apprehension or describes a sensation that the shoulder is beginning to slide out of position, the test result is considered positive.

Sulcus sign

What it tests: Inferior glenohumeral laxity.

What to do: Ask the patient to stand with the arms resting comfortably at the sides. Stand beside the patient. With one hand, grasp the elbow and apply steady downward force to the arm. With the other hand, stabilize the shoulder girdle by resting the hand over the clavicle, scapular spine, and trapezius.

Normal: Less than 1 cm of inferior translation of the humerus, with no visible "sulcus" (ie, indentation of the skin). It is critical to note any asymmetry from side to side.

Interpretation: This maneuver creates inferior translation of the humeral head on the glenoid fossa. If the humeral head translates down to or below the rim of the glenoid fossa, a sulcus may be seen between the lateral aspect of the acromion and the humeral head.

Comment: If this test result is positive, check other joints (ie, passive hyperextension of the fingers) for generalized laxity, which may suggest a collagen disorder.

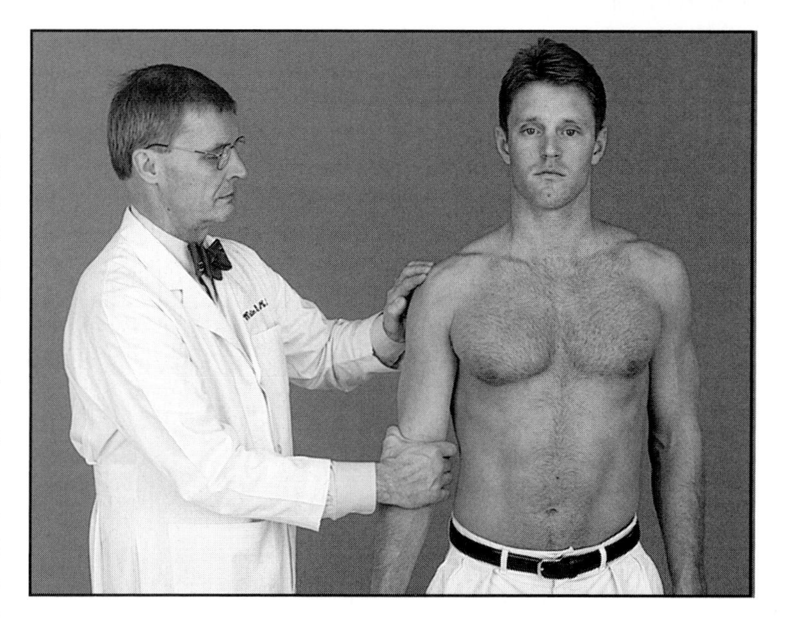

Cross-body adduction

What it tests: Because cross-body adduction loads the AC joint, pain with this maneuver suggests pathology in that joint.

What to do: Locate the AC joint with one hand. With the other hand, elevate the patient's arm to 90° in the sagittal plane. Then, gradually adduct the arm across the patient's body.

Normal: No pain at the AC joint.

Interpretation: Discomfort will occur when inflammation or arthrosis is present because this maneuver compresses the distal end of the clavicle against the medial aspect of the acromion at the AC joint. Reproduction of symptoms at the AC joint is a positive test result.

Comment: It is critical to localize the discomfort to the AC joint. For example, a tight capsule may create pain behind the shoulder rather than at the AC joint.

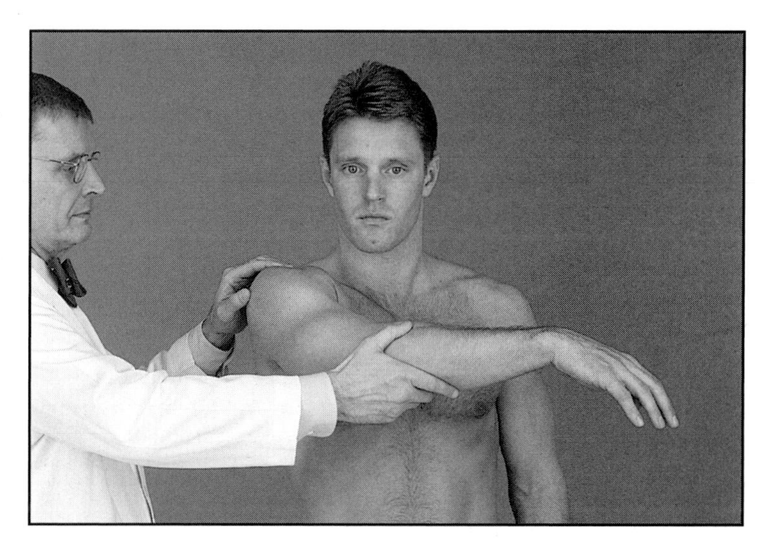

Elbow and Forearm

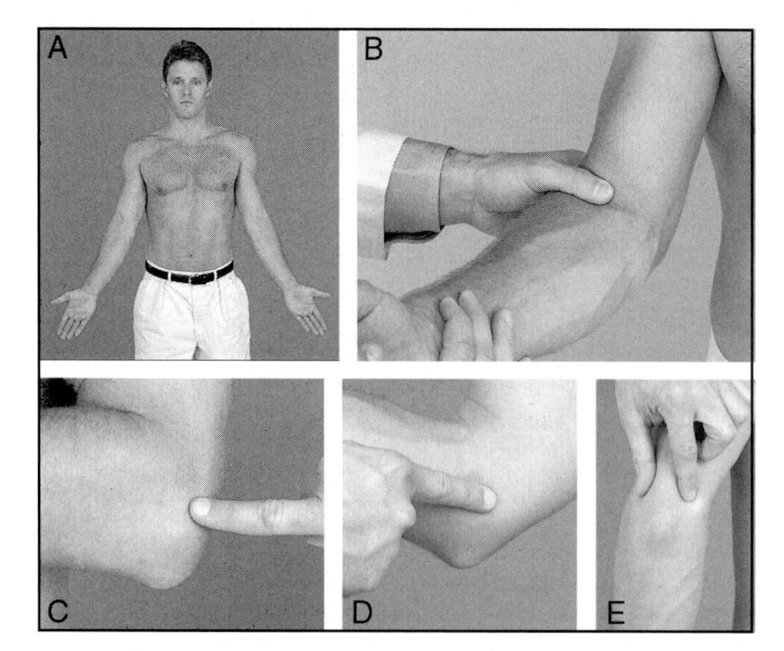

Inspection

Look for swelling and redness, especially at the olecranon bursa. Observe the "carrying angle," **(A)** made by the humeral shaft and the forearm (normally about 10° as shown here). In addition, look for any sign of trauma, such as ecchymosis in the antecubital fossa **(B)** (as seen with distal biceps rupture). Scarring of the skin may be responsible for joint contractures.

Palpation

Palpation Point	Significance
Lateral epicondyle **(C)**	Tenderness coupled with pain with resisted wrist extension suggests "tennis elbow," a tendinitis of the wrist extensor muscles that originate there.
Medial epicondyle **(D)**	Tenderness suggests tendinitis or medial collateral ligament sprain.
Biceps tendon	Palpable gap where the tendon should be suggests a distal rupture, especially if flexion and supination are weak.
Radial head	Tenderness after a fall on the outstretched hand suggests a radial head fracture. If the patient is an adolescent, suspect osteochondritis dissecans of the capitellum.
Triceps insertion **(E)**	Tenderness suggests tendinitis.
Brachial artery pulse	Located medial to the biceps. It may be difficult to palpate. A common location to auscultate when measuring blood pressure.
Ulnar nerve at the elbow	Abnormal sensation on palpation (Tinel's sign) suggests cubital tunnel syndrome (ie, irritation of the ulnar nerve).
Olecranon bursa	Tenderness is best assessed by pinching rather than pushing because pushing may evoke tenderness in the bone below.

Elbow flexion

What to do: Ask the patient to face you with the arms resting comfortably at the side. Grasp the patient's distal forearm. Ask the patient to maximally flex the elbow while you resist that motion.

Normal: Ability to flex to 135° with good strength and no pain.

Interpretation: If the patient cannot do this, you must exclude a mechanical block to motion by flexing the patient's elbow. If passive motion is normal but active motion is not, the cause may be a ruptured musculotendinous unit or a neurologic injury.

Comment: The brachialis powers elbow flexion in tandem with the biceps. Therefore, even with a distal biceps rupture, flexion is maintained.

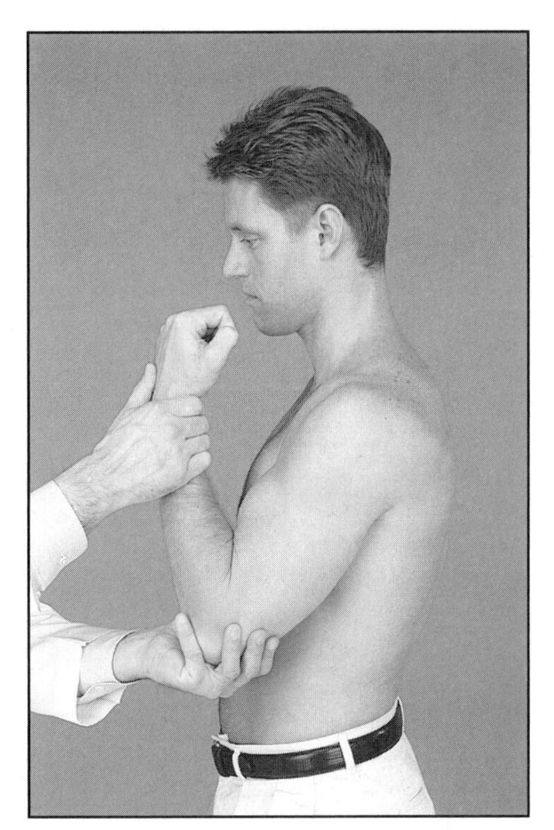

Elbow extension

What to do: Ask the patient to hold the elbow flexed to 90°. Then grasp the patient's forearm and ask the patient to extend the elbow while resisting this motion.

Normal: Extension to 0° with good strength and no pain.

Interpretation: A normal examination implies the presence of an intact triceps muscle and tendon and an intact radial nerve. The triceps is the sole elbow extensor, powered by the radial nerve. The inability to extend beyond 7° of flexion may be considered normal if present bilaterally.

Comment: In cases of profound weakness, extension should be tested with the arm elevated to remove any effect of gravity.

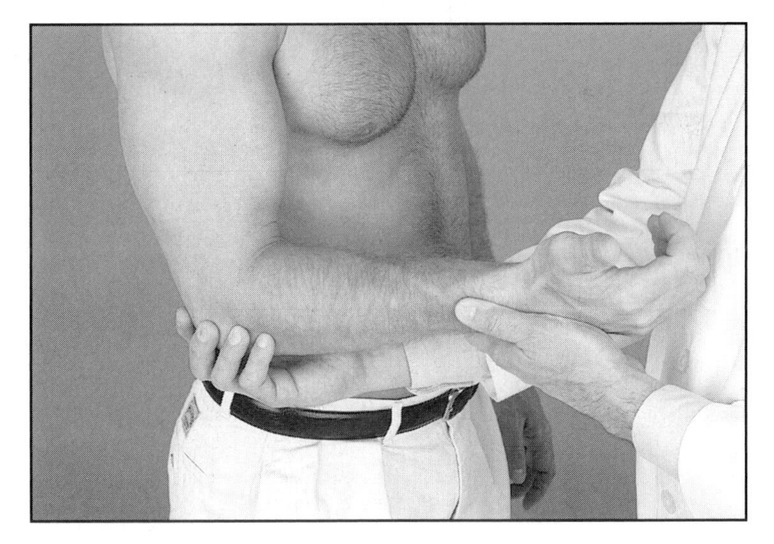

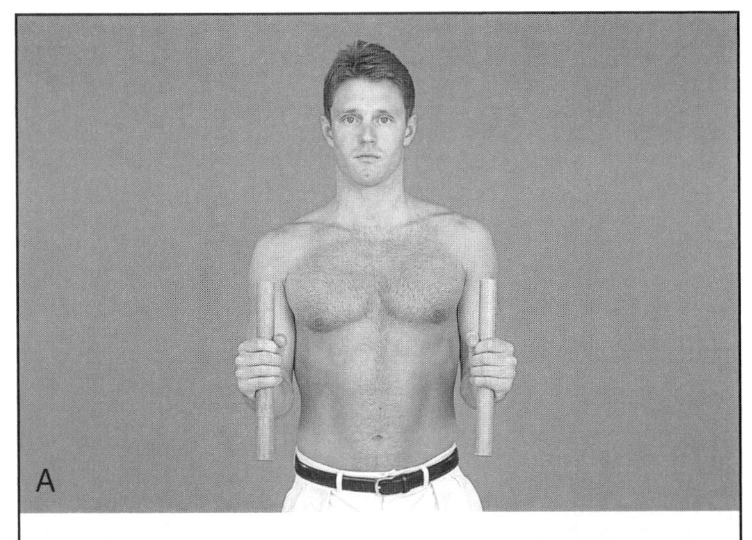

A

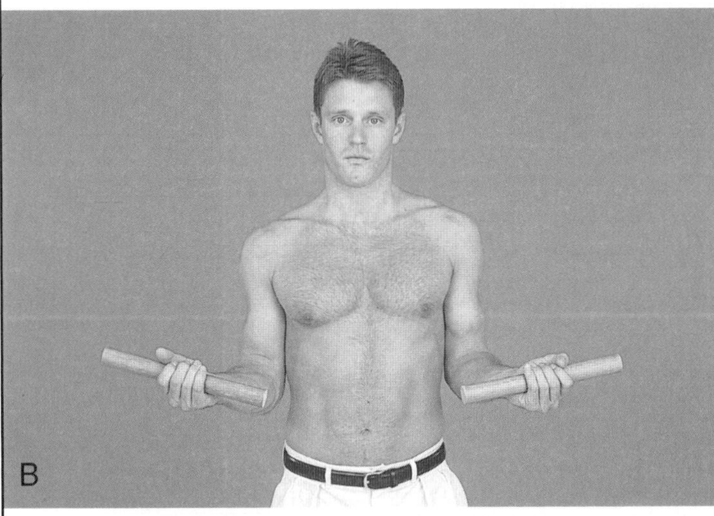

B

C

Forearm supination

What to do: Ask the patient to face you with the arms held at the side and the elbows flexed **(A)**. Then ask the patient to rotate the palms to face the ceiling **(B)**. This can be tested with and without resistance.

Normal: Nearly 90° of motion (ie, the palms should fully face the ceiling).

Interpretation: This motion is powered by the biceps, since it inserts on the radius. A rupture of the distal biceps will substantially weaken supination, even when the supinator muscle is intact.

Comment: Supination and pronation take place at the joint between the distal humerus (capitellum) and the radius, as well as between the proximal and distal radioulnar joints.

Forearm pronation

What to do: Ask the patient to face you with the arms held at the side and the elbows flexed **(A)**. Then ask the patient to rotate the palms to face the floor **(C)**. This can be tested with and without resistance.

Normal: Nearly 90° of motion (ie, the palms should fully face the floor).

Interpretation: This motion is powered by the pronator teres and assisted by the pronator quadratus, both of which are supplied by the median nerve.

Comment: Abnormal passive motion can occur following fractures of the elbow, forearm, or wrist.

Special Tests

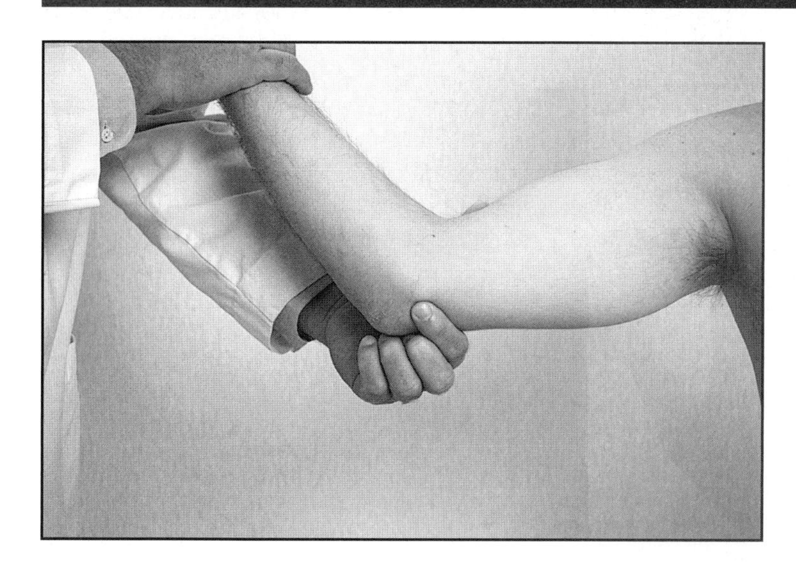

Tinel's sign at the elbow

What it tests: Irritation of the ulnar nerve at the elbow (cubital tunnel syndrome).

What to do: Gently tap the patient's elbow near the medial border of the proximal ulna, just distal to the elbow joint over the ulnar nerve.

Normal: Patient should not report any abnormal sensation ("electric shocks") in the hand, although mild discomfort localized to the elbow is normal.

Interpretation: Paresthesias radiating to the ulnar side of the hand suggest irritation of the ulnar nerve at the cubital tunnel.

Hand and Wrist

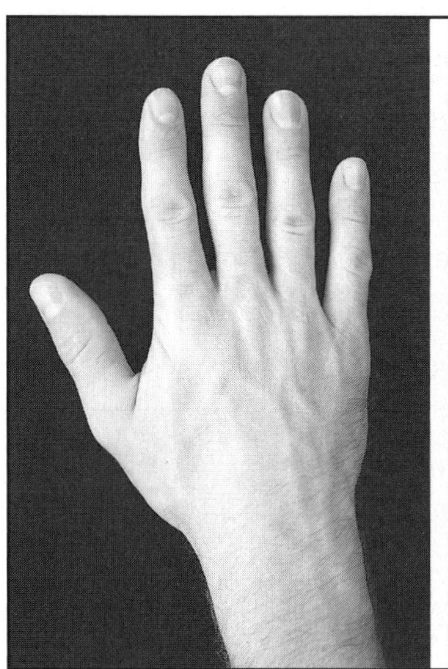

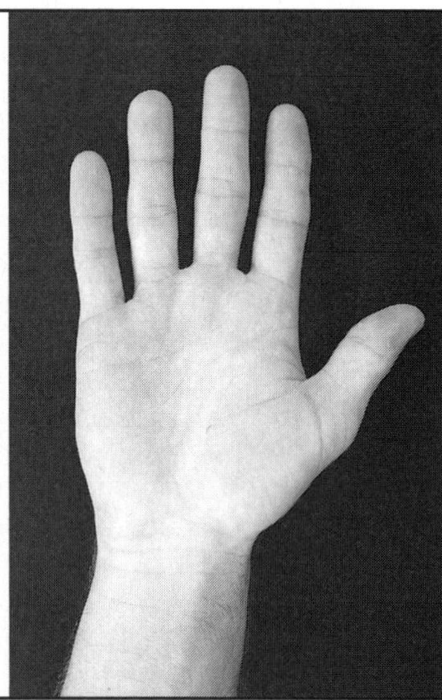

Inspection

Begin inspection by simply observing the patient using the hands. The alignment of the fingers should be noted, as well as any pitting or discoloration of the nails. Atrophy of the thenar muscles (innervated by the median nerve) or the intrinsics between the metacarpals (innervated by the ulnar nerve) should also be noted. While in a resting position, all of the fingers should be slightly flexed. Check for angular and rotational deformities at the joints.

Palpation

Palpation Point	Significance
"Anatomic snuff box"	Tenderness is associated with fracture of the scaphoid (see Chapter 16, Figure 8).
Phalangeal joints	Bony swelling (nodes) suggests arthritis; when occurring at the distal interphalangeal (DIP) joints, it suggests osteoarthritis; when occurring at the metacarpophalangeal (MCP) joints, it suggests rheumatoid arthritis.
Distal ulna	Tenderness suggests injury to the triangular fibrocartilage complex (TFCC).
Thenar muscles	Atrophy suggests median nerve lesion.
Radial artery pulse	Occlusion should not cut off the blood supply to the hand if the ulnar artery and arterial arches in the palm are patent. Can be injured in wrist or forearm fractures.
Palmar surface of the tip of the index finger	The sensory region of the median nerve. Sensation is decreased in injuries to the median nerve (ie, carpal tunnel syndrome).
Palmar surface of the tip of the little finger	The sensory region of the ulnar nerve. Sensation is decreased in injuries to the ulnar nerve (ie, cubital tunnel syndrome).
Hook of the hamate	Tenderness after trauma suggests a fracture, which may be apparent only on CT and missed on standard radiographs.
Base of the thumb (metacarpal)	Tenderness suggests carpometacarpal (CMC) arthritis.
Finger mass	Firm mass on volar surface associated with trigger finger.
Palmar mass	Soft mass suggests lipoma.
Nailbed	Check capillary refill.

Wrist flexion

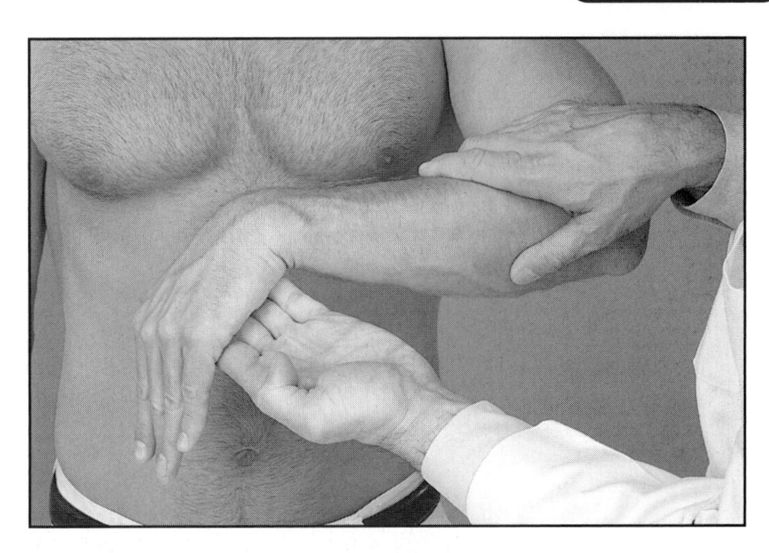

What to do: Support the patient's forearm and place one hand on the patient's palm; the patient's wrist should be in neutral position. Ask the patient to flex the wrist as you apply resistance to the palm. For passive motion testing, there should be no resistance. Instead, with the patient relaxed, bend the wrist by applying force across the dorsum of the hand.

Normal: Nearly 90° of flexion, with good strength and no deviation to the radial or ulnar side.

Interpretation: Flexion is powered by the flexor carpi radialis (median nerve) and the flexor carpi ulnaris (ulnar nerve). Pain at the humeral medial epicondyle, the origin of both of these muscles, suggests medial epicondylitis (ie, golfer's elbow). Flexion loss may indicate intra-articular pathology (ie, carpal instability). Diminished wrist flexion is usually well tolerated.

Comment: The patient's fingers should be extended to avoid some wrist flexion by the finger flexor muscles.

Wrist extension

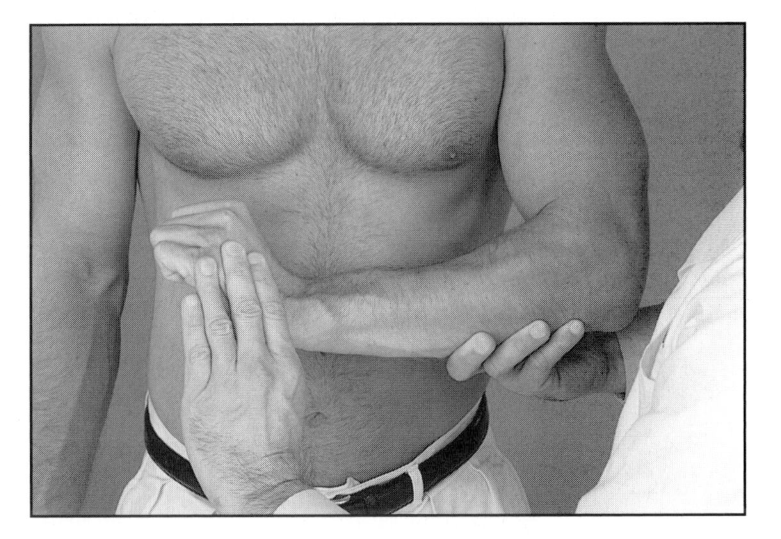

What to do: Support the patient's forearm and place one hand on the dorsum of the hand; the patient's wrist should be in neutral position. Ask the patient to extend the wrist as you apply resistance to the hand. A passive motion test is performed by extending the wrist by placing force on the palm.

Normal: 70° of extension.

Interpretation: Extension is powered by the extensor carpi radialis longus and the extensor carpi radialis brevis, which are innervated by the radial nerve, primarily at the C6 level. Pain at the lateral epicondyle, the origin of the extensor muscles, suggests lateral epicondylitis, (ie, tennis elbow). As with flexion loss, loss of extension may indicate an intra-articular pathology. However, loss of extension is less well tolerated for normal wrist function.

Comment: The fingers should be held in flexion during this maneuver.

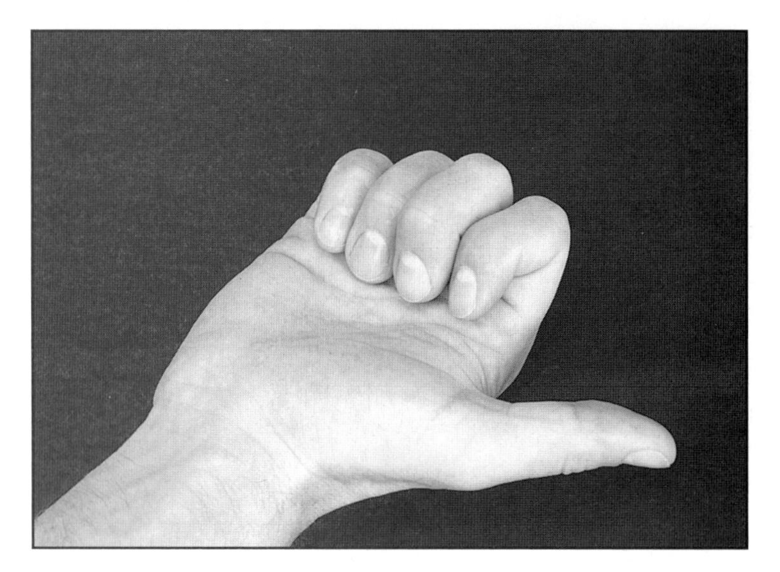

Finger flexion and extension

What to do: With the patient's hand supinated, ask the patient to make a fist by flexing the fingers to the palm and holding the thumb abducted.

Normal: The fingertips should touch the distal palmar crease. In addition, the patient should be able to extend all fingers completely straight (0° of extension).

Interpretation: Any lack of flexion or extension should be checked both actively and passively. Passive restriction of motion is usually caused by arthritis of the joint or joint contracture; active restriction can be caused by muscle weakness or neurologic impairment. Trauma to a flexor or extensor tendon with stiffness can also produce active and passive restriction in motion.

Comment: Finger flexion and extension is a composite movement of the MCP, PIP, and DIP joints. If any finger lacks full flexion or extension, each joint should be examined to identify the affected joint.

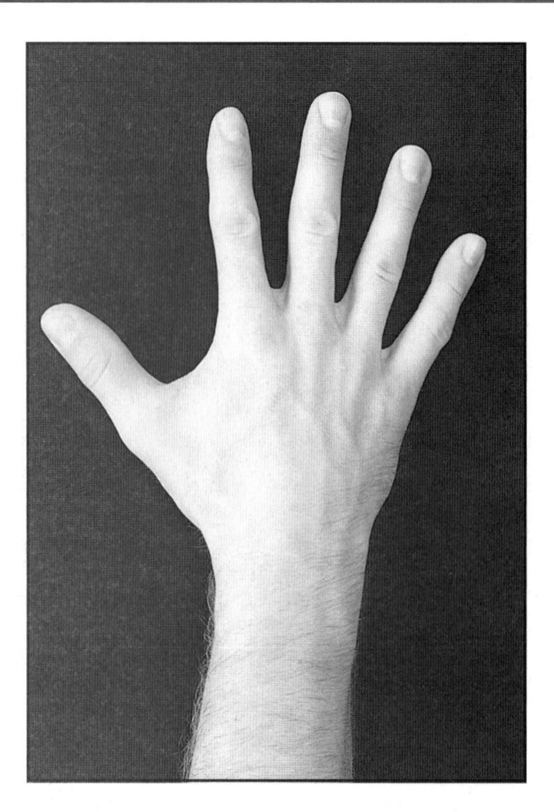

Finger abduction and adduction

What to do: With the patient's fingers extended, ask the patient to fully abduct and adduct the fingers.

Normal: The fingers should fully adduct together and abduct (ie, separate) approximately 20°.

Interpretation: Restricted abduction is usually caused by intrinsic muscle weakness.

Thumb opposition

What to do: Ask the patient to bring the thumb across the palm toward the base of the little finger.

Normal: The patient should be able to bring the thumb to the MCP joint of the little finger.

Interpretation: Restricted opposition can be caused by pathology of the CMC, MCP, or interphalangeal (IP) joint of the thumb. It can also be secondary to muscle weakness.

Comment: Thumb opposition is a composite movement of the CMC, MCP, and IP joints of the thumb. It is powered by the thenar muscles (ie, the abductor pollicis brevis, opponens pollicis, and flexor pollicis brevis).

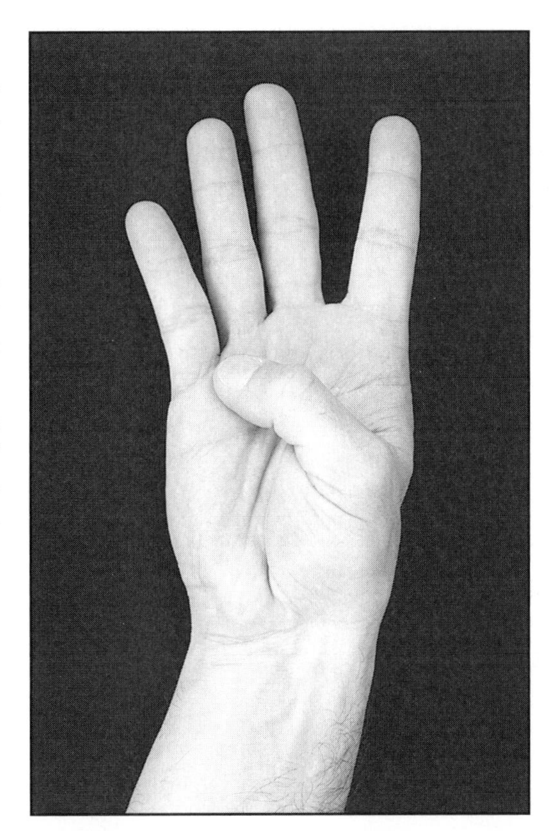

Thumb flexion and extension

What to do: Ask the patient to bend and extend the thumb in the plane of the palm.

Normal: Extension should be 0° at the MCP joint and 5° to 10° at the IP joint **(A)**. Flexion at the MCP joint **(B)** is usually 50° to 60°, and flexion at the IP joint is usually 55° to 75° **(C)**.

Interpretation: Restricted flexion and extension are most commonly caused by stenosing tenosynovitis (trigger thumb). Patients may demonstrate "locking" or "catching" during active motion.

Comment: A trigger thumb may be tender at the palmar base of the thumb. Pain may limit flexion, and locked flexion may intermittently occur. This occurs when a nodule forms on the flexor tendon and catches within the pulleys that hold the tendon close to the bone.

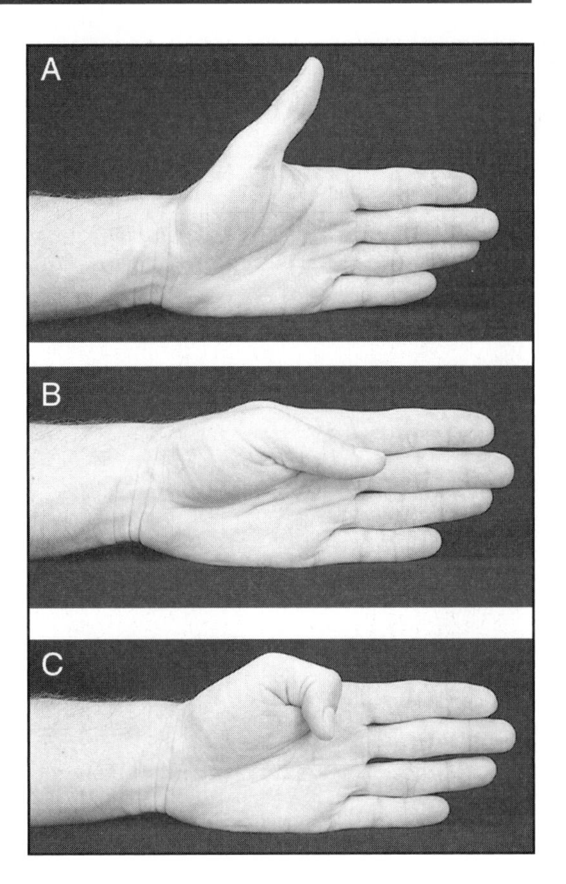

Muscle Testing

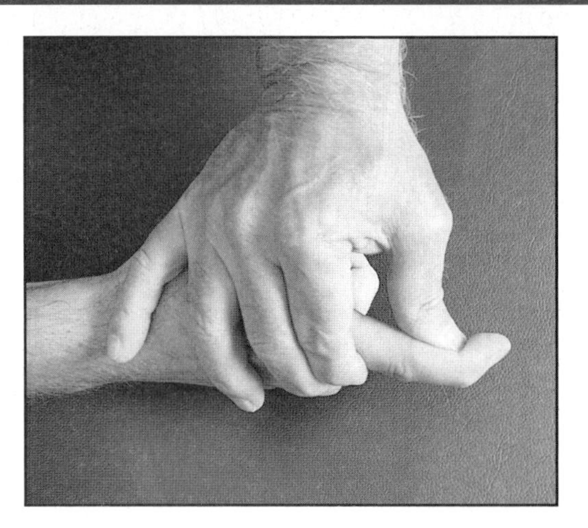

Finger/DIP flexion (flexor digitorum profundus)

What to do: Hold the finger with the PIP joint in extension, and ask the patient to bend the finger.

Normal: 90°.

Interpretation: Inability to flex only one finger's DIP joint suggests flexor digitorum profundus injury or bony avulsion.

Comment: This test is done when trauma or laceration is suspected. Assess routine DIP and PIP flexion by asking the patient to make a fist.

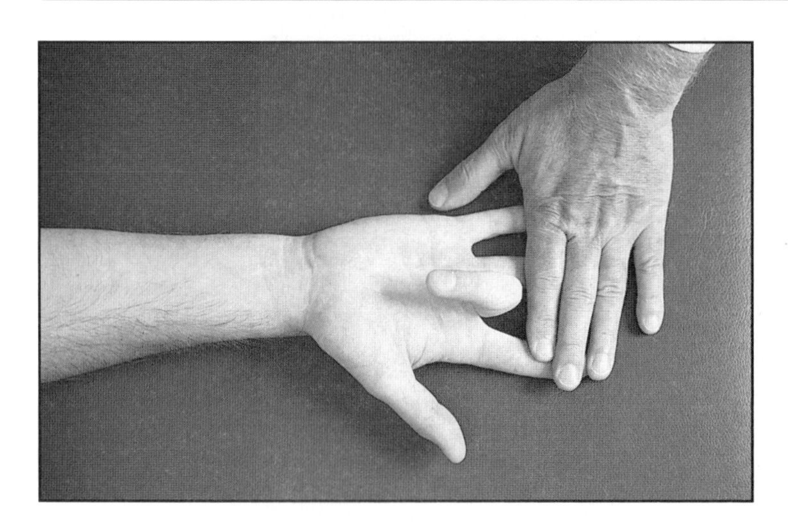

Finger/PIP flexion (flexor digitorum superficialis)

What to do: Grasp those fingers not being tested and hold those PIP and DIP joints in extension. Ask the patient to flex the finger not being held.

Normal: 100°.

Interpretation: Inability to flex suggests flexor digitorum superficialis tendon injury.

Comment: The flexor digitorum superficialis is specific for PIP flexion, but the flexor digitorum profundus (FDP) can also flex the PIP. Holding the other fingers in extension prohibits contraction of the FDP, which has one muscle belly attached to all four fingers.

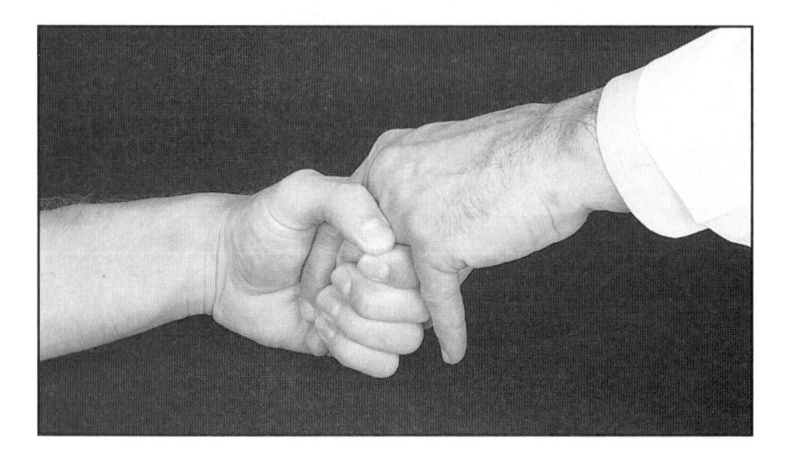

Grip strength

What to do: Ask the patient to squeeze your fingers as hard as possible.

Normal: Painless motion and symmetrical strength should be noted.

Interpretation: Power grip is primarily mediated by the ring and little fingers, as well as the intrinsic muscles of the hand. These are innervated by the ulnar nerve. Precision grip or pinch is mediated by the median nerve. Normal grip also requires good strength of the intrinsic muscles of the hand, which are also (for the most part) innervated by the ulnar nerve. Grip strength should be compared with the opposite side.

Palmar abduction of the thumb

What to do: Ask the patient to raise the thumb from the plane of the palm. Resist the patient's effort.

Normal: Good strength compared with the opposite side.

Interpretation: Poor strength usually indicates weakness of the abductor pollicis brevis. This is most commonly related to the motor nerve effects of carpal tunnel syndrome.

Comment: Atrophy of the thenar eminence may be present in severe carpal tunnel syndrome.

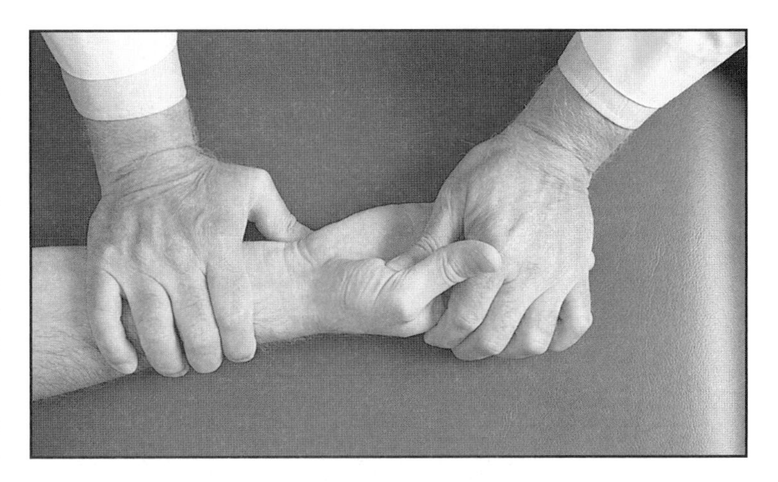

Special Tests

Phalen's test

What it tests: Compression of the median nerve at the wrist (ie, carpal tunnel syndrome).

What to do: Ask the patient to sit comfortably or stand with the wrists flexed maximally for 60 seconds.

Normal: No symptoms produced in the fingers.

Interpretation: Paresthesias in the median nerve distribution suggest carpal tunnel syndrome.

Comment: This test should be combined with a positive history and other tests to confirm the diagnosis.

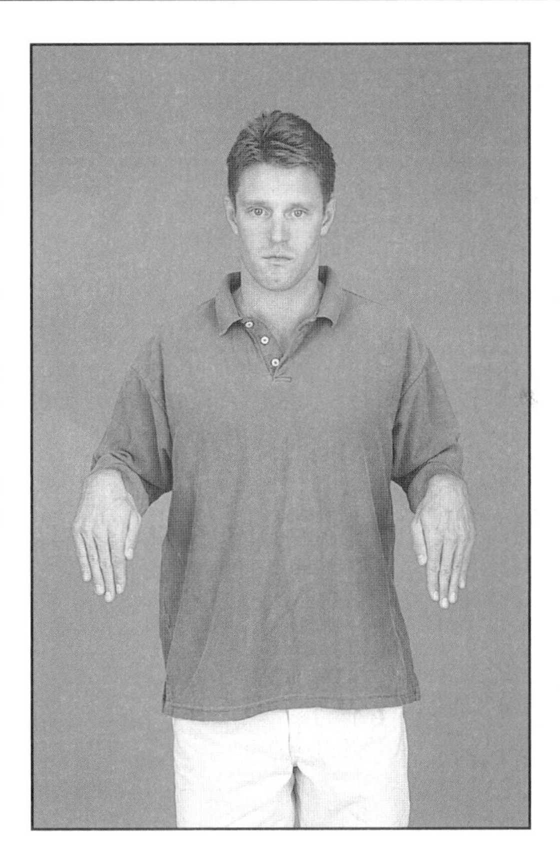

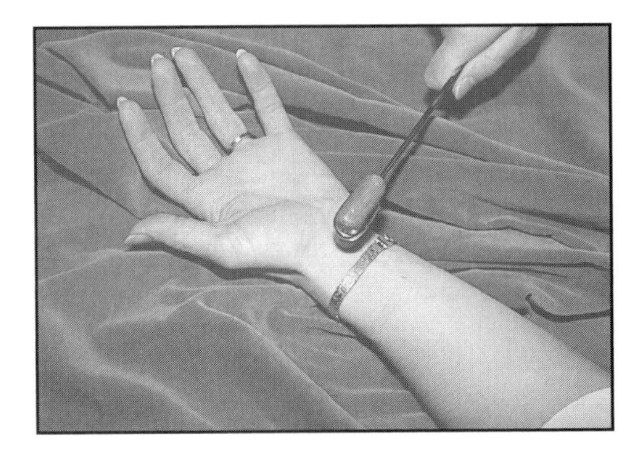

Carpal tunnel percussion (Tinel's sign at the wrist)

What it tests: Compression of the median nerve at the wrist (ie, carpal tunnel syndrome).

What to do: Place the patient's hand on the table with the palm side up. Tap on the carpal tunnel using a reflex hammer or your index finger.

Normal: No symptoms produced in the fingers.

Interpretation: Paresthesias in the index and long fingers (and, less commonly, the thumb) suggest irritation of the median nerve at the carpal tunnel.

Comment: This test should be combined with a positive history and other tests to confirm the diagnosis. The presence of other signs and symptoms of carpal tunnel syndrome may help make the diagnosis.

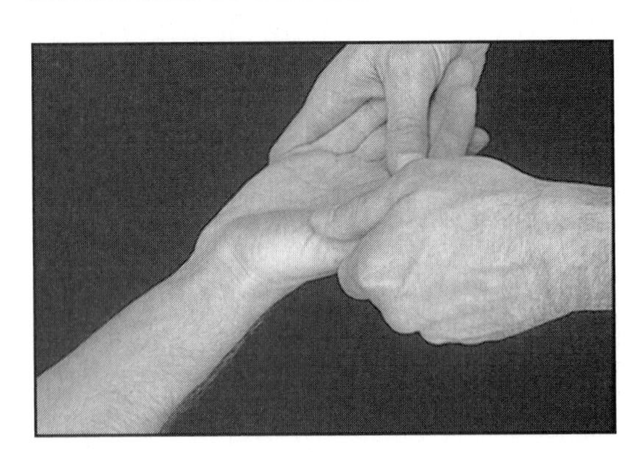

Thumb CMC stress test

What it tests: Arthritis at the CMC joint of the thumb.

What to do: Hold the thumb metacarpal. Place axial compression across the CMC joint and gently rotate the metacarpal to create a gentle grind.

Normal: Mild discomfort or no discomfort.

Interpretation: Pain at the base of the thumb suggests osteoarthritis.

Comment: The CMC joint of the thumb is the most common site of osteoarthritis in the hand.

Allen's test

What it tests: Collateral circulation of the hand from both the radial and ulnar arteries.

What to do: Ask the patient to rapidly open and close the hand three times and then make a fist; this will exsanguinate the hand. Compress the radial and ulnar arteries by placing your fingers on their pulse. Release pressure on the ulnar artery, but maintain pressure on the radial artery. Observe return of normal perfusion (ie, color) to the hand. Repeat by releasing pressure on the radial artery and compressing the ulnar artery.

Normal: The hand should become pink, signifying adequate perfusion, even with one artery compressed by digital pressure.

Interpretation: This test helps determine if circulation in either artery is solely sufficient to perfuse the entire hand. If one artery is slow or sluggish in reperfusing the hand, it should be noted.

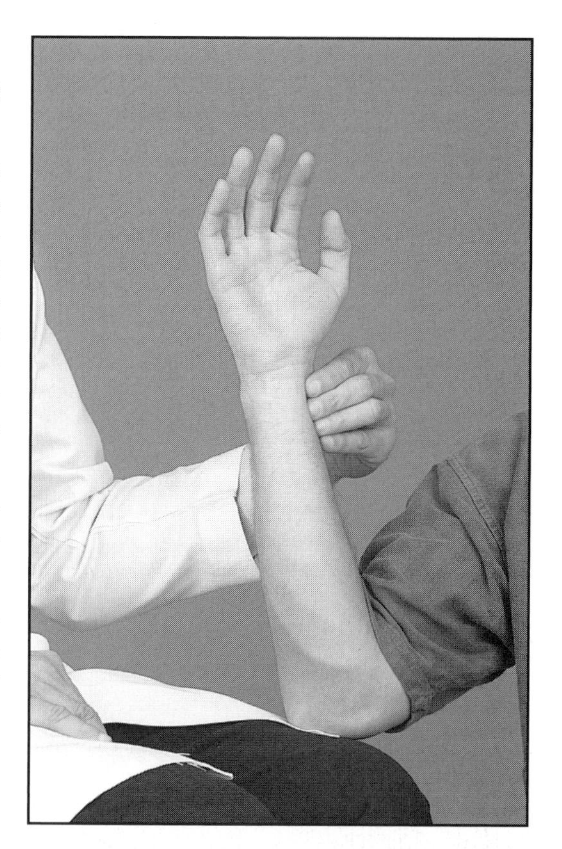

Finkelstein's test

What it tests: Painless function of the abductor pollicis longus and extensor pollicis brevis tendons (located in the first dorsal compartment of the wrist).

What to do: Flex and ulnarly deviate the wrist, then push the thumb into flexion.

Normal: Mild discomfort.

Interpretation: Sharp pain on the radial border of the wrist suggests tenosynovitis of the abductor pollicis longus or extensor brevis tendons.

Comment: This test can be very painful in affected patients.

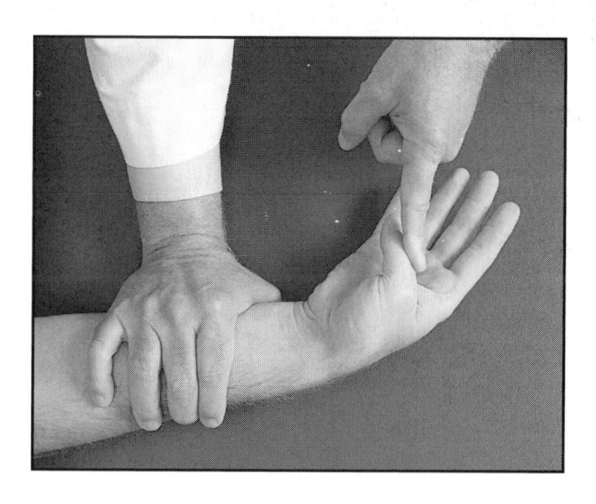

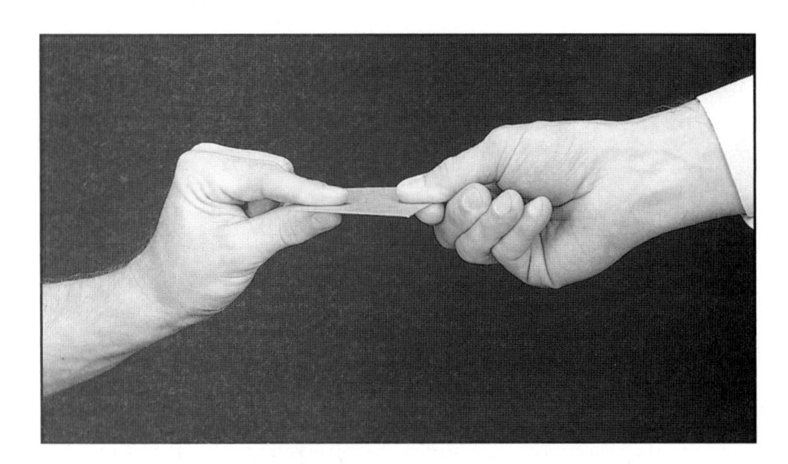

Froment's test

What it tests: Function of the adductor pollicis (ie, the ulnar nerve to the thumb).

What to do: Ask the patient to pinch a piece of paper between the thumb and the tip of the index finger. Then attempt to pull the piece of paper from the patient.

Normal: The patient will be able to pinch the piece of paper with good strength and without flexing the IP joint of the thumb.

Interpretation: Froment's sign is flexion of the IP joint of the thumb while pinching. IP flexion is powered by the median nerve. Adduction of the thumb is powered by the ulnar nerve. When pinching, IP flexion may be used instead of thumb adduction if there is an ulnar nerve injury.

Hip and Thigh

Inspection

Look for atrophy of the muscles. Assess overall alignment of the limbs and tilt (obliquity) of the pelvis **(A)**. The iliac crest **(B)** is a useful landmark for assessing pelvic tilt.

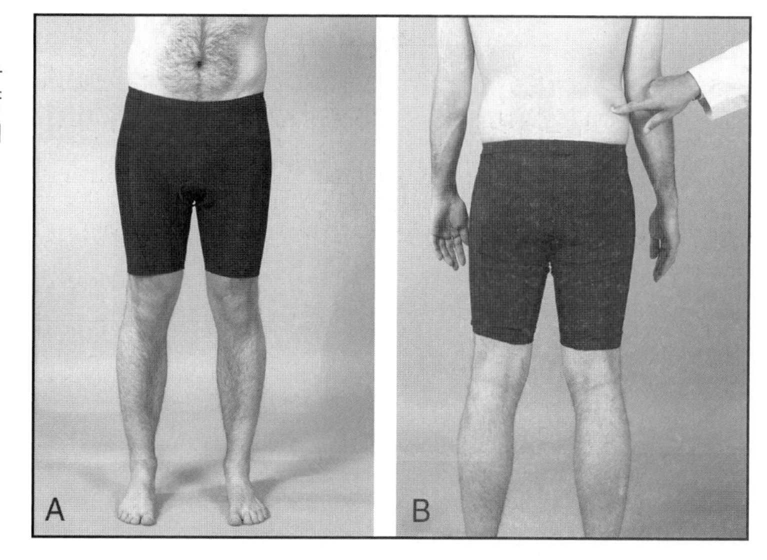

Palpation

Palpation Point	Significance
Anterior-superior iliac spine	Tenderness near the anterior superior iliac spine suggests an avulsion of the sartorius.
Greater trochanter	Tenderness suggests trochanteric bursitis.
Ischial tuberosity	Tenderness suggests contusions or hamstring avulsion.
Inguinal lymph nodes	Enlargement of lymph nodes suggests infection or neoplastic process.
Femoral pulse	Abnormal femoral pulse suggests vascular disease.

Hip flexion

What to do: Ask the patient to lie in the supine position. Passively flex the hip maximally toward the chest. Record the angle of the femur and horizontal line parallel to the examining table as the degree of hip flexion.

Normal: 0° to 130°.

Interpretation: Lack of normal hip flexion is most commonly secondary to osteoarthritis.

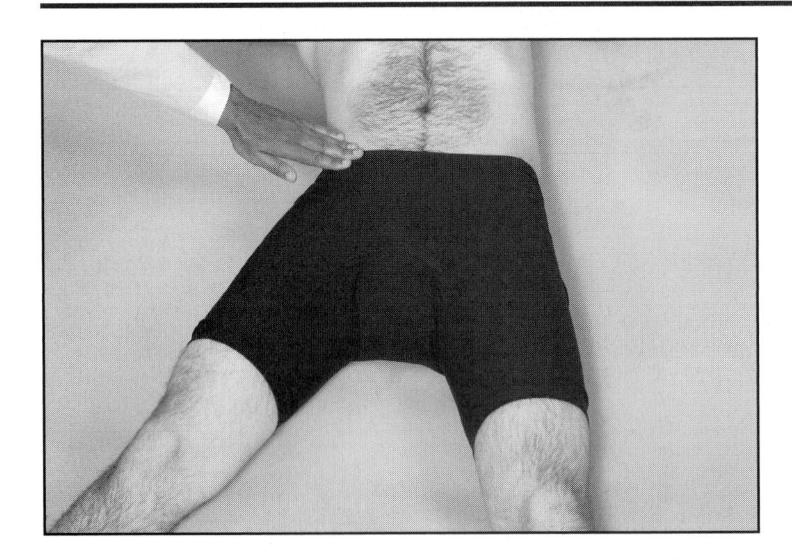

Hip abduction

What to do: Ask the patient to lie in the supine position. Stabilize the patient's pelvis and ask the patient to abduct the affected extremity. Alternatively, with the patient's perineum covered with an examination gown, ask the patient to separate the legs simultaneously and maximally (contralateral hip abduction simultaneously stabilizes the pelvis). Record the angle of abduction.

Normal: 0° to 45°.

Interpretation: Restricted abduction is a late manifestation of hip osteoarthritis.

Hip adduction

What to do: With the patient in the supine position, passively extend the entire leg across the midline. The contralateral leg remains on the examination table while the tested leg passes above the elevated leg. Record the extent to which the extremity angularly crosses the midline.

Normal: 20° to 30°.

Interpretation: Restriction in adduction is a late manifestation of hip osteoarthritis.

Internal and external rotation of the hip

What to do: With the patient in the supine position, flex the patient's hip to 90°. Passively rotate the tibia away from the midline to assess internal rotation. Rotate the tibia toward the midline to assess external rotation.

Normal: 0° to 35° for internal rotation **(A)** and 0° to 45° for external rotation **(B)**.

Interpretation: Loss of hip rotation is commonly the first sign of osteoarthritis. In children, loss of rotation may suggest a slipped capital femoral epiphysis.

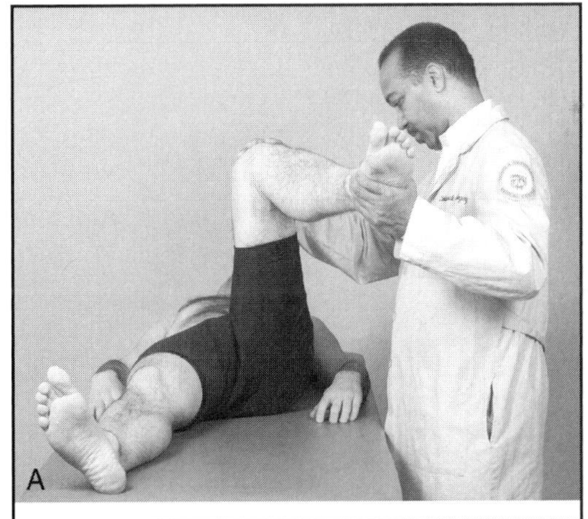

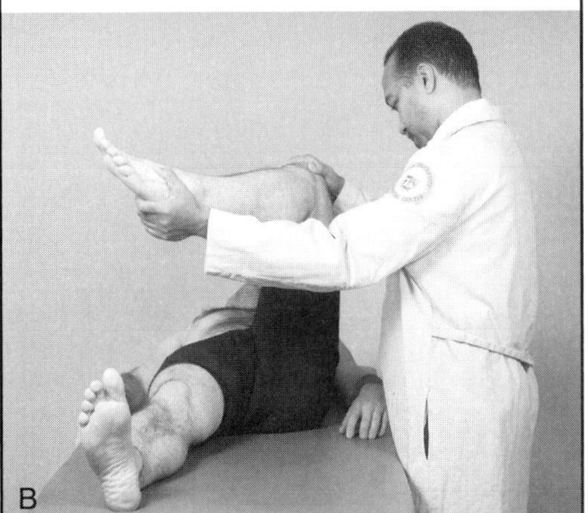

Muscle Testing

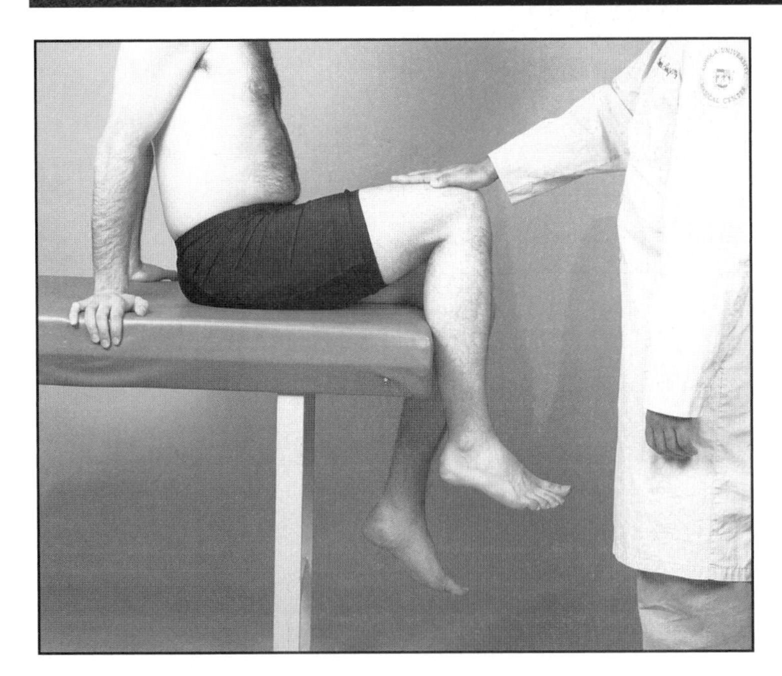

Hip flexion

What to do: With the patient in the seated position, place your hand on the thigh just above the knee. Ask the patient to flex the hip against resistance.

Normal: Normal muscle strength.

Interpretation: Absence of normal muscle strength may be secondary to pain or may indicate muscular or neurologic disease. Focal processes near the iliopsoas and high in the pelvis may cause weakness of hip flexion.

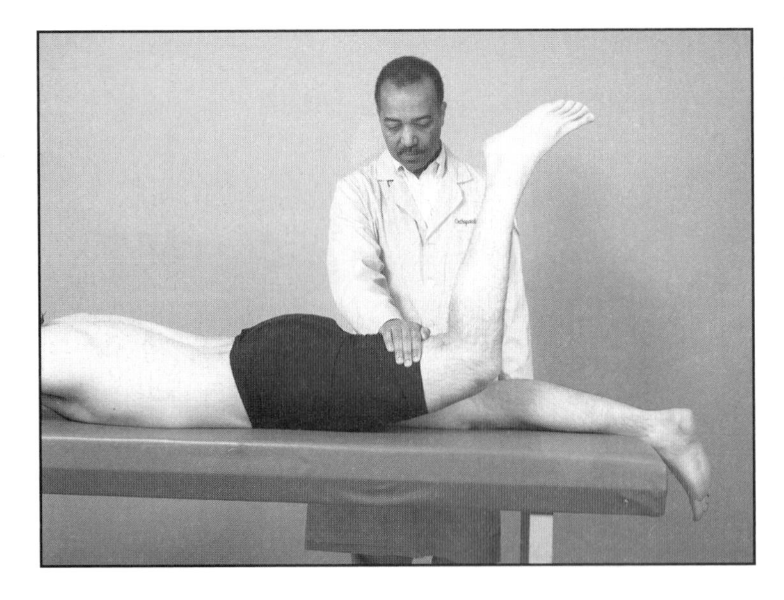

Hip extension

What to do: With the patient in the prone position, place the knee in 90° of flexion with your hand on the posterior thigh above the knee. Ask the patient to extend the hip against resistance.

Normal: Normal muscle strength.

Interpretation: Absence of normal muscle strength may be secondary to pain or may indicate muscular or neurologic disease.

Comment: A hip flexion contracture may limit the ability to test hip extension in this position; if this is the case, hip extension should be tested with the patient in the supine position.

Active adduction

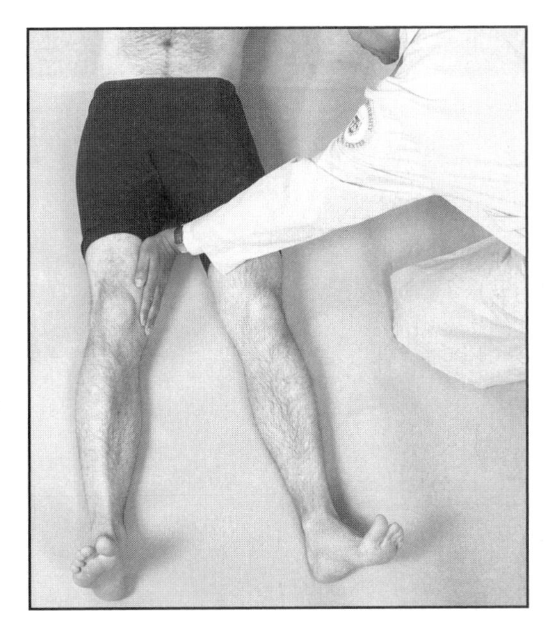

What to do: With the patient in the supine position and the legs slightly apart, place one hand on the medial border of the thigh or leg and ask the patient to adduct the hip against manual resistance by bringing the legs together.

Normal: Normal muscle strength and absence of pain.

Interpretation: An adductor strain (also known as a pulled groin) causes pain-induced weakness.

Active abduction

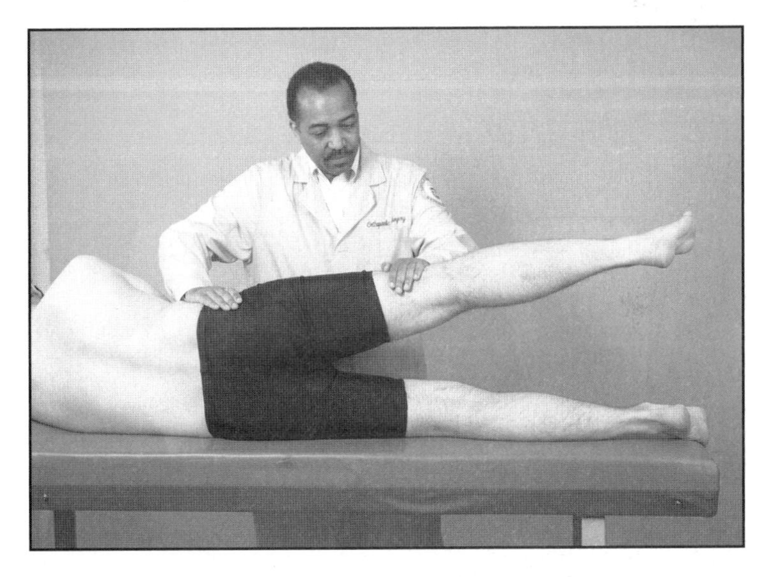

What to do: With the patient lying on the unaffected side, place one hand on the lateral aspect of the thigh. Ask the patient to abduct the hip against resistance.

Normal: Normal muscle strength.

Interpretation: Absence of normal muscle strength may be secondary to pain or may indicate muscular or neurologic disease. Muscle atrophy and weakness commonly occur with chronic hip diseases, such as osteoarthritis.

Comment: Hip abductor strength can also be assessed with the Trendelenburg test.

Special Tests

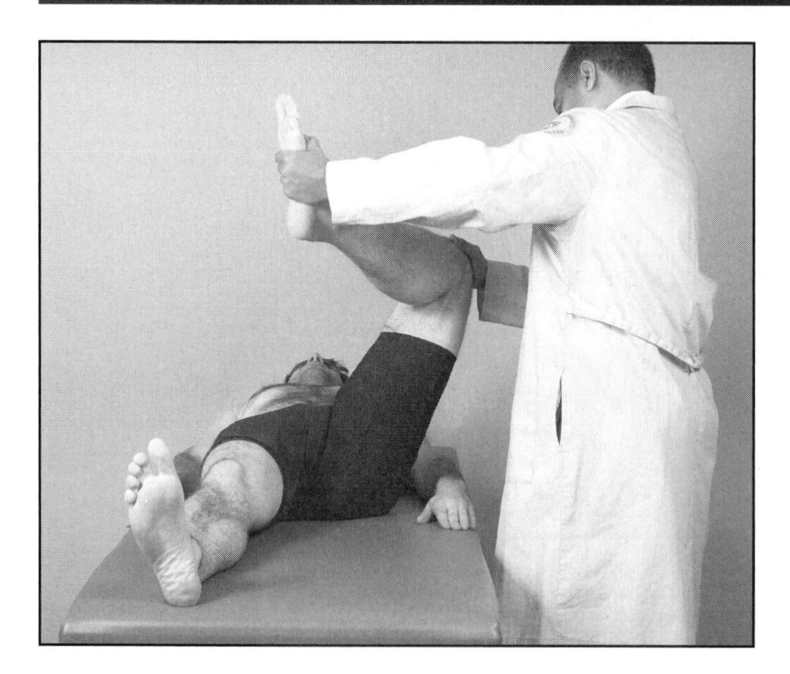

FABER test

What it tests: The FABER test (flexion-abduction-external rotation) is a stress maneuver to detect pathology at the sacroiliac joint.

What to do: With the patient in the supine position, place the patient's affected hip in flexion, abduction, and external rotation.

Normal: Absence of pain.

Interpretation: If pain is produced, confirm that it emanates from the sacroiliac joint by palpating it.

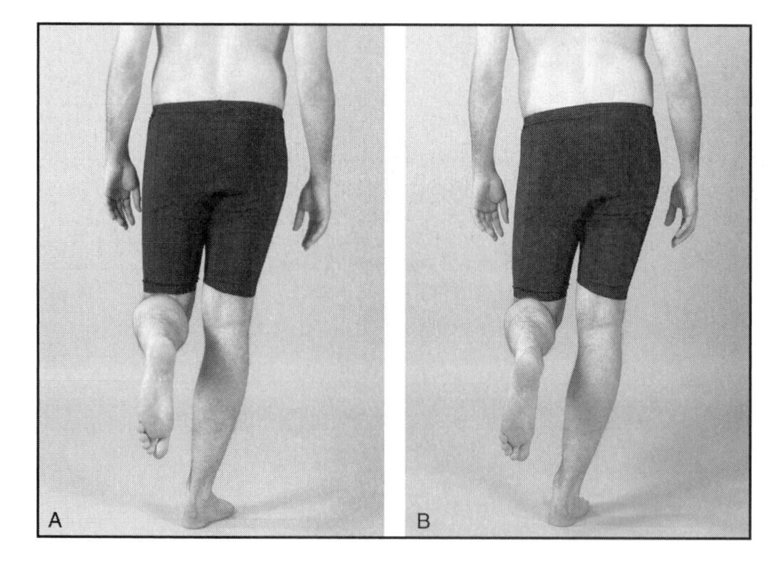

Trendelenburg test

What it tests: Abductor muscle strength of the hip.

What to do: With the patient standing, ask the patient to lift one foot off the ground and note whether the pelvis remains level or is slightly inclined toward the side of the supporting extremity.

Normal: Pelvis remains level **(A)**.

Interpretation: If the pelvis dips toward the side of the foot being lifted off the ground and the trunk inclines toward the supporting side, the test result is positive **(B)**.

Comment: Positive test results can be noted in patients with hip pain because abductor muscle contraction creates compression of the hip joint (reactive joint force).

Thomas test

What it tests: Hip flexion contracture (lack of extension).

What to do: With the patient in the supine position, passively flex both of the patient's hips into the chest. Then extend the affected hip while keeping the opposite hip in the flexed position.

Normal: The affected hip should extend completely to the examination table **(A)**.

Interpretation: As this maneuver is being performed, the pelvis will be brought forward and flexion will occur at the lumbosacral articulations and eliminate lumbar lordosis. If the hip being tested cannot extend to the examination table, the test result is positive, in which case a fixed flexion contracture of the affected hip is diagnosed and measured **(B)**.

Comment: Apparent "full extension" can be achieved even if there is a contracture (ie, false-negative test result) when the patient accentuates lumbar lordosis. By stabilizing the pelvis with the contralateral extremity flexed, the patient cannot compensate, and a true flexion contracture can be detected.

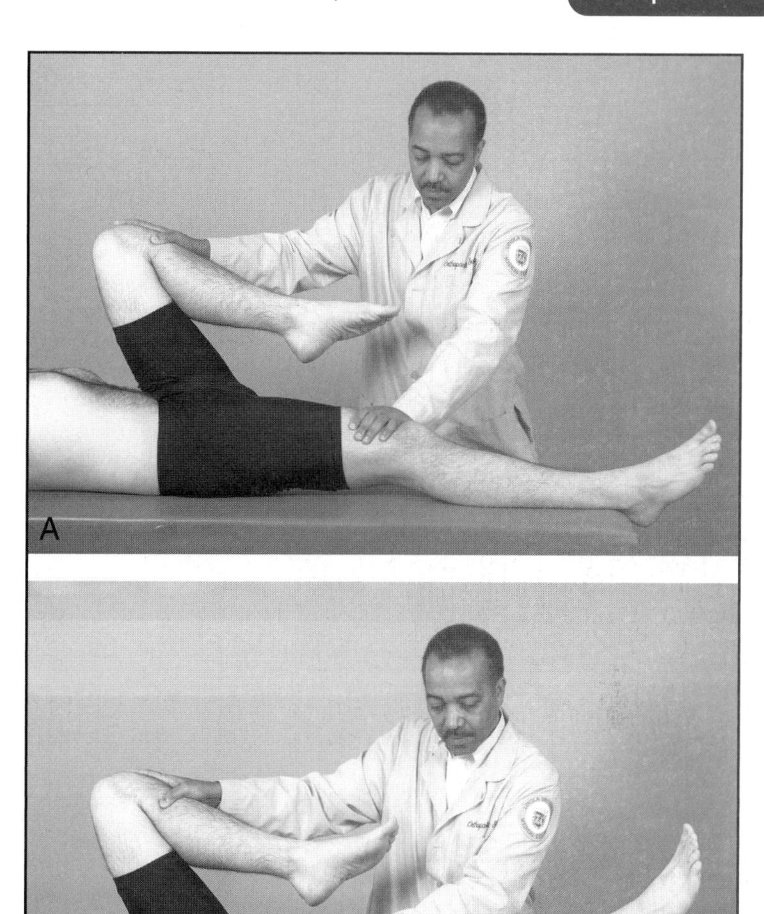

Knee and Leg

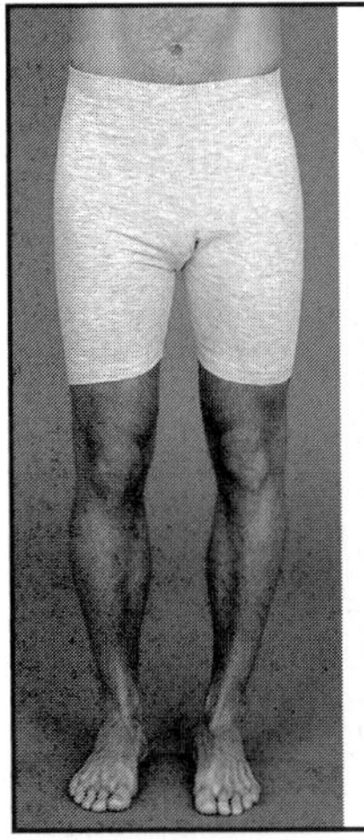

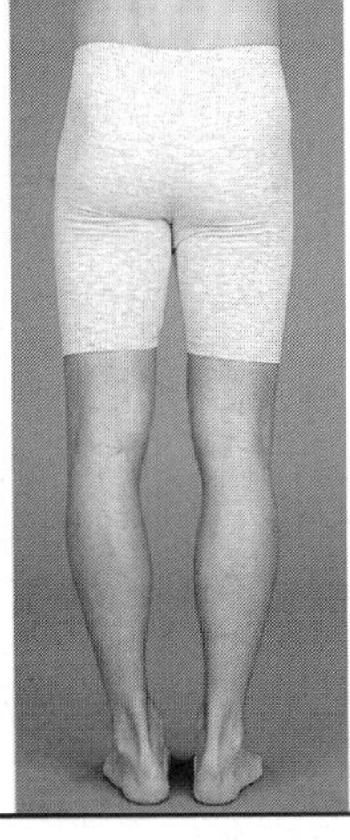

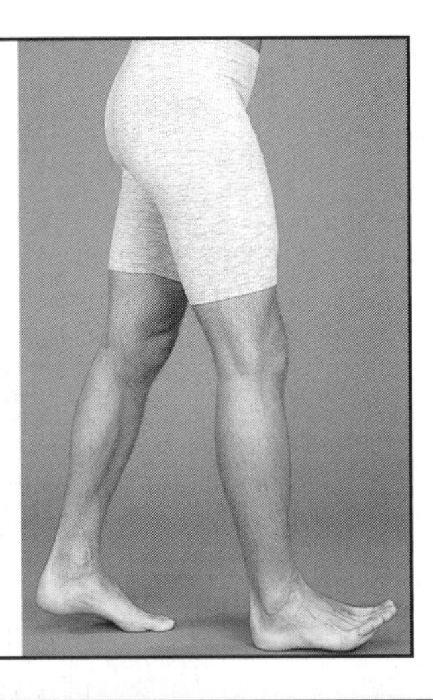

Inspection

Inspect the weight-bearing knees for overall alignment. The normal knee should be in slight valgus (knock-knees, or about 5° to 7°). Note any muscle atrophy, swelling, or malalignment, which are most evident on side-to-side inspection.

Palpation

Palpation Point	Significance
Patella	Tenderness at the superior or inferior poles of the patella suggests tendinitis.
Prepatellar bursa	Tenderness with gentle pinching suggests prepatellar bursitis.
Medial joint line	Tenderness suggests medial compartment osteoarthritis, meniscus tear, or medial collateral ligament injury.
Lateral joint line	Tenderness suggests lateral compartment osteoarthritis, meniscus tear, or lateral collateral ligament injury.
Medial and lateral epicondyles	Tenderness suggests injury to the origin of the medial or lateral collateral ligament.
Pes anserine bursa	Tenderness or swelling suggests pes anserine bursitis.
Tibial tubercle	Tenderness or prominence suggests Osgood-Schlatter's disease.
Popliteal fossa	Fullness suggests the presence of a large popliteal cyst. The popliteal pulse is typically not easily palpable.
Fibular head	Tenderness suggests a fracture or injury to the lateral collateral ligament.

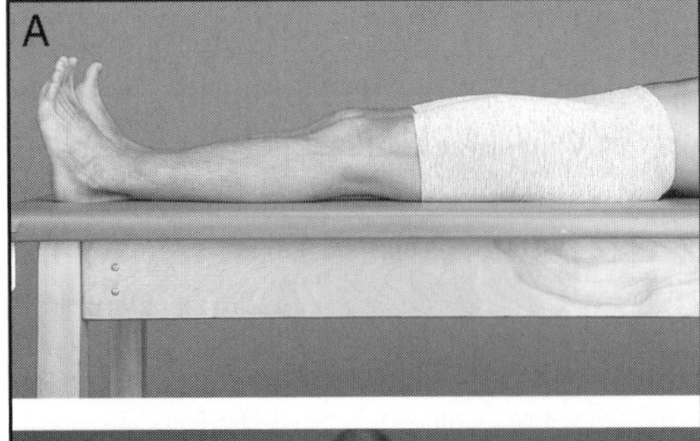

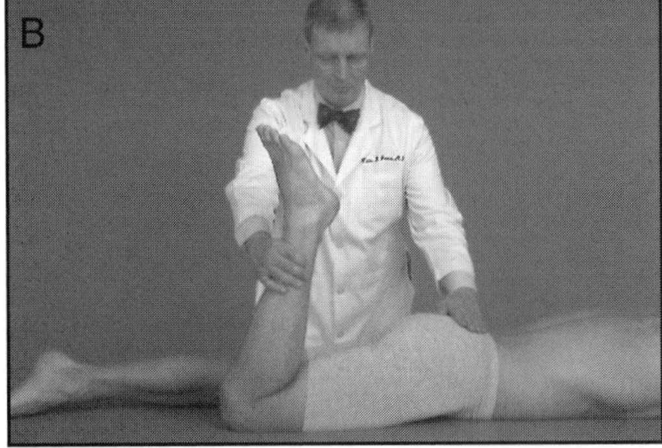

Knee flexion and extension

What to do: With the patient in the supine position, fully extend the knee by pulling the heel off of the examination table **(A)**. Fully flex the knee by bringing the foot toward the buttock **(B)**.

Normal: Extension to 0° and flexion to 135°.

Interpretation: Reduced range of motion suggests osteoarthritis, acute hemarthrosis, or capsular contracture (arthrofibrosis).

Muscle Testing

Quadriceps

What to do: With the patient in the seated position with the legs off of the examining table, ask the patient to extend the knee against resistance.

Normal: Good strength throughout a full range of motion (0° to 135°).

Interpretation: Complete inability to extend the knee indicates disruption of the extensor mechanism, including a possible quadriceps tendon tear, patellar tendon tear, or patella fracture. Weakness can be secondary to a muscular or neurologic cause. Chronic knee pain causes disuse atrophy of the quadriceps muscle, leading to subtle weakness.

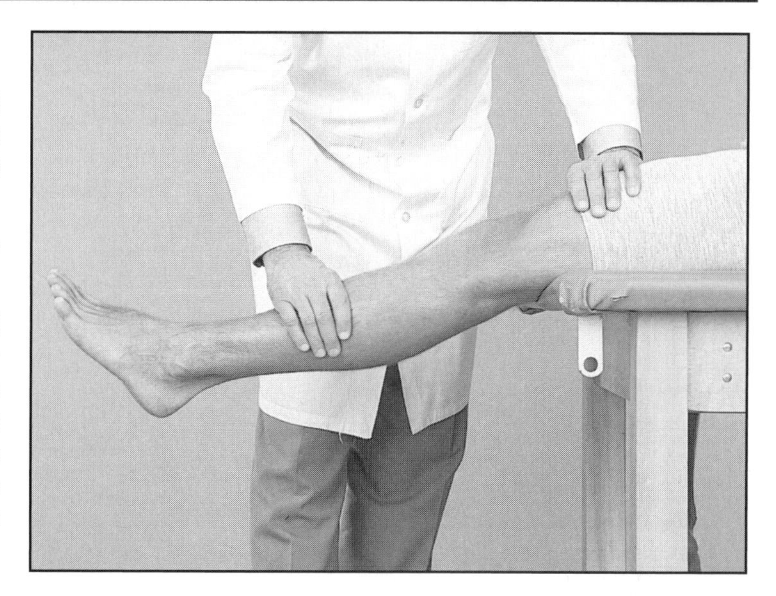

Hamstring

What to do: With the patient in the supine position with the hip flexed, bring the heel toward the buttock against resistance by placing the hand behind the ankle.

Normal: Good strength throughout a full range of motion.

Interpretation: Weakness can be secondary to pain or a muscular or neurologic cause (including disk herniation).

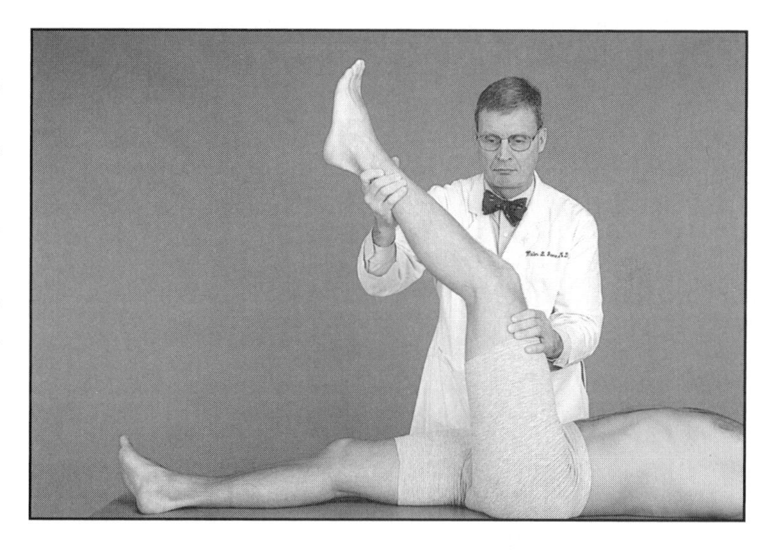

Special Tests

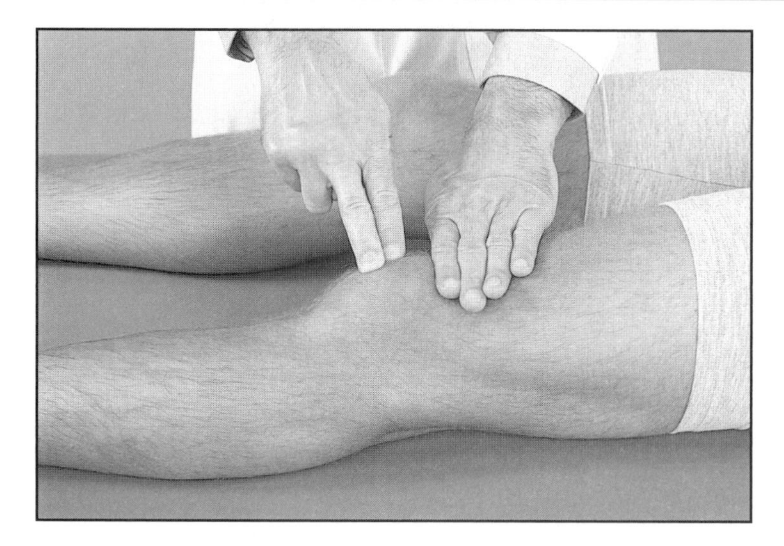

Patellar ballottement

What it tests: Effusions in the knee joint.

What to do: With the patient in the supine position, use one hand to compress the suprapatellar pouch, pushing whatever fluid may be in the pouch distally beneath the patella. Then use the other hand to tap the patella.

Normal: No effusions and the patella remains stable.

Interpretation: A positive result occurs when there is fluid in the knee joint that causes the patella to ballotte (bounce) as the fluid displaces with tapping. An effusion can be caused by acute trauma, inflammation, or infection.

Comment: The ballottement test may fail to detect small effusions if the fluid is not adequately compressed from the suprapatellar pouch.

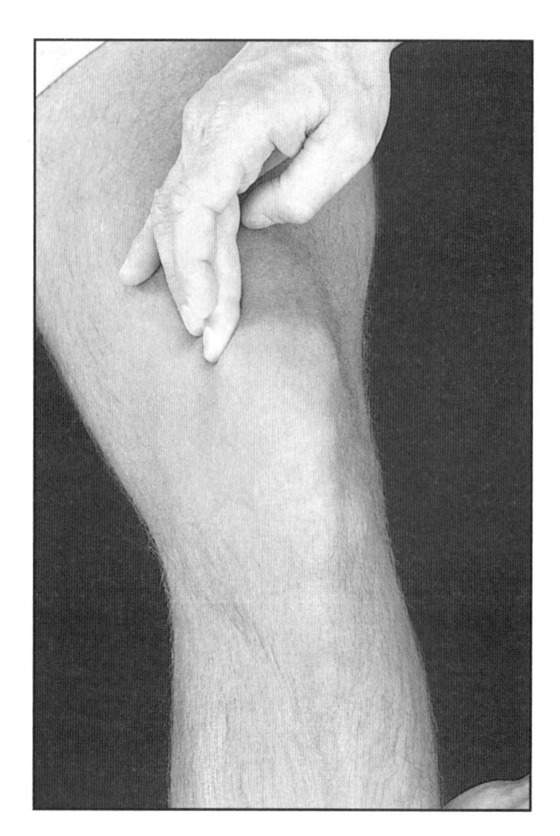

Patellar apprehension test

What it tests: Patellar stability.

What to do: With the patient in the supine position, ask the patient to relax the quadriceps muscles. Attempt to laterally translate the patella by pushing on the medial aspect of the patella with a medial to lateral force, noting any apprehension or distress in the patient's face.

Normal: No apprehension or distress by the patient.

Interpretation: A positive test result occurs when the patient reacts with apprehension or distress to the lateral translation of the patella. A patient with ligamentous laxity and instability may not react as dramatically as will a patient with traumatic dislocation.

Patellar inhibition test

What it tests: Function of the patellofemoral joint.

What to do: With the patient in the supine position, the quadriceps muscles relaxed, and the knee in extension, push the patella distally in the trochlear groove and then ask the patient to contract the quadriceps muscles. Apply resistance to the proximal excursion of the patella as the patient contracts the muscles.

Normal: No pain; patient is able to actively contract quadriceps.

Interpretation: A positive test result usually indicates arthrosis of the patellofemoral joint; patients will inhibit the quadriceps because the contraction causes pain.

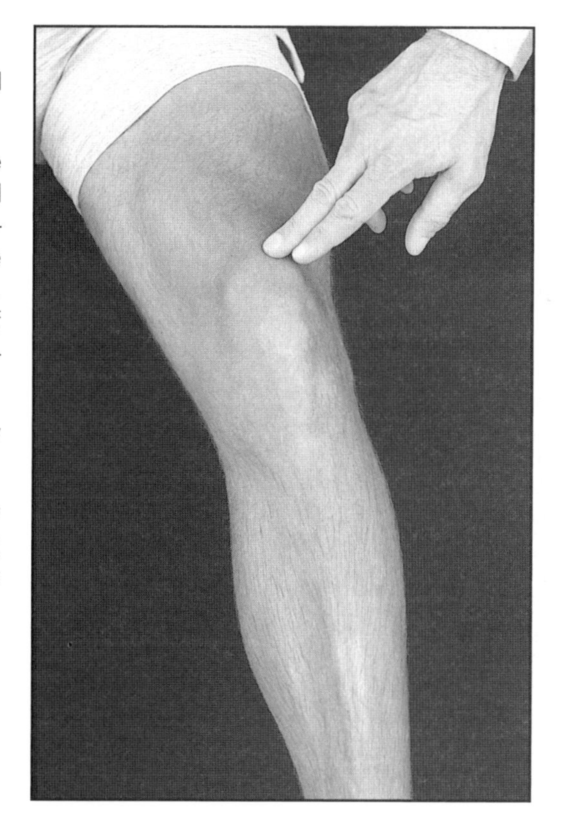

Valgus stress test

What it tests: Integrity of the medial collateral ligament.

What to do: With the patient lying supine on the examination table, hold the leg in one hand while applying valgus stress to the knee with the other. Perform this test twice: first with the patient's knee fully extended and again with the patient's knee at 30° of flexion.

Normal: No significant opening of the medial joint space.

Interpretation: The test result is positive when there is more than 5° of opening of the medial joint space in response to the valgus stress. When the test result is positive with the knee at 30° of flexion but negative at 0°, isolated injury of the medial collateral ligament is suspected. When the test results are positive at both 0° and 30°, capsular damage is indicated in addition to a medial collateral ligament injury.

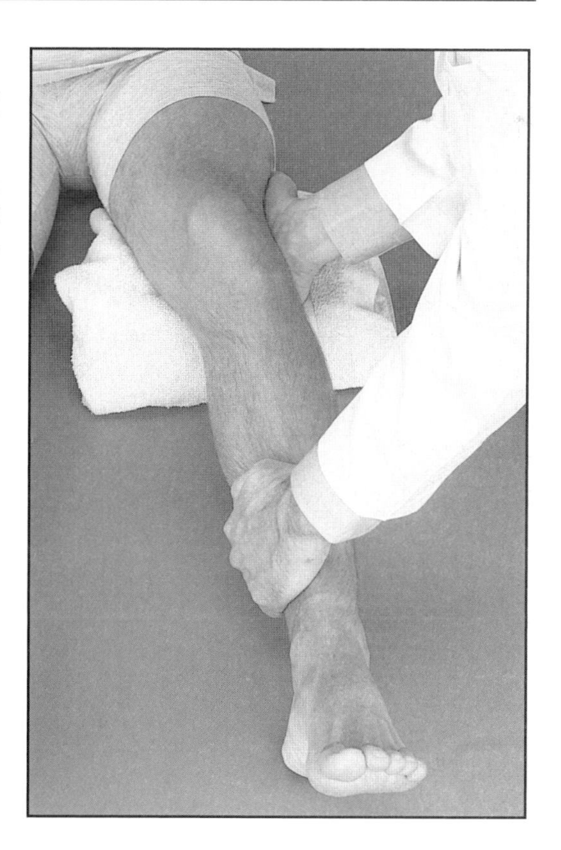

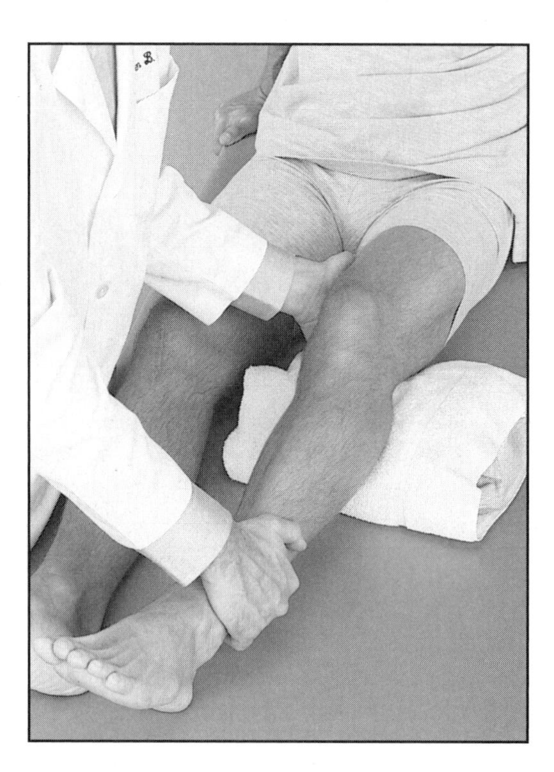

Varus stress test (lateral collateral ligament)

What it tests: Integrity of the lateral collateral ligament.

What to do: With the patient lying supine, hold the leg in one hand while applying varus stress to the knee with the other hand. As with the valgus stress test, perform this test twice: first with the patient's knee fully extended (0°) and again with the patient's knee at 30° of flexion.

Normal: No significant opening of the lateral joint space. It is normal to have slightly more lateral gapping than medial.

Interpretation: The test result is positive when there is more than 7° of opening of the lateral joint space in response to varus stress. When the test result is positive with the knee at 30° of flexion but negative at 0°, isolated injury to the lateral collateral ligament is suspected; however, when the test results are positive at both 0° and 30°, capsular damage is indicated in addition to a lateral collateral ligament injury.

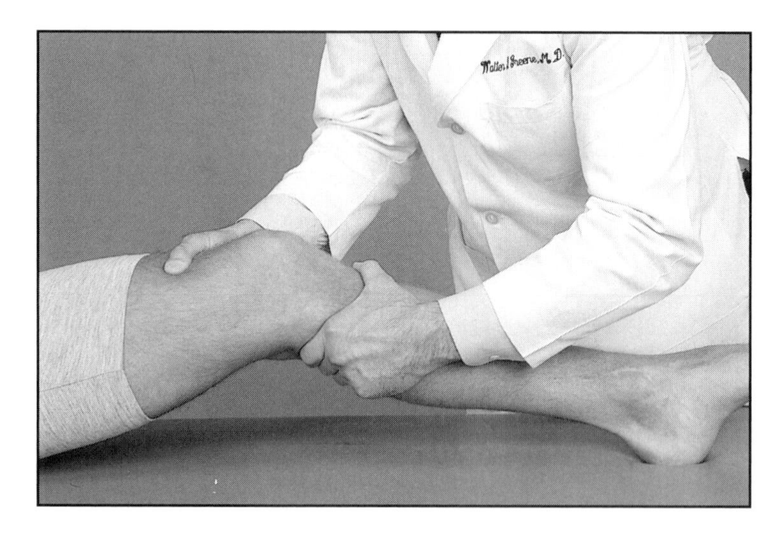

Lachman test

What it tests: Integrity of the anterior cruciate ligament (ACL).

What to do: With the patient lying supine with the knee at 20° to 30° of flexion, stabilize the distal femur in one hand while grasping the proximal tibia with the other hand. The thumb of the hand grasping the tibia should be placed just below the joint line. Rapidly apply an anterior force to the tibia (a jerking motion). Grade the quality of the end point (taut versus soft) and the amount of anterior translation of the tibia on the femur compared with the normal side.

Normal: A firm end point implies continuity of the ACL.

Interpretation: Increased translation of the tibia or a soft end point indicates an ACL tear.

Comment: This test requires an experienced examiner and a relaxed patient.

Anterior and posterior drawer test

What it tests: Integrity of the ACL and posterior cruciate ligament (PCL).

What to do: With the patient lying supine on the examination table, the knee flexed to 90°, and the hip flexed to 45°, rest your buttock on the patient's foot to maintain the flexion and stabilize the leg. Then attempt to displace the tibial plateau first anteriorly and then posteriorly by pulling and pushing on the proximal tibia.

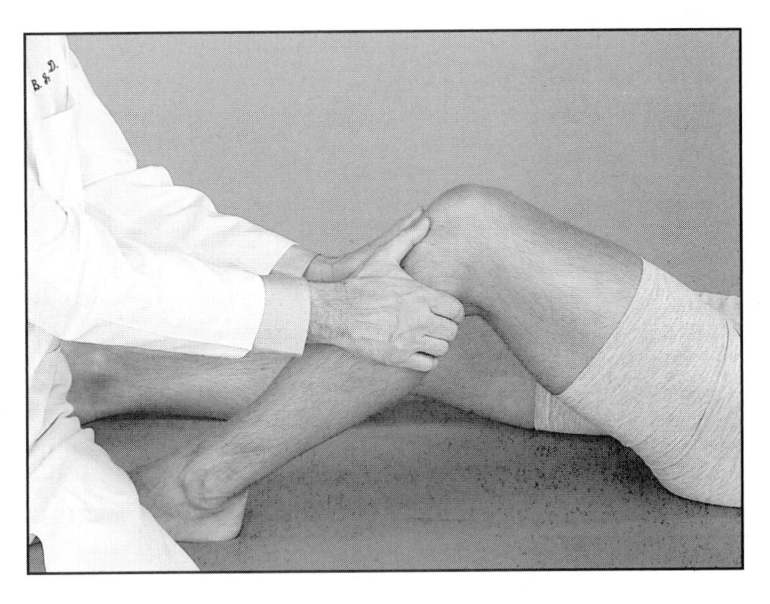

Normal: No excessive translation of the tibia on the femur; a firm end point is felt in each direction.

Interpretation: A positive test result for ACL insufficiency consists of tibial plateau subluxation anteriorly relative to the femoral condyles. This can be seen and felt as the force is applied. Excessive posterior translation is positive for a PCL injury.

Comment: The anterior drawer test is less accurate than the Lachman test in detecting ACL injury because some subluxation occurs even with an intact ACL. Nevertheless, the test has some utility when the patient's body habitus or lack of relaxation makes it difficult to perform the Lachman test. If the test result is inconclusive and an effusion is present, aspiration may help make the diagnosis. The posterior drawer test is the most accurate test for detecting PCL injuries because the PCL is the tightest in 90° of knee flexion.

Foot and Ankle

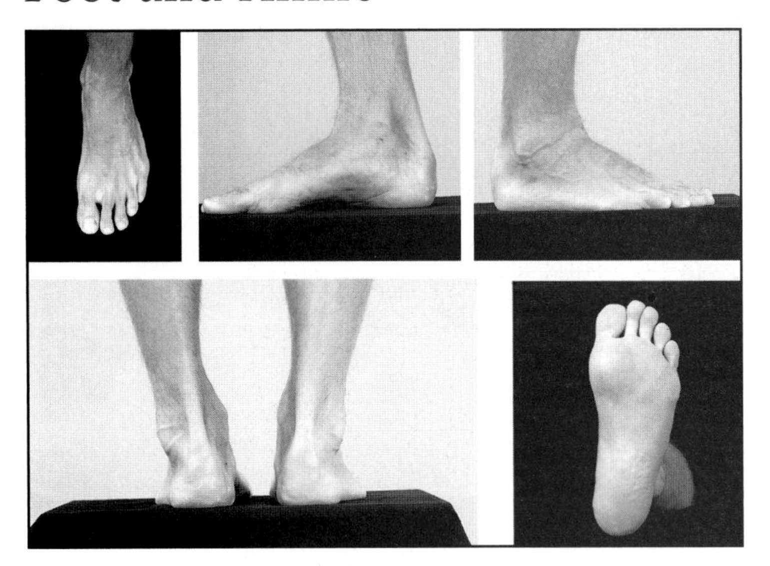

Inspection

Inspect the feet and ankles using a combination of standing, ambulating, and non–weight-bearing positions. All surfaces of the foot and ankle should be inspected. Gait should also be inspected.

Palpation

Palpation Point	Significance
First metatarsophalangeal (MTP) joint	Pain or deformity suggests hallux valgus (bunion deformity) or hallux rigidus.
Tarsometatarsal (Lisfranc) joint	Tenderness suggests acute sprain or dislocation.
Dorsalis pedis artery	Impaired vascularity can contribute to ulcers or other foot and ankle problems, especially in those with diabetes.
Syndesmosis of ankle	Tenderness between the tibia and fibula above the ankle joint line suggests a "high" ankle sprain.
Medial malleolus	Ecchymosis and swelling suggest fracture or sprain.
Deltoid ligament	Commonly injured in ankle sprains or fractures.
Posterior tibial tendon	Tenderness suggests posterior tibial tendinitis or rupture (causing flat feet).
Posterior tibial artery	The pulse may be harder to detect here than at the dorsalis pedis artery. Consider using Doppler ultrasonography.
Lateral malleolus	Tenderness or ecchymosis suggests fracture.
Lateral ankle ligaments	Tenderness suggests lateral ankle sprain.
Base of fifth metatarsal	Tenderness suggests fracture (including stress fracture) and peroneus brevis overuse.
Achilles tendon	Tendon defect suggests Achilles tendon tear; nodule or tenderness suggests degeneration.
Calcaneus	Tenderness suggests plantar fasciitis or stress fracture.
Sesamoids	Tenderness suggests sesamoiditis or sesamoid fracture.

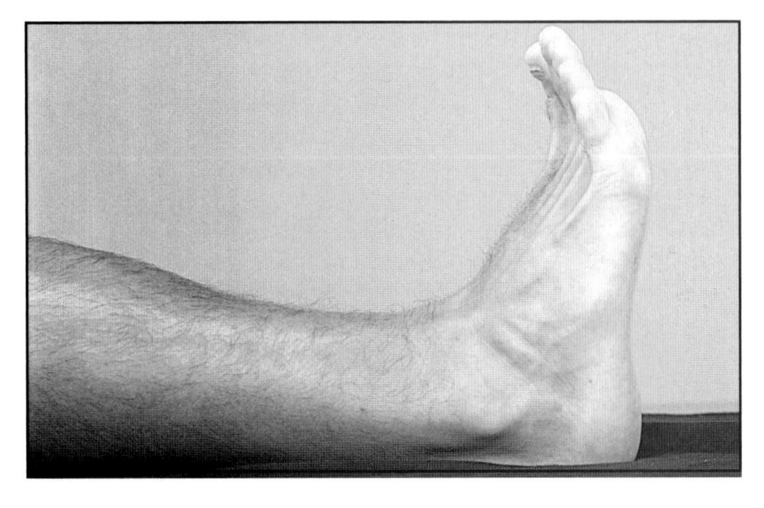

Ankle dorsiflexion and plantar flexion

What to do: With the patient seated and the heel stabilized, grasp the foot in the midfoot position and maximally dorsiflex and plantar flex the ankle joint.

Normal: Dorsiflexion of 0° to 20° and plantar flexion of 0° to 50°.

Interpretation: Loss of ankle range of motion is usually caused by arthritis or a capsular contracture.

Comment: Ankle extension is termed dorsiflexion by convention.

Subtalar inversion and eversion

What to do: With the patient seated, grasp the heel and apply gentle force in inversion (turn the heel in) and eversion (turn the heel out).

Normal: While considerable variability exists, average inversion is approximately 5° to 20°; average eversion is approximately 5° to 20°.

Interpretation: Restricted motion of the subtalar joint may be seen in patients with subtalar arthritis, end-stage posterior tibial tendon dysfunction, or tarsal coalition.

Comment: History of pain resulting while walking on uneven terrain suggests subtalar pathology.

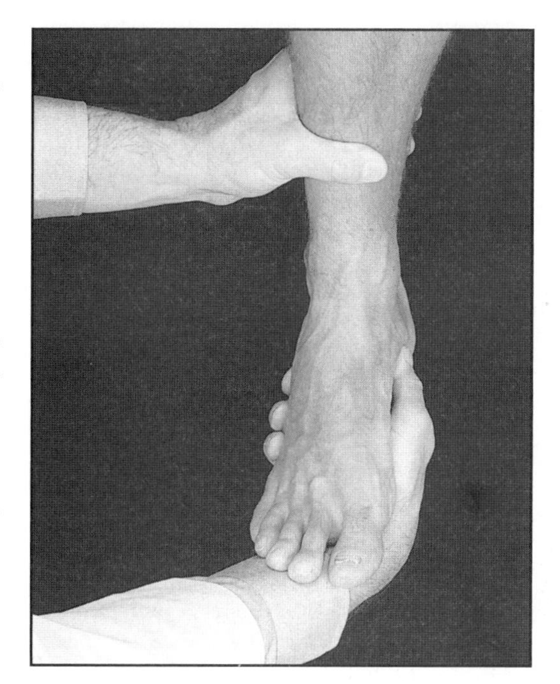

Supination and pronation

What to do: With the patient in the seated position, ask the patient to rotate the foot in (supination) **(A)** and out (pronation) **(B)**. Supination includes inversion of the heel, as well as adduction and plantar flexion of the midfoot. Pronation is the opposite motion and includes eversion of the heel and abduction and dorsiflexion of the midfoot.

Normal: Because the degree of supination and pronation is difficult to quantify, compare with the opposite side.

Interpretation: Restricted motion most commonly occurs secondary to midfoot arthritis; excess motion can be caused by injury to the ligaments of the midfoot joints.

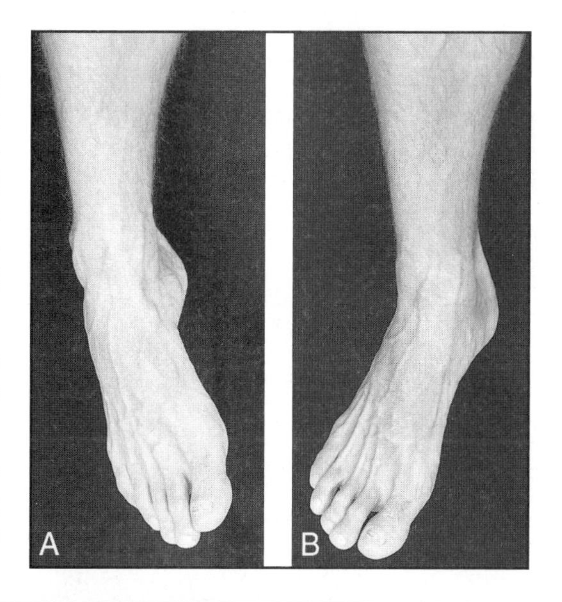

Great toe dorsiflexion and plantar flexion

What to do: Ask the patient to maximally dorsiflex and plantar flex the MTP joint.

Normal: Extension 0° to 70°, flexion 0° to 45°.

Interpretation: Restricted motion, often associated with pain, is commonly associated with hallux rigidus or arthritic change in the MTP joint.

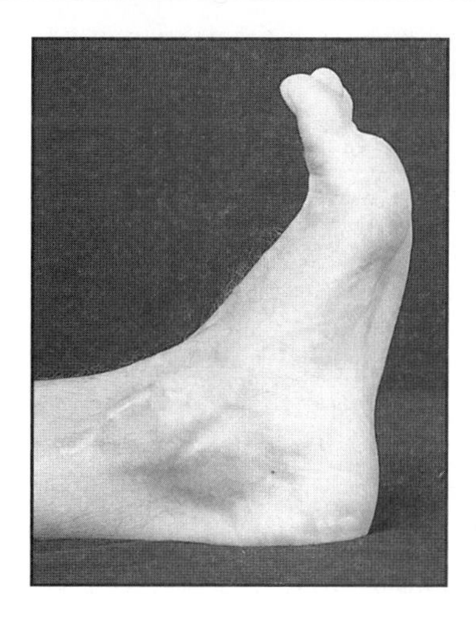

Muscle Testing

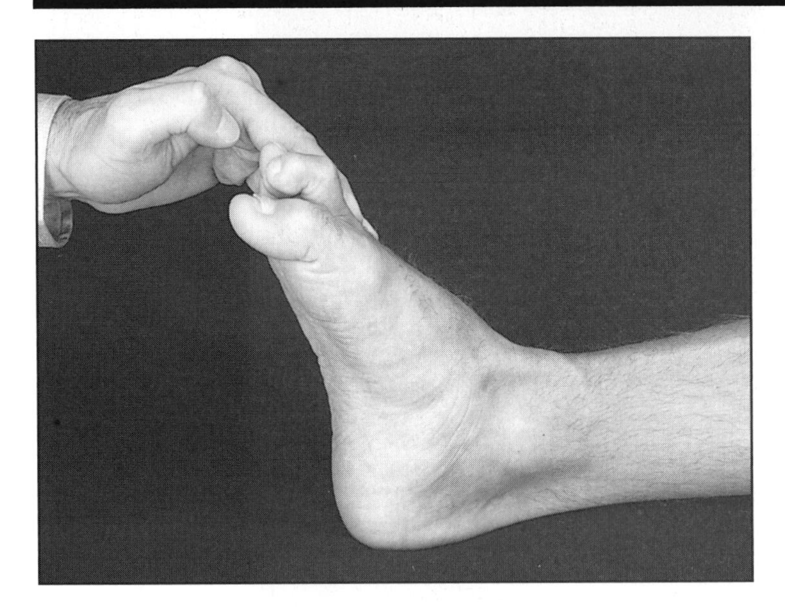

Tibialis anterior

What to do: With the patient in the supine position, ask the patient to maximally dorsiflex the ankle and maintain that position while you attempt to push the ankle into plantar flexion.

Normal: Normal strength against resistance (compare with contralateral side).

Interpretation: Weakness or loss of movement in the tibialis anterior leads to a "foot drop" during gait, which is commonly caused by peroneal nerve injury, or less commonly, by isolated L4 radiculopathy.

Comment: Because the extensor digitorum longus, extensor hallucis longus, and peroneus tertius are also ankle dorsiflexors, they may assist and mask isolated tibialis anterior weakness.

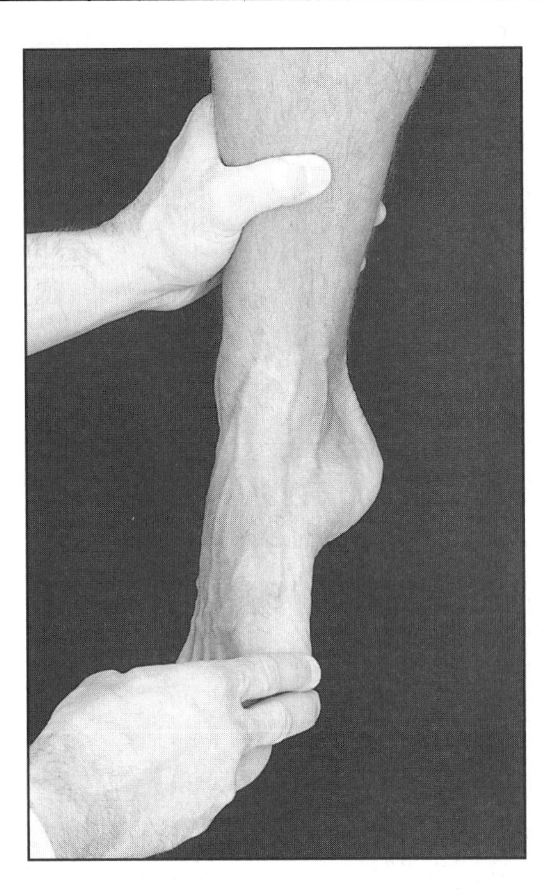

Tibialis posterior

What to do: Place the patient's ankle in a plantar flexed and inverted position. Then ask the patient to resist your attempt at eversion.

Normal: Normal strength against resistance; the posterior tibialis tendon should visibly tighten.

Interpretation: Weakness can be associated with tendinitis or rupture and warrants further investigation; with the patient standing, you should evaluate the medial longitudinal arch for collapse and perform the single heel rise test.

Peroneus longus and brevis

What to do: Ask the patient to plantar flex and evert the ankle. Then ask the patient to maintain this position as you attempt to invert the ankle with finger pressure on the lateral border of the foot.

Normal: Normal strength against resistance.

Interpretation: Inability to evert the foot and ankle may indicate tendinitis, instability of the peroneal tendons, Charcot-Marie-Tooth disease, superficial peroneal nerve palsy, or lumbosacral radiculopathy (S1).

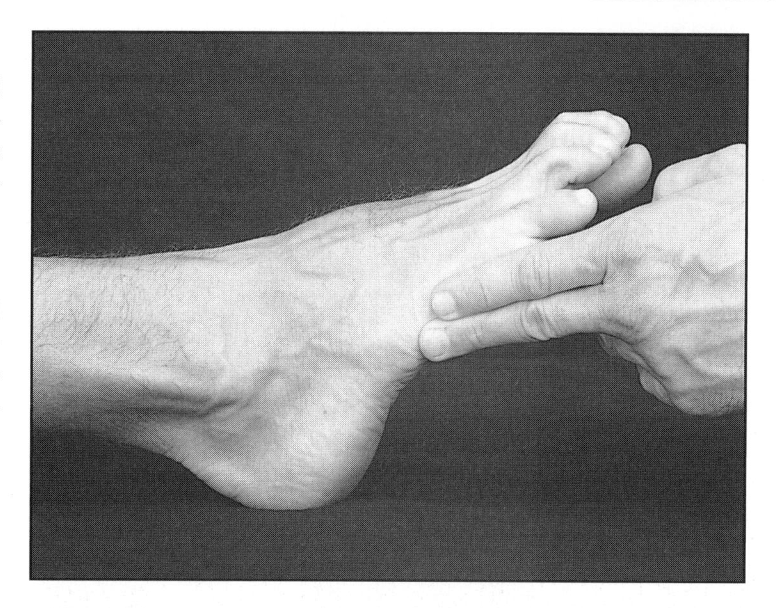

Extensor hallucis longus

What to do: While supporting the plantar forefoot, ask the patient to extend the great toe and resist your attempts to push the toe into flexion.

Normal: Normal strength against resistance.

Interpretation: Weakness or the inability to extend the great toe indicates injury to the extensor hallucis longus, the deep peroneal nerve, or, most commonly, the L5 nerve root.

Comment: Testing of the extensor digitorum longus muscle may be performed in an identical manner by isolating each of the remaining four toes.

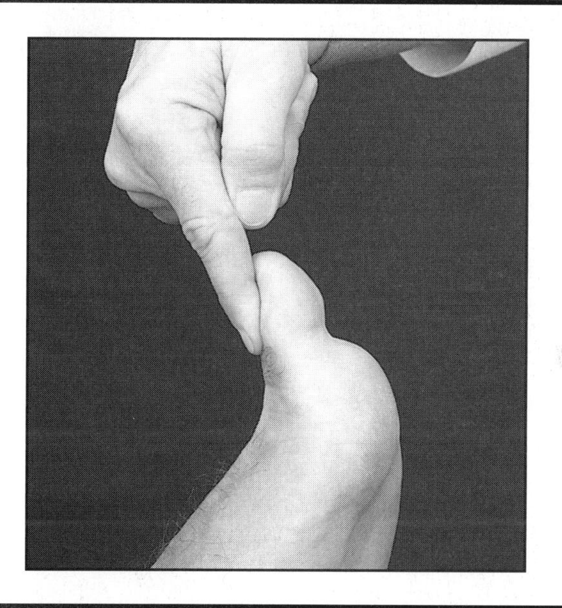

Flexor hallucis longus

What to do: With the patient's ankle in the neutral position, ask the patient to flex the great toe against resistance to the distal phalanx.

Normal: Normal strength against resistance.

Interpretation: Weakness can indicate tendinitis or a tendon tear of the flexor hallucis longus. In addition, flexor hallucis longus weakness can indicate dysfunction of the S1 nerve root.

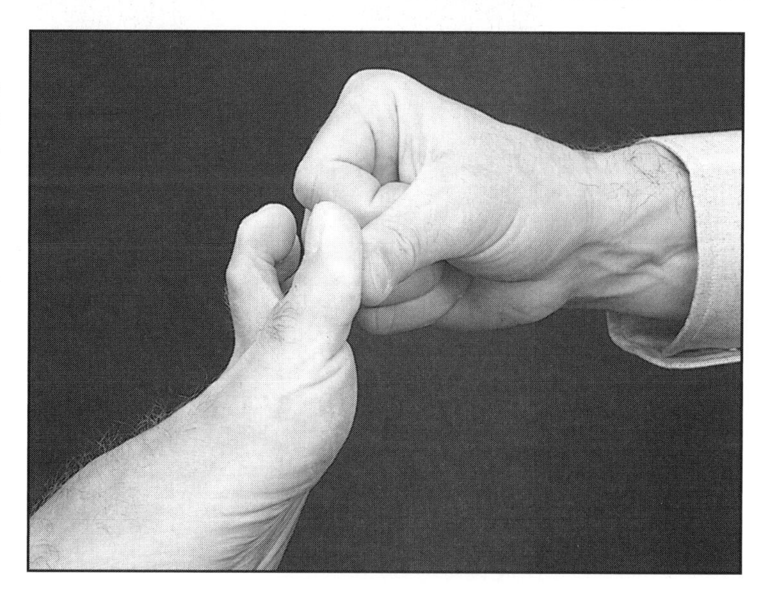

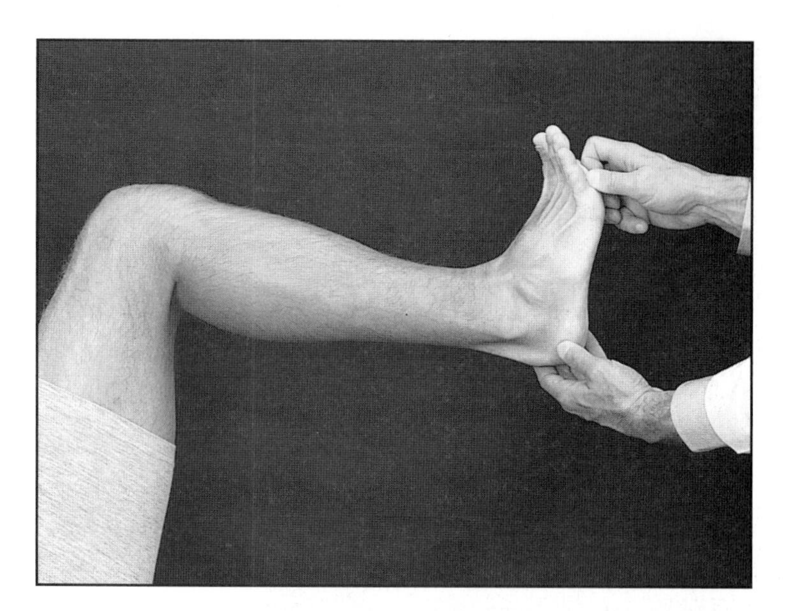

Gastrocnemius-soleus complex

What to do: Ask the patient to plantar flex the ankle against resistance.

Normal: Full plantar flexion.

Interpretation: Weakness or the inability to plantar flex against resistance indicates possible Achilles tendon rupture, sciatic and tibial nerve injury, or lumbosacral radiculopathy (S1 and S2). The ability to plantar flex against resistance does not rule out a ruptured Achilles tendon (see Thompson's test).

Special Tests

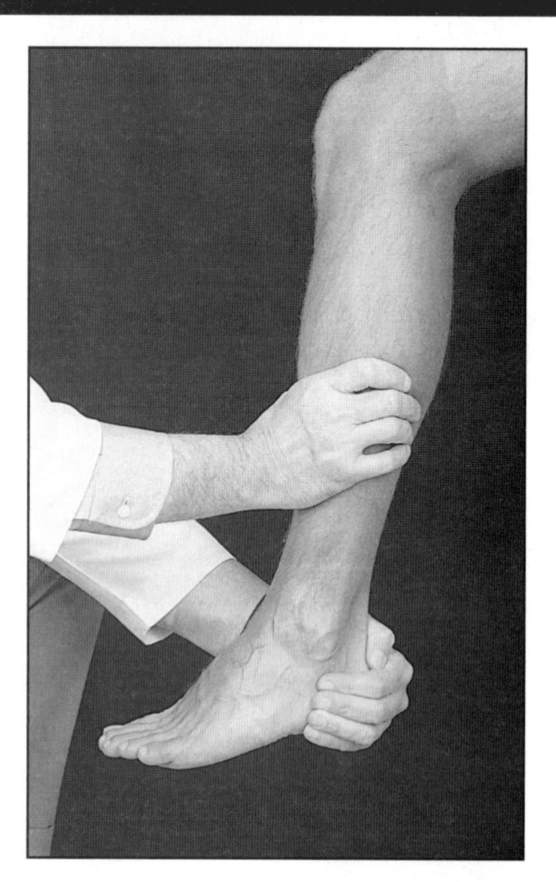

Anterior drawer test

What it tests: Integrity of the anterior talofibular ligament.

What to do: With the patient seated and the ankle slightly plantar flexed, stabilize the distal tibia with one hand and cup the heel in the other hand. Rapidly apply an anterior force using the hand cupping the heel. Assess for anterior translation of the talus on the tibia.

Normal: 3 to 5 mm of anterior talus translation followed by a firm end point.

Interpretation: Translation of greater than 3 to 5 mm difference from the contralateral ankle or a "soft" or absent end point, suggests insufficiency (disruption) of the anterior talofibular ligament.

Comment: The same test performed with the ankle in dorsiflexion tests the calcaneofibular ligament.

Inversion stress test

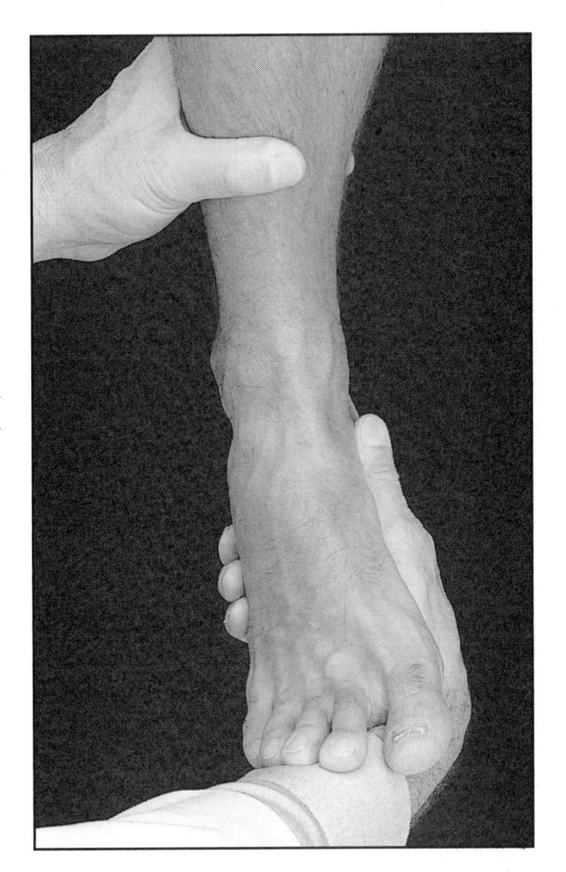

What it tests: Integrity of the calcaneofibular ligament.

What to do: With the patient seated, maximally dorsiflex the ankle by applying upward force on the forefoot. Then apply a varus movement to the ankle by cupping the heel medially and inverting it.

Normal: No significant motion of the talus.

Interpretation: With disruption of the calcaneofibular ligament, the talus can be felt rocking into inversion.

Thompson's test

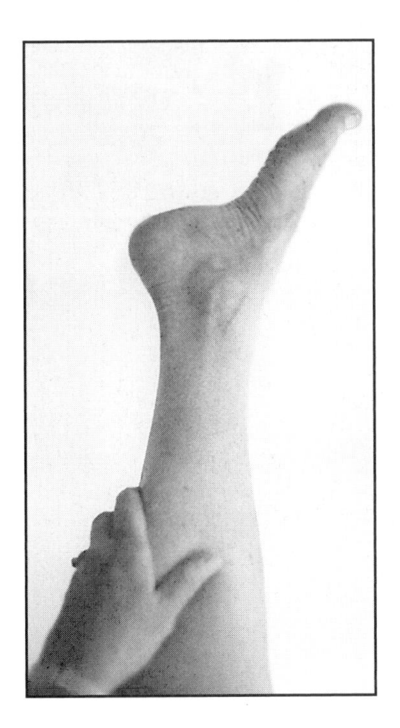

What it tests: The continuity of the gastrocnemius-soleus complex in cases of possible Achilles tendon rupture.

What to do: With the patient in the prone position, gently squeeze the calf using one or both hands.

Normal: Passive plantar flexion of the ankle is produced as the calf is compressed.

Interpretation: With a fully ruptured Achilles tendon, there should be essentially no motion at the ankle. If there is only partial disruption of the Achilles tendon complex, slight plantar flexion may still occur.

Comment: Active plantar flexion does not rule out a torn Achilles tendon because the toe flexors and tibialis posterior can provide some plantar flexion strength; therefore, Thompson's test is necessary in the assessment of the continuity of the gastrocnemius-soleus complex in cases of possible Achilles tendon rupture.

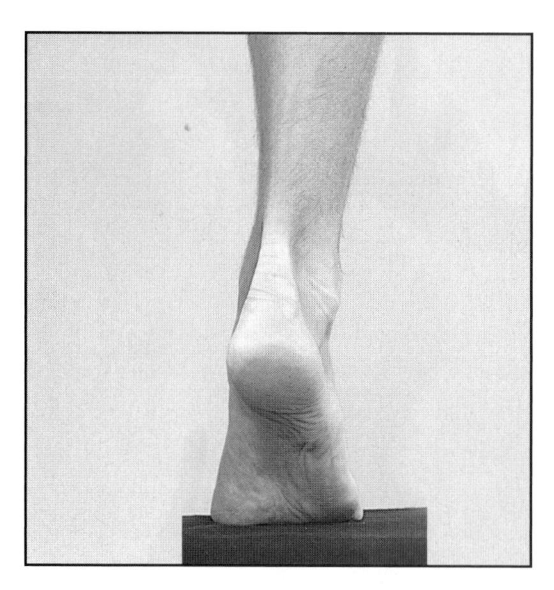

Single heel rise test

What it tests: Competence of the posterior tibial tendon.

What to do: Ask the patient to stand and face a wall using the hands for stabilization. Then ask the patient to lift the unaffected foot off the ground and stand up on the toes of the affected side. Observe the heel for movement.

Normal: Ability to stand up on toes; heel should tilt into an inverted (varus) position (medial side higher).

Interpretation: A patient with posterior tibial tendon dysfunction will not be able to perform the single heel rise and the heel will not move into the normal varus position. This can indicate posterior tibial tendinitis or a posterior tibial tendon tear.

Comment: It is crucial to begin the test with the unaffected leg off the ground. If the patient elevates onto the toes using both feet, the Achilles tendon can hold the heel in varus despite a dysfunctional posterior tibial tendon.

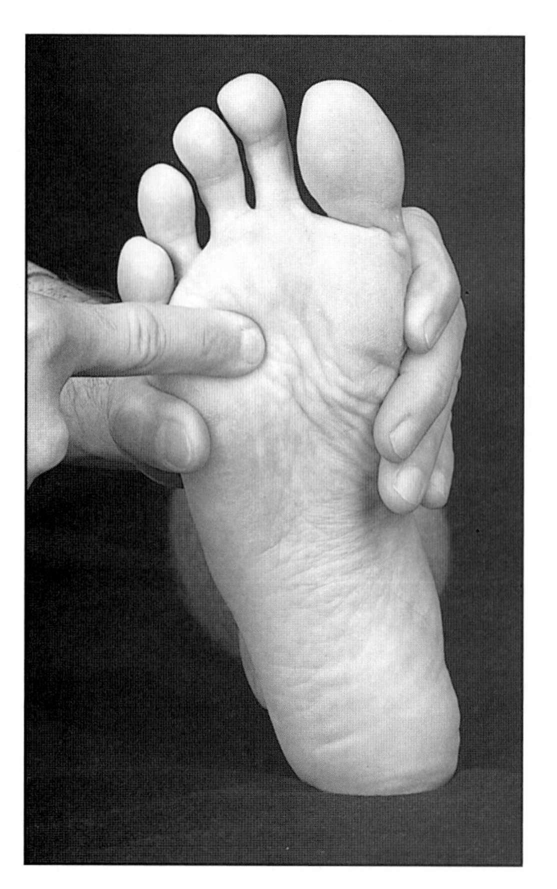

Interdigital neuroma (Morton's) test

What it tests: The presence of interdigital neuroma as the source of web-space pain.

What to do: Grab the heads of the first and fifth metatarsals and squeeze them, capturing the other metatarsal heads between them.

Normal: Absence of pain.

Interpretation: The test result is positive when tenderness occurs with compression of interdigital neuromas between metatarsals. When a palpable click (Mulder's click) is felt during compression, a large neuroma may extrude plantarly from between the metatarsal heads.

Spine

Inspection

With the patient standing, inspect for deviations in the normal spinal curves. In the lateral position, inspect for lordosis and kyphosis. Lordosis (a concave curve) should be apparent at the cervical and lumbar spine segments. Kyphosis (a mild convex curve) should be apparent in the thoracic spine. Deviation from these normal curves can occur with painful conditions such as acute sprains, fractures, infections, or neoplastic processes. Abnormalities also can be congenital. With the patient standing and facing away from you, view the alignment of the spine. Moderate to severe scoliosis (lateral "S" or "C" curvature) will be obvious, if present. A lumbar "list," which causes the patient to lean to one side to alleviate nerve root compression, may also be present in association with a herniated disk or other conditions. Look for any atrophy of the gluteus muscles and posterior thigh musculature.

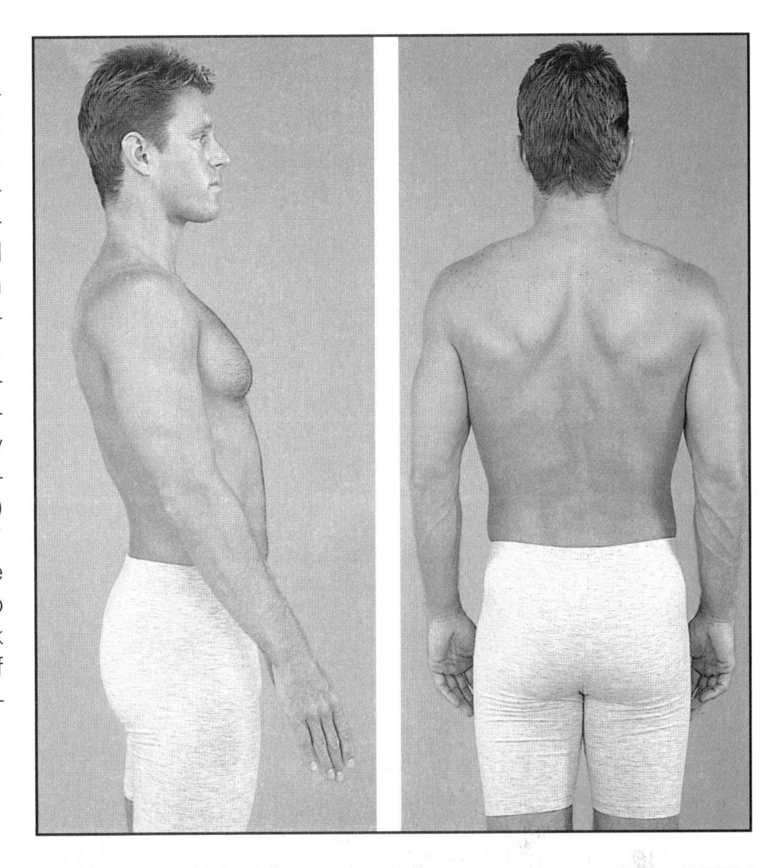

Palpation

Palpation Point	Significance
Cervical spinous processes	C7 is most prominent in the cervical spine. Bony tenderness suggests possible fracture or other bony lesion.
Thoracic spinous processes	Scoliosis (lateral curvature) of the thoracic spine can suggest acute sprains, fractures, infections, or neoplastic processes.
Lumbar spinous processes	Bony tenderness suggests possible fracture or other bony lesion.
Paraspinal musculature	An area of tenderness is most commonly associated with muscle strain.
Cervical lymph nodes	Swelling suggests the presence of an infectious or inflammatory process.

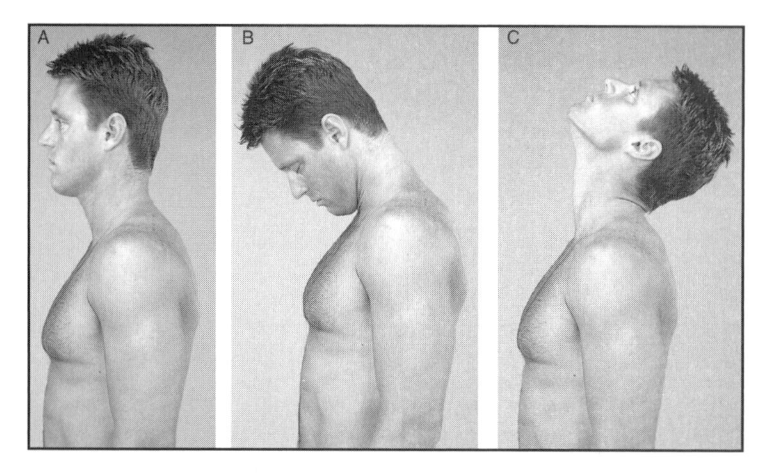

Neck flexion and extension

What to do: With the patient standing, align the neck with the trunk **(A)**. Ask the patient to flex the neck by bending the head forward and touching the chin to the chest **(B)**. Then ask the patient to extend the neck by looking up at the ceiling **(C)**.

Normal: Flexion: able to touch the chin to the chest. Extension: able to look directly at the ceiling.

Interpretation: Limited range of motion or pain with this maneuver can indicate cervical spinal pathology, disk disease, or a muscle strain.

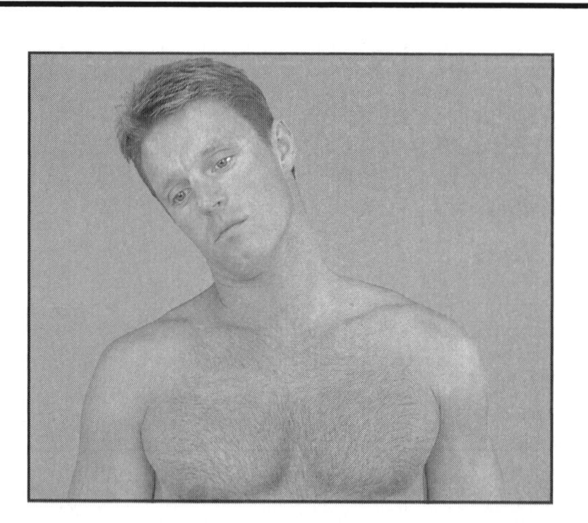

Lateral flexion of the neck

What to do: With the patient's neck aligned with the trunk, ask the patient to touch the ear to one shoulder and then the other shoulder.

Normal: Able to bend to 45° the shoulder.

Interpretation: Limited range of motion or pain with this maneuver can indicate cervical spinal pathology, disk disease, or a muscle strain.

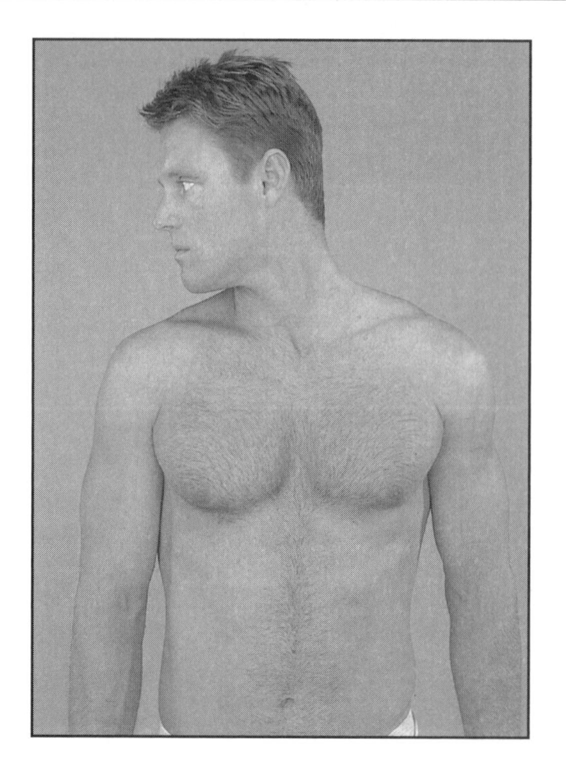

Neck rotation

What to do: With the patient's neck aligned with the trunk, ask the patient to turn the head to the right as far as possible. Then ask the patient to repeat this motion to the left.

Normal: Able to place the chin in line with each shoulder (approximately 90°).

Interpretation: Limited range of motion or pain with this maneuver can indicate cervical spinal pathology, disk disease, or a muscle strain.

Lumbar flexion and extension

What to do: Ask the patient to stand straight with the feet comfortably apart for balance. Then ask the patient to flex the lower back by bending forward to touch the toes **(A)**. Measure how close the fingers come to touching the floor. Ask the patient to return to standing. Then ask the patient to extend the lower back by leaning backward as far as possible **(B)**. Visually note the angle of extension during this test.

Normal: For normal lumbar flexion, the fingertips should reach within 10 cm of the floor. For normal lumbar extension, the patient should be able to lean backward 15° to 30° from standing straight.

Interpretation: If lumbar flexion is less than normal, lumbar radiculopathy or nonorganic pathology may be present. Pain with extension in young patients can be caused by spondylolysis.

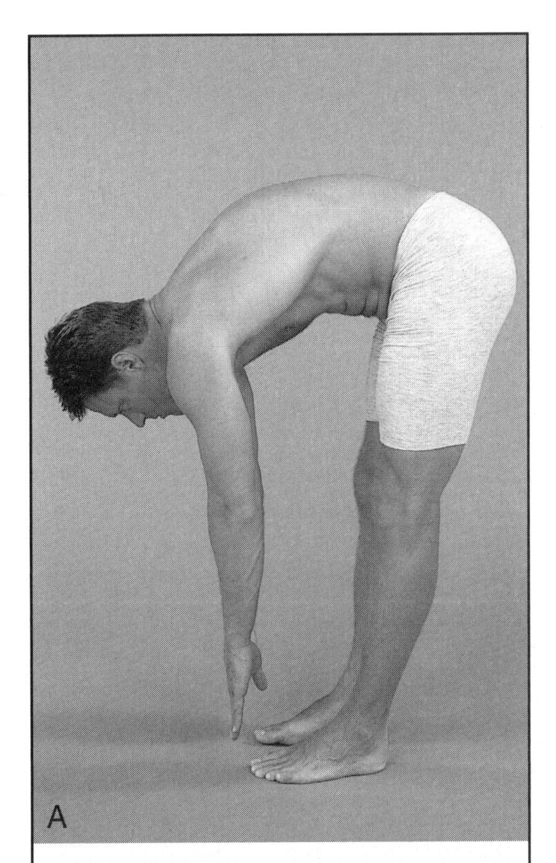

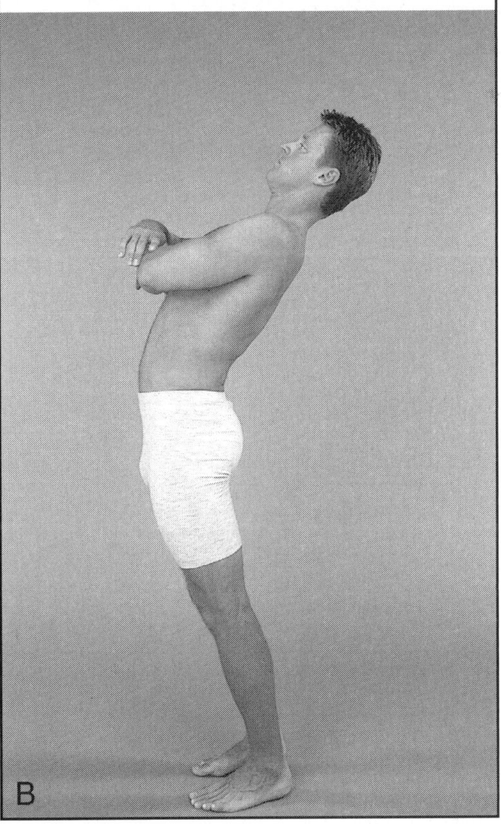

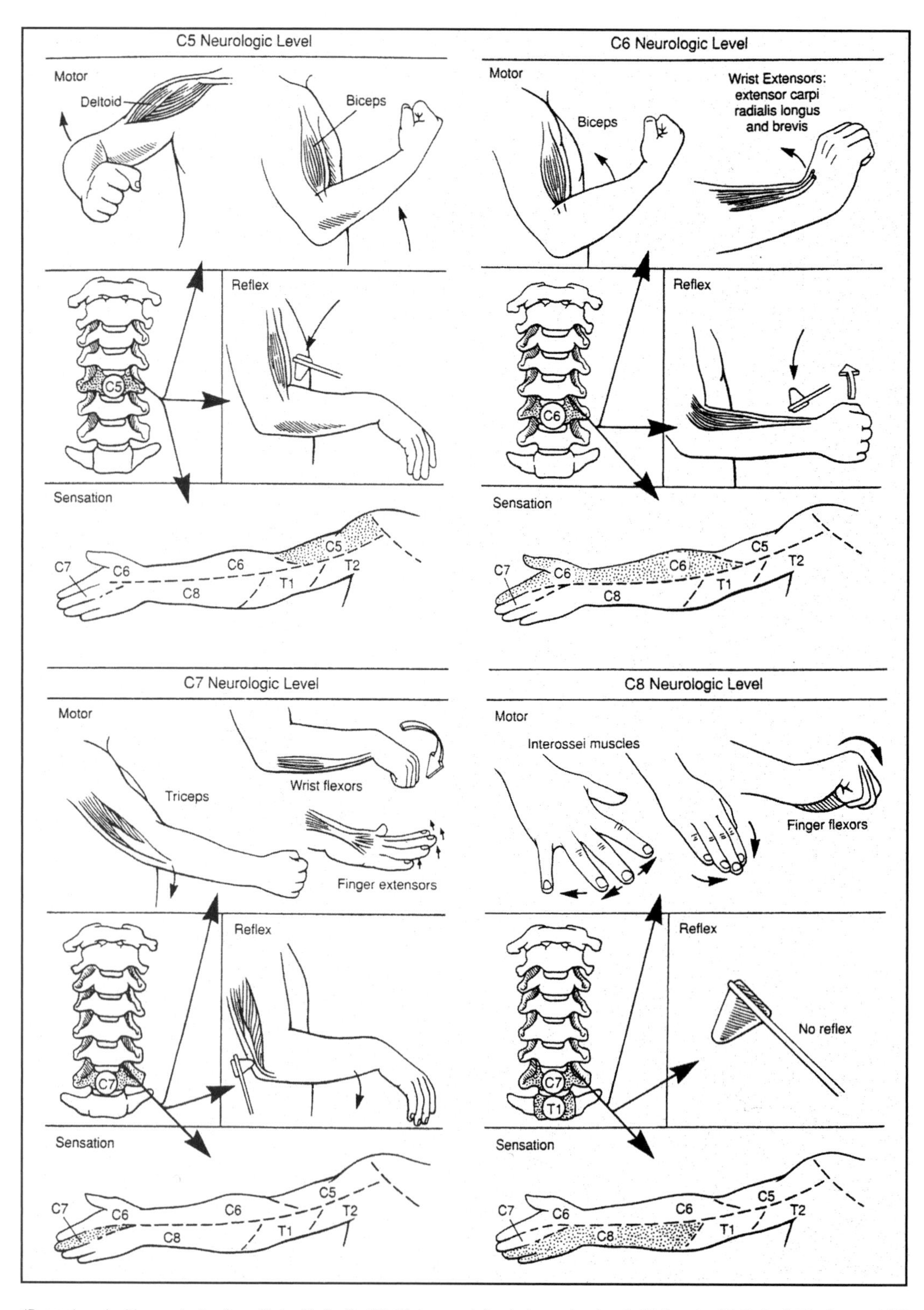

(Reproduced with permission from Klein JD, Garfin SR: History and physical examination, in Weinstein JN, Rydevik BL, Somtag VKH (eds): *Essentials of the Spine*. New York, NY, Raven Press, 1995, pp 71-95.)

Muscle Testing and Neurologic Examination of the Spine and Pelvis

The spinal cord cannot be examined directly. Therefore, specific tests designed to assess the function of each nerve root can be used to determine whether a given nerve is intact. These functions include muscle strength, inducible reflexes, and sensation. For the purposes of conducting a thorough physical examination of the spine, these tests are grouped by the nerve root whose function they assess. Table 1 is a scale for grading muscle strength. Table 2 provides a grading system for reflexes.

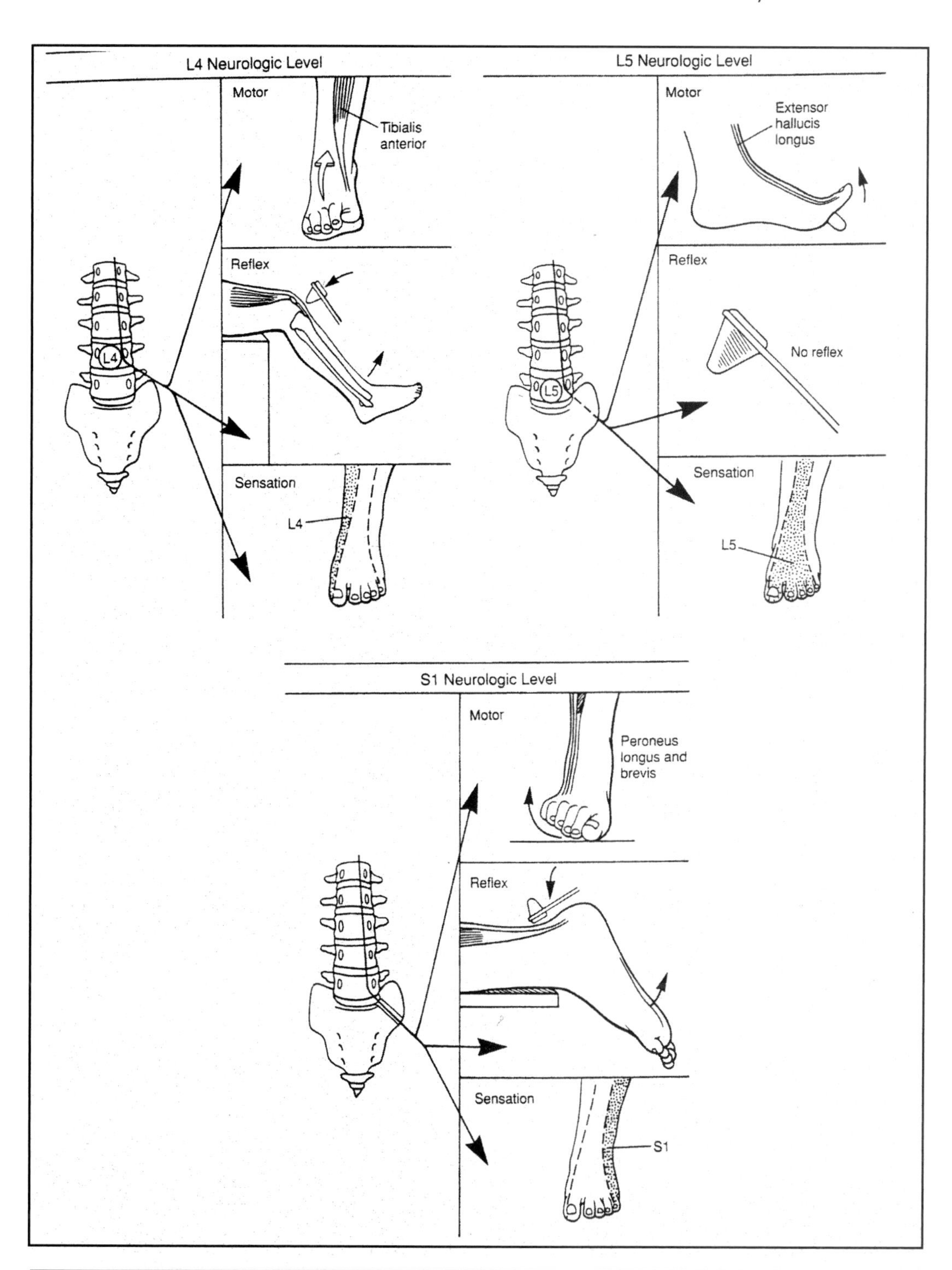

Table 1
Grading Muscle Strength

Grade	Description
0	Total paralysis
1	Palpable or visible contraction but no motion
2	Active movement with gravity eliminated
3	Active movement against gravity
4	Active movement with some weakness
5	Normal movement against full resistance

Table 2
Grading Reflexes

Grade	Description
4+	Very brisk, hyperactive
3+	Brisker than average
2+	Normal
1+	Somewhat diminished
0	No response

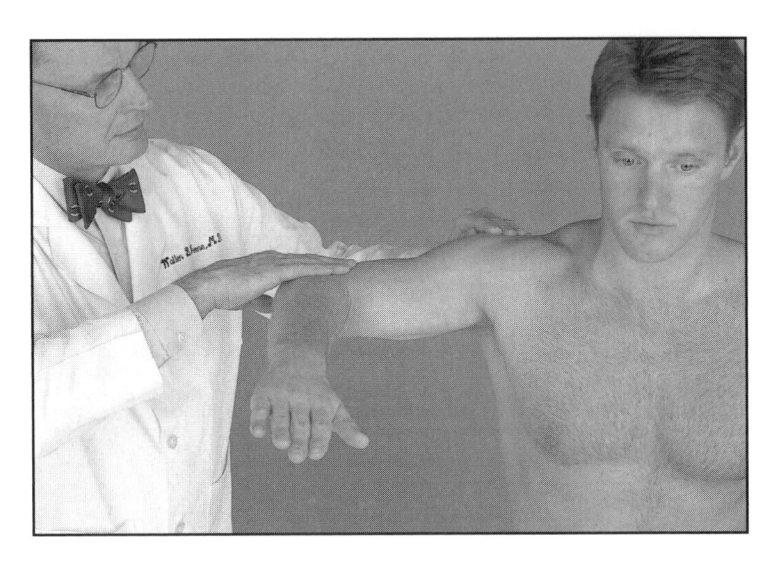

C5—Deltoid muscle

What to do: With the patient seated, abduct the shoulder to 90°. Push down on the arm to resist activity of the deltoid while the patient abducts the shoulder.

Normal: Grade 5 strength.

Interpretation: Weakness indicates possible nerve compression at the spinal level. Complete absence of function is likely caused by peripheral nerve injury.

Comment: The deltoid muscle is innervated by the axillary nerve and is composed of C5 and C6 spinal cord segments.

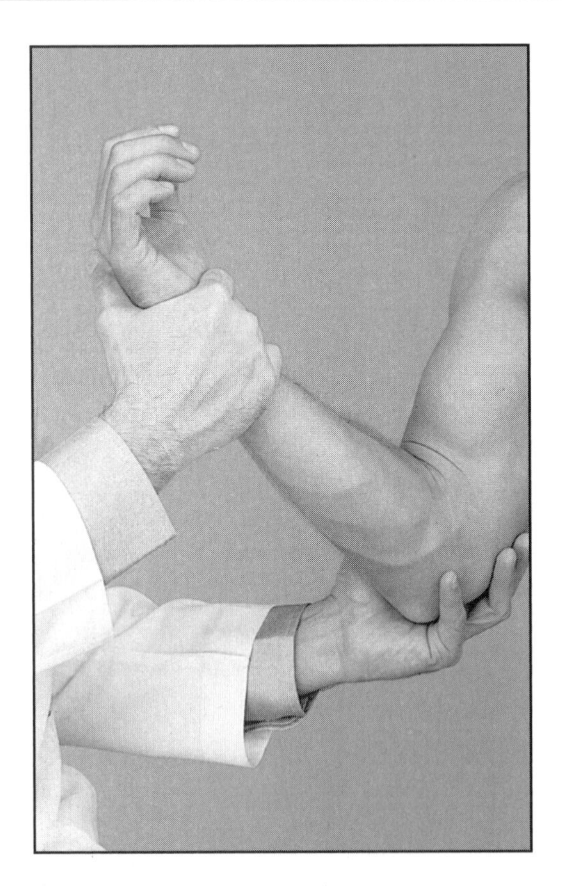

C5—Biceps

What to do: With the patient seated, ask the patient to "make a muscle" with the arm. To accomplish this, the patient must flex the elbow in a supinated position as you apply resistance.

Normal: Grade 5 strength.

Interpretation: Although partially innervated by C6, if C5 is intact, biceps strength should be at least grade 3.

Comment: The biceps muscle is innervated by the musculocutaneous nerve and is composed of C5 and C6 spinal cord segments.

C6—Radial wrist extensors

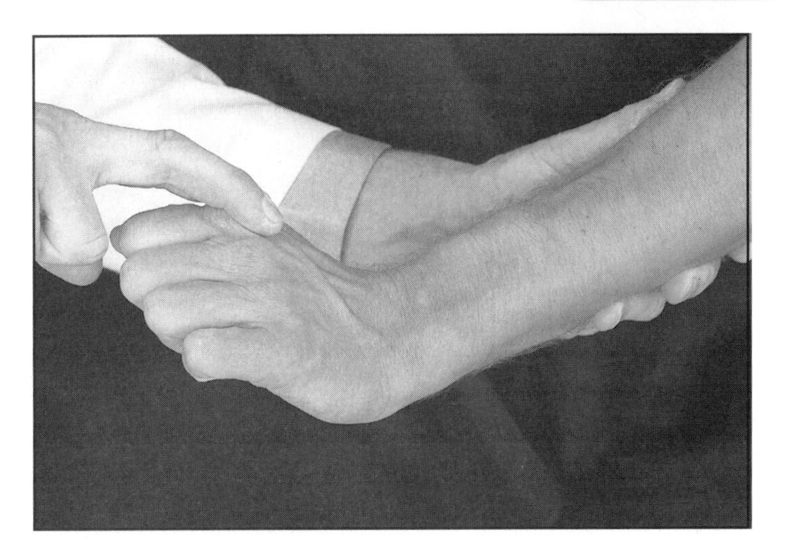

What to do: Begin with the patient's fingers flexed to eliminate wrist extension activity by the finger extensor muscles. With the forearm pronated, ask the patient to extend the wrist in an upward direction as you apply resistance.

Normal: Grade 5 strength.

Interpretation: Weakness indicates possible nerve compression at the spinal level. Complete absence of function is likely caused by a radial nerve injury.

Comment: The radial wrist extensors include the extensor carpi radialis longus and brevis muscles. They are innervated by the radial nerve from the C6 and C7 spinal cord segments.

C7—Triceps

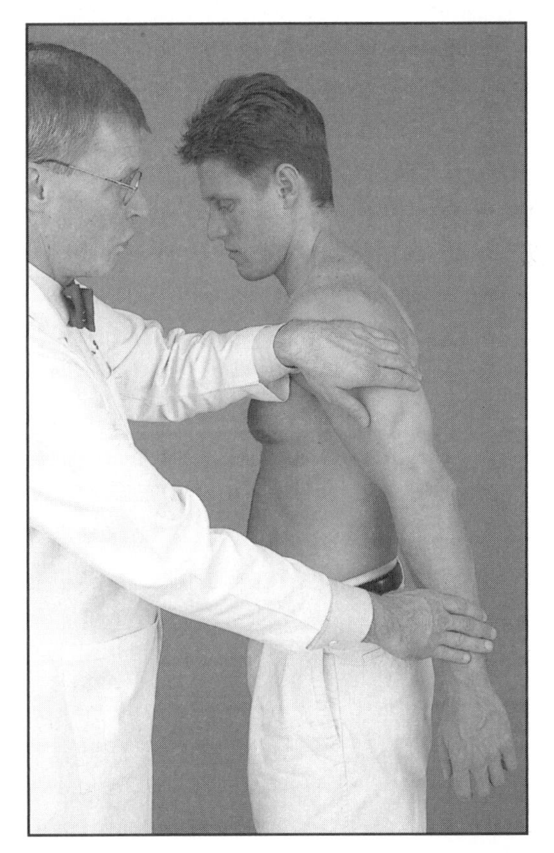

What to do: Ask the patient to extend the elbow as you apply resistance.

Normal: Grade 5 strength.

Interpretation: Weakness indicates possible nerve compression at the spinal level. Complete absence of function is likely caused by a radial nerve injury.

Comment: The triceps muscle is innervated by the radial nerve and is composed of C6, C7, and C8 spinal cord segments.

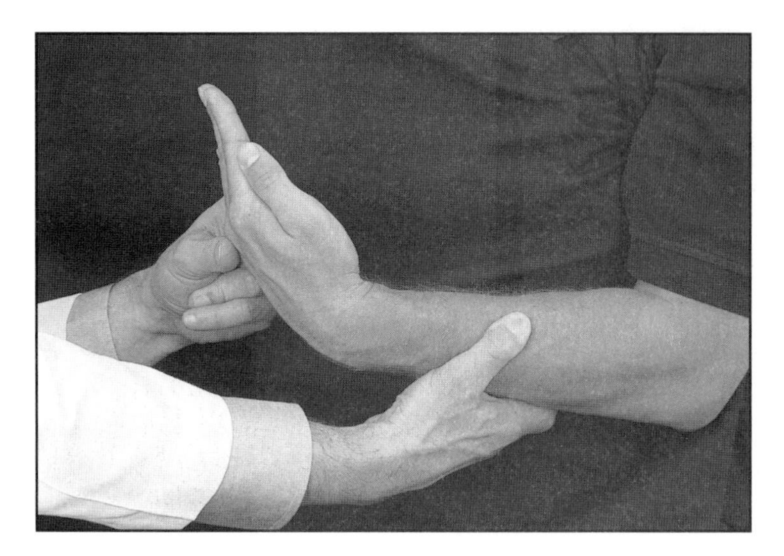

C7—Flexor carpi radialis

What to do: With the patient seated, place the patient's fingers in extension to eliminate wrist flexor activity by the finger flexor muscles. Ask the patient to flex the wrist in the radial direction as you apply resistance.

Normal: Grade 5 strength.

Interpretation: Weakness indicates possible nerve compression at the spinal level. Complete absence of function is likely caused by a median nerve injury.

Comment: The flexor carpi radialis muscle is innervated by the median nerve and is composed of C6 and C7 spinal cord segments.

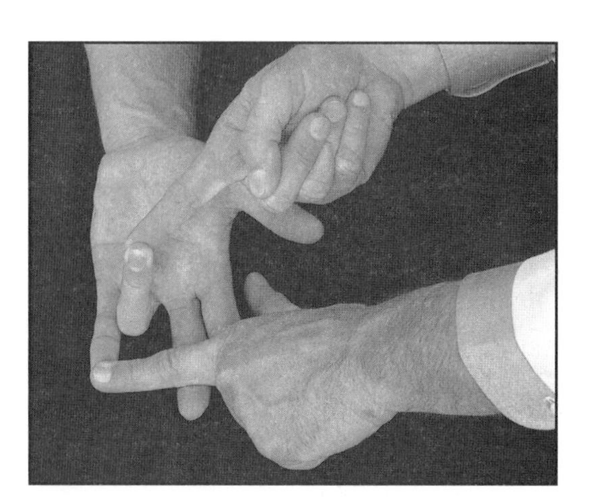

C8—Flexor digitorum superficialis

What to do: While stabilizing the long, index, and little fingers in extension, ask the patient to flex the fingers as you apply resistance.

Normal: Grade 5 strength.

Interpretation: Weakness indicates possible nerve compression at the spinal level. Complete absence of function is likely caused by a median nerve injury.

Comment: The flexor digitorum superficialis muscle is innervated by the median nerve and is composed of C7, C8, and T1 spinal cord segments.

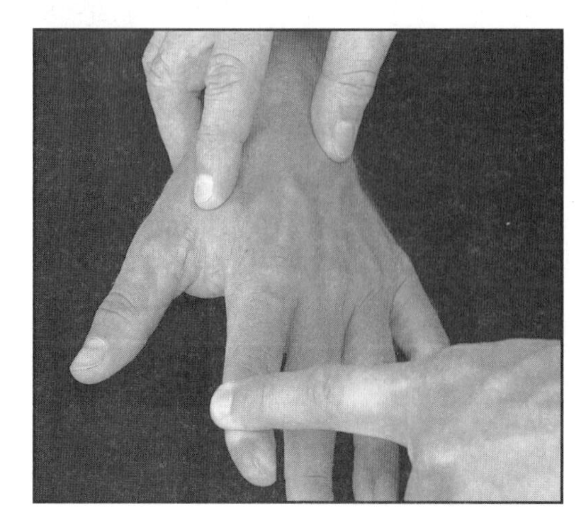

T1—First dorsal interossei

What to do: Ask the patient to abduct the index finger as you apply resistance. Palpate the muscle belly of the first dorsal interossei to confirm activity of the muscle.

Normal: Grade 5 strength.

Interpretation: Weakness indicates possible nerve compression at the spinal level. Complete absence of function is likely caused by an ulnar nerve injury.

Comment: The dorsal interossei muscles are innervated by the deep ulnar nerve and are composed of C8 and T1 spinal cord segments.

L1—Hip flexors

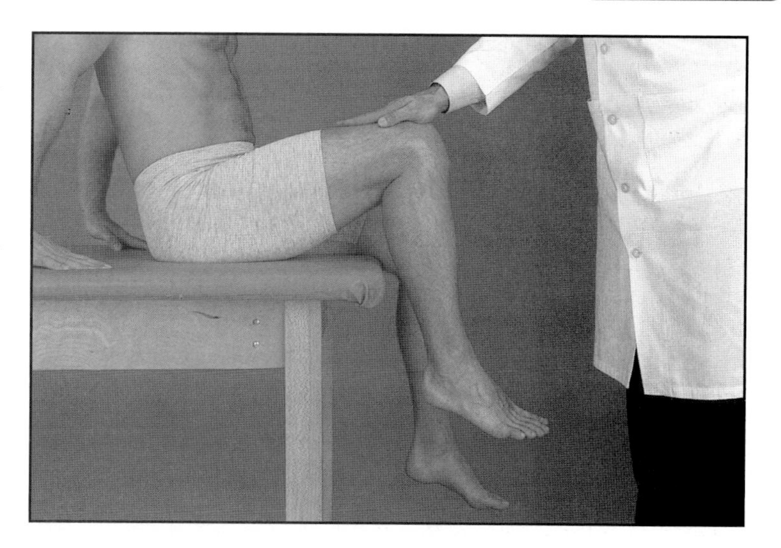

What to do: With the patient seated, ask the patient to flex the hip as you apply resistance.

Normal: Grade 5 strength.

Interpretation: Weakness indicates possible nerve compression at the spinal level. Complete absence of function is likely caused by a femoral nerve injury.

Comment: Although many muscles aid in the flexion of the hip, the iliopsoas muscle is the predominant muscle. It is innervated in part by the femoral nerve and is composed of L1, L2, and L3 spinal cord segments. It is the only hip flexor with an L1 component.

L2—Hip adductors

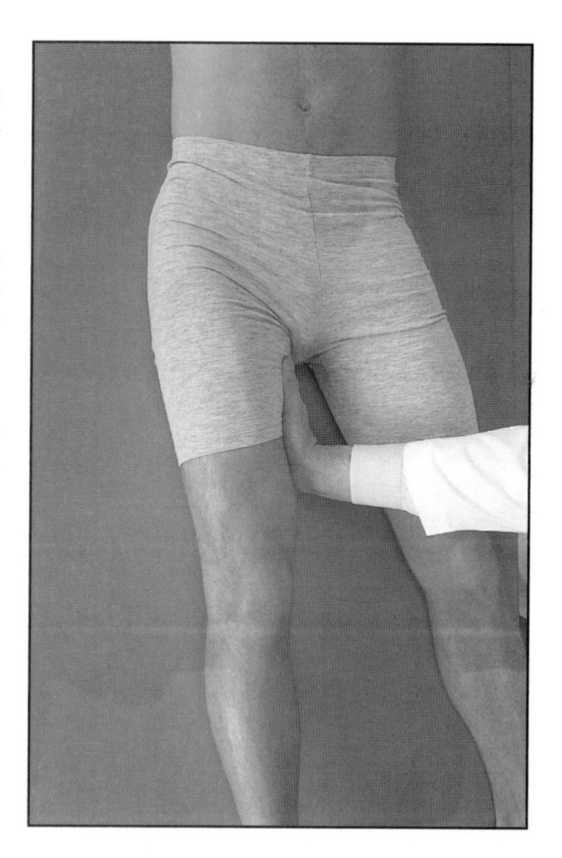

What to do: With the patient supine, ask the patient to adduct the thigh as you apply resistance.

Normal: Grade 5 strength.

Interpretation: Weakness indicates possible nerve compression at the spinal level. Complete absence of function is likely caused by an obturator nerve injury.

Comment: Multiple muscles act to adduct the thigh, but the major innervation to these muscles is the obturator nerve. This nerve is composed of L2, L3, and L4 spinal cord segments.

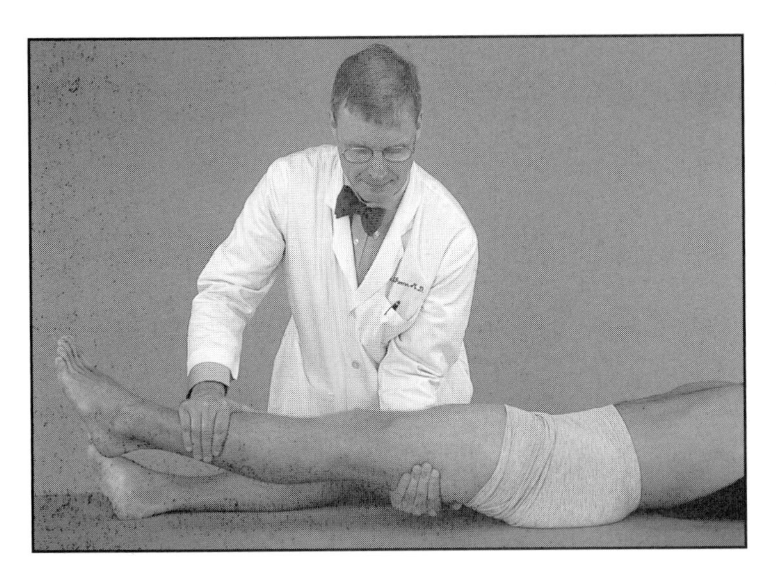

L3—Quadriceps

What to do: With the patient supine, ask the patient to extend the knee as you apply resistance.

Normal: Grade 5 strength.

Interpretation: Weakness indicates possible nerve compression at the spinal level. Complete absence of function is likely caused by a femoral nerve injury.

Comment: The quadriceps consists of the rectus femoris, vastus lateralis, vastus medialis, and vastus intermedius muscles. They are innervated by the femoral nerve, which is composed of L2, L3, and L4 spinal cord segments.

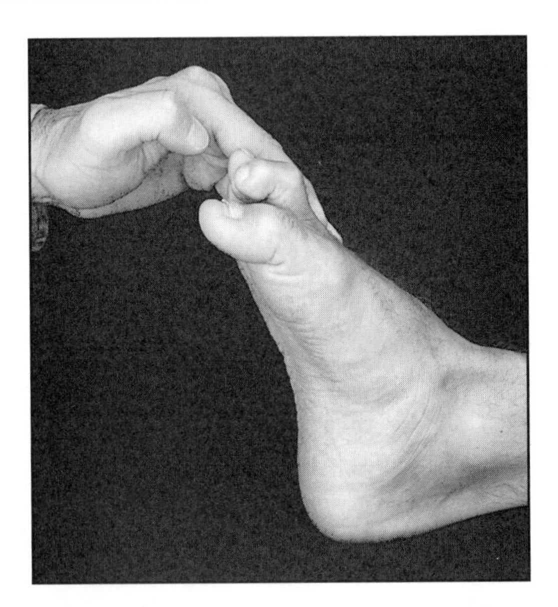

L4—Tibialis anterior

What to do: With the patient seated, ask the patient to flex the toes to eliminate dorsiflexor activity by the toe extensor muscles. Then ask the patient to dorsiflex the ankle as you apply resistance.

Normal: Grade 5 strength.

Interpretation: Weakness indicates possible nerve compression at the spinal level. Complete absence of function is likely caused by a peroneal nerve injury.

Comment: The tibialis anterior is innervated by the deep peroneal nerve and is composed of L4, L5, and S1 spinal cord segments.

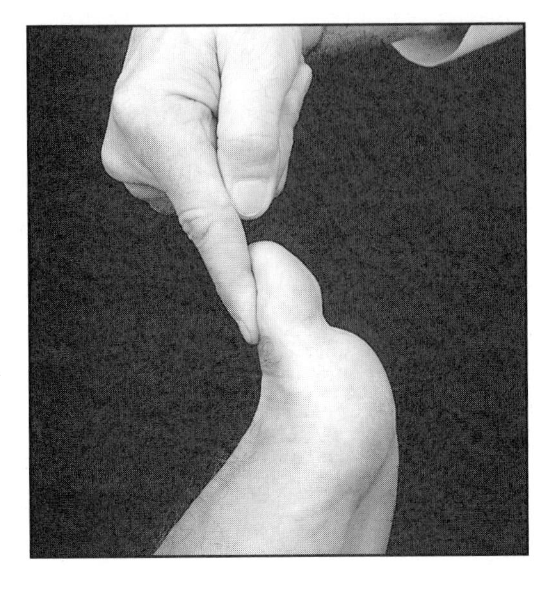

L5—Extensor hallucis longus

What to do: With the patient supine or seated, stabilize the foot in a neutral position with one hand and ask the patient to extend the great toe as you apply resistance.

Normal: Grade 5 strength.

Interpretation: Weakness indicates possible nerve compression at the spinal level. Complete absence of function is likely caused by a peroneal nerve injury.

Comment: The extensor hallucis longus muscle is innervated by the deep peroneal nerve and is composed of L4, L5, and S1 spinal cord segments.

S1—Flexor hallucis longus

What to do: With the patient supine or seated, stabilize the foot in a neutral position with one hand and ask the patient to flex the great toe as you apply resistance.

Normal: Grade 5 strength.

Interpretation: Weakness indicates possible nerve compression at the spinal level. Complete absence of function is likely caused by a tibial nerve injury.

Comment: The flexor hallucis longus muscle is innervated by the tibial nerve and is composed of L5, S1, and S2 spinal cord segments.

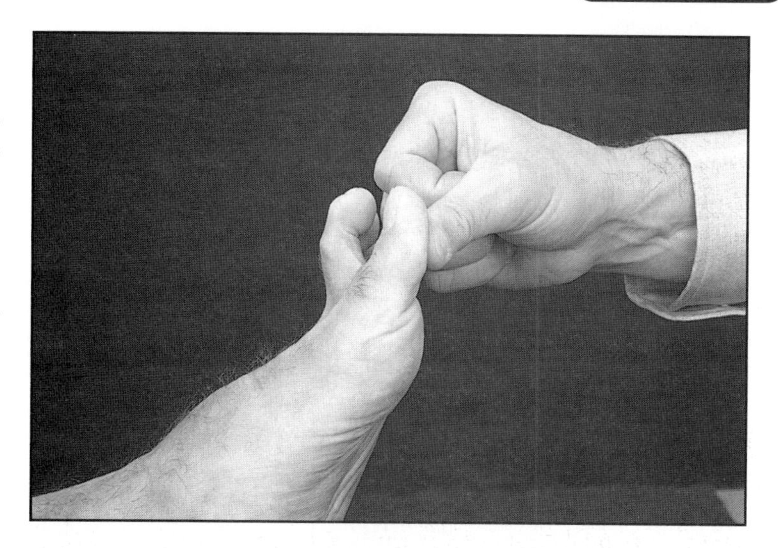

Special Tests

Hoffmann reflex

What it tests: Cervical myelopathy.

What to do: With the patient seated and the patient's relaxed hand cradled in yours, repeatedly flick the long fingernail.

Normal: No flexion of the index finger and thumb.

Interpretation: Flexion of the index finger and thumb is a sign of long-tract spinal cord involvement.

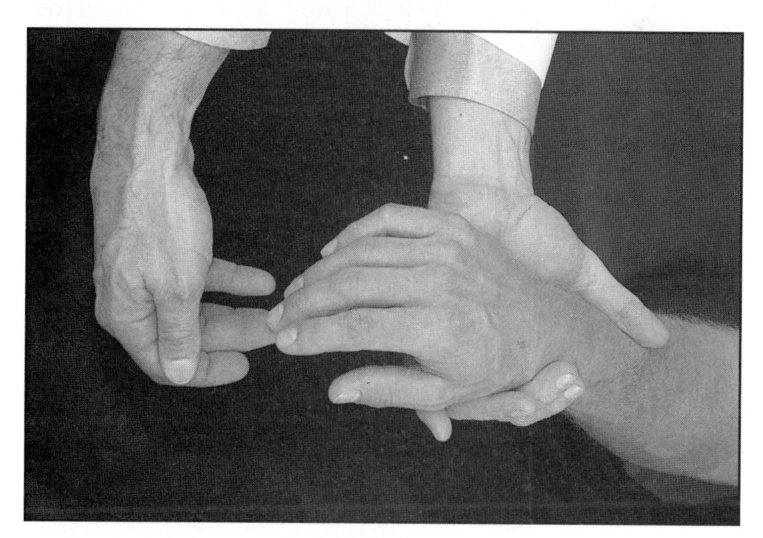

Reverse straight-leg raising

What it tests: L1-L4 nerve root impingement.

What to do: With the patient in the prone position, lift the hip into extension while keeping the knee straight.

Normal: No pain.

Interpretation: Pain with this maneuver suggests compression of the upper lumbar nerve roots.

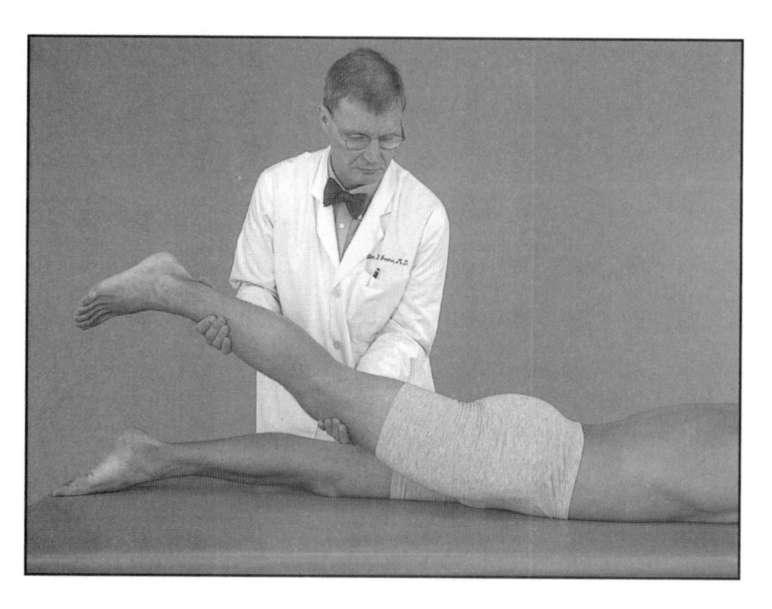

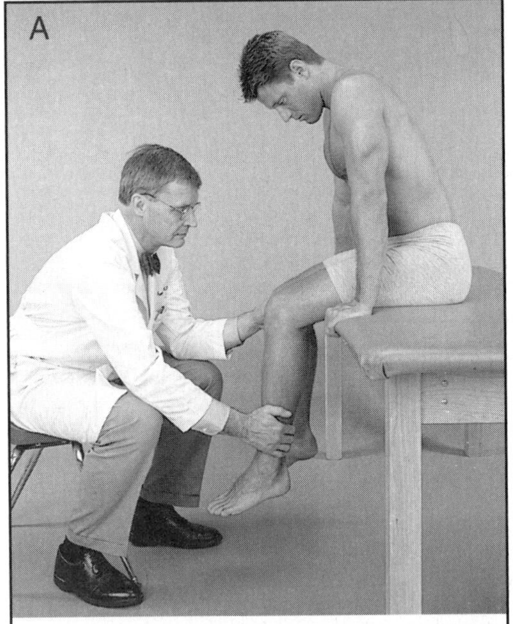

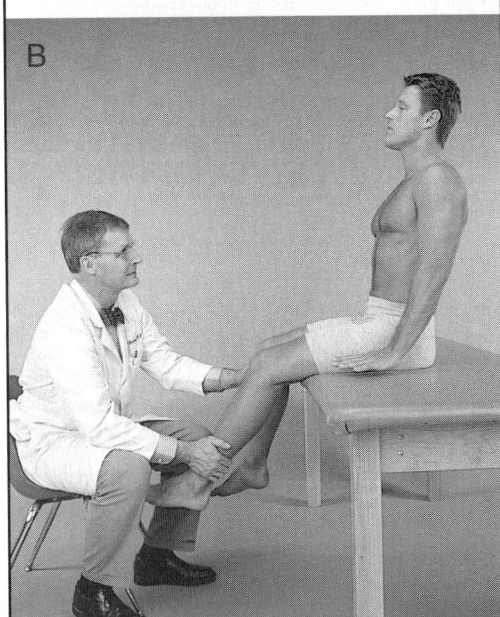

Flip sign

What it tests: Lumbar nerve root impingement.

What to do: With the patient seated and hands resting on the edge of the examination table, examine the patient's reflexes, extensor hallucis longus, and quadriceps strength. Next, distract the patient by asking about knee symptoms **(A)**. During the distraction, dorsiflex the ankle and extend the knee, noting the degree of knee flexion when back pain occurs.

Normal: No back pain with knee extension.

Interpretation: This test correlates with the straight-leg raise in the supine position. Patients with sciatic tension will have significant pain with extension of the knee and may "flip" backward to avoid tension on the nerve **(B)**.

Spurling test

What it tests: Cervical spinal and cervical disk pathology.

What to do: Ask the patient to extend the neck while you passively tilt the head to one side with approximately 5 lb of pressure and hold for 10 to 30 seconds.

Normal: No radicular pain or paresthesias.

Interpretation: This maneuver narrows the neural foramen and will increase or reproduce radicular arm pain associated with cervical disk herniations or cervical spondylosis.

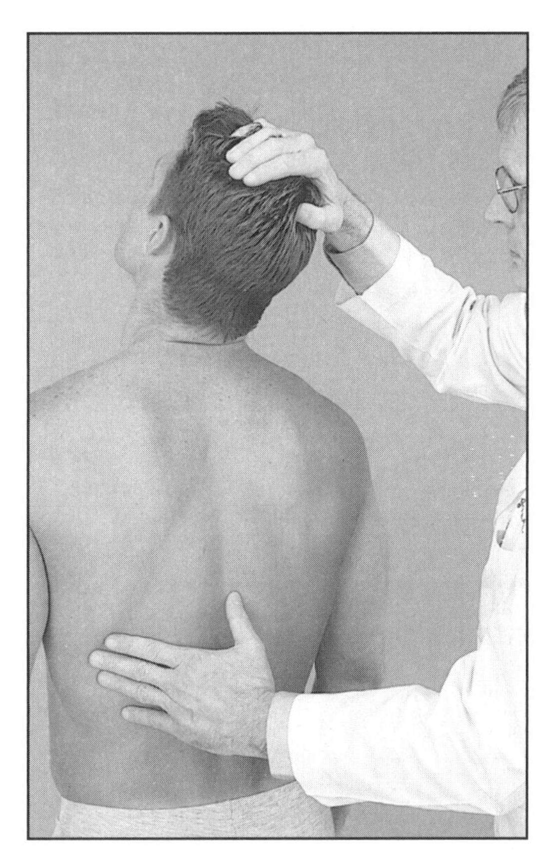

Axial loading

What it tests: Cervical disk pathology.

What to do: With the patient standing, push down on the patient's head.

Normal: No pain.

Interpretation: This maneuver may elicit neck pain in some patients with disk pathology. Increased low back pain is usually a nonorganic finding.

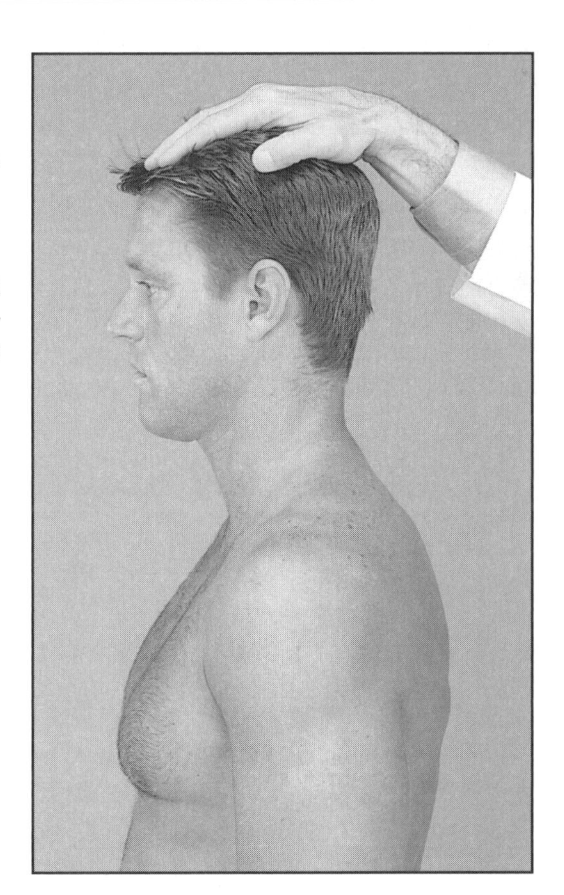

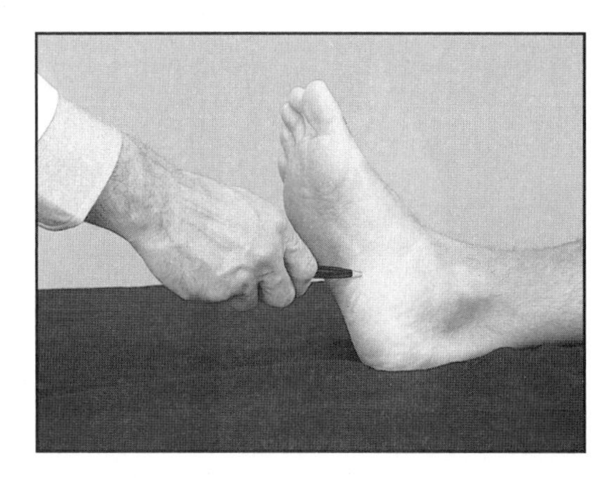

Babinski sign

What it tests: Upper motor neuron disease caused by spinal cord injury to the lower back; also called the flexor plantar response.

What to do: With the patient supine, stroke lightly upward on the plantar surface of the foot and look for great toe extension (withdrawal response) and fanning of the lesser toes.

Normal: A neurologically intact person will flex the toes or have no reaction at all.

Interpretation: Fanning of the toes is a positive test result and indicates long-tract spinal cord involvement.

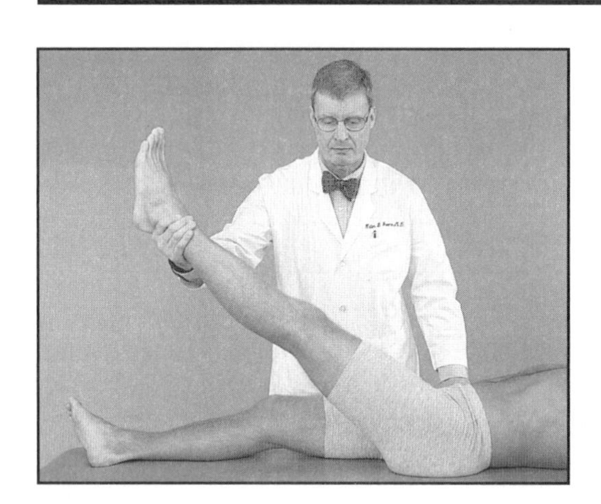

Straight-leg raising

What it tests: The sciatic nerve.

What to do: With the patient in the supine position and relaxed, elevate the leg by flexing the hip until either the knee begins to bend or the patient reports severe pain in the buttock or back. Record the degree of elevation at which pain occurs. Then dorsiflex the ankle to determine whether this motion increases pain (ankle dorsiflexion further stretches the L5 and S1 nerve roots).

Normal: No pain radiating down the leg.

Interpretation: Radicular pain down the back of the leg produced by this maneuver is evidence of sciatic nerve irritation, usually resulting from a herniated disk. Pain in the lower back with this maneuver is common with lumbar muscle strain; pain radiation is essential for a test result to be considered positive. Repeating this test with the patient seated may increase the specificity.

Chapter 27

Radiography

Plain radiographs are medical images that are obtained by transmitting high-energy electromagnetic radiation through a subject onto a radiographic plate or film. The rays that strike the film blacken it.[1,2] In the human body, differences in tissue density affect the degree to which the radiation is able to pass through the tissue. On the radiographic film, the parts of the body that are dense or thick appear bright because less radiation is able to pass through and blacken the film.[1,3] A dark or black region indicates tissues of low density or slight thickness.[2]

Radiolucency refers to matter that is easily penetrated by x-rays, whereas radiopacity indicates matter that absorbs x-rays.[2] Anatomic tissues and substances can be arranged in a scale ranging from the most radiolucent to the most radiopaque. According to this classification, gas is very radiolucent; fatty tissue is moderately radiolucent; and connective tissue, muscle, blood, and cartilage are intermediately radiolucent. Bones and calcium salts are moderately radiopaque, and heavy metals are very radiopaque.[2]

Because plain radiography offers excellent spatial resolution, it is typically the imaging modality of choice for patients with suspected bone lesions, such as fractures, dislocations, arthritis, and neoplasms. Readily available and relatively inexpensive, radiography may be used in screening and follow-up, and radiographs often serve as a guide for the correct interpretation of advanced imaging studies. For example, oblique radiographs of the cervical spine may reveal osteophytes that cannot be seen using MRI.

Radiography, however, does have limitations. For example, it cannot assess the loss of bone mineral content very well: loss of even 50% of the bone mass can elude detection on radiographs.[4,5] Radiography is also of limited value in the evaluation of soft-tissue abnormalities such as rotator cuff tears and intervertebral disk displacement. Radiography cannot accurately differentiate among edema, cellulitis, and soft-tissue abscesses. Moreover, radiography cannot detect osteomyelitis with sensitivity because its earliest radiographic signs (ie, soft-tissue swelling and mild bone mineral loss) may be too subtle to be seen. The permeative destruction caused by infection may be mistakenly attributed to a malignancy. The late osseous changes of osteomyelitis may simulate radiographic findings of neuropathic osteoarthropathy or healing fractures.[4,6] Given these limitations, the specificity of radiography in the diagnosis of infection is likewise far from perfect.[7,8]

Another limitation of plain radiography is that it provides a two-dimensional view of a three-dimensional structure. The radiographic image of an object is a composite of multiple sections of the object superimposed over one another.[2] As a projectional method, radiography can be of limited value in the diagnosis and precise localization of musculoskeletal abnormalities. On a single radiographic view, it is impossible to define which structures are located toward the front and which are located toward the back. For instance, an AP view of a dislocated shoulder does not identify whether the humeral head is posterior or anterior to the glenoid.

A particular anatomic part must be imaged from at least two sides to provide a three-dimensional perspective.[2] For routine radiographic examination, AP and lateral views are obtained. Occasionally, oblique views are needed to help visually separate overlapping structures.[2] CT and MRI may complement radiography when a three-dimensional perspective is needed.

Regional Radiographic Anatomy

The following section introduces the radiographic anatomy by region. The radiographs shown are not an exhaustive set. Instead, they are offered to provide an introduction to radiographic anatomy. Correlation of these images with surface anatomy and deep soft-tissue structures will be helpful.

Shoulder and Arm

Figure 1

Adequate radiographic evaluation of the glenohumeral joint requires AP and lateral views. Images obtained with internal and external rotation of the arm are needed for complete evaluation of suspected calcification in the rotator cuff. A radiographic series for a suspected dislocation must include an axillary view. Stress radiographs, obtained with a weight placed in the patient's hand, are used for evaluating acromioclavicular (AC) joint separation.

Elbow and Forearm

Figure 2

AP and lateral radiographs are routinely obtained to evaluate the elbow. Oblique views may provide additional information. The lateral view obtained with the elbow in flexion allows for visualization of the fat pads about the elbow, which may help detect an occult fracture. In the case of fracture, blood from the fracture will elevate the fat pad from the bone and will be seen as a distinct shadow.

Hand and Wrist

Figure 3

The radiographic examination of the hand includes AP, lateral, and oblique views. Special AP and lateral views obtained with the metacarpophalangeal (MCP) joints flexed may be required to evaluate injuries of these joints and the dorsum of the hand.[9] AP and lateral views of the wrist are considered routine. Oblique radiographs are useful for identifying arthritic changes of the wrist. Specialized views of the scaphoid may be required to detect a fracture; however, fractures can still be missed even with specialized views. Thus, MRI or bone scans may be needed. The carpal tunnel view is best for depicting the osseous structures and soft tissues of the carpal canal.

Pelvis

Figure 4

AP radiographs of the pelvis are obtained with the patient supine. Because of the normal anteversion of the hip, the feet are placed in 15° of internal rotation to show a true AP view of the femoral neck.[9]

Knee and Leg

Figure 5

Several radiographic views, including the AP and lateral views, are used to evaluate the knee joint. Weight-bearing radiographs permit assessment of the articular space and are particularly useful in the evaluation of degenerative joint disease; when cartilage is lost, the bones can touch each other. Radiographs obtained with side-to-side stress applied to the knee can help diagnose instability of the joint. A sunrise view to visualize the patellofemoral joint can be obtained with the knee flexed and the beam aimed parallel to the direction of the leg.

Foot and Ankle

Figures 6 and 7

AP, lateral, and medial oblique views are obtained in the routine evaluation of the foot. An axial radiograph provides visualization of the posterior subtalar joint and part of the anterior talocalcaneonavicular joint. Oblique views also provide visualization of these joints. A lateral view as well as an angulated AP view offers optimal visualization of the calcaneus.

The radiographic evaluation of the ankle requires AP and lateral views. An AP radiograph obtained with 15° to 20° of internal rotation of the foot (the mortise view) allows better delineation of the ankle joint because the fibula is slightly behind the center of the joint and would overlap with the tibia on a straight AP view if this rotation is not made.[9]

Spine

Cervical Spine

Figure 8

Plain radiographs are useful for assessing abnormalities of the cervical spine. The standard radiographic examination consists of multiple views, including AP and lateral views. The lateral view, however, may be supplemented with flexion-extension radiographs. Right and left 45° oblique views are helpful in the evaluation of the intervertebral foramina. Because the cervical apophyseal joints are oriented in a horizontal articular plane of about 30° to 45°, AP views obtained with neck extension (pillar views) are useful in examining the articular processes and the vertebral arches.[9] Open-mouth odontoid views provide visualization of the atlas and axis. For patients with significant neck trauma, the screening examination of the cervical spine should include cross-table

AP and lateral views. (Note that the radiograph in Figure 8 does not include the complete cervical spine—the junction between C7 and T1 is not shown; therefore, this particular radiograph would be inadequate for a trauma evaluation.)

Thoracic Spine
Figure 9
The standard radiographic series for assessing abnormalities of the thoracic spine includes AP and lateral radiographs. The swimmer's view (so named because the patient is positioned as if swimming the crawl stroke, with one arm up and one arm down) allows visualization of the cervicothoracic region.[9] In patients with scoliosis, AP and PA views are used to measure the angle of curvature.

Lumbar Spine
Figure 10
A radiographic series of the lumbar spine includes AP, lateral, and oblique views. The AP radiograph is obtained with the patient standing or supine. For the supine AP radiograph, the hips and knees are flexed, which reduces the lumbar lordosis and improves visualization of the vertebral bodies and intervertebral disks.[9] The lateral radiograph is taken with slight flexion of the hips and knees. Oblique radiographs are useful in the evaluation of the posterior elements of the lumbar spine.

Common Pathologies

This section presents a sampling of musculoskeletal conditions that are clearly delineated on radiographs. Most radiographs seen in clinical practice will be normal and are taken as part of a screening examination. Other radiographs may demonstrate pathology, but this pathology will not always be clinically significant. Imaging studies, therefore, must be considered as a complement to the patient's history and physical examination.

Fracture

Figures 11, 12, and 13
Characterization of the location and type of fracture—its position, alignment, and presence or absence of rotation between fragments—is necessary in the evaluation and treatment of injury.[10] The radiographic examination of extremity fractures should include, at the minimum, AP and lateral views. In the case of a possible midshaft fracture, the joints above and below the suspected fracture site must be included on the radiograph to help identify any associated joint injuries (Fig. 11). Radiographic evaluation of suspected metaphyseal or epiphyseal fractures usually includes imaging of only the joint adjacent to the fracture site. Comminuted fractures involve more than two fracture fragments (Fig. 12). In most instances, a callus develops as the fracture heals (Fig. 13).

Dislocations

Figure 14
In a joint dislocation, complete loss of contact between the articular surfaces occurs; in a subluxation, the loss of contact is only partial. Dislocations caused by trauma also may be associated with the fracture of an adjacent bone (fracture-dislocation). Diastasis refers to separation of a joint that is normally only slightly moveable (eg, the sacroiliac joint and the symphysis pubis).

Dislocations and subluxations result from shifting of a bone from the normal anatomic position. Imaging methods useful for assessing these injuries include conventional radiography, CT scanning, and MRI. The evaluation of a patient with possible dislocation requires multiple radiographic projections (Fig. 14). Stress radiographs with or without weight bearing frequently are necessary to diagnose dislocation. Additionally, comparison of radiographs of the side with the suspected dislocation with radiographs showing the unaffected side may be needed.

Cervical Spine Trauma

Figure 15
Radiographic evaluation of the cervical spine is mandatory in all instances of suspected trauma. The objectives of the initial radiographic evaluation of the injured cervical spine include the identification of all injuries, detection of neural canal compromise, and assessment of potential mechanical instability. Because the shoulders obscure the lower cervical vertebrae, traction on the arms may be necessary to visualize the entire cervical spine. An AP open-mouth odontoid view allows visualization of the atlas and axis. CT is often needed to fully visualize the injury.

Degenerative Changes of the Cervical Spine

Figure 16
Degenerative processes in the cervical spine are common in adults. Degenerative changes

usually become evident in multiple spinal levels and predominate in the lower cervical spine. Degenerative alterations of the intervertebral disks most frequently occur at the C5-C6 and C6-C7 levels. Osteoarthritis shows a predilection for the middle and lower cervical spine. With increasing degeneration of the intervertebral disks, progressive loss of height of the disk space occurs. Exuberant osteophytosis may cause narrowing of the intervertebral foramina, most frequently affecting the C3-C6 levels. Compromise of the adjacent nerve roots may develop. However, the relationship of radiographic abnormalities of degeneration in the cervical spine to clinical symptoms and signs is not clear.[11]

Osteoarthritis of the Hand

Figure 17

In the hand, primary osteoarthritis is manifested as bony enlargements about the finger joints. Those affecting the distal interphalangeal (DIP) joints are termed Heberden's nodes; similar enlargement about the proximal interphalangeal (PIP) joints are termed Bouchard's nodes. Typically, abnormalities in the DIP and PIP joints are more prominent than changes in the MCP joints. Radiographs of involved interphalangeal joints reveal loss of joint space and prominent osteophytes, causing close apposition of adjacent articular surfaces.[12]

Rheumatoid Arthritis of the Hand

Figure 18

In rheumatoid arthritis of the hand, the MCP and PIP joints are most commonly affected; the DIP joints are less frequently affected.[13] Characteristic radiographic findings include symmetric soft-tissue swelling, periarticular osteoporosis, concentric joint space loss, and marginal erosions.[14-18] Superficial surface-bone resorption may become apparent in the diaphyses and metaphyses of the phalanges.

In advanced stages of disease, severe narrowing or obliteration of the joint space occurs. Subluxation of the small joints of the hand, producing boutonnière and swan-neck deformities, are common complications.[13,19] Malalignment of the MCP joints may result in ulnar deviation of the fingers and volar subluxation and flexion of the MCP joint.

Osteonecrosis of the Hip

Figure 19

Osteonecrosis of the femoral head can lead to the collapse of the articular surface and severe structural joint deformity.[20-22] The radiographic findings associated with osteonecrosis of the hip vary with the stage of the disease.[23] Initial findings may be subtle and include mottled, poorly defined, radiolucent lesions or scattered, patchy, radiodense areas. Patchy, radiolucent areas may be surrounded by a peripheral rim of sclerosis (Fig. 19, *A*). With progression of disease, arc-like, subchondral, radiolucent lesions and flattening of the femoral head become apparent. In advanced stages, the articular surface collapses and osteoarthritis develops[23] (Fig. 19, *B*).

Knee Osteoarthritis

Figure 20

Osteoarthritis most commonly affects the knee joint. Radiographic changes usually predominate in the medial femorotibial compartment.[24] In patients with knee osteoarthritis, radiographs may reveal joint space narrowing, sclerosis of subchondral bone, subchondral cysts, and osteophytes, particularly at the margins of the joint. In patients with patellofemoral osteoarthritis, radiographic abnormalities may include joint space narrowing, sclerosis, and osteophytes, particularly on the patellar side of the space. Other findings include angulation and subluxation, joint malalignment, and the presence of intra-articular bodies, joint effusions, and synovial cysts. Varus deformity is more common than valgus deformity.[12]

Osteomyelitis

Figure 21

Osteomyelitis is an infection of bone and marrow that is generally characterized by acute, subacute, and chronic clinical stages. In patients with osteomyelitis, the earliest radiographic abnormalities may be subtle, consisting of soft-tissue swelling. Chronic infection, however, results in bone resorption. Therefore, as the infection progresses, poorly defined, radiolucent lesions develop in the metaphysis of tubular bones and, in children, may extend to the growth plate. Cortical resorption and periostitis also may be present. In the subacute and chronic stages of hematogenous osteomyelitis, bone abscesses (Brodie's abscesses) may appear in the metaphyseal region of the long bones (Fig. 21). On radiographs, a Brodie's abscess appears as a sharply delineated, elongated, radiolucent lesion surrounded by bone sclerosis. In patients with subacute and chronic osteomyelitis, characteristic radiographic findings include a sequestrum (necrotic bone fragmentation) and an involucrum (a zone of living bone surrounding sequestered bone).[4,5]

Osteoporosis

Figure 22

Osteoporosis most commonly affects the axial skeleton and proximal portions of the long bones. In the diagnosis of early bone loss, radiographs are ineffective because even significant loss of bone mass may elude detection.[25] Therefore, other imaging modalities (eg, dual-energy x-ray absorptiometry) are used to better assess osteoporosis.[26-32] Radiographic findings in patients with osteoporosis include increased radiolucency of the bones, changes in the trabecular pattern and shape of vertebral bodies (eg, wedge-shaped vertebrae, compressed vertebrae, and fish vertebrae), and acute and insufficiency fractures[33,34] (Fig. 22, *A*). The Singh index characterizes the severity of osteoporosis on radiographs according to patterns of trabecular loss and cortical thinning in the hip.[35] As osteoporosis progresses, trabecular groups become obliterated in sequence; with severe disease, only the compressive trabeculae remain (Fig. 22, *B*). (This process may be appreciated better by contrasting the appearance of the femoral neck shown in Figure 22, *B* with that shown in Figure 4.)

Osteomalacia and Rickets

Figure 23

Defective mineralization of the adult skeleton is termed osteomalacia; it is called rickets when found in the immature skeleton.[36] Radiographic manifestations of rickets include retarded bone growth and osteopenia.[37,38] Alterations of the growth plate are characteristic of the disease, and widening at the physis represents the earliest specific radiographic finding.[39] Decreased bone density on the metaphyseal side of the growth plate becomes evident as the disease progresses, and progressive widening and irregularity of the growth plate also may be observed. In the metaphyseal region, disorganization and "fraying" of the spongy bone occurs, with widening and cupping of the metaphysis.[37,38] In the epiphysis, changes consist of deossification and blunting of the ossified periphery.[40] In the long bones, bulky growth plates and bowing deformities are typical manifestations of rickets.

Tumors

Hemangioma

Figure 24

Hemangioma of bone is a benign lesion composed of newly formed capillary, cavernous, or venous blood vessels.[41] Vertebral hemangiomas are usually asymptomatic lesions that can be incidental findings on radiographs. Although these tumors involve primarily the vertebral body, the pedicles, laminae, and transverse or spinous processes also can be affected. Spinal hemangiomas assume a characteristic coarse, vertical trabecular pattern; a honeycomb or cartwheel configuration also may be seen.[42] Bone scans can be useful in the diagnosis of hemangiomas. CT and MRI, particularly when used with contrast material, can provide additional information regarding the characterization of osseous hemangiomas.

Osteosarcoma

Figure 25

Osteosarcoma is a malignant tumor of bone in which the proliferating tumor cells produce osteoid.[43] In general, the radiographic appearance of osteosarcomas depends on three parameters: (1) the degree of destruction of cortical or medullary bone (osteolysis), (2) the amount of bone production and calcification (osteosclerosis), and (3) the presence of periosteal bone formation.[43] In tubular bones, conventional osteosarcomas appear as poorly defined, destructive, intramedullary lesions that violate the cortex and extend to soft tissue.

Skeletal Metastases

Figure 26

One of the basic characteristics of all malignant tumors is the capacity to metastasize. The primary tumors involved in skeletal metastases most frequently are carcinoma of the breast, prostate, and lung. The most common sites of metastasis are the spine, pelvis, ribs, sternum, femoral and humeral diaphyses, and skull. In general, metastatic lesions demonstrate either bone resorption (Fig. 26, *A*) or bone formation (Fig. 26, *B*). These patterns are termed osteolytic and osteoblastic, respectively. Mixed osteolytic-osteoblastic patterns can be seen as well. Findings associated with skeletal metastases include soft-tissue masses, soft-tissue ossification, and pathologic fractures.[44]

References

1. Parker DL: X-ray: X-ray attenuation: X-ray contrast: X-ray image: X-ray interaction with matter, in von Schulthess GK, Smith HJ (eds): *The Encyclopaedia of Medical Imaging Vol. I: Physics, Techniques and Procedures*. Oslo, Norway, the NICER Institute, 1998, pp 454-459.
2. Meschan I: Fundamental background for radiologic anatomy, in Meschan I (ed): *An Atlas of Anatomy Basic to Radiology*. Philadelphia, PA, WB Saunders, 1975, pp 1-23.
3. Kiuru A: Radiographic contrast, in von Schulthess GK, Smith HJ (eds): *The Encyclopaedia of Medical Imaging Vol. I: Physics, Techniques and Procedures*. Oslo, Norway, the NICER Institute, 1998, p 394.
4. Resnick D, Niwayama G: Osteomyelitis, septic arthritis, and soft tissue infection: Mechanisms and situations, in Resnick D (ed): *Diagnosis of Bone and Joint Disorders*, ed 3. Philadelphia, PA, WB Saunders, 1995, vol 4, pp 2325-2418.
5. Theodorou DJ, Theodorou SJ, Kakitsubata Y, Sartoris DJ, Resnick D: Imaging characteristics and epidemiologic features of atypical mycobacterial infections involving the musculoskeletal system. *AJR Am J Roentgenol* 2001;176:341-349.
6. Resnick D, Niwayama G: Osteomyelitis, septic arthritis, and soft tissue infection: Organisms, in Resnick D (ed): *Diagnosis of Bone and Joint Disorders*, ed 3. Philadelphia, PA, WB Saunders, 1995, vol 4, pp 2448-2558.
7. Larcos G, Brown ML, Sutton RT: Diagnosis of osteomyelitis of the foot in diabetic patients: Value of 111In-leukocyte scintigraphy. *AJR Am J Roentgenol* 1991;157:527-531.
8. Tehranzadeh J, Wang F, Mesqarzadeh M: Magnetic resonance imaging of osteomyelitis. *Crit Rev Diagn Imaging* 1992;33:495-534.
9. Sartoris DJ, Resnick D: Plain film radiography: Routine techniques, in Resnick D (ed) Bone and Joint Imaging: ed 2. Philadelphia, PA, WB Saunders, 1996, pp 19-41.
10. Weissman B: General principles, in Weissman BN, Sledge CB (eds): *Orthopedic Radiology*. Philadelphia, PA, WB Saunders, 1986, pp 1-69.
11. Lestini WF, Wiesel SW: The pathogenesis of cervical spondylosis. *Clin Orthop* 1989;239:69-93.
12. Resnick D: Degenerative disease of extraspinal locations, in Resnick D (ed): *Bone and Joint Imaging*, ed 2. Philadelphia, PA, WB Saunders, 1996, pp 321-354.
13. Resnick D, Niwayama G: Rheumatoid arthritis, in Resnick D (ed): *Diagnosis of Bone and Joint Disorders*, ed 3. Philadelphia, PA, WB Saunders, 1995, vol 2, pp 866-970.
14. Alenfeld FE, Diessel E, Brezger M, Sieper J, Felsenberg D, Braun J: Detailed analyses of periarticular osteoporosis in rheumatoid arthritis. *Osteoporos Int* 2000;11:400-407.
15. Winalski CS, Palmer WE, Rosenthal DI, Weissman BN: Magnetic resonance imaging of rheumatoid arthritis. *Radiol Clin North Am* 1996;34:243-258.
16. Sharp JT, Gardner JC, Bennett EM: Computer-based methods for measuring joint space and estimating erosion volume in the finger and wrist joints of patients with rheumatoid arthritis. *Arthritis Rheum* 2000;43:1378-1386.
17. van der Heijde D, Boers M, Lassere M: Methodological issues in radiographic scoring methods in rheumatoid arthritis. *J Rheumatol* 1999;26:726-730.
18. Theodorou DJ, Theodorou SJ, Resnick D: Imaging of rheumatoid arthritis, in Hochberg M, Silman A, Smolen J, Weinblatt M, Weisman M (eds): *Rheumatology*, ed 3. London, England, Harcourt, 2003, pp 51-60.
19. Rosen A, Weiland AJ: Rheumatoid arthritis of the wrist and hand. *Rheum Dis Clin North Am* 1998;24:101-128.
20. Ohzono K, Saito M, Takaoka K, et al: Natural history of nontraumatic avascular necrosis of the femoral head. *J Bone Joint Surg Br* 1991;73:68-72.
21. Takatori Y, Kokubo T, Ninomiya S, Nakamura S, Morimoto S, Kusaba I: Avascular necrosis of the femoral head: Natural history and magnetic resonance imaging. *J Bone Joint Surg Br* 1993;75:217-221.
22. Koo KH, Kim R, Ko GH, Song HR, Jeong ST, Cho SH: Preventing collapse in early osteonecrosis of the femoral head: A randomised clinical trial of core decompression. *J Bone Joint Surg Br* 1995;77:870-874.
23. Resnick D, Niwayama G: Osteonecrosis: Diagnostic techniques, specific situations, and complications, in Resnick D (ed): *Diagnosis of Bone and Joint Disorders*, ed 3. Philadelphia, PA, WB Saunders, 1995, vol 5, pp 3495-3558.
24. Barrett JP Jr, Rashkoff E, Sirna EC, Wilson A: Correlation of roentgenographic patterns and clinical manifestations of symptomatic idiopathic osteoarthritis of the knee. *Clin Orthop* 1990;253:179-183.
25. Finsen V, Anda S: Accuracy of visually estimated bone mineralization in routine radiographs of the lower extremity. *Skeletal Radiol* 1988;17:270-275.
26. Theodorou DJ, Theodorou SJ, Andre MP, Kubota D, Weigert JM, Sartoris DJ: Quantitative computed tomography of spine: Comparison of three-dimensional and two-dimensional imaging approaches in clinical practice. *J Clin Densitom* 2001;4:57-62.
27. Sartoris DJ: Quantitative bone mineral analysis, in Resnick D (ed): *Bone and Joint Imaging*, ed 2. Philadelphia, PA, WB Saunders, 1996, pp 154-164.
28. Theodorou DJ, Theodorou SJ, Sartoris DJ: Treatment of osteoporosis: Current status and recent advances. *Compr Ther* 2002;28:109-122.
29. Theodorou DJ, Theodorou SJ: Dual-energy x-ray absorptiometry in clinical practice: Application and interpretation of scans beyond the numbers. *Clin Imaging* 2002;26:43-49.
30. Theodorou DJ, Theodorou SJ, Sartoris DJ: Osteoporosis: Prevention and diagnostic work-up. *Hosp Med* 2002;63:396-400.
31. Theodorou DJ, Theodorou SJ, Sartoris DJ: Dual-energy x-ray absorptiometry in diagnosis of osteoporosis: Basic principles, indications, and scan interpretation. *Compr Ther* 2002;28:190-200.
32. Theodorou DJ, Theodorou SJ, Sartoris D: Imaging modalities in the assessment of osteoporosis. *Compr Ther* 2002;28:189-199.
33. Theodorou DJ, Theodorou SJ, Duncan TD, Garfin SR, Wong WH: Percutaneous balloon kyphoplasty for the correction of spinal deformity in painful vertebral body compression fractures. *Clin Imaging* 2002;26:1-5.
34. Resnick D: Osteoporosis, in Resnick D, Boutin RD (eds): *The Encyclopaedia of Medical ImagingVol. III: 1 Musculoskeletal and Soft Tissue Imaging*. Oslo, Norway, the NICER Institute, 1999, pp 322-323.

35. Weissman BN: The hip, in Weissman BN, Sledge CB (eds): *Orthopedic Radiology*. Philadelphia, PA, WB Saunders, 1986, pp 385-495.
36. Resnick D, Boutin RD: Rickets, in Resnick D, Boutin RD (eds): *The Encyclopaedia of Medical Imaging Vol. III: 1 Musculoskeletal and Soft Tissue Imaging*. Oslo, Norway, the NICER Institute, 1999, pp 371-372.
37. Pitt MJ: Rickets and osteomalacia, in Resnick D (ed): *Diagnosis of Bone and Joint Disorders*, ed 3. Philadelphia, PA, WB Saunders, 1995, vol 4, pp 1885-1922.
38. Theodorou DJ, Theodorou SJ, Resnick D: The hand in endocrine disorders, in Guglielmi G, van Kuijk C, Genant HK (eds): *Fundamentals of Hand and Wrist Imaging*. Berlin, Germany, Springer-Verlag, 2001, pp 203-229.
39. Steinbach HL, Noetzli M: Roentgen appearance of the skeleton in osteomalacia and rickets. *AJR Am J Roentgenol* 1964;91:955-972.
40. Pitt MJ: Rickets and osteomalacia, in Resnick D (ed): *Bone and Joint Imaging*, ed 2. Philadelphia, PA, WB Saunders, 1996, pp 511-524.
41. Huvos AG: Hemangioma of bone: Lymphangioma of bone: Diffuse skeletal angiomatosis and lymphangiomatosis: Glomus tumor: "Disappearing bone disease", in Huvos AG (ed): *Bone Tumors: Diagnosis, Treatment, and Prognosis*, ed 2. Philadelphia, PA, WB Saunders, 1991, pp 553-578.
42. Resnick D, Greenway GD: Tumors and tumor-like lesions of bone: Imaging and pathology of specific lesions, in Resnick D (ed): *Bone and Joint Imaging*, ed 2. Philadelphia, PA, WB Saunders, 1996, pp 991-1063.
43. Huvos AG: Osteogenic sarcoma, in Huvos AG (ed): *Bone Tumors: Diagnosis, Treatment, and Prognosis*. Philadelphia, PA, WB Saunders, 1991, pp 85-155.
44. Resnick D: Metastasis, in Resnick D, Boutin RD (eds): *The Encyclopaedia of Medical Imaging Vol. III: 1 Musculoskeletal and Soft Tissue Imaging*. Oslo, Norway, the NICER Institute, 1999, pp 280-281.

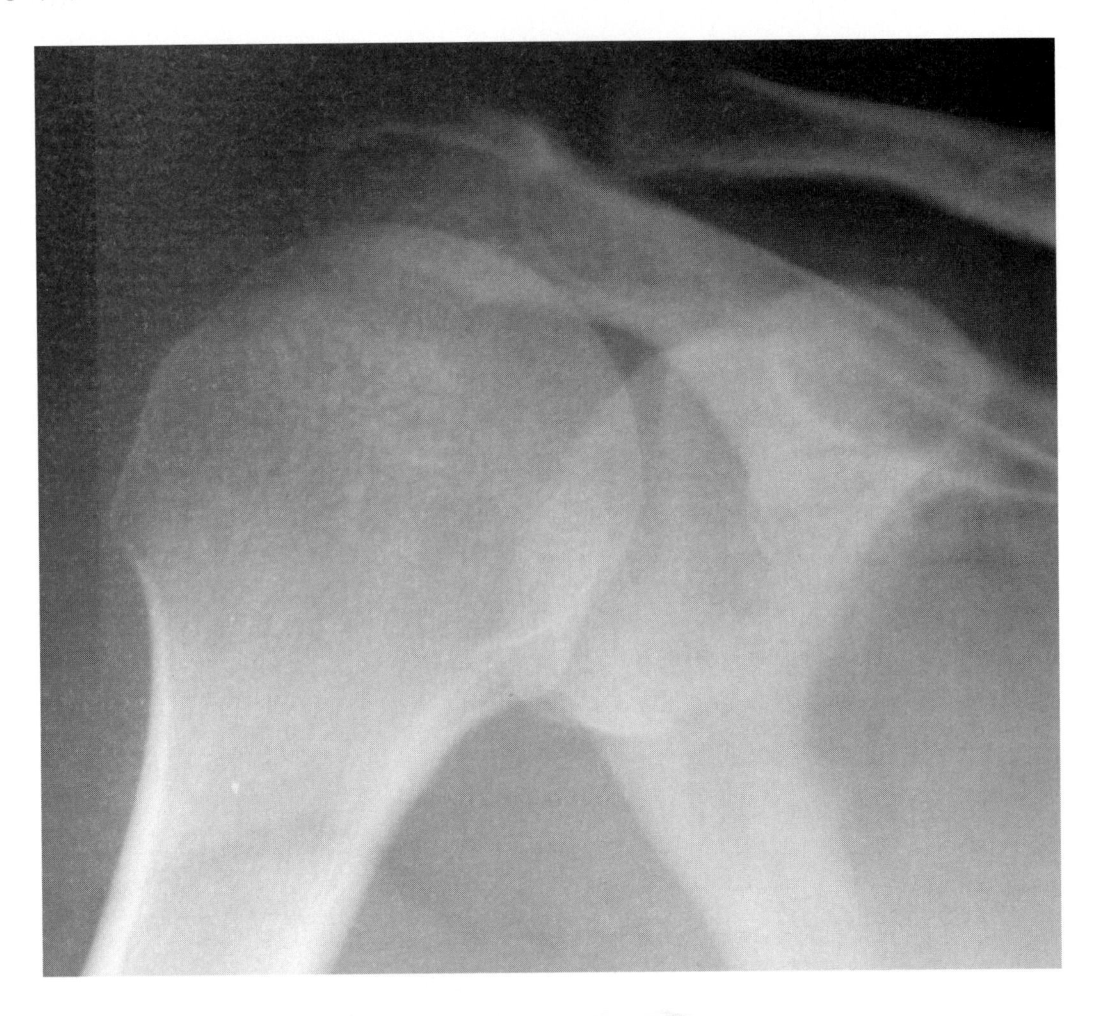

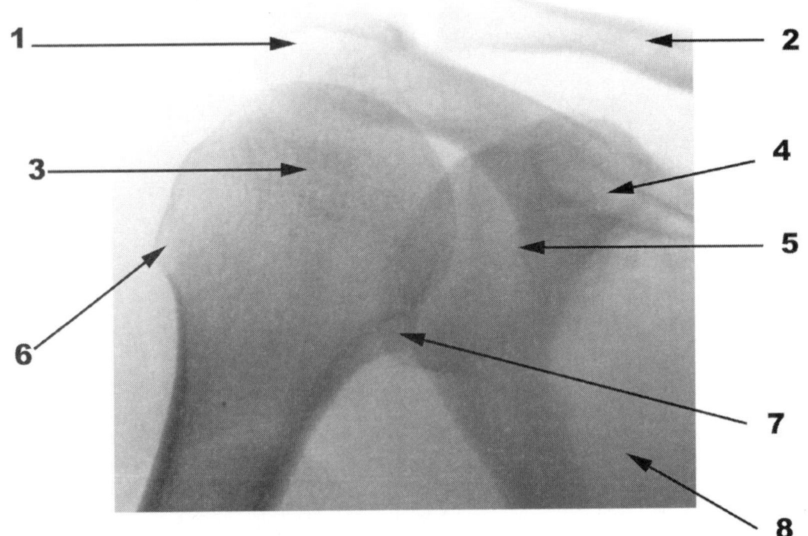

Figure 1
AP view of the shoulder (internal rotation). 1 = Acromion, 2 = Clavicle, 3 = Humeral head, 4 = Coracoid, 5 = Glenoid, 6 = Greater tuberosity, 7 = Lesser tuberosity, 8 = Scapula.

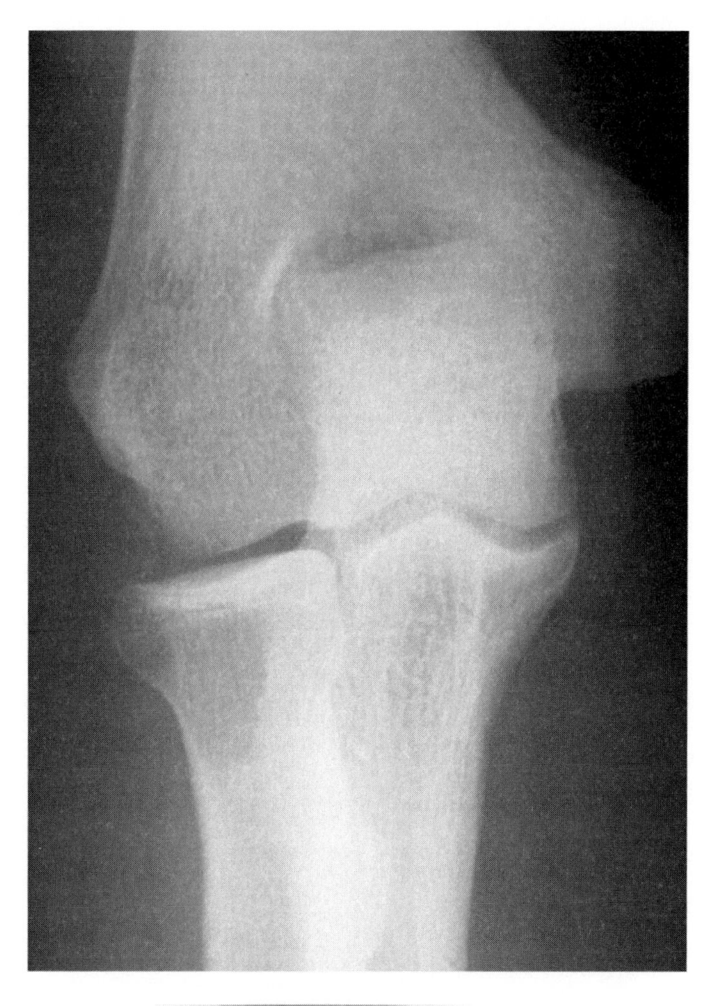

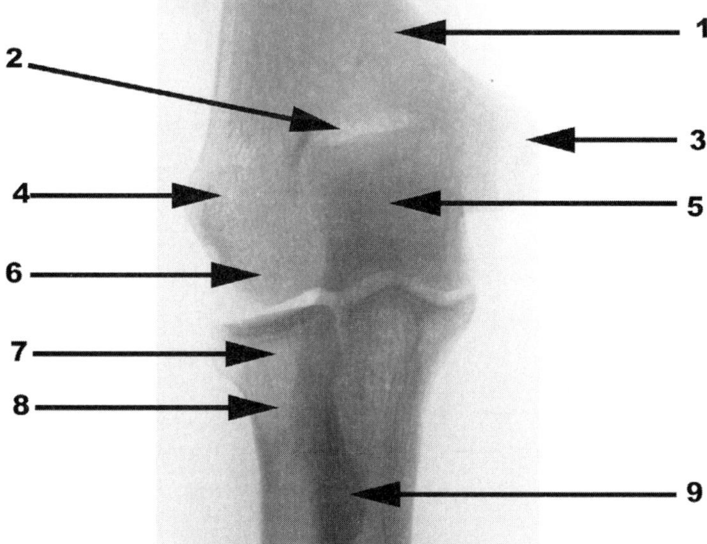

Figure 2
AP view of the elbow. 1 = Humerus, 2 = Olecranon fossa, 3 = Medial epicondyle, 4 = Lateral epicondyle , 5 = Olecranon, 6 = Capitulum of humerus, 7 = Radial head, 8 = Radial neck, 9 = Radial tuberosity.

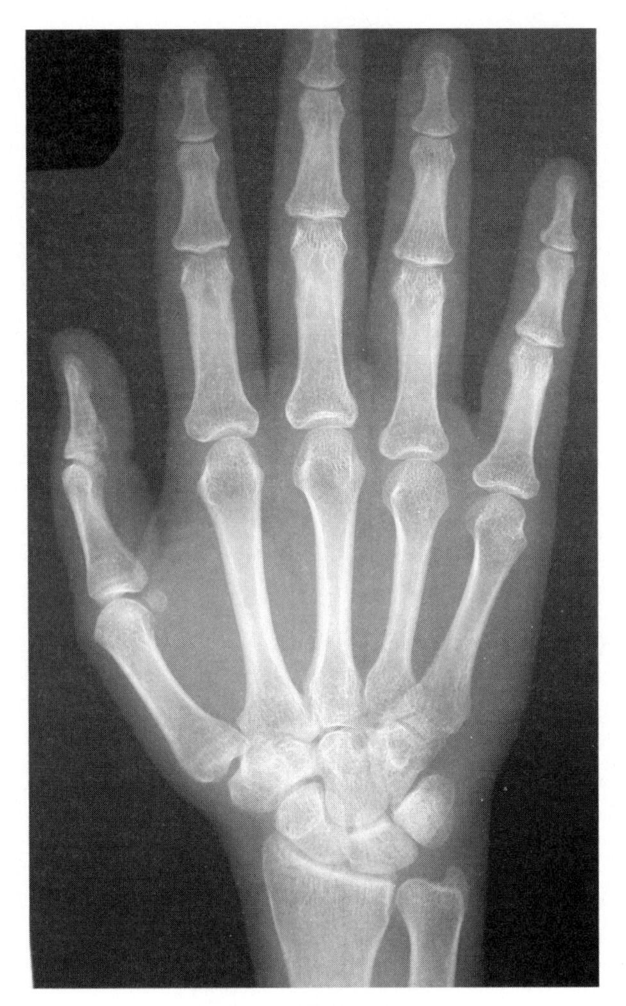

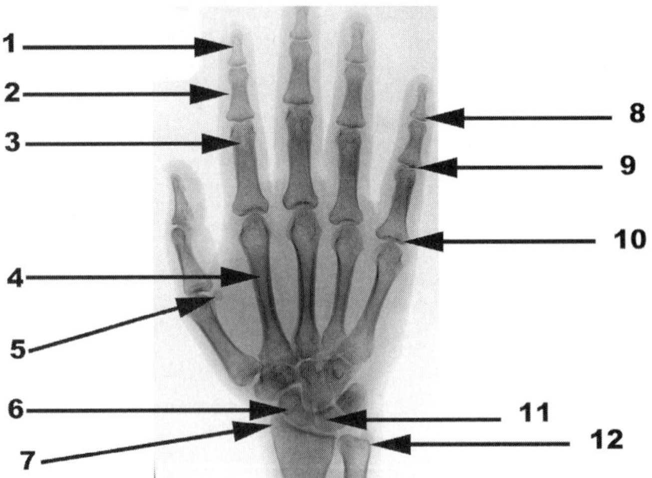

Figure 3
AP view of the hand. 1 = Distal phalanx, 2 = Middle phalanx, 3 = Proximal phalanx, 4 = Metacarpal bone, 5 = Sesamoid, 6 = Scaphoid, 7 = Styloid process of radius, 8 = Distal interphalangeal joint, 9 = Proximal interphalangeal joint, 10 = Metacarpophalangeal joint, 11 = Lunate, 12 = Styloid process of ulna.

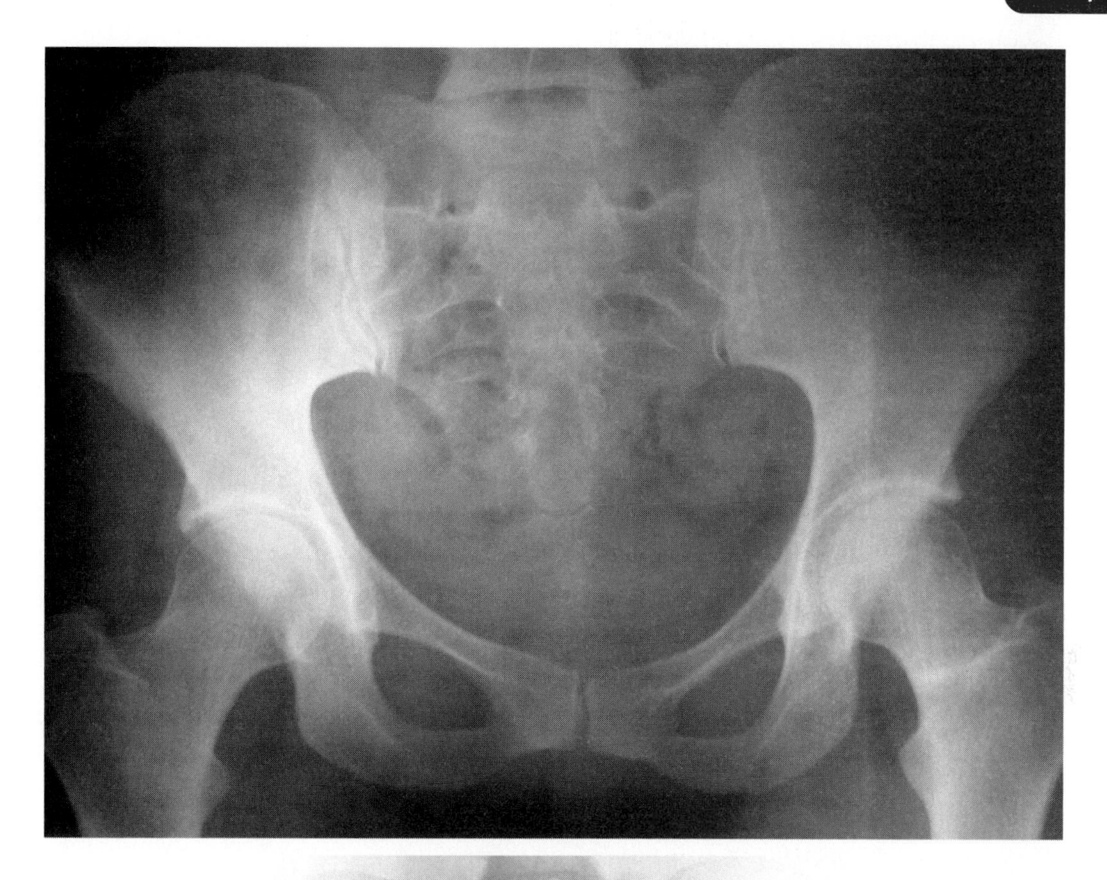

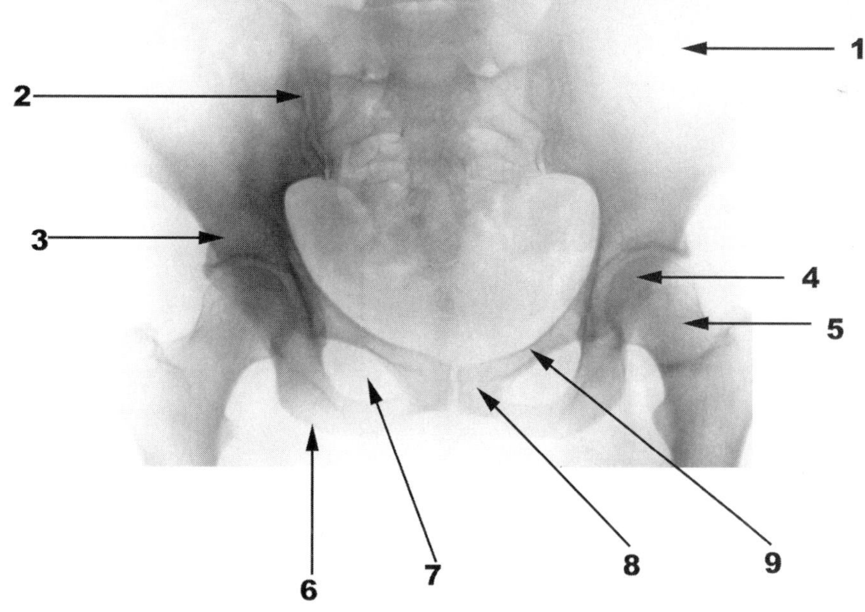

Figure 4
AP view of the pelvis. 1 = Ilium, 2 = Sacroiliac joint, 3 = Acetabulum, 4 = Femoral head, 5 = Femoral neck, 6 = Inferior pubic ramus, 7 = Obturator foramen, 8 = Pubis, 9 = Superior pubic ramus.

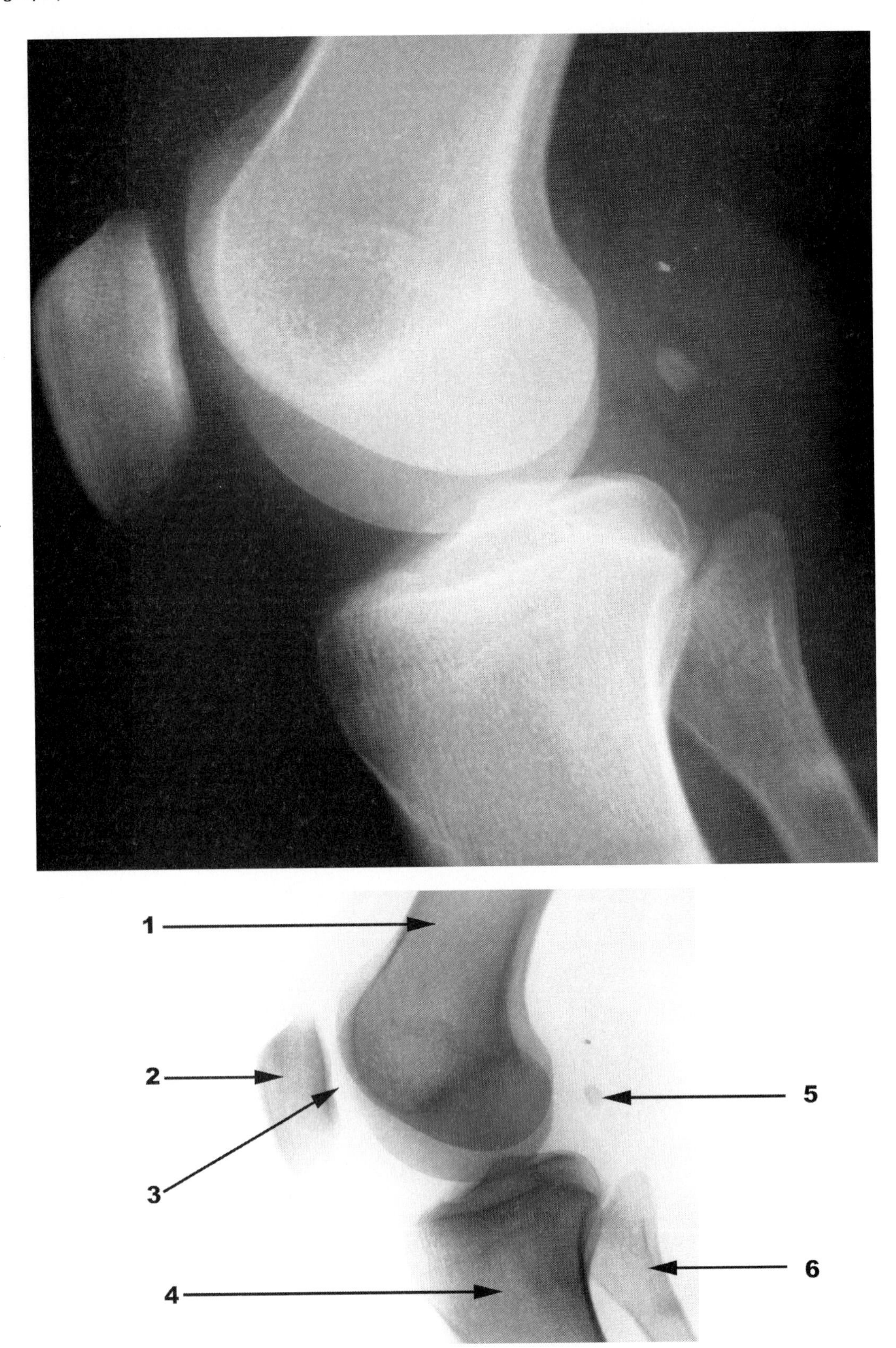

Figure 5
Lateral view of the knee. 1 = Femur, 2 = Patella, 3 = Patellofemoral joint, 4 = Tibia, 5 = Fabella, 6 = Fibula.

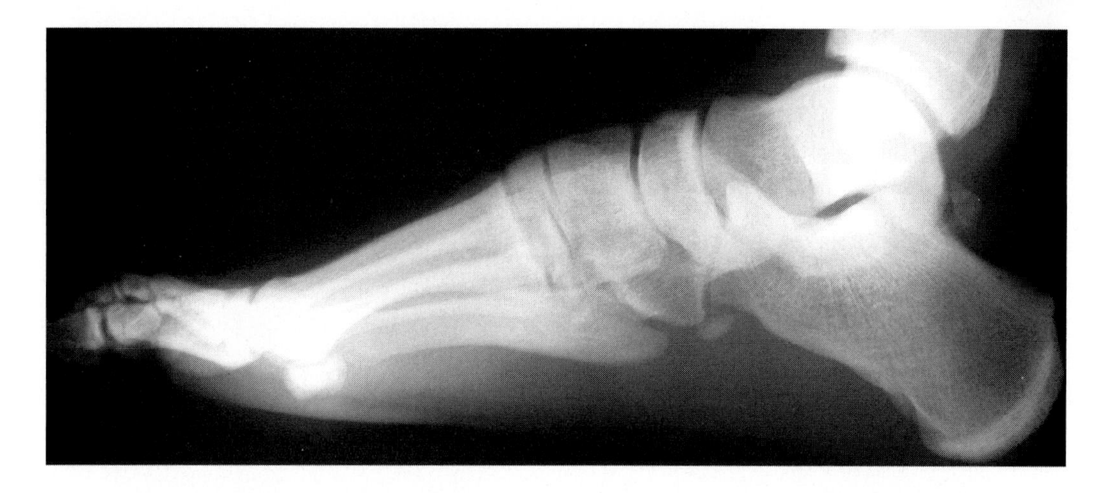

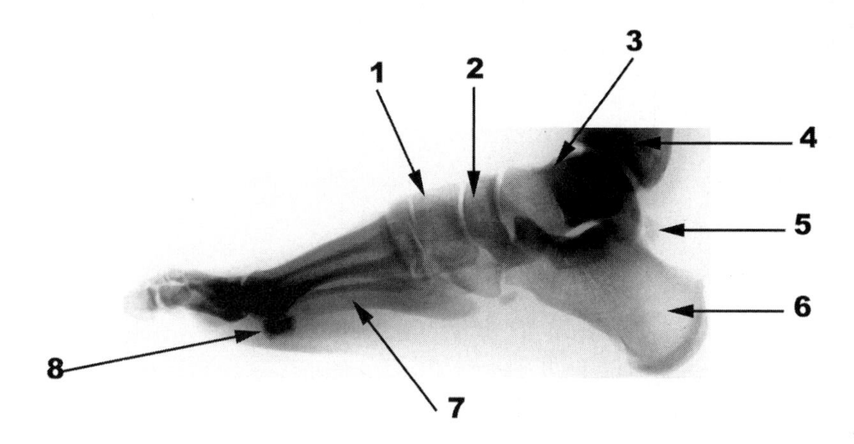

Figure 6
Lateral view of the foot. 1 = Cuneiform bones, 2 = Navicular, 3 = Talus, 4 = Tibia, 5 = Os trigonum, 6 = Calcaneus, 7 = Metatarsal bone, 8 = Sesamoid.

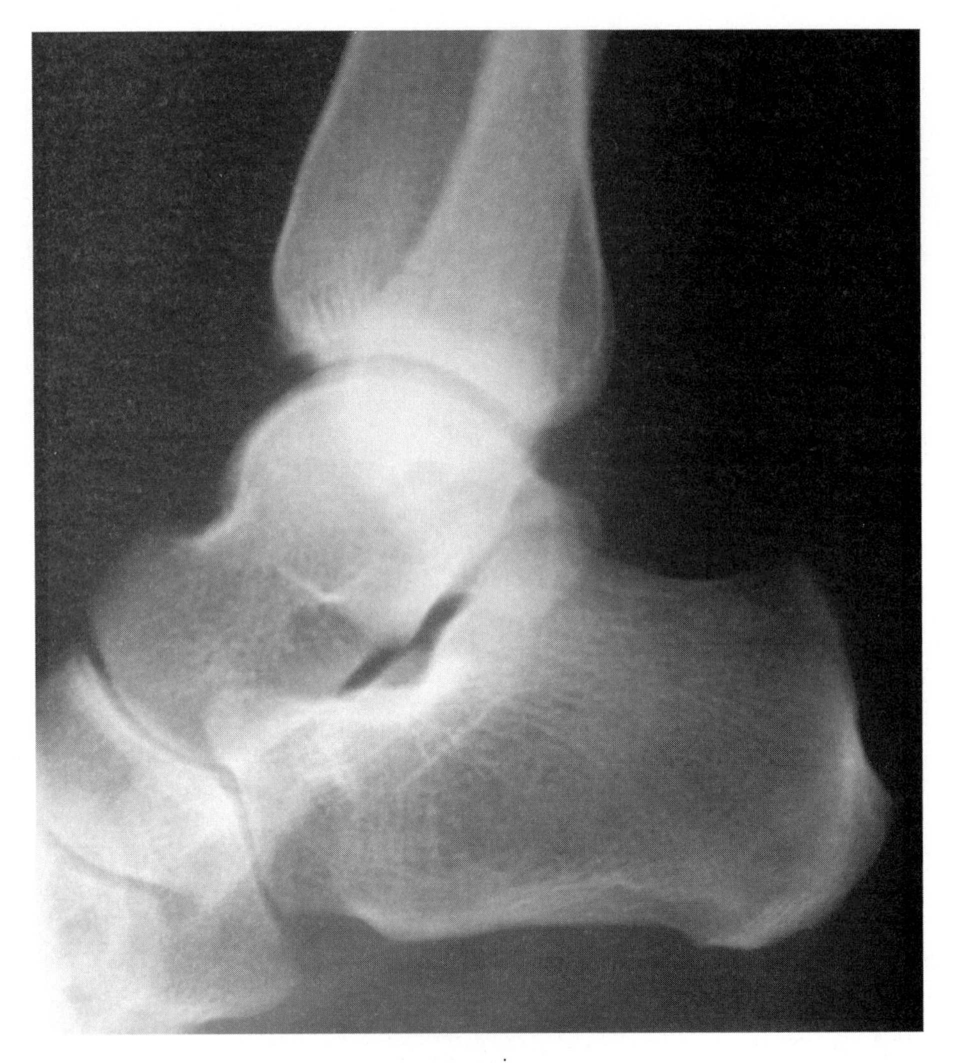

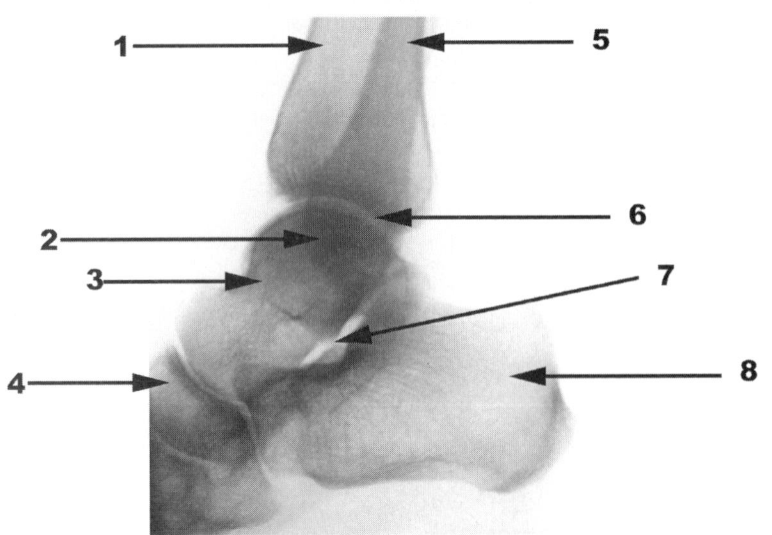

Figure 7

Lateral view of the ankle. 1 = Tibia, 2 = Talus, 3 = Talar neck, 4 = Navicular, 5 = Fibula, 6 = Tibiotalar joint, 7 = Subtalar joint, 8 = Calcaneus.

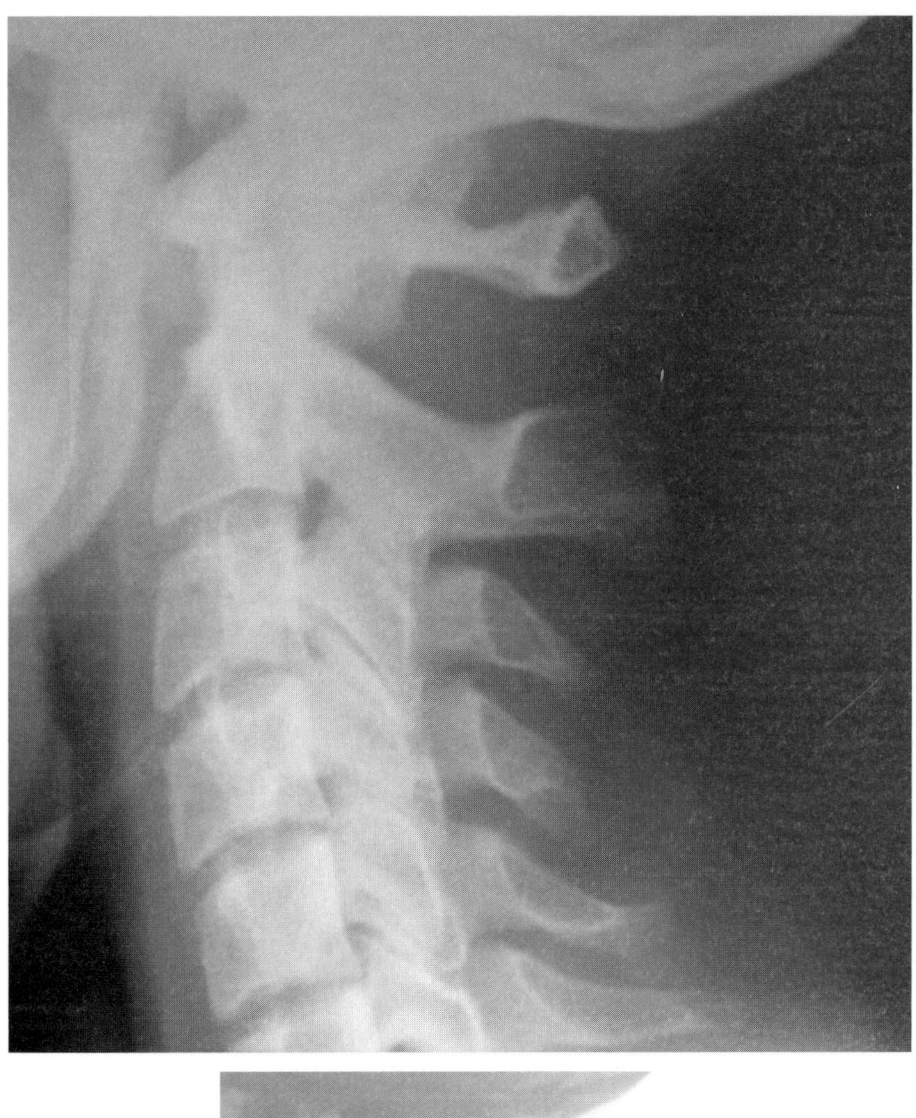

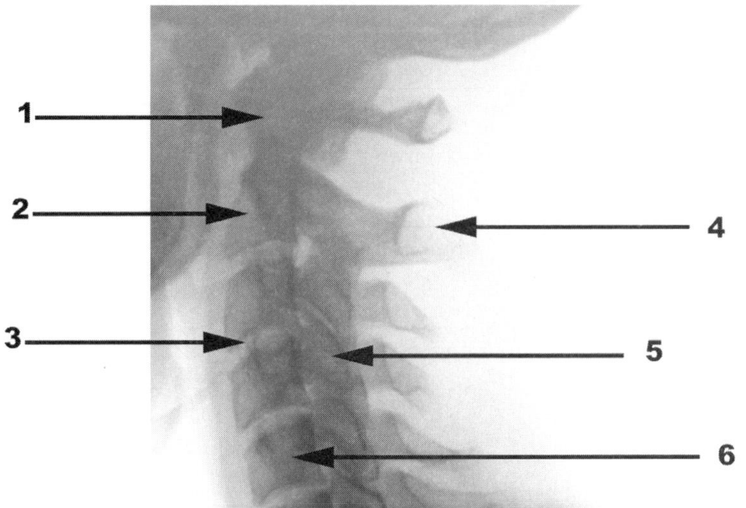

Figure 8
Lateral view of cervical spine. 1 = Dens of axis, 2 = Axis, 3 = Intervertebral disk, 4 = Spinous process, 5 = Posterior elements, 6 = Vertebral body.

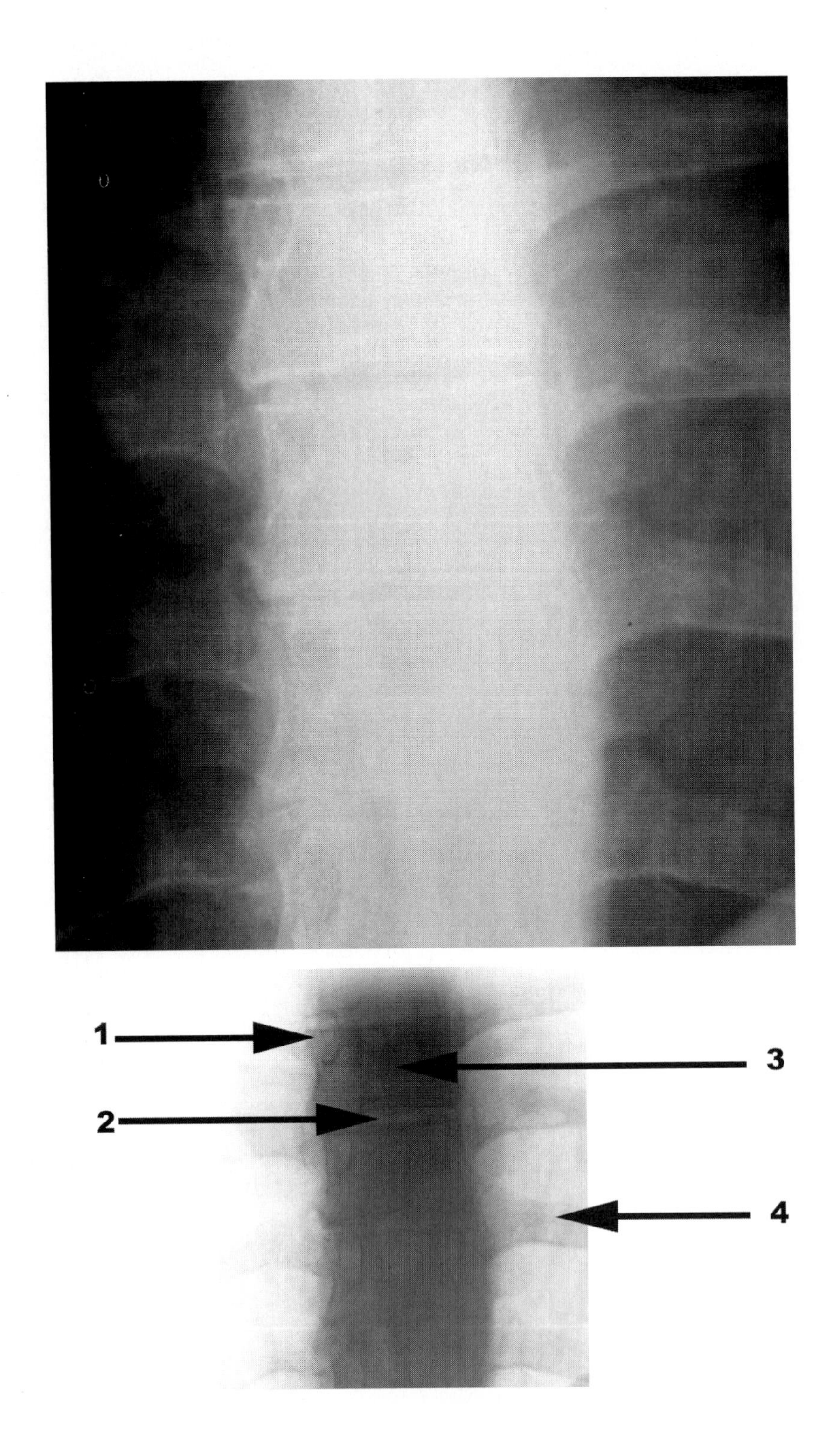

Figure 9
AP view of the thoracic spine. 1 = Pedicle, 2 = Intervertebral disk, 3 = Vertebral body, 4 = Rib.

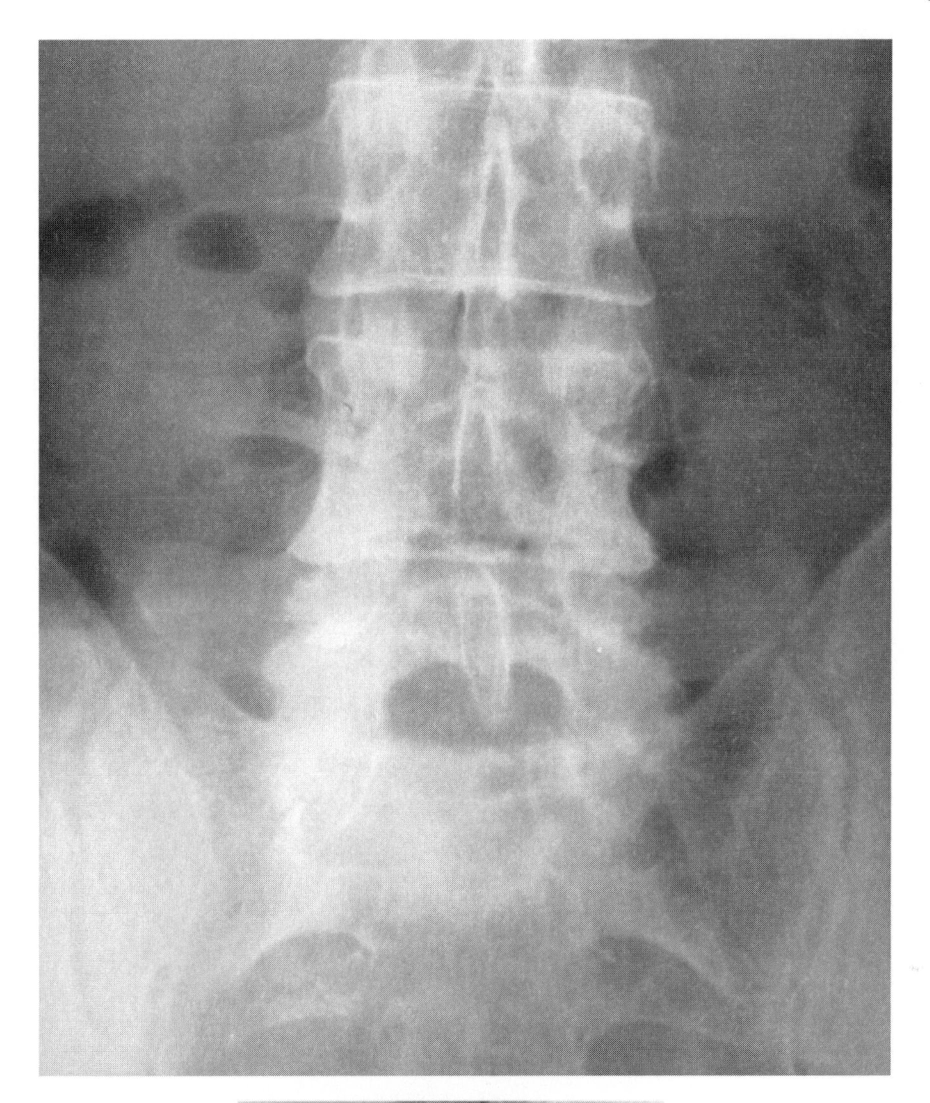

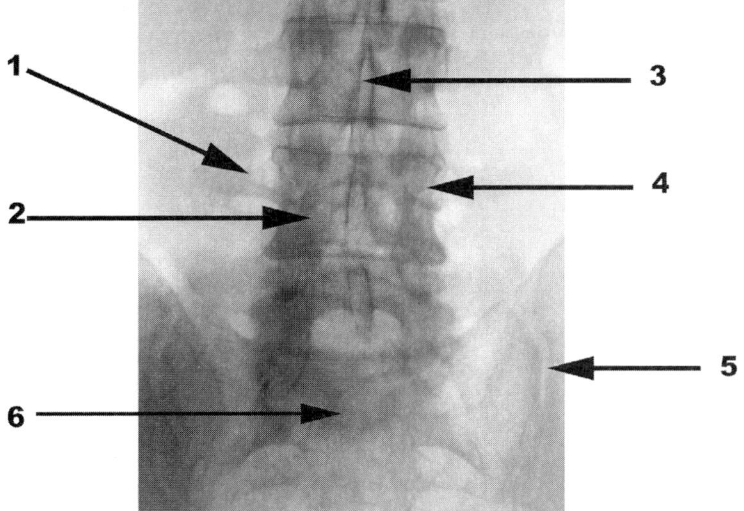

Figure 10
AP view of the lumbar spine. 1 = Transverse process, 2 = Vertebral body, 3 = Spinous process, 4 = Pedicle, 5 = Sacroiliac joint, 6 = Sacrum.

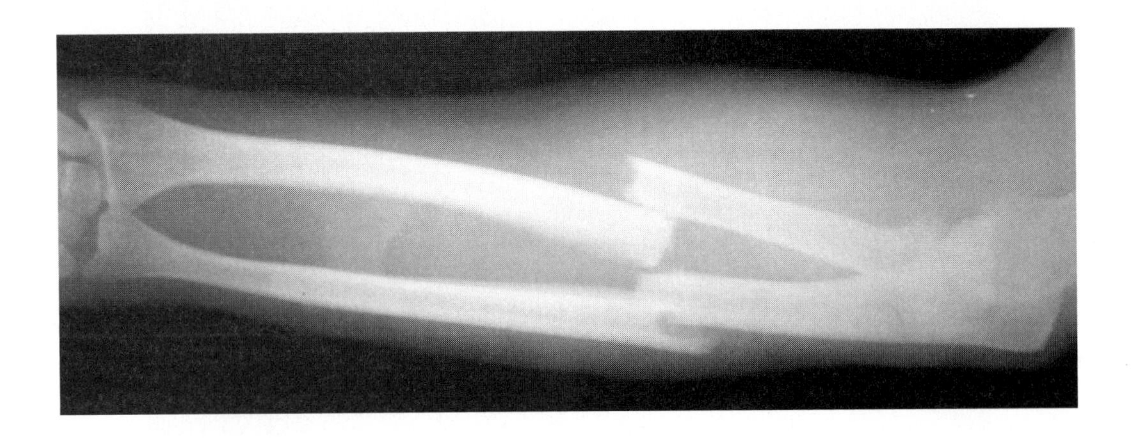

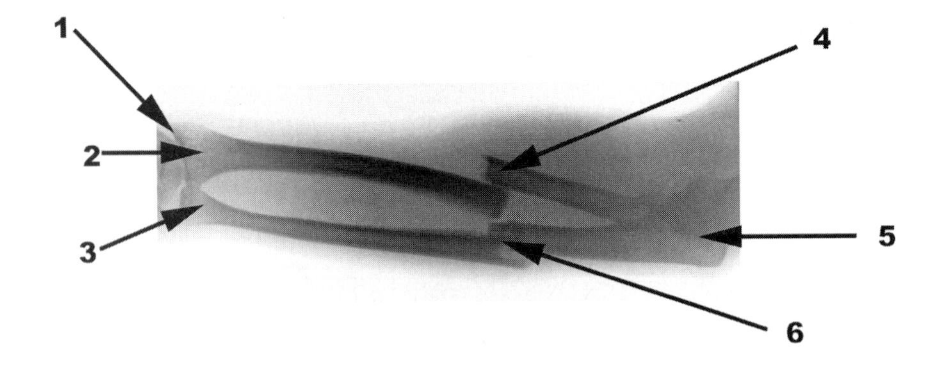

Figure 11
AP view of the forearm demonstrates midshaft radial and ulnar fractures with overriding and displaced fragments. 1 = Radiocarpal joint, 2 = Radius, 3 = Ulna, 4 = Radial fracture, 5 = Elbow joint, 6 = Ulnar fracture.

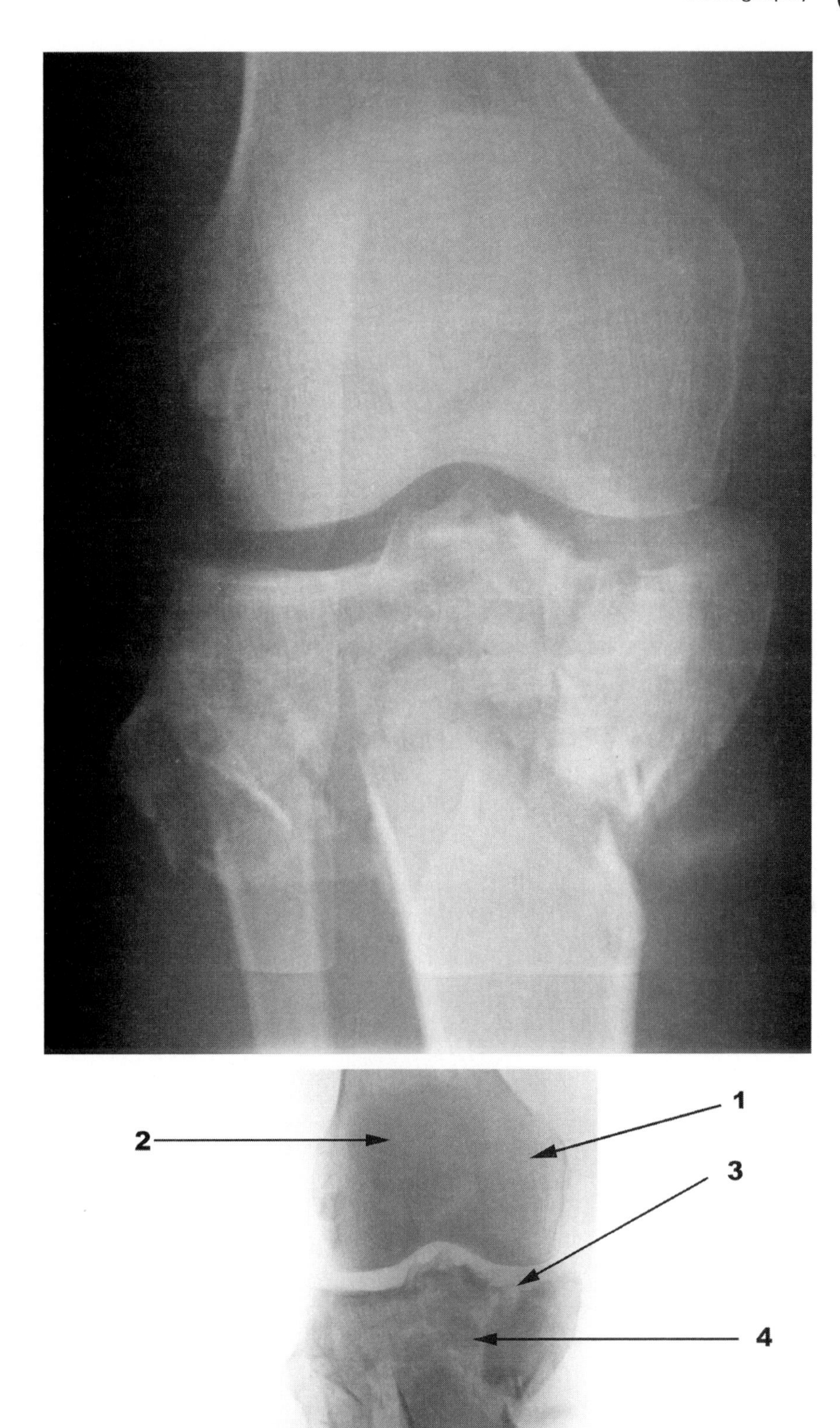

Figure 12
AP view of the knee shows a highly comminuted fracture of the proximal portion of the tibia (tibial plateau). The fracture involves the articular surface on both the medial and lateral sides of the tibia. A portion of the medial tibial articular surface is depressed. There is also a proximal fibular fracture. 1 = Femur, 2 = Patella, 3 = Intra-articular extension of fracture, 4 = Comminuted fracture, 5 = Fibula, 6 = Tibia.

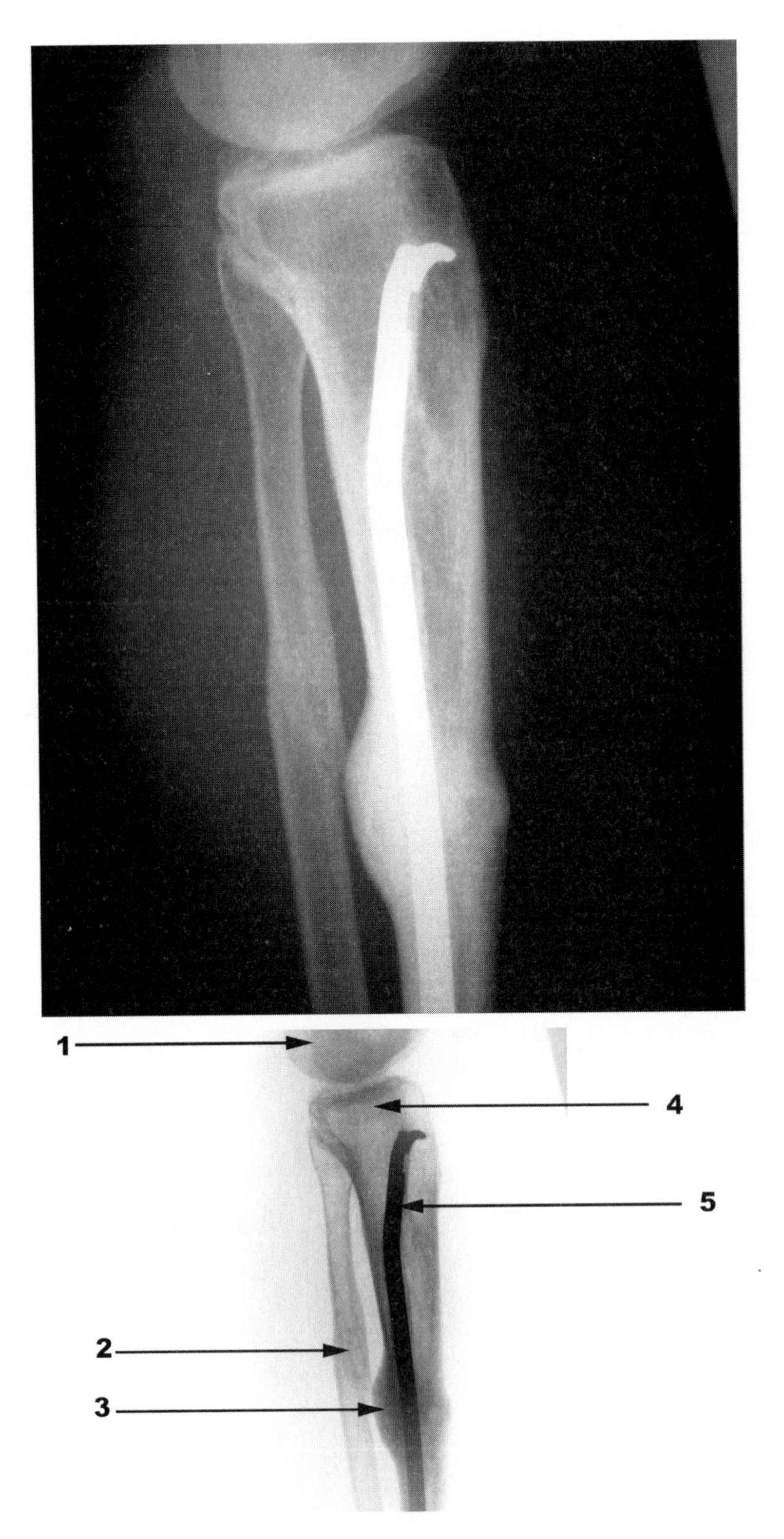

Figure 13
Lateral view of the leg in a patient with fractures of the tibia and fibula. An intramedullary rod was placed in the tibia. Callus at the fracture site is seen. 1 = Femur, 2 = Healed fibula fracture with callus, 3 = Healed tibia fracture with callus, 4 = Tibia, 5 = Intramedullary rod.

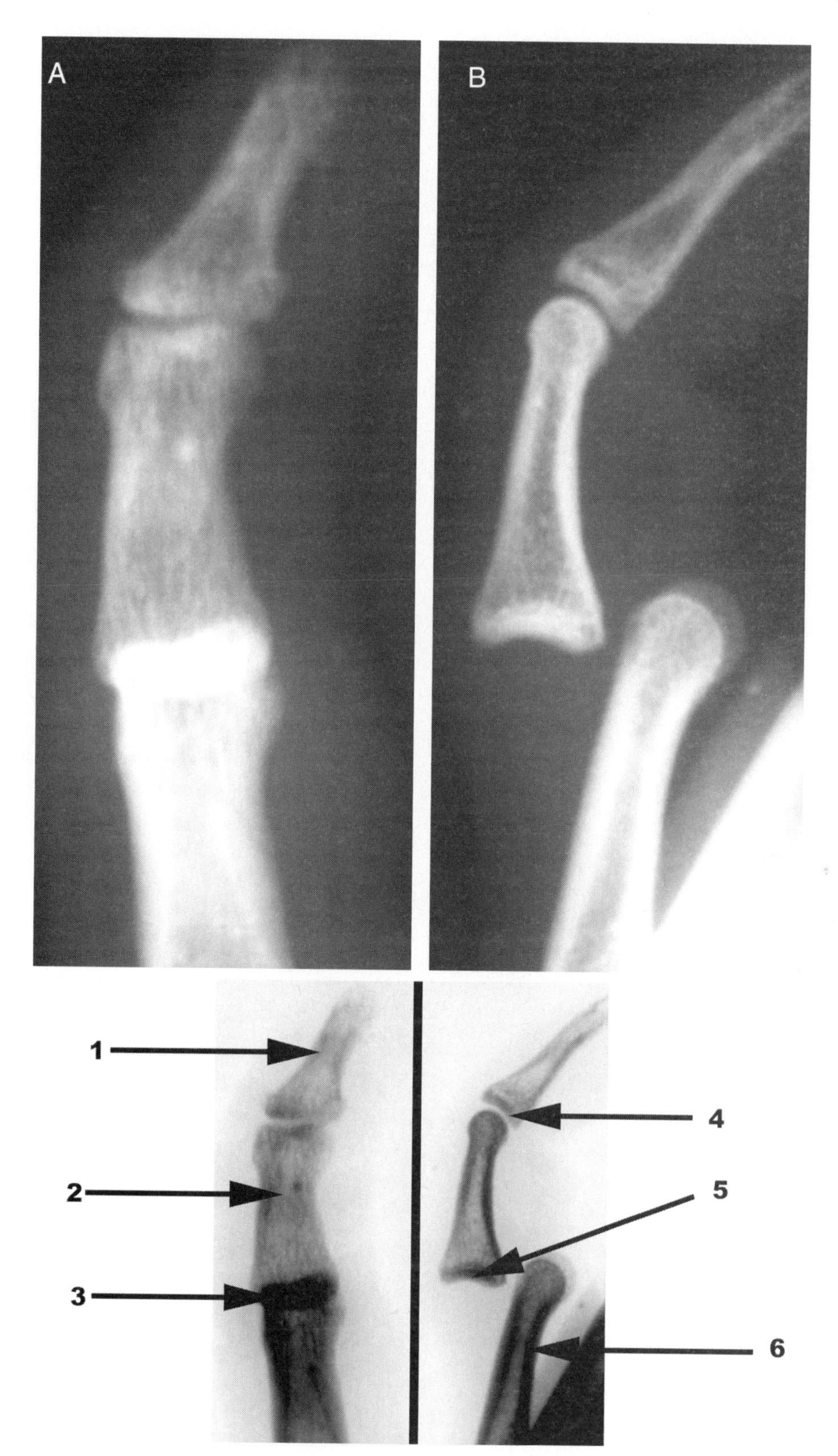

Figure 14
AP (**A**) and oblique (**B**) views of the little finger demonstrate dorsal dislocation of the fifth MCP joint. Although an abnormality may be inferred by the position of the bones on the AP view, the presence of the dislocation and its direction is best appreciated on the oblique view. This pair of radiographs demonstrates the need for two views. 1 = Distal phalanx, 2 = Middle phalanx, 3 = Abnormal proximal interphalangeal joint, 4 = Distal interphalangeal joint, 5 = Dislocated proximal interphalangeal joint, 6 = Proximal phalanx.

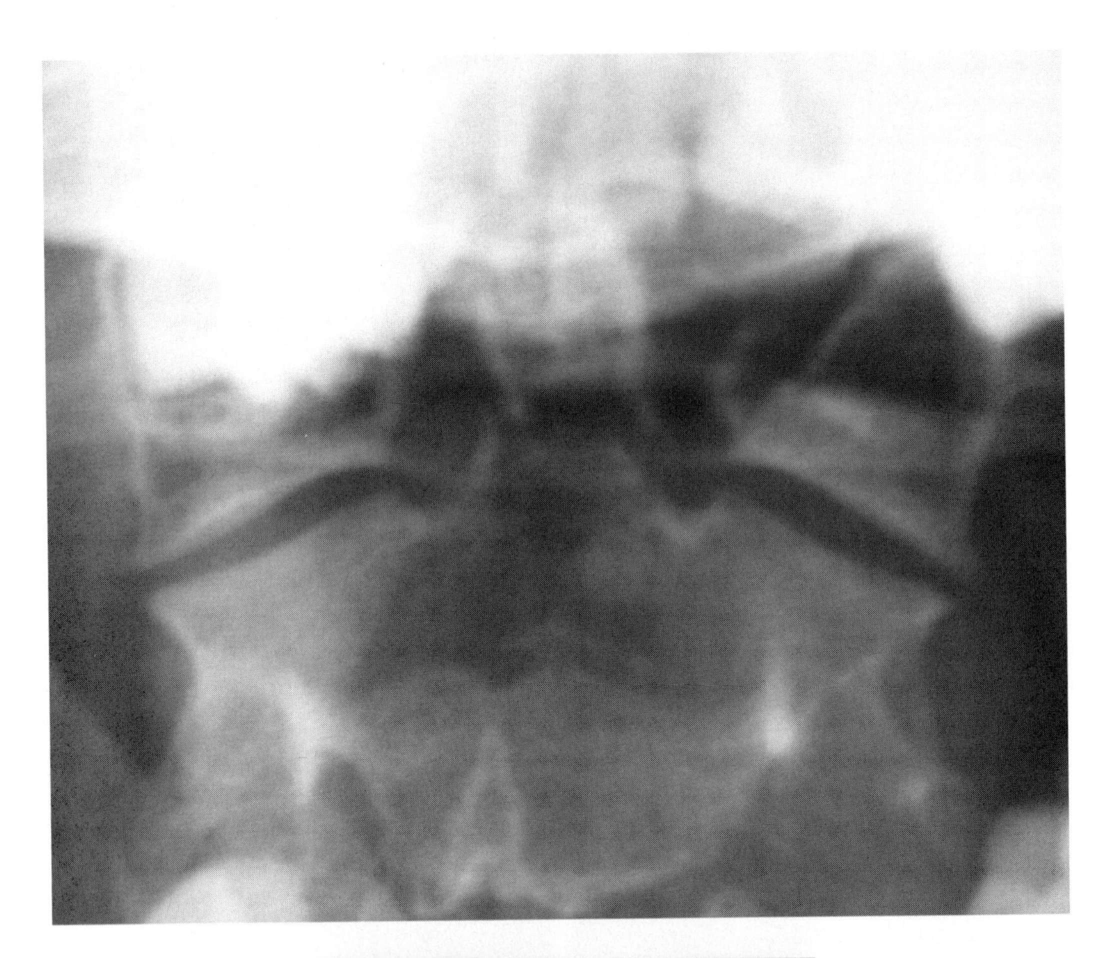

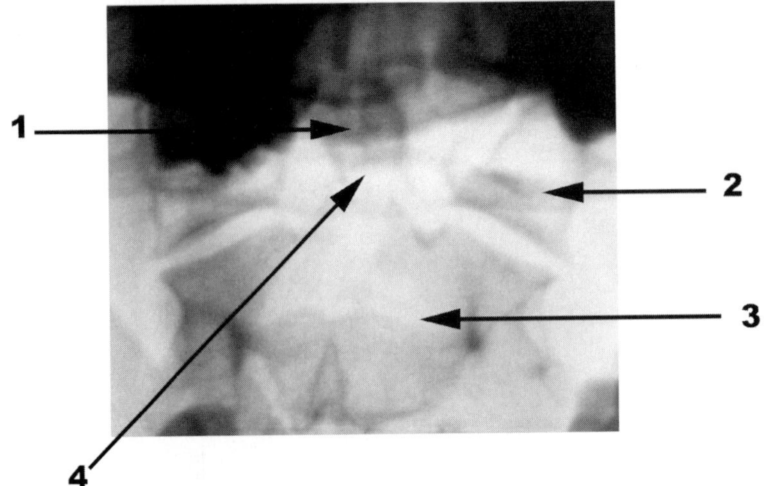

Figure 15
AP open-mouth view of the odontoid process (dens) shows a transverse fracture at the base of the dens at the same level as the joints of the C1-C2 lateral masses. 1 = Dens, 2 = Lateral mass, 3 = Vertebral body, 4 = Fracture.

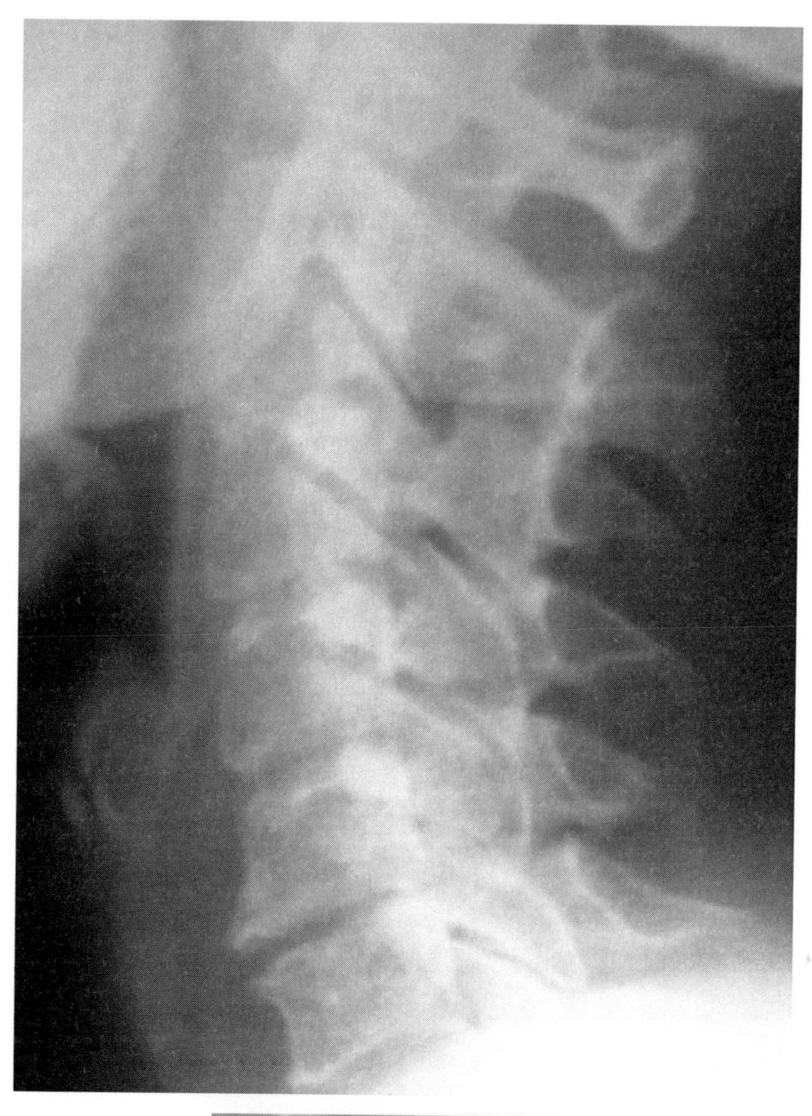

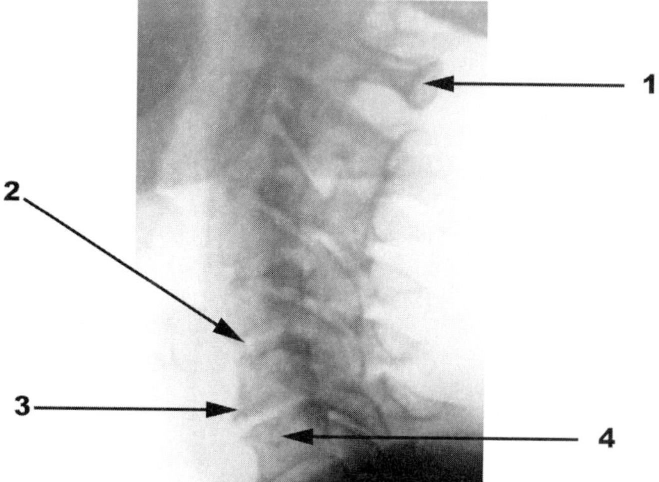

Figure 16
Lateral view of the cervical spine shows degenerative changes that include disk space narrowing, bone sclerosis of adjacent vertebral surfaces, and osteophytes. 1 = Spinous process, 2 = Intervertebral disk space narrowing, 3 = Osteophyte, 4 = Vertebral body.

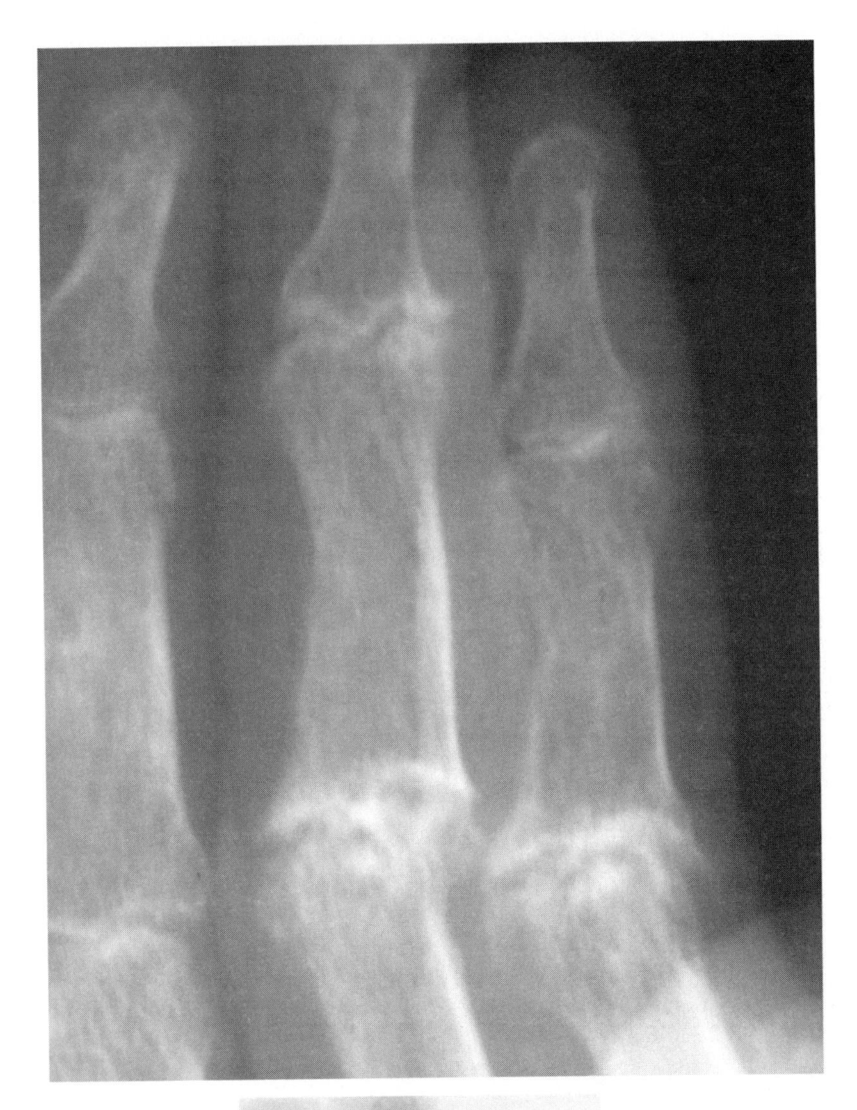

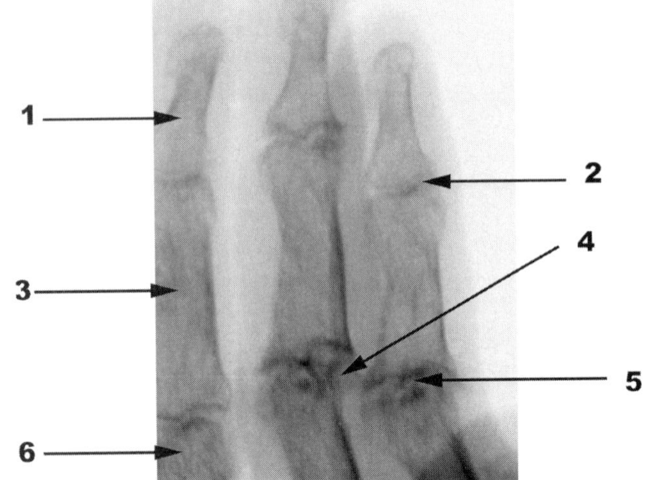

Figure 17
Oblique view of the fingers shows degenerative changes of the proximal interphalangeal and distal interphalangeal joints, including loss of joint space, subchondral sclerosis, and marginal osteophytes. 1 = Distal phalanx, 2 = Joint space narrowing, 3 = Middle phalanx, 4 = Osteophytes, 5 = Subchondral sclerosis, 6 = Proximal phalanx.

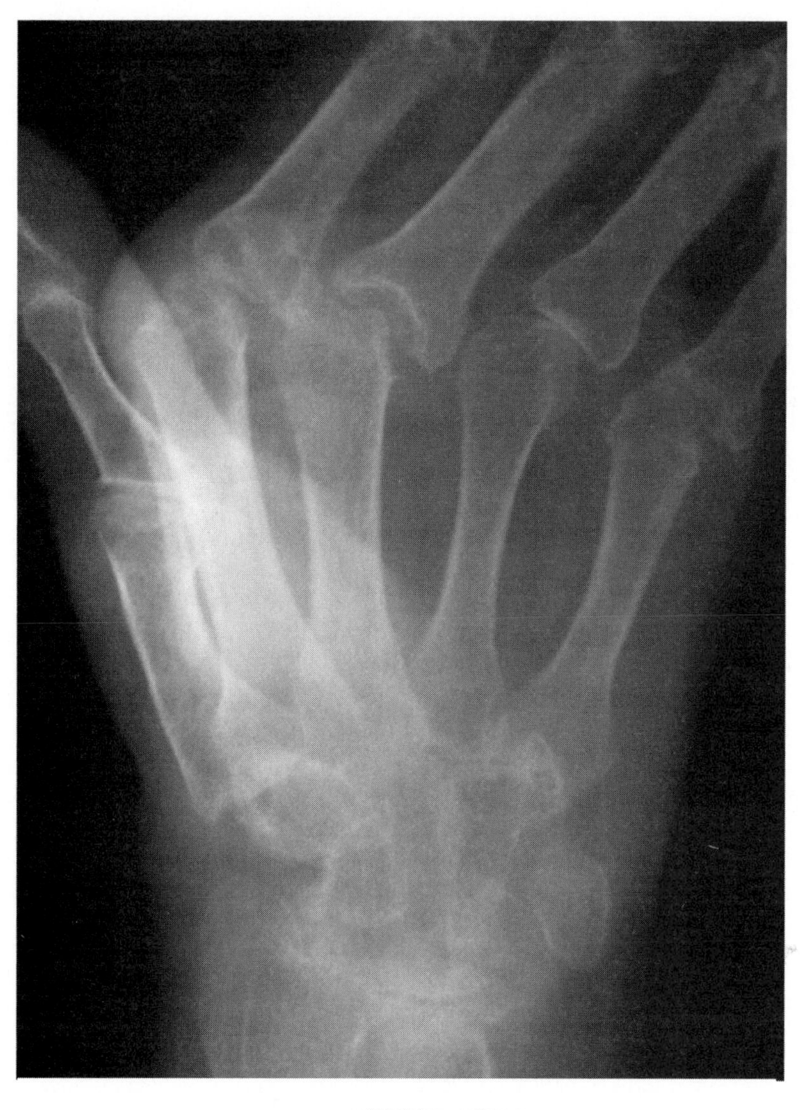

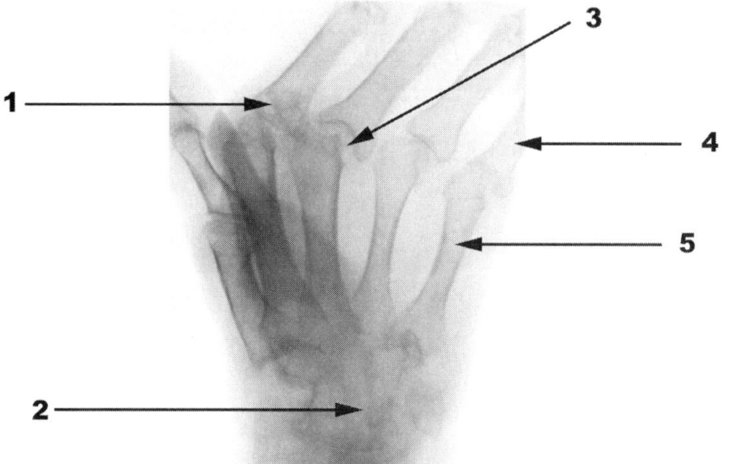

Figure 18
AP view of the hand demonstrates abnormalities of advanced rheumatoid arthritis characterized by obliteration of articular space at the metacarpophalangeal (MCP), carpometacarpal, midcarpal, and radiocarpal joints. Large osseous erosions and osseous defects are predominant at the MCP joints. Ulnar deviation and flexion at the MCP joints is seen. The thumb also is subluxated. Widespread abnormalities of rheumatoid arthritis assuming pancompartmental distribution are apparent in the wrist. Changes in the distal ulna include bony resorption and sclerosis. 1 = Subchondral cysts, 2 = Carpal arthritis and osteopenia, 3 = Subluxated MCP joint, 4 = Proximal phalanx, 5 = Metacarpal bone.

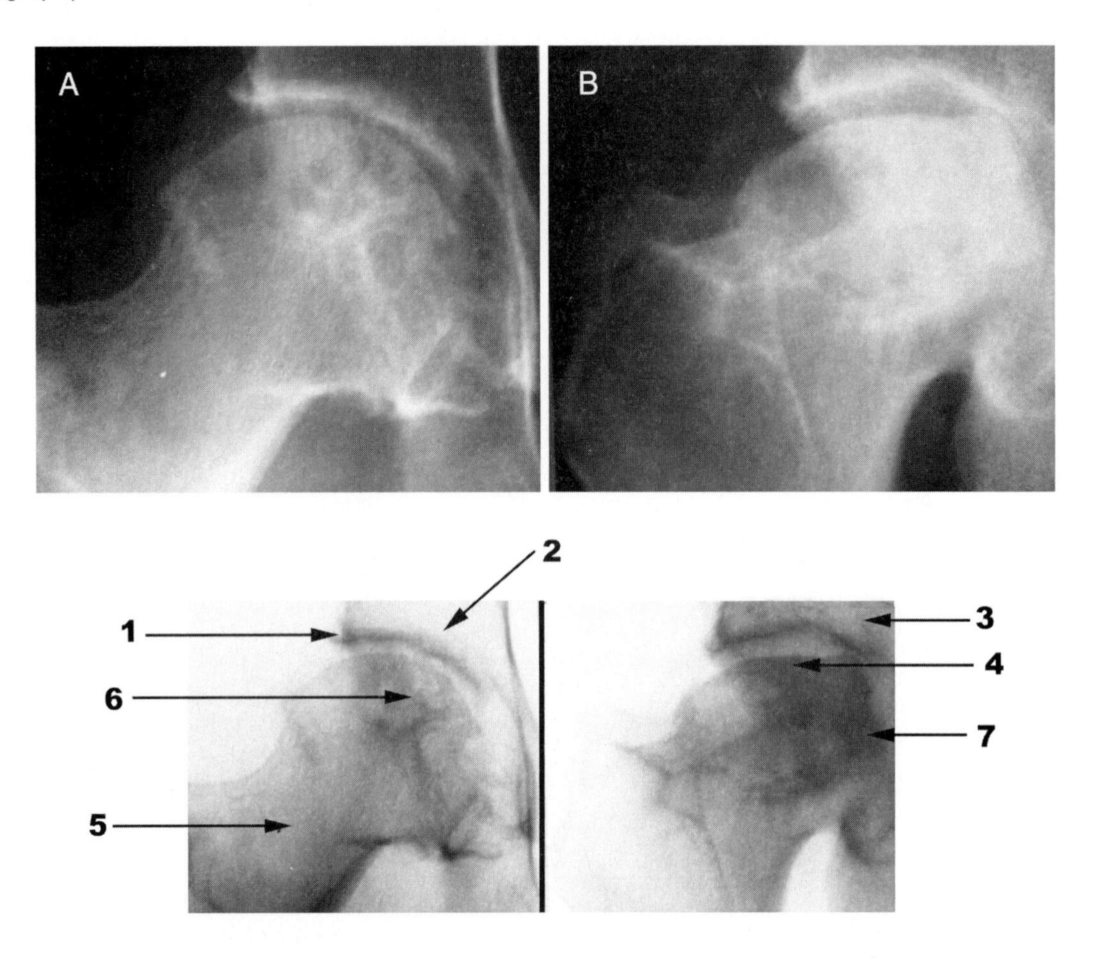

Figure 19

A, AP view of the hip shows changes of osteonecrosis that consist of cystic lucent areas and patchy sclerosis. No collapse of the articular surface is seen. The changes of osteoarthritis are apparent as well and include symmetric loss of joint space, osteophytosis in the femoral head and acetabular region, and sclerosis in the acetabular rim. **B**, In a different patient with more advanced disease, significant collapse of the superolateral aspect of the femoral head is seen. The joint space is not yet narrowed. 1 = Osteophyte, 2 = Sclerosis (acetabulum), 3 = Acetabulum, 4 = Collapse of femoral head, 5 = Femoral neck, 6 = Cyst in femoral head, 7 = Sclerosis (femoral head).

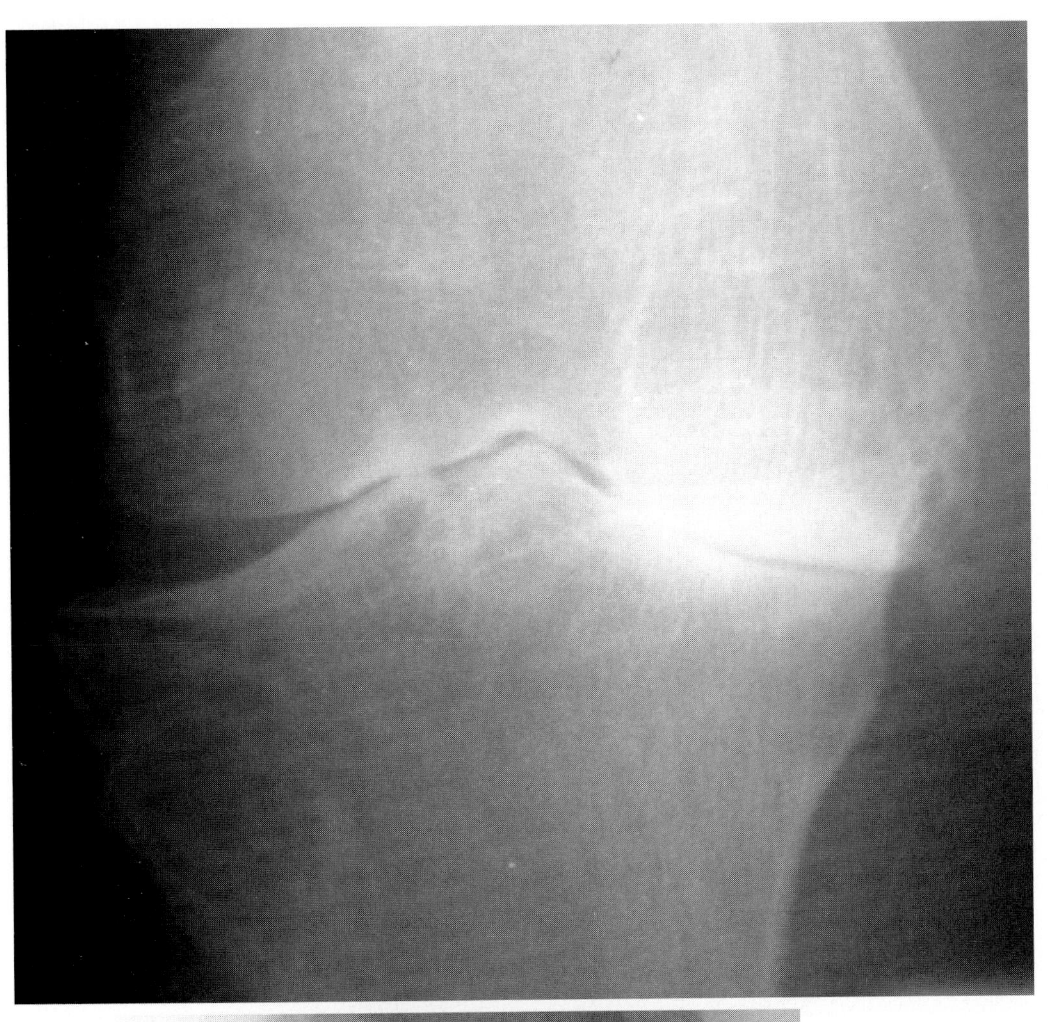

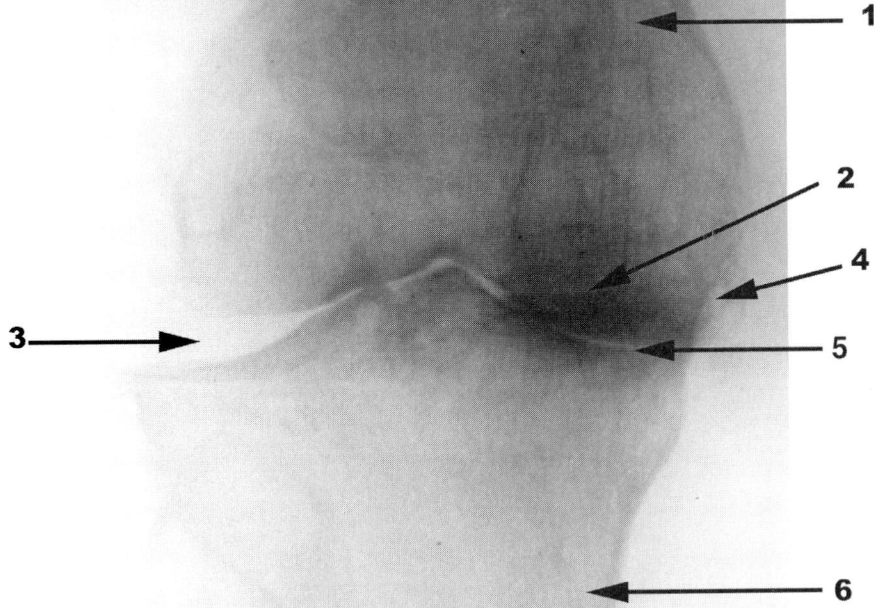

Figure 20
AP view of the knee reveals obliteration of joint space in the medial compartment. Additional findings of osteoarthritis include bone sclerosis and osteophytosis. The fourth cardinal sign of osteoarthritis, subchondral cyst formation, is not seen well on this radiograph. 1 = Femur, 2 = Subchondral sclerosis, 3 = Lateral joint space widening, 4 = Osteophytes, 5 = Medial joint space narrowing, 6 = Tibia.

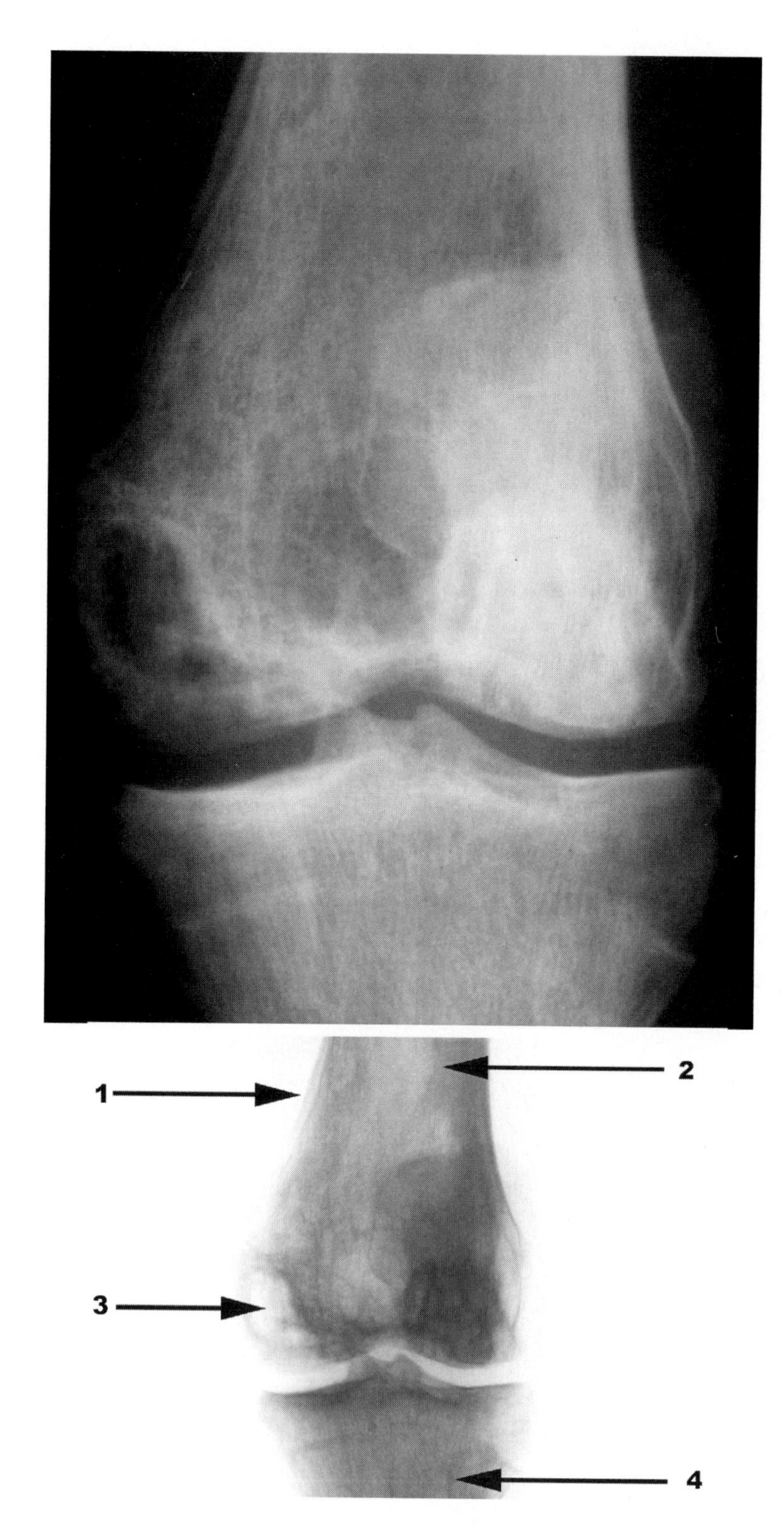

Figure 21
AP view of the knee in a patient with chronic osteomyelitis shows a Brodie's abscess of the medial femoral condyle. A metaphyseal, well-defined radiolucent lesion surrounded by a sclerotic margin with periosteal reaction is seen. 1 = Periosteal reaction, 2 = Femur, 3 = Abscess, 4 = Tibia.

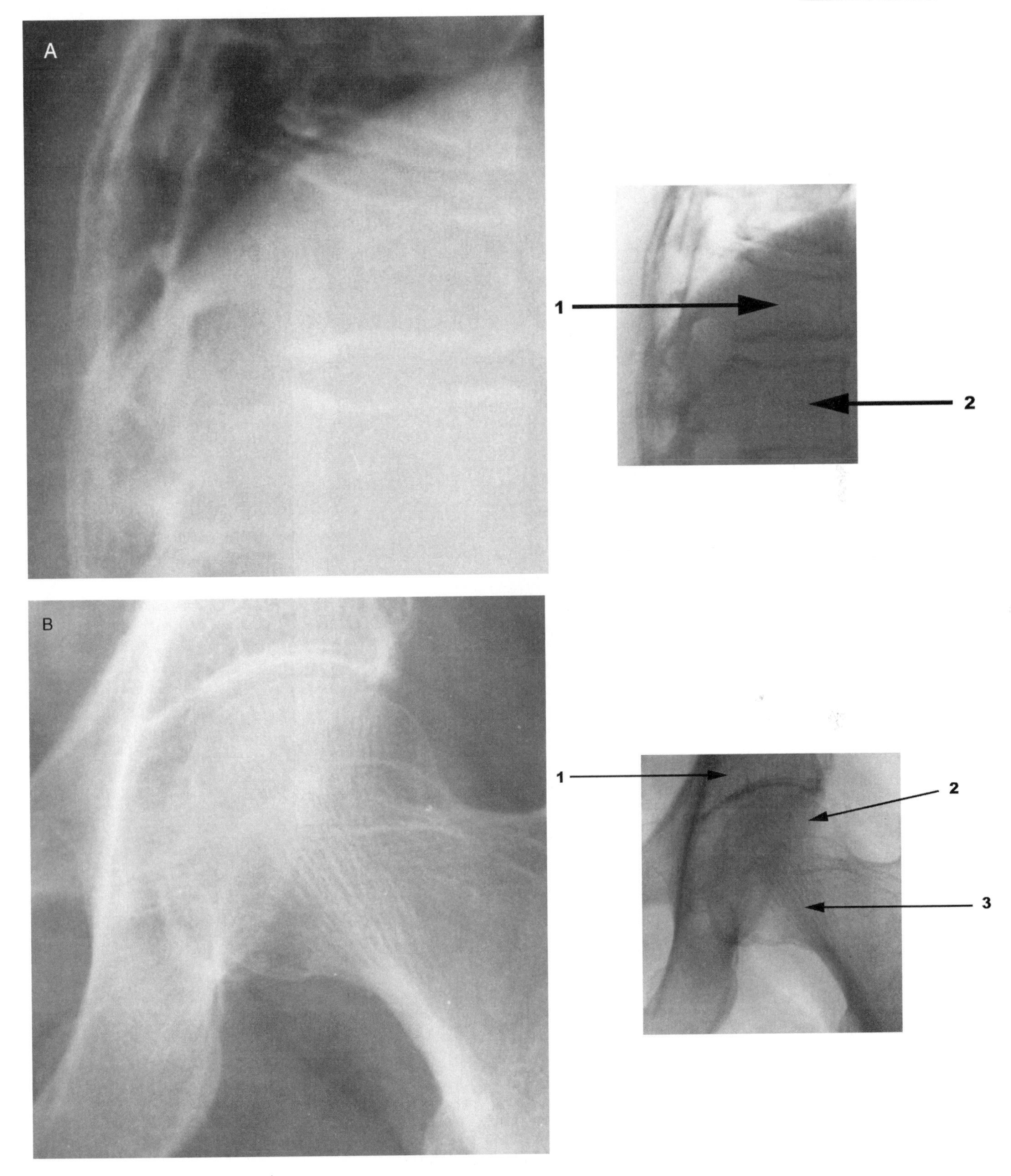

Figure 22
A, A wedge-shaped deformity associated with collapse of the anterior aspect of the vertebral body is seen in the T12 vertebra. 1 = Collapsed vertebral body, 2 = Comparatively normal vertebral body. **B**, AP view of the hip shows osteopenia of the femoral head and neck. The principal compressive trabecular group appears accentuated because all others are lost. 1 = Acetabulum, 2 = Femoral head, 3 = Compressive trabecular group.

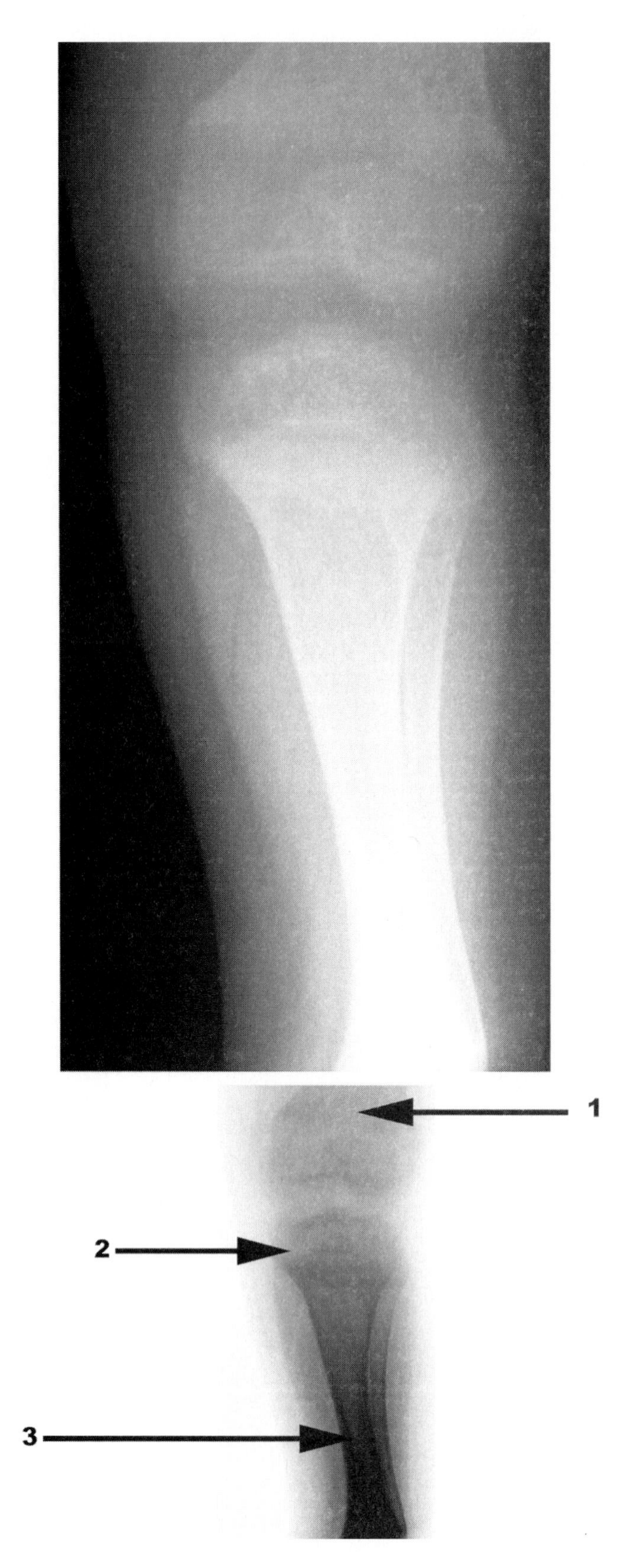

Figure 23

AP view displays bowing deformity in the lower extremity as a result of rickets. Additional findings include widening and cupping of the metaphyses, widening of growth plates, and osteopenia. 1 = Femoral physis, 2 = Abnormal tibial growth plate, 3 = Bowed tibia.

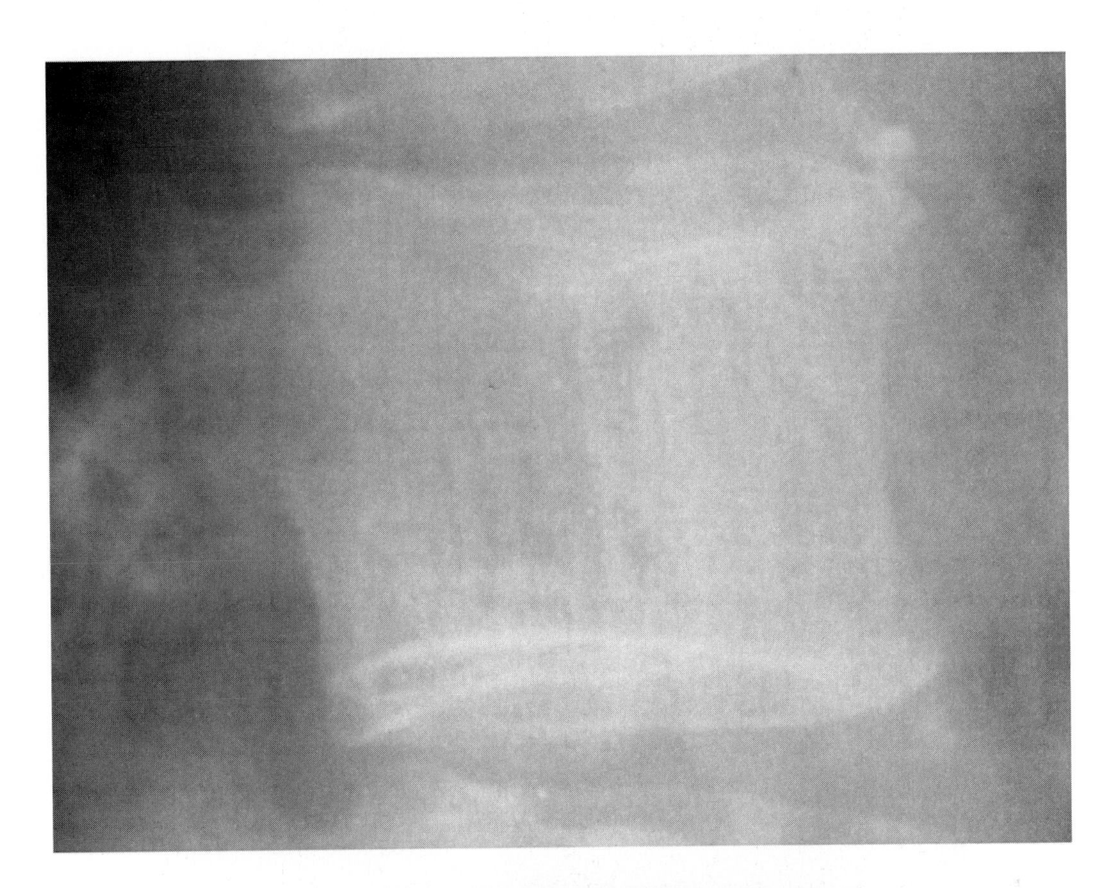

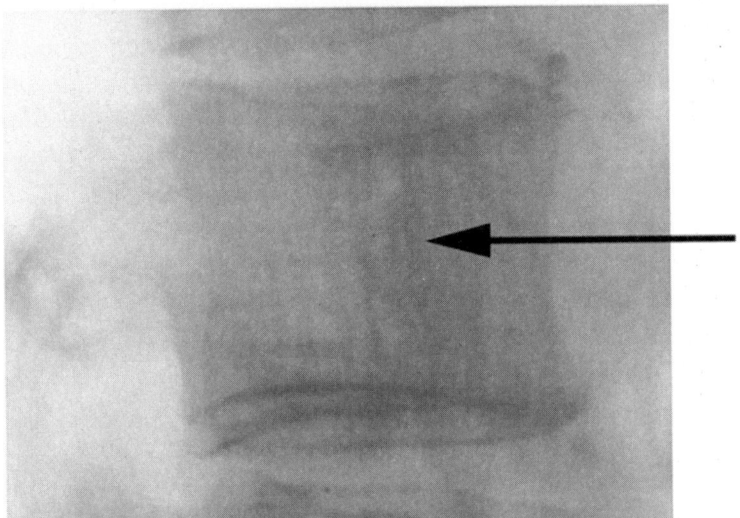

Figure 24
Lateral view shows osteopenia of the second lumbar vertebral body and the coarse trabecular pattern (arrow) characteristic of hemangioma.

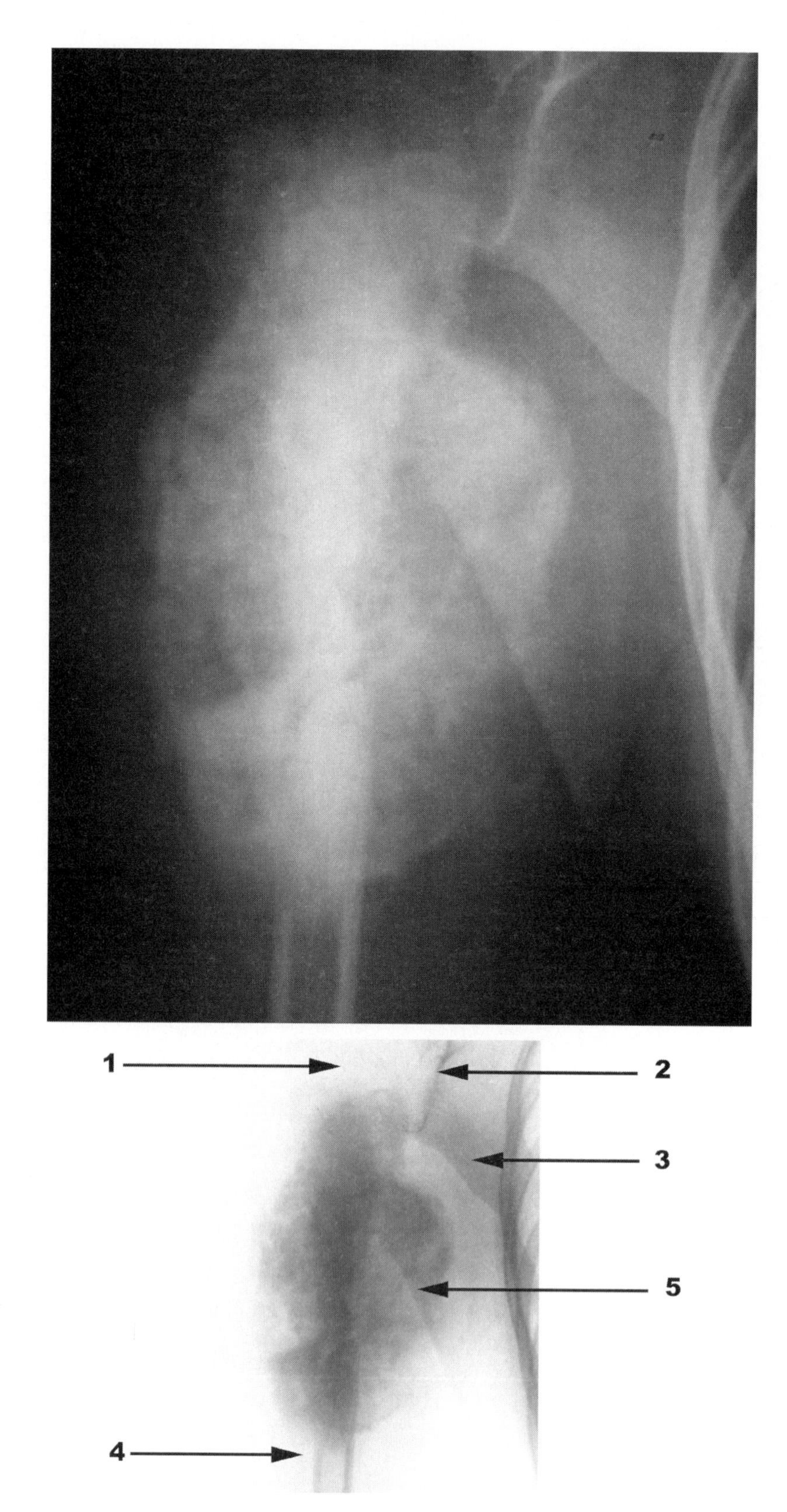

Figure 25

Osteosarcoma. AP view shows a large, mixed osteolytic and osteosclerotic lesion involving the proximal metaphysis and diaphysis of the humerus. 1 = Humeral head, 2 = Glenoid, 3 = Scapula, 4 = Periosteal reaction, 5 = Mass.

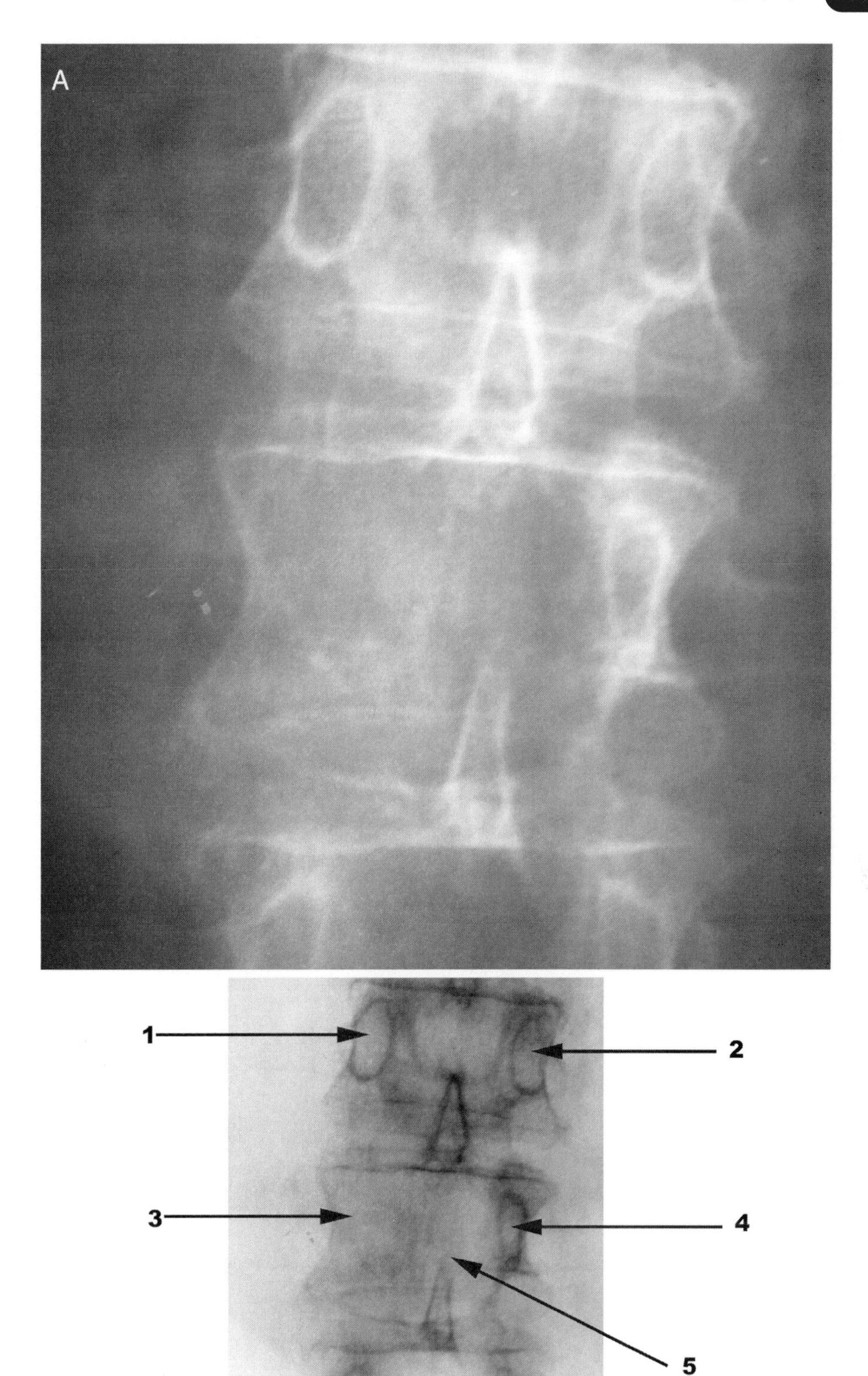

Figure 26

A, AP view reveals osteolytic lesion involving the pedicle, transverse process, lamina, and inferior articulating processes of L2. 1 = Right pedicle, 2 = Left pedicle, 3 = Absent right pedicle, 4 = Left pedicle, 5 = Vertebral body.

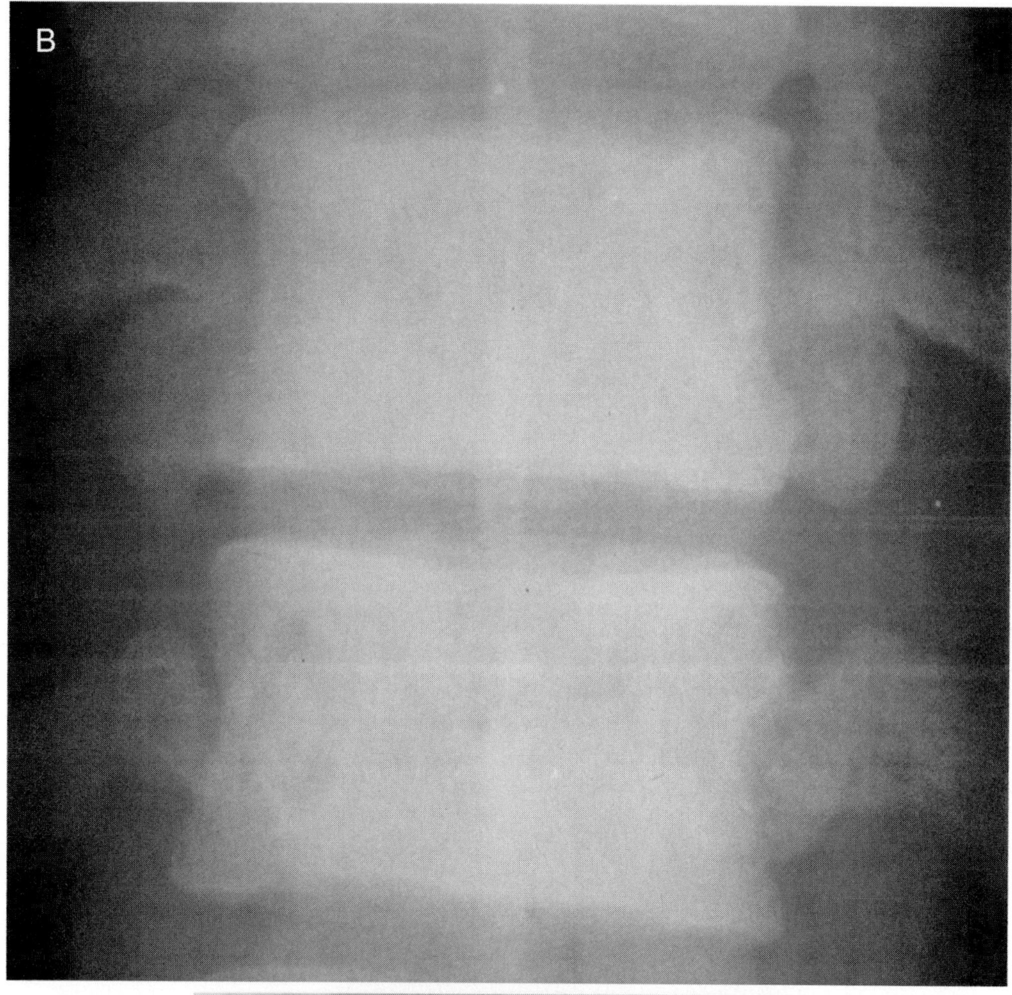

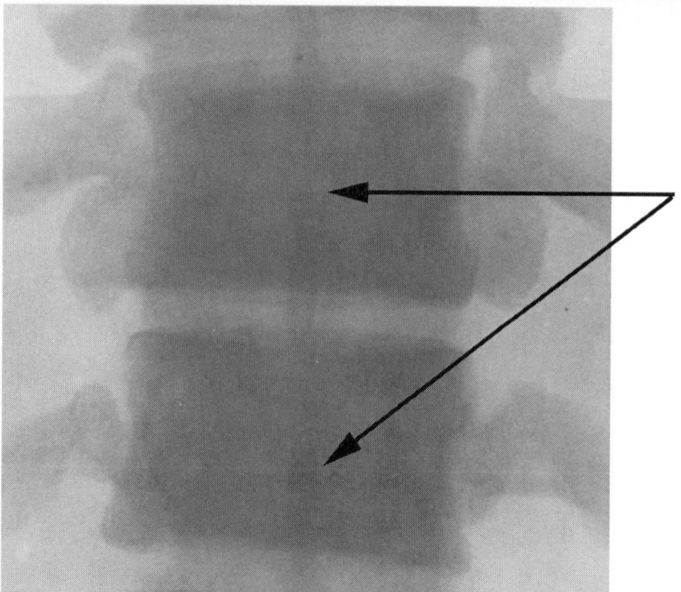

Figure 26
B, AP view shows uniformly increased radiodensity (arrows) of T10 and T11 related to metastases from carcinoma of the prostate. This appearance is known as the "ivory vertebral body."

Chapter 28

Special Imaging Studies

For the initial evaluation of many pathologic conditions involving the musculoskeletal system, radiography remains the mainstay.[1,2] However, special imaging studies such as MRI, CT, and ultrasonography are essential components of the diagnostic workup of patients with both complex disease processes such as bone tumors and simple ones such as cartilage tears.[3,4] Special imaging is also valuable for the evaluation of infections and comminuted fractures. Cross-sectional imaging studies may supplement radiography, providing more detailed information regarding the identification,localization, and characterization of abnormalities.[5-12] Bone scanning, ultrasonography, and interventional procedures can provide diagnostic information that may help evaluate bone and joint disorders or postoperative complications.[13,14] Many types of diagnostic imaging tools are available, but the capabilities, limitations, and potential pitfalls of the different imaging methods vary among them.

The purpose of this chapter is to present the basic concepts and principles of different special diagnostic imaging methods to help physicians select the appropriate imaging studies for their patients. To this end, a sample of imaging studies for common musculoskeletal conditions is presented and described to provide practical information regarding diagnosis and treatment of these problems.

Magnetic Resonance Imaging

MRI is a useful tool for evaluating neoplastic, traumatic, and inflammatory disorders of the musculoskeletal system, and it is exquisitely sensitive to the detection of abnormalities in bone marrow.[8-10] For example, MRI can detect subtle changes in osteonecrosis, early infections, and occult fractures.[10] Because MRI most clearly delineates the cartilage, menisci, ligaments, and soft tissues, it has become the standard method to evaluate these structures.[15-17] MRI can also assess the spinal canal and nerve roots.[18-21] In addition, because MRI detects water easily, blood flow in the major blood vessels can be clearly visualized without the administration of contrast agents.[4]

The advantages of MRI include its high sensitivity and superior contrast resolution, which, together with its multiplanar imaging capabilities, make it the method of choice for the precise delineation of anatomy and the characterization of a wide array of pathologic conditions (Figs. 1 through 15).[4-12] Advances in surface-coil technology have improved the spatial resolution of MRI and increased considerably its usefulness in the diagnosis of the internal derangements of joints.[6,11,22] Furthermore, the injection of contrast agents at the time of obtaining the MRI images—magnetic resonance (MR) arthrography—has significantly enhanced physicians' ability to visualize the joint capsule and small intra-articular structures.[5,23] Although MRI provides clear visualization of the various osseous and musculotendinous structures, its diagnostic utility may be diminished following internal fixation or joint replacement: metal produces artifacts that corrupt the image. The high cost and limited capability of MRI to detect small calcifications may also be considered among its drawbacks.[4]

Although analysis of the technical aspects of MRI is beyond the scope of this chapter, the general principles that apply in musculoskeletal imaging will be briefly discussed. The first idea is that MRI, unlike radiography or CT, does not use ionizing radiation. Rather, the machine detects electromagnetic signals released from a body part placed within a magnet and subjected to radio waves. MRI scans thus reflect the chemical properties of tissue. The signal characteristics of a tissue (which affect how bright the

Table 1
Relative MR Signal Intensity of Normal Tissues and Fluids

Tissue	T1-Weighted	T2-Weighted
Yellow marrow	High	Intermediate
Red marrow	Low	Intermediate
Cortical bone/calcification	Very low	Very low
Muscle	Intermediate	Low
Tendon/ligament	Low	Low
Anulus fibrosus	Low	Low
Nucleus pulposus	Intermediate	High
Articular cartilage	Intermediate	Low
Nerve	Low	Low
Vessels/cerebrospinal fluid	Low	High
Fat	High	Intermediate

tissue appears) depend on both the nature of the tissue and the settings of the machine. Tissue parameters include hydrogen (water) density and two magnetic relaxation constants: the longitudinal constant (T1) and the transverse constant (T2). Each tissue is characterized by its own hydrogen density and T1-weighted and T2-weighted constants and therefore has its own unique fingerprint (Table 1). Instrument parameters that are varied by the technician include the pulse-repetition time (TR) and the echo-delay time (TE). Spin-echo sequences with short TR and TE are said to be T1-weighted. These can be obtained quickly, providing excellent anatomic detail. Sequences with long TR and TE are said to be T2-weighted and are helpful in the assessment of hemorrhages and changes in water content of the muscles, tendons, ligaments, and other soft tissues. As a rule, fatty marrow shows high signal intensity on T1-weighted sequences, and hematopoietic marrow shows much lower signal intensity on T1-weighted sequences.[4]

T1- and T2-weighted spin-echo sequences are widely used for imaging of the musculoskeletal system. Other combinations such as the gradient echo, short tau inversion recovery (STIR), and fast spin-echo sequences have increased the sensitivity of MRI for demonstrating alterations in bone marrow.[4] Because the STIR sequence does not report the signal derived from fat and highlights any abnormal signal derived from other sources, marrow edema is readily depicted on the STIR sequence. Fat suppression techniques and the use of intravascular contrast agents can increase tissue contrast, improve image quality, and enhance the sensitivity of MRI in the detection of abnormalities affecting bones and soft tissues.[4]

Computed Tomography

CT is a planar, transaxial imaging method that can provide useful diagnostic information regarding a variety of disorders of the musculoskeletal system.[3] CT scanning involves a highly focused (collimated) x-ray source, efficient detectors, and a data processing system.[24] The x-ray source and detectors are mounted on a frame (gantry) that is positioned around the patient who moves through the unit. The x-ray tube, in turn, moves in spiral fashion around the patient during the acquisition of a single CT slice. As the tube rotates, the x-ray beams are emitted from the source, and the amount of radiation transmitted through the tissue is detected. Data are processed by a computer, which uses an algorithm to reconstruct the image. A description of technical details of CT is beyond the scope of this chapter, but a nice summary appears at http://www.nobel.se/medicine/laureates/1979/press.html, the press release announcing the 1979 Nobel Prize awarded to Cormack and Hounsfield for the development of CT.

In a CT image, each picture element (pixel), or square, represents a piece of tissue—specifically that tissue's ability to attenuate the x-ray beam's passage through it. Each square is assigned a CT number (Hounsfield unit) that corresponds to the attenuation coefficient. CT numbers provide characterization of the nature of the tissues imaged[3] (Table 2).

Because of its cross-sectional display, excellent contrast resolution, and ability to

Table 2
CT Numbers for Various Tissues

Tissue	CT Number (HU)*
Bone	1,000
Blood	40
Muscle	10 to 40
Cerebrospinal fluid	15
Water	0
Fat	−50 to −100
Air	−1,000

*HU = Hounsfield Units

Table 3
Comparison of MRI and CT Sensitivity

MRI more sensitive
Bone marrow disease
Disk disease
Extent of bone tumors
Extent of soft-tissue tumors
Ligament and tendon injuries
Articular cartilage lesions
Metastatic disease
Osteomyelitis
Septic arthritis
Soft-tissue infection
Joint effusion
Pigmented villonodular synovitis
Postoperative scarring
Spinal metastases
Subacute/chronic hematoma
Osteonecrosis
Occult fracture
Compression neuropathy
Compartment syndrome
Brachial plexopathy
CT more sensitive
Osteolysis
Small calcified lesions
Degenerative bony abnormalities
Complex fracture
Intra-articular bodies
MRI and CT approximately equivalent
Advanced lumbar disk disease
Spinal stenosis

measure specific attenuation values, CT can define alterations in soft tissue and bone that may not be visualized on radiographs. The inherent high resolution of CT allows fine discrimination of subtle changes in cortical bone. Because CT can detect density so well, it is considered superior to MRI in the assessment of calcification, ossification, cortical destruction, and endosteal or periosteal reaction (Table 3).

Because CT images are produced in computers, it is possible to reformat the transaxial data into coronal, sagittal, or even three-dimensional images.[3] In the musculoskeletal system, three-dimensional display is particularly useful in imaging regions of complex anatomy, such as the pelvis, spine, shoulder, wrist, knee, midfoot, and hindfoot.[25] Detailed delineation of anatomy aids in surgical planning. CT is also a valuable diagnostic tool for evaluating fractures and dislocations; intra-articular bodies; infection, such as osteomyelitis and septic arthritis; neoplasms; joint disease; vascular lesions, such as aneurysms and arterial entrapment syndromes; congenital or metabolic disease; low back pain; and complex fractures and dislocations of the spine[3,25,26] (Figs. 16 through 19). In patients with dislocation of the sternoclavicular and glenohumeral joints, for instance, CT can provide a perspective that cannot be obtained using standard radiography.[27] In patients with musculoskeletal neoplasms, CT is used in surgical planning.[3]

The intravenous, intraspinal, or intradiskal administration of a radiopaque contrast agent or the injection of a radiopaque contrast agent or air into a joint coupled with CT can be helpful in the assessment of the vascularity of soft tissue or osseous lesions; the identification of degenerative or recurrent herniated intervertebral disks; and the evaluation of the glenoid labrum, patellar cartilage, synovial plicae, and cruciate ligaments of the knee.[3] In the postoperative setting, the quality of CT images may be compromised because of the artifacts created by metal components, screws, and other orthopaedic implants.[28] However, valuable information can still be obtained regarding bone stock, marrow, and soft tissue. CT scanners can also generate digital radiographs by sampling at intervals from the x-ray detectors. From these digital images, accurate measurements can be obtained that are useful in the assessment of conditions such as leg-length discrepancy.[29,30] CT is also capable of performing quantitative measurements of bone mineral content, a test used for the evaluation of metabolic bone diseases.[31,32]

Nuclear Medicine

Radionuclide bone scanning is an imaging method in which bone scans are produced using radiopharmaceutical agents that accumulate in areas of pathology. After intravenous injection of a radionuclide agent, imaging data are collected with a gamma camera. Bone scanning is based on accumulation of the radioactive tracer at the sites of abnormalities.[13] The increased concentration of tracer can be seen in a variety of conditions that lead to bone production and resorption, such as bone tumors, infections, neuropathic osteoarthropathies, fractures, and postoperative changes.[13,33] In this regard, bone scanning is considered highly sensitive but relatively nonspecific for evaluating a multitude of pathologic processes.

In the musculoskeletal system, radionuclide bone scanning is particularly useful for detecting and localizing metastatic disease, diagnosing osteomyelitis in its early stages (when it may be confused with cellulitis), assessing whether joint prostheses are loose or infected, evaluating peripheral vascular disease, and defining the age of fractures.[13] Chronic fractures, such as osteoporosis-related compression fractures, do not need aggressive treatment and thus the age of the fracture—acute versus chronic—is important. In addition, bone scanning can assist in the diagnosis of conditions such as child abuse, metabolic bone disease (eg, osteomalacia and renal osteodystrophy), Paget's disease, osteonecrosis, arthritis, complex regional pain syndrome, heterotopic ossification, and various soft-tissue lesions.[10] Other uses of radionuclide bone scanning include the evaluation of stress fractures and the assessment of fracture healing and nonunion[13] (Figs. 20 and 21).

In most patients with metastatic bone disease, foci of increased accumulation of the radionuclide ("hot" lesions) are observed on radionuclide bone scans. In whole-body imaging of these patients, the scintigraphic pattern of a "superscan" is seen when the metastatic bone disease is widespread; marked increased radionuclide uptake by all the bones in the body is evident.[13,33] In some instances, however, skeletal metastases may appear as "cold" (ie, photon-deficient) areas on radionuclide bone scans.[13]

A more recent advance in radionuclide bone imaging is the single-photon emission computed tomography (SPECT) camera, which displays tomographic slice images. The camera rotates 360° around the patient, acquiring data from which images are generated. Images can then be viewed in coronal, sagittal, and transaxial projections.[13] SPECT has been shown to provide useful diagnostic information in the evaluation of regions with complex anatomy, such as the spine, knee, and facial bones.

Arthrography and Magnetic Resonance Arthrography

Arthrography is the broad category of medical imaging in which radiography is performed after the injection of opaque contrast material into the joint cavities. It comprises not only radiography but also fluoroscopy, conventional tomography and CT (arthrotomography), and digital radiography and MRI. During single-contrast technique, either radiopaque (dye) or radiolucent (air) contrast material can be used. Double-contrast technique refers to imaging obtained after using both radiopaque and radiolucent contrast materials.[5]

Arthrography is used to evaluate the shoulder, hip, knee, ankle, and wrist. In the glenohumeral joint, arthrography is useful in diagnosing rotator cuff tears, adhesive capsulitis, bicipital tendon abnormalities, and ligament disruption from previous dislocation.[5] In the hip, arthrography is used to assess developmental dysplasia of the hip, septic arthritis in infants, Legg-Calvé-Perthes disease, trauma, pigmented villonodular synovitis, and synovial osteochondromatosis.[34] In the knee, arthrography is used to evaluate meniscal abnormalities, ligamentous injuries, osteochondral fractures (osteochondritis dissecans), synovial plicae and cysts, different types of arthritis, and synovial osteochondromatosis. In the ankle, arthrography is used to detect ligamentous injuries, transchondral fractures, and adhesive capsulitis.[23] Arthrography of the wrist is used to demonstrate degenerative changes and tears of the triangular fibrocartilage, detect osteocartilaginous fragments and ganglion cysts, evaluate the changes of rheumatoid arthritis, and determine intercompartmental communications.

MR arthrography combines the direct intra-articular injection of a gadolinium compound with MRI. MR arthrography has been used in the glenohumeral joint for identifying rotator cuff tears. Other potential diagnostic applications of MR arthrography of the shoulder include diagnosing impingement syndrome, analyzing the articular cartilage of the glenohumeral joint, and evaluating labral abnormalities and postoperative changes.[5] Although the role for MR arthrography of the elbow is not defined, the technique allows full distention of the joint, thereby facilitating the delineation of anatomy.[35] MR arthrography also may be useful in the identification of intra-articular osteocartilaginous bodies, chondral defects, and partial deep ligamentous tears. The main advantages of MR arthrography of the wrist consist of distention of the joint capsule and delineation of the joint space. MR arthrography has been used for investigating chronic wrist pain and showing the intrinsic and extrinsic ligaments of the wrist. In the hip, MR arthrography is useful in the assessment of acetabular labral tears and loose bodies.[36] In the knee, MR arthrography may be helpful in detecting and characterizing meniscal tears, showing articular cartilage abnormalities, and evaluating the postoperative menisci for recurrent lesions[37] (Fig. 22). In the ankle, MR arthrography is used for delineating ligamentous injuries, assessing abnormalities of the tarsal sinus, and evaluating cartilage abnormalities.[23]

Diskography and Magnetic Resonance Diskography

Diskography is a diagnostic imaging technique in which an iodinated contrast agent is injected in the intervertebral disk while the physician or technician views the injection by fluoroscopy (dynamic, real time radiography) (Fig. 23). Because diskography allows for morphologic assessment of the intervertebral disk, it has been used for the evaluation of cervical or lumbar diskogenic pain, especially after other examinations have yielded questionable results.[14,38] Diskography also can be used as a provocative test, to determine if the patient's pain is reproduced when the seemingly abnormal spinal level is injected.[14,39-42] Detractors of this technique argue that diskography is painful, invasive, relatively expensive, and time-consuming.[41,43] MRI following the intradiskal injection of gadolinium-based contrast material (MR diskography) reportedly has proved helpful in the detection of tears of the anulus fibrosus of the spinal disk.[44]

Ultrasonography

Musculoskeletal ultrasonography is a rapidly evolving imaging method that has proved helpful in the diagnosis of joint and soft-tissue disorders. New developments in technology with higher frequency transducers and power Doppler ultrasonographic machines have improved the diagnostic capabilities of this imaging method. Advantages of ultrasonography over existing imaging methods include accessibility and ease of application, noninvasiveness, quick scan time, low cost, absence of ionizing radiation, and the ability to perform dynamic imaging with contralateral comparison. With real-time ultrasonography, anatomic structures such as tendons can be followed in motion. In addition, power Doppler ultrasonography may allow detection of active inflammatory disease and tumor growth. Furthermore, ultrasound-guided procedures, such as aspiration of fluid collections, are commonly used in clinical practice.[45] Ultrasonography can be used to study patients with metallic pacemakers or cochlear implants or patients for whom MRI is contraindicated. Sound waves are also not distorted by ferromagnetic implants. Diagnostic power is limited by the talent of the examiner.

In the shoulder, the primary indication for sonography is assessment of the integrity and abnormalities of the rotator cuff tendons. Ultrasonography is useful in the detection of joint effusion and loose bodies in the elbow. In the hand and wrist, ultrasonography is used to demonstrate ganglion cysts, evaluate tenosynovitis around the wrist, and determine compression of the median nerve in carpal tunnel syndrome. Ultrasonography allows evaluation of developmental dysplasia of the hip in the neonate and permits rapid detection of a joint effusion or bursitis.[46] In the knee, meniscal and synovial cysts, joint effusion, and bursitis are easily evaluated on the ultrasonographic images. In addition, ultrasonography may identify tears of the patellar retinacula and abnormalities of the patellar tendon.[46] With regard to the ankle and foot, ultrasonography has proved useful in the assessment of joint effusion, tenosynovitis, tendinosis, and plantar fasciitis.[6,8,46]

References

1. Weissman BN, Pathria MN, Garfin SR: Imaging after bone and spine surgery, in Resnick D (ed): *Bone and Joint Imaging*, ed 2. Philadelphia, PA, WB Saunders, 1996, pp 165-182.
2. Weissman BN: Imaging of joint replacement, in Resnick D (ed): *Bone and Joint Imaging*, ed 2. Philadelphia, PA, WB Saunders, 1996, pp 183-193.
3. André M, Resnick D: Computed tomography, in Resnick D (ed): *Bone and Joint Imaging*, ed 2. Philadelphia, PA, WB Saunders, 1996, pp 70-83.
4. McEnery KW, Murphy WA Jr: Magnetic resonance imaging, in Resnick D (ed): *Bone and Joint Imaging*, ed 2. Philadelphia, PA, WB Saunders, 1996, pp 84-93.
5. Rafii M, Minkoff J: Advanced arthrography of the shoulder with CT and MR imaging. *Radiol Clin North Am* 1998;36:609-633.
6. Theodorou DJ, Theodorou SJ, Kakitsubata Y, et al: Plantar fasciitis and fascial rupture: MR imaging findings in 26 patients supplemented with anatomic data in cadavers. *Radiographics* 2000;20:S181-S197.
7. de la Puente R, Boutin RD, Theodorou DJ, Hooper A, Schweitzer M, Resnick D: Post-traumatic and stress-induced osteolysis of the distal clavicle: MR imaging findings in 17 patients. *Skeletal Radiol* 1999;28:202-208.
8. Theodorou DJ, Theodorou SJ, Farooki S, Kakitsubata Y, Resnick D: Disorders of the plantar aponeurosis: A spectrum of MR imaging findings. *AJR Am J Roentgenol* 2001;176:97-104.
9. Kakitsubata Y, Boutin RD, Theodorou DJ, et al: Calcium pyrophosphate dihydrate crystal deposition in and around the atlantoaxial joint: Association with type 2 odontoid fractures in nine patients. *Radiology* 2000;216:213-219.
10. Theodorou DJ, Malizos KN, Beris AE, Theodorou SJ, Soucacos PN: Multimodal imaging quantitation of the lesion size in osteonecrosis of the femoral head. *Clin Orthop* 2001;386:54-63.
11. Farooki S, Sokoloff RM, Theodorou DJ, Trudell D, Resnick D: MR imaging of the midfoot: Normal anatomy with cadaveric correlation. *Magn Reson Imaging Clin N Am* 2001;9:419-434.
12. Theodorou DJ: The intravertebral vacuum cleft sign. *Radiology* 2001;221:787-788.
13. Alazraki N: Radionuclide techniques, in Resnick D (ed): *Bone and Joint Imaging*, ed 2. Philadelphia, PA, WB Saunders, 1996, pp 141-150.
14. Haughton V: Imaging techniques in intraspinal diseases, in Resnick D (ed): *Bone and Joint Imaging*, ed 2. Philadelphia, PA, WB Saunders, 1996, pp 100-112.
15. Byers GE, Berquist TH: MR imaging techniques for soft-tissue lesions. *Magn Reson Imaging Clin N Am* 1995;3:563-576.
16. Sims RE, Genant HK: Magnetic resonance imaging of joint disease. *Radiol Clin North Am* 1986;24:179-188.
17. Li DK, Adams ME, McConkey JP: Magnetic resonance imaging of the ligaments and menisci of the knee. *Radiol Clin North Am* 1986;24:209-227.
18. Bates D, Ruggieri P: Imaging modalities for evaluation of the spine. *Radiol Clin North Am* 1991;29:675-690.
19. Modic MT, Masaryk TJ, Ross JS, Carter JR: Imaging of degenerative disk disease. *Radiology* 1988;168:177-186.
20. Pech P, Haughton VM: Lumbar intervertebral disk: Correlative MR and anatomic study. *Radiology* 1985;156:699-701.
21. Schiebler ML, Grenier N, Fallon M, Camerino V, Zlatkin M, Kressel HY: Normal and degenerated intervertebral disk: In vivo and in vitro MR imaging with histopathologic correlation. *AJR Am J Roentgenol* 1991;157:93-97.
22. Peterfy CG, Roberts T, Genant HK: Dedicated extremity MR imaging: An emerging technology. *Radiol Clin North Am* 1997;35:1-20.
23. Helgason JW, Chandnani VP: MR arthrography of the ankle. *Radiol Clin North Am* 1998;36:729-738.
24. Parker DL: Computed tomography (CT), in von Schulthess GK, Smith HJ (eds): *The Encyclopaedia of Medical Imaging: Vol. I: Physics, Techniques and Procedures*. Oslo, Norway, the NICER Institute, 1998, pp 82-83.
25. Hunter JC, Brandser EA, Tran KA: Pelvic and acetabular trauma. *Radiol Clin North Am* 1997;35:559-590.
26. Rabassa AE, Guinto FC Jr, Crow WN, Chaljub G, Wright GD, Storey GS: CT of the spine: Value of reformatted images. *AJR Am J Roentgenol* 1993;161:1223-1227.
27. Burnstein MI, Pozniak MA: Computed tomography with stress maneuver to demonstrate sternoclavicular joint dislocation. *J Comput Assist Tomogr* 1990;14:159-160.
28. Young SW, Muller HH, Marshall WH: Computed tomography: Beam hardening and environmental density artifact. *Radiology* 1983;148:279-283.
29. Aaron A, Weinstein D, Thickman D, Eilert R: Comparison of orthoroentgenography and computed tomography in the measurement of limb-length discrepancy. *J Bone Joint Surg Am* 1992;74:897-902.
30. Helms CA, McCarthy S: CT scanograms for measuring leg length discrepancy. *Radiology* 1984;151:802.
31. Liu CC, Theodorou DJ, Theodorou SJ, et al: Quantitative computed tomography in the evaluation of spinal osteoporosis following spinal cord injury. *Osteoporos Int* 2000;11:889-896.
32. Theodorou DJ, Theodorou SJ, Andre MP, Kubota D, Weigert JM, Sartoris DJ: Quantitative computed tomography of spine: Comparison of three-dimensional and two-dimensional imaging approaches in clinical practice. *J Clin Densitom* 2001;4:57-62.
33. Brown ML: Bone scintigraphy in benign and malignant tumors. *Radiol Clin North Am* 1993;31:731-738.
34. Aliabadi P, Baker ND, Jaramillo D: Hip arthrography, aspiration, block, and bursography. *Radiol Clin North Am* 1998;36:673-690.
35. Steinbach LS, Schwartz M: Elbow arthrography. *Radiol Clin North Am* 1998;36:635-649.
36. Haims A, Katz LD, Busconi B: MR arthrography of the hip. *Radiol Clin North Am* 1998;36:691-702.
37. Resnick D: Knee, in Resnick D, Kang HS (eds): *Internal Derangements of Joints: Emphasis on MR Imaging*. Philadelphia, PA, WB Saunders, 1997, pp 555-785.
38. Schellhas KP, Smith MD, Gundry CR, Pollei SR: Cervical discogenic pain: Prospective correlation of magnetic resonance imaging and discography in asymptomatic subjects and pain sufferers. *Spine* 1996;21:300-312.

39. Tehranzadeh J: Discography 2000. *Radiol Clin North Am* 1998;36:463-495.
40. Adams MA, Dolan P, Hutton WC: The stages of disc degeneration as revealed by discograms. *J Bone Joint Surg Br* 1986;68:36-41.
41. Bogduk N, Modic MT: Controversy: Lumbar discography. *Spine* 1996;21:402-404.
42. Osti OL, Fraser RD: MRI and discography of annular tears and intervertebral disc degeneration: A prospective clinical comparison. *J Bone Joint Surg Br* 1992;74:431-435.
43. Nachemson A: Lumbar discography: Where are we today? *Spine* 1989;14:555-557.
44. Kakitsubata Y, Theodorou DJ, Theodorou SJ, et al: Normal and abnormal lumbar intervertebral disk: Cadaveric study correlating MR imaging and MR discography. *AJR Am J Roentgenol* 2001;176(suppl 3):S71.
45. Craig JG: Infection: Ultrasound-guided procedures. *Radiol Clin North Am* 1999;37:669-678.
46. Mack L, Scheible W: Diagnostic ultrasonography, in Resnick D (ed): *Bone and Joint Imaging*. Philadelphia, PA, WB Saunders, 1996, pp 94-99.

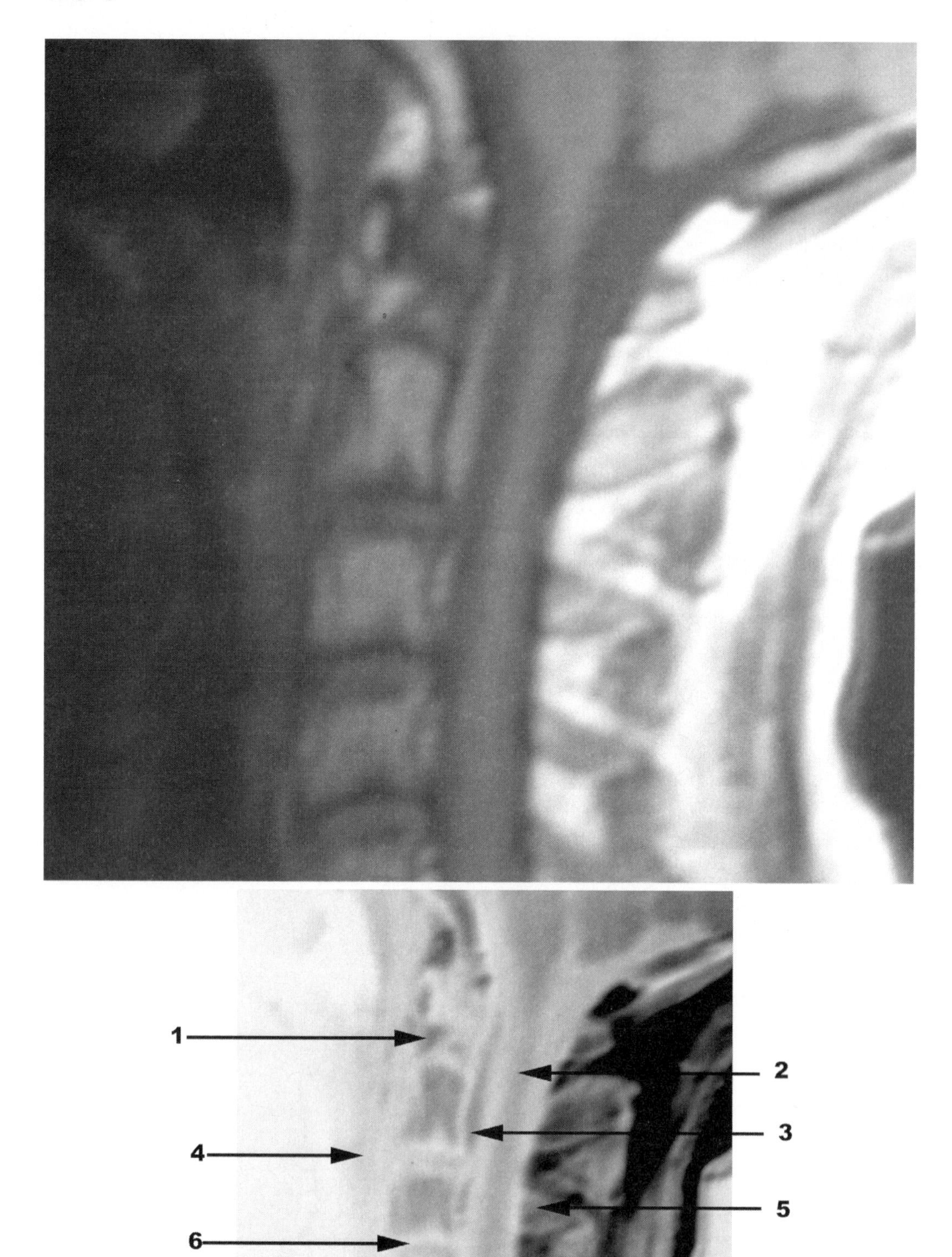

Figure 1

Sagittal plane T1-weighted MRI of the cervical spine. 1 = Axis, 2 = Spinal cord, 3 = Posterior longitudinal ligament, 4 = Anterior longitudinal ligament, 5 = Spinous process, 6 = Intervertebral disk, 7 = Vertebral body.

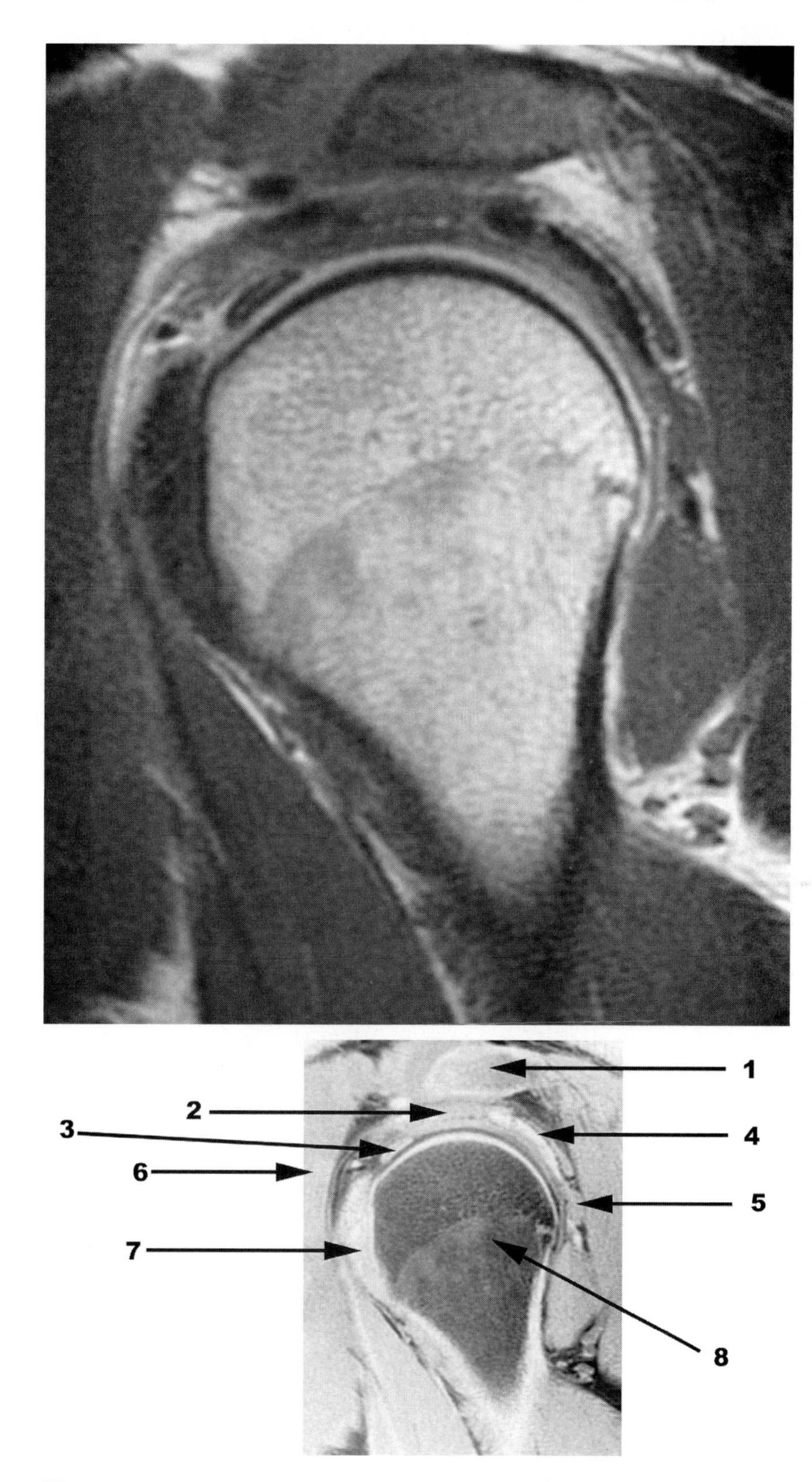

Figure 2
Sagittal plane T1-weighted MRI of the shoulder. 1 = Acromion, 2 = Supraspinatus, 3 = Biceps tendon, 4 = Infraspinatus, 5 = Teres minor, 6 = Deltoid, 7 = Subscapularis, 8 = Humerus.

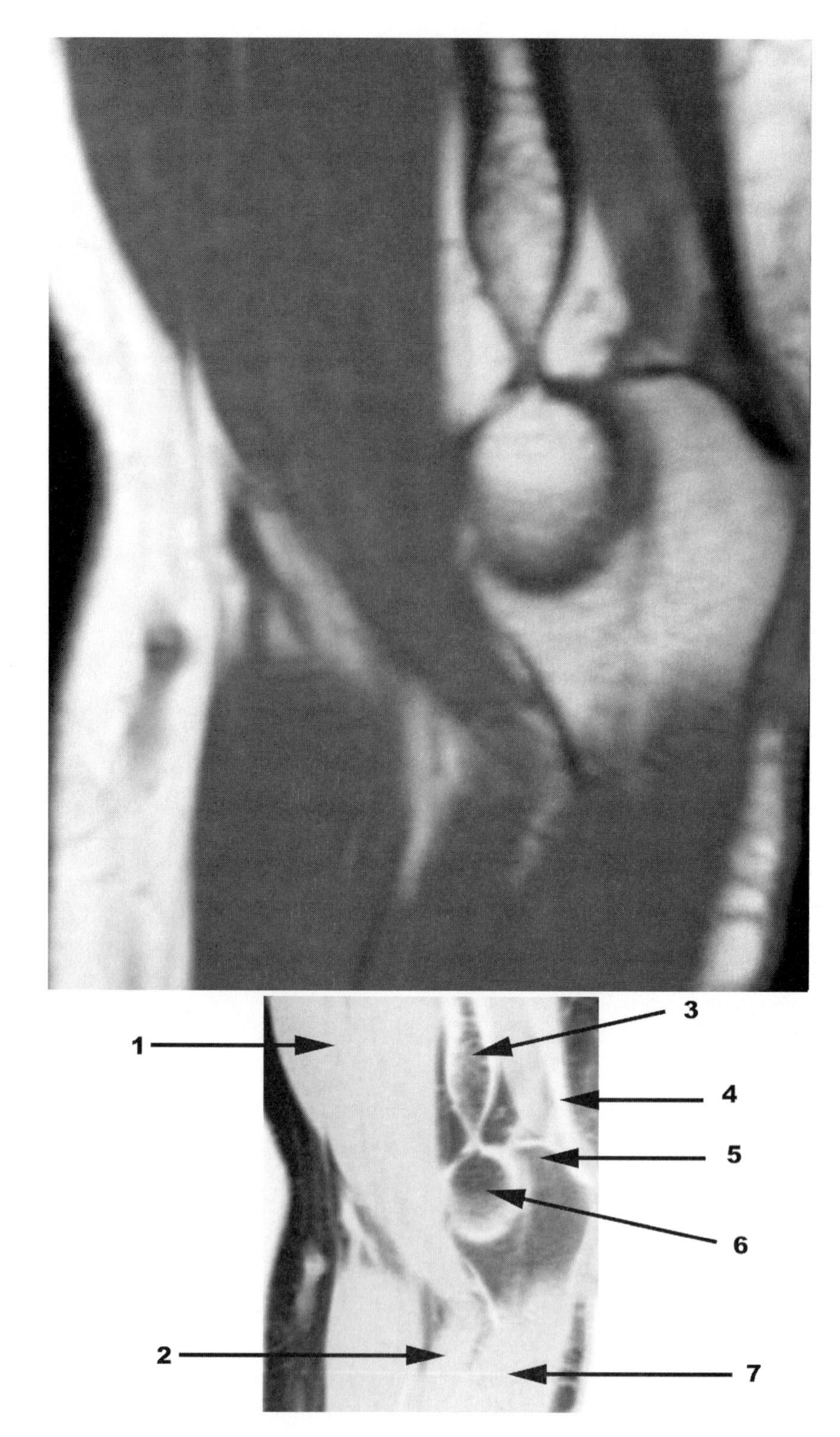

Figure 3
Sagittal plane T1-weighted MRI of the elbow. 1 = Brachialis, 2 = Flexor digitorum superficialis, 3 = Humerus, 4 = Triceps, 5 = Olecranon, 6 = Trochlea of humerus, 7 = Flexor digitorum profundus.

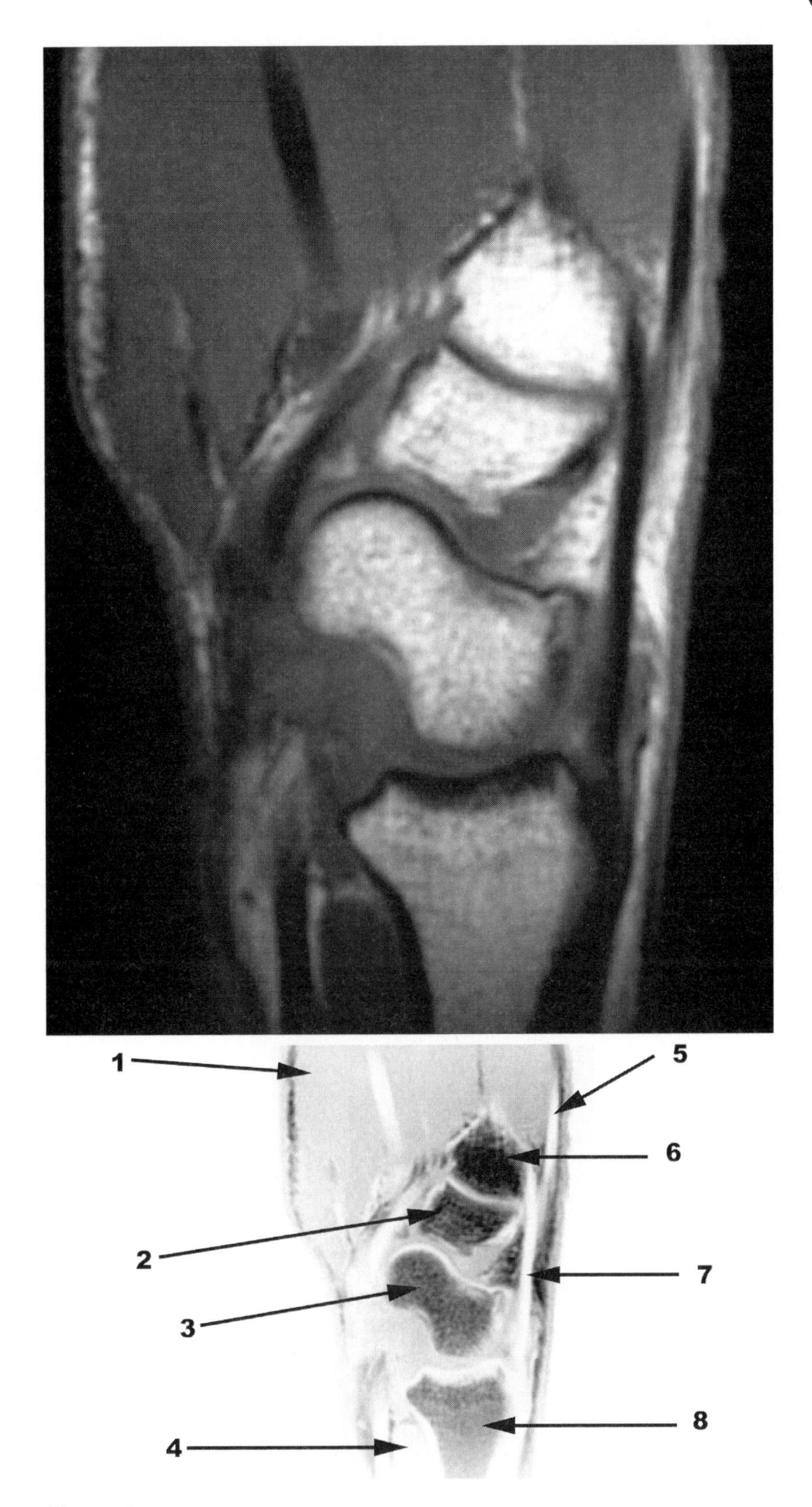

Figure 4
Sagittal plane T1-weighted MRI of the wrist. 1 = Thenar muscles, 2 = Trapezoid, 3 = Scaphoid, 4 = Pronator quadratus, 5 = Tendon of extensor digitorum, 6 = Second metacarpal, 7 = Tendon of extensor carpi radialis brevis, 8 = Radius.

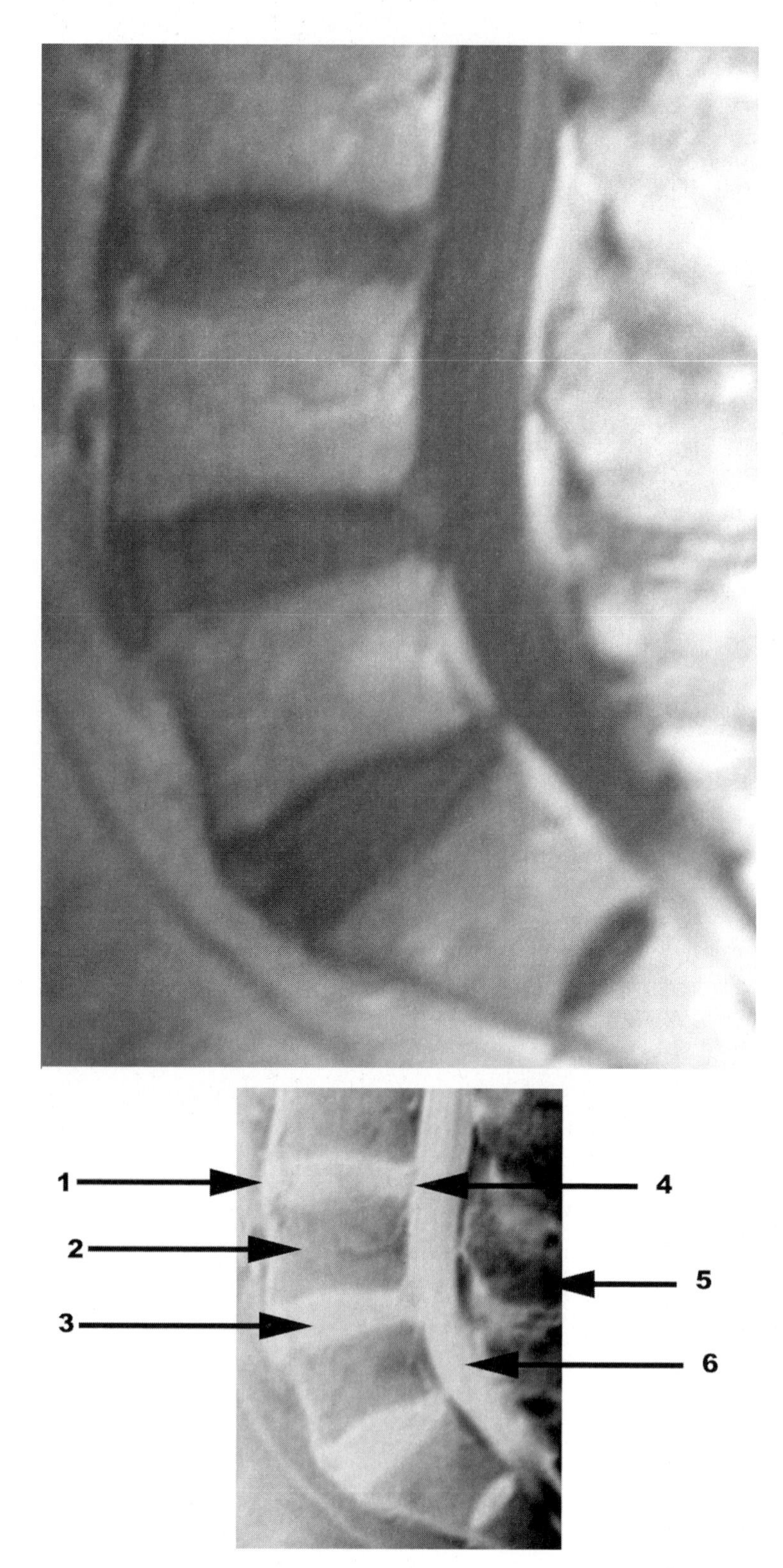

Figure 5
Sagittal plane T1-weighted MRI of the lumbar spine. 1 = Anterior longitudinal ligament, 2 = Vertebral body, 3 = Invertebral disk, 4 = Posterior longitudinal ligament, 5 = Spinous process, 6 = Cauda equina.

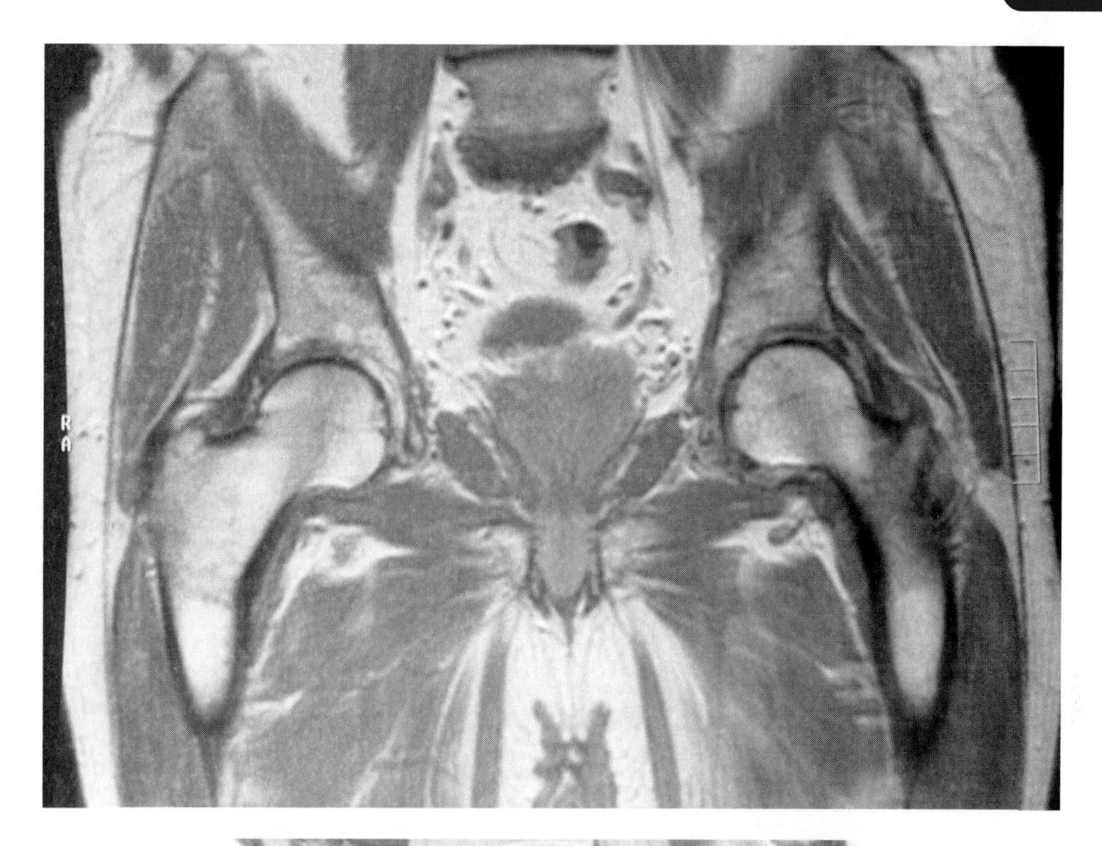

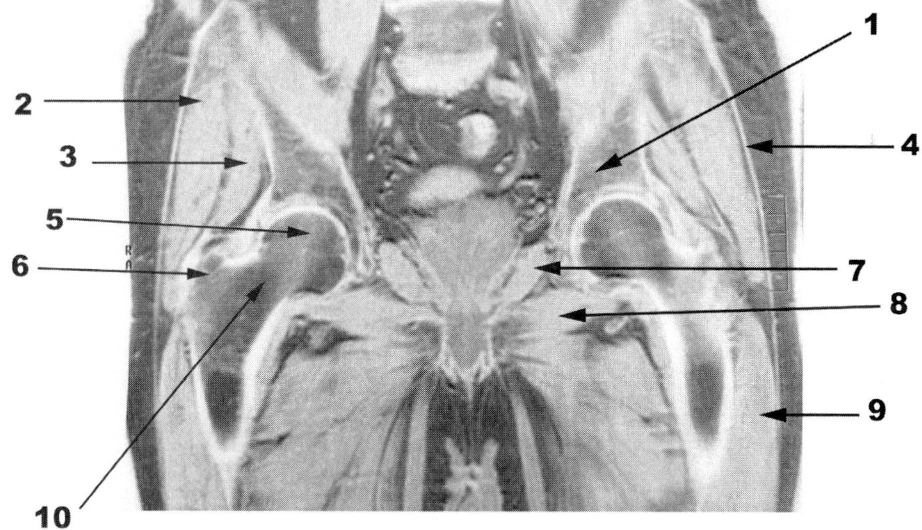

Figure 6
Coronal plane T1-weighted MRI of the pelvis. 1 = Acetabulum, 2 = Gluteus medius, 3 = Gluteus minimus, 4 = Iliotibial band, 5 = Femoral head, 6 = Greater trochanter, 7 = Obturator internus, 8 = Obturator externus, 9 = Vastus lateralis, 10 = Femoral neck.

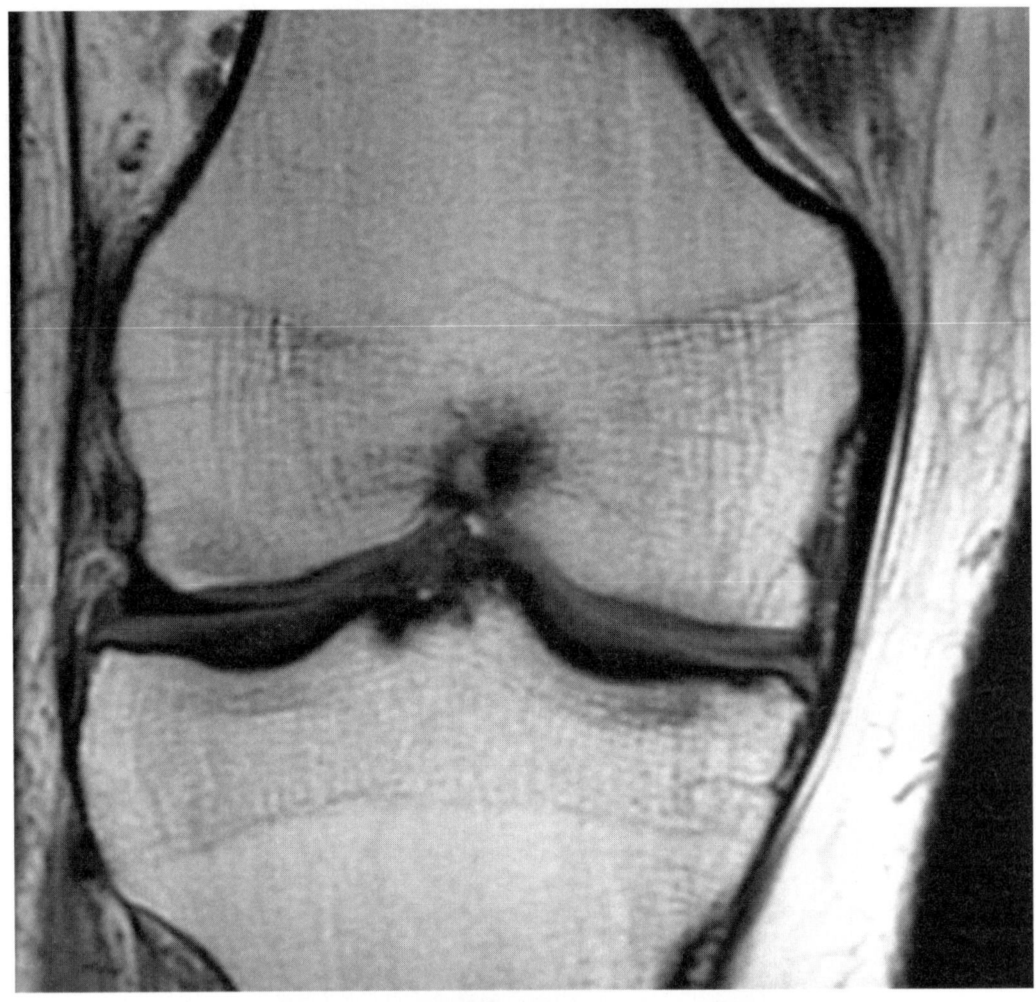

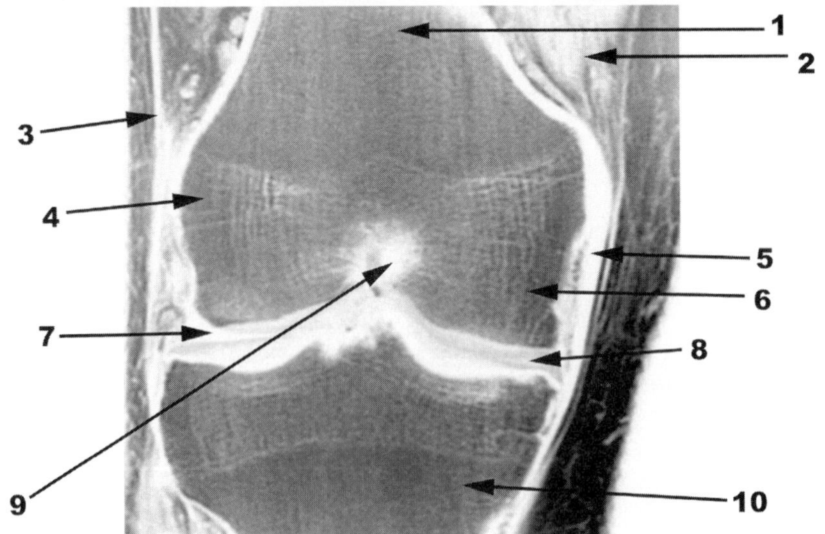

Figure 7
Coronal plane T1-weighted MRI of the knee. 1 = Femur, 2 = Vastus medialis, 3 = Iliotibial band, 4 = Lateral femoral condyle, 5 = Medial collateral ligament, 6 = Medial femoral condyle, 7 = Lateral meniscus, 8 = Medial meniscus, 9 = Posterior cruciate ligament, 10 = Tibia.

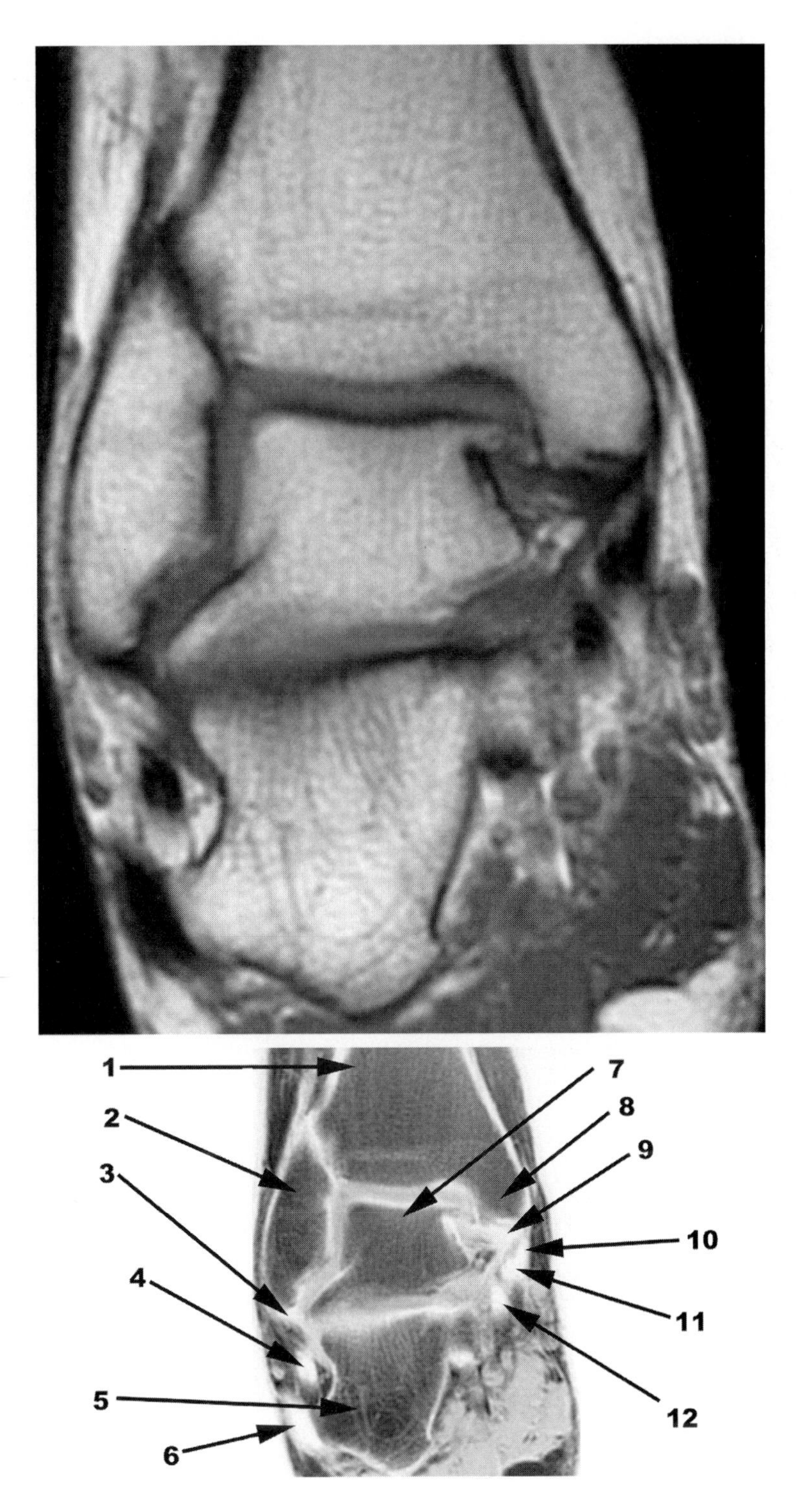

Figure 8

Coronal plane T1-weighted MRI of the ankle. 1 = Tibia, 2 = Fibula, 3 = Calcaneofibular ligament, 4 = Tendon of peroneus brevis, 5 = Calcaneus, 6 = Tendon of peroneus longus, 7 = Talus, 8 = Medial malleolus, 9 = Deltoid ligament, 10 = Tendon of tibialis posterior, 11 = Tendon of flexor digitorum longus, 12 = Tendon of flexor hallucis longus.

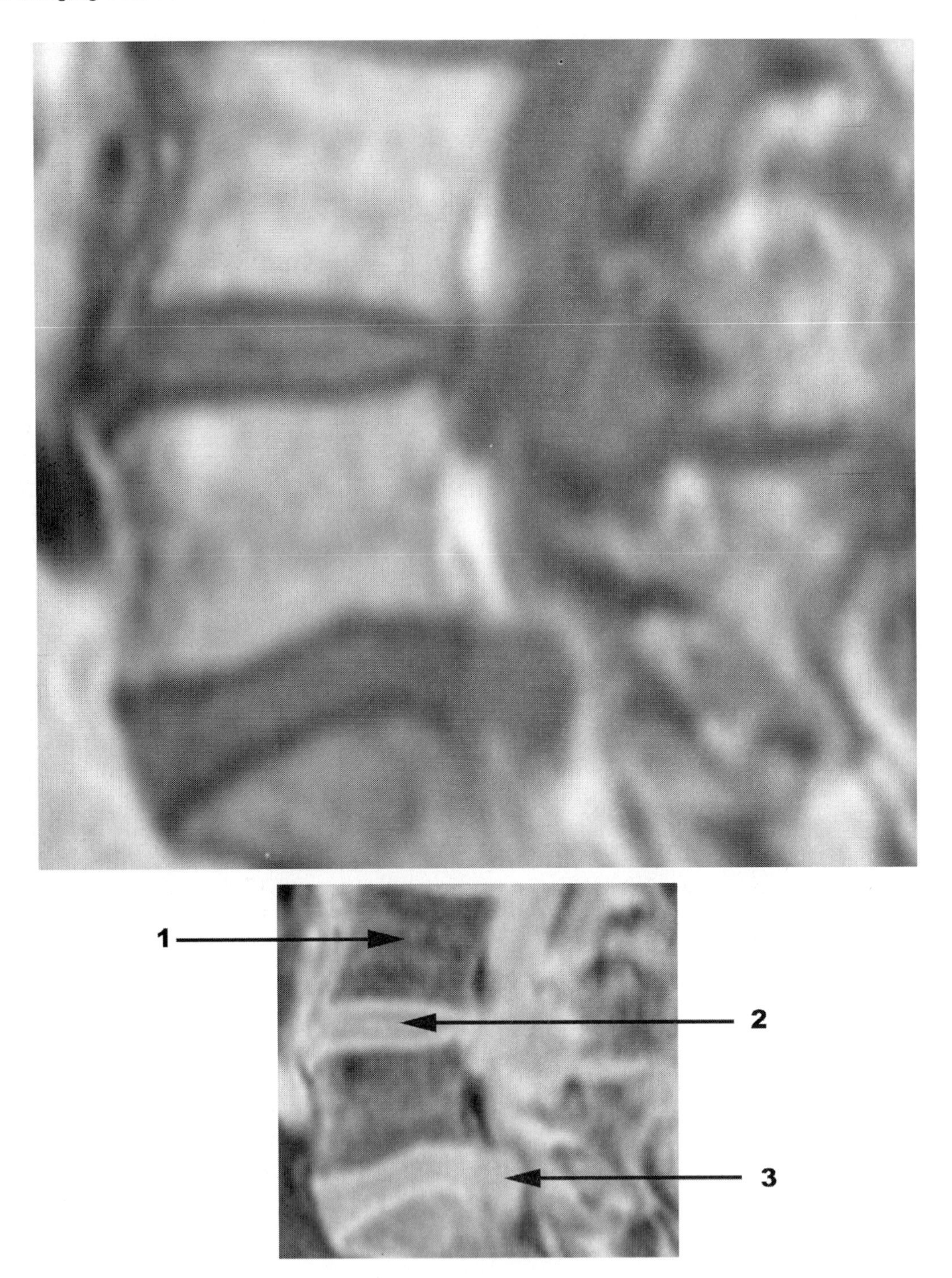

Figure 9
Sagittal T1-weighted, spin-echo MRI of the lumbar spine shows a large posterior herniated L5-S1 intervertebral disk. 1 = Vertebral body, 2 = Normal intervertebral disk, 3 = Herniated nucleus pulposus.

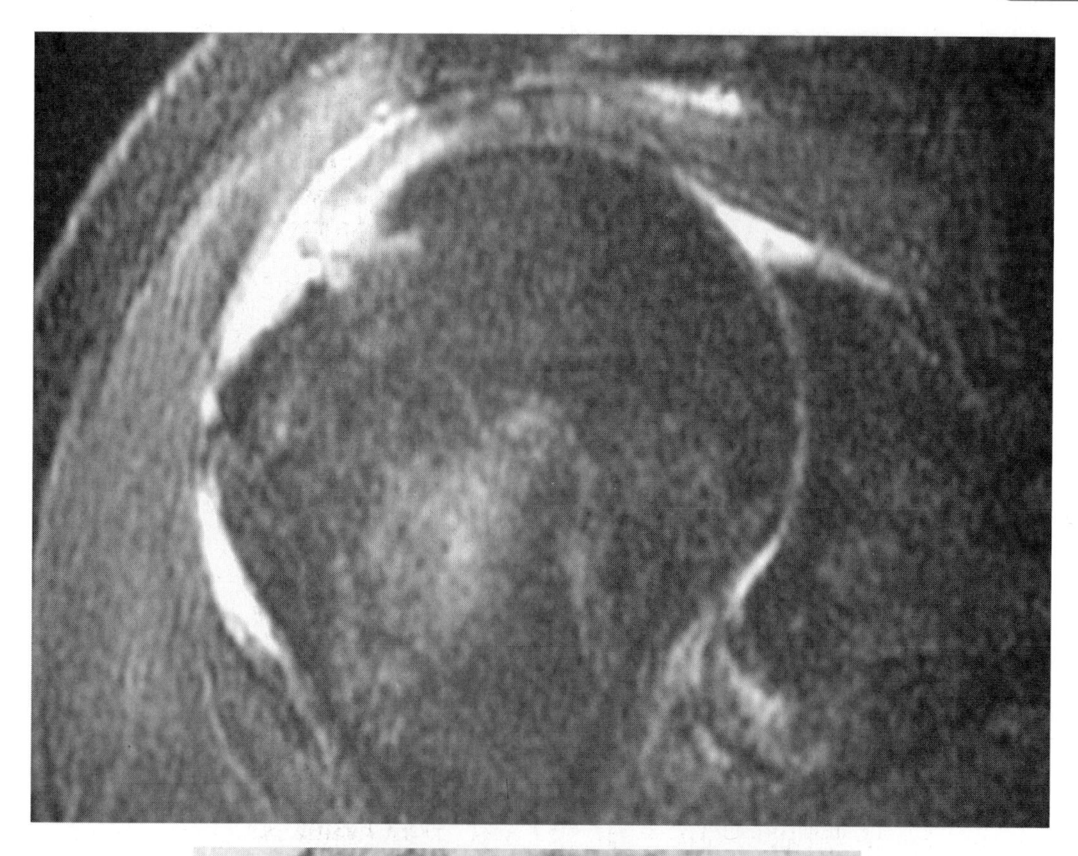

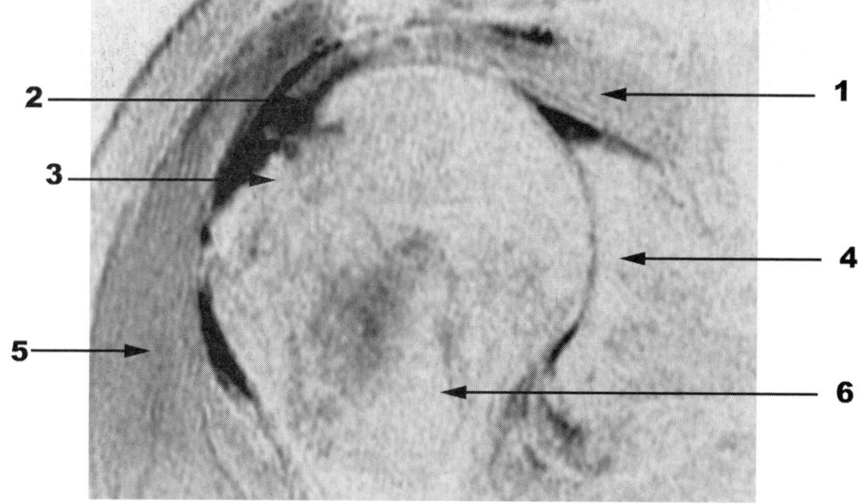

Figure 10
Coronal oblique T2-weighted, spin-echo MRI of the shoulder shows a tear of the supraspinatus muscle-tendon unit from its normal attachment on the greater tuberosity. 1 = Supraspinatus, 2 = Torn edge of the supraspinatus tendon, 3 = Greater tuberosity, 4 = Glenoid, 5 = Deltoid, 6 = Humerus.

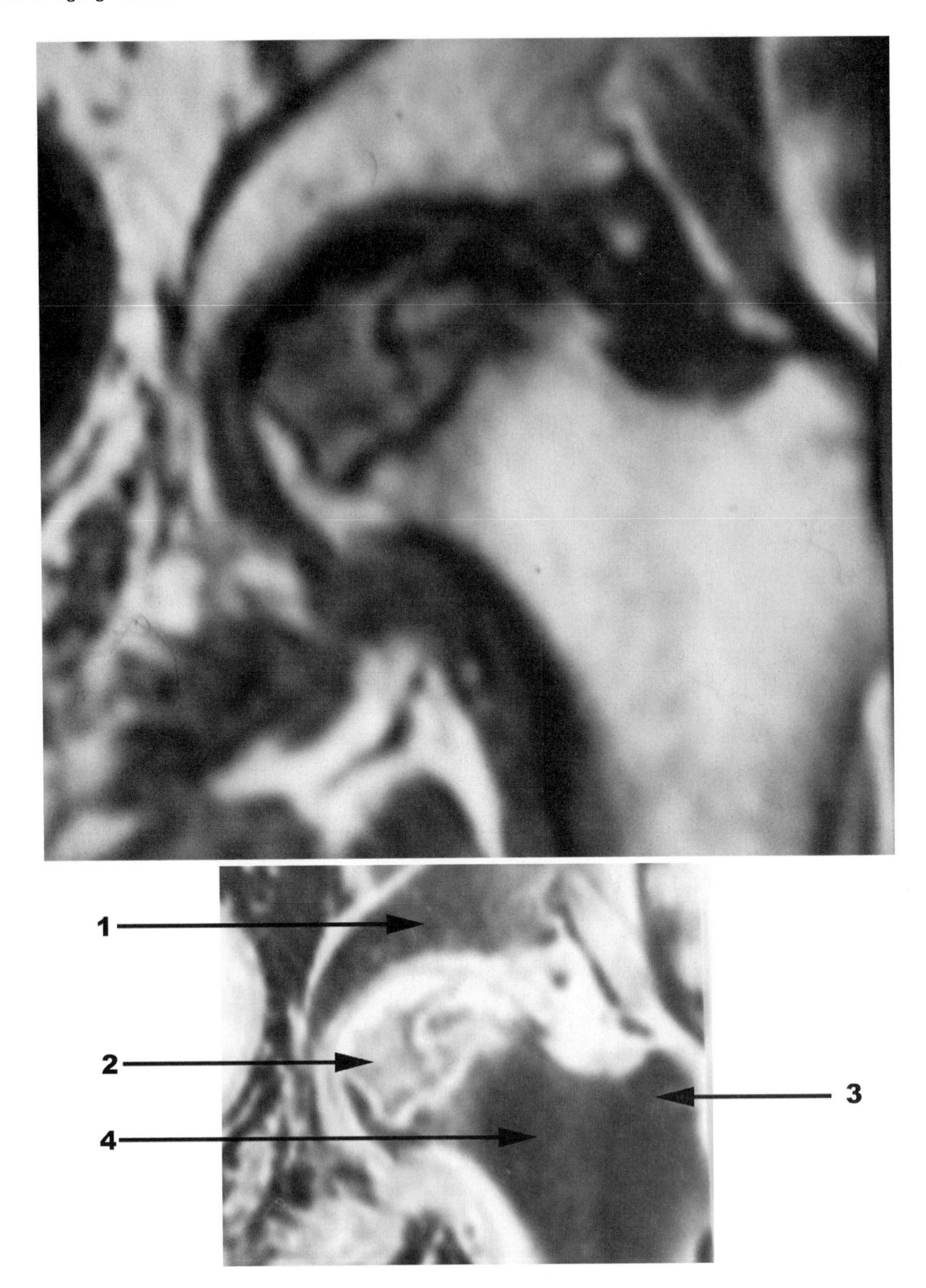

Figure 11
Coronal T1-weighted, spin-echo MRI of the hip reveals a large area of osteonecrosis in the femoral head. 1 = Acetabulum, 2 = Osteonecrosis, 3 = Greater trochanter, 4 = Femoral neck.

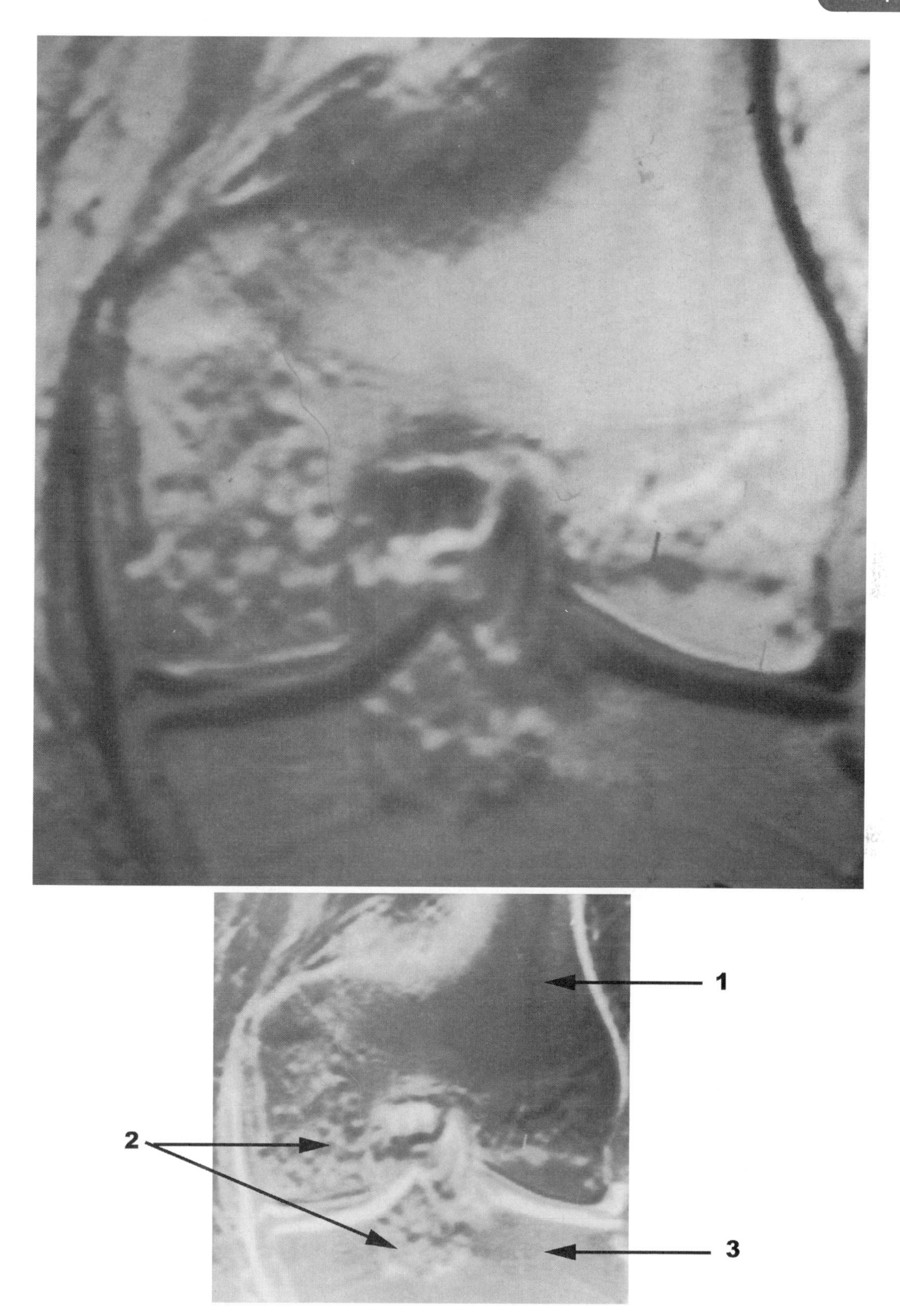

Figure 12
Coronal intermediate-weighted spin-echo MRI of the knee shows diffusely abnormal signal intensity of bone marrow related to osteomyelitis. (Courtesy of C. Beaulieu, Stanford, CA.) 1 = Femur, 2 = Osteomyelitis, 3 = Tibia.

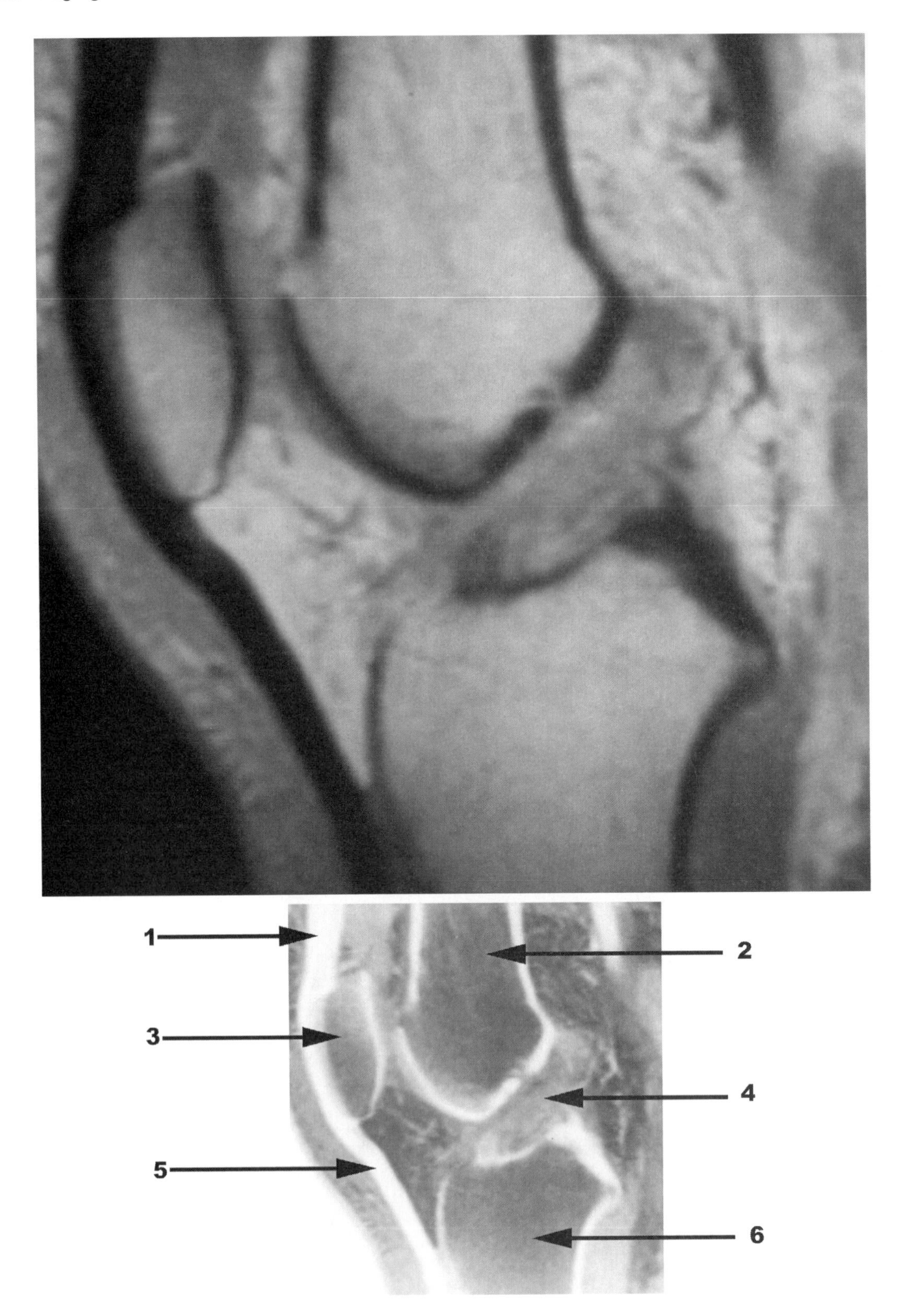

Figure 13
Sagittal intermediate-weighted, spin-echo MRI of the knee shows an acute tear of the anterior cruciate ligament. A normal anterior cruciate ligament should appear as discrete line of low-signal intensity (similar to the patellar and quadriceps tendons). 1 = Quadriceps tendon, 2 = Femur, 3 = Patella, 4 = Torn anterior cruciate ligament, 5 = Patellar tendon, 6 = Tibia.

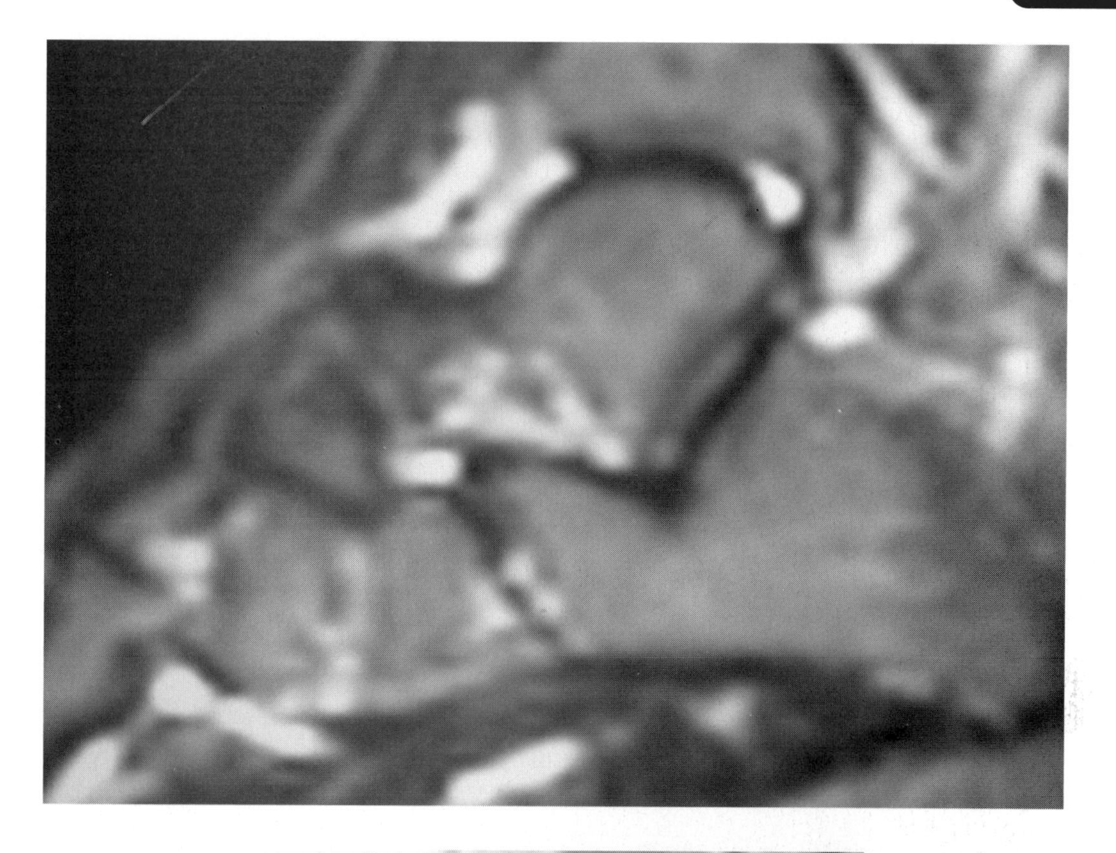

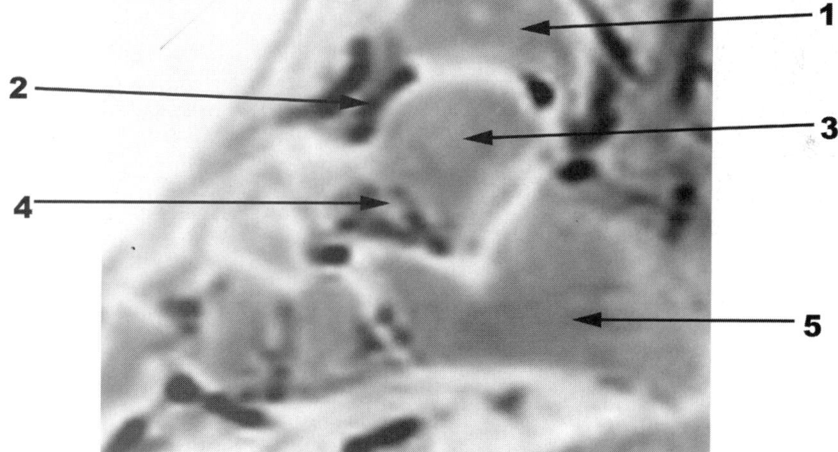

Figure 14
Sagittal T2-weighted, spin-echo MRI of the ankle reveals an osteoid osteoma in the neck of the talus. The nidus within the tumor is of low-signal intensity. A joint effusion is also present. 1 = Tibia, 2 = Effusion, 3 = Talus, 4 = Nidus within osteoid osteoma, 5 = Calcaneus.

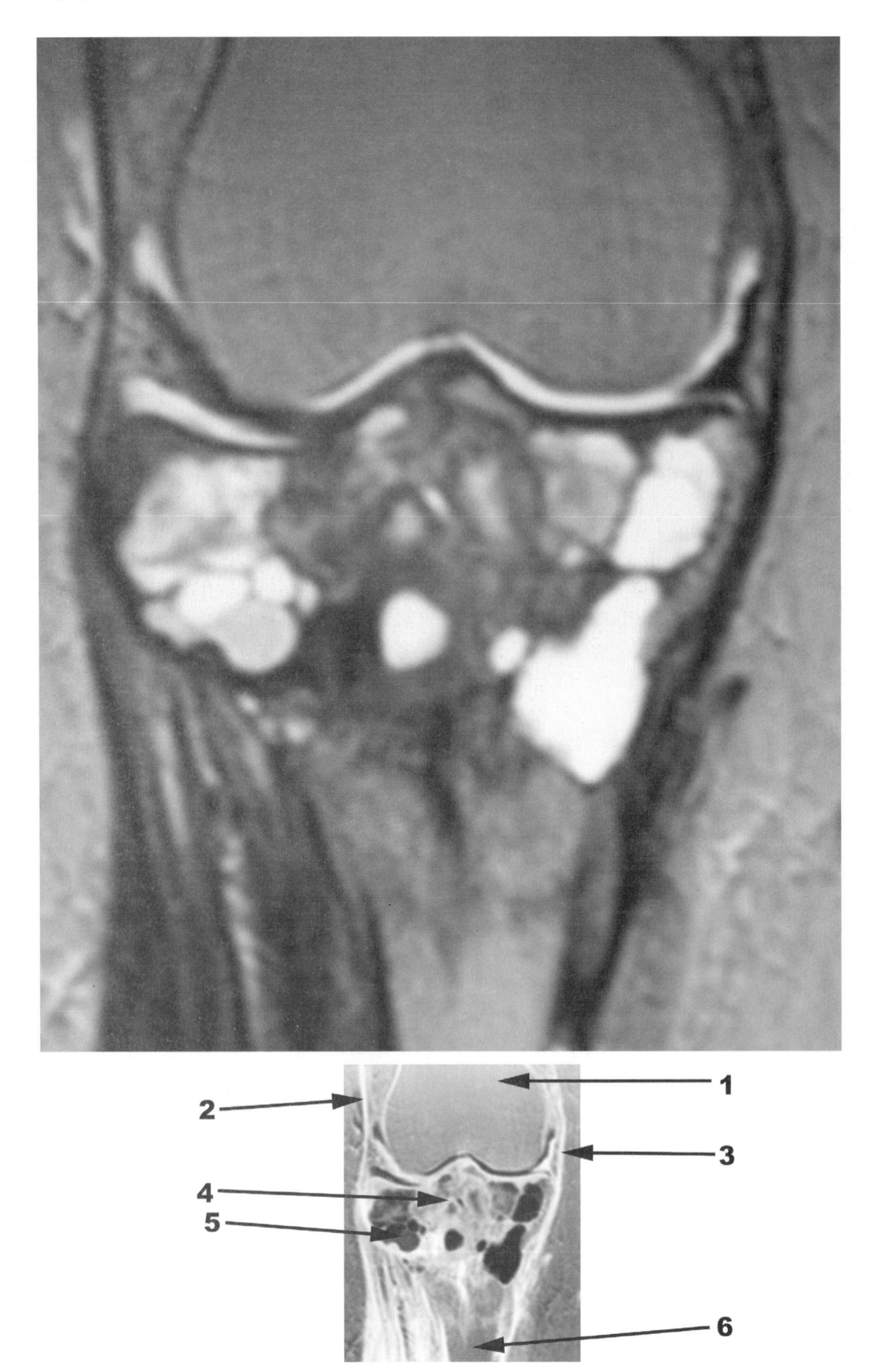

Figure 15

Coronal T2-weighted, spin-echo MRI of the knee shows a giant cell tumor involving the proximal tibia. The tumor lacks homogeneous signal intensity; there is high-signal intensity peripherally and mainly low-signal intensity centrally. 1 = Femur, 2 = Iliotibial band, 3 = Medial collateral ligament, 4 = Giant cell tumor (area of low-signal intensity), 5 = Giant cell tumor (area of high-signal intensity), 6 = Tibia.

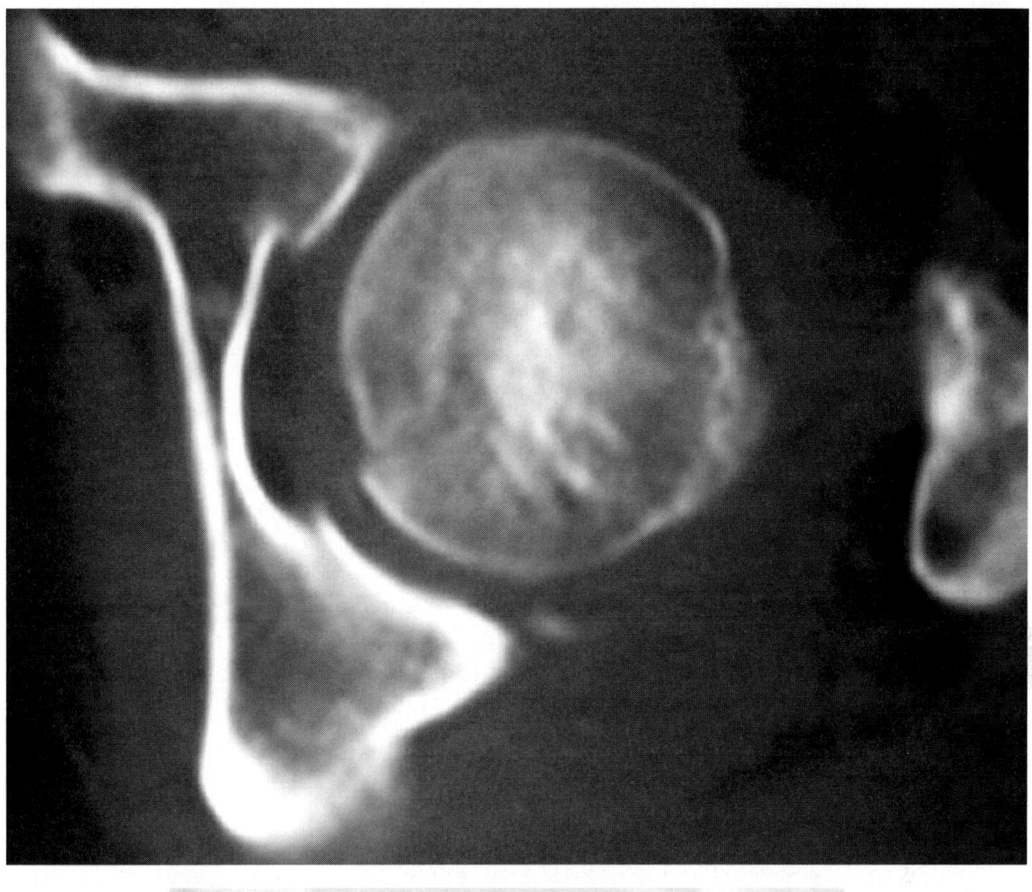

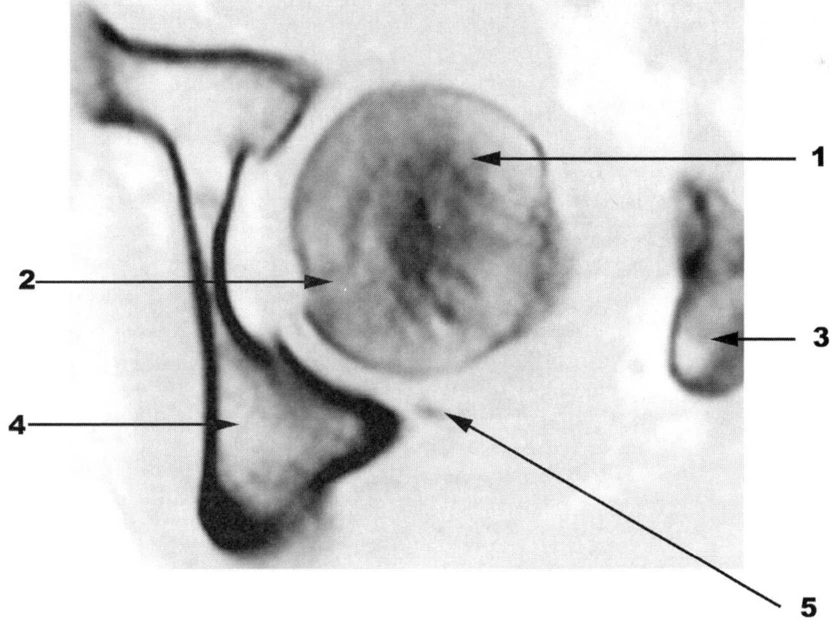

Figure 16
Transaxial CT scan of the hip shows fracture of the femoral head and intra-articular fracture fragments. 1 = Femoral head, 2 = Fracture, 3 = Greater trochanter, 4 = Acetabulum, 5 = Loose body.

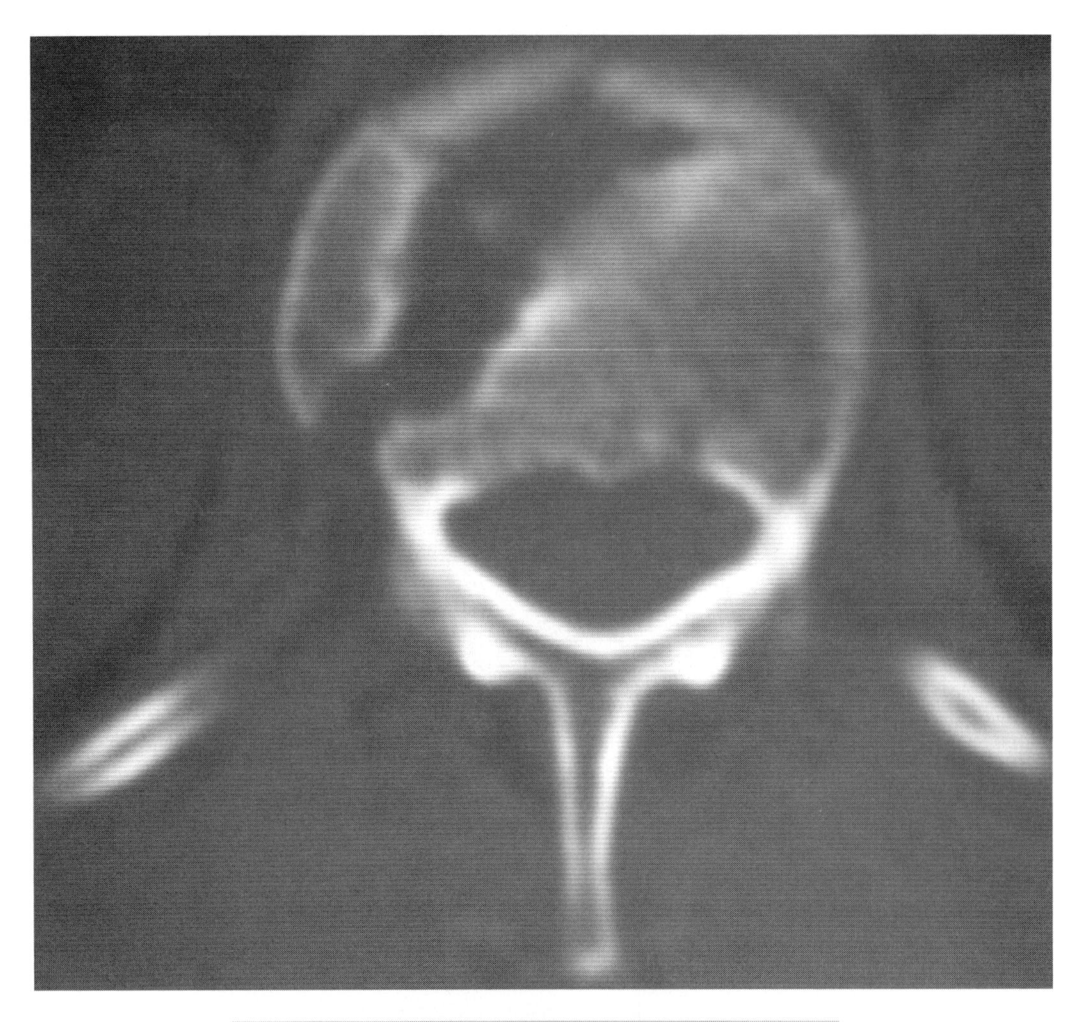

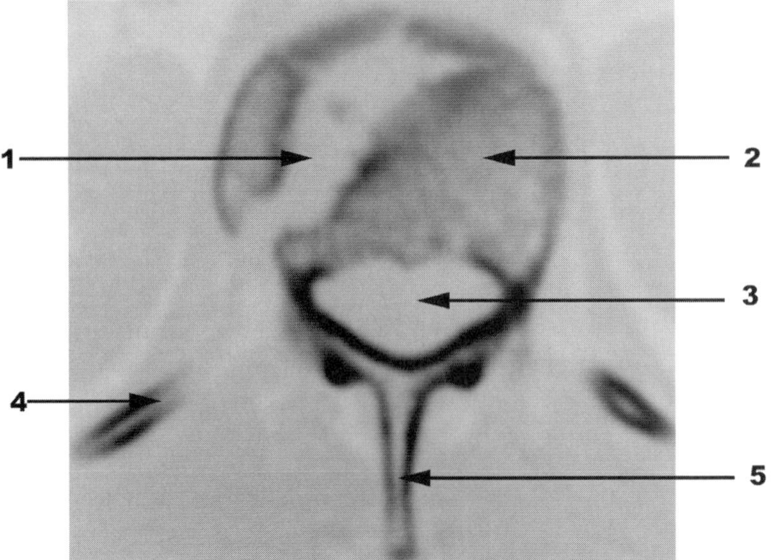

Figure 17
Transaxial CT scan of T12 shows a comminuted fracture of the vertebral body. No fragments within the spinal canal are identified. 1 = Fracture, 2 = Vertebral body, 3 = Spinal canal, 4 = Transverse process, 5 = Spinous process.

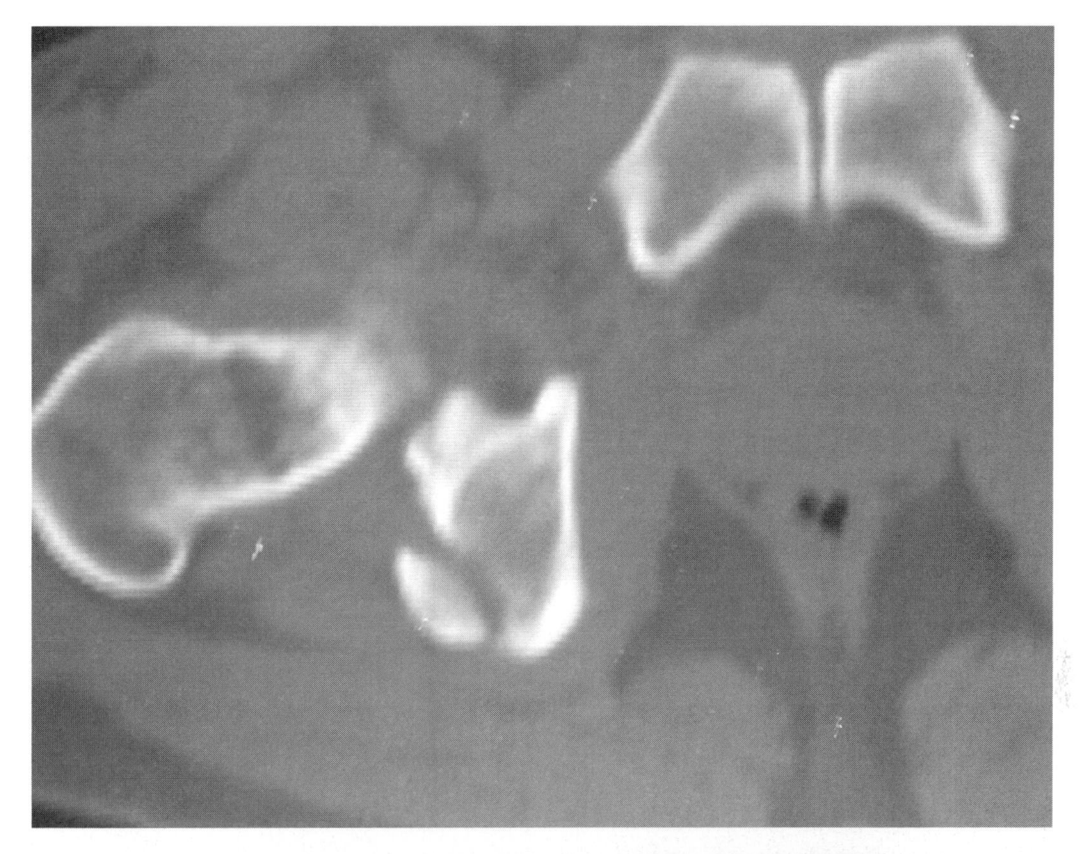

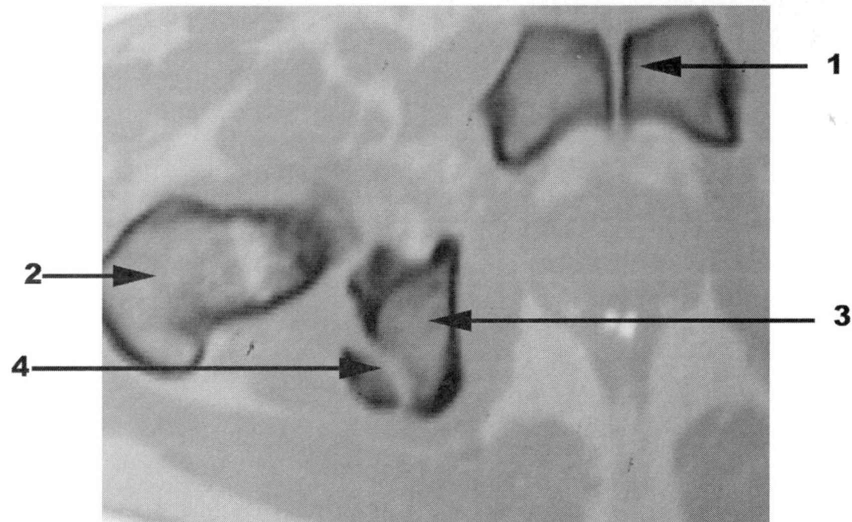

Figure 18
Transaxial CT scan of the pelvis reveals a fracture of the ischial tuberosity. 1 = Pubic symphysis, 2 = Femur, 3 = Ischial tuberosity, 4 = Fracture.

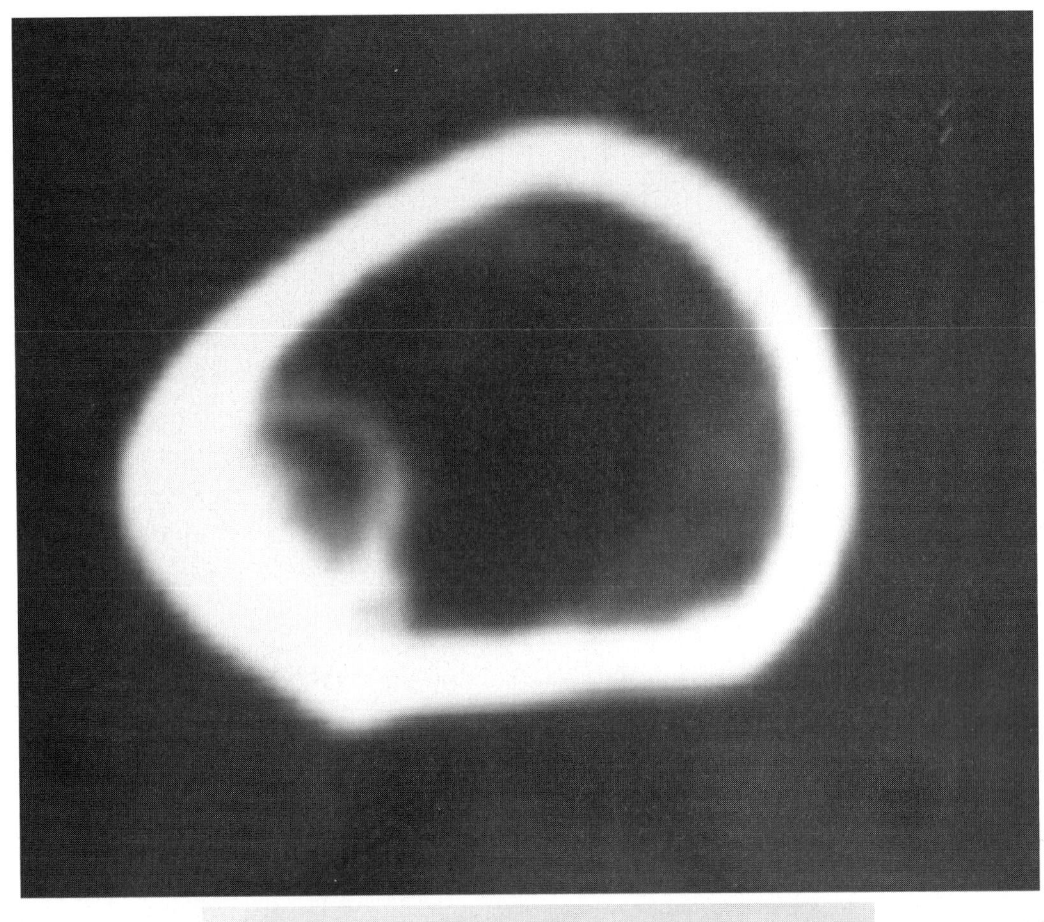

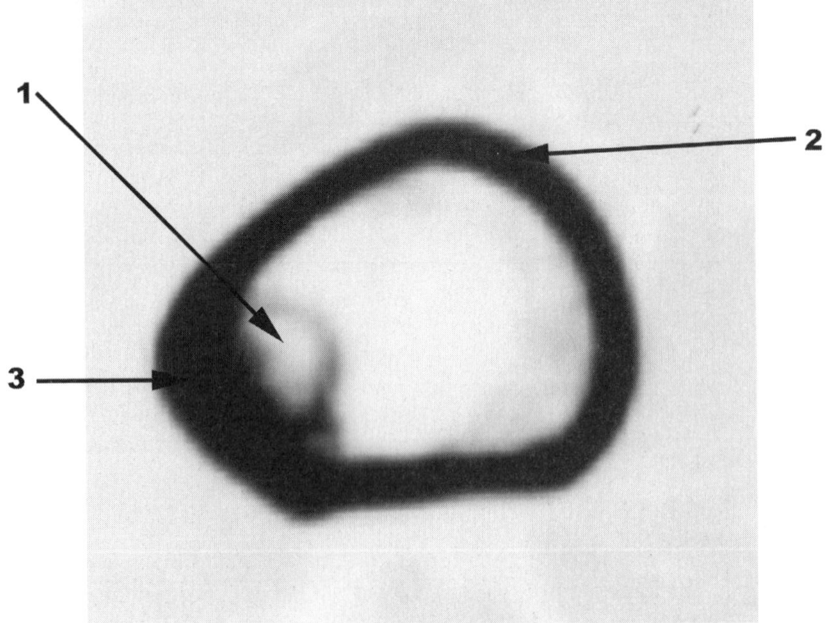

Figure 19
Transaxial CT scan of the femoral diaphysis reveals the radiolucent nidus of an osteoid osteoma. Note thickening of adjacent cortical bone. 1 = Nidus of osteoid osteoma, 2 = Normal femoral cortex, 3 = Thickened femoral cortex.

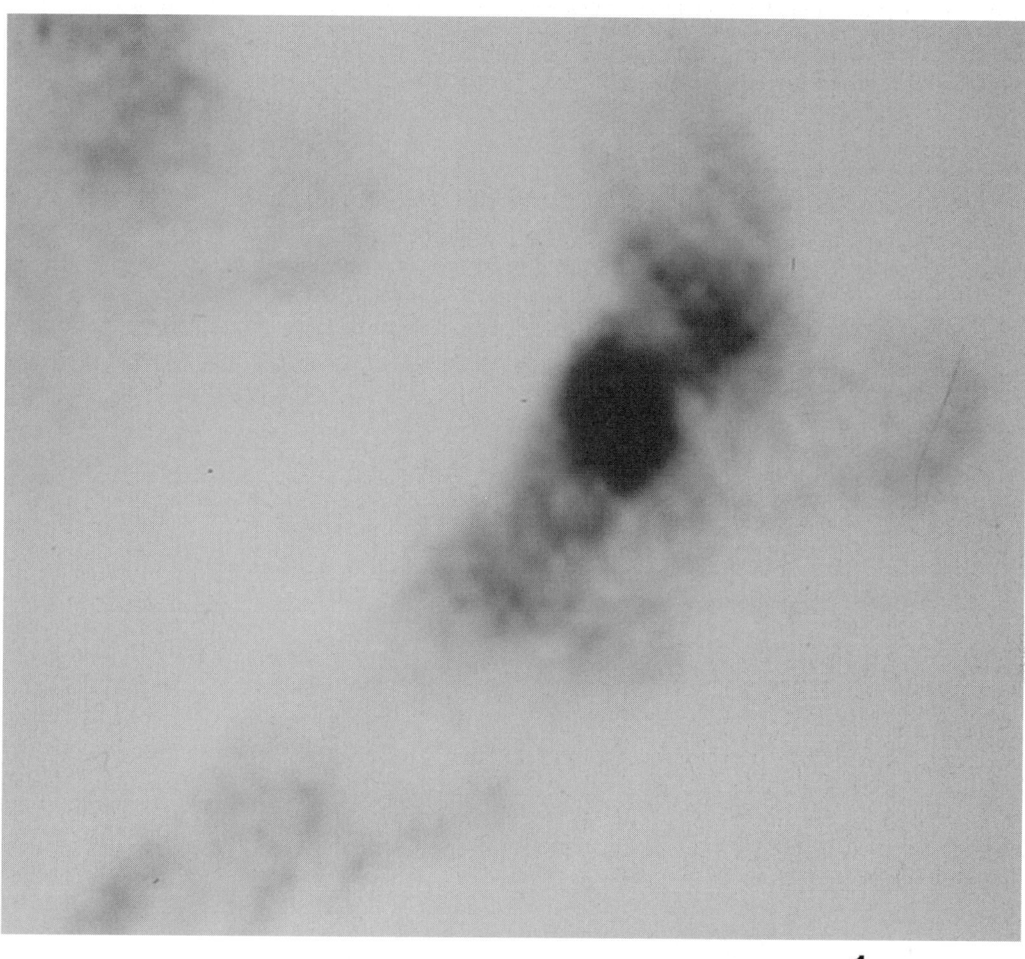

Figure 20
Bone scan of the foot shows accumulation of the radionuclide in the talar neck, consistent with an occult fracture. 1 = Increased signal (fracture), 2 = Tibia, 3 = Talus, 4 = Calcaneus, 5 = Metatarsals.

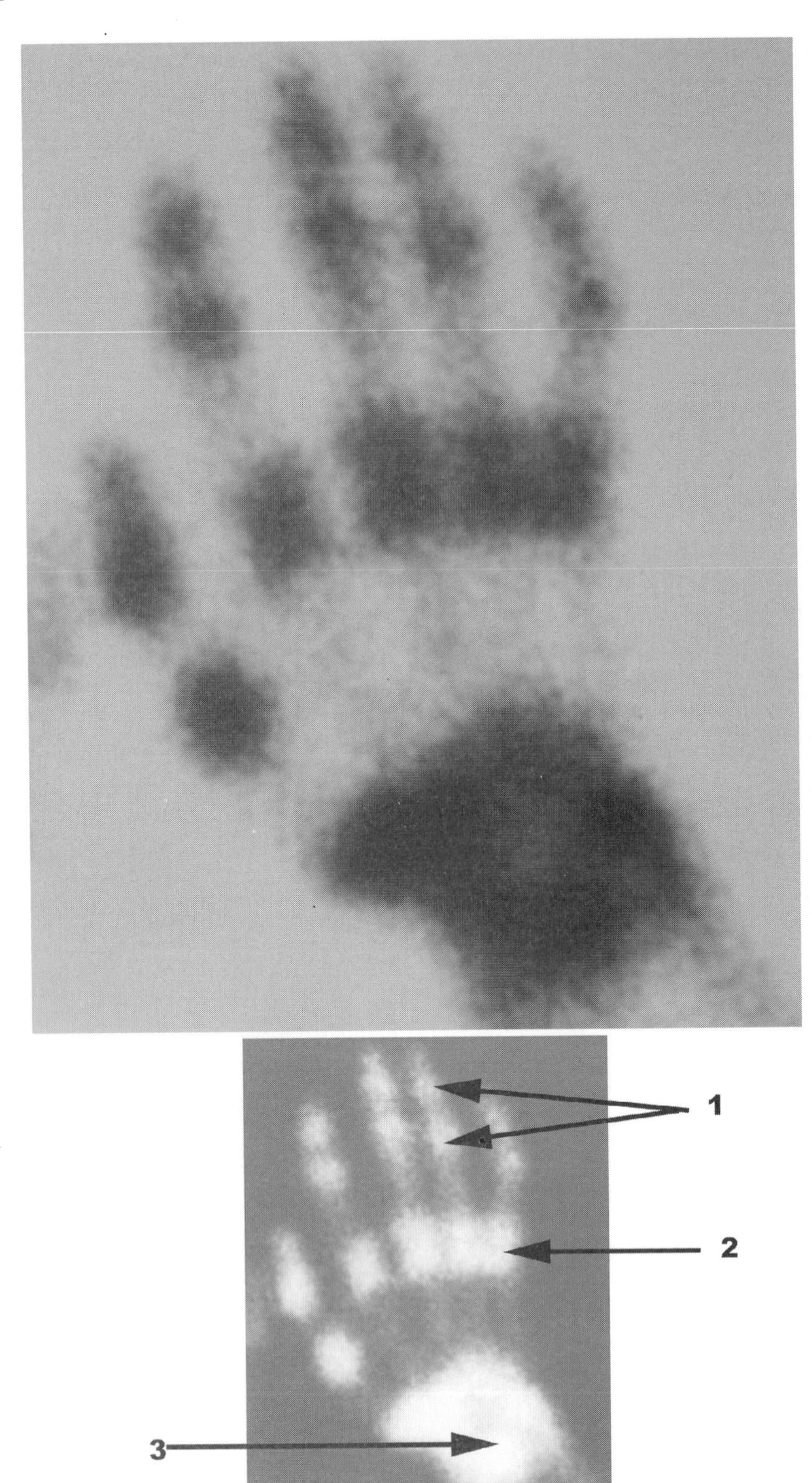

Figure 21
This patient had severe pain following a Colles' fracture that healed. The bone scan shows abnormal radionuclide accumulation in the bones about the wrist and in the hand, suggestive of complex regional pain syndrome. 1 = Interphalangeal joints, 2 = Metacarpophalangeal joint, 3 = Midcarpal and radiocarpal joints (wrist).

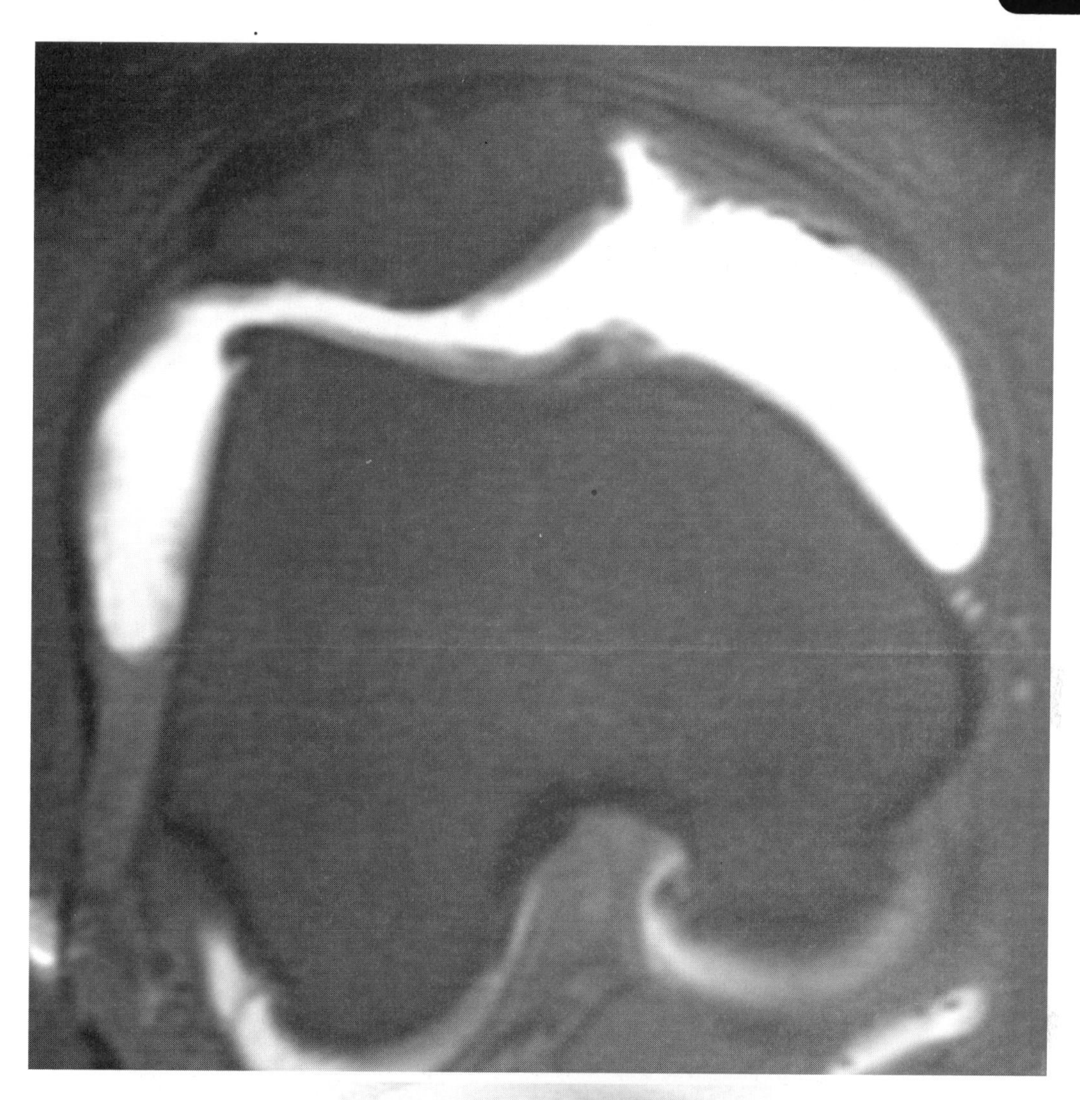

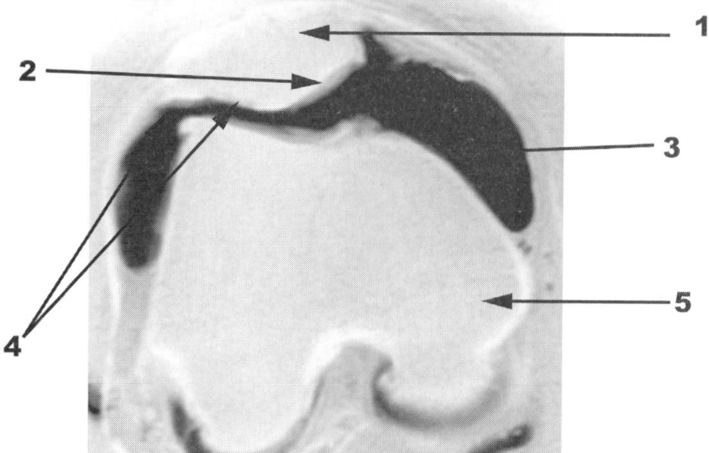

Figure 22

MR arthrography of the knee. An axial fat suppressed T1-weighted spin-echo MRI obtained following the intra-articular injection of a gadolinium compound shows surface irregularity of the patellar articular cartilage. The articular cartilage of the patella should appear as a gray band. On the lateral side of the patella (left of figure), this cartilage is focally absent or severely thinned. 1 = Patella, 2 = Normal articular cartilage, 3 = Effusion, 4 = Articular lesion, 5 = Femur.

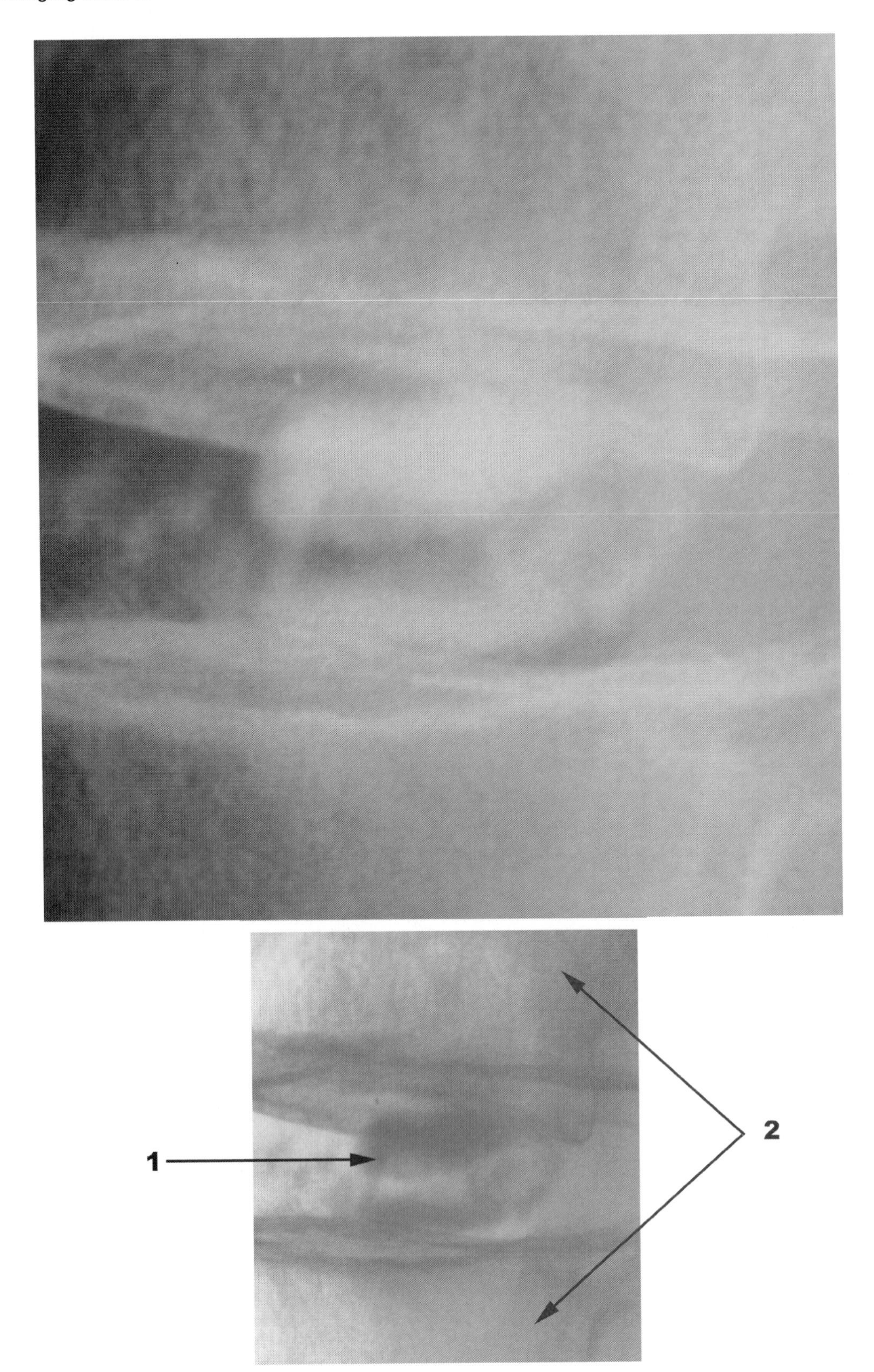

Figure 23
Diskogram displays a normal pattern (bilobular) of contrast opacification of the intervertebral disk at the L4-5 spinal level. 1 = Intervertebral disk, 2 = Vertebral body.

Chapter 29

Electrodiagnostic Studies

Electrodiagnostic studies are useful diagnostic tools for localizing and characterizing pathology in nerve or muscle. Among the conditions amenable to electrodiagnostic evaluation are focal neuropathies, radiculopathies, or plexopathies; polyneuropathies; and other neuromuscular lesions (eg, myopathies and central lesions). Information rendered by electrodiagnostic studies includes the type of pathology, its location, and its severity.

A thorough history and physical examination is the prerequisite to an accurate electrodiagnostic workup. Particular attention must be placed on recognizing patterns of weakness, sensory abnormalities, or pain. Patterns that must be observed include location (proximal versus distal or symmetric versus asymmetric involvement) and type (motor versus sensory). In this sense, electrodiagnostic testing is an extension of the history and physical examination, and the results must be interpreted in the context of the foundation established by the clinical data.

It is not the intention of this chapter to empower the reader to successfully interpret the raw data of electrodiagnostic studies; rather, it is to provide an appreciation of what these studies can reveal and to introduce and review the basic science that underlies the observed responses.

A complete electrodiagnostic evaluation consists of nerve conduction studies (NCSs) (also known as nerve conduction velocity tests) and electromyography (EMG). Each part of the evaluation offers distinct information. Unless there are specific contraindications, a complete electrodiagnostic study should include both NCSs and EMG. The combined testing is commonly referred to as EMG.

Nerve Conduction Studies

Nerve conduction studies test the speed by which motor, sensory, or mixed (combined motor and sensory) nerves transmit impulses. The nerves most often studied are the median, ulnar, and radial nerves in the arm and the tibial, peroneal, and sural nerves in the leg. In this noninvasive and essentially painless test, two pairs of surface electrodes are placed on the skin: one pair proximal and one pair more distal. The general format is as follows: the proximal pair stimulates the nerve with a very mild electrical impulse, and the distal pair records the resulting electrical activity. The time it takes for electrical impulses to travel between the electrodes is noted; this value is then divided by the known distance between the electrodes to yield the nerve conduction velocity.

Motor nerve conduction studies are performed by placing recording electrodes over a muscle and stimulating the nerve that innervates that muscle (Fig. 1). The recording electrodes consist of an active electrode placed over the muscle and a reference electrode placed distally over its tendon. The nerve innervating the muscle is then stimulated, usually at a proximal and distal site. The response recorded over the muscle is called a compound muscle action potential (CMAP). The CMAP can be described in terms of latency, amplitude, and conduction velocity (Fig. 2). Latency denotes the time required to note the initial response of the muscle after the application of the stimulus. Amplitude reflects the magnitude of the response and is proportional to the total number of nerve fibers activated.

If the appearance of the CMAP is considered to be the "response" to stimulation, the

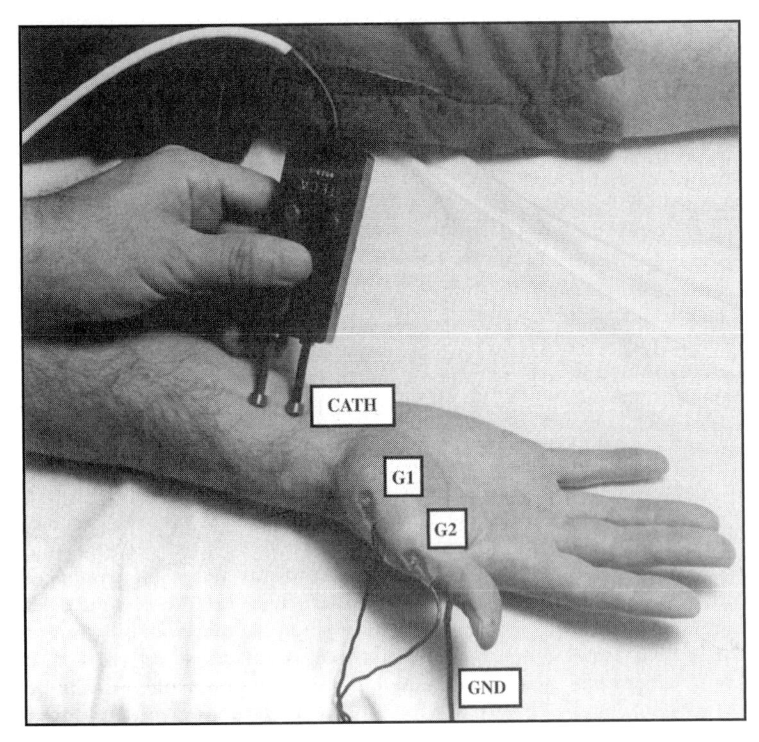

Figure 1
Motor nerve conduction study setup. The active recording electrode (G1) is placed on the center of the muscle, and the reference electrode (G2) is placed distally over the tendon. CATH = stimulator cathode, GND = ground.

*(Reproduced with permission from Preston DC, Shapiro BE: **Electromyography and Neuromuscular Disorders: Clinical-Electrophysiologic Correlations**. Boston, MA, Butterworth-Heinemann, 1998, p 26.)*

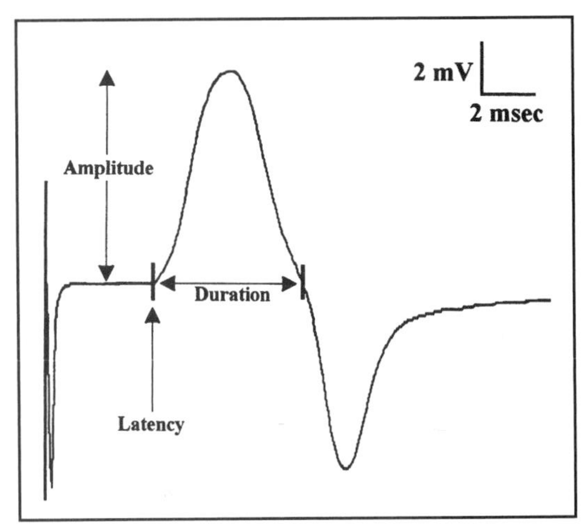

Figure 2
Compound muscle action potential (CMAP). Latency measures the delay between stimulus and response.

*(Reproduced with permission from Preston DC, Shapiro BE: **Electromyography and Neuromuscular Disorders: Clinical-Electrophysiologic Correlations**. Boston, MA, Butterworth-Heinemann, 1998, p 27.)*

conduction velocity cannot be calculated unless the nerve is stimulated at both a proximal and distal site. With such dual stimulation, the time required for the impulse to cross the neuromuscular junction (NMJ) and depolarize the muscle can be subtracted, yielding the nerve conduction time alone.

Sensory nerve conduction studies are performed by placing a recording and reference electrode over the nerve to be examined and then stimulating the nerve proximal or distal to the recording site (Fig. 3). The response is called a sensory nerve action potential (SNAP) and, as with CMAP, is quantified by latency, amplitude, and conduction velocity (Fig. 4). Because the SNAP is stimulated and recorded over the nerve, no NMJ or muscle depolarization time has to be accounted for, and conduction velocity can be calculated with one stimulation site.

NCSs also can be used to assess F waves in addition to the CMAP response. F waves represent delayed muscular stimulation that occurs after and as a result of the initial motor nerve stimulation. F waves appear as a "backfiring" or echo—the very same nerve impulse that leads to muscle contraction also travels away from the muscle toward the spinal cord, where it can stimulate the motor neuron in the spinal cord. This sends a recurrent antegrade nerve impulse that results in a second, delayed motor response. F-wave measurements reflect conduction along the entire nerve and are useful in the study of general polyneuropathies and Guillain-Barré syndrome.

As the F wave is routinely measured from a distal muscle, the latency may be prolonged in distal entrapments as well as in diseases of the proximal neuron. It may be more useful to think of the F wave as a nonspecific test of the entire nerve circuit.[1]

Electromyography

Electromyography (EMG) is a test that measures the electrical response of muscle contraction; in this respect, it is similar to an electrocardiogram, which is a test that measures the electrical effects of heart muscle contraction. EMG (or needle electromyographic examination) involves inserting a needle into a muscle and evaluating spontaneous and voluntary activity as detected by an electrode (Fig. 5). This test is considered by many to be mildly uncomfortable and is, by its nature, invasive.

EMG is an integral part of the electrodiagnostic study. Abnormal activity may be seen in disease states ranging from motor neuron disease to myopathy. EMG is very important in determining not only the severity of pathology but also the chronicity. It can be used for evaluating proximal muscles that are not easily studied with NCSs and often yields important information about the nervous system. It can be helpful, for example, in differentiating between primary nerve and muscle dysfunction and partial and complete nerve dysfunction.

After the resolution of the insertional activity (electrical activity in the muscle elicited by mechanical stimulation), there should be a period of electrical silence. A normal muscle will have no activity at rest. Common findings in acute denervation include fibrillation potentials (Fig. 6) or positive sharp waves, which represent the spontaneous depolarization of a single muscle fiber. Fasciculations result from spontaneous activity of groups of muscle fibers. These are common in amyotrophic lateral sclerosis (ALS). This first part of the EMG can help differentiate between primary muscle disease and apparent muscle dysfunction caused by denervation; in the latter, the muscle may be seen to fire at rest.

Voluntary activity is then evaluated by asking the patient to contract the muscle being examined. Motor unit action potential (MUAP) morphology is evaluated for amplitude, duration, and the pattern of activation and recruitment of muscle fibers. In general, neuropathies with adequate time for reinnervation can demonstrate large amplitude, duration, and/or polyphasic MUAPs. Recruitment of MUAPs is decreased with axonal loss or with demyelination and conduction block.

Variables that affect NCS results are numerous and include temperature, age, height, anomalous innervations, and timing of the electrodiagnostic study relative to the onset of injury.[1,2] The recognition of these variables and patterns is essential to avoid making mistakes in diagnosis. Cool temperatures slow conduction velocity, prolong latencies, and increase CMAP and SNAP amplitudes. Skin temperatures should be monitored and maintained, ideally, between 33°C (91.4°F) and 34°C (93.2°F) during the electrophysiologic examination.[1] Conduction velocities and CMAP and SNAP amplitudes decrease with advancing age. Decreased nerve conduction

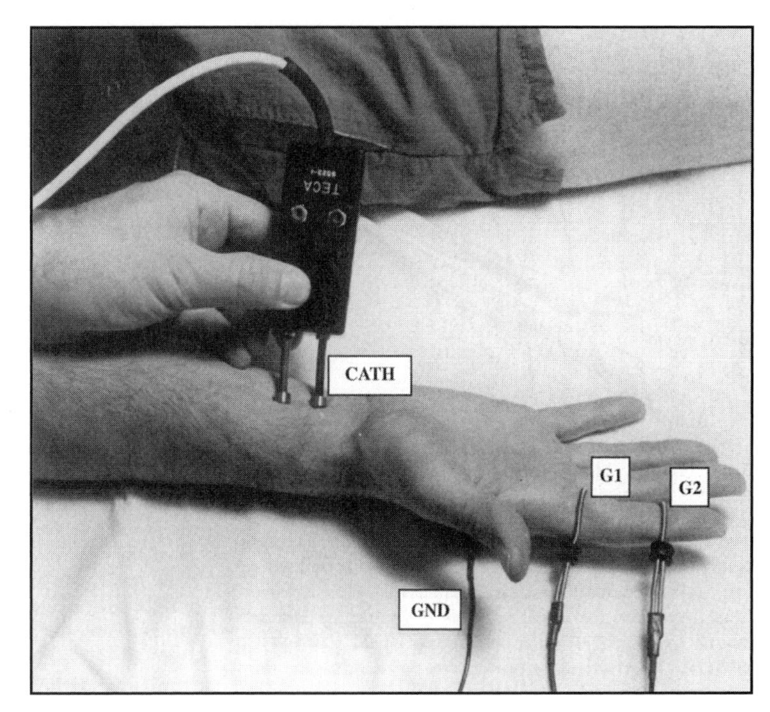

Figure 3

Sensory nerve conduction study setup. Ring electrodes are placed over the index finger, 3 to 4 cm apart. The active recording electrode (G1) is placed more proximally, closest to the stimulator. CATH = stimulator cathode, GND = ground, G2 = reference electrode.

*(Reproduced with permission from Preston DC, Shapiro BE: **Electromyography and Neuromuscular Disorders: Clinical-Electrophysiologic Correlations**. Boston, MA, Butterworth-Heinemann, 1998, p 29.)*

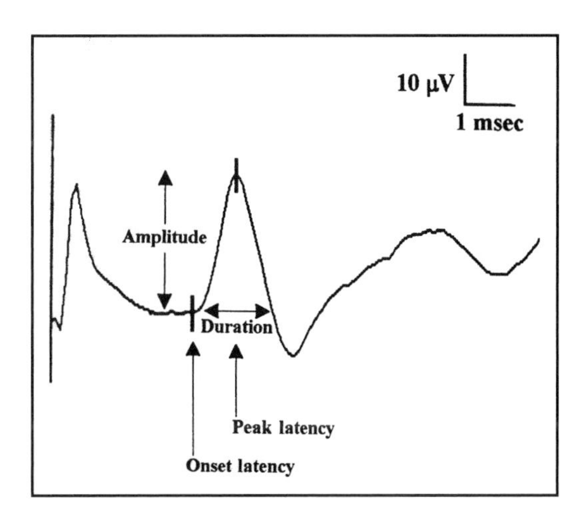

Figure 4

Sensory nerve action potential (SNAP). Onset latency measures the interval between the stimulus and the onset of response; peak latency measures the interval between stimulus and maximal repsonse.

*(Reproduced with permission from Preston DC, Shapiro BE: **Electromyography and Neuromuscular Disorders: Clinical-Electrophysiologic Correlations**. Boston, MA, Butterworth-Heinemann, 1998, p 30.)*

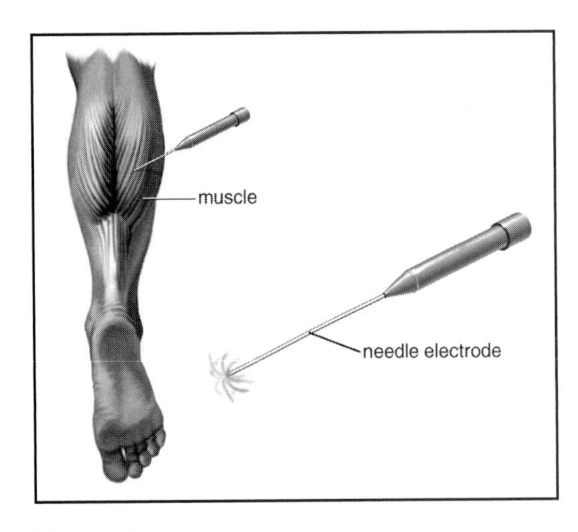

Figure 5
During EMG, a needle electrode is inserted into the muscle. The oscilloscope (not shown) registers the shape, size, and presence of the waveform of the action potential of each contracting muscle fiber, providing data regarding the muscle's response to stimulation of the nerves more proximally (not shown).

(© A.D.A.M., Inc.)

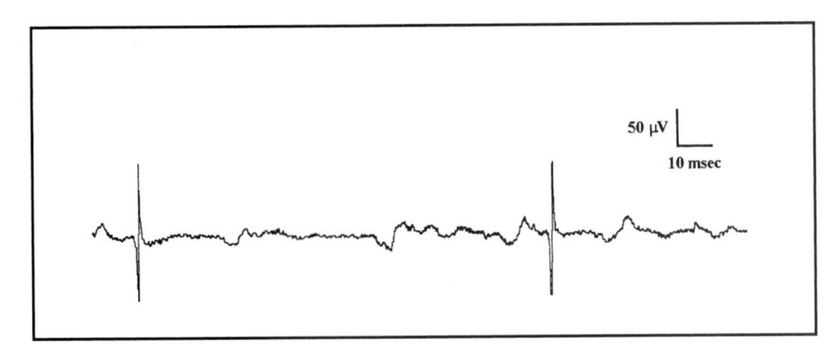

Figure 6
Fibrillation potential. Spontaneous depolarization of a single muscle fiber. This pattern may indicate acute denervation.

*(Reproduced with permission from Preston DC, Shapiro BE: **Electromyography and Neuromuscular Disorders: Clinical-Electrophysiologic Correlations**. Boston, MA, Butterworth-Heinemann, 1998, p 184.)*

velocities are seen in tall patients. Knowledge of anomalous innervations and their subsequent NCS patterns is also important to avoid mistaking a normal anatomic variant for nerve pathology.

Appreciation of the timing of the electrodiagnostic study relative to the time of the injury, if any, is very important because injured nerves undergo wallerian degeneration, a process that affects the nerve beyond the immediate location of injury. Wallerian degeneration is the means by which injured axons are broken down. It is characterized by Schwann cell degradation of the myelin sheath—a process that clears a path, so to speak, for the regenerating axons. The degeneration extends proximally from the injury to the preceding node of Ranvier and distally to the axon terminus. This process is not complete until approximately 7 days after injury.

Because of the timing of axonal degeneration, an NCS with stimulation and recording done distal to the lesion may produce normal results if performed too soon after injury; the distal axon functions normally. Conduction block (with demyelinating lesions) may be the only abnormality seen before wallerian degeneration occurs. EMG abnormalities occur in a proximal to distal temporal pattern. For example, in a proximal lesion (radiculopathy), the paraspinal muscles are affected first, and then more distal muscles are affected. Fibrillations or positive sharp waves obtained during EMG may take 1 to 2 weeks to occur in proximal muscles and 5 to 6 weeks to occur in more distal muscles. Reinnervated-appearing MUAPs (large amplitude, duration, or polyphasia) will also occur in a proximal to distal fashion, with early changes seen within 2 to 3 months after the injury. Knowledge of the acuity of the lesion and the time in which potential abnormalities may develop is essential to interpret results and recommend proper timing of the study.

Clinical Electrodiagnostic Testing

The purpose and scope of this section is not to extensively review all of the described neuromuscular disorders; rather, it is to provide an overview of commonly encountered disorders with a specific focus on their anatomy, presentation, examination findings, and related electrodiagnostic findings.

Motor Neuron Diseases

Motor neuron diseases are a group of disorders characterized by progressive deterioration of motor neurons. The most common form is ALS. This is a progressive disorder affecting both upper and lower motor neurons, usually beginning in a segmental distribution (ie, hand or leg), with symmetric or asymmetric involvement. Weakness, muscle wasting, and fasciculations are common findings on presentation. Speech and/or swallowing difficulties and respiratory compromise may occur. Physical examination findings include mixed upper and lower motor neuron signs, such as atrophy, hyperreflexia, and weakness. Sensation is spared.

Electrodiagnostic studies are useful in the workup of motor neuron diseases, but the diagnosis remains a clinical one.[1] Electrodiagnostic studies can also be very useful to evaluate other potential diagnoses, such as spinal stenosis. NCSs should include motor and sensory studies in the upper and lower extremities of the affected side. Motor NCS results may be normal or have decreased CMAP amplitudes.[3,4] Sensory NCS results are normal, but the presence of sensory abnormalities may suggest coexisting neuropathies and does not rule out motor neuron disease.[1,3]

Radiculopathies

Electrodiagnostic evaluation is an important tool in combination with imaging studies in the evaluation of cervical and lumbar **radiculopathies**. These conditions are often caused by nerve root compression, disk impingement, or central or neuroforaminal stenosis. Imaging studies, such as MRI or CT, can demonstrate the structure of the nerve root and possible impingement, but these studies do not evaluate nerve function. Additionally, MRI has a substantial clinical false-positive rate; disk degeneration, herniation, and spondylosis are seen often in asymptomatic individuals.[5]

Electrodiagnostic studies can be helpful to evaluate radiculopathy when imaging study results are normal and there is a high degree of clinical suspicion or when there are nonspecific abnormalities on the imaging study that do not correlate with the clinical presentation.[6]

Electrodiagnostic studies can exclude more distal lesions, such as plexopathies, polyneuropathies, and mononeuropathies, and localize the level of involvement in multilevel degeneration or disease. For example, a patient who presents with thumb numbness, nocturnal paresthesias, neck pain, and cervical spondylosis may have a C6 radiculopathy, a median neuropathy at the wrist (carpal tunnel syndrome), or both. If MRI of the neck shows nonspecific mild to moderate multilevel disk degeneration with neuroforaminal stenosis, electrodiagnostic studies will be helpful to evaluate the presence or absence of a distal entrapment neuropathy or a coexisting cervical radiculopathy. Radiculopathies typically present with radiating arm or leg pain and are usually but not always associated with neck and back pain, paresthesias, or weakness.

Motor NCS results are usually normal in cases of radiculopathy but may show decreased CMAP amplitudes if there is significant axonal loss. Decreased CMAP amplitudes are less commonly seen in cervical radiculopathies. This is because the motor NCSs for the median and ulnar nerves are recorded over the muscles that are innervated by the C8 and T1 nerve roots (ie, the abductor pollicis brevis and abductor digiti minimi). These more caudal roots are less commonly involved in cervical radiculopathies.

As radiculopathies usually occur from compression of the nerve root proximal to the dorsal root ganglion, sensory NCS results are normal in these lesions. Therefore, sensory NCSs are very useful in localizing lesions proximal (radiculopathy) and distal (plexopathy, polyneuropathy, and mononeuropathy) to the dorsal root ganglion.

On EMG, abnormal spontaneous and voluntary activity can be seen in proximal and distal muscles of the affected extremity as well as the paraspinal muscles. The distribution of abnormalities should be in multiple peripheral nerves with a specific myotomal distribution. If muscles examined are limited to those innervated by a single nerve, a distinction cannot be made on the basis of EMG alone between a mononeuropathy and a radiculopathy. Paraspinal muscle involvement is very helpful in localizing the nerve lesion at or proximal to the nerve roots because the paraspinal muscles are innervated by the dorsal rami that arise directly from the spinal nerves.[1] The absence of paraspinal involvement cannot, however, rule out a radiculopathy.

Plexopathies

The causes of brachial and lumbosacral **plexopathies** may be idiopathic, iatrogenic, traumatic, or neoplastic. Clinical findings on presentation include numbness and weakness with or without pain, depending on the etiology of the lesion. Weakness, altered sensation, and/or decreased reflexes in multiple peripheral nerve and nerve root distributions are also common. Electrodiagnostic testing can be very helpful in localizing and characterizing these complex lesions.

Routine motor NCSs may show decreased CMAP amplitude; conduction velocity may be decreased or normal, depending on the involved region of the plexus. Sensory NCSs are useful in localizing plexopathies because there are common sensory NCSs that can be performed to assess the C5 through T1 dermatomes. The results should be abnormal in plexopathies because the lesion is distal to the dorsal root ganglion. The distribution of abnormalities should be in the nerves supplied by the involved region of the plexus.

Polyneuropathies

Because etiologies of polyneuropathies are so numerous, electrodiagnostic evaluation is an important tool not only in establishing the diagnosis of polyneuropathy but also in characterizing specific polyneuropathies. Distinctions such as demyelinating or axonal and hereditary or acquired can help to narrow the differential diagnosis.

Polyneuropathies classically present as distal (stocking-glove) numbness and weakness. Pain may be present. On physical examination, a proximal to distal gradient of normal to abnormal symmetric findings is present. Polyneuropathies tend to appear distally first, with lower extremity involvement occurring before upper extremity involvement. Polyneuropathies usually affect both motor and sensory fibers, but there are predominately motor or sensory forms of polyneuropathies.

NCSs should be performed on both the upper and lower extremities. Motor and sensory NCS results will be abnormal, with more severe involvement in the lower extremities. Sensory NCS results may be more affected than motor NCS results. Limited comparison of the contralateral extremity will demonstrate symmetry of findings. If significant asymmetry is demonstrated, other diagnoses (such as plexopathy) should be considered. The presence of demyelination can be determined by the degree of latency prolongation and conduction velocity slowing. Conduction block may be present.

EMG should also show symmetric abnormalities in multiple peripheral nerve and nerve root distributions in a proximal to distal gradient. Lower extremity muscles should have greater involvement than upper extremity muscles. Proximal extremity or paraspinal muscles also should be examined; the presence of abnormalities in these muscles suggests more proximal pathology (ie, plexopathy and radiculopathy).

Mononeuropathies

An understanding of the anatomy of the upper and lower extremity will aid in identifying various mononeuropathies and differentiating these lesions from radiculopathies, plexopathies, polyneuropathies, and other lesions. As the electrodiagnostic studies are an extension of the physical examination, patterns of injuries will be similar, with abnormalities below the level of injury and normal function above the level of injury. Mononeuropathies of the upper extremities are commonly seen in the median, radial, and ulnar nerves and are often the result of direct compression.

The median nerve is the most commonly affected nerve in the arm, with the site of compression typically occurring at the carpal tunnel. Electrodiagnostic testing for carpal tunnel syndrome includes testing for median and ulnar motor and sensory responses. Specific comparison of the results of testing the median nerve with those of the ulnar nerve increases the electrodiagnostic sensitivity. Furthermore, two sequential comparison studies should be performed for patients with mild carpal tunnel syndrome to decrease the possibility of false-positive results in this population.[7] EMG is performed to assess severity (axonal loss) as well as to exclude proximal median neuropathies, brachial plexopathies, and cervical radiculopathies.

Radial nerve compression can be a source of pain, weakness, and sensory disturbances. The radial nerve is commonly injured at the spiral groove of the humerus secondary to trauma or fracture.[8] Radial neuropathies may occur in the axilla secondary to compression (as may be caused by using poorly fitting crutches, for example).[1] When there is a radial nerve injury at the axilla, patients will have elbow extension (triceps) weakness. Sensory abnormalities may affect the posterior arm and forearm (posterior cutaneous nerve of arm and forearm). EMG can reveal triceps abnormalities; however, distinguishing a high radial neuropathy from a plexopathy (posterior cord) or radiculopathy requires evaluation of nonradial muscles (eg, deltoid and paraspinals). Furthermore, tests of the median and ulnar nerves and nonradial innervated muscles are performed to exclude radiculopathy, brachial plexopathy, or other possible lesions.

The posterior interosseous nerve (a

branch of the radial nerve) can become entrapped near the elbow. Patients with such compression may have weakness on extension of the fingers and part of the wrist, but there should be no weakness proximal to the elbow. Sensation is preserved.

Ulnar neuropathies at the elbow are the second most commonly encountered nerve compression syndromes in the arm after carpal tunnel syndrome.[9] It is more difficult to localize an ulnar neuropathy at the elbow than it is to localize a median neuropathy at the wrist using electrodiagnostic studies.[1] Injuries of the lower trunk of the brachial plexus or the C8 and T1 nerve roots may also appear similar to an ulnar neuropathy, and electrodiagnostic studies will be helpful in localizing the injury. Entrapment of the ulnar nerve at the wrist occurs less commonly than at the elbow and may be associated with increased pressure over the ulnar surface of the palm, trauma, or compression.[1]

Femoral neuropathies have been described in association with surgical procedures (eg, hysterectomy, hip arthroplasty, and renal transplantation) as well as with hyperextension injuries at the hip, tumors, diabetes mellitus, and iliopsoas hematomas.[10] Depending on the etiology, patients may present with quadriceps weakness (knee buckling), decreased patellar reflexes, abnormal sensation or inguinal pain, swelling, or ecchymoses. A routine lower extremity electrodiagnostic examination will include testing of the tibial, peroneal, and sural nerves to differentiate plexopathies from polyneuropathies and saphenous and femoral motor studies to evaluate the femoral nerve. A screening examination of muscles innervated by the femoral, obturator, tibial, peroneal, and other nerves is performed to exclude a more generalized process.

Tibial nerve injuries are more common distally, where the nerve is more superficial.[10] Proximal tibial neuropathies are described secondary to trauma (eg, penetrating injuries and knee dislocations), tumors, cysts, or entrapment at the soleus. Distally, the nerve can become compressed posterior to the medial malleolus, where it travels under the flexor retinaculum (tarsal tunnel syndrome). The clinical presentation may include paresthesias of the plantar foot, posterior calf numbness, plantar flexion weakness, abnormal Achilles reflexes, intrinsic foot muscle atrophy, and clawing of the toes. The differential diagnosis includes incomplete neuropathies, sacral plexopathies, and S1-2 radiculopathies.

The peroneal nerve is a continuation of the sciatic nerve, with motor and sensory fibers from the L4-S2 nerve roots. The nerve divides into superficial and deep branches posterolaterally below the knee. The superficial peroneal nerve provides motor branches to the peroneus longus and brevis muscles, as well as sensation to the dorsum of the foot. The deep peroneal nerve innervates the tibialis anterior, peroneus tertius, and toe extensor muscles and also provides sensation for the first web space on the dorsum of the foot. Injuries can occur at the common nerve or at either the deep or superficial branch.

The peroneal nerve can be injured by trauma (eg, knee injuries, ankle sprains, impact at the fibular head, and fractures), masses, compartment syndromes, entrapment, external compression, vascular compromise, and other conditions.[10]

On physical examination, patients typically have dorsiflexion weakness (footdrop), eversion weakness (perceived as ankle instability), numbness, tenderness, and localized pain. The differential diagnosis includes L5 radiculopathies, lumbosacral plexopathies, sciatic neuropathies, and peripheral neuropathies.

Routine electrodiagnostic studies of the lower extremity include peroneal motor responses to the extensor digitorum brevis and are supplemented by peroneal motor studies to the tibialis anterior and superficial peroneal sensory responses. On EMG, both deep and superficial peroneal nerve innervated muscles are examined in addition to the short head of the biceps femoris, which is innervated by the common peroneal nerve. Muscles innervated by the tibial nerve are also examined, and more proximal muscles may need to be examined to differentiate a mononeuropathy from radiculopathies, plexopathies, and polyneuropathies, depending on the findings.

Neuromuscular Junction Disorders

Neuromuscular junction disorders are uncommon and usually present as symmetric proximal weakness and fatigue without sensory involvement that can be mistaken for myopathy. Eye findings, such as extraocular muscle weakness and ptosis, are common and may be asymmetric. Bulbar involvement may affect speech and swallowing. Physical examination findings usually include fatigability and proximal weakness, with preserved sensation and preserved reflexes. NMJ disorders can be classified by the site of dysfunction and include Lambert-Eaton myasthenic syndrome, tetanus and botulism (presynaptic), organophosphate poisoning (synaptic), and myasthenia gravis (postsynaptic).[11] Routine electrodiagnostic testing is important to evaluate and rule out myopathies or other neuromuscular disorders with predominately motor involvement. Specialized electrodiagnostic testing using repetitive nerve stimulation can help make the diagnosis of an NMJ disorder and characterize the site of the pathophysiology.

Motor NCS results are usually normal in patients with myasthenia gravis; the CMAP amplitudes may be diffusely low in patients with Lambert-Eaton myasthenic syndrome.[11] Sensory NCS results are usually normal, and there is usually no abnormal spontaneous activity on EMG, with the exception of botulism, which may have fibrillation potentials.[1,11] Voluntary MUAPs may appear myopathic, with small amplitudes, short durations, and polyphasia. Repetitive nerve stimulation at a low frequency (3 Hz) can show a decreasing amplitude response in presynaptic and postsynaptic disorders. Repetitive nerve stimulation at a high frequency (50 Hz) will show an increasing amplitude response in presynaptic disorders.

Myopathies

Electrodiagnostic evaluation is a useful tool in combination with other diagnostic tests, such as muscle biopsy, to diagnose and characterize myopathic disorders. As with polyneuropathies, myopathies represent a wide and varied group of disorders that cannot be comprehensively covered within the scope of this chapter. Myopathies classically present with symmetric proximal weakness involving the upper extremities, lower extremities, or both. They may be acquired or hereditary, with weakness and functional impairments ranging from mild to severe. Sensation is not affected. Exceptions exist, and some myopathies may be asymmetric (eg, inclusion body myositis and facioscapulohumeral muscular dystrophy), distal (eg, myotonic dystrophy and inclusion body myositis), or painful (eg, dermatomyositis, polymyositis, and toxic myopathies).[1,12] Physical examination findings include proximal weakness and atrophy, spared sensation, and usually normal reflexes.

Motor NCS results are usually normal because routine upper and lower extremity studies (ie, of the median, ulnar, tibial, and peroneal nerves) use recording electrodes over distal muscles, whereas myopathies typically affect proximal muscles. In severe myopathies or those with distal involvement, CMAP amplitudes may be low. Sensory NCS results will be normal. F waves are usually normal. NCSs are done to evaluate other disorders that may present with symptoms similar to those of myopathies (eg, motor neuron diseases, motor neuropathies, and NMJ disorders).[1]

EMG is the most important part of the electrodiagnostic study in the workup of myopathy. Spontaneous activity may be absent or present and can help to narrow the differential diagnosis. Voluntary activity in myopathy is usually seen as small amplitude and duration, polyphasic MUAPs secondary to muscle fiber dysfunction, or dropout.[7] However, this type of MUAP can be seen in a variety of disorders, including NMJ disorders, early reinnervation (nascent motor units), and periodic paralysis.[1]

Recruitment is often described as "early" and can only be assessed by the examiner resisting the force generated by the patient during active muscle contraction. As the activated MUAPs have fewer muscle fibers, a smaller amount of force is generated than would typically be seen for the same amount of MUAPs being activated.

Key Terms

Compound muscle action potential (CMAP) The latency, amplitude, and conduction velocity recorded over muscle in motor nerve conduction studies

Electromyography (EMG) A test that measures the electrical response of muscle contraction

Mononeuropathies A group of disorders characterized by disease of a single nerve

Motor nerve conduction studies Studies that test the speed by which motor nerves transmit impulses by placing recording electrodes over a muscle and stimulating the nerve that innervates that muscle

Motor neuron diseases A group of disorders characterized by progressive deterioration of motor neurons

Myopathies A wide and varied group of primary muscle disorders characterized by weakness

Nerve conduction studies Studies that test the speed by which motor, sensory, or mixed (combined motor and sensory) nerves transmit impulses

Neuromuscular junction disorders A group of disorders affecting the neuromuscular junction characterized by symmetric proximal weakness without sensory involvement

Plexopathies A group of disorders characterized by numbness and weakness with or without pain and caused by injury to a collection of nerves, such as the brachial plexus

Polyneuropathies Neuropathies of several peripheral nerves simultaneously

Radiculopathies A group of disorders often caused by nerve root compression or central or neuroforaminal stenosis

Sensory nerve action potential (SNAP) The latency, amplitude, and conduction velocity recorded in sensory nerve conduction studies

Sensory nerve conduction studies Studies that test the speed by which sensory nerves transmit impulses by placing a recording and reference electrode over the nerve to be examined and then stimulating the nerve proximal or distal to the recording site

References

1. Preston DC, Shapiro BE: (eds): *Electromyography and Neuromuscular Disorders: Clinical-Electrophysiologic Correlations*. Boston, MA, Butterworth-Heinemann, 1998.
2. Kimura J: (ed): *Electrodiagnosis in Diseases of Nerve and Muscle: Principles and Practice*, ed 2. Philadelphia, PA, FA Davis, 1989.
3. Dumitru D: (ed): *Electrodiagnostic Medicine*. Philadelphia, PA, Hanley & Belfus, 1995.
4. Krivickas LS: Electrodiagnosis in neuromuscular diseases. *Phys Med Rehabil Clin North Am* 1998;9: 83-114.
5. Boden SD, McCowin PR, Davis DO, et al: Abnormal magnetic-resonance scans of the cervical spine in asymptomatic subjects: A prospective investigation. *J Bone Joint Surg Am* 1990;72:1178-1184.
6. Haig AJ, LeBreck DB, Powley SG: Paraspinal mapping: Quantified needle electromyography of the paraspinal muscles in persons without low back pain. *Spine* 1995;20:715-721.
7. Preston DC, Ross MH, Kothari MJ, et al: The median-ulnar latency difference studies are comparable in mild carpal tunnel syndrome. *Muscle Nerve* 1994;17:1469-1471.
8. Barton NJ: Radial nerve lesions. *Hand* 1973;5: 200-208.
9. Bradshaw DY, Shefner JM: Ulnar neuropathy at the elbow. *Neurol Clin* 1999;17:447-461.
10. Stewart JD: (ed): *Focal Peripheral Neuropathies*. New York, NY, Elsevier, 1987.
11. Strommen J: Electrodiagnosis in neuromuscular junction disorders. *Phys Med Rehabil* 1999;13: 281-306.
12. Krivickas L, Carlson N: Myopathies. *Phys Med Rehabil* 1999;13:307-332.

Section Editor
Marshall L. Balk, MD

Section 5

Management

Contributing Authors
Jaimo Ahn, MD, PhD
Marshall L. Balk, MD
Joseph Bernstein, MD, MS
Nicholas A. DiNubile, MD
Sarah Yuxing Fan, MD
Kevin M. Guskiewicz, PhD, ATC-L
David Hill, PT
Seth S. Leopold, MD
James A. Onate, MA, ATC-L
Brigham B. Redd, MD
Jeffrey D. Stone, MD
A.J. Yates, Jr, MD

Management

The physician's job, said Voltaire, is to amuse the patient until nature heals the disease. Of course, Voltaire lived in the eighteenth century when patients would have been lucky if their physicians attempted no more than amusement. This was the age before antibiotics, anesthesia, and other innovations that define modern medicine. Today, the art of medicine is much more.

Modern physicians have three responsibilities: to make a diagnosis, to offer a prognosis, and to create and implement a treatment plan. The last task, treatment, seems to be the essential one—the one that defines the physician's therapeutic role. Nevertheless, this textbook deliberately relegates the discussion of management to a small section at the end. The primary reason for this is that *Musculoskeletal Medicine* is written for medical students, physicians-in-training who are charged more with cultivating a basic understanding of disease processes than with learning the fundamentals of management. The nuances of treatment will be covered in residency training and are, of course, the subject of many other fine books.

The details of treatment are also glossed over, one may add, in an attempt to challenge the student to deduce them from basic principles. Training resident physicians by forcing them to deduce what is already known is certainly impractical, but demanding hard thinking from students is an extremely effective means of enhancing insight. Just as the jogger jogs for the sake of improving fitness and not to get to a specific destination, so too the student may attempt to deduce treatment approaches as a form of mental exercise and not practical training.

The last reason for limiting the discussion of therapeutics is because this category is the one most likely to need revision. When I entered medical school, the dean greeted us with the following jocular warning: "Half of what we are to teach you is not true. The problem is we don't know which half." The dean was wrong; he did know which half, or at least he could have made a good guess. If anything we were taught was wrong, these inaccuracies were almost certainly to be found in the realm of therapeutics. The history of medicine is littered with examples of well-intentioned and well-reasoned therapies that were, in retrospect, simply off-base. This process of discovering our missteps is not ancient history either—the surgical resection of gastric ulcers was a mainstay of my third year surgery clerkship not too long ago. Accordingly, the discussion of therapeutics does not dominate this book simply because the information in this section is least likely to endure.

Another way of looking at this, though, is that the section on therapeutics comes last as congratulatory coda to *Musculoskeletal Medicine*. As wrongheaded as our therapies may be proven some day in the future, they are still extremely effective in mitigating human suffering now. The modern treatment of musculoskeletal disease (however flawed) has essentially obliterated the word "cripple" from the contemporary vernacular. This section is worth studying in detail because it describes treatment methods that are extremely effective. That these methods are imperfect is no secret and no source of shame. Rather, their imperfection should stand as an invitation to the next generation to improve them.

—Joseph Bernstein, MD, MS

Chapter 30

Nonsurgical Therapy

The successful management of musculoskeletal injuries begins with an initial evaluation and ends with the return to normal activities. For many injuries, there are several important steps along the way to achieving a positive outcome. The goals of management include: (1) protecting damaged tissues, (2) controlling swelling, (3) reducing pain, and (4) limiting loss of function. Meeting these initial goals can be facilitated using a combination of protection, rest, ice, compression, and elevation, better known as the RICE method.

The healing process will be delayed if the injured structure is not rested initially. Rest or restricted activity following injury is very important if the healing process is to begin in a timely fashion. The duration of rest depends on the severity of the injury and the amount of tissue damage; in most cases, at least 48 hours is needed before rehabilitation is begun. Protecting the injured area using a splint, crutches, or sling or by taping or wrapping also should be considered during this period of rest.

Types of Nonsurgical Therapy

Cryotherapy

Cryotherapy, the therapeutic use of cold, should be the initial treatment of choice for musculoskeletal injuries because it minimizes secondary cell death from hypoxia. It also reduces pain.[1] Cooling of tissues decreases metabolic activity, thereby reducing oxygen demand and allowing injured cells in the area to survive. Pain relief is most likely the result of slowed nerve conduction velocity and reduction of the reflex muscle spasm. Following musculoskeletal injury, the application of ice is most effective at 20- to 30-minute intervals three to four times within the first several hours of injury, followed by at least three applications daily thereafter. Cold can be applied in many ways, including wrapping a bag of crushed ice to the injury site, massaging the area with an ice cup, or submerging the body part into a cold whirlpool. Cryotherapy should continue until the signs and symptoms of inflammation have disappeared or no further benefits of applying cold are evident.

Compression and Elevation

Compression and elevation are arguably more helpful in reducing initial swelling than is applying ice. Compression reduces swelling by mechanically reducing the amount of space available for fluid to accumulate.[1-3] The best compression technique involves applying an elastic bandage directly over the injury. The bandage should be wrapped in a distal to proximal direction, with pressure greatest at the most distal point and gradually decreasing as it moves proximally (Fig. 1). In some regions, especially around bony structures such as the ankle, horseshoes and donuts made with felt padding can increase the effectiveness of external compression. Elevating the injured body part will eliminate the effects of gravity and assist in returning the body's fluids back to the central circulatory system. When combined, compression and elevation decrease the intravascular hydrostatic pressure and increase the extravascular hydrostatic pressure, thus promoting lymphatic drainage.[1]

Thermotherapy

Other therapeutic modalities, such as superficial heating, ultrasound, and electrical stimulation are also effective tools that can be used in the rehabilitation process. Thermotherapy, whether using a hydroculator pack or warm whirlpool, is a universal treatment for pain and discomfort. Externally applied heat penetrates to a tissue depth of approximately 2 cm.[2] However, heat must not be introduced too soon after injury, as it increases capillary blood pressure and cellular permeability, which can result in additional swelling and edema.[2] Heat modalities should

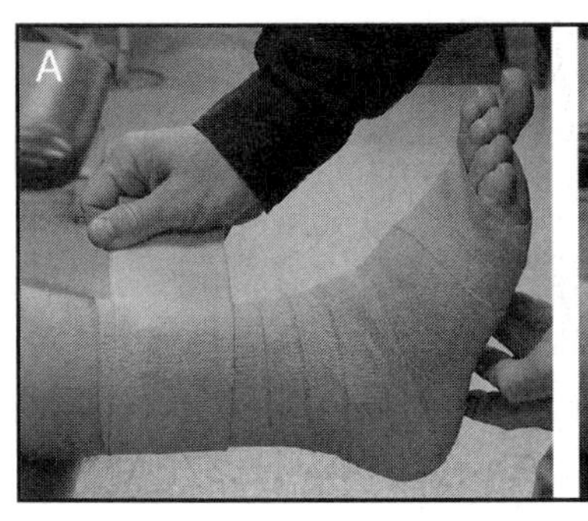
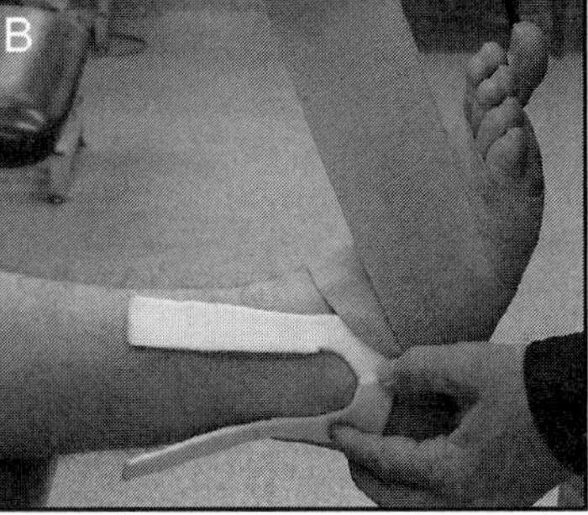

Figure 1
A, Elastic wrap using graduated compression from distal to proximal.
B, Elastic wrap using a felt horseshoe for focal compression.

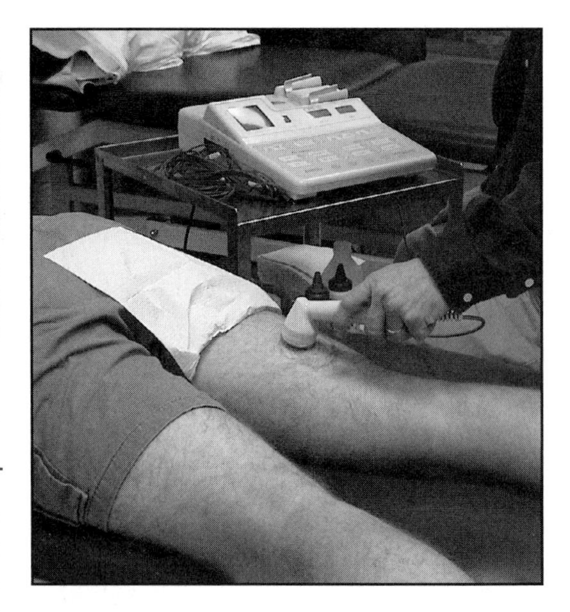

Figure 2
Therapeutic ultrasound.

not be introduced until signs and symptoms associated with the inflammatory phase have resolved. The following questions should be asked to help determine when to move from cryotherapy to thermotherapy: (1) Is the swelling controlled? (2) Is range of motion limited by stiffness rather than swelling? (3) Is skin temperature normal? (4) Has progress with cold treatments reached a plateau?

The primary goals of thermotherapy include increasing blood flow and muscle temperature to stimulate analgesia; stimulating cellular metabolism to increase the elasticity of muscle, tendon, and ligamentous tissue; increasing nutrition to the cellular level; and increasing lymphatic drainage for the removal of metabolites and other products of the inflammatory process.[2]

Therapeutic Ultrasound

Therapeutic ultrasound, which delivers inaudible, high-frequency acoustic vibrations to a tissue depth of up to 5 cm, is classified as a deep-heating modality.[4,5] The clinical effects of ultrasound are similar to those of superficial heating modalities; however, ultrasound can be used for either thermal or nonthermal therapy. Thermal effects include increasing collagen extensibility and blood flow, while decreasing pain, muscle spasm, and joint stiffness. Nonthermal effects of ultrasound include increasing interstitial fluid flow around an area (cavitation) and altering the permeability of cell membranes to sodium and calcium ions (microstreaming), both of which can help during the healing process. Ultrasound is especially effective in muscle tissue because of muscle's high concentration of protein, which allows for an increased absorption rate. Thus, ultrasound is very effective in the treatment of strains and contusions (Fig. 2). Phonophoresis is the transdermal introduction of a topically applied medication, usually either an anti-inflammatory or analgesic, into soft tissue using the ultrasound's acoustic waves. Although ultrasound can be used for many purposes, it is used most often to enhance soft-tissue healing, to release scar tissue and joint contractures, and to reduce chronic inflammation.

Electrical Stimulation

Electrical stimulation is one of the most frequently used treatment modalities for musculoskeletal injuries because the electrical current passing through tissue can trigger physiologic, chemical, or thermal effects. The type and extent of the response to electrical stimulation depend on the type of tissue and its physiologic response characteristics and the parameters of the electrical current applied. The parameters of electrical stimulation include waveform, current intensity, current duration, current frequency, and polarity. Perhaps the most common indication for electrical stimulation is pain modulation, whereby the sensory nerves are stimulated to alter a patient's perception of a painful stimulus.

Various types of stimulators and electrode configurations are available for administering electrical stimulation. Standard units offer a variety of waveforms and treatment

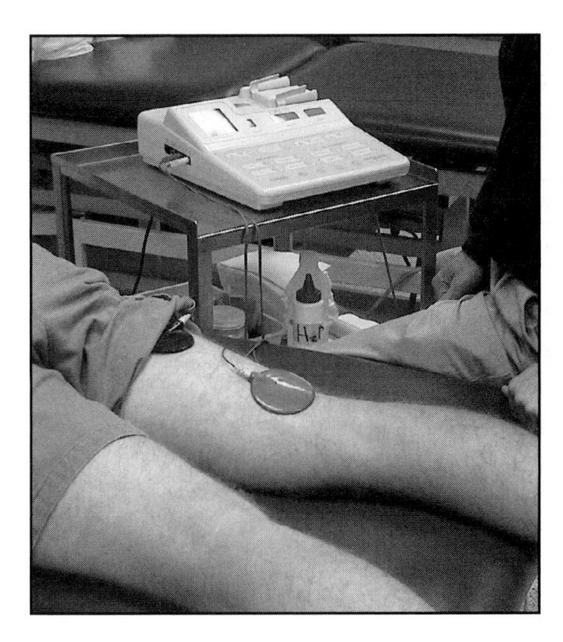

Figure 3
Electrical stimulation.

combinations, including an electrical stimulation and ultrasound treatment option (Fig. 3). Another commonly used treatment option is a portable transcutaneous electrical nerve stimulation (TENS) unit (Fig. 4), which is useful for home treatments because patients can easily be trained in its use. Various pain modulation techniques can be administered using these units, and in some instances, patients can exercise while using them.

Joint Mobilization

Joint mobilization is a manually administered treatment modality that is useful for improving flexibility and decreasing pain. Other goals of joint mobilization include decreasing muscle guarding, stretching or lengthening tissue surrounding a joint (especially capsular and ligamentous tissue), stimulating reflexogenic effects that can inhibit or facilitate the stretch reflex, and improving proprioceptive awareness about a joint. Appropriately administered joint mobilization requires a clinician with an understanding of joint anatomy and the types of movement that govern the motion of each joint. Joint movement is categorized by physiologic motion and accessory motion. Physiologic motion refers to traditional cardinal plane movement as a result of contracting muscles, whereas accessory motion refers to the manner in which one articulating joint surface moves relative to another. Accessory motions

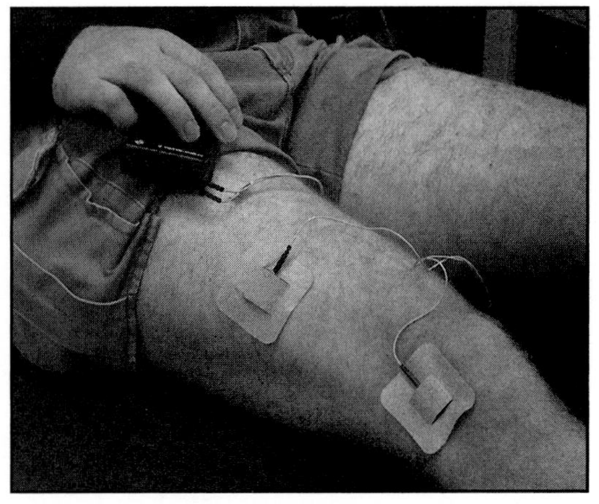

Figure 4
Portable TENS unit.

are also known as joint arthrokinematics, which include spin, roll, and glide.[3]

In a contracted joint, there is an abnormal point of limitation. This point is short of the anatomic limit because of pain, spasm, or tissue resistance. In such instances, joint mobilization can improve accessory movement, but the clinician using this type of therapy must have a clear understanding of the convex/concave rules for joint mobilization. In the case of shoulder capsulitis, the convex surface of the humeral head can be oscillated with gentle pushing and pulling in an inferior direction on the concave surface of the glenoid fossa (Fig. 5). This movement will improve shoulder abduction when performed in the coronal plane and will improve shoulder flexion when performed in the sagittal plane. Joints that are hypomobile and have restricted movement should be treated three to four times per week, with motion exercises performed at home on the other days. Typical mobilization of a joint may involve a series of three to six sets of oscillations lasting between 20 and 60 seconds each, with one to three oscillations per second.[3] Mobilizations can be performed on any joint with range-of-motion restrictions in the subacute and chronic injury phases. However, the clinician must avoid aggressive movement of joints that still show signs of increased pain and/or inflammation following the initial treatments.

Bracing, Taping, and Orthotics

Bracing

With ligamentous injuries to the knee, full or partial immobilization of the joint following the injury or subsequent surgery is often required. A knee immobilizer, which locks the

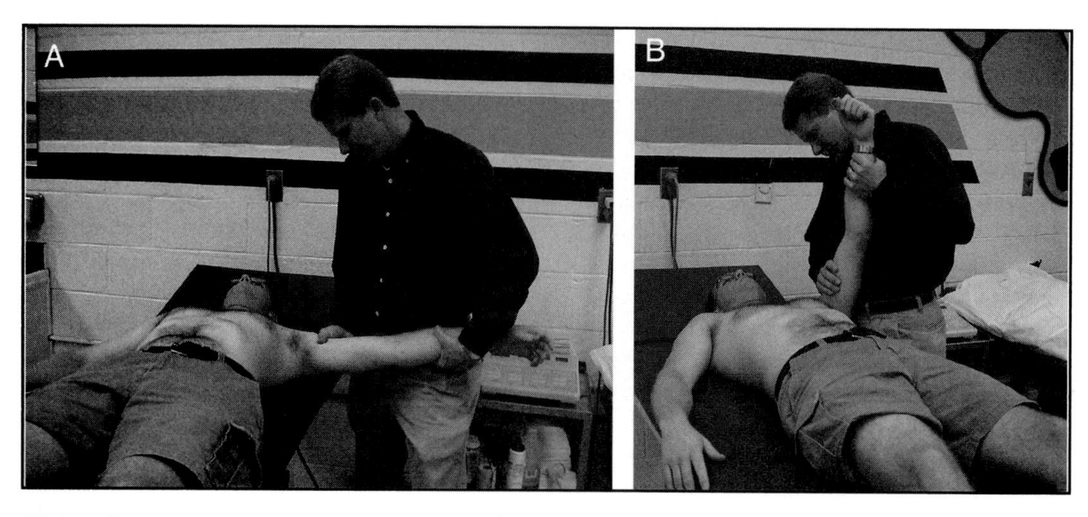

Figure 5
Inferior glide of the humerus in the frontal plane (**A**) and the sagittal plane (**B**).

knee in 0° extension, should be applied initially, followed by a rehabilitative or functional knee brace that can be set to various ranges of motion (Fig. 6). When range of motion is tightly restricted, crutches are necessary for ambulation. As pain subsides and range of motion improves, the rehabilitative brace is most effective for allowing joint movement within a safe range of motion, thereby limiting the potential for disuse atrophy while protecting the joint from further injury. These braces are important for rehabilitating the joint while progressing to strengthening exercises. Although not as popular as those used to treat lower extremity injuries, immobilizers and rehabilitative braces are also available for the treatment of elbow and wrist injuries.

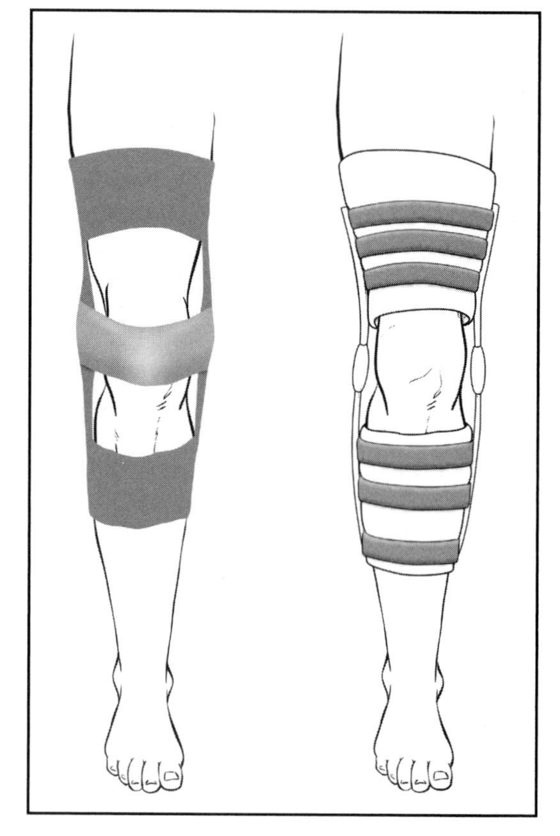

Figure 6
A, Knee immobilizer. **B**, Rehabilitative hinged knee brace.

(Adapted from France EP, Paulos LE: Knee bracing. ***J Am Acad Orthop Surg*** *1994;2:281-287.)*

Taping

Protective taping and wrapping can be effective for increasing stability, especially if the patient intends to be physically active. The purpose of taping a joint is to limit excessive motion. Taping should be done as an adjunct to other therapeutic modalities and rehabilitation strengthening techniques. Relying on protective taping alone may lead to inherent muscle and ligament weakness. Ankle taping is commonly used, but ankle bracing is becoming more popular because of logistical and cost considerations.

Orthotics

Orthotic devices can be used to correct certain lower extremity malalignments, such as pes planus, pes cavus, or hypermobility at the subtalar joint.[3] Secondary injuries resulting from these malalignments, including stress fracture, tibial stress reaction, plantar fasciitis, and patellofemoral pain, can often be successfully managed using orthotics because small modifications in the subtalar joint position can affect the stresses placed at the ankle, knee, hip, lumbar spine, or any of various points along the kinetic chain. Orthotic devices are available as pads or felt supports (soft and flexible, semirigid, or rigid) and may be either custom-molded or premolded.[3] Because many different materials and methods are used to fabricate orthotic devices, clinicians should thoroughly investigate which style and brand is best for each patient.

Functional Progression in Rehabilitation

Range of Motion

Range of motion is defined as the amount of movement available at a joint. After an injury, loss of range of motion may occur as a result of contracture of connective tissue or resistance of the musculotendinous unit to stretch. The amount of range of motion available at an injured joint often determines how well the patient can move the joint following an injury. If the injured body part cannot move through the normal range of motion, then use of the injured part will be limited, resulting in alterations in movement strategies (eg, altered gait). Reestablishing complete, pain-free range of motion is crucial in the functional progression of injury rehabilitation.[6] The clinical evaluation of range of motion can be conducted objectively by assessing muscle tension. Postoperatively, a goniometer is often used to measure the precise amount of joint motion that can be obtained by a patient both actively and passively. To track progress in the rehabilitation plan, the amount of motion available at a joint should be assessed during the early, intermediate, and late stages of injury.

Injury prevention and rehabilitation is dependent on the ability of a joint to complete a nonrestricted, pain-free motion while a person performs an intended movement.[6,7] Therefore, aggressive rehabilitation must be deferred until motion is normal. Certain physical activities require increased levels of flexibility to achieve desired motion patterns. Flexibility or range-of-motion exercises can be active, active-assistive, or passive. Active range-of-motion exercises are those in which the patient moves a joint through a range of motion via muscle contraction. Active-assistive range-of-motion exercises are those in which the injured individual performs a muscle contraction to move a joint and the clinician assists the injured individual to move the joint further through the range of motion. Passive range-of-motion exercises are those in which the injured limb is moved through a range of motion by the clinician without any muscle contraction elicited from the injured individual.

Flexibility

Flexibility can be increased using various modalities and training techniques. Stretching techniques include static, ballistic, and proprioceptive neuromuscular facilitation (PNF). Static stretching, used by active individuals before exercising, is the most commonly used method of increasing flexibility. It involves passively stretching a given antagonist muscle by placing it in a maximal position of stretch and holding it there for an extended period. Static stretching is a safe technique for individuals beginning a flexibility routine or during the initial stages of injury rehabilitation. Ballistic stretching, also known as dynamic stretching, consists of repetitive ("bouncing") contractions of the agonist muscle to stretch the antagonist muscle. An advanced form of flexibility training, ballistic stretching should be used cautiously. It should not be used for rehabilitation because the uncontrolled forces that are created when ballistically stretching a muscle may exceed the extensibility limits of the muscle fiber and cause microtears.[6,7]

Proprioceptive neuromuscular facilitation (PNF) is an advanced stretching technique that incorporates the body's reflexes to facilitate stretching of the affected body part. A number of different PNF techniques can increase flexibility, including contract-relax, hold-relax, and slow-reversal-hold-relax. All PNF techniques are based on the underlying principle that combinations of alternating isometric or isotonic contractions and relaxations of both agonist and antagonist muscles can increase flexibility. Usually performed with a rehabilitation specialist applying resistance or assistance, PNF can also be done alone with the aid of some external force, such as a wall or doorway.[6]

Strength

Strength is important in the functional progression of injury rehabilitation.[6,8] Restoring strength is based on the overload theory, which holds that if a muscle is forced to work harder than it is accustomed to working, then gradually it will adapt to the imposed demands. Resistive training exercises involve two different types of muscle contractions: concentric or eccentric.[8] Concentric exercises are those in which the muscle shortens while contracting against resistance. Eccentric exercises are those in which the muscle lengthens despite resisting a force, as in slowly lowering a weight. For example, during a biceps curl, the concentric phase occurs while flexing the elbow, and the eccentric phase occurs while extending the elbow. Each type of muscle contraction is important in strength training and should be emphasized throughout the rehabilitation process.[8]

Three basic types of resistance training techniques can be used: isometric, isotonic, and isokinetic. Isometric exercise involves a muscle contraction in which the muscle length remains constant while tension develops within the muscle. Isometric exercise is useful in the early stages of rehabilitation because it activates injured body parts without placing harmful stress by moving a joint throughout a range of motion. Since isometric exercise increases blood pressure and provides beneficial effects only at isolated angles of contraction, it should be used cautiously; however, it should not be avoided entirely during the initial stages of rehabilitation.

Isotonic exercise increases strength by muscle shortening (concentric phase) and lengthening (eccentric phase) throughout the full range of motion using a constant load at a variable speed.[7,8] Barbells and most manual-resistance weight machines can provide isotonic exercise.

Isokinetic exercise, performed with the aid of specially designed machines, restricts the amount of force applied against a resistance so that a muscle can be moved only at a fixed speed. In other words, the speed of movement in an isokinetic exercise will remain constant despite the amount of force that is being externally applied. Isokinetic exercise devices are costly, but they can provide objective, quantifiable data for evaluating an injured patient's rehabilitative progress. Data regarding an individual's force output, power, work, and endurance throughout a selected range of motion can be obtained for comparison with reciprocal muscle groups or the uninvolved side.

Balance

Musculoskeletal injuries to the lower extremity, such as sprains, result in decreased proprioception. Joint proprioceptors are believed to be damaged during both complete and incomplete rupture of the ligaments because the joint receptor fibers possess less tensile strength than the ligament fibers.[9,10] The damage to the joint receptors is believed to diminish the supply of messages from the injured joint up to the brain.[10] For instance, when the anterior cruciate ligament (ACL) in the knee is torn or stretched, proprioception from the knee joint is decreased, and a patient's ability to balance on the ACL-injured leg may be decreased, even following surgical reconstruction of the knee.[11-13]

Rehabilitation protocols for the lower extremity should emphasize the importance of balance and proprioceptive exercises once the patient has regained range of motion and some degree of overall strength.

A close relationship exists between proprioception and balance. Therefore, many of the exercises proposed for kinesthetic training indirectly enhance balance. A variety of activities can be used to improve balance, but clinicians should ensure that the exercises are safe yet challenging; stress multiple planes of motion; incorporate a multisensory approach; begin with static, bilateral, and stable surfaces and progress to dynamic, unilateral, and unstable surfaces; and progress to more functional exercises.

Balance exercises should be performed in an open area, where the patient will not be injured in the event of a fall. It is best to perform exercises with an assistive device, such as a chair, railing, table, or wall, within reach, especially during the initial phase of rehabilitation. When considering exercise duration for balance exercises, either sets and repetitions or a time-based protocol can be recommended. The patient can perform two to three sets of 15 repetitions and progress to 30 repetitions as tolerated, or perform 10 of the exercises for a 15-second period and progress to 30-second periods later in the program. Figure 7 represents single-leg balance activities performed on foam and a biomedical ankle platform system, or BAPS board.[14]

Aquatic Therapy

Aquatic therapy is gaining popularity within the musculoskeletal rehabilitation setting be-

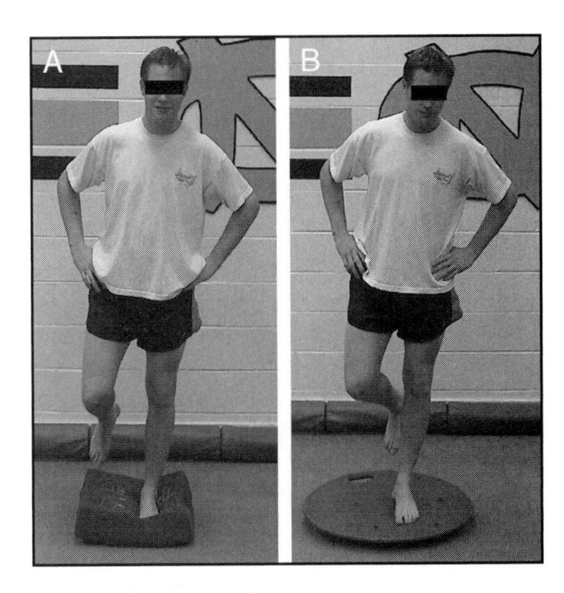

Figure 7
Balance training on medium-density foam (**A**) and a BAPS board (**B**).

cause gravitational forces are reduced in water due to the buoyancy of the human body. Thus, patients with lower extremity injuries can resume walking or running while reducing the amount of load placed on the affected body part. Water also provides resistance to movement.[15] With an overload program designed to strengthen certain muscles, movements that oppose buoyancy (such as an upward thrust of the legs) will be resisted by the water and help strengthen the affected body part. Safe range-of-motion and strengthening exercises can be achieved by understanding the potential effects of aquatic therapy and can be manipulated according to height of water level, speed of movement, and direction of movement.[15] Aquatic therapy can be particularly helpful for obese patients for whom land-based exercise is painful because of overload.

Occupational Therapy

Occupational therapy, as defined by the American Occupational Therapy Association, is the discipline that helps people "regain, develop, and build skills that are essential for independent functioning, health, and well being." Occupational therapy is particularly useful after injuries to the hand. The details of occupational therapy are beyond the scope of this book but may be found at http://www.aota.org.

Prescription Procedures

A physical therapist first evaluates a patient's condition by reading the physician's prescription. It is important, therefore, that each particular diagnosis and condition is clearly written and followed by the words "evaluate and treat." Although prescriptions often specify the frequency and duration of therapy, the optimal duration is best determined by the patient's response to therapy and the ongoing communication between the physician and the physical therapist. Physicians can specify particular treatments by listing protocols or procedures for a particular condition. There are also many textbooks with protocols listed for specific conditions.[16-18]

Key Terms

Ballistic stretching Repetitive, bouncing contractions of an agonist muscle to stretch its antagonist muscle

Concentric exercises Exercises in which the muscle shortens while contracting against resistance

Cryotherapy The therapeutic use of cold

Eccentric exercises Exercises in which the muscle lengthens despite resisting a force, as in slowly lowering a weight

Isokinetic Literally "same speed"; when applied to muscle action, it implies constant velocity of shortening

Isometric Literally "same length"; when applied to muscle action, it implies that the muscle length is held constant

Isotonic When applied to muscle action, it implies that the load is constant

Joint mobilization A manually administered treatment modality in which joints are manipulated to improve flexibility and decrease pain

Phonophoresis The transdermal introduction of a topically applied medication (usually either an anti-inflammatory or analgesic) into soft tissue using ultrasound

Physiologic motion Traditional cardinal plane movement as a result of contracting muscles

Portable transcutaneous electrical nerve stimulation (TENS) unit A portable therapeutic modality that uses electrical stimulation to attempt to modulate pain, strengthen muscles, and enhance soft-tissue healing

Proprioceptive neuromuscular facilitation (PNF) An advanced stretching technique that incorporates the body's reflexes to facilitate stretching of the affected body part

Range of motion The amount of movement available at a joint

RICE method A method of treatment of acute injury that is used to counteract the body's initial response to injury; RICE is an acronym for rest, ice, compression, and elevation

Static stretching The passive stretching of a given antagonist muscle by placing it in a position of maximal stretch and holding it there for an extended period

Therapeutic ultrasound The therapeutic use of mechanical radiant energy

Thermotherapy The therapeutic use of heat

References

1. Knight KL: (ed): *Cryotherapy in Sport Injury Management*. Champaign, IL, Human Kinetics, 1995.
2. Bell GW, Prentice WE: Chapter 9: Infrared modalities: Therapeutic heat and cold, in Prentice WE (ed), *Therapeutic Modalities in Sports Medicine*, ed 4. Boston, MA, WCB/McGraw-Hill, 1999, pp 173-206.
3. Prentice WE: Principles of rehabilitation, in, Almekinders LC (ed): *Soft Tissue Injuries in Sports Medicine*. Cambridge, MA, Blackwell Science, 1996, pp 62-106.
4. Castel C, Draper D, Castel D: Rate of temperature increase during ultrasound treatments: Are traditional times long enough? *J Athletic Training* 1994;29:156-161.
5. Draper DO, Castel JC, Castel D: Rate of temperature increase in human muscle during 1 MHz and 3 MHz continuous ultrasound. *J Orthop Sports Phys Ther* 1995;22:142-150.
6. Tippett SR, Voight ML: (eds): *Functional Progressions for Sport Rehabilitation*. Champaign, IL, Human Kinetics, 1995.
7. Clark MA: (ed): *Integrated Training for the New Millennium*. Thousand Oaks, CA, National Academy of Sports Medicine, 2000.
8. Baechle TR: (ed): *Essentials of Strength Training and Conditioning*. Champaign, IL, Human Kinetics, 1994.
9. Freeman MA: Instability of the foot after injuries to the lateral ligament of the ankle. *J Bone Joint Surg Br* 1965;47:669-677.
10. Freeman MA, Dean MR, Hanham IW: The etiology and prevention of functional instability of the foot. *J Bone Joint Surg Br* 1965;47:678-685.
11. Gauffin H, Tropp H: Altered movement and muscular-activation patterns during the one-legged jump in patients with an old anterior cruciate ligament rupture. *Am J Sports Med* 1992;20:182-192.
12. Noyes FR, Barber SD, Mangine RE: Abnormal lower limb symmetry determined by function hop tests after anterior cruciate ligament rupture. *Am J Sports Med* 1991;19:513-518.
13. Barrett DS: Proprioception and function after anterior cruciate reconstruction. *J Bone Joint Surg Br* 1991;73:833-837.
14. Guskiewicz KM: Regaining balance and postural

equilibrium, in Prentice WE (ed): *Rehabilitation Techniques in Sports Medicine*, ed 3. Boston, MA, WCB/McGraw-Hill, 1999, pp 107-131.

15. Koury JM: (ed): *Aquatic Therapy Programming: Guidelines for Orthopedic Rehabilitation*. Champaign, IL, Human Kinetics, 1996.
16. Andrews JR, Harrelson GL, Wilk KE: (eds): *Physical Rehabilitation of the Injured Athlete*, ed 2. Philadelphia, PA, WB Saunders 1998.
17. Brotzman SB: (ed): *Clinical Orthopaedic Rehabilitation*. St Louis, Mosby-Year Book, 1996.
18. Kisner C, Colby LA: (eds): *Therapeutic Exercise: Foundations and Techniques*, ed 2. Philadelphia, PA, FA Davis Company, 1990.

Exercise Prescription and Promotion

Prevention

In this era of health care reform, any serious attempt to control costs must include urging people to take more responsibility for their own health and urging health care professionals to focus on prevention. In the United States, only a scant amount of health care spending goes toward prevention services. This may have been appropriate at the turn of the century, when people lived only into their 40s and tuberculosis, pneumonia, and gastrointestinal infections were essentially resistant to treatment. However, health care has changed considerably since then. More years have been added to the average life expectancy in the past century than in the preceding 5,000 years. On average, people in the United States now live to age 76 years (mid-80s in Japan), and the current major causes of death, heart disease, cancer, and stroke, all have strong links to lifestyle.

To underscore the importance of prevention, the US Public Health Service published *Healthy People 2000*, which listed 300 objectives for our nation's health for the year 2000, with a strong emphasis on prevention.[1] In this manual for our nation's health, exercise promotion was listed as a cornerstone of a comprehensive health promotion and disease prevention effort. The newest national objectives, published in *Healthy People 2010*, continue to emphasize the key role of exercise.[2]

Physician Involvement

Among the objectives published in *Healthy People 2000* and *Healthy People 2010* is a directive that physicians and other health care professionals become more involved in the promotion and prescription of exercise. Currently, fewer than 30% of physicians in the United States actually discuss exercise with their patients.[3] Two national surveys, the Prevention Index-89 and a study by the President's Council on Physical Fitness and Sports, were conducted to determine which factors influence people to become involved in regular exercise programs. Survey respondents reported that the most influential factor was their physician's recommendation for exercise. Therefore, physicians and other health care professionals must assume a leadership role in promoting exercise.

The promotion of exercise as part of health care is not a new concept. Hippocrates, the father of medicine, routinely prescribed exercise for patients with a wide variety of ailments, and scientific data substantiate that Hippocrates was wise to do so. So why don't physicians prescribe exercise more often?

Exercise as Medicine

Physicians should think of exercise as a medicine. The *Dorland's Illustrated Medical Dictionary, 29th Edition,* defines medicine as "any drug or remedy." Although the relationship between exercise and disease has not been fully defined, data continue to suggest that enormous benefits are derived when exercise is prescribed for health promotion and disease prevention and treatment.

As is the case with certain medications, exercise not only can be used to prevent and treat many diseases, but regular use also results in relatively predictable, specific changes in the human body. These adaptations occur both centrally and peripherally and include structural, hormonal, and biochemical changes. In addition, as is also the case with medications, a dose-response curve should be considered when developing safe, sensible, and effective exercise programs. As ongoing scientific research attempts to define the optimal dose ranges for a variety of exercise-related effects, it is becoming more apparent that the quantity and quality of exercise or activity that is required for certain health-related outcomes may actually differ from what is needed for fitness benefits. Interestingly, exercise has been linked to allergy (eg, exercise-induced urticaria and exercise-induced anaphylaxis) and addiction (eg, exercise addiction and withdrawal), making the "exercise as medicine" concept even stronger.

Exercise Prescription

Once physicians are convinced that exercise plays a significant role in the prevention and treatment of many ailments, the next step is to identify how to most effectively prescribe exercise. How much exercise should be prescribed? How often should patients exercise? What types of exercise should be prescribed?

The ideal prescription should include a specific exercise program for the individual based on his or her goals, health and/or fitness needs, level of physical conditioning, and past or present illness or injuries. The ideal prescription should also specify the frequency, intensity, duration, and type of exercise and include advice for graduated progression. Physicians also need to be prepared to modify exercise programs and routines for individuals with certain ailments.[4] This is especially true for those with musculoskeletal conditions such as arthritis, tendinitis, back pain, osteoporosis, and other bone and joint problems that limit the ability to be optimally active.[5]

The prescription of exercise may seem complicated, but it is actually simple. The key is to individualize the program and identify activities the patient will enjoy and continue for life. For years, the American College of Sports Medicine (ACSM) has provided excellent guidelines regarding exercise prescription.[6] Its 1978 position statement on exercise focused on moderate-to-vigorous aerobic exercise as it relates to fitness. In a 1990 update, the ACSM refined its aerobic exercise recommendations and added the development of muscular strength and endurance as a major objective.[7] Thus, any balanced fitness program for adults should include aerobic or cardiovascular exercise and resistance or strength training.

More recent scientific data have confirmed that significant health-related benefits can be derived from exercise and activity at levels lower than those recommended for fitness purposes in the ACSM guidelines.[8] These benefits, including protection against coronary artery disease, type II diabetes mellitus, hypertension, certain cancers, and osteoporosis, can be achieved with relatively moderate activity programs, such as walking, cycling, or gardening.

Studies comparing three groups of individuals with different activity levels (sedentary, moderate activity, or vigorous activity) have demonstrated reduced mortality rates in the active groups versus the sedentary group.[8] Results included lower death rates not only from cardiovascular disease, but also from cancer, stroke, and all other causes of death. The most dramatic improvement was apparent when comparing the sedentary with the moderate activity group. Because of findings such as these, sedentary behavior is now considered a major coronary risk factor that, in terms of its potential damaging effects, is equal to smoking, a high total cholesterol level, and hypertension.[3] As a result, a new effort is underway to shift the almost 30% of the US population that is totally sedentary into the moderate activity category. In 1993, the ACSM, in conjunction with the Centers for Disease Control and Prevention and the President's Council on Physical Fitness and Sports, recommended that every American adult should spend 30 minutes per day or more engaged in moderately intense physical activity over the course of most days of the week. New guidelines from the Institute of Medicine recommend that people exercise for 60 minutes every day to avoid weight gain as they age.

Program Design: Health Protection Versus Fitness

In recommending exercise or activity programs for patients, identifying the lifestyles of individuals as either sedentary or active has practical applications for program design. In addition, the spectrum of activities of each person should be considered on a continuum from sedentary (no activity) to activity to exercise to fitness. If the individual's lifestyle is sedentary, the major emphasis should be to activate that person. To do so, discussing the new information regarding the benefits of moderate activity should be motivating. People who were once intimidated by vigorous exercise or stopped because of the level of difficulty of high-intensity exercise programs or musculoskeletal ailments can now be easily activated by convincing them that even a minor increase in their activity levels can produce beneficial results.

An individual activity program can prescribe exercise sessions or recommend new daily routines, such as taking the stairs at work, parking farther away in the parking lot, or walking to the store. Individuals are more likely to make these adjustments a regular part of their lives if they are reminded to do so, if family or friends are included in their programs, if they keep activity logs, and if they receive positive support and feedback. If physicians can motivate patients to this level, they have done them a tremendous favor.

The activation phase should be seen as a hook. Once patients are activated, it is not as hard to move patients along the activity continuum into more balanced exercise and fitness programs. In addition to the health benefits, each patient can enjoy the benefits of a stronger, more fit, and more functional body. More comprehensive fitness programs should be prescribed for those who are already involved in exercise programs or, as noted above, those who can be moved further along in the fitness continuum. Every balanced fitness program should include three basic components: cardiovascular or aerobic exercise, resistance or strength training, and flexibility exercise.

Aerobic Exercise

Aerobic exercise strengthens the most important muscle in the body—the heart. In addition to improving cardiac function, the metabolic effects of aerobic exercise increase caloric consumption, which is important in weight control and fat loss. Aerobic exercises include walking, hiking, cycling, running, stair climbing, aerobic dance, and cross-country skiing. For a training effect, aerobic exercise should be performed within a target heart rate range and sustained at that level for at least 20 or 30 minutes, three times per week. To calculate their target heart rate zone, patients should be instructed to first determine their maximum heart rate by subtracting their age from 220; the target heart rate range is 50% to 75% of the maximum heart rate. Gradually increasing the intensity and duration of aerobic exercise will increase the training effect. Perceived exertion scales, such as the Borg Scale of Perceived Exertion, can also be used. These scales allow individuals to accurately determine their degree of cardiovascular effort during exercise based on their perceived subjective work effort.

Strength Training

Exercise for muscular strength and endurance involves the use of resistance exercise to build muscle tone and strength.[9] This can be accomplished with free weights or weight machines. Progressively overloading muscle tissue increases strength. This structural response not only affects muscle but also bone and the surrounding ligaments and tendons. Strength training programs should include all major muscle groups of the upper and lower extremities as well as the lower back and abdomen.

Strength training is finally receiving the recognition it has long deserved. Once only used by football players and other select athletes, its indications have significantly broadened.[9] Strength training is equally important to men and women. No longer synonymous with body building, it is now considered a method for making muscle and bone stronger. Although the effect of strength training on bone is often overlooked, it strengthens bone tissue and is therefore useful in osteoporosis prevention and treatment. Moreover, age is no barrier. Provided that appropriate precautions are taken, strength training exercise is as safe for children as it is for adults.

The greatest potential benefits of strength training, however, may be for older people. Studies have documented that strength training exercise resulted in improved strength and function even among frail nursing-home residents age 90 years or older.[4,10] For aging adults trying to maintain functional independence, this effect is extremely important. Many of the physiologic changes often attributed to aging are, in large part, the result of inactivity rather than aging and are preventable and reversible to some degree. A 60-year-old who exercises can have better functional capacity than an inactive 30-year-old. Therefore, exercise may be the closest thing to the fountain of youth that we have.

Additional strength training benefits include injury prevention, especially among athletes, and weight and fat control. Along with aerobic exercise and dietary modifications, strength-training exercises should be an integral part of any weight-control or weight-reduction program.

Flexibility

Stretching improves muscle and joint flexibility, which in turn reduces the likelihood of muscle strain and injury. In addition, stretching helps prevent muscle soreness associated with exercise or activity and helps maintain mobility and function in arthritic joints when used in conjunction with range-of-motion and strengthening exercises. When done following a brief aerobic warm-up, stretching can also improve muscle elasticity.

All of the major muscle groups of the upper and lower extremities should be stretched. Typical adult problem areas include the anterior shoulder, lower back, hamstrings, and calves. Slow, static stretches with no bouncy or ballistic movements should be used, and the stretch should be held for 15 to 20 seconds. When stretching correctly, a slight pulling sensation (not pain) should be felt. Stretches should be repeated several times per exercise session and attempts should be made to gradually improve flexibility with each session. Stretching can safely be done every day.

Before Starting

Certain individuals require medical clearance and/or exercise testing before initiating an exercise program, for which the ACSM has issued excellent guidelines.[6] Variables for clinicians to consider include the presence of risk factors or known disease, the intended level of activity, and the age and sex of the individual. Most healthy, previously sedentary people can safely start a moderate activity program, such as walking. For people with musculoskeletal conditions, the American Academy of Orthopaedic Surgeons (http://www.aaos.org) has developed sample exercise programs. This concept of modified exercise programs will become increasingly important with the aging population, especially the baby boomers. Nearly all people, including pregnant women; older people; and people with chronic, degenerative, or handicapping conditions, can benefit from a well-designed, individualized exercise program.

References

1. *Healthy People 2000: National Health Promotion and Disease Prevention Objectives, Full Report, With Commentary.* Washington, DC, US Department of Health and Human Services, Public Health Service, 1990. DHHS Publication No. (PHS)91-50212.
2. *Healthy People 2010: With Understanding and Improving Health and Objectives for Improving Health.* Washington, DC, US Government Printing Office, November 2000.
3. Physical Activity and Health: *A Report of the Surgeon General.* Altanta, GA, US Department of Health and Human Services, Centers for Disease Control and Prevention, National Center for Chronic Disease Prevention and Health Promotion, 1996.
4. DiNubile NA: (ed): *The Exercise Prescription.* Philadelphia, PA, WB Saunders, 1991.
5. DiNubile NA: The role of exercise in the treatment of osteoarthritis. *Am J Sports Med* 1999;1:188-200.
6. Kenney WL, Humphrey RH, Bryant CX, et al: (eds): *ACSM's Guidelines for Exercise Testing and Prescription*, ed 5. Baltimore, MD, Williams & Wilkins, 1995.
7. American College of Sports Medicine: Position stand: The recommended quantity and quality of exercise for developing and maintaining cardiorespiratory and muscular fitness in healthy adults. *Med Sci Sports Exerc* 1978;10:vii-x.
8. Blair SN, Kohn WH III, Paffenbarger RS Jr., et al: Physical fitness and all-cause mortality: A prospective study of healthy men and women. *JAMA* 1989;262:2395-2401.
9. DiNubile NA: Strength training. *Clin Sports Med* 1991;10:33-62.
10. Fiatarone MA, O'Neill EF, Ryan ND, et al: Exercise training and nutritional supplementation for physical frailty in very elderly people. *N Engl J Med* 1994;330:1769-1775.

Chapter 31

Medications

This chapter provides a clinically based introduction to the pharmacologic therapies most commonly used in musculoskeletal medicine. The types of medicine discussed include anti-inflammatory pain medications, viscosupplements for arthritis, disease-modifying drugs for rheumatoid arthritis, medications for increasing bone density and preventing blood clots, and antibiotics.

Analgesia

The appropriate and skillful management of pain is one of the most important interventions physicians can perform. Although physicians should rarely want to obliterate pain, as pain itself is a persuasive indication to change behavior, there is convincing evidence that pain is inadequately treated by physicians of all disciplines.[1] Although most disciplines within medicine address analgesia (the relief of pain), orthopaedic surgeons are required to treat pain more frequently than most doctors because, for a majority of musculoskeletal conditions, pain is the chief complaint. Many musculoskeletal complaints,however, can be treated successfully with one or more pharmacologic modalities without the need for surgery. Given the large assortment of analgesics in numerous classes that are available and the great variation of patient responses to pain and pain therapy, providing effective analgesia is often challenging. But if time is taken to learn about the broad spectrum of options and to individually tailor management to each patient, effective pain control can be achieved in nearly all cases. Acute, severe pain—such as postoperative pain—may be effectively managed with opioids (ie, narcotic analgesics). For less severe pain, the use of nonopioid analgesics, such as acetaminophen and tramadol, or nonsteroidal anti-inflammatory drugs (NSAIDs) may be indicated.

Opioids

The oldest class of medicines used to control postoperative pain is the opioids. Opioids are frequently used to provide postoperative analgesia and analgesia for painful emergency department procedures, such as fracture reduction.

Morphine was the first opioid analgesic used in medicine and remains the most commonly prescribed. It is the standard against which all other opioids are compared. Although a number of other opioids are commonly used today, the primary difference between these opioids and morphine is one of dose potency. In addition, patients may experience adverse effects from one opioid but not another. Among the more commonly used opioid analgesics are codeine, hydrocodone, oxycodone, meperidine, and fentanyl. Combination analgesics, which may include acetaminophen or an NSAID in addition to an opioid, also are available in a variety of generic and brand-name formulations.

Opioid analgesics can be delivered by several routes—orally, rectally, intramuscularly, or intravenously. Their pharmacokinetics vary by the route of administration. More recently, transcutaneous patches and sublingual "lollipops" have been made available. The latter is useful when a young child requires analgesia.

On-demand intramuscular opioid delivery (in which a patient must report pain and request medication) has been the traditional approach to postoperative pain control. More recently, however, patient-controlled analgesia (PCA) has become popular. It can deliver any of several intravenous or epidural opioids via a pump that is controlled by the patient. The pump can be regulated to minimize the likelihood of overdose or respiratory depression. More importantly, this method of management provides patients with a sense of control over their pain management. Allowing patients to provide themselves with appropriate doses of opioid analgesics when needed helps minimize anxiety and improves pain control. PCA is most appropriate for patients who are able to understand and use the device correctly.[2] The very young, the very old, or those with difficulty comprehending or physically performing what is required of them cannot use PCA. Consultation with a pain management specialist is typically indicated for epidural PCA; for patients with complex, severe, or atypical pain patterns; and for patients in whom drug interactions are of particular concern.

Once severe postoperative pain has diminished (typically in 2 or 3 days) patients can be switched to orally administered opioids. Often, combinations of opioids and acetaminophen or an NSAID can be used for synergy. This permits the use of lower opioid doses. Patients sometimes will require parenteral dosing of opioids given on an as-needed basis to control "breakthrough" pain. If a patient is expected to have significant pain at a continuous level for several days, a longer-acting oral opioid can be administered to permit once- or twice-daily dosing. This may minimize the likelihood of analgesics reaching subtherapeutic concentrations as a result of missed or delayed doses and improve overall pain control.

The most severe potential adverse effect of opioids is respiratory depression, which is dose related. Other undesirable effects of these medications include nausea, urinary retention, pruritus, and constipation. When opioids are appropriately used for control of acute or postoperative pain, the likelihood of addiction is very low.[3]

Nonopioid Analgesics

Acetaminophen has been used for its analgesic properties for more than 100 years. It became a favored choice for reducing fever and pain in children in the 1980s when the use of aspirin was linked to the onset of Reye's syndrome, an encephalitic disorder associated with seizures. While acetaminophen has analgesic and antipyretic properties, it has no proven anti-inflammatory effects and does not reduce platelet aggregation as do NSAIDs. It has a favorable safety profile and is a good first-line analgesic for mild to moderate pain. The mechanism by which it works as an analgesic is poorly understood; its antipyretic properties act directly on the hypothalamic heat-regulating center. Acetaminophen is metabolized by the liver. In therapeutic doses, serious adverse effects are rare; however, overdoses can cause permanent liver damage. Patients should be advised not to take additional acetaminophen if they are already taking a combination drug containing acetaminophen.

Tramadol is a synthetic opioid agonist. It also inhibits the re-uptake of norepinephrine and serotonin. Tramadol appears to have a lower propensity to cause physical dependence than opioids. Tramadol appears to effectively control pain resulting from postoperative surgical trauma and is roughly equivalent to acetaminophen-codeine combination agents. It also has been used for many chronic malignant and nonmalignant disease states. Tramadol can be administered concomitantly with other analgesics, particularly those with peripheral action. Drugs that depress central nervous system function may enhance the sedative effect of tramadol and should be combined with caution. Common dose-related adverse effects of tramadol include dizziness, nausea, vomiting, dry mouth, and/or drowsiness. Other adverse effects of tramadol are generally similar to those of opioids, although they are usually less severe and can include respiratory depression, dysphoria, and constipation. Tramadol should not be administered to patients receiving monoamine oxidase inhibitors or tricyclic antidepressant drugs.[4]

Nonsteroidal Anti-inflammatory Drugs

Salicylates were the first widely used class of NSAIDs. In 1763, Reverend Edmond Stone described the analgesic and anti-inflammatory effects of willow bark, which had been used for centuries to relieve pain. In 1860, salicylic acid was synthesized from this natural product. Nearly 40 years later, the compound acetylsalicylic acid (aspirin) was produced. Since then, aspirin has been a mainstay in the treatment of the arthritides, other inflammatory conditions, and numer-

Table 1
Types of NSAIDs

Drug	Strength (mg)	Trade Name	Typical Dosage
Salicylates			
Aspirin	300, 325, 600, 650	Several	325 mg qid
Choline magnesium trisalicylate	500, 750, 1,000	Trilisate	1,500 mg bid
Salsalate	500, 750	Disalcid	1,000 mg tid, 1,500 mg bid
Diflunisal	500	Dolobid	250 mg bid or tid, 500 mg bid
Propionic Acids			
Naproxen sodium	220	Aleve	220 mg bid
	375, 500	Naprelan	750-1,000 mg/day
	275, DS 550	Anaprox	275 mg bid or tid, 550 mg bid
Naproxen	250, 375, 500	Naprosyn	250, 375, or 500 mg bid
Flurbiprofen	50, 100	Ansaid	100 mg bid
Oxaprozin	600	Daypro	600-1,200 mg/day
Ibuprofen	400, 600, 800	Motrin	400-800 mg bid or tid
Ibuprofen (OTC)	200	Motrin IB, Advil, Rufin, Nuprin	200-400 mg bid or tid
Ketoprofen	25, 50, 75	Orudis	50 mg qid, 75 mg tid
Ketoprofen extended release	100, 150, 200	Oruvail	200 mg/day
Ketorolac tromethamine	10	Toradol Oral	10 mg qid; 5 days max of oral and intramuscular combined
Indolacetic Acids and Related Compounds			
Sulindac	150, 200	Clinoril	150-200 mg bid
Indomethacin	25, 50	Indocin	25-50 mg tid
Indomethacin sustained release	75	Indocin SR	75 mg/day
Etodolac	200-500	Lodine	200-400 mg tid, 500 mg bid
COX-2 Inhibitors			
Celecoxib	100, 200	Celebrex	100-200 mg/day
Rofecoxib	12.5, 25	Vioxx	12.5-25 mg/day
Valdecoxib	10, 20	Bextra	10 mg/day

ous noninflammatory but painful conditions. Despite its long history, new therapeutic uses for acetylsalicylic acid continue to be found, including the prevention and treatment of myocardial infarction and noncoronary vascular disease and the prevention of thromboembolic disease in high-risk surgical patients.

Dozens of nonaspirin NSAIDs are commercially available, and NSAIDs are thought to be the most widely prescribed class of drugs in the world (Table 1). In addition to providing analgesia, NSAIDs have both antipyretic and anti-inflammatory effects. All medications in this class inhibit prostaglandin formation by blocking cyclooxygenase. There are two forms of the cyclooxygenase (COX) enzyme, COX-1 and COX-2. COX-1 is present in most bodily tissues (including platelets and gastrointestinal mucosal tissues) and serves as a "housekeeping" enzyme to form protective prostaglandins. COX-2, on the other hand, is not present in most tissues unless induced in response to inflammation. It is responsible for the formation of prostaglandins that contribute to pain and inflammation. Unselective cyclooxygenase inhibition (of both COX-1 and COX-2) is the mechanism of most older NSAIDs and is responsible both for the beneficial and most of the harmful effects of salicylates and other NSAIDs.[5] COX-2–selective NSAIDs (including rofecoxib, valdecoxib, and celecoxib) irreversibly inhibit COX-2 in peripheral tissues and thereby interfere only with prostaglandin formation in inflamed joints and other body tissues.

Early studies have shown that COX-2 inhibitors are as effective as nonselective NSAIDs and have an improved safety profile because they seem to spare the COX-1 enzyme in the gastrointestinal tract. Specifically, the reported frequency of gastrointestinal ulcers caused by use of these medications has been similar to that of placebo in early studies.[6,7] However, as with all new medications, caution is warranted until our experience with these medications grows and follow-up studies can be performed.

The most common severe adverse effect of salicylates and NSAIDs is gastrointestinal irritation, leading sometimes to ulceration and hemorrhage. Because some of the prescribers of these medications (orthopaedic surgeons, for example) do not themselves treat the medical complications that arise from the use of NSAIDs, they may have an overly optimistic perception of the safety profile of these drugs. It is estimated that the incidence of gastric or duodenal ulcers among users of nonselective NSAIDs is between 14.0% and 44.0% and that the risk for hospitalization from serious complications is between 0.2% and 4.0%. Because these drugs are so frequently prescribed, even infrequent complications can lead to significant morbidity and mortality.[8]

Nonselective NSAIDs also disrupt platelet function and hemostasis and thereby prolong bleeding time. This effect is sometimes used therapeutically for prophylaxis against thromboembolic disease. Other adverse effects include hypersensitivity reactions; nasal polyps; and renal, hepatic, and central nervous system toxicity. Hepatic and renal toxicity resulting from the use of these medications initially can be silent, and it is prudent to screen patients who are going to take NSAIDs for prolonged periods with regularly scheduled kidney and liver function tests. NSAIDs also can adversely affect blood pressure, sometimes through fluid retention, or can be a primary cause of hypertension because of adverse effects on the kidneys.[9] COX-2 inhibitors do not offer increased protection against renal adverse effects. Some patients are at particular risk for NSAID-related complications and should therefore be observed closely while taking these medications. Well-defined risk factors include the following: being older than 65 years, a history of gastrointestinal symptoms with NSAID use, taking high doses of NSAIDs, and concomitant use of oral corticosteroids or oral anticoagulants.

Other Medications Used to Control Pain

In addition to pure analgesics or anti-inflammatories, a large number of medications are also used to control pain. These are generally used for atypical pain patterns, neuropathic pain, or painful refractory conditions, such as complex regional pain syndrome. These medications are most often prescribed by health care providers with a subspecialty interest in the management of chronic pain. Antidepressants—including tricyclics and monoamine oxidase inhibitors—are sometimes used and seem to have analgesic effects in addition to their function as antidepressants. Several anticonvulsants have also been found to be helpful in treating pain that is neurogenic in nature, presumably through a mechanism that is similar to their anticonvulsant properties. Anticonvulsants may be more effective for treating intermittent lancinating neuropathic dysesthesias than for continuous pain. Numerous anxiolytics are used alongside analgesics to alleviate the anxiety often associated with painful conditions or surgery.

Viscosupplementation

In osteoarthritic joints, the concentration of hyaluronic acid, the major component of synovial fluid, is decreased. This change may have a detrimental mechanical effect on the already weakened articular cartilage. Injected hyaluronic acid was first used as treatment for arthritic knees of racehorses in the 1970s, and studies examining its use in humans were conducted later that decade in Europe. In the United States, intra-articular hyaluronic acid products, sometimes called viscosupplements, have been used to treat pain associated with osteoarthritis of the knee for only a short time.

The mechanism of action of intra-articular viscosupplementation is not completely understood. A direct mechanical action was first postulated—hence the term viscosupplementation—but the observed benefits cannot be attributed solely to the mechanical effects of hyaluronic acid injections. In one study, patients with osteoarthritis of the knee who were treated with viscosupplements (high-molecular-weight sodium hyaluronate [hyalectin]) experienced decreased symptoms for 6 to 12 months.[10] However, another study found the half-life of the injected hyaluronic acid molecule, which may have anti-inflammatory properties, to be only 20 hours in normal joints and less than 12 hours in inflamed joints.[11] The overall role of this relatively expensive treatment in the routine management of knee arthritis is not clear at this time, and not every study has confirmed a beneficial effect of treatment.[12,13] Adverse reactions from intra-articular hyaluronic acid injections are uncommon and generally of limited severity. However, an acute inflammatory reaction that mimics septic arthritis can occur in people who receive injections. Hyaluronic acid injections have not yet been approved by the US Food and Drug Administration (FDA) for use in joints other than the knee at this time.

Corticosteroid Injections

Corticosteroids have been used for decades to decrease inflammation in arthritic joints. Despite over 50 years of clinical use, the literature on intra-articular and periarticular corticosteroid injections is remarkably scant, and few studies are of high quality. Every diarthrodial joint, however, is a potential injection site for either diagnostic or therapeutic purposes. From a diagnostic standpoint, pain relief from an intra-articular injection—even if only temporary—strongly suggests that the painful pathology is intra-articular rather than peri- or extra-articular. The therapeutic effect of these injections is variable, depending on the joint, the etiology of disease, and associated rehabilitative interventions. Most corticosteroid injections are given in conjunction with short- and long-acting local anesthetics. These adjuvants provide immediate relief and minimize the burning sensation associated with pure corticosteroid injections. They also can inform the physician that the mixture was delivered to the correct location.

Among the large joints, arthritis develops most commonly in the knee. As a subcutaneous joint, the knee is also the most common joint to be treated with intra-articular injections. Most studies on the effect of intra-articular injections for the treatment of arthritis reveal short-term benefits in terms of decreased pain. Pain relief can last from days to months, and there is no reliable way to predict which patients will benefit most from this intervention.

Periarticular injections for the treatment of bursitis also are performed commonly. The subacromial bursa (above the rotator cuff), the greater trochanteric bursa at the hip, and the medial epicondyle of the elbow often are injected for symptomatic pain relief and for diagnostic purposes. These interventions should be combined with physical therapy or similar rehabilitative interventions to maximize their effects.

Tendon sheath and paratenon injections are performed most commonly in the hand and wrist. Such injections are widely considered risky when done in the analogous, weight-bearing tendons of the foot. Achilles tendon and anterior tibial tendon ruptures have been reported after corticosteroid injections to those structures.[14] Apart from tendon rupture, the most worrisome complication of intra-articular and periarticular injections is infection. When careful aseptic technique is used, the risk of causing joint sepsis is very low, estimated to be 1 in 14,000 to 50,000.[15] Skin depigmentation in darkly pigmented patients and fat necrosis, especially in the heel, can occur as well. Although animal studies have raised questions about the possibility of injury to cartilage from intra-articular corticosteroids, there are no satisfactory data demonstrating this problem in humans. Although infections from single injections certainly occur, most complications result from frequent injections, technical errors, or errors in judgment. As with any medical or surgical procedure, the patient should always be advised about the risks prior to the intervention, and informed consent should always be obtained.

Corticosteroids are also used in the emergency setting to decrease the swelling associated with spinal cord trauma. If given within 8 hours of a spinal cord injury, high-dose intravenous steroids have been shown to help lessen the severity of injury. They are even more effective if given within 3 hours of injury. Even then, however, the benefit obtained is not universal, nor is it miraculous. A reasonable goal of treatment is an improvement in the sensory or motor function of one or two nerve root levels. Even injury to one nerve root level, however, can significantly impair function; therefore, these medications are an essential component in the treatment of acute spinal cord trauma.[16]

Management

Rheumatoid Arthritis

The management of patients with rheumatoid arthritis (RA) and other inflammatory arthritides is complex and changes frequently. Broadly speaking, the medications used to treat RA are divided into two classes, symptom-modifying antirheumatic drugs (SMARDs) and disease-modifying antirheumatic drugs (DMARDs). Which type of therapy is most appropriate is an individual decision, but most patients with RA take, or have taken, at least one medication from each class. Therapy for RA is the topic of literally thousands of peer-reviewed journal articles each year. This section will provide only a brief introduction to that ever-growing body of work.

SMARDs consist primarily of NSAIDs and oral steroids. This class of drugs is used to decrease pain and inflammation quickly and to suppress chronic inflammation. Oral corticosteroids decrease inflammation through a variety of pathways, including inhibiting prostaglandin and leukotriene production, opsonizing antigens, blocking cytokines, and impairing adhesion and migration of inflammatory cells through vascular endothelium. They have a host of adverse effects, including interference with insulin function and lipogenesis, which can result in hyperglycemia, centripetal fat deposition, and hyperlipidemia; impairment of bone metabolism and promotion of osteoporosis; edema; fluid retention; hypertension; and hypokalemia. Skin problems include poor wound healing from decreased collagen production and fragile capillaries, which causes easy bruising, acne, hirsuitism, and striae. Additional adverse effects can include cataracts, glaucoma, gastrointestinal ulcers with bleeding and perforation, pancreatitis, myopathies, osteonecrosis, and premature atherosclerosis. Finally, because oral corticosteroids blunt the immune response, patients taking these medications over the long term are more susceptible to bacterial and other opportunistic infections. Nonetheless, in many cases the cost-benefit ratio favors their use.

SMARDs, as the name implies, do not affect disease progression or avert the long-term destructive sequelae of RA. Because patients with extra-articular manifestations of RA have a life expectancy that is 10 to 15 years shorter than age-matched controls, the use of DMARDs is often indicated.[17] DMARDs comprise a class of drugs that includes a number of agents with dissimilar mechanisms of action and includes many of the drugs formerly classified as "slow-acting," "second-line," or "remittive agents" for the treatment of RA. DMARDs work to blunt immune response and slow or halt the destructive effects of inflammatory arthritis. In the past, drugs now classified as DMARDs were typically thought of as fairly toxic. As a group, they vary greatly in their relative efficacy and adverse effect profiles.[18]

Many DMARDs have a safety profile comparable to NSAIDs and are preferable to prednisone for the treatment of RA.[19] The older DMARDs include minocycline, chloroquine, hydroxychloroquine, gold compounds, methotrexate, D-penicillamine, azathioprine, cyclophosphamide, and cyclosporine. The latter three are still considered to be quite toxic and should be used only when safer agents have not been effective. Newer agents include leflunomide (a dihydroorotate dehydrogenase inhibitor) and etanercept (a tumor-necrosis factor inhibitor); long-term efficacy and safety profiles of these medications remain in question, but they appear promising. Methotrexate remains a mainstay of therapy because of its rapid onset (3 to 4 weeks), its ability to provide long-term therapeutic benefit to individual patients, its long track record, and its intermediate-to-low toxicity at doses commonly used to treat RA.

Osteoporosis and Fracture Prevention

Common areas of osteoporotic fracture include the spine, hip, and wrist. Because the 1-year mortality rate after a hip fracture in elderly patients can be 20% or higher, pathologic fractures arising from osteoporosis represent a significant public health risk and result in enormous costs to society. Preventing this disease, however, is easier and safer than treating fractures. Osteoporosis prevention is thus the responsibility of all physicians who treat musculoskeletal conditions.

Primary prevention of osteoporosis is nonpharmacologic and includes weight-bearing exercise, smoking cessation, good nutrition (especially sufficient nutritional intake of calcium and vitamin D), and avoidance of excessive alcohol intake. However, in many individuals—especially postmenopausal women—these measures are insufficient to prevent osteoporosis. Moreover, cal-

Table 2
Dietary Reference Intake Values for Calcium by Life-Stage Group

Life-Stage Group*	Adequate Intake (mg/day)	Pregnancy	Adequate Intake (mg/day)	Lactation	Adequate intake (mg/day)
0 to 6 months	210	≤18 years	1,300	≤18 years	1,300
6 to 12 months	270	19 through 50 years	1,000	19 through 50 years	1,000
1 through 3 years	500				
4 through 8 years	800				
9 through 13 years	1,300				
14 through 18 years	1,300				
19 through 30 years	1,000				
31 through 50 years	1,000				
51 through 70 years	1,200				
>70 years	1,200				

*All groups except Pregnancy and Lactation are both males and females
Adapted with permission from the National Osteoporosis Foundation: ***Physician's Guide to the Prevention and Treatment of Osteoporosis.*** Washington, DC, National Osteoporosis Foundation, 1998.

cium supplementation, although essential, is probably not enough on its own to prevent fractures in patients at risk for osteoporosis (Table 2). Therefore, other therapies have been advocated to help prevent osteoporosis, including hormone replacement therapy (HRT), raloxifene, calcitonin, and bisphosphonates.

Hormone Replacement Therapy and Selective Estrogen Receptor Modulators

In women who are within 5 years of menopause, estrogen deficiency is primarily responsible for bone loss. In postmenopausal women, HRT improves bone density in the spine and hip and may decrease the risk of fracture by up to 50%, provided that treatment is initiated within 5 years of menopause. It is thought to act by directly binding estrogen receptors on osteoblasts or by limiting the action of cytokines that promote bone resorption. Estrogen is often given with progesterone to offset the higher risk for endometrial cancer, breast cancer, and thromboembolic disease that was observed in women taking estrogen alone. The decision to use estrogen to help prevent bone loss must be made on an individual basis after considering a woman's personal and family history of cancer and thromboembolic disease. However, a well-designed, prospective, randomized trial has called into question the widespread use of HRT for prophylaxis against osteoporosis.[20]

Raloxifene hydrochloride is a selective estrogen receptor modulator (SERM). Its action is not well understood, but it decreases bone resorption without apparent increased risk of endometrial or breast cancer. Recent studies have shown a modest improvement in bone density in the hip and spine and a 30% to 50% decrease in the incidence of vertebral fractures. The effect of raloxifene on hip fracture incidence is not known. Its adverse effects include primarily vasomotor symptoms, hot flashes, and leg cramping. Tamoxifen, another SERM, is sometimes used to treat breast cancer and is being investigated for potential use in the prevention of osteoporosis.

Calcitonin

Calcitonin is a peptide produced by the thyroid parafollicular cells. It inhibits bone resorption through a direct inhibition of osteoclasts, which have a high affinity for calcitonin. It was approved by the FDA in 1984 for use in an injectable form and in 1995 for use as a nasal spray. It has been shown to prevent bone loss in the spine and forearm in postmenopausal women with osteoporosis, and a recent study demonstrated a decrease in vertebral fractures in women using intranasal calcitonin.[21] Adverse effects of intranasal calcitonin generally are minor and include occasional rhinitis, nausea, and flushing.

Bisphosphonates

Bisphosphonates are drugs that are adsorbed onto bone hydroxyapatite and inhibit bone resorption by interfering with osteoclast-ruffled border membranes; however, the mechanism of action of this class of drugs is not entirely known. Etidronate was the first medicine in this class to be used to treat osteoporosis. It has been shown to increase bone density and decrease fractures of the spine in older women with osteoporosis. Alendronate appears to be more effective in the prevention and treatment of osteoporosis and also has been proven to reduce the risk of hip and spine fractures. Risedronate is the most recent drug in this class to be promoted for fracture prevention. A recent study showed that patients treated with risedronate had a 40% reduction in fracture risk.[22] Adverse effects of bisphosphonates are most commonly related to gastrointestinal irritation and include erosive gastritis and esophagitis. Patients are told not to lie supine for at least 30 minutes after taking alendronate to help minimize the likelihood of these problems occurring.

Prevention of Deep Venous Thrombosis and Pulmonary Embolism

Patients undergoing major orthopaedic surgery on the lower extremities have a higher risk for the development of deep venous thrombosis (DVT) or pulmonary embolism (PE) than any other group of surgical patients, and those undergoing joint replacement or surgery for pelvic trauma are at greatest risk. Nonpharmacologic approaches to thromboprophylaxis include early mobilization; surgical techniques minimizing the time the limb is positioned in flexion; mechanical interventions, such as compressive stockings and sequential compression devices placed on the feet or legs; and possibly alterations in anesthesia technique (ie, using hypotensive epidural anesthesia) and preoperative collection of autologous blood.[23]

Controversy exists, however, regarding whether and how pharmacologic thromboprophylaxis should be used. A recent meta-analysis of the English-language literature concluded that pharmacologic thromboprophylaxis does not decrease the risk of fatal PE after total hip replacement.[24] On the other hand, there is ample evidence that the DVT rate can be significantly decreased with thromboprophylactic agents. Nevertheless, it has not been shown whether the prevention of DVT justifies the assumption of risk for complications. Since many DVTs are benign, the treatment may be worse than the disease. All thromboprophylactic agents that have been approved by the FDA are anticoagulants. As such, they all share the potential risk of bleeding complications, ranging from minor wound complications to fatal hemorrhages. The consensus in the United States is to use routine pharmacologic thromboprophylaxis during routine joint replacement surgery. The norm in Great Britain and much of Europe, however, is not to use these agents at all.

Warfarin

Warfarin is commonly used as a thromboprophylactic agent during joint replacement surgery and is the only oral anticoagulant currently approved by the FDA for this indication. It works by reducing the synthesis of vitamin K–dependent clotting factors II, VII, IX, and X in the liver. As such, it does not have an immediate effect on clotting (as does heparin); the body's supply of these factors are not depleted for at least 8 hours. Its effects can be monitored by testing the prothrombin time or international normalized ratio (INR). A large number of warfarin regimens have been advocated, including minidose therapy (ie, fixed doses of 2 mg or less, which are not designed to elevate prothrombin time or INR) to adjusted-dose treatments (ie, doses designed to more than double the INR). There is no consensus regarding duration of effectiveness of warfarin therapy, which may range from several days to several months. The variation in practice stems from the absence of well-designed, evidence-based clinical trials.

Unfractionated Heparin

Unfractionated heparin, usually given as 5,000 U subcutaneously either twice or three times daily, has not been shown to be effective in patients undergoing joint replacement and therefore is no longer preferred for that procedure. It is sometimes used to treat musculoskeletal trauma patients because it presents a relatively low risk of causing serious bleeding and has a short half-life.

Low-Molecular-Weight Heparins

Fractionated, low-molecular-weight heparins (LMWHs) are effective anticoagulants that have been commonly used as thromboprophylactic agents for the last decade. They work by a mechanism similar to that of other heparin products, namely, by binding to antithrombin-III and catalyzing its inactivation of factor Xa. Unlike unfractionated heparins, LMWHs cause less inhibition of thrombin and platelet aggregation; are less bound to plasma proteins and, therefore, are more bioavailable; and do not require that patients undergo routine blood testing for management of coagulation parameters. They have also been shown to be more effective thromboprophylactic agents than unfractionated heparin for most types of orthopaedic surgery. When the efficacy of warfarin and LMWHs were compared, most studies found LMWHs to be more effective in preventing DVT. With respect to PE and fatal PE, however, the data are not conclusive. In general, bleeding complications and transfusion requirements have been more common in patients treated with LMWHs than in those treated with warfarin, although this has not been the case in all series. LMWHs are more expensive than warfarin, although this difference in cost may be offset by a lower incidence of DVT and by the cost of routine blood testing required of patients managed with warfarin.

Aspirin

Aspirin was used frequently in the past to decrease the incidence of perioperative DVT. Aspirin works as a platelet inhibitor, by irreversibly inactivating cyclooxygenase. It has a fairly good safety profile, and it poses minimal risk of wound or remote-site bleeding complications. Subsequently, as physicians realized that thromboembolic disease causes significant morbidity, more aggressive agents—such as warfarin and LMWHs—have become more popular choices for perioperative use. However, aspirin remains a useful agent, and recent research has shown it to be effective, particularly when combined with hypotensive epidural anesthesia.[25] A recent meta-analysis also found aspirin to be an effective thromboprophylaxis for patients with hip fractures.[26]

Other Agents

Dextran, a polysaccharide solution, decreases platelet adhesion and aggregation. However, it is inconvenient to use, and it has an unacceptable adverse effect profile, including increased bleeding, allergic reaction in 1% of patients, and pulmonary edema from volume overload in patients with cardiac problems. Recombinant hirudin, an agent that has been newly developed for DVT prophylaxis, inactivates both free and clot-bound thrombin in a highly potent and specific manner. Preliminary data have shown it to be effective and safe, but more studies are necessary.

Antibiotics

A large proportion of the antibiotics used in hospitals are for routine prophylaxis to prevent infection—not to treat illness. Many surgeons insist on prophylaxis for every surgical case, and most use antibiotics whenever hardware is implanted. Some have argued that prophylaxis is unnecessary, costly, and contributes to the proliferation of drug-resistant organisms. However, infected implants, especially infected artificial joint prostheses, result in such severe morbidity that even small decreases in the rate of infection provide a very meaningful advantage.

Early studies examining the effectiveness of perioperative antibiotic prophylaxis failed to show a decrease in infection rates. This was because the drugs were only given postoperatively. In the last 2 decades, numerous studies have made it clear that prophylactic antibiotics significantly decrease the rate of infection during surgery in which hardware is implanted and especially during joint replacement surgery.[27] More recently, it has been shown that prophylactic antibiotics given to patients more than 2 hours before surgery or intraoperatively do not decrease the infection rate compared with those patients to whom no antibiotic was administered.[28] The 2-hour period before surgery is the critical window. When administered too early, the concentration of antibiotics in the blood will be too low when the surgery begins. When administered intraoperatively, the blood supply to the surgical field may be compromised by incision and dissection; when a tourniquet is used, no antibiotics will reach the site after it is inflated.

For "clean" cases (ie, those without prior infection or skin breakdown), most surgeons choose first-generation cephalosporins for prophylaxis. These drugs are particularly effective against staphylococci, which most commonly cause infection during surgery. Among the first-generation cephalosporins, cefazolin sodium is known to reach high therapeutic levels in muscle and hematoma. It also has a long half-life and is relatively inexpensive.[29] Its action blocks the transpeptidation of peptidoglycan and thereby inhibits bacterial cell-wall synthesis. It also activates autolytic cell-wall enzymes, resulting in bacterial death. Because cephalosporins are structurally related to penicillin, there is a small chance of an allergic reaction occurring among patients known to be allergic to penicillin.

The appropriate duration of perioperative antibiotic prophylaxis remains somewhat controversial, but the consensus at present is that for joint replacements and procedures in which bone or joint is exposed, antibiotics should be started preoperatively and continued for 24 to 48 hours. Longer durations of treatment have not been shown to be more effective and probably increase the likelihood of promoting drug resistance among bacteria or cause adverse drug reactions. Colitis caused by *Clostridium difficile*, for example, is more likely to occur when more protracted courses of antibiotics are used because the antibiotics kill some of the normal gastrointestinal flora and allow this organism, which is not normally found in great amounts, to grow without competition and upset the ecologic balance of the colon.

Key Terms

Acetaminophen A nonprescription pain medication comparable in potency to NSAIDs but with absent or weak anti-inflammatory effects

Analgesia The relief of pain

Aspirin A nonprescription NSAID that is used for its pain-relieving and antiplatelet effects

Bisphosphonates Drugs that are absorbed into bone hydroxyapatite and inhibit bone resorption by interfering with osteoclast-ruffled border membranes

Calcitonin A peptide produced by the thyroid parafollicular cells that inhibits bone resorption through a direct inhibition of osteoclasts

Corticosteroids Hormones that affect carbohydrate, fat, and protein metabolism or affect the regulation of water and electrolyte balance; used to decrease inflammation but has significant adverse effects

COX-1 Cyclooxygenase-1 enzyme; an enzyme that is present in most bodily tissues (including platelets and gastrointestinal mucosal tissues) and serves as a "housekeeping" enzyme to form prostaglandins

COX-2 Cyclooxygenase-2 enzyme; an enzyme that is thought to be present in the body only when induced in response to injury and is responsible for the formation of prostaglandins that mediate pain and inflammation

Dextran A polysaccharide solution that decreases platelet adhesiveness and aggregation

Disease-modifying antirheumatic drugs (DMARDs) A class of drugs used to blunt immune response and slow or halt the destructive effects of inflammatory arthritis

Hormone replacement therapy (HRT) Medications that are administered to treat estrogen deficiency and improve bone density

Low-molecular-weight heparins (LMWHs) Anticoagulants that work by binding to antithrombin-III and catalyze its inactivation of factor Xa

Morphine A commonly prescribed opioid analgesic

Nonsteroidal anti-inflammatory drugs (NSAIDs) Inhibitors of cyclooxygenase and therefore prostaglandin synthesis

Patient-controlled analgesia The intravenous or epidural delivery of opioids via a pump that is controlled by the patient

Selective estrogen receptor modulator (SERM) A class of drugs that is thought to provide the beneficial effects of hormone replacement therapy without some of its adverse effects

Symptom-modifying antirheumatic drugs (SMARDs) A class of drugs that decreases pain and inflammation in rheumatic disease but does not prevent disease progression

Tramadol A synthetic opioid agonist

Viscosupplements Intra-articular hyaluronic acid preparations commonly used to treat osteoarthritis; thought to increase joint lubrication

Warfarin A drug that reduces the synthesis of vitamin K-dependent clotting factors II, VII, IX, and X in the liver

References

1. Starck PL, Sherwood GD, Adams-McNeill J, et al: Identifying and addressing medical errors in pain mismanagement. *Jt Comm J Qual Improv* 2001;27:191-199.
2. Ballantyne JC, Carr DB, Chalmers TC, et al: Postoperative patient-controlled analgesia: Meta-analyses of initial randomized control trials. *J Clin Anesth* 1993;5:182-193.
3. Porter J, Jick H: Letter: Addiction rare in patients treated with narcotics. *N Engl J Med* 1980;302:123.
4. Dayer P, Desmeules J, Collart L: Pharmacology of tramadol. *Drugs* 1997;53(suppl 2):18-24.
5. Fries JF, Williams CA, Bloch DA: The relative toxicity of nonsteroidal anti-inflammatory drugs. *Arthritis Rheum* 1991;34:1353-1360.
6. Silverstein FE, Faich G, Goldstein JL, et al: Gastrointestinal toxicity with celecoxib vs nonsteroidal antiinflammatory drugs for osteoarthritis and rheumatoid arthritis: The CLASS study: A randomized controlled trial: Celecoxib Long-Term Arthritis Safety Study. *JAMA* 2000;284:1247-1255.
7. Bombardier C, Laine L, Reicin A, et al: Comparison of upper gastrointestinal toxicity of rofecoxib and naproxen in patients with rheumatoid arthritis: VIGOR Study Group. *N Engl J Med* 2000;343:1520-1528.
8. Gabriel SE, Jaakkimainen L, Bombardier C: Risk for serious gastrointestinal complications related to use of nonsteroidal anti-inflammatory drugs: A meta-analysis. *Ann Intern Med* 1991;115:787-796.
9. Pope JE, Anderson JJ, Felson DT: A meta-analysis of the effects of nonsteroidal anti-inflammatory drugs on blood pressure. *Arch Intern Med* 1993;153:477-484.
10. Dougados M, Nguyen M, Listrat V, et al: High molecular weight sodium hyaluronate (hyalectin) in osteoarthritis of the knee: A 1 year placebo-controlled trial. *Osteoarthritis Cartilage* 1993;1:97-103.
11. Fraser JR, Kimpton WG, Pierscionek BK, Cahill RN: The kinetics of hyaluronan in normal and acutely inflamed synovial joints: Observations with experimental arthritis in sheep. *Semin Arthritis Rheum* 1993:22(suppl 1):9-17.
12. Lohmander LS, Dalen N, Englund G, et al: Intra-articular hyaluronan injections in the treatment of osteoarthritis of the knee: A randomised, double blind, placebo controlled multicentre trial: Hyaluronan Multicentre Trial Group. *Ann Rheum Dis* 1996;55:424-431.
13. Altman RD, Moskowitz R: Intraarticular sodium hyaluronate (Hyalgan) in the treatment of patients with osteoarthritis of the knee: A randomized clinical trial: Hyalgan Study Group. *J Rheumatol* 1998;25:2203-2212.
14. Ford LT, DeBender J: Tendon rupture after local steroid injection. *South Med J* 1979;72:827-830.
15. Gray RG, Gottlieb NL: Intra-articular corticosteroids: An updated assessment. *Clin Orthop* 1983;177:235-263.
16. Bracken MB, Shepard MJ, Collins WF, et al: A randomized controlled trial of methylprednisolone or naloxone in the treatment of acute spinal-cord injury: Results of the Second National Acute Spinal Cord Injury Study. *N Engl J Med* 1990;322:1405-1411.
17. Wolfe F, Mitchell DM, Sibley JT, et al: The mortality of rheumatoid arthritis. *Arthritis Rheum* 1994;37:481-494.
18. Felson DT, Anderson JJ, Meenan RF: The comparative efficacy and toxicity of second-line drugs in rheumatoid arthritis: Results of two meta-analyses. *Arthritis Rheum* 1990;33:1449-1461.
19. Fries JF, Williams CA, Bloch DA: The relative toxicity of nonsteroidal anti-inflammatory drugs. *Arthritis Rheum* 1991;34:1353-1360.
20. Nelson HD, Humphrey LL, Nygren P, et al: Postmenopausal hormone replacement therapy: Scientific review. *JAMA* 2002;288:872-881.
21. Chesnut CH III, Silverman S, Andriano K, et al: A randomized trial of nasal spray salmon calcitonin in postmenopausal women with established osteoporosis: The prevent recurrence of osteoporotic fractures study: PROOF Study Group. *Am J Med* 2000;109:267-276.
22. Harris ST, Watts NB, Genant HK, et al: Effects of risedronate treatment on vertebral and nonvertebral fractures in women with postmenopausal osteoporosis: A randomized controlled trial: Vertebral Efficacy With Risedronate Therapy (VERT) Study Group. *JAMA* 1999;282:1344-1352.
23. Salvati EA, Pellegrini VD Jr., Sharrock NE, et al: Recent advances in venous thromboembolic prophylaxis during and after total hip replacement. *J Bone Joint Surg Am* 2000;82:252-270.
24. Murray DW, Britton AR, Bulstrode CJ: Thromboprophylaxis and death after total hip replacement. *J Bone Joint Surg Br* 1996;78:863-870.
25. Westrich GH, Farrell C, Bono JV, et al: The incidence of venous thromboembolism after total hip arthroplasty: A specific hypotensive epidural anesthesia protocol. *J Arthroplasty* 1999;14:456-463.
26. Rodgers A: Letter: Low-dose aspirin prevented deep venous thrombosis and pulmonary embolism after surgery for hip fracture: Pulmonary Embolism Prevention (PEP) Trial Collaborative Group. *J Bone Joint Surg Am* 2000;82:1807.
27. Bodoky A, Neff U, Heberer M, et al: Antibiotic prophylaxis with two doses of cephalosporin in patients managed with internal fixation for a fracture of the hip. *J Bone Joint Surg Am* 1993;75:61-65.
28. Classen DC, Evans RS, Pestotnik SL, et al: The timing of prophylactic administration of antibiotics and the risk of surgical-wound infection. *N Engl J Med* 1992;326:281-286.
29. Jones S, DiPiro JT, Nix DE, et al: Cephalosporins for prophylaxis in operative repair of femoral fractures: Levels in serum, muscle, and hematoma. *J Bone Joint Surg Am* 1985;67:921-924.

Chapter 32

Alternative Medicine

Alternative medicine defies easy definition. Its broadest definition comprises that which is not taught in a traditional medical school curriculum. This definition, of course, makes it impossible for a conventional medium such as this text to address the topic because by simply including a discussion of a given remedy here, it might no longer be considered an alternative approach. Still, it is important that alternative medicine be given as precise a meaning as possible, especially if value judgments are assigned to various treatment modalities based on their designation—as is often the case.

Alternative medicine can be defined only in the context of cultural norms. Three hundred years ago, for example, an individual with heart failure was instructed by doctors to chew on the flowers, leaves, stems, and seeds of the foxglove plant. Today, by contrast, botanical medicine is considered an alternative remedy. Yet we now know that the foxglove plant contains digitalis. Should that knowledge compel us to reconsider the labels?

Alternative medicine includes a wide spectrum of treatments—many finding support from collective anecdotal evidence—that is not considered standard therapy because of a lack of a scientific rationale, clinical evidence, or a favorable historic tradition. A more rigorous distinction between alternative and traditional medicine would involve considering a functional classification of all treatments based on their modes of action, in which case the distinction between alternative medicine and its mainstream counterpart becomes less important.

Classification of Treatments

Regardless of their labeling, therapies may be effective or perceived to be effective for one of the following reasons: (1) direct homeostasis; (2) biologic effects by stimulation of physiologic homeostasis or interference with abnormal physiology; (3) analgesic effects, either direct or indirect; and (4) the placebo effect.

Direct Homeostasis

Some therapies directly help return the body to the normal healthy state—that is, they do not rely on the body's response to achieve their ends. Direct homeostatic therapies include treatments such as the removal of a foreign body or the administration of antibiotics. Here, the end point of the treatment is reached without specifically invoking a biologic process; the treatment itself produces the result. Not all antibiotics have a purely direct effect: bacteriostatic drugs, which only inhibit bacterial growth, are clinically helpful to the extent that they control the infection as the body mounts its response.

Biologic Effects

Biologic therapies come in two types: those that stimulate the body to produce the desired physiologic effect and those that interfere with pathophysiologic processes. In both categories, the reaction of the living organism to the applied therapy produces the end product of therapy.

Stimulation of Physiologic Homeostasis

Therapies that stimulate the body's homeostatic mechanisms are those that create the appropriate setting for the body to achieve the best result from its own healing process. A musculoskeletal example of this is fixation of a fracture. Although a surgically implanted rod helps align the bone, creates the right mechanical environment for healing, and brings together fragments that would otherwise not unite, placing a rod is not tantamount to "curing" the fracture. Rather, the rod implantation optimizes the body's own processes, which eventually "cure" the fracture.

Blocking Pathophysiologic Processes

An often-quoted pharmacologic aphorism states that "all medicines are poisons." Although this statement is not entirely true, neither is it totally false; rather, it is based on the notion that many medications are administered to interfere with biologic processes. Therapies that block biologic processes are often chosen because the process is either abnormal or excessive, such as with inflammation. In other instances, the process is entirely normal but not desirable. For example, the process that a beta-blocker blocks is entirely normal, yet the adrenergic effect of that response increases blood pressure and myocardial oxygen consumption. Thus, it may be desirable to block those end results.

Analgesic Effects

Analgesic therapies are used simply to decrease pain. Pain is certainly the primary reason that people seek medical attention for musculoskeletal complaints. Patients may be seeking a "cure," but it is their pain that signals that something is wrong. Pain can be decreased by treating the underlying condition. In musculoskeletal medicine, for instance, immobilizing a fracture helps the body begin the process of repairing the damage, but it also immediately relieves pain. Pain relief for its own sake is often a necessary component of treating disease as well. This may be achieved via two mechanisms: direct and indirect analgesic effects.

Direct Analgesic Effect

Pain is a central nervous system response to noxious stimulants. Any intervention that decreases the signal to the brain or modulates its interpretation can decrease pain. This category runs the gamut from general anesthesia to acetaminophen.

Indirect Analgesic Effect

The body produces its own natural painkillers: endorphins. Any intervention that stimulates the production of these chemicals will have an analgesic effect. It is also known that the interpretation of pain can be modulated. For example, simply rubbing the skin over a bruised area may make it feel better, probably because the thalamus, which integrates pain signals in the brain, is "distracted." Processes that modulate the interpretation of pain signals, without actually modifying them, are also analgesic.

Placebo Effect

The placebo effect is one of the most misunderstood concepts in medicine. Hrobjartsson and Gotzsche[1] actually questioned the existence of this effect. These authors suggested that researchers who compare an active drug with a sugar pill may be fooled into thinking that the sugar pill "helped" when all that happened was a natural return to health (ie, both groups got well, but they would have done so anyway without *any* intervention). Alternatively, the process of observation may have produced the effect. This may be seen, for instance, in a study in which patients are given a placebo for weight loss but are monitored weekly by a physician. It is the monitoring that induces compliance with a diet that may be the cause of the observed weight loss. In order to demonstrate a true placebo effect, then, a study must have three arms: the active treatment, the placebo treatment, and no treatment at all.

For the purpose of this classification of therapies, a placebo effect will be defined as a therapy that, for unknown reasons, stimulates the mind to help affect either healing or analgesia that would not have occurred in its absence. Psychologic processes can produce biologic effects. For example, embarrassment leads to blushing because at some level the higher cognitive message of shame stimulates a histamine response in the cheeks.

Alternative Medicine in a Scientific Context

Given the previous discussion, there should be no distinction between alternative and traditional medicine. A treatment either works by one of the mechanisms listed above—or it does not work. For example, unless there is definitive proof that qi (vital energy, pronounced "chee") or "water memory" are false concepts, there is the possibility that a therapy based on qi or "water memory" may be effective but not by its putative mechanism. Over time, traditional medicine assimilates those alternative therapies that have scientifically proven effectiveness—foxglove (and its digitalis), for instance. Thus, at any one time, alternative medicine will include those treatments that are ineffective (as many effective treatments will be co-opted by mainstream medicine), as well those that are unproven but may still be effective. It is easy to deride this latter category, especially if the theory underlying it sounds questionable to western ears. Still, any physician with an eye toward the past (and a historical appreciation of the quackery that once was conventional medicine) will approach alternative medicine with a healthy dose of humility. From the treatment of scurvy with lemons and oranges to the discovery that penicillins grow from molds, what was once considered absurd becomes an accepted principle of modern medicine through scientific experimentation and the accumulation of supporting evidence.

Understanding Scientific Investigation

Modern medical knowledge is almost always derived from a process of statistical sampling. The researcher does not study all possible subjects, but only a sample of an underlying population. Statistical tests are used to determine whether it is reasonable to draw inferences from that sampling. Thus, uncertainty enters the process, and statements are made in terms of probability. Sometimes probability is so close to 100% that it would be foolish to consider an inference to be false. In other instances, scientific studies report a P value. When the P value of a study is less than 0.05 (by convention), the results are considered statistically significant. That means that there is less than a 0.05 (1/20) chance that two samples that were previously seen as different are not from different underlying populations. An incorrect assumption that two similar groups are different constitutes a type I error.

The Importance of Correct Interpretation

The use of statistical testing has two important implications. First, even if the study is statistically significant, there is a measurable chance that its results are wrong—that a type I error has occurred. This risk is usually low, but it is not zero. Second, statistical significance says nothing about clinical significance—the magnitude of the effect may be nothing worthwhile, even if its existence is a statistical fact.

Readers of journal articles often do not need to worry about type I error; editors reject papers in which the results are not statistically significant. The problem posed to readers is that they are able to read only what is published, resulting in a publication bias, and only those studies that show a positive effect in statistically significant fashion are accepted for publication in the peer-reviewed literature.

For example, consider a perfectly executed, prospective, randomized controlled trial showing that chanting a mantra decreases the pain of arthritis ($P = 0.04$). What do these results mean? This may be hard for the reader to determine, as the reader does not have the full body of literature written on the subject at hand. The effectiveness of chanting is certainly called into question if there are 35 other perfectly executed, prospective, randomized controlled trials that show that chanting may *not* decrease pain; however, the reader does not have access to these studies if they were deemed either "not interesting" enough (to the author or the editor) or not statistically significant enough to be published. Alternatively, a researcher who believes that chanting is effective may simply repeat the experiment again and again until the desired results are found; all other runs are discarded because of "obvious flaws" in their execution.

The fact that studies with "negative" results were not published reflects a bias that may be impossible to detect. The reader cannot evaluate what may or may not exist. The possibility of publication bias means that a study or collection of studies with positive results does not necessarily prove a fact. Alternatively, when study results controvert what we know to be true, it is likely that these results are statistical quirks.

Analysis of Alternative Medicine

In the discussion that follows, an overview of some of the alternative medicines used in musculoskeletal medicine will be presented. These therapies are listed here as a means of introduction but they are not referenced in any other section of the book. There is certainly no disparagement intended by simply labeling a treatment as "alternative." Claims that defy our concepts of biology will be explicitly noted. Also, no disparagement is directed at those therapies not selected for discussion.

Classically, the delivery of healthcare has been divided into medical and surgical domains: interventions that use pharmaceuticals and those that use a hands-on approach (note that the etymology of the word "surgery" is the Greek word meaning "hand"). Our initial discussion of specific alternative medicines follows that division, with two examples chosen from each category: acupuncture and chiropractic care (alternative hands-on interventions) and glucosamine and chondroitin therapy and homeopathy (alternative pharmaceutical interventions). A third category, patient-directed interventions or self-healing, then discusses meditation, yoga, qigong, and tai chi. And finally, answering to modern, high-tech sensibilities, a fourth category, alternative medical devices, discusses magnets and copper bracelets.

Hands-On Interventions

Acupuncture

Background and Theory

Acupuncture, one of the oldest forms of therapy, has been used as a medical modality for over 3,000 years in China. It was first developed as a naturalistic and human-centered therapeutic replacement for the supernatural shamanistic rituals of earlier periods. Developed in the same era as Confucianism and Taoism, the practice of acupuncture is deeply rooted in ordinary human sensory awareness. Even so, the conceptual underpinnings of acupuncture such as yin-yang (polar opposing forces, representing female/passive and male/active) and qi are difficult for westerners to immediately appreciate.

Traditional Chinese medicine theorizes that any manifestation of disease is a sign of imbalance between the yin and yang forces in the body. There are 12 main and eight secondary pathways called meridians that are connected by over 2,000 acupuncture points on the human body, and each has either yin or yang characteristics. These meridians conduct qi, which regulates a person's mental and physical balance, and it can be affected by the imbalances of yin-yang. By inserting needles into precisely defined points on the body, acupuncture theoretically realigns imbalances of yin-yang and qi to bring about harmony to the "climate" of the individual. The resulting balance will maintain the normal flow of energy and maintain or restore health to the body and mind.[2]

Western scientists tend to think of meridians as a nebulous concept because they do not directly correspond to nerve or blood circulation pathways. Some researchers believe that meridians are located throughout the body's connective tissue; others do not believe that qi exists at all.[3,4] Such differences of opinion have made acupuncture an area of scientific controversy. Still, the effectiveness of acupuncture can be explained in the framework of western medicine as follows:

Gate control theory This theory maintains that the brain can process only so much information at one time. Thus, one type of sensory input, such as low back pain, can be temporarily inhibited in the central nervous system by another pain stimulus—acupuncture needling.

Activation of opioid systems Some evidence points to the possibility that acupuncture stimulates the production of endorphin, serotonin, and acetylcholine in the central nervous system, enhancing analgesia. Opioid antagonists, such as naloxone, reverse the analgesic effect, which further strengthens this hypothesis.[5]

Diffuse noxious inhibitory control (DNIC) Noxious stimulation of other areas of the body alters the pain sensation where an individual feels the original pain.

Vascular permeability Acupuncture may affect vascular permeability and, therefore, inflammation.

Indications and Use

In Eastern cultures, acupuncture has been used to treat a wide range of human ailments. As an adjunct to modern musculoskeletal medical care, acupuncture has been evaluated and used for the treatment of pain (eg, neck and back pain or pain associated with osteoarthritis) and for the control of postoperative nausea and vomiting. Although its mechanism of action remains unclear, acupuncture has gained considerable popularity in the United States and recognition by the US Food and Drug Administration (FDA). In 1996, the FDA reclassified acupuncture needles from a class III (experimental) to class II (nonexperimental but regulated) medical device.[5]

The National Institutes of Health (NIH) established a Consensus Development Panel (NIHCDP) in 1997 to evaluate the clinical efficacy of acupuncture. After reviewing the literature, the NIHCDP concluded in their report that there is definitive evidence of acupuncture's efficacy in postoperative and chemotherapy-induced nausea and vomiting, as well as in treating postoperative dental pain. They also listed the following conditions in which acupuncture may be effective as an adjunctive treatment or an alternative therapy: fibromyalgia, low back pain, lateral epicondylitis, alcohol addiction, stroke rehabilitation, headache, menstrual cramps, myofascial pain, osteoarthritis, carpal tunnel syndrome, and asthma.[6]

The incidence of adverse events in the practice of acupuncture has been documented to be extremely low, although a life-threatening event such as pneumothorax has occurred on rare occasions. Therefore, it is important that appropriate safeguards for protecting patients be in place. Since acupuncture needles are under FDA regulation, they are expected to be sterile, single-use needles. If a patient is under the care of an acupuncturist and a physician, both practitioners should be informed so that important medical problems are not overlooked.

Chiropractic Therapy

Background and Theory

The practice of modern chiropractic care was founded by Daniel Palmer in 1895, but the general practice of spinal manipulation as a healing method dates as far back as Hippocrates and Galen. The foundation of Palmer's therapeutic philosophy is based on the "innate intelligence" of the body to heal itself and the belief that medications and surgeries were unnatural invasions. Chiropractic care tenets are based on the assumption that many diseases are caused or exacerbated by disruption of the flow of vital force through the spinal cord by subluxation (that is, small malalignment) of the vertebrae, which causes impingement of the nerves as they exit the neural foramen. Chiropractic therapy focuses on identification of these subluxations and restoration of the proper vertebral alignment via spinal manipulation.

The practice consists of procedures involving the application of force to a specific tissue (usually associated with the spine) with therapeutic intent. The biologic rationale for chiropractic therapy maintains that spinal subluxation and manipulation cause, in addition to anatomic shifts, physiologic changes in neuronal activities, facet joint kinematics, proprioception, and endorphin elaboration.[7]

At least five mechanical and neurologic mechanisms have been proposed to explain the benefits of spinal manipulation.

1. It releases entrapped facet joint inclusions that are heavily innervated.
2. It repositions fragments of posterior annular material from the intervertebral disk.
3. It relaxes hypertonic muscle by sudden stretching.
4. It disrupts articular or periarticular adhesions from prior injury.
5. It inhibits excess reflex activity in intrinsic spinal musculature.

Although each of these mechanisms of action are plausible, no clear evidence yet exists to support the anatomic or physiologic underpinnings of chiropractic manipulations.

Indications and Use

Patients who consult a chiropractor for musculoskeletal problems most often do so for low back pain (60%); other common complaints include head, neck, and extremity symptoms. Approximately half of these patients have chronic symptoms. Randomized clinical trials involving spinal manipulation abound in the current literature. While publication bias may be suspected, most published trials either reported equivocal findings or beneficial effects of manipulation.[7] Currently, there are over 70 randomized clinical trials on this topic, and most were conducted on patients with low back, neck, and head pain, and a few have examined other conditions. The clinical trials include placebo-controlled comparisons, comparisons with other treatments, and pragmatic comparisons of chiropractic management with common medical management.

A majority of the randomized trials on treatment of low back pain showed beneficial effects of spinal manipulation over conventional therapy in at least one subgroup of patients. Although the positive effect appeared to be statistically significant from these trials, the improvement was not dramatic. Meta-analyses in the early 1990s made cautiously positive or equivocal statements regarding the effectiveness of spinal manipulations for low back pain.[8,9] However, a systematic review in 1997 found sufficient evidence that manipulation is helpful for chronic but not acute back pain.[10] The most recent systematic review found insufficient evidence to define the role of chiropractic manipulation in treating sciatica.[11]

A smaller number of trials were conducted for neck pain and headache. The trials examining chiropractic manipulations for neck pain compared it with both sham manipulations as well as conventional therapy. Chiropractic manipulations were shown to be superior in both settings.[7] Likewise, the published reports of a few trials examining the efficacy of chiropractic therapy for headache showed a benefit for this approach. The efficacy of manipulation for other conditions, such as sciatica and carpal tunnel syndrome, has also been examined but results are inconclusive.

Spinal manipulation has been a controversial topic. Some rather minor adverse effects include transient, localized discomfort, headache, or fatigue. The more serious complications reported include cauda equina syndrome from lumbar manipulation and cerebrovascular artery dissection from cervical manipulation. Although the rate of serious complications is unknown, estimates vary from 1 per 400,000 to between 3 and 6 per 10 million.[9] Needless to say, even "benign" western treatments for back pain (using nonsteroidal anti-inflammatory drugs (NSAIDs), for example) are not without their complications either.

Pharmaceutical Interventions

Glucosamine and Chondroitin

Background and Theory

The molecular constituents of the articular cartilage matrix are composed primarily of collagen and proteoglycans. These molecules are critical to conferring the structural integrity and compressive resistance characteristic of cartilage. The rationale for the use of glucosamine sulfate and chondroitin sulfate for joint pain, specifically that caused by osteoarthritis, is to promote cartilage repair and synthesis and to prevent enzymatic degradation of the building blocks of joint cartilage, respectively. Chondroitin is also believed to promote water retention and elasticity in cartilage.

Glucosamine is a fundamental component in the synthesis of both hyaluronic acid (the backbone molecule in proteoglycans) and chondroitin (an important class of glycosaminoglycans in articular proteoglycans). Thus, the rationale for using these compounds as a dietary supplement is to provide biosynthetic substrate for cartilage repair and synthesis. There is also limited evidence in animal and explant tissue culture systems that dietary supplementation may stimulate synthesis of glycosaminoglycans and proteoglycans and suppress cartilage degradation.[12] However, a direct and compelling mechanism of action has not been elucidated nor have the histologic and biochemical results of these studies been substantiated in human studies.[13]

Indications and Use

The primary musculoskeletal indication for glucosamine and chondroitin is to prevent and treat osteoarthritis. A variety of studies (ranging from small, uncontrolled studies to moderately sized, controlled trials) have examined the effectiveness of glucosamine and chondroitin in the treatment of osteoarthritis and have reported conflicting results. A meta-analysis of placebo-controlled trials (most with industry financial support) involving osteoarthritis of the knee and hip revealed some improvement in short-term outcomes (including symptoms and function) with treatment, but many of the studies this analysis examined were flawed.[14] More recently, a randomized, double-blind, placebo-controlled trial assessing the long-term effects (up to 3 years) of oral glucosamine on the progression of osteoarthritis of the knee suggested a cartilage protective effect and revealed small but significant improvements in symptoms and function.[15]

Both supplements also have some anti-inflammatory effects that may account for pain relief. If glucosamine or chondroitin prove to have significant anti-inflammatory effects without the gastrointestinal adverse effects of NSAIDs, then these supplements may become very useful in treating arthritis on that basis alone. Scientifically, further studies are necessary to construct a biochemical and physiologic or, as some have suggested, psychologic basis for the efficacy of glucosamine and chondroitin.

Clinically, the issue of industry support, which has been considerable in trials to date, as a potential cause of publication bias needs to be adequately addressed.[16] In addition, the effectiveness of glucosamine and chondroitin, which currently appears small and variable, as an alternative to or as concurrent treatment with other modalities (eg, cyclooxygenase-2 inhibitors) should be further studied in randomized controlled trials. In March 2000, the NIH announced that it would initiate a large, multicenter study to provide definitive answers about the effectiveness of glucosamine and chondroitin in treating osteoarthritis. This study is to be a nine-center effort, enrolling more than 1,000 patients. The data from this study should be available in the middle of the decade.

The risk of using glucosamine and chondroitin is minimal, as excess intake is apparently filtered out in the urine. However, they may interact with certain medications (causing the need for higher doses of blood pressure medications among certain patients); therefore, patients should inform their physicians if they are taking glucosamine and chondroitin.

Homeopathy

Background and Theory

The term homeopathy is derived from the Greek *homoios* (similar) and *pathos* (suffering) and was initially conceived and developed by the German physician Samuel Hahnemann (1755-1843). Hahnemann, not burdened by the regulatory environment of the 21st century, assembled a library of "symptoms pictures" by administering various noxious substances to healthy individuals and recording the resultant effects. Homeopathy, then, is a system of therapeutics in which disorders characterized by a particular symptom are treated with a substance that causes that symptom in a healthy individual. In an attempt to minimize adverse effects of his medications, Hahnemann found that a stepwise dilution with vigorous shaking paradoxically improved the efficacy of his treatments—the greater the dilutions, the greater the efficacy.[17]

Dating back to the time of Hippocrates (fifth century BC), the idea of healing "like with like" was met and continues to be met with considerable skepticism from the established medical community. Note that the process of vaccination, though superficially similar, does not propose treatment with the agent, but rather immune stimulation. Vaccines are administered to healthy individuals as a preventive measure prior to exposure.

Given that homeopathy is based on the theory that decreasing the concentration increases the potency of the solution, it is no surprise that this is one of the least accepted alternative medical approaches. The notion of dilution increasing potency runs counter not only to conventional medicine but also to the basic science of chemistry. In some typical doses, a homeopathic medication may not have even a single molecule of the active pharmacologic substance. A number of theories that attempt to explain the mechanism of action in the absence of the original active ingredient (ie, the water "remembers" the encounter with the substance) remain controversial and unsubstantiated by scientific methods. Nonetheless, homeopathic practitioners and millions of consumers maintain that empiric results and observations validate the practice.

Indications and Use

A variety of musculoskeletal conditions are treated with homeopathic remedies, including acute sprains, hemarthosis, rheumatoid arthritis, osteoarthritis, muscle cramps, and myalgia. In addition to Hahnemann's original studies and other observational reports, there has been a plethora of randomized placebo-controlled trials examining homeopathic treatments of musculoskeletal conditions. One large meta-analysis concluded that the overall clinical improvement with homeopathic treatment, although small in magnitude, is unlikely to be the result of placebo effects alone.[18] However, the study also indicated that homeopathy was not clearly efficacious in the treatment of any particular clinical condition, including musculoskeletal conditions. Another meta-analysis offered similar conclusions but further noted poor methodology in many trials and that trials of high quality were more likely to show negative efficacy.[19] To date, the effectiveness of homeopathy in treating musculoskeletal conditions has not been clearly proven scientifically.

In one large, randomized, double-blind, placebo-controlled, crossover study, 112 patients with stable rheumatoid arthritis who were taking NSAIDs but not steroids in the past 12 months were selected to receive either a placebo or a homeopathic arthritis medicine.[20] At 3 months, the medications for patients in each group were switched. In this study, the homeopathic medicine did not provide pain relief for patients; in fact, the placebo resulted in better pain relief.

The best case to be made for homeopathy is that it is almost certainly benign. Homeopathic medicines, devoid of active ingredients, are unlikely to inflict harm. Arguably, if homeopathy were to assuage concerned patients to the extent that it helps them resist taking more toxic alternative medications, it may have some beneficial role after all.

Patient-Directed Interventions

Meditation

Background and Theory

Meditation is an ancient practice thought to promote relief for stress, headaches, anxiety, and emotional disorders. In addition, it is now being investigated as a means for patients to cope with pain associated with chronic illness. Meditation involves slow, rhythmic breathing while the mind is focused on a symbol, such as a candle, a mental image, or a mantra, which is a word or phrase that is rhythmically repeated. The practice of meditation, once a cultural phenomenon, is now an accepted therapy for many conditions. Rather than costly equipment, drugs, or surgery, meditation requires only a quiet environment, comfortable clothing, and a position in which the spine is vertical. Consequently, many insurance plans are allowing for reimbursement of meditation lessons as a treatment option.

Scientists believe that meditation influences how the brain functions and that thoughts can then influence the brain and body. Meditation may have the ability to lower stress hormone levels, improve circulation, lower blood pressure, and moderate the immune system.

Indications and Use

Several studies have been launched to investigate the impact of meditation programs on various illnesses. Because stress may cause flares in arthritis, systemic lupus erythematosus, rheumatoid arthritis, and fibromyalgia, it is believed that meditation, by lowering stress, can reduce the incidence of flares and thereby alleviate disease-associated pain. Singh and associates[21] studied the effect of behavioral therapy in 28 patients with fibromyalgia over the course of 8 weeks. During weekly sessions of 2½ hours each, patients were given a lesson focusing on the mind-body connection, a portion focusing on relaxation response mechanisms, and a movement therapy session. Standard outcome measures reported significant reduction in pain, fatigue, and sleeplessness, along with improved function, mood state, and general health. The authors concluded that meditation techniques, patient education, and movement therapy appear to be an effective therapy for patients with fibromyalgia.

In a study by Kaplan and associates,[22] 77 patients meeting the criteria of the American College of Rheumatology for fibromyalgia took part in a 10-week group outpatient program to evaluate the effectiveness of a meditation-based, stress-reduction program to treat their illness. Patients were evaluated before and after the program. Outcome measures included visual analog scales to measure global well-being, pain, sleep, fatigue, and feeling refreshed in the morning. They also completed a medical symptom checklist and other standard questionnaires. The mean scores of all patients completing the program showed improvement, and 51% showed moderate to marked improvement. The authors concluded that a meditation-based, stress-reduction program is effective for patients with fibromyalgia.

Although these preliminary findings appear to be promising, controlled trials with a larger patient population are necessary to establish meditation as an effective treatment for chronic illnesses. Even though there are no evident adverse effects from the practice of meditation, it is important for patients to maintain their conventional treatments. Patients must be aware that meditation will not take away their pain but may allow them to better cope with their illness.

Yoga, Qigong, and Tai Chi

Background and Theory

Yoga, qigong, and tai chi belong to a family of movement therapies designed to reduce stress and anxiety, while improving circulation, range of motion, flexibility, and overall physical fitness.

The practice of yoga is rooted in Hindu religion and dates back 5,000 years. It refers to a variety of disciplines designed to bring practitioners into union with mankind and a higher God or life force. Deep breathing is important because the life force, or *prana*, is believed to enter the body through the breath. Some yoga postures have been developed as stretching and strengthening exercises to improve posture and the musculoskeletal system, relax nerves, and enhance the body. No special equipment is needed. Patients who have a back injury, recently had surgery, or are pregnant should avoid yoga; and those with arthritis, heart disease, or high blood pressure may need to avoid certain yoga postures as a precaution. Health advocates promote yoga as a method to lower blood pressure as well as a method to increase strength and flexibility. It has been used as a complementary therapy for cancer, arthritis, weight loss, anxiety, and diabetes.

Qigong (pronounced *chee-gong*) is a form of gentle exercises in which the flow of qi is stimulated along the meridians throughout the body. Its practice in Chinese medicine dates back more than 2,000 years. The philosophy of qigong considers illness to be the result of an imbalance of qi and suggests that regulating this flow of energy will restore balance within the body and return a diseased part of the body to health. There are thousands of variations of qigong exercises, ranging from simple to complex. Proponents believe the physical benefits of qigong include lowering the heart rate and blood pressure, increasing circulation, reducing pain, and increasing the flow within the lymphatic system, which in turn, improves the immune system. Although proponents credit it with curing AIDS, arthritis, asthma, cancer, diabetes, and heart disease and alleviating pain, there is no definitive proof that qigong has these curative powers.

Tai chi is a low-impact exercise program that is suitable for almost anyone to practice, including the sick, the young, and the elderly. It is derived from the martial arts in China and also founded in accordance with the belief that chi (chi and qi are transliterations of the same word) is a vital life force flowing through the body. All tai chi exercises are pairs of opposite movements, such as left and right, forward and backward. Exercises are correctly performed slowly and gracefully, with the knees slightly bent through each sequence and all movements originating from the waist, or tantien, the area of the body located beneath the navel and considered to be the center of the body's qi. Practitioners believe that by breaking up blockages in the flow of qi, the balance of the life force within the body will be restored. Advocates report that tai chi improves health, promotes concentration, improves posture and balance, and increases energy, flexibility, and muscle tone. It is often recommended for older adults as a means to prevent osteoporosis.

Indications and Use

Creamer and associates[23] conducted a pilot study to examine the practicality of delivering a package of nonpharmacologic, behavioral-based treatment to patients with fibromyalgia. For 8 weeks, 28 patients were given formal meditation-relaxation training and instruction in qigong. Outcome measurements included Fibromyalgia Impact Questionnaire scores and a range of other outcome measures, including tender points and pain thresholds. Improvement was sustained for 4 months after the end of the intervention. The authors concluded that nonpharmacologic, behavioral-based treatment appears to produce sustained benefit in a range of outcomes for patients with fibromyalgia.

In a 1999 nonrandomized, noncontrolled pilot study, Husted and associates[24] conducted an 8-week tai chi program to explore the psychosocial and physical benefits for patients with multiple sclerosis. Nineteen patients participated, and outcome measures included walking speed (distance, 25 ft), hamstring flexibility, and Medical Outcomes Study 36-Item Short Form Health Survey scores. Results showed that walking speed increased by 21% and hamstring flexibility increased by 28%. Patients experienced improvements in vitality, social functioning, mental health, and the ability to carry out physical and emotional roles. The results of this pilot program led to the implementation of several tai chi classes for people with multiple sclerosis. Advocates cite this study for using tai chi as a means for improving the quality of life for people with chronic disabling conditions.

For rheumatic and musculoskeletal diseases, stress, pain, and mobility are closely related. Therefore, treatments that decrease stress may arguably reduce pain and alter the course of these diseases. However, there are no large-scale, scientifically organized clinical trials to prove the claims that these mind-body methods cure disease.

Devices

Magnets

Background and Theory

The use of magnets for pain relief can be traced back to the days of Cleopatra, who slept on a magnet stone to prevent aging. An Austrian physician, Franz Mesmer, popularized magnet therapy in the late 1800s by achieving "miracle cures" of many different illnesses—illnesses that were mainly self-limiting. This may have been a form of hypnosis: the word "mesmerize," after all, is derived from Mesmer's name.

Two major types of magnet therapy are currently in use: pulsating electromagnet field (PEMF) therapy and static magnet therapy. Although numerous studies have shown the efficacy of PEMF magnets in treating various conditions, such as enhancing bone and wound healing, similar evidence regarding the efficacy of static magnets is less abundant. Static magnets, however, have gained popularity as an alternative therapy for musculoskeletal complaints.

Static magnet therapy involves taping a small disk of magnet to the body over the painful area. The magnets used for this therapy should generate a magnetic field that is 10 times the strength of a typical refrigerator magnet, although without regulation, refrigerator magnets themselves may be sold as therapeutic devices. There are two types of static magnets that are commercially available: unipolar and bipolar. The unipolar magnet has only one pole on the surface that is applied to the skin. Most manufacturers designate it "N." With the bipolar magnet, there is an alternating north and south pole generating a concentric magnetic pattern or a grid on the surface that is applied to the skin. Manufacturers of each type of magnet claim that their type of magnet is superior. Working from the assumption that increased blood flow is the mechanism by which magnets control pain, proponents of the bipolar magnet claim that their magnet promotes blood flow in skin, muscle, and tendon, despite the low surface intensity of the magnetic field. Proponents of the unipolar magnet believe that their magnet has a stronger magnetic field. Field strength and tissue penetration are presumably the most important properties that account for a magnet's therapeutic effects.

The exact mechanism of the interaction between magnetic field and biologic tissue is unknown. Various hypotheses proposed by different researchers include the following:[25]

1. *Solid-state theory of cellular function* According to this theory, small regions of tissue and molecules in the body are sensitive to the effects of external magnetic fields, regardless of strength. Consequently, such interactions lead to a cascade of cellular events, such as altered enzymatic activity and cellular potential.
2. *Theory of closed electric circuits* In this theory, the endothelial lining of the cardiovascular system serves as a large conduit to conduct the electrical current that is generated by the magnetic field.
3. *Association-induction hypothesis* This theory offers multiple explanations to account for the biologic effect of a magnetic field, one of which is similar to that of the solid-state theory and proposes that a magnetic field induces changes in the small foci of sensitive molecules, which triggers a cascading event that alters the physiologic function of adjacent tissues. Another possible explanation is that when a sensitive molecule within the bloodstream is exposed to a magnetic field, the affected molecule can have physiologic effects that are distant from the original site of magnet application.
4. *Ion parametric resonance theory* In this model, it is thought that magnetic effects on an ion within a molecular complex (such as an enzyme binding site) can alter the conformation and in turn affect the biologic properties of the molecular complex.

Indications and Use

Unlike its counterpart, PEMF, for which there is an abundance of clinical trials evaluating its efficacy in various indications, only a handful of clinical trials evaluate the efficacy of static magnets, largely for pain reduction in various conditions. One of the first clinical trials was a randomized, double-blinded clinical trial of 50 patients with post-polio syndrome.[26] Patients were asked to rate their pain on a scale of 0 to 10 after wearing either an active or inactive bipolar magnet at their pain trigger point for 45 minutes. The magnets had a field strength of between 300 and 500 gauss. Patients who wore active magnets reported more pain relief than their counterparts who wore the inactive magnets; however, one limitation of this study is that researchers did not compare efficacy of unipolar and bipolar magnets. And although there are no studies of the long-term effects of magnets on pain relief, this study became widely quoted by magnet manufacturers to demonstrate the efficacy of static magnets.

Another study of patients with diabetic polyneuropathy showed a statistically significant reduction of neuropathic pain in patients who wore magnetic foot pads for a total of 12 weeks.[27] The effects of bipolar permanent magnets in treating chronic low back pain were also recently examined in a yearlong randomized, double-blinded, placebo-controlled, crossover study.[28] All 24 patients alternated between 1 week of active bipolar magnet therapy (magnet strength, 300 gauss) and 1 week of wearing inactive magnets, with 1 week of washout in between. All patients wore the magnet for 6 hours a day, 3 days a week, and rated their pain at the end of each week. In this study, there was no statistically significant pain reduction in either group.

Because the strength of the commercially available static magnets is relatively low (between 300 and 500 gauss), the risks of use are thought to be low as well. There are no known contraindications to using small magnets on the extremities.

Copper

Background and Theory

The use of copper as a therapeutic agent dates back to 1550 BC. For thousands of years, people have worn copper bracelets and anklets to reduce pain and inflammation associated with joint and connective tissue problems. Although the anti-inflammatory property of copper has been long recognized, the exact mechanism is still under investigation. The proponents of copper therapy argue that most people currently do not consume enough trace minerals, including copper, from their regular diets. A copper bracelet or anklet, when worn next to the skin, allows for direct absorption of trace amounts of dissolved copper into the body on a daily basis. This absorption process, proponents argue, bypasses the digestive route, thus avoiding the potential gastrointestinal adverse effects produced by conventional NSAIDs. Currently, copper bracelets are promoted to be beneficial in the treatment of arthritis, tendinitis, bursitis, osteoporosis, sports injuries, and lateral epicondylitis.

Copper dissolves in human sweat. Over a 7-week period, a pair of copper bracelets worn around the wrist or ankles will lose about 80 to 90 mg of mass.[29] Given the skin's natural permeability for copper, it is possible that dissolved copper can penetrate through the dermis and reach the local tissues and the systemic circulation to offer some therapeutic effect.

It has been long known that patients with rheumatoid arthritis have elevated levels of serum copper. Initially thought to be a consequence of inflammation, it is now known that the increased serum level of copper in these patients is an acute-phase response. Moreover, copper seems to have both pro- and anti-inflammatory potentials, depending on the ligands to which it is bound.[30] Although the exact anti-inflammatory mechanism is unknown, there is ample evidence suggesting that copper is involved in several different biologic pathways.

1. *Oxygen free radical metabolism* Copper is known to be an essential cofactor for enzyme cytoplasmic superoxide dismutase, an important enzyme that protects cells from oxidative stress.
2. *Prostaglandin biosynthesis* When a copper ion is bound to a nonsteroidal anti-inflammatory ligand, the new complex is more potent than the parent drug.
3. *Neutrophil adhesions* A recent study evaluated copper's effect on neutrophils during the inflammatory process using a rat model. It was shown that dietary copper seems to inhibit neutrophil adhesion, a crucial step in the development of inflammation. This may be one of the mechanisms by which copper induces its anti-inflammatory effect.

Indications and Use

There are scant references to studies assessing the effectiveness of copper bracelets for musculoskeletal complaints. In one study, using questionnaires and psychological parameters as outcome measures, a comparison was made among patients with self-diagnosed arthritis or rheumatism—hardly rigorous inclusion criteria—who wore copper bracelets, anodized aluminum bracelets resembling copper, or no bracelets at all.[31] The study obtained a 93% response rate from 323 patients and noted that those who wore copper bracelets felt worse when not wearing them ($P < 0.02$). A significant number of patients also noted greater symptomatic improvement with the copper bracelets (copper versus aluminum, $P < 0.01$).

In another clinical trial, the pain-relieving effect of a copper-salicylate gel in osteoarthritis was studied.[32] This randomized controlled trial recruited 116 patients with osteoarthritis of the hip and/or knee who were asked to apply copper-salicylate gel to their forearms. Patients had to apply either the active or placebo gel twice daily for 4 weeks and then complete a self-assessment questionnaire comparing pain before and after the trial. The result of the study showed no significant difference between the two groups in terms of pain reduction. However, significantly more patients in the treatment group reported adverse skin reactions than did the placebo group (83% versus 52%, respectively).

Although the physiologic effects of trace elements such as copper may be reasonably established, their therapeutic effectiveness remains less compelling. The only noted adverse effect of wearing copper bracelets is allergic skin reaction; therefore, they are relatively safe to wear.

Alternative Therapy Versus Traditional Medicine

Millions of Americans suffer from chronic illnesses. Unhappy, perhaps, with the results obtained from conventional medicine, an increasing number of patients are turning to alternative therapies to improve their quality of life. Because so many patients are using alternative therapies, it is important for physicians to be familiar with these remedies and to be able to answer questions from patients regarding their efficacy. The more popular alternative therapies include not only those discussed in this chapter, but others such as aromatherapy, music therapy, biofeedback, reflexology, imagery, naturopathy, and many more.

Much information regarding mainstream and alternative medicine is readily available on the Internet. How do physicians help the healthcare consumer make sense of it all? Patients should first and foremost be encouraged to always obtain a medical diagnosis from their physicians if they are concerned about a particular condition. Patients should also be cautioned to never discontinue a medication or treatment of conventional medicine in favor of an unproven complementary therapy without at least discussing this with their doctor. Regarding consumer healthcare information, there are numerous medical resources available; however, patients should be warned to be wary of information regarding products that result in miracle cures or contain secret formulas. When product information reports results based on research studies, it is important for the physician and the healthcare consumer to examine the study, look for valid citations and references, and verify that the study is based on human subjects rather than animals.

The NIH supports clinical trials of numerous therapies and treatments that are currently considered alternative or complementary. Information about these studies can be found on the NIH website at http://www.nccam.nih.gov. The American Academy of Orthopaedic Surgeons launched its own website for Complementary and Alternative Medicine at http://www3.aaos.org/courses/cam/camtoc.htm, where it offers useful links, information on herb-drug interactions (Table 1), and relevant articles on complementary and alternative medicine. Just because a therapy is labeled "alternative" does not mean it does not work—or works any less well than conventional treatments. Nonetheless, alternative medicine includes many untested (and therefore possibly dangerous) treatments. Thus, a cautious approach, such as asking "What is the evidence and what does it show?" is necessary to adhere to medicine's golden rule: first do no harm. Once it is established that the risk of harm is not too high, alternative therapies may be fine choices, especially for those patients not helped by traditional approaches.

Table 1
Complementary and Alternative Medicine Herb/Drug Interactions

Herbal Supplement	Common Uses	Potential Problems	Potential Interactions
Dong Quai (*Angelica*)	To treat menopausal symptoms, premenstrual syndrome, dysmenorrhea	Enhances bleeding	Anticoagulants
Echinacea	To treat colds, flu, and mild infections, especially upper respiratory infections	Hepatotoxicity; intestinal upset	Other hepatotoxic drugs; anabolic steroids; methotrexate
Ephedra (*Ma Huang, Ephedrine, Pseudo-ephedrine*)	To treat asthma, cough, and to induce weight loss	Seizures; adverse cardiovascular events; hypertension; death	Cardiac glycosides; general anesthesia; monoamine oxidase inhibitors; decongestants, stimulants
Garlic	To decrease cholesterol and blood clot formation	Enhances bleeding	Anticoagulants
Ginger	To relieve nausea	Enhances bleeding; central nervous system depression; hypotension; cardiac arrhythmia; hypoglycemia	Anticoagulants; enhances the effects of barbiturates; antihypertensives; cardiac drugs; hypoglycemic drugs
Ginkgo Biloba	To improve circulation, especially to brain; for memory loss, dizziness, and headache	Enhances bleeding; cramps, muscle spasms	Anticoagulants
Ginseng	To increase energy and reduce stress	Enhances bleeding; tachycardia and hypertension; mania	Anticoagulants; stimulants; antihypertensives; antidepressants/phenelzine; digoxin; potentiates the effects of corticosteroids and estrogens
Goldenseal	Used as a mild antibiotic to treat sore throats and upper respiratory infections	Increases fluid retention; hypertension; nausea; nervousness	Diuretics; antihypertensives
Kava Kava	To treat anxiety, nervousness, and insomnia	Upset stomach; allergic skin reaction, yellow discoloration of skin; central nervous system depression, liver toxicity	Potentiates the effects of antidepressants, barbiturates, and benzodiazepines; skeletal muscle relaxants; anesthetics
Licorice	To treat hepatitis and peptic ulcers	Hypertension; hypokalemia; edema	Antihypertensives; potentiates the effects of corticosteroids
SAM-e (*S-adenosyl-L-methionine*)	To treat depression or osteoarthritis	Mimics serotonin; nausea, upset stomach	Drugs that can increase or mimic serotonin, such as antidepressants
St. John's Wort	To treat mild depression, anxiety, seasonal affective disorder	Enhances bleeding; hastens metabolic breakdown of drugs; contraindicated for organ transplant recipients	Anticoagulants; antidepressants; decreases the effectiveness of cyclosporine, antiviral drugs; digoxin; dextromethorphan; prolongs the effects of general anesthetics; monoamine oxidase inhibitors
Valerian	To treat insomnia, anxiety	Sedation; digestion problems	Potentiates the effects of barbiturates

Key Terms

Acupuncture The insertion of needles into precisely defined points on the body; thought to realign imbalances of yin-yang and qi and thereby bring harmony to the "climate" of an individual

Alternative medicine A wide spectrum of treatments—many finding support from collective anecdotal evidence—that is not considered standard therapy because of the lack of a scientific rationale, clinical evidence, or a favorable historic tradition

Analgesic therapies Treatments used to decrease pain

Biologic therapies Treatments that make use of the body's response to the applied therapy to either stimulate the body to produce the desired physiologic effect or interfere with the body's pathophysiologic processes

Chiropractic therapy The identification and restoration of proper vertebral alignment via spinal manipulation

Chondroitin sulfate An important class of glycosaminoglycans in articular proteoglycans; the oral form is thought to prevent degradation of joint cartilage and relieve symptoms

Copper therapy The application of any of a number of copper devices to the body to reduce pain and inflammation associated with joint and connective tissue problems

Direct homeostatic therapies Treatments that help return the body to the normal healthy state without relying on the body's response to achieve their ends

Glucosamine sulfate A fundamental component in the synthesis of both hyaluronic acid and chondroitin that is thought to promote cartilage repair and synthesis; the oral form is taken as a dietary supplement to treat arthritis

Homeopathy A system of therapeutics in which disorders characterized by a particular symptom are treated with a substance that causes that symptom in a healthy individual

Meditation Slow, rhythmic breathing while the mind is focused on a symbol, a mental image, or a mantra that is thought to promote relief for stress, headaches, anxiety, and emotional disorders

Placebo effect The beneficial effect of an inert therapy; thought to stimulate the mind to help produce either healing or analgesia that would not have occurred in its absence

Qigong A form of gentle exercises in which the flow of qi is stimulated along the meridians throughout the body

Static magnet therapy Application of a small disk of magnet to the body to relieve pain

Tai chi A low-impact exercise program that is derived from the martial arts in China and was founded in accordance with the belief that chi is a vital life force flowing through the body

Yoga A variety of disciplines designed to bring practitioners into union with mankind and a higher God or life force

References

1. Hrobjartsson A, Gotzsche PC: Is the placebo powerless? An analysis of clinical trials comparing placebo with no treatment. *N Engl J Med* 2001;344:1594-1602.
2. Kaptchuk TJ: Acupuncture: Theory, efficacy, and practice. *Ann Intern Med* 2002;136:374-383.
3. Brown D: Three Generations of Alternative Medicine: Behavioral Medicine, Integrated Medicine, and Energy Medicine. *Boston University School of Medicine Alumni Report.* Fall 1996.
4. Raso J, Barrett S: (eds): *Alternative Health Care: A Comprehensive Guide: Natural Medicine, "Hands-On" Healing, Spiritualism, Occultism, and Much More.* Amherst, NY, Prometheus Books, 1994.
5. Han JS: Acupuncture Activates Endogenous Systems of Analgesia. *NIH Consensus Development Conference on Acupuncture: Program & Abstracts.* Bethesda, MD, National Institutes of Health, 1997.
6. National Institutes of Health: *NIH Consensus Development Conference on Acupuncture: Program and Abstracts.* Bethesda, MD, National Institutes of Health (US), 1997, pp 1-38.
7. Meeker WC, Haldeman S: Chiropractic: A profession at the crossroads of mainstream and alternative medicine. *Ann Intern Med* 2002;136:216-227.
8. Koes BW, Assendelft WJ, van der Heijden GJ, et al: Spinal manipulation for low back pain: An updated systematic review of randomized clinical trials. *Spine* 1996;21:2860-2873.

9. Hurwitz EL, Aker PD, Adams AH, et al: Manipulation and mobilization of the cervical spine: A systematic review of the literature. *Spine* 1996;21:1746-1760.
10. van Tulder MW, Koes BW, Bouter LM: Conservative treatment of acute and chronic nonspecific low back pain: A systematic review of randomized controlled trials of the most common interventions. *Spine* 1997;22:2128-2156.
11. Bronfort G: Spinal manipulation: Current state of research and its indications. *Neurol Clin* 1999;17:91-111.
12. Lippiello L, Woodward J, Karpman R, et al: In vivo chondroprotection and metabolic synergy of glucosamine and chondroitin sulfate. *Clin Orthop* 2000;381:229-240.
13. Buckwalter JA, Callaghan JJ, Rosier RN: From oranges and lemons to glucosamine and chondroitin sulfate: Clinical observations stimulate basic research. *J Bone Joint Surg Am* 2001;83:1266-1268.
14. McAlindon TE, LaValley MP, Gulin JP, et al: Glucosamine and chondroitin for treatment of osteoarthritis: A systematic quality assessment and meta-analysis. *JAMA* 2000;283:1469-1475.
15. Reginster JY, Deroisy R, Rovati LC, et al: Long-term effects of glucosamine sulphate on osteoarthritis progression: A randomised, placebo-controlled clinical trial. *Lancet* 2001;357:251-256.
16. Thornton A, Lee P: Publication bias in meta-analysis: Its causes and consequences. *J Clin Epidemiol* 2000;53:207-216.
17. Haehl R, Clarke JH, Wheeler FJ: (eds): *Samuel Hahnemann: His Life and Work: based on Recently Discovered State Papers, Documents, Letters, and etc.* London, England, Homoeopathic Publishing, 1922.
18. Linde K, Clausius N, Ramirez G, et al: Are the clinical effects of homeopathy placebo effects? A meta-analysis of placebo-controlled trials. *Lancet* 1997;350:834-843.
19. Cucherat M, Haugh MC, Gooch M, et al: Evidence of clinical efficacy of homeopathy: A meta-analysis of clinical trials: HMRAG: Homeopathic Medicines Research Advisory Group. *Eur J Clin Pharmacol* 2000;56:27-33.
20. Fisher P, Scott DL: A randomized controlled trial of homeopathy in rheumatoid arthritis. *Rheumatology* 2001;40:1052-1055.
21. Singh BB, Berman BM, Hadhazy VA, et al: A pilot study of cognitive behavioral therapy in fibromyalgia. *Altern Ther Health Med* 1998;4:67-70.
22. Kaplan KH, Goldenberg DL, Galvin-Nadeau M: The impact of a meditation-based stress reduction program on fibromyalgia. *Gen Hosp Psychiatry* 1993;15:284-289.
23. Creamer P, Singh BB, Hochberg MC, et al: Sustained improvement produced by nonpharmacologic intervention in fibromyalgia: Results of a pilot study. *Arthritis Care Res* 2000;13:198-204.
24. Husted C, Pham L, Hekking A, et al: Improving quality of life for people with chronic conditions: The example of t'ai chi and multiple sclerosis. *Altern Ther Health Med* 1999;5:70-74.
25. Vallbona C, Richards T: Evolution of magnetic therapy from alternative to traditional medicine. *Phys Med Rehabil Clin North Am* 1999;10:729-754.
26. Vallbona C, Hazlewood CF, Jurida G: Response of pain to static magnetic fields in postpolio patients: A double-blind pilot study. *Arch Phys Med Rehabil* 1997;78:1200-1203.
27. Weinstraub M: Magnetic biostimulation in painful diabetic peripheral neuropathy: A novel intervention: A randomized, double-placebo crossover study. *Am J Pain Manage* 1999;9:8-17.
28. Collacott EA, Zimmerman JT, White DW, et al: Bipolar permanent magnets for the treatment of chronic low back pain: A pilot study. *JAMA* 2000;283:1322-1325.
29. Walker WR, Keats DM: An investigation of the therapeutic value of the "copper bracelet": Dermal assimilation of copper in arthritic/rheumatoid conditions. *Agents Actions* 1976;6:454-459.
30. Berthon G: Is copper pro- or anti-inflammatory? A reconciling view and a novel approach for the use of copper in the control of inflammation. *Agents Actions* 1993;39:210-217.
31. Walker WR, Beveridge SJ, Whitehouse MW: Dermal copper drugs: The copper bracelet and Cu(II) salicylate complexes. *Agents Actions* 1981;(suppl 8):359-367.
32. Shackel NA, Day RO, Kellett B, et al: Copper-salicylate gel for pain relief in osteoarthritis: A randomised controlled trial. *Med J Aust* 1997;167:134-136.

Chapter 33

Surgical Treatments

Many musculoskeletal conditions can be treated nonsurgically. In most instances, surgical treatment is reserved for conditions that do not respond to nonsurgical measures. However, some conditions, such as open fractures, demand urgent surgical attention. The principal goals of surgical treatment may include the restoration of normal anatomic relationships and functions, the promotion of healing, the correction of deformity, the removal of abnormal tissue, or the replacement of structures that cannot be repaired.

Conditions

Infection

In musculoskeletal medicine, three forms of infection are encountered: soft-tissue infection, joint infection (septic arthritis), and bone infection (osteomyelitis). Many soft-tissue infections do not require surgical treatment. When the body's defenses are overwhelmed, however, removal of the infected tissue, often devitalized and necrotic, is necessary to restore homeostasis and allow healthy granulation tissue to form.

An infection within a joint usually requires some type of surgical treatment because the joint space is, to some extent, inaccessible to the circulatory system. Surgical drainage is also used to preempt the body's own response because the cellular machinery built to fight infection can also destroy articular cartilage. It is imperative, therefore, to treat the infection before this response is mounted.

Surgical drainage of a joint infection does not always have to be performed in an operating room with the patient under anesthesia. Aspiration of many superficial joints, such as the knee, can be performed with a needle at the bedside. Aspiration of deeper joints, such as the hip, may require at least radiographic guidance.

Lavage, also known as irrigation, is another surgical procedure that is used to treat infected joints. Lavage is often done with an arthroscope, especially when treating infections in the large joints. In this procedure, the joint is thoroughly washed with a high-volume saline solution flush to remove the offending organism and prevent the host response to it. For example, articular cartilage is extremely sensitive to the degradative enzymes that are released by white blood cells to combat bacteria. Thus, the response to the bacteria may result in a Pyrrhic victory because the body will win the battle against the bacteria but lose the war in that the articular surface will be destroyed in the process.[1,2] In this example, lavage removes bacteria but helps preserve joint surfaces.

Because the deep core of the bone is relatively inaccessible to circulating antibiotics, infected bone is typically treated with surgical drainage as well. Moreover, bone responds to infection by forming new bone called the involucrum that creates a barrier around the infected nidus. The infected area isolated by the involucrum is called the sequestrum. To drain the sequestrum surgically, the surrounding involucrum must be removed.[3] Bone drainage procedures are often performed in combination with soft-tissue coverage operations, such as skin grafts and muscle flaps. This is because soft-tissue defects are commonly seen in conjunction with osteomyelitis and because both can be caused by an open fracture.

Fractures

Although many fractures that are treated surgically would heal without surgical intervention, surgery may nevertheless be needed to preserve the anatomic alignment of the bone or to allow the bone to heal without excessive immobilization. In addition, surgical treatment is sometimes needed for fractures that would not heal well otherwise. Open fractures, for example, require wound débri-

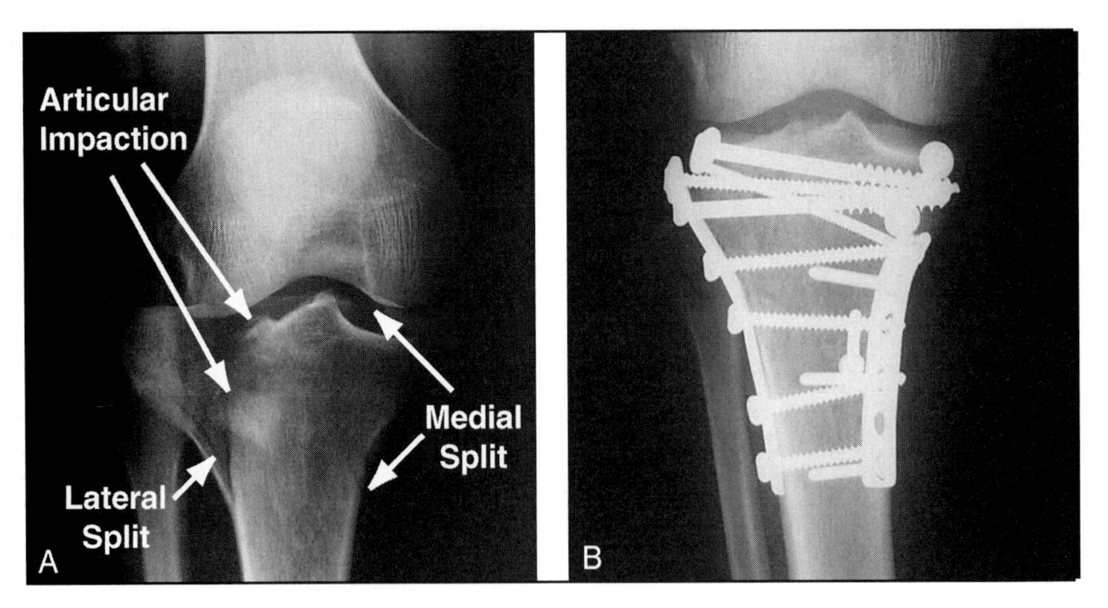

Figure 1
A, Radiograph of an extensively comminuted fracture of the tibial plateau with disruption of the joint surface. **B**, Radiograph showing restitution of the joint achieved with open reduction and internal fixation using medial and lateral plates and interfragmentary (free) screws.

(Reproduced from Lang GJ: Knee and leg: Bone trauma, in Koval KJ (ed): ***Orthopaedic Knowledge Update 7****. Rosemont, IL, American Academy of Orthopaedic Surgeons, 2002, pp 479-488.)*

dement and restoration of the soft-tissue envelope. Such a fracture may also need to have additional blood supply brought to it in the form of a vascularized muscle flap, a procedure called revascularization. Additionally, some fractures associated with great bone loss require bone grafting to augment healing.

Surgical treatment of fractures can lead to either primary or secondary bone healing. Primary bone healing occurs when the bone ends are anatomically opposed and held together rigidly. This is similar to the healing seen in sutured skin, but no scar is formed in bone. Primary healing occurs only when there is no motion between the fracture ends. In that case, the bone heals by direct formation of new bone, and no fracture callus is formed. Primary bone healing is also known as osteonal healing because osteons cut across the fracture site to allow the formation of new bone in the gap. Secondary bone healing is characterized by the formation of fracture callus. A fracture callus is initially formed from the hematoma that develops from the bleeding edges of bone. This callus then forms a cartilage mass that is remodeled into mature bone.[4,5]

Types of Surgery

Open Reduction and Internal Fixation

Open reduction and internal fixation, often abbreviated as ORIF, is a procedure that involves incising the skin and soft tissue to repair a fracture under direct visualization. Once the incision is made, the fracture fragments are placed into anatomic alignment (hence, open reduction) and rigidly held in this position by means of screws, plates, or wires (internal fixation).

Open reduction and internal fixation is typically used to mend fractures for which perfect anatomic alignment is necessary. Typically, these fractures involve or are in close proximity to the joint line (Fig. 1). An advantage of this surgical approach is that anatomic alignment can be restored; however, a disadvantage is that it requires surgical dissection near the site of the fracture, which disrupts the biologic envelope (ie, the surrounding soft tissue and periosteal sleeve). This disruption increases the risk of infection and may also increase the risk of fracture nonunion.[6]

Open reduction and internal fixation can fail if there is too much motion at the fracture site, which can occur when the hardware itself fails or when the screws pull out of the bone. In both situations, another surgical procedure would be required to repair the problems.

Intramedullary Nailing

Another method of surgical fixation of fractures is intramedullary nailing or rodding. (The terms are interchangeable.) This method is best employed for diaphyseal (shaft) fractures of the long bones. This is deemed to be an indirect method of fixation because the fracture site itself is not surgically exposed; rather, a rod is inserted into the intramedullary canal of the bone at one of its two ends (Fig. 2). For example, an intramedullary nail can be inserted into the tibia through a hole made near the tibial tubercle, and the femoral shaft can accept an intramedullary rod either through the hip or the knee.

An intramedullary nail functions as an internal splint to stabilize a fracture. Even when the proximal and distal ends of the nail are secured to the bone with screws, there is some small degree of motion at the fracture site. This leads to secondary bone healing through callus formation. The proximal and distal locking screws provide rotational stability to the fracture site, which allows for early mobilization of the fractured limb and possibly early weight bearing. This method also allows motion of the joint above and below the fracture site. By way of contrast, it should be noted that the cast needed to control rotation of a tibial fracture spans from foot to groin and may produce knee or ankle stiffness.

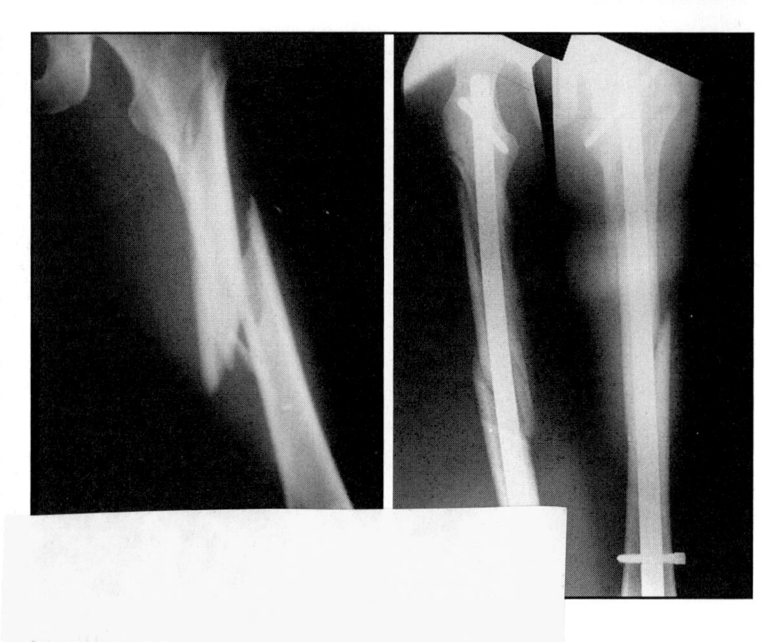

A, Midshaft femur fracture with comminution. **B**, AP and lateral view of this femur after fixation with a locked intramedullary nail. Note that the fracture fragments are not perfectly interdigitated and will unite only if callus forms across this gap.

(Reproduced from Johnson KD: Femur trauma, in Poss R, Bucholz RW, Frymoyer JW, Gelberman RH, Hensinger RN, Morrey BF (eds): ***Orthopaedic Knowledge Update 3****. Rosemont, IL, American Academy of Orthopaedic Surgeons, 1990, pp 513-527.)*

Intramedullary rods have a number of advantages and disadvantages. Compared with plating, intramedullary fixation does not require any stripping of the soft tissue or periosteum near the fracture site. Also, the length of incision typically needed to insert the rod is considerably smaller than that needed to expose a fracture site. However, there is a risk of fat embolism associated with placing a rod in the intramedullary canal.[7,8] Intramedullary nailing does not allow for as precise an anatomic alignment as that achieved with a plate and, of course, cannot be used for articular fractures.

One special case for which intramedullary nailing is particularly well suited is in the treatment of an impending pathologic fracture, in which a metastatic lesion within a long bone weakens the bone and predisposes it to fracture. In many instances, pain develops in a long bone in which there is a metastatic lesion, even though the bone has not yet fractured. For patients with metastatic bone lesions, a true fracture is avoided and much of their pain alleviated when the "intact" bone receives an intramedullary nail.[9-11] By spanning the pathologic site, the intramedullary nail is able to accept some of the forces of weight bearing and prevent the bone from breaking.

External Fixation

External fixation is a method of immobilization that uses pins inserted through the skin to attach fragments of fractured bone to external stabilizing frames (Fig. 3). These frames are placed outside the body at a distance from the zone of injury. Malgaine used the first external fixation device in 1853 when he placed a claw-like clamp through the skin to compress and immobilize a fractured patella. Fixation devices historically have spanned at least one joint, but some of the newer models use wires inserted at the ends of the bone such that the fixator does not cross the adjacent joints. This can help preserve the motion of these joints.

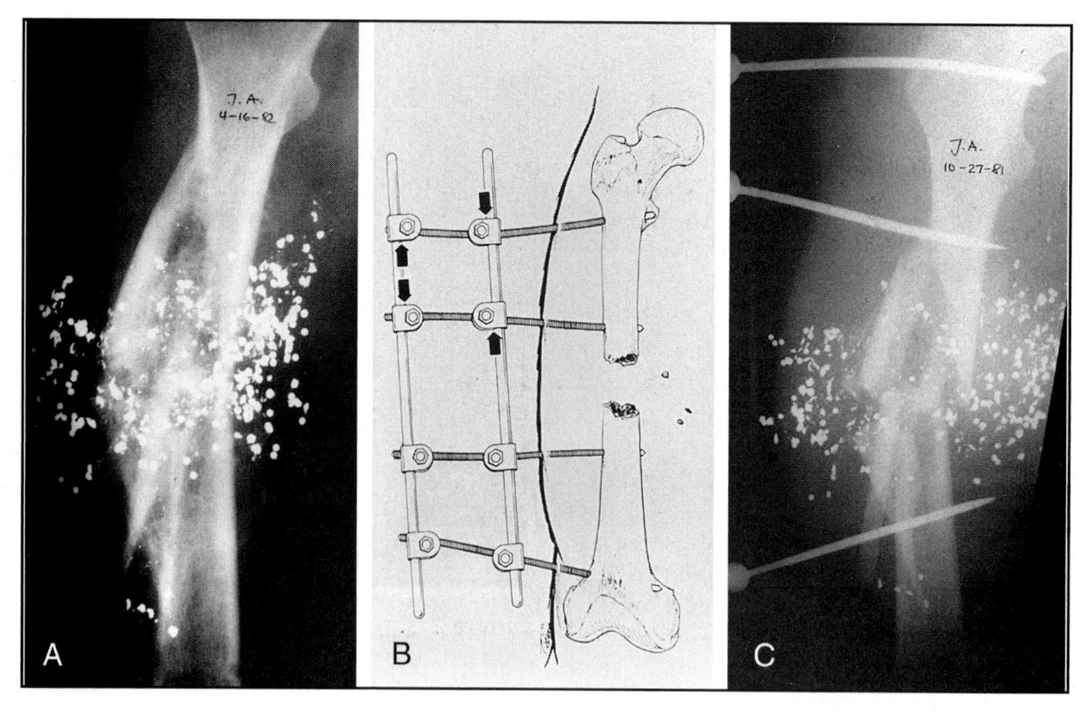

Figure 3
A, Highly comminuted femur fracture caused by a shotgun blast inflicted at close range. There was extensive soft-tissue damage as well. **B**, Schematic representation of external fixation for this fracture pattern. **C**, Radiograph of this fracture after the application of the fixator.

(Reproduced from Johnson KD: Femur trauma, in Poss R, Bucholz RW, Frymoyer JW, Gelberman RH, Hensinger RN, Morrey BF (eds): ***Orthopaedic Knowledge Update 3****. Rosemont, IL, American Academy of Orthopaedic Surgeons, 1990, pp 513-527.)*

Several situations call for the use of external fixation devices. Highly comminuted (shattered) fractures, for example, frequently cannot be treated with plates and screws because there is simply not enough solid bone left to accept the screws. Open fractures with severe soft-tissue injury may be best treated with devices that can temporarily stabilize the bone as the soft-tissue envelope and blood supply are being restored.[12]

External fixation is now also used for limb lengthening. In the Ilizarov technique, an external fixation device is placed across a bone, which is then cut transversely.[13] Gradual tension across this man-made fracture site allows for gradual lengthening of the bone—a process called distraction osteogenesis. Such an approach is also useful to correct angular deformities.

The major limitation of external fixation is that it requires an open channel of communication to the outside world in the form of pin tracts, which are subject to infection. What may begin as a superficial pin tract infection can actually extend deep into the bone, causing osteomyelitis. Therefore, it is not unusual for an external fixation device to be exchanged for another means of immobilization once the specific goals of external fixation are accomplished. For example, once the skin wound of an open fracture heals, the fixator may be replaced with a rod or cast.

Total Joint Arthroplasty

The mainstay of surgical treatment for arthritis of the hip and knee is joint replacement, a procedure known asarthroplasty . The goal of arthroplasty is to replace a destroyed, painful joint with an artificial joint that is functional and pain free. During the procedure, the diseased ends of the bone with their irregular articular surfaces are removed and replaced with man-made materials.

When both sides of a joint are replaced, the procedure is known as a total joint replacement. When only one side is replaced, the procedure is called a partial joint replacement, or hemiarthroplasty. Hemiarthroplasty is most often used in the hip to treat a fracture that is unlikely to heal (Fig. 4). In this procedure, the femoral head is removed and an artificial one is placed on a stem that extends into the shaft of the femur. This artificial head can then articulate with the native acetabulum. In a total hip replacement, there is an acetabular component added as well. With a total knee replacement, both sides of all three compartments (medial, lateral, and patellofemoral) are replaced. A partial knee replacement, also called a unicompartmental arthroplasty, involves the replacement of only one compartment, typically on the medial side.

Arthroplasty is major surgery. Thus, there is a measurable risk of perioperative complications. Beyond the medical risks associated with major surgery in general, with joint replacement there is a specific risk for deep venous thrombosis and subsequent pulmonary embolism; however, with the use of chemical or mechanical prophylaxis, the risk of deep venous thrombosis is reduced.[14] Other risks include infection or dislocation of the prosthesis. Finally, even if these complications are avoided, there is the risk of mechanical (aseptic) loosening of the prosthesis.[15] Aseptic loosening refers to a process that exceeds the normal wear and tear of the prosthesis. It is a biologic process in which small particles from the plastic surface of the artificial joint are generated from motion and incite an inflammatory response. This inflammation ultimately loosens the prosthesis. Because of these limitations, joint replacement is a less appealing alternative for young people; it is likely that they will outlive their prosthetic joints and thus require one or more revision surgeries, each with a diminishing chance of good results.

Despite its limitations, total joint arthroplasty is an extremely effective operation. It has nearly obliterated the word "cripple" from the English language. However, it is hoped that arthroplasty will be replaced by more biologically oriented procedures in the future.

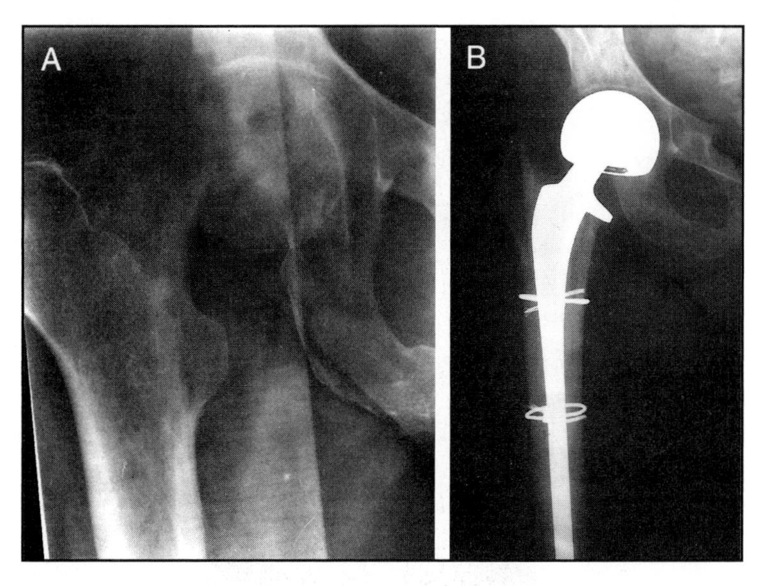

Figure 4
A, Preoperative AP view of the right hip demonstrating a femoral neck fracture caused by tumor metastasis. **B**, Postoperative AP view of a partial hip replacement (hemiarthroplasty) in which the femoral head and neck were removed and long-stem femoral components were placed. The "ball" of this joint sits in the native acetabulum.

(Reproduced from Damron TA, Sim FH: Surgical treatment for metastatic disease of the pelvis and the proximal end of the femur. ***Instr Course Lect*** *2000;49:461-470.)*

Resection Arthroplasty
An arthritic joint can be treated with other surgical approaches as well. Some joints are amenable to removal, or resection arthroplasty. A resection arthroplasty involves simply cutting out the surfaces of the diseased joint and allowing a fibrocartilage scar to grow in its place. Resection arthroplasty is often the best initial surgical option for joints that have little function. For example, surgical treatment of a painful acromioclavicular joint can be successful by simply removing the contacting surfaces of the clavicle and acromion. Even major joints can be treated with resection arthroplasty, although this is rarely a first-choice solution. Resection arthroplasty of the hip was originally developed to treat tuberculosis. In this operation, the bones of the hip joint are removed so that a fibrous union can form between the pelvis and lower extremity. Although this shortens the leg considerably and produces a floppy joint, this is the ideal operation for some patients because it alleviates joint pain.

Arthrodesis
A painless joint also can be achieved by arthrodesis, or fusion. With arthrodesis, the painful edges of the joint are removed, and the two ends of raw bone are allowed to unite as if they were the edges of a fracture (Fig. 5). This radical operation is used only when other options are not feasible. For example, a joint that has been destroyed by infection may require fusion and not replacement when there is suspicion that some of the infection still lingers. With fusion, all motion across the joint is eliminated. This typically eliminates pain permanently, but motion is likewise permanently lost. This loss of motion may limit function substantially. The glenohumeral joint of the shoulder tolerates

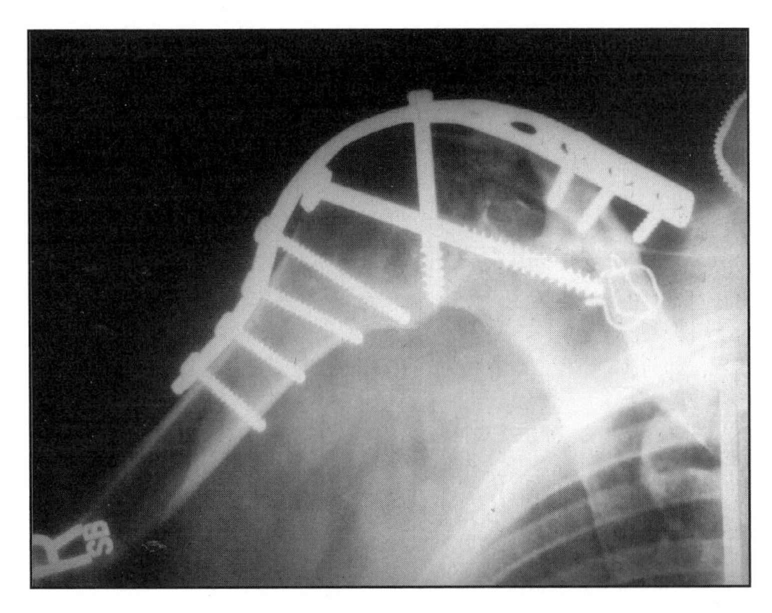

Figure 5
Radiograph of fusion of the glenohumeral joint using a plate that spans from the humeral shaft to the scapula. A separate long screw was used to compress the humerus into the glenoid socket.

(Reproduced from Bennett JB, Allan CH: Tendon transfers about the shoulder and elbow in obstetrical brachial plexus palsy. ***Instr Course Lect*** *2000;49:319-332.)*

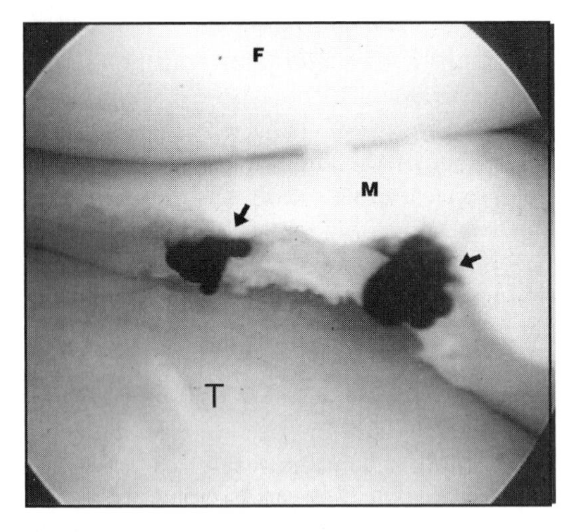

Figure 6
Arthroscopic view of the knee joint demonstrates meniscal repair. Note the two suture knots in the meniscus (arrows). The femur (F) is above the meniscus (M). The tibial plateau (T) is below.

(Reproduced from Rodeo S: Arthroscopic meniscal repair with use of the outside-in technique. ***Instr Course Lect*** *2000;49:195-206.)*

fusion better than many other major joints because the nearby scapulothoracic joint can substitute some of the lost motion. When the nearby joint is not able to provide replacement motion, great stresses are placed on it. This, in turn, places adjacent joints at increased risk for early arthritis.

Arthroscopy

Arthroscopy is a form of minimally invasive surgery in which a fiberoptic camera, the arthroscope, is introduced into a joint through a small incision. This allows the surgeon to observe the inner environment of the joint without making large incisions and disrupting the normal soft tissue surrounding it.

Arthroscopy was originally used as a diagnostic tool. Although MRI has obviated the need for many diagnostic surgeries, arthroscopy still has some role in making diagnoses. For the most part, though, arthroscopy is used as a therapeutic tool. By making additional small incisions, the surgeon can introduce specially designed surgical instruments that allow for removal or repair of abnormal tissue. Common arthroscopic procedures include the removal of a torn meniscus (partial meniscectomy); the flattening of a bone spur extending from the acromion (acromioplasty); and various surgical repairs and reconstructions, including meniscal repair (Fig. 6), anterior cruciate ligament reconstruction, and rotator cuff repair.[16-18]

As surgeons' arthroscopic skills have improved over the years, the number of conditions now amenable to arthroscopic treatment has increased. Arthroscopy of the wrist, elbow, and ankle has now become routine. The major limitation of arthroscopy is that it can address only surface phenomena in the joint; diseases within the soft tissue or the bone are beyond its reach. In addition, arthroscopy can be used only in joints that have enough room for the arthroscopic equipment to function without hitting (and possibly damaging) the articular surfaces.

Surgical Decompression

When clinical symptoms result from pressure on vital structures, surgical treatment may consist of simply eliminating the pressure by removing the offending tissue. This is a broad category of procedures that include carpal tunnel release, in which the transverse carpal ligament of the hand is simply cut to allow more space in a crowded carpal tunnel; **laminectomy** and **diskectomy**, in which the bone and disk pressing on a nerve root in the spine are removed (Fig. 7); and **acromioplasty**, in which a bone spur extending from the acromion and pressing on the rotator cuff is flattened.

The common theme of these operations is a controlled injury: without precision control, tissues may be damaged or an exuberant healing response may be induced such that even greater compression results.

Amputation

The goal of surgery on the musculoskeletal system is to restore function. In some cases, however, a better outcome can be achieved with amputation, the surgical removal of a damaged limb, as opposed to repeated surgical attempts to fix a problem that may not be fixable. For example, if a patient has an open fracture of the leg that would require extensive vascular and nerve repair, multiple débridements, and a muscle flap for soft-tissue coverage, it may be better to simply amputate distal to the injury and fit the patient with a prosthesis.[19] This is often a complex and difficult decision; however, it should be noted that there have been patients with above-knee amputations who were able to compete successfully in marathon races.

Amputation can be used to address traumatic injuries or it can be performed as a planned treatment for pathologic conditions, such as vascular insufficiency, life-threatening infections, malignant tumors, or severe degenerative or acquired deformities. Traumatic amputations can be complete or partial. Complete amputation can result from sharp transection (eg, from saw blades or knives), crushing injuries (eg, from hydraulic presses), or avulsion injuries (eg, from motor vehicle accidents). The amputated part is often too severely damaged to attempt replantation, in which case the remaining stump is treated with débridement of severely contaminated tissue and closure of the wound after appropriate decontamination. The remaining bone ends may require shortening to obtain adequate soft-tissue and/or skin coverage for the remaining stump. Skin grafts and soft-tissue flaps also may be used to maintain the length of the amputated limb. Preservation of limb length is important because it allows for a more efficient prosthetic or artificial limb fitting and function.

Replantation

Traumatic amputation that results from a clean transection of the limb or body part without a large zone of crush injury is generally amenable to attempted replantation. Replantation is often a lengthy and tedious surgical procedure for which the skill of the surgical team and the age and overall health status of the patient must be considered. For example, an elderly man who has severe coronary artery disease and cardiac insufficiency may suffer grave complications during an extended replantation procedure; therefore, he may not be a good candidate because the risks outweigh the benefits. Severely contaminated or damaged partial amputations or injuries with a large zone of devastating damage to the nerves, blood vessels, bone, and soft tissues are also frequently treated by amputation. This is known as completion amputation.

Microsurgical techniques have been developed that allow surgeons to anastamose (reconnect) arteries, veins, and nerves in conjunction with repair of the tendons and muscles. Microsurgical techniques have also been used in conjunction with fracture fixation and soft-tissue débridement and coverage to enable surgeons to reattach fingers, hands, arms, feet, and even legs under certain circumstances. For example, the great toe may be attached to the hand to establish a functioning thumb after traumatic loss of the thumb.

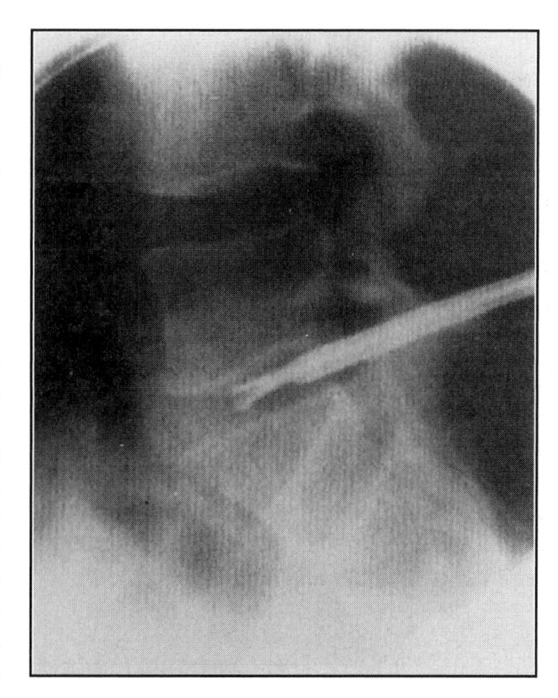

Figure 7
Intraoperative radiograph showing forceps in the L5-S1 disk space removing a herniated spinal disk.

(Reproduced with permission from Kambin P, Schaffer JL: Percutaneous lumbar discectomy: Review of 100 patients and current practice. ***Clin Orthop*** *1989;238:24-34.)*

Key Terms

Acromioplasty A surgical decompression procedure in which a bone spur extending from the acromion and pressing on the rotator cuff is removed

Arthrodesis A procedure in which the edges of a joint are removed and the two ends of raw bone are allowed to unite as if they were the edges of a fracture; also known as joint fusion

Arthroplasty A procedure in which the diseased ends of bone with their irregular articular surfaces are removed and replaced with man-made materials or scar tissue

Arthroscopy A form of minimally invasive surgery in which a fiberoptic camera, the arthroscope, is introduced into a joint through a small incision

Aspiration Removal of fluids from a body cavity; often done to obtain specimens for analysis

Diskectomy A surgical decompression procedure in which an intervertebral disk is removed

External fixation A method of immobilization that uses pins inserted through the skin to attach fragments of fractured bone to external stabilizing frames

Fracture callus Structure developed after a fracture; initially formed from a hematoma at the bleeding edges of bone, it eventually forms a cartilage mass that is remodeled into mature bone

Intramedullary nailing or rodding A procedure for the fixation of fractures in which a nail or rod is inserted into the intramedullary canal of the bone at one of its two ends

Involucrum In osteomyelitis, the barrier of new bone formed around an infested nidius (ie, the sequestrum)

Laminectomy A surgical decompression procedure in which part of the posterior arch of a vertebra is removed; allows access to the disk

Lavage The irrigation or thorough washing of an infected joint with high-volume saline solution

Open reduction and internal fixation Often abbreviated ORIF, a procedure that involves incising the skin and soft tissue to repair a fracture under direct visualization

Primary bone healing The end-to-end repair process that occurs when the bone ends are anatomically opposed and held together rigidly; no callus forms

Resection arthroplasty A procedure in which the surfaces of diseased bone are excised and fibrocartilage is allowed to grow in its place

Revascularization A procedure to provide an additional blood supply to fractured bone

Secondary bone healing The repair process that is characterized by the formation of fracture callus, which then remodels to form new bone

Sequestrum The infected area of bone that is walled off by the involucrum

Surgical drainage A procedure in which fluid is removed from an infected body part, often to treat infection

References

1. Goldenberg DL: Septic arthritis. *Lancet* 1998;351:197-202.
2. Deng GM, Nilsson IM, Verdrengh M, et al: Intra-articularly localized bacterial DNA containing CpG motifs induces arthritis. *Nat Med* 1999;5:702-705.
3. Mader JT, Calhoun J: Osteomyelitis, in Mandell GL, Bennett JE, Dolin R (eds): *Mandell, Douglas and Bennett's Principles and Practice of Infectious Diseases*, ed 4. New York, NY, Churchill Livingstone, 1995, pp 1039-1051.
4. McKibbin B: The biology of fracture healing in long bones. *J Bone Joint Surg Br* 1978;60:150-162.
5. Brighton CT: Principles of fracture healing: Part I. The biology of fracture repair. *Instr Course Lect* 1984;33:60-82.
6. Leunig M, Hertel R, Siebenrock KA, et al: *Clin Orthop* 2000;375:7-14.
7. Wenda K, Runkel M, Degreif J, et al: Pathogenesis and clinical relevance of bone marrow embolism in medullary nailing: Demonstrated by intraoperative echocardiography. *Injury* 1993;24(suppl 3):S73-S81.
8. Kropfl A, Berger U, Neureiter H, et al: Intramedullary pressure and bone marrow fat intravasation in unreamed femoral nailing. *J Trauma* 1997;42:946-954.
9. Harrington KD: Orthopedic surgical management of skeletal complications of malignancy. *Cancer* 1997;80(suppl 8):1614-1627.
10. van der Hulst RR, van den Wildenberg FA, Vroemen JP, et al: Intramedullary nailing of (impending) pathologic fractures. *J Trauma* 1994;36:211-215.
11. Ryan JR, Rowe DE, Salciccioli GG: Prophylactic internal fixation of the femur for neoplastic lesions. *J Bone Joint Surg Am* 1976;58:1071-1074.
12. Browner BD, Alberta FG, Mastella DJ: A new era in orthopedic trauma care. *Surg Clin North Am* 1999;79:1431-1448.
13. Shtarker H, Volpin G, Lerner A, et al: Ilizarov reconstructive surgery: A solution for complex problems of the musculoskeletal system. *Harefuah* 1999;136:182-256.
14. Geerts WH, Heit JA, Clagett GP, et al: Prevention of venous thromboembolism. *Chest* 2001;119(suppl 1):S132-S175.
15. Goldstein J, Zuckerman JD: Selected orthopedic problems in the elderly. *Rheum Dis Clin North Am* 2000;26:593-616.
16. McGinity JB, Geuss LF, Marvin RA: Partial or total meniscectomy: A comparative analysis. *J Bone Joint Surg Am* 1977;59:763-766.
17. DeHaven KE: Meniscus repair. *Am J Sports Med* 1999;27:242-250.
18. Gartsman GM: All arthroscopic rotator cuff repairs. *Orthop Clin North Am* 2001;32:501-510.
19. Sudkamp N, Haas N, Flory PJ, et al: Criteria for amputation, reconstruction and replantation of extremities in multiple trauma patients. *Chirurg* 1989;60:774-781.

USMLE Musculoskeletal System Content Outline

Step 1

Musculoskeletal System

Normal Processes

- Embryonic development, fetal maturation, and perinatal changes
- Organ structure and function
- Cell/tissue structure and function
  - biology of bones, joints, tendons, skeletal muscle
  - exercise and physical conditioning
- Repair, regeneration, and changes associated with stage of life

Abnormal Processes

- Infectious, inflammatory, and immunologic disorders
  - infectious disorders (eg, septic arthritis, Lyme disease, osteomyelitis)
  - inflammatory disorders (eg, fibrositis, synovitis, tenosynovitis)
  - immunologic disorders (eg, rheumatoid arthritis, ankylosing spondylitis, polymyositis, systemic lupus erythematosus, dermatomyositis, polymyalgia rheumatica)
- Traumatic and mechanical disorders (eg, fractures, sprains, strains, dislocations, repetitive motion injuries)
- Neoplastic disorders (eg, osteosarcoma, metastatic disease)
- Metabolic, regulatory, and structural disorders (eg, dwarfism, osteogenesis imperfecta, osteomalacia, osteoporosis, osteodystrophy, gout)
- Vascular disorders (eg, polyarteritis nodosa, bone infarcts)
- Systemic disorders affecting the musculoskeletal system (eg, diabetes mellitus)
- Idiopathic disorders (eg, Dupuytren's contracture, scoliosis, Paget's disease)
- Degenerative disorders (eg, disk disease, osteoarthritis)

Principles of Therapeutics

- Mechanism of action, use, and adverse effects of drugs for treatment of disorders of the musculoskeletal system
  - nonsteroidal anti-inflammatory drugs
  - muscle relaxants
  - antigout therapy (eg, allopurinol, colchicine, uricosuric drugs)
  - immunosuppressive drugs (eg, glucocorticoids, gold, cytotoxic agents)
  - drugs affecting bone mineralization (eg, diphosphonates, calcitonin, estrogen analogs)
- Other theraupeutic modalities (eg, radiation, surgery, casts, rehabilitation)

Gender, Ethnic, and Behavioral Considerations Affecting Disease Treatment and Prevention (Including Psychosocial, Cultural, Occupational, and Environmental)

- Emotional and behavioral factors (eg, diet, exercise, seat belts, bicycle helmets)
- Influence on person, family, and society (eg, osteoporosis, fractures in elderly, alcohol abuse, fractures)
- Occupational and other environmental risk factors (eg, athletes, musicians)
- Gender and ethnic factors (eg, bone mass)

Step 2

Diseases of the Musculoskeletal System and Connective Tissue

Health and Health Maintenance

- Epidemiology, impact, and prevention of degenerative joint and disk disease
- Prevention of disability due to musculoskeletal disorders or infection (eg, osteomyelitis, septic arthritis, Lyme disease, gonococcal tenosynovitis)

Mechanism of Disease

- Infections
- Nerve compressions and degenerative, metabolic, and nutritional disorders
- Inherited, congenital, or development disorders
- Inflammatory of immunologic disorders

Diagnosis

- Infections (eg, osteomyelitis, septic arthritis, Lyme disease, gonococcal tenosynovitis)
- Degenerative, metabolic, and nutritional disorders (eg, degenerative joint disease, degenerative disk disease, gout, rickets)
- Inherited, congenital, or developmental disorders (eg, congenital hip dysplasia, phocomelia, osteochondritis, slipped capital femoral epiphysis, scoliosis, syringomyelia, dislocated hip in infantile spinal muscular atrophy)
- Inflammatory, immunologic, and other disorders (eg, polymyalgia rheumatica, lupus arthritis, polymyositis-dermatomyositis, rheumatoid arthritis, ankylosing spondylitis, bursitis, tendinitis, myofascial pain, fibromyalgia, shoulder-hand syndrome, Dupuytren's contracture, Paget's disease)
- Neoplasm (eg, osteosarcoma, metastases to bone, pulmonary osteoarthropathy)
- Traumatic injury and nerve compression and injury (eg, fractures, sprains, dislocations, carpal tunnel syndrome, cauda equina syndrome, low back pain)

Principles of Management (With Emphasis on Topics Covered in Diagnosis)

- Pharmacotherapy only
- Management decision (treatment/diagnosis steps)
- Treatment only

Step 3

Disease/Disorders of the Musculoskeletal System

General Musculoskeletal Problems

- Internal derangement of knee
- Effusion of joint
- Affections of shoulder region
- Synovitis/tenosynovitis
- Bursitis
- Ganglion and cyst of synovium/tendon/bursa
- Muscular wasting and disuse atrophy
- Infective myositis
- Myositis ossificans
- Myalgia/myositis
- Disorders of bone and cartilage
- Developmental problems
- Congenital anomalies of limbs
- Muscular dystrophy

Disorders of Back/Spine

- Spinal enthesopathy
- Intravertebral disk disorders
- Disk displacement
- Spinal stenosis
- Lumbago, sciatica
- Kyphoscoliosis and scoliosis
- Sprains and strains of back
- Contusion of back
- Cervicalgia

Malignancy

- Secondary malignancy
- Neoplasm of bone/marrow
- Pathologic fracture

Contusions

Fractures/Dislocations

- Closed fracture of facial bones
- Fracture of vertebral column
- Fracture of ribs
- Closed fracture, upper extremity
- Fracture of neck of femur
- Various fractures, lower limbs
- Dislocations/separations

Sprains/Strains

- Temporomandibular joint disorders
- Rotator cuff syndrome
- Enthesopathy of elbow or ankle
- Various sprains and strains

Rheumatoid Conditions

- Collagen diseases
- Rheumatoid arthritis
- Nonarticular rheumatism

Osteoarthritis and Arthropathies

- Infective arthritis
- Osteoarthritis
- Monarthritis
- Arthropathy
- Polyarthritis

Ill-defined Symptoms, Musculoskeletal System

- Pain in joint
- Stiffness of joint
- Pain in limb
- Cramp, swelling

INDEX

B

C

D

E

F

G

H

I

J

K

L

M

N

O

P

Q

R

S